Biology and Ecology
of Crayfish

Biology and Ecology of Crayfish

Editors

Matt Longshaw
Product Development Manager
Benchmark Animal Health Ltd.
Edinburgh
Scotland
UK

Paul Stebbing
Aquatic non-native species specialist
Centre for Environment Fisheries and
Aquaculture Science (Cefas)
Barrack Road, The Nothe
Weymouth, Dorset
UK

CRC Press
Taylor & Francis Group
Boca Raton London New York

CRC Press is an imprint of the
Taylor & Francis Group, an **informa** business
A SCIENCE PUBLISHERS BOOK

CRC Press
Taylor & Francis Group
6000 Broken Sound Parkway NW, Suite 300
Boca Raton, FL 33487-2742

First issued in paperback 2021

© 2016 by Taylor & Francis Group, LLC
CRC Press is an imprint of Taylor & Francis Group, an Informa business

No claim to original U.S. Government works

Version Date: 20160311

ISBN 13: 978-0-367-78298-6 (pbk)
ISBN 13: 978-1-4987-6732-3 (hbk)

Visit the Taylor & Francis Web site at
http://www.taylorandfrancis.com

and the CRC Press Web site at
http://www.crcpress.com

Dedication

This book is dedicated to the memory of Dr. Francesca Gherardi. Her influence on our collective understanding of the biology and ecology of crayfish is outstanding; without her collaborations and her publications we would know much less about the subject. Her impact on the subject as both an outstanding scientist and an enthusiastic individual will be sorely missed. The world is a quieter place without her.

Preface

It would be fair to say that anybody that has worked with, or written about crayfish, has probably read the seminal work of Huxley (1880) "The crayfish. An introduction to the study of zoology". His preface ably sums up why many of us work with these animals when he states "…how the careful study of one of the commonest and most insignificant animals, leads us, step by step, from every-day knowledge to the widest generalizations and the most difficult problems of zoology; and, indeed, of biological science in general…". Whilst at the time that Huxley wrote his book in 1880 crayfish may have been viewed as "insignificant animals", they are now seen in a much different light, with the importance of crayfish now being seen on a global scale. Crayfish are now extensively used in aquaculture and wild harvest for human consumption, in the aquarium trade as pets, in scientific studies as model organisms, recognised as key components of freshwater ecosystems, and are some of the most widely spread and damaging invasive species. We wonder if Huxley would have been surprised at how important his "insignificant animals" have now become.

Huxley eludes to the point that crayfish are an ideal model organism for study. As this book will hopefully demonstrate, we agree that crayfish are an excellent model as some species are readily available and are identifiable through morphological and molecular methods (Chapters 1 and 2), methods for their capture and holding are established or can be adapted from known methods (Chapter 8), they can be maintained in the lab (Chapter 9), and their inter- and intra-specific interactions as well as their drivers for individual and population success are beginning to be known to us (Chapters 3, 4, 5 and 7).

Of course, we recognise that subsequent to Huxley's book there have been a number of books on the subject of crayfish biology, culture and ecology; for example, those published by David Holdich, and the excellent works published by some of our co-authors and others. We don't expect to compete with these, rather we consider our book as complimentary to them. When we set out to edit this book, it was our intention to have a combination of reviews of the current state of knowledge in the respective disciplines that was balanced with providing some practical hints and tips that could be used on a daily basis when working with crayfish. We each recognise our own inherent biases and interests; as editors, we allowed those biases to come through in each chapter. This has meant that there is some overlap in the topics in some chapter, albeit with a different viewpoint, depending on the overall subject being discussed. However, we feel that this minor repetition helps to reinforce some key concepts throughout the book. Simply put, crayfish biology is an amalgam of a number of disciplines; none of them should be seen as isolated from any other and the use of integrated studies to fully understand crayfish in totality is of paramount importance to us. In pulling together the list of authors, we tried to find like-minded individuals—those who reflect our views of taking a holistic view of the world, those who are active in the field of crustacean biology and, to misquote Huxley, to avoid a book that was "a treatise upon our English crayfish", hence our international authorship.

The first chapter of this book covers the latest information on the taxonomy, phylogeny and global distribution of crayfish by Catherine Souty-Grosset and James Fetzner Jr. from France and the USA respectively, which is followed by a detailed chapter on the population genetics of a range of species by Catherine, emphasising the importance of understanding genetics for protection of crayfish species. The next chapter by Colin McLay from New Zealand and Anneke van den Brink from The Netherlands looks at perhaps the two most important facets of any animals' biology—growth and reproduction (Chapter 3). Without these, there would be no progeny and no more crayfish. Linked with this chapter is

Chapter 4 on behavioural aspects of crayfish biology by Ana M. Jurcak, Sara E. Lahman, Sarah J. Wofford, and Paul A. Moore from the USA. Behaviour is covered from the perspective of crayfish as predators and as prey and how they compete for space and mates, covering the range of mating behaviours that allow them to successfully grow and breed. Explanations for some of the observed behaviours are further explored in the chapter on chemical ecology (Chapter 5) by Thomas Breithaupt from the UK, Francesca Gherardi (deceased) and Laura Aquiloni and Elena Tricarico from Italy. The pivotal role of these chemicals in how crayfish make sense of their world is explored through descriptions of the morphology of the various sensory systems and the role of "infochemicals" in predator/prey interactions, social interactions, reproduction and progeny interactions, and considers the role of pollutants as "info-disrupters". In a shift towards factors affecting individuals and populations of crayfish, Matt Longshaw from the UK, provides an up to date review and listing of the estimated 900 disease agents, parasites and commensals of crayfish across their global range (Chapter 6). Touching briefly on disease as a driver for population success, Ed Willis Jones from the UK, Michelle Jackson from South Africa and Jonathan Grey from the UK, begin to tie together the various environmental drivers for population success by examining population biology and community dynamics. They consider factors such as hydrography and habitat quality as well as abiotic factors (broadly termed by us as water quality). Recruitment, a population biology measure of successful reproduction is explored in relation to those environmental factors that impact reproduction, successful dispersal and survival. Community dynamics and species interactions, again from a population driver perspective, are further explored along with the ecosystem function of crayfish in food webs and in their wider, non-trophic, interactions. Eric Larson and Julian Olden from the USA, ask the question "why do we study crayfish?" at the start of their chapter on field sampling (Chapter 8) which they use to springboard into a review of the diverse methods available for assessing population numbers. The potential biases and limitations with different methods are highlighted, including the impact of trap choice on population estimates. Suggestions for addressing the variations in population estimates through a variety of methods including tagging studies as well as statistical approaches are covered. The next chapter on laboratory methods by both of us emphasises the need for integrated studies to better understand crayfish in natural and artificial habitats. We consider methods for transporting and holding crayfish followed by a proposed methodology for tissue sampling—whether that be for disease screening, assessing reproductive state, for collection of material for molecular/genetic studies or to describe new species. The final chapter by Paul Stebbing addresses how crayfish as invasive species can be managed, a subject close to the heart of Francesca Gherardi, to whom this book is dedicated, and who contributed so much to this specific subject area.

So, here it is. Our book. We entrust it to you to use as you see fit. Delve into it for the bits that you work on or take a risk and read a chapter on an area outside your comfort zone. Either way, we hope it's useful. One of us (Paul) has been actively involved in crayfish research for a number of years with a particular interest in their management and control. The other, Matt, has been involved in the periphery of crayfish research, focusing on the parasites and diseases of indigenous and non-native crayfish in the United Kingdom. For both of us we have found the exercise of editing the book exhilarating, debilitating, frightening, frustrating, educational, enlightening and, above all fun.

Finally, we wish to extend our thanks to all the authors involved in writing this book—every editor says the same… without the support of the authors, etc… but in this case, that is true. Halfway through the process of developing the book, one of us (Matt) changed jobs, moving from a research position in the British Government to a commercially focused role in the pharmaceutical industry. The hiatus caused by the need to focus on a whole new set of skills was partly responsible for a delay in getting the book to press. In addition, during the production of the book Paul's wife gave birth to two children, also resulting in the learning of and the need to focus on a whole new set of skills, causing further delay. We are pleased to say that each and every author was understanding, for which we are eternally grateful! We thank them for their patience in getting their book out there for you all to see. We hope you appreciate their hard work and patience as much as we do; this book is a testament to their skill and knowledge.

December 15th 2015

Matt Longshaw, Edinburgh
Paul Stebbing, Weymouth

Contents

Taxonomy and Identification

Catherine Souty-Grosset[1],* and *James W. Fetzner Jr.*[2]

"My purpose is to exemplify the general truths respecting the development of zoological science which have just been stated by the study of a special case; and, to this end, I have selected an animal, the Common Crayfish, which, taking it altogether, is better fitted for my purpose than any other".

(T.H. Huxley 1880)

Introduction

According to Huxley (1880), the origin of the common name, "crayfish" involves some interesting questions of etymology, and indeed, of history. It might readily be supposed that the word "cray" had a meaning of its own, and qualified with the substantive "fish", but this is not certain. The old English method of writing the word was "crevis" or "crevice", and "cray" was simply a phonetic spelling, with the word "fish" added to reinforce our perception of it as an aquatic animal. The term "crevis" has two distinct meanings. Swahn (2004) suggests that, according to the French, that the English were the first *Astacus* eaters (there is a historical reference to people eating crayfish in England from the tenth century onwards), and as in many other cases, they accepted not only the food but also the old French name for it. The French word "(é)crevisse" was modified and the new word "cray-fish" created. In the United States, crayfish are commonly known as crawfish, crawdads, or mudbugs and constitute a diverse and important component of freshwater aquatic and semi-aquatic ecosystems around the world (Taylor and Schuster 2004). The etymology of these terms is less clear.

Similar to crabs, shrimps and lobsters, the freshwater crayfish belong to the phylum Arthropoda, subphylum Crustacea, class Malacostraca, which contains about 25,000 species with a standard segmented body plan of 20 segments within the Subclass Eumalacostraca and the Superorder Eucarida. They are decapods (the Order Decapoda contains about 14,335 species, De Grave et al. 2009) because they have ten legs, including 8 pairs of thoracic limbs, but only 5 pairs are ambulatory (pereiopods), giving the group its name. The head has a compound eye, usually stalked, two pairs of sensory antennae and three pairs of

[1] Université de Poitiers, Laboratoire Ecologie & Biologie des Interactions - UMR CNRS 7267, Equipe Ecologie Evolution Symbiose - Bât B8, 40, avenue du Recteur Pineau, F-86022 POITIERS Cedex FRANCE.
[2] Carnegie Museum of Natural History, Section of Invertebrate Zoology, 4400 Forbes Avenue, Pittsburgh, PA 15213-4080. Email: FetznerJ@CarnegieMNH.Org
* Corresponding author: Catherine.souty@univ-poitiers.fr

mouthparts. A further three pairs of thoracic limbs (the maxillipeds) are incorporated into the mouthparts. The six abdominal segments each have a pair of swimming limbs, the last pair (the uropods) expanded into a tail fan in crawling and swimming forms. They are most closely related to marine lobsters (Crandall et al. 2000) and differ from those organisms by possessing a direct juvenile development rather than a dimorphic larval stage.

Among decapods, freshwater crayfish are represented by over 640 species (Crandall and Buhay 2008), with the southeastern United States being one of the epicenters of diversity. Three hundred sixty-three species are represented in the United States (Taylor et al. 2007) and according to De Grave et al. (2009), freshwater crayfish are widely distributed across the globe, mainly in temperate and subtropical water bodies and wetlands.

Systematics of the freshwater crayfish: Infraorder Astacidea Latreille 1802

The basic taxonomy of the Infraorder Astacidea has been summarized by C. Souty-Grosset in the Treatise on Zoology - Anatomy, Taxonomy, Biology (cf. Gherardi et al. 2010: Chapter Infraorder Astacidea Latreille, 1802 Volume 9). The infraorder is further subdivided into two superfamilies, the Astacoidea and Parastacoidea. The taxonomy of crayfish was extensively studied by Hobbs between 1974 and 1994. According to Hobbs "The nephropoids, ancestors to the modern lobsters, initiated a line that was the most conservative. Not only have the descendants remained in the sea, basically an environment in which their ancestors came into existence, but also many of the characteristics that constitute the lobster facies, and the release of young as larvae by modern descendants, suggest a more generalized condition than that which exists in current derivatives of the early astacoid and parastacoid stocks". Hobbs (1988), goes on to say that "the generally more morphologically divergent and venturesome astacoids and parastacoids, forebearers of modern crayfish, were destined to invade and, for the most part, to become restricted to the freshwaters of the Northern and Southern Hemispheres, respectively, having successfully negotiated the transition from the sea to freshwater, an environment which, in the late Jurassic, seems to have been discovered by few, if any, other decapods". The diagnosis of the two superfamilies of Astacoidea and Parastacoidea is based on the description of the carapace, the form of sternal plates and podobranchia, the branchial formula, and the differences between the first pleopods of males and females. Both superfamilies lack a dorsomedian longitudinal suture or a ridge in the cardiac and posterior gastric regions of the carapace and the sternal plate between the fifth pereiopods is not fused with the sternal complex anteriorly.

Throughout his career, Horton Hobbs, Jr. described many new taxa, including one new family (Cambaridae), 38 new genera and subgenera and 286 species, all of which were based on morphological characteristics alone. His most recent taxonomic summary of species was published in 1989 and was entitled *An Illustrated Checklist of the American Crayfishes (Decapoda: Astacidae, Cambaridae and Parastacidae).*

A detailed treatise of the taxonomy is given below.

Superfamily ASTACOIDEA Latreille, 1802

Articles of the lateral ramus of antennules bear two clusters of aesthetascs (except in Cambaroidinae Villalobos, 1955, in which there is only one); branchial formula is 16 + ep; 17 + ep; 18 + 2r + ep; or 18 + 3r + ep (ep: epipod; r: rudimentary), podobranchiae of the first three pereiopods not differentiated into branchial and epipodite portions; males have first pleopods with a single sperm groove, groove may be present or absent in females, second pleopods of males show a spiral element frequently borne on a subtriangular lobe; telson divided by a transverse suture almost always, and usually completely. Species live in fresh waters but some migrate into salt waters for part of their life cycle.

Family Astacidae Latreille, 1802

Some articles of the lateral rami of antennules bear 2 clusters of aesthetascs; branchial formula 18 + 2r + ep or 18 + 3r + ep; ischia of male pereiopods lack hooks; females lack first pleopods and *annulus ventralis* (sclerites present but lack sinus and fossa); males never exhibit cyclic dimorphism, distal portion of the male first pleopods rolled to form a cylinder, distal most part contracted to form either a tube or produced into 2 simple spoon-like lobes.

Family Cambaridae Hobbs, 1942

Some articles of lateral rami of antennules bear 1 or 2 clusters of aesthetascs; branchial formula 18 + 3r + ep; 17 + ep or 16 + ep; ischia of one or more of second-fourth pereiopods with hooks; first pleopods and *annulus ventralis* may be present or absent; males exhibit cyclic dimorphism, male first pleopods either medially bear shallow sperm grooves or distal portions tightly folded with distal end of sperm groove opening on one of 2–4 terminal elements.

Family †Cricoidoscelosidae Taylor, Schram and Shen, 1999

Rostrum with rounded base and lateral spines; blade-like scaphocerite; no ischial hooks on pereiopods; rounded pleomeral pleua; first pleopod styliform, remaining pleopods annulate; telson not divided by a transverse suture.

A single extinct species, *Cricoidoscelosus aethus*, originated from the Jurassic, Jehol Group of northeastern China.

Superfamily PARASTACOIDEA Huxley, 1880

Articles of the lateral ramus of antennules never bear more than one cluster of aesthetascs; Branchial formula ranges between 12 + epr + 5r and 21 + ep (epr: rudimentary epipod), epipodite of the first maxillipeds usually have branchial filaments, podobranchiae of the first three pereiopods differentiated into branchial and epipodite portions; first pleopods absent, second pleopods of males similar to third; telson never completely divided by a transverse suture.

Family Parastacidae Huxley, 1880

Diagnosis is the same as in the superfamily.

After Hobbs, crayfish taxonomy has mainly been updated at the generic level (e.g., Fitzpatrick 1983, Fetzner and Crandall 2002, Taylor 2002). Initial efforts examined allozyme variation (Fetzner 1996, Horwitz and Adams 2000), and 16S DNA sequences from the mitochondrial genome (Pedraza-Olvera et al. 2004, Sinclair et al. 2004). In particular, Crandall and Fitzpatrick (1996) gave new insights into the molecular systematics of crayfish by using a combination of procedures. Molecular studies have also elucidated a wealth of cryptic species that likely represent units of evolution. Their identification is thus highly relevant for conservation purposes (Crandall et al. 2000, Fetzner and Crandall 2003).

Origin of crayfish and fossil taxa

Hobbs (1988) provided insightful discussions of the known fossil crayfish taxa and discussed in detail their presumed evolutionary history. There are several fossil representatives included within the Astacidea, and a few are closely related to extant crayfish species (based on morphological evidence). However, several recent discoveries may suggest an alternative evolutionary history involving crayfish. Crayfish fossils and burrows have been found in the Triassic formations of North Carolina (Olsen 1977), of Arizona (Miller and Ash 1988) and of Utah (Hasiotis 1999) dating back 225 million years (Early Carboniferous during the formation of the Pangean supercontinent). These trace and body fossils confirm that crayfish were established across a variety of ecological settings ranging from fully terrestrial to fully aquatic. The Erymidae were marine representatives and were most likely the progenitors to the clawed lobster and freshwater crayfish lines. Members of this group first appear in the fossil record some 245 million years ago and disappear around 75 million years ago.

After the break-up of Pangaea into a northern and a southern continent, Laurasia and Gondwana respectively, the differences are believed to have evolved between the northern hemisphere Astacoidea and the southern hemisphere Parastacoidea. The monophyly of the two crayfish superfamilies Parastacoidea (southern hemisphere) and Astacoidea (northern hemisphere) (Crandall et al. 2000) is consistent with the break-up of Pangaea. Subsequently, the Parastacidae have radiated in Australasia, New Zealand, South America and Madagascar. Molecular genetic studies support the monophyly of the continental subgroups (Sinclair et al. 2004), but the relationships between them remain unresolved. However, Riek (1972) suggested that members of *Astacoides* from Madagascar appear closer to *Astacopsis* from Tasmania than they are to the South American species.

The fossil record appears older for Astacoidea than for Parastacoidea (Scholtz 2002) and is supported by fossil evidence of burrows. The centre of origin of the Astacoidea is suggested to be eastern Asia, from where the cambarid ancestors could have migrated via the Bering land bridge to their current position in eastern modern day North America, while most of the Astacidae dispersed westwards into Europe, with the oldest known *Austropotamobius* appearing there in the early Cretaceous (Souty-Grosset et al. 2006: Box 1). In Europe, a petrified specimen found in the Jurassic limestone of Solnhofen, Bavaria, from the same place where *Archaeopteryx* was found, is dated to 135–145 million years ago (MYA) and has been assigned to either *Aeger tippularius* (Schlothuis, 1822), *A. bronni* (Oppel, 1862) or *A. antumpsos speciosus* (Münster, 1839). A specimen of this fossil is currently used as the insignia for the President of the International Association of Astacology (IAA) (Picture 1).

Box 1. Summary of "Geological times" and "crayfish events" (Souty-Grosset et al. 2006).

ERA	System	Epoch (Million yrs)	Major crayfish events
"Quaternary"		(ca 3.000-1000 yrs BC) = *Littorina* period	
		Holocene (0.01)	post-glacial colonizations
		Pleistocene (1.8)	actual sub-sp differentiations?
	Neogene	Pliocene (5)	actual species differentiations?
CENOZOIC		Miocene (23)	Messinian crisis (ca 5.5 MY)
"Tertiary"			
		Oligocene (33.9)	
	Paleogene	Eocene (55.8)	
		Paleocene	
		-(65.5)-	
	Cretaceous		oldest known *Austropotamobius*
		-(145)-	
MESOZOIC	Jurassic		oldest known Astacidae
"Secondary"		-(200)-	
	Triassic		differentiation Astacoids/Parastacoids
		-(251)-	
PALEOZOIC	Permian		from seawater to freshwater
			"crayfish" in Antarctica
"Primary"			Nephropoid ancestors
		-(299)-	

American astacids of the genus *Pacifastacus* must have dispersed eastward after the cambarids; they are considered to be the most primitive of this family (Scholtz 2002). Ice ages following the break-up of Pangaea would have extinguished crayfish from Siberia and central Asia, although this does not explain their absence from Africa and India. Either they never got there, or were eliminated by some process, with competitive exclusion by freshwater crabs being one postulated scenario. However, elsewhere, representatives of these two groups coexist today in southern Europe, Turkey, Madagascar, Australia and New Guinea (Scholtz 2002).

Breinholt et al. (2009) presented a recent analysis of the timing of the diversification of the freshwater crayfishes by calibrating the times with multiple fossils, including a newly discovered Parastacoid fossil from Australia. With such a narrow taxonomic focus, they were able to increase accuracy and provide divergence estimates that were more specific to freshwater crayfish. Their molecular time estimates support a late Permian to early Triassic divergence from Nephropoidea, with a subsequent radiation and dispersal before the breakup of Pangaea, as well as later speciation and radiation prior to, or directly associated with, the breakup of Gondwana and Laurasia. The breakup of Gondwana and Laurasia resulted in the separation of the Parastacoidea and Astacoidea during the Jurassic period. The hypothesized divergence and radiation of these two superfamilies is also supported by their molecular time estimates. For the three

Picture 1. Fossil of IAA insignia.

families of crayfish, they estimate the Astacidae radiated at 153 MYA, the Cambaridae at 90 MYA, and the diversification of the Parastacidae at 161 MYA.

Phylogeny and genera

The freshwater crayfish are a well-established monophyletic group (Scholtz and Richter 1995, Crandall et al. 2000). Recent analyses support the sister relationship between clawed lobsters and freshwater crayfishes (Crandall et al. 2000, Dixon et al. 2003, Ahyong and O'Meally 2004, Porter et al. 2005, Bracken et al. 2009, Breinholt et al. 2009), lending support to the continued recognition of Astacidea. Within the freshwater crayfish, both of the superfamilies, Parastacoidea and Astacoidea, are monophyletic groups. The generic level taxonomy of parastacoids was recently revised with the splitting of the genus *Parastacoides* Clark 1936, into two new genera, *Spinastacoides* and *Ombrastacoides* (Hansen and Richardson 2006). De Grave et al. (2009) followed the conventional higher level taxonomy outlined by Hobbs (1974), with the adjustments proposed by Hansen and Richardson (2006). There is some debate about the monophyletic status of several genera in the family Cambaridae (Fetzner 1996; Crandall and Fitzpatrick 1996, Breinholt et al. 2009). Moreover, there have also been recent additions at both the family (Taylor et al. 1999) and generic levels (Martin et al. 2008, Feldmann et al. 2011) for fossil crayfish. Breinholt et al. (2012) suggest that convergent evolution has impacted the morphological features used to delimit the subgenera of *Cambarus*, as relationships based on chelae and carapace morphology are incongruent with estimated phylogenetic relationships. Many of the current systematic relationships within the Cambaridae are based on first form male gonopod morphology. Several features suggest that subgeneric morphological diagnoses used in traditional cambarid crayfish taxonomy (form one male gonopods in combination with chela and carapace characters) might be confounded by convergent evolution across all cambarids. The use of molecular-based phylogenies may be useful in evaluating synapomorphic morphological characters that reflect evolutionary relationships that are less affected by convergent evolution. While one goal of systematic studies is to revise taxonomy to reflect evolutionary history, for *Cambarus*, this task seems unwise without complete taxon sampling. Future work in this genus specifically needs to obtain complete taxon sampling as well as increased sampling throughout the geographic range of each species. Additional studies have used extensive sampling of species from the genera *Orconectes*, *Procambarus* and *Cambarus* and have found significant population structure and cryptic diversity (Buhay and Crandall 2008, 2009). Breinholt et al. (2012) concluded that extensive sampling within species is critically important for all cambarid crayfish before inferring meaningful evolutionary hypotheses or when making taxonomic changes.

Present distribution of the families of crayfish

As explained above, freshwater crayfishes are taxonomically distributed among three families; two Northern Hemisphere families, Astacidae and Cambaridae, and one Southern Hemisphere family, Parastacidae. There are two centres of species diversity for freshwater crayfishes. The first is located in the Southeastern United States where some 80% of the cambarid species can be found. The second centre of diversity is in Victoria, Australia; which contains a large proportion of the parastacid species. Freshwater crayfishes naturally occur on all of the continents except Africa and Antarctica (Fig. 1). The Astacidae are distributed in Europe and also west of the Rocky Mountains in the Northwestern United States and extending into southern British Columbia, Canada. The Cambaridae are found in the Eastern United States and south through Mexico, with members of the genus *Cambaroides* having a disjunct distribution in Southeastern Russia, Japan and the Korean Peninsula. The Parastacidae are distributed in Australia, New Guinea, New Zealand, South America, and Madagascar. Crayfish are naturally absent from the Antarctic continent, continental Africa, the Indian subcontinent, and much of Asia.

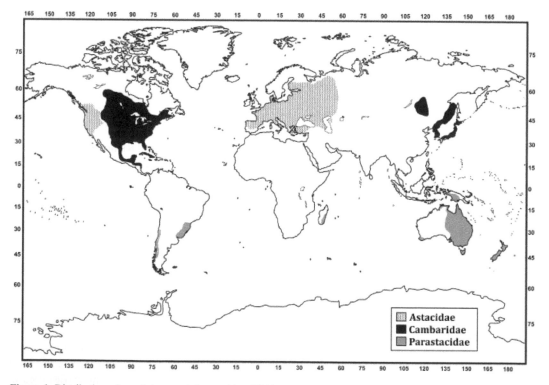

Figure 1. Distribution of crayfish around the world, exhibiting two centers of diversity in North America and Australia, respectively [Figure adapted from the Freshwater Crayfish and Lobster Taxonomy Browser: http://iz.carnegiemnh.org/crayfish/NewAstacidea/].

De Grave et al. (2009) listed below a recent classification that included both living and fossil genera of the Astacidea, in which they gave a comprehensive catalogue of crustaceans, including an examination of over 400 papers on decapod fossils. This compendium listed the current state of knowledge of the number of species of Decapoda in the Infraorder Astacidea Latreille, 1802, including 653 extant species, 5 extant species also known as fossils, and 124 exclusively fossil species (Table 1 below), although subsequent new species descriptions over the last five years have modified these numbers somewhat.

Table 1. According to De Grave et al. 2009: Taxa that are either exclusively living (extant) or exclusively fossil (the latter indicated by ††), have only one tally associated with them. In contrast, taxa that are known to include both extant and fossil species (indicated by †) are provided with three counts, e.g., "Family Parastacidae † Huxley, 1880 (164, 1, 3)". The first number represents exclusively extant species, the second number indicates extant species also represented in the fossil record, and the third is the number of exclusively fossil species. The total number of extant species is the sum of the first two numbers; in this case there are 165 known species of living Parastacidae. The total number of fossil species is the sum of the second and third numbers: there are 4 known species of fossil Parastacidae. The total number of known species (extant, fossil, or both) is the sum of all three numbers: there are 168 known species in Parastacidae.

+ indicates the number of new species described since De Grave et al. (2009).

SUPERFAMILY ASTACOIDEA † Latreille, 1802 (428, 0, 9) + 34

Family Astacidae † Latreille, 1802 (11, 0, 6)
Astacus † Fabricius, 1775 (3, 0, 4)
Austropotamobius † Skorikow, 1907 (3, 0, 1)
Pacifastacus † Bott, 1950 (5, 0, 1)

Family Cambaridae † Hobbs, 1942 (417, 0, 1) + 34
Barbicambarus Hobbs, 1969 (1) +1
Bouchardina Hobbs, 1977 (1)
Cambarellus Ortmann, 1905b (17) +1
Cambaroides Faxon, 1884 (7)
Cambarus Erichson, 1846 (100) +10
Distocambarus Hobbs, 1981 (5)
Fallicambarus Hobbs, 1969 (18) +1
Faxonella Creaser, 1933 (4)
Hobbseus Fitzpatrick and Payne, 1968 (7)
Orconectes Cope, 1872 (91) + 8
Palaeocambarus †† Taylor, Schram and Shen, 1999 (1)
Procambarus † Ortmann, 1905a (165, 0, 1) + 13
Troglocambarus Hobbs, 1942 (1)

Family Cricoidoscelosldae †† Taylor, Schram and Shen, 1999 (1)
Cricoidoscelosus †† Taylor, Schram and Shen, 1999 (1)

SUPERFAMILY PARASTACOIDEA † Huxley, 1880 (164, 1, 3) + 22

Family Parastacidae † Huxley, 1880 (164, 1, 3) + 22
Aenigmastacus †† Feldmann, Schweitzer and Leahy, 2011 (1) +1
Astacopsis † Huxley, 1880 (2, 1, 0) + 1
Astacoides Guérin-Méneville, 1839 (7)
Cherax Erichson, 1846 (34) + 15
Engaeus Erichson, 1846 (35)
Engaewa Riek, 1967 (5)
Euastacus E.M. Clark, 1936 (49) + 3
Geocharax E.M. Clark, 1936 (2)
Gramastacus Riek, 1972 (1) + 1
Lammuastacus †† Aguirre-Urreta, 1992 (1)
Ombrastacoides Hansen and Richardson, 2006 (11)
Palaeoechinastacus †† Martin, Rich, Poore, Schultz, Austin, Kool and Vickers-Rich, 2008 (1)
Paranephrops † White, 1842 (2, 0, 1) + 1
Parastacus Huxley, 1880 (8)
Samastacus Riek, 1971 (1)
Spinastacoides Hansen and Richardson, 2006 (3)
Tenuibranchiurus Riek, 1951 (1)
Virilastacus Hobbs, 1991 (3) + 1

The authors specified that subspecies were not counted, which is understandable given the fact that some may be species complexes or represent cryptic species. For example, among the Astacidae, the taxon *Austropotamobius pallipes* (Leboullet 1858) is a species complex (see Chapter 2 on Genetics) and *Pacifastacus leniusculus* (Dana 1852) currently contains three subspecies recognised in North America: *Pacifastacus l. leniusculus*, *P. l. trowbridgii* and *P. l. klamathensis* (Miller 1960). Among the Cambaridae, *Orconectes virilis* also corresponds to a cryptic species complex with several species known from North America (Mathews et al. 2008, Mathews and Warren 2008, Filipova et al. 2009); another example includes the White River Crayfish, *Procambarus zonangulus* (Hobbs and Hobbs 1990), which was part of a species complex formerly known as *P. acutus acutus* (Girard 1852), which still includes at least 3 species of crayfishes in the eastern United States (Hobbs and Hobbs 1990, Huner and Barr 1991, Huner 2002, Taylor et al. 1996).

The number of described crayfish species is subject to change, and since De Grave et al. (2009), numerous new species have been described every year. Feldmann et al. (2011) added a new monotypic fossil genus *Aenigmastacus* with *A. crandalli* designated as the type species. The most recent estimates suggest there are currently 180 extant described Parastacid species (Fetzner, personal communication). Some examples of the species described since 2007 include:

- 2007: *Procambarus maya* was described from a salt marsh 1 km from the coast, within Sian Ka'an Nature Reserve, Municipio de Felipe Carrillo Puerto, Quintana Roo, Mexico by Alvarez et al. (2007).
- 2008: Lukhaup and Herbert (2008) described a new species of crayfish, *Cherax (Cherax) peknyi*, from the Fly River drainage, in the western province region of Papua New Guinea. This species differs from all others in its subgenus by the shape of the rostrum, and chelae, and in coloration.
- 2008: In the United States, a new species, *Orconectes taylori* (common name: Crescent Crayfish), was described by Schuster (2008) from tributaries of the North Fork Obion River in western Tennessee. It occurs in small to medium size sandy bottom streams, and is found in leaf litter and woody debris along the banks. It belongs to the subgenus *Trisellescens* Bouchard and Bouchard, 1995 and can be distinguished from other species in the group by a combination of the length and curvature of the central projection of the form I gonopod, carina on the rostrum, appressed tubercles on the margin of the palm of the chela, and width of the areola.
- from 2008 to 2011, Johnson described two new crayfishes from southeastern Texas, in the United States, within the genus *Fallicambarus*, four new crayfishes from the genus *Orconectes*, a new burrowing crayfish from eastern Texas, *Fallicambarus (F.) wallsi*, and *Procambarus (Ortmannicus) luxus* from the southern part of the state (Johnson 2008, 2010, 2011a, 2011b).
- 2009: Thoma and Stocker discovered a new species of crayfish named *Orconectes (Procericambarus) raymondi* from south-central Ohio, North America. Of the recognized members of the subgenus, it is morphologically most similar to *Orconectes (P.) putnami*, found in Kentucky and Tennessee and is easily separated from it by the presence of a strong rostral carina. It is distinguished from other recognized members of the subgenus by the rostral carina, mandible structure, and a first form male gonopod having a central projection approximately 50% of total gonopod length.
- 2010: Cooper and Price described *Cambarus (Puncticambarus) aldermanorum*, a new species of crayfish that appears to be endemic to the lower Catawba and Saluda river basins in the Piedmont Plateau of South Carolina in the United States. Morphologically, it is most similar to *C. (P.) hobbsorum* and *C. (P.) hystricosus*. It differs from both species in having a long, narrow, lanceolate rostrum, and in lacking a proximomesial tubercle or spine on the ventral surface of the carpus. It further differs from *C. (P.) hobbsorum* in having hepatic spines, in other aspects of spination, and in having a broader areola. *Cambarus (P.) spicatus* of the Broad River basin is another very spinose crayfish that bears some resemblances to *C. (P.) aldermanorum*, from which it differs in having a broader rostrum with a very short acumen that is delineated at its base by marginal spines or tubercles, and a much broader, more punctate areola. Adams et al. (2010) also gave the description of three new crayfish species in the Tennessee River basin in Mississippi, and the first drainage-specific distributional information in the state for a fourth. The species—*Cambarus girardianus*, *Cambarus rusticiformis*, *Orconectes spinosus* and *Orconectes wrighti*—are also known from Alabama. They discussed taxonomic issues involving *C. girardianus*

Table 2. MS Crayfish Database (Adams and Henderson 2009). Species in bold are new state records and those with asterisks have new distributional information; ^Undescribed species.

Species	Subgenus	Authority	Common name
Cambarus diogenes	*Lacunicambarus*	Girard	Devil Crawfish
C. girardianus	**Hiaticambarus**	Faxon	Tanback Crayfish
C. ludovicianus	*Lacunicambarus*	Faxon	Painted Devil Crayfish
C. rusticiformis	**Erebicambarus**	Rhoades	Depression Crayfish
C. striatus	*Depressicambarus*	Hay	Ambiguous Crayfish
Orconectes compressus	*Gremicambarus*	(Faxon)	Slender Crayfish
O. etnieri	*Trisellescens*	Bouchard and Bouchard	Ets Crayfish
Orconectes sp. *A*	*Trisellescens*		
O. spinosus	**Procericambarus**	Bundy	Coosa River Spiny Crayfish
*O. wrighti**	*Faxonius*	Hobbs	Hardin Crayfish
Procambarus ablusus	*Pennides*	Penn	Hatchie River Crayfish
P. acutus	*Ortmannicus*	(Girard)	White River Crawfish
*P. viaeviridis**	*Ortmannicus*	(Faxon)	Vernal Crayfish

and *O. spinosus*. Based on their distributions in neighboring states, they think that several other species may occur in the Mississippi portion of the basin. According to Adams et al. (2010), the Table 2 below summarizes the crayfish species known from the Tennessee River basin in Mississippi.

- 2011. Taylor and Schuster gave a description of a new crayfish of the genus *Barbicambarus* Hobbs, 1969 discovered in only two locations of the Tennessee River drainage using both morphological characters and molecular data. The new species differs from the type species in possessing a median carina, less dense setae on the antennae, a less angular central projection, a spine at the dorsodistal margin of the merus of the cheliped, and a high level of divergence in the COI gene region. They gave the name *Barbicambarus simmonsi* for this giant crayfish, which is considered native to Tennessee.
- 2011 *Cambarus (Puncticambarus) smilax*, a new species of crayfish (Cambaridae), was discovered by Loughman et al. (2011) in the Greenbrier River of West Virginia. The authors estimated there are approximately 20 to 30 undescribed species of crayfish in the state. The new species is morphologically most similar to *C. (P.) robustus*, from which it can be distinguished by a combination of the following characters: adult palm length comprising 73–76% of palm width as opposed to 63–70% in *C. (P.) robustus*; ventral surface of chela of cheliped with 0–2 subpalmar tubercles compared to 3–6 subpalmar tubercles in *C. (P.) robustus*; lack of tubercles on the dorsal surface of chela; longer, more tapering, less rectangular rostrum (47–52% rostrum width/length ratio) compared to *C. (P.) robustus* shorter, less tapering rectangular rostrum (54–63% rostrum width/length ratio); and the central projection of the form-I male gonopod curved ≤ 90 degrees to the shaft.
- 2011 In Australia, *Euastacus morgani* sp. n. was described by Coughran and McCormack (2011) from a highland, rainforest site in Bindarri National Park, in eastern New South Wales. *Euastacus morgani* is found living sympatrically with two more common species, *Euastacus dangadi* Morgan, 1997 and *Euastacus neohirsutus* Riek, 1956. Systematically, the species belongs in the 'simplex' complex of the genus that includes *Euastacus simplex* Riek, 1956, *Euastacus clarkae* Morgan, 1997, *Euastacus maccai* McCormack and Coughran 2008 and *E. morgani* Coughran and McCormack 2011. This new species differs from its nearest congenor, *Euastacus simplex* Riek 1956, in having three mesial carpal spines. We give here the example of one type of the keys the authors give each time they discover a new species in order to specify morphological and relevant characters. Coughran and McCormack (2011) gave a key to the 'simplex' complex of the genus *Euastacus* detailed in Box 2.

Box 2. Example identification key to the *Euastacus simplex* complex.

1	Chelae with elongate, tapered fingers. Apart from one or two large molars, development of teeth on cutting edges of chelae distinctly reduced. Gape between fingers distinctly broad and lanceolate in shape	*Euastacus maccai* McCormack and Coughran, 2008
1'	Chelae with stout fingers, without distinctive gape between fingers. Lesser cutting edge teeth of moderate size	2
2	Cheliped with 3 mesial carpal spines	*Euastacus morgani* sp. n.
2'	Cheliped with 2 mesial carpal spines	3
3	Dorsal apical propodal spines present. Suborbital spine medium to large	*Euastacus clarkae* Morgan, 1997
3'	Dorsal apical propodal spines absent. Suborbital spine barely discernible to small	*Euastacus simplex* Riek, 1956

- 2012 Rudolph and Crandall discovered a new species of burrowing crayfish, *Virilastacus jarai* (Parastacidae) in the south central part of Chile. This is the fourth species of *Virilastacus*, a genus endemic to Chile, to be described to date. Features that distinguish *V. jarai* from its congeneric species are: (1) rostral carina, short, slightly prominent and widely separated from the orbital margin; (2) pilous dorsal side of the opposable margin of the P1 propodus, as is the basal zone of the ventral side, 11 to 22 teeth on its opposable margin; (3) dorsal surface of the P1 dactylus close to the opposable border, hirsute; external distal border of the ischiopodite of the third maxilliped with a large extension that ends in the form of a right angle; (4) precervical cephalothorax with dorsal ridges absent, or with two or four; (5) areola, wide and extended; (6) telson with small, but sharp, lateral spines. Morphologically, this new species is similar to *Virilastacus araucanius* and *V. retamali*, with whom it shares 14 of the 27 morphological attributes analyzed, nine of which are common to these three species. These same attributes (13 of 14) differentiate *V. jarai* from *V. rucapihuelensis*, with whom it only shares seven morphological traits. The morphological similarity of *V. jarai* with *V. araucanius* and *V. retamali* contrasts with the degree of genetic divergence that exists between these species.
- 2013 Furse et al. discovered two new species of the crayfish genus *Euastacus*, described from the Gondwana Rainforests on the Queensland—New South Wales border region of Australia—*Euastacus binzayedi* and *Euastacus angustus*. Both are small, poorly spinose species that are broadly similar in appearance and coloration to *Euastacus dalagarbe* Coughran, from the same region. Both species can be readily distinguished from *E. dalagarbe*; *Euastacus binzayedi* by the numerous bumps and protrusions on the dorsal and ventral surfaces of its chelae, and *Euastacus angustus* by its unusual, laterally compressed body shape, and the large ventromesial carpal spine. Cytochrome oxidase I divergence estimates from the most closely related species were high for both *Euastacus binzayedi* (4.8%), and *Euastacus angustus* (8.7%). Morphologically, both of these new species belong in a clearly defined, poorly spinose group, and both appear to be exceptionally rare, each known from a single locality. That same year, Loughman et al. (2013a) discovered *Cambarus (Puncticambarus) theepiensis*, a stream-dwelling crayfish that appears to be endemic to the junction of the Cumberland Mountains with the Appalachian Plateau in West Virginia and Kentucky. The new species is morphologically most similar to *Cambarus robustus* and *Cambarus sciotensis*. Moreover, Loughman et al. (2013b) described the new species *Cambarus (Cambarus) hatfieldi*, a stream-dwelling crayfish that appears to be endemic to the Tug Fork River system of West Virginia, Virginia, and Kentucky. The new species is morphologically most similar to *Cambarus sciotensis* and *Cambarus angularis*.
- 2014 Simon and McMurray (2014a) described *Orconectes alluvius* (detrital crayfish) from southwestern Indiana while Simon and Morris (2014b) described *Cambarus erythrodactylus* (warpaint mudbug) from Alabama and Mississippi, which was formerly part of the *Cambarus diogenes* complex. In addition, Thoma et al. (2014) described *Cambarus callainus* (Big Sandy crayfish), from the Big Sandy River drainage system in Kentucky, Virginia and West Virginia. These populations were previously considered to be *Cambarus veteranus* (now restricted to the Guyandotte River drainage), but both morphological and genetic data suggests these are separate taxa, both of which are being considered for listing under the U.S. Endangered Species Act.

- 2015 Several new *Cherax* species have been described from West Papua, Indonesia, including *Cherax pulcher* (Lukhaup 2015) and *Cherax gherardii* (Patoka et al. 2015), both with very restricted distributions. In the United States, Thoma and Fetzner (2015) described *Cambarus magerae* from Big Stone Gap in Virginia while Loughman et al. (2015) described *Cambarus pauleyi* from two adjacent counties in West Virginia. And finally, Pedraza-Lara and Doadrio (2015) have described Cambarellus zacapuensis, which is known from only a single locality in central México.

Insights on the families and genera

Taylor et al. (2007) provided a list of all crayfish (families Astacidae and Cambaridae) in the United States and Canada. The two families occur natively in North America and it is here that crayfish reach their highest level of diversity. In Europe, indigenous crayfish species (ICS) are only represented by members of the Astacidae, with the cambarids being non-indigenous crayfish species (NICS) recently introduced into Europe. As previously explained, the family Parastacidae contains all freshwater crayfish found naturally occurring in the southern hemisphere.

Family Astacidae Latreille, 1802

The Family Astacidae (three genera, 16 species according to Hobbs, 1989) are distributed both in Europe and west of the Rocky Mountains in the northwestern United States and northward into southern British Columbia, Canada.

Astacidae in Europe

The *Atlas of Crayfish in Europe* (Souty-Grosset et al. 2006) shows that only five crayfish species, all belonging to the family Astacidae, are native to Europe, according to the taxonomy adopted by Holdich (2002), with three from the genus *Astacus* and two from *Austropotamobius*. For *Astacus*, these include the noble crayfish, *Astacus astacus* (Linnaeus 1758), the narrow-clawed crayfish, *A. leptodactylus* (Eschscholtz 1823), and the thick-clawed crayfish, *A. pachypus* (Rathke 1837). The genus *Austropotamobius* includes the white-clawed crayfish, *A. pallipes*, and the stone crayfish, *A. torrentium* (Schrank 1803). The present distributions of these ICS are the result of both natural events that occurred during the Pleistocene up until recent historical times, and translocations attributable to human activities. Identification keys are given in order to identify ICS and NICS in Europe; Box 3 below is extracted from the crayfish guide of Romania (Pârvulescu 2010) illustrating how to distinguish *Astacus astacus* from *Astacus leptodactylus*, *Austropotamobius torrentium* and the introduced North American species, *Orconectes limosus* (Rafinesque 1817), the spinycheek crayfish.

Recent molecular studies by Filipova et al. (2011) were aimed at verifying the taxonomic status of European crayfish through DNA barcoding. They compared sequences obtained from the cytochrome *c* oxidase subunit I (COI) gene fragment from sampled American populations with populations now present in European waters. They demonstrated that DNA barcoding is useful for the rapid and accurate identification of exotic crayfish in Europe, and also provided insights into overall variation within these species.

The genus *Astacus*

The genus *Astacus* must have been formed during the Paleogene in response to a changing regime of inland waters. The genus dispersed widely in Europe during Neogene times, but the Pliocene cooling of the climate is believed to have divided a single species (*A. astacus*) of the genus into three species or subspecies: *A. a. colchicus* is the most archaic of them, having survived in Western Transcaucasia; *A. a. balcanicus* must have emerged in the Balkan peninsula, while *A. astacus*, the most advanced species, occupied all the northern parts of Europe.

Box 3. Key to crayfish in Europe (courtesy of Lucian Pârvulescu).

1. Two pairs of post-orbital ridges: 3

    3a. First post-orbital ridges more prominent and ending atypically with a spine, the second post-orbital ridges are blunt. Strong rostrum with parallel edges more or less sharp apex. Sides of the cephalothorax and of the cervical groove with spines: *Astacus astacus*

    3b. Both post-orbital ridges visible and with one apical spine each. Strong rostrum with parallel edges, sharp apex. On the sides of the cephalothorax and of the cervical groove 1–3 prominent spines and several tubercles or small spines: *Astacus leptodactylus*

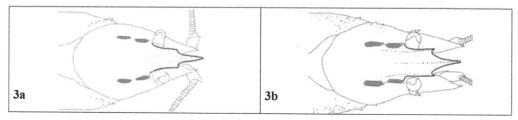

2. One pair of post-orbital ridges: 4

    4a. The post-orbital ridges as a crease, triangular shaped rostrum with a less obvious apex and without median carina: *Austropotamobius torrentium*

    4b. Post-orbital ridges prominent, ending atypically in obvious spine, rostrum with parallel edges and sharp apex. Many hepatic spines on the sides of the cephalothorax: *Orconectes limosus*

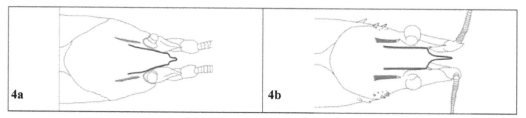

Astacus astacus (Linnaeus 1758) (noble crayfish):

In the genus *Astacus*, the noble crayfish, *Astacus astacus* (Linnaeus 1758), is widely distributed in Europe, from France in the southwest to Russia in the east, and from Italy, Albania and Greece in the south to Scandinavia in the north (Cukerzis 1988, Skurdal and Taugbøl 2002) and has been stocked into numerous new localities, especially in marginal areas, so that this crayfish now has a larger distribution than it originally had, and is currently found in 39 territories across most of northern Europe. Although its range was much greater before the onset of the crayfish plague, a fungal infection which is responsible for the widespread reduction of the number of crayfish populations throughout Europe (Holdich et al. 2009). The species is widely harvested, and many countries have national and federal regulations governing trapping seasons and size restrictions (Skurdal and Taugbøl 2002). The most abundant populations exist in Nordic and Baltic countries (e.g., it is the dominant crayfish in Latvian waters, having been found in 220 out of 258 crayfish localities) (Taugbøl et al. 2004, Arens and Taugbøl 2005). *Astacus astacus* was introduced into Cyprus from Denmark in the 1970s, and although the original stock has disappeared, it apparently occurs by the Lefkara dam (Stephanou 1987).

According to the recent IUCN assessment (Edsman et al. 2010), although the noble crayfish is a widespread species, it has undergone significant declines in population numbers due to unfavourable interactions with non-indigenous species, crayfish plague, habitat loss and over-harvesting. Estimates of the rate of decline in Sweden, Finland and Norway are as high as 78%, ~ 20% and 61%, respectively, over a 3 generation period. Similar rates of decline are being seen within a number of other countries. Globally, this species is estimated to be declining at a rate of 50–70%, however, in some parts of its range, numbers are stable and there have been some successful re-stocking programs, so the true rate of decline may in fact be slightly lower.

Astacus leptodactylus Eschscholtz, 1823 (narrow-clawed crayfish)—species complex:
According to Holdich et al. (2009), the narrow-clawed crayfish, *Astacus leptodactylus* Eschscholtz, 1823, has a southeastern European range and is indigenous to Russia and the Ponto-Caspian area. It was originally distributed over an area corresponding more or less to Turkey, the Ukraine, Turkmenistan and southwestern Russia, but is also found in Iran, Kazakstan, Georgia, Belarus, Bulgaria, Romania, Hungary and Slovakia. The Caspian Sea, the Black Sea and the lower and middle Danube are in its original distribution area, along with the lower reaches of the rivers Don, Dniester and Volga and their tributaries (Köksal 1988, Holdich 2002). However, this crayfish has been widely introduced into many countries as a replacement for noble crayfish populations lost due to crayfish plague. The species is currently found in 32 territories across most of Europe, with the exception of Scandinavia and the Baltic Peninsula, although its range was probably greater before the onset of the crayfish plague (Holdich et al. 2009). The systematics of this species complex is in a state of flux, with populations from Western and Central Europe referred to as *A. leptodactylus*, whilst in Eastern Europe, a number of species are recognized within a separate genus *Pontastacus* (Starobogatov 1995, Śmietana et al. 2006). *Astacus leptodactylus* is considered a species complex (see IUCN assessment: Gherardi and Souty-Grosset 2010a). In the 1950s, this species was believed to belong to the subgenus *Astacus* (*Pontastacus*), and included *A. (P.) pachypus, A. (P.) pyzlowi* and *A. (P.) kessleri*. The following four subspecies were assigned to *A. (P.) leptodactylus: eichwaldi, cubanicus, salinus,* and *leptodactylus*. Karaman (1962, 1963), however, did not acknowledge *A. (P.) cubanicus* as a subspecies. In the 1970s, *Pontastacus* was raised to the generic level. In the 1980s, Brodskij made a number of revisions within *Pontastacus*, but the number of taxa varied with each paper. In the mid-1990s, Starobogatov (1995) split *Pontastacus* into two genera including *Pontastacus*, which contained *P. angulosus* (Rathke, 1837), *P. cubanicus* (Birstein and Winogradow 1934), *P. danubialis* (Brodskij, 1967), *P. eichwaldi* (Bott, 1950), *P. intermedius* (Bott, 1950), *P. kessleri* (Schimkewitsch, 1886), *P. pyzlowi* (Skorikov, 1911) and *P. salinus* (Nordmann, 1942). The second genus, *Caspiastacus*, contained two species (*C. pachypus* Rathke, 1837 and *C. daucinus* (Brodsky, 1981)). However, there is a great deal of criticism over recent taxonomic revisions made by Ukranian and Russian taxonomists, as they appear to be based on little evidence. The most abundant populations are found in Eastern Europe and the Near East. Large commercial stocks of *A. leptodactylus* exist in Belarus (Alekhnovich 2006). This species is prone to the effects of crayfish plague and in recent times there have been reports of introduced populations being affected in England (Environment Agency 2007), and Switzerland (Hefti and Stucki 2006). Simić et al. (2008) report that although *A. leptodactylus* is spreading in some regions of Serbia, in others, their numbers are being reduced by the presence of *O. limosus*. Although crayfish plague devastated populations in Turkey in the 1980s, resulting in very low harvests (down from 5000 tonnes in 1984 to 320 tonnes in 1991), a partial recovery has been recorded in the 2000s (i.e., 2317 tonnes in 2004), and previously plague-infected lakes are productive again (Harlioğlu 2008). Perdikaris et al. (2007) have also reported *A. leptodactylus* from the River Evros in Greece, where it was probably deliberately introduced. Tertyshny and Panchishny (2009) have reported large-scale mortalities amongst the stocks from aquaculture of *A. leptodactylus* in the Ukraine, which are partly attributable to disease.

Astacus pachypus Rathke, 1837 (thick-clawed crayfish), occurs in the Caspian Sea and in the brackish waters of the estuaries of the Dniester and the Bug, and is recorded from two European countries, as well as some western Asian countries. At the present time, this species is indigenous to Russia, Ukraine, Azerbaijan, Turkmenistan and Kazakhstan (Machino and Holdich 2006, Holdich et al. 2009). In Azerbaijan, it is known from the coastal waters off Baku (Holdich 2002); in Kazakhastan, this species is known from the coastal waters of the Caspian Sea (Sokolsky et al. 1999); in Turkmenistan, it is known from coastal waters (Cherkashina 1999a). In the Ukraine, this species is known from the Dneiper-Bug Lagoon of the Azov-Black Sea Basin (Cherkashina 1999b). In the 1950s, this species was thought to belong to a different subgenus from *Astacus* (i.e., was thought to belong to *A.* (*Pontastacus*) *pachypus*, along with the species *pylzowi, kessleri* and *leptodactylus*). The subgenus *Pontastacus* has since been elevated to generic level, and subsequently to a new genus (*Caspiastacus*) in 1995 by Starobogatov (Souty-Grosset et al. 2006). The taxonomy of Eurasian crayfish is questionable, as there appears to be little validity for revisions to the existing taxonomy. There is considerable morphological variation across the Eurasian species, and it is thought that there is hybridization between *A. pachypus* and *A. leptodactylus*. The only way to truly delineate the precise range of each species is to conduct comparative molecular genetic studies (see IUCN assessment by Gherardi and Souty-Grosset 2010b).

The genus *Austropotamobius*

Among the European crayfish, the genus *Austropotamobius* Skorikov, 1907, is widely distributed throughout west and central Europe, from the Iberian Peninsula in the west and the British Isles in the north to Italy and the Balkan Peninsula in the south and east (Holdich 2002). It comprises two species, the circum-alpine stone crayfish, *Austropotamobius torrentium* (Schrank, 1803), and the white-clawed crayfish, *Austropotamobius pallipes* (Lereboullet, 1858). However, the situation within each species is more complicated, especially for the white-clawed crayfish (*A. pallipes*). Its taxonomy is still under debate in spite of well-advanced research, particularly in genetics (see details in Chapter 2 of this volume).

According to molecular data, the historical events leading to the main splits in the genus took place during the second half of the Miocene. At that time, the landmass of the Adriatic microplate separated the Parathetys from the paleo-Mediterranean sea (Dercourt et al. 1986). The resulting two major drainages thus might have formed the basis for the split of the ancestral *Austropotamobius* into the *pallipes* and *torrentium* lineages (Souty-Grosset et al. 2006).

Austropotamobius torrentium (Schrank, 1803) (stone crayfish): three subspecies are known: *Austropotamobius torrentium torrentium, A. t. danubicus,* and *A. t. macedonicus.* This species is mainly confined to Central Europe where it is known from France and western Germany in the west of its range, to Turkey in the east (Füreder et al. 2006). Indeed, the species is currently known from 20 countries, but with a restricted range in Central and Southeastern Europe, where it was most likely more widely distributed in the past. It is the most southerly of the European ICS, extending as far as Bulgaria and Romania, and has recently been found in Turkey (Harlioğlu 2007). Perdikaris et al. (2007) have confirmed the presence of *A. torrentium* in the River Evros in Greece for the first time in 112 years. While this species is relatively widespread across Europe, it is undergoing significant declines throughout much of its range (IUCN assessment, Füreder et al. 2010a).

Austropotamobius pallipes (Lereboullet, 1858) (white-clawed crayfish—species complex) has a wide distribution throughout Europe. It was previously thought that the western limit of the species range was in Portugal (although it is now thought to be extinct there), but is now northwestern Spain. Montenegro is the eastern limit, whilst Spain and Scotland are the southern and northern limits, respectively. Its distribution is restricted in Austria, Corsica, Germany, Lichtenstein and Montenegro (Souty-Grosset et al. 2006). This species, currently found in 17 countries, has a narrower range than those of *A. astacus* and *A. leptodactylus*, and is more centred in Western, Central and Southern Europe. Machino et al. (2004) have catalogued the many introductions of *A. pallipes* that have been made throughout Europe. There is a wide genetic diversity within the second taxon, so that some authors have suggested dividing it into two phylogenetic species: *Austropotamobius pallipes* and *A. italicus* (Grandjean et al. 2000, Zaccara et al. 2004, Fratini et al. 2005). If the division of the species into *A. italicus* (Faxon, 1814) and *A. pallipes* is accepted, then the distribution map will have to be redrawn (Fratini et al. 2005, Bertocchi et al. 2008). In northern Europe, these two species can be clearly defined at a molecular level, but farther south, several subspecies have been recognized. Allopatric speciation of the two taxa led to *A. pallipes* being distributed in west-central Europe (France, Great Britain and Northern Italy) and *A. italicus* in Switzerland, Austria, Italy, the Balkans and Spain (Grandjean et al. 2002a,b). Phylogeographic studies confirmed the presence in Italy of both *A. pallipes*, confined to North-Western Italy, and *A. italicus*, distributed across the peninsula (Fratini et al. 2005). However, research is still in a state of flux, so that the general consensus is to define the taxon as a species complex, with a number of distinct genetic strains related to their recent history, but which are not distinguishable morphologically (Souty-Grosset et al. 2006).

Some consider *Austropotamobius pallipes* as a species complex comprised of two genetically distinct species; *A. pallipes* and an Italian species for which the name is being discussed. The Italian species is thought to be comprised of a number of subspecies, though this depends on the author. Both the Italian form and *A. pallipes* can be found in Spain, France, Italy and Switzerland. It is also suggested that there are two subspecies of *A. pallipes*: *A. pallipes pallipes*, which exists in France, the British Isles, Spain, Switzerland, and Germany, and *A. p.* subsp. which is known from Liguria in Italy and the Alpes Maritimes region of France. There still exists some debate as to whether the Italian form should be raised to species level, though recent genetic work (Grandjean et al. 2000a, Fratini et al. 2005, Bertocchi et al. 2008) would support a separate species, *Austropotamobius italicus* with 4 subspecies. The White-clawed crayfish has been assessed as Endangered under criterion A2ce. In the last ten years, this species is suspected to have undergone a decline of somewhere between 50–80% based on presence/absence data available for England, France and Italy (IUCN assessment, Füreder et al. 2010b).

Astacidae in North and Central America

The family Astacidae and the particular case of the genus *Pacifastacus* (example of *Pacifastacus leniusculus*, ICS in North America and NICS in Europe).

Pacifastacus leniusculus Dana 1852 (signal crayfish): three subspecies have been historically recognised in North America including: *Pacifastacus l. leniusculus, P. l. trowbridgii* and *P. l. klamathensis* (Miller 1960). These sub-species are difficult to distinguish because both their morphological characters and their distribution range overlap. Sonntag (2006) examined mtDNA variation in signal crayfish populations from the Klamath River Basin in California and Oregon in North America and was able to distinguish the three subspecies using this DNA marker. In Europe, the first studies were based on the RFLP analysis of mtDNA (Grandjean and Souty-Grosset 1997) suggested that the high variation found in three French signal crayfish populations could reflect the presence of more subspecies in Europe. However, the recent study by Filipova (2012) used an mtDNA analysis of signal crayfish sampled from 17 European countries and showed that only the lineage corresponding to *P. l. leniusculus* seems to have been introduced into Europe. A recent study by Larson et al. (2012) found substantial cryptic diversity across the range of the species, with three main groups that were highly distinct from *P. leniusculus*, each being found in discrete geographic regions. In North America, *Pacifastacus leniusculus leniusculus* is distributed in southern British Columbia in Canada, and in California, Idaho, Oregon, and Washington in the USA. *Pacifastacus leniusculus klamathensis* is distributed in British Columbia in Canada, Idaho and south to central California in the USA. *Pacifastacus leniusculus trowbridgii* ranges from British Columbia in Canada to California, Idaho, Oregon and Washington in the USA, and has been introduced into California and Nevada in the USA, and also introduced into Japan. Furthermore, this subspecies is also known to occur in Greece (Koutrakis et al. 2007). *Pacifastacus leniusculus* has been introduced into many countries throughout Europe, as well as to California, Nevada and Utah in the USA. This species was introduced during the 1970s and 1980s, is widely cultivated, and is established in the wild, from where it is harvested (Harlioğlu and Holdich 2001).

Family Cambaridae

The family Cambaridae is distributed in North America east of the Rocky Mountains, from southern Canada in the north to Central America in the south, and with one genus (*Cambaroides*) being disjunct and restricted to eastern Asia. This family contains the most described freshwater crayfish species, with 444 species currently distributed among 12 genera (see Table 1).

In the family Cambaridae, there are three highly specious genera that account for roughly 86% of the known species and these include: *Procambarus* (178 species, 40.1%), *Cambarus* (106, 23.9%) and *Orconectes* (97, 21.9%). The remaining nine genera contain fewer species and include *Barbicambarus* (2 taxa), *Bouchardina* (1), *Cambarellus* (17), *Cambaroides* (7), *Distocambarus* (5), *Fallicambarus* (19), *Faxonella* (4), *Hobbseus* (7) and *Troglocambarus* (1).

North and Central America

Members of the family Cambaridae occur natively in North America and it is here that crayfish reach their highest level of diversity. Reasons for this high level of biodiversity include isolation from glacial advances and geological and topographic diversity. Approximately 68% (444 species and subspecies) of the world's known species occur in North America (Taylor 2002), with the overwhelming majority of this continent's crayfish fauna (98%) assigned to the family Cambaridae (the remainder are from the family Astacidae, see above). With over two-thirds of its species endemic to the southeastern United States, the distribution of crayfish diversity in North America closely follows that observed in other freshwater aquatic taxa, such as fishes (Warren and Burr 1994) and mussels (Williams et al. 1993).

The state of Alabama in the USA is home to one of the most diverse crayfish faunas in the world, with a current count of 89 species, all found in an area of roughly 135,000 km^2 (Schuster et al. 2008). These species are from seven different genera, of which only the virile crayfish, *Orconectes virilis*, is considered to be non-native. Eleven of these species are endemic to the state, and thus are found nowhere else. The state of Alabama lists almost one third (28) of these as species of greatest conservation need (Wildlife and Freshwater Fisheries Division 2005). Several other states from the southeastern U.S. have similarly high levels of diversity and these include: Tennessee (84 species), Georgia (72), Mississippi (63), Arkansas (60), Florida (56) and Kentucky (52). Unfortunately, there is still very little known about many of these species, especially the limits of their distributions and detailed life histories (Moore et al. 2013).

The number of crayfish species described or reported from North America declines from south to north. For example, in Canada only 11 species in total are known (Hamr 1998). Most (nine crayfish species) are found in Ontario, including five species of *Orconectes*, two *Cambarus* and one *Fallicambarus*, plus the northern clear water crayfish, *Orconectes propinquus* (Girard, 1852) which is dispersing up the St. Lawrence River from Quebec. Ontario and Quebec (eight species) have the highest species richness of crayfishes in Canada. Of the 11 Canadian crayfishes, the only two that do not occur in Ontario are the spinycheek crayfish (*Orconectes limosus*), which is found in Quebec, New Brunswick and Nova Scotia, and the signal crayfish, which is found in British Columbia. Two provinces (Newfoundland and Labrador and Prince Edward Island) and the three Canadian territories lack crayfish faunas due to their extreme northern latitudes.

The genus *Procambarus*

The genus *Procambarus* contains the largest number of species of any genus of freshwater crayfish worldwide. Currently, there are 170 described species and 16 taxa listed as subspecies (Fetzner 2005). They are distinguished from other genera by having a male gonopod with four terminal elements. The native range of the genus is in North America, ranging along the eastern seaboard and the coastal regions of the Gulf of Mexico, up the Mississippi River drainage as far as southern Wisconsin, and south through Texas and Mexico to Honduras (Hobbs 1972). The genus was divided into 16 different subgenera by Hobbs (1972) and these include: *Acucauda* (1 species), *Austrocambarus* (24), *Capillicambarus* (3), *Girardiella* (22), *Hagenides* (10), *Leconticambarus* (14), *Lonnbergius* (2), *Mexicambarus* (1), *Ortmannicus* (59), *Paracambarus* (2), *Pennides* (20), *Procambarus* (1), *Scapulicambarus* (6), *Tenuicambarus* (1) and *Villalobosus* (13). The last subgenus, *Remoticambarus*, which contained the monotypic cave-adapted species, *Procambarus* (*R.*) *pecki*, was recently found to be most closely related to members of the *Cambarus* subgenus *Aviticambarus*, and is now considered a member of that group (Buhay and Crandall 2009).

At least one species, the red swamp crayfish (*Procambarus clarkii*, Girard 1852), now has what could be considered a worldwide distribution, after being introduced into many countries either intentionally via the aquaculture industry, or accidently via the pet trade. In many areas where it has been introduced, it has had severe adverse impacts on native crayfish, either by direct competition or through the spread of the crayfish plague, and has been implicated in the declines of other native species of aquatic flora and fauna.

Mexico and Central America together comprise another "hot spot", which contains 55 native cambarid species from two genera (*Cambarellus* and *Procambarus*), whereas only four species have been described from Guatemala, two from Belize, one in Costa Rica (probably introduced) and the Dominican Republic (also introduced) and four from Cuba. According to Mejía-Ortíz et al. (2003), the Mexican crayfish fauna is restricted to the two previously mentioned native genera, of which only members of the genus *Procambarus* have been recorded from underground habitats (Reddell 1981, Hobbs 1989). In Mexico, *Procambarus* is represented by members from nine of the 16 recognized subgenera, with *Austrocambarus* having the greatest representation with 16 species and subspecies (Villalobos et al. 1993, Rojas et al. 1999,

2000), and possibly several as yet undescribed species (Allegrucci et al. 1992). In Mexico, 15 species of *Procambarus* have been previously recorded from hypogean waters. Ten of these are stygophiles (those that can be found in caves but that lack the adaptations to cave life), and in most cases they are also known from epigean waters. Only five are considered true stygobites (those being found exclusively in caves and with clear adaptations to cave habitats). These species include *Procambarus* (*Ortmannicus*) *xilitlae* (Hobbs and Grubbs 1982), distributed to the north of the Cordillera, and *Procambarus* (*Austrocambarus*) *rodriguezi* (Hobbs 1943), *Procambarus* (*Austrocambarus*) *oaxacae oaxacae* (Hobbs 1973), *Procambarus* (*Austrocambarus*) *oaxacae reddelli* (Hobbs 1973) and *Procambarus* (*Austrocambarus*) sp. 2 (Allegrucci et al. 1992), all from south of the Cordillera. Mejía-Ortíz et al. (2003) described a new stygobitic species of *Procambarus* (*Austrocambarus*) inhabiting Gabriel Cave near Buenos Aires, Oaxaca, Mexico and discuss its affinities with other members of the subgenus. They also reviewed the distribution of stygobitic and stygophilic species of *Procambarus* in Mexico.

Several recent attempts have been made, not only by Mexican scientists but by many other organizations, to identify all Mexican crayfish species. For example, *Procambarus regiomontanus* was only found in the state of Nuevo Leon and this endemic species is now endangered due to the introduction of *Procambarus clarkii* into streams in this region.

The genus *Cambarus*

The genus *Cambarus* is the second largest crayfish genus in the Northern Hemisphere and it currently contains 12 subgenera (*Aviticambarus* (6 species), *Cambarus* (11), *Depressicambarus* (17), *Erebicambarus* (5), *Exilicambarus* (1), *Glareocola* (3), *Hiaticambarus* (11), *Jugicambarus* (26), *Lacunicambarus* (4), *Puncticambarus* (23), *Tubericambarus* (4), *Veticambarus* (1)) and 111 species. Members of this genus can be distinguished by the presence of two terminal elements on the male form I gonopod that are curved at an angle of roughly 90° from the main axis of the appendage. *Cambarus* ranges from the coastal region of New Brunswick, Canada, south to the Florida panhandle, west to Texas, and northward to Minnesota and southern Ontario, Canada (Hobbs 1969). The genus has its center of diversity in the Southern Appalachian Mountains of the eastern United States (Hobbs 1969).

Members of the subgenus *Jugicambarus* are a variable group, with some forms occupying diverse habitats and ecological niches, such as rivers and streams, lakes, burrows and caves. The burrowing crayfish, such as *C.* (*J.*) *dubius*, *C.* (*J.*) *monongalensis* and *C.* (*J.*) *carolinus* (among others), can have quite striking color variations, ranging from deep red, orange, and royal blue, and multiple combinations thereof. Most of these species are primary burrowers (Hobbs 1969, 1989), and spend the majority of their lives underground in the burrows they construct.

Another large subgenus, *Puncticambarus*, contains large crayfish (> 15 cm) that mostly inhabit bigger river systems. Species such as *C.* (*P.*) *robustus*, *C.* (*P.*) *cumberlandensis,* and the highly imperiled *Cambarus* (*P.*) *veteranus* are all part of this group. *Cambarus robustus* has a rather large distribution, ranging from southern Ontario and New York to North Carolina and Tennessee in the south, and then to Illinois in the west. However, Hobbs (1989) and others have considered this to be a large species complex for quite some time, and several new species have already been described from this complex, such as *Cambarus smilax* (Loughman et al. 2011). *Cambarus veteranus* is limited in its distribution to the upper tributaries of the Guyandotte River drainage of West Virginia. This species is quite rare and is being adversely impacted by human land use practices, such as logging operations and mountain top removal mining, which dump large quantities of sediment into rivers, making them uninhabitable for crayfish and other aquatic organisms.

The genus *Orconectes*

The genus *Orconectes* is comprised of 11 subgenera, 93 species and 11 subspecies. Members of the genus also have a male gonopod with two terminal elements, but rather than being curved at 90° like in *Cambarus*, they are usually longer and straight to only slightly curved. The distribution of the genus ranges from the eastern side of the Rocky Mountains to the east coast and from southern Canada southward to the Gulf coast, but is mostly absent from the core southeastern states of Alabama, Georgia, South Carolina and Florida.

The eleven subgenera include the monotypic *Billecambarus* (1 species), *Buannulifictus* (8), *Crockerinus* (16), *Faxonius* (3), *Gremicambarus* (7), *Hespericambarus* (8), *Orconectes* (10), *Procericambarus* (31), *Rhoadesius* (2), monotypic *Tragulicambarus* (1) and *Trisellescens* (10).

One of the most frequently mentioned species, *Orconectes virilis* (Hagen, 1870), commonly called the northern or virile crayfish, and a member of the subgenus *Gremicambarus*, grows on average to 10–12 cm in length, excluding its antennae and large chelipeds. The color of the body and abdomen are brownish-red, dappled with dark brown spots. The chelae, or the palm of the large chelipeds, are wide, flattened and possess a straight dactyl margin. The chelae and legs have a bluish tint with yellow tubercles (Hamr 2002). *O. virilis* has a wide natural range from Alberta to Quebec, Canada, throughout more than half of the United States from Texas to Maine, and Chihuahua, Mexico (Hamr 2002). But in Massachusetts, *O. virilis* is believed to be an invasive species (Hobbs 1989), and is listed as such by the Global Invasive Species Database (http://www.issg.org/database/). Its habitats include rivers, streams, lakes, ponds, and marshes.

Another species from the genus that has received considerable attention, especially as an invasive species is *Orconectes rusticus* (Girard, 1852), or the rusty crayfish. This species is a member of the large subgenus *Procericambarus* and has a native range in parts of Indiana, Michigan and Ohio, but has been widely introduced into other areas of the United States (e.g., Illinois, Wisconsin, Pennsylvania, Maryland, and others), and has caused significant declines or local extinctions of other native crayfish species, and have also impacted other aquatic flora and fauna. These crayfish typically grow larger and are able to outcompete other crayfish for food, shelter and other resources. Introgression also appears to be a common method that this species employs to displace other native crayfish from the habitats it invades (Perry et al. 2001a,b).

The spinycheek crayfish, *Orconectes limosus* (Rafinesque, 1817), a member of the subgenus *Faxonius*, is native to the northeastern states, and has also been introduced into many parts of Europe, where it has dispersed widely from its original sites of introduction, either through natural dispersal or human-aided translocations. The species also carries the crayfish plague, and has adversely impacted many of the native European species.

In Missouri, the long pincered crayfish, *Orconectes longidigitus* (Faxon 1898), is native to the White River drainage, and is one of the largest species in North America. In more recent years, are creational fishery in the state has become increasingly popular. Another species, *Orconectes meeki* (Faxon, 1989) is found in the upper White River drainage of Missouri and Arkansas. It is listed as critically imperiled and is among the rarest crayfish in the state. Additionally, the coldwater crayfish, *Orconectes eupunctus* (Williams, 1952), is also critically imperiled and is restricted to only three river drainages in Missouri and Arkansas. The species is typically associated with cold spring-fed rivers in the region and is being impacted by the recent introduction of another crayfish.

Japan and Southeast Asia

The taxonomy of the Asian cambarid genus *Cambaroides*, known from southeastern Russia, Mongolia, North and South Korea, China and Japan, still remains unresolved (e.g., Starabogatov 1995, Kawai et al. 2003, Braband et al. 2006, Kawai 2012). The taxa from Mongolia, Russia and Japan are considered endangered. Starobogatov (1995) suggested that there were 7 species, but recent studies by Kawai and workers suggest that there may only be four.

Cambaroides japonicus (De Haan, 1841) is the only crayfish native to Japan and is restricted to Hokkaido (Kawai 1996). The invaders, such as *Procambarus clarkii* and *Pacifastacus leniusculus* (Dana, 1852), are now also present. The same native genus is encountered in Korea, with the named species *Cambaroides similis* (Koelbel, 1892) and *C. wladiwostokensis*, and in central China, with two species *Cambaroides koshevonikowi*. Here again, *P. clarkii* is present, and farmed on a large scale (production exceeding 100000 tonnes per year).

Family Parastacidae

The family Parastacidae comprises 14 extant genera (~180 species) of which 10 are found in Australia, New Guinea and New Zealand, three in South America and one in Madagascar (Crandall and Buhay 2008, Toon et al. 2010).

South America

According to the review of Almerão et al. (2015), in South America, the first observations of crayfish were made by von Martens (1869) with the description of *Astacus pilimanus* and *A. brasiliensis*, collected in Porto Alegre and Santa Cruz do Sul (Brazil). Following this work, there were many other works on the taxonomy and systematics of South American crayfish: including those by Huxley (1880), Faxon (1898, 1914), Ortmann (1902), Riek (1969, 1972), Buckup and Rossi (1980), Crandall et al. (2000), and Rudolph and Crandall (2005, 2007). Currently, there are 13 species (aforementioned) and for a few, some remarks are necessary. *Parastacus saffordi* was described by Faxon (1898) based on the examination of one individual collected in Montevideo (Uruguay). Buckup and Rossi (1980) identified two specimens of *P. saffordi* from Siderópolis (state of Santa Catarina) and another from the collection of the National Museum of Rio de Janeiro (Cubatão River, state of Santa Catarina). Over a twenty-year period, Buckup performed numerous sampling campaigns in both states (Santa Catarina and Rio Grande do Sul), but never found a specimen with the morphological features originally ascribed to *P. saffordi*. Investigations of the collections from Museo de Historia Natural and from Facultad de Ciencias de la Universidad de la Republica in Montevideo did not reveal the presence of this species. Moreover, *P. saffordi* shows some morphological similarities with *P. varicosus* and thus it is probable that the two species are conspecific. Another taxonomic problem concerns *P. laevigatus* that was described based on individuals deposited in National Museum of Rio de Janeiro (NMRJ) (Buckup and Rossi 1980). Unfortunately, these individuals were lost and never found. In 1990, Buckup collected only one individual of *P. laevigatus* further south (Laguna, state of Santa Catarina) from the type locality (Joinville, state of Santa Catarina). However, it was not possible to confirm it was *P. laevigatus* because the type material deposited in NMRJ was lost. All of these taxonomic uncertainties are being investigated further (Buckup, pers. comm.).

The thirteen South American crayfish species all belong to three genera—*Parastacus* (8 species), *Virilastacus* (4) and *Samastacus* (1)—and they are distributed in Southern Brazil, Uruguay, central to southern Chile and in Southern Argentina (Crandall et al. 2000). This group forms a well-supported clade within the larger Parastacid phylogeny, with supported subclades representing the three genera (Crandall et al. 2000, Sinclair et al. 2004, Toon et al. 2010).

The first collections of freshwater crayfish in South America were made in the 18th century (Buckup 1998). Since then, populations have been identified in several localities in Brazil, Uruguay, Argentina and Chile. It has been postulated that this distribution pattern has been influenced by marine water permanence during the transgressions that occurred from the Cretaceous to the Middle Paleogene periods (Collins et al. 2011). The *Parastacus* group seems to have a disjunct distribution, in which two species (*P. brasiliensis* and *P. laevigatus*) are supposed to be endemic to southern Brazil, two (*P. pugnax* and *P. nicoleti*) are endemic to Chile, and the other four species (*P. saffordi*, *P. varicosus*, *P. defossus* and *P. pilimanus*) are distributed in Southern Brazil and Uruguay (Buckup 1999). The monotypic genus *Samastacus* (type species, *S. spinifrons*) occurs in Chile and Argentina (Rudolph 2010), while all the species of the *Virilastacus* group (*V. araucanius*, *V. retamali*, *V. rucapihuelensis* and *V. jarai*) are endemic to Chile (Rudolph 2010).

Australia

Australia, including Tasmania, holds the world's largest crayfish species, which includes several good examples of flagship species in conservation terms. Within Australia, freshwater crayfish are distributed in all states and territories, but mainly in costal temperate regions of southwestern, southeastern and eastern Australia, and they occupy a variety of different aquatic habitats (Taylor 2002). As summarized by Beatty (2005), two genera (*Astacopsis* and *Parastacoides*) are endemic to Tasmania, but the latter genus was subsequently revised (see Hansen and Richardson 2006). For example, the Tasmanian giant freshwater crayfish, *Astacopsis gouldi* (Clark, 1936), is found in the rivers of northern Tasmania. While two genera (*Engaeus* and *Geocharax*) are recorded in southeastern Australia and Tasmania.

In New South Wales, the Murray River crayfish, *Euastacus armatus* (Von Martens, 1866), is the most commonly known species, and is the world's second largest freshwater crayfish, endemic to the streams and tributaries of the Murray-Darling catchment where it plays a vital role in ecosystem processes and is an important tourist attraction. However, population numbers have been declining due to habitat modification and overfishing (IUCN: Alves et al. 2010).

Several studies addressing the taxonomy of *Cherax* species in Australia have described new species or synonymised others, which have resulted in some confusion and disagreement regarding the status of certain taxa. *Cherax* species from different regions can often be quite different in appearance. There is even evidence that specimens from within the same waterway can look quite different. However, based on past revisions and new descriptions, classifications are based largely after Riek (1969), Austin (1996) and Munasinghe et al. (2004). Several member of the genus *Cherax* have been studied extensively and include the marron (*C. tenuimanus* Smith 1912), the red-claw crayfish (*C. quadricarinatus* Von Martens 1868), the western yabby and also the koonac (*C. preissii*, Erichson, 1846), and the yabby (*C. destructor*, Clark, 1936), that latter which supports a large aquaculture industry and aquarium pet trade. In northeastern New South Wales, two endemic species of *Cherax*, *C. cuspidatus* and *C. leckii*, were recently discovered (Coughran 2006). In Western Australia the hairy marron, *C. tenuimanus*, is endemic to the Margaret River and is under threat of extinction due to its rapid replacement following the introduction of the widespread smooth marron, *C. cainii* (Austin and Bunn 2010). This species is indigenous to southwestern Western Australia between Harvey and Albany (Kent River), and is considered a good biological indicator of water quality for the rivers in the region. Yabbies, *Cherax destructor*, are native to the eastern states of Australia and are considered invasive in Western Australia, where they compete with the native marron (*C. tenuimanus*). *C. destructor* is of special interest because the species is the most widespread and abundant of all Australian freshwater crayfish, with a natural distribution covering over two million square kilometers, from South Australia and the southern portion of the Northern Territory in central Australia, to the Great Dividing Range in the east (Nguyen et al. 2004). *Cherax quadricarinatus* is indigenous to the rivers of northwestern Queensland and the northern territory in tropical Australia, and also extends into the catchments of southeastern Papua New Guinea. According to Beatty (2005), 22 species of *Cherax* have been described and are native to Australia.

Tasmania has a rich freshwater crayfish fauna, with about 37 species from four genera, which is relatively high in the context of the total Australian fauna (Whiting et al. 2000). They range from the world's largest crayfish, *Astacopsis gouldi*, which are found in the northern part of the island, to the tiny burrowing crayfish from the genus *Engaeus* that are found throughout the island. Within *Engaeus* there are 15 species known, 13 of which occur only in Tasmania, and two shared with Victoria on the Australian mainland. Areas of high diversity are in the northeast (*Engaeus* spp.) and the central west (*Engaeus* and *Parastacoides* spp.) (Richardson et al. 2006).

Finally, several genera are restricted to certain regions. For example, the genus *Tenuibranchiurus* is only present in southeastern Queensland, *Gramastacus* in western Victoria, and *Engaewa* in the southwestern part of Australia. In southwestern Australia, the burrowing freshwater crayfish genus *Engaewa* is a Gondwanian relict restricted to the high rainfall zone. Of five species of *Engaewa* recognized in the genus, three are of conservation concern.

In Victoria, the Grampians National Park harbours seven species from six different genera (*Euastacus bispinosus* (Clark, 1936), *Cherax destructor*, *Geocharax falcata* Clark, 1936, *Gramastacus insolitus* Riek, 1972 (the smallest Western swamp crayfish) and *Engaeus lyelli* Clark, 1936) all of which occur in sympatry. This region is considered a "hot spot" for crayfish diversity in Queensland, and here the crayfish *Euastacus sulcatus* is quite abundant. This species is a keystone species and functions as an ecosystem engineer (Furse 2010).

In New Zealand, the family Parastacidae is also present, with just two endemic species of *Paranephrops* on the main islands, and no introduced species.

Madagascar

Freshwater crayfish of the genus *Astacoides* are endemic to the highlands of eastern Madagascar, with six uniquely tropical species listed as of 2005: *Astacoides madagascariensis* (Milne Edwards and Audouin, 1839), *A. caldwelli* (Bate, 1865), *A. betsileoensis* (Petit, 1923), *A. granulimanus* (Monot and Petit, 1929), *A. crosnieri* (Hobbs 1987) and *A. petiti* (Hobbs 1987). Growth rates for *Astacoides granulimanu*s and

A. crosnieri are among the slowest known of any crayfish. *Astacoides madagascariensis* is endemic to Madagascar, and extends a little further north than that of any other Malagasy crayfish. The distribution of this species lies at latitudes 18° to 21° S, longitudes 47° to 49° E. Type specimens were probably collected in the vicinity of Tananarive (Hobbs 1987). This species is found in the Toamasina and Antananarivo provinces (Boyko et al. 2005). These authors also described a new species commemorating Hobbs and named it *Astacoides hobbsi*. Madagascar's freshwater habitats have great significance for global biodiversity, yet conservation efforts, as in so much of the world, has focused on terrestrial ecosystems. Jones et al. (2007) call for more attention to be paid to Madagascar's exceptional, yet understudied, freshwater biodiversity which is now coming under increasing threat.

Conclusions

If marine crustaceans are economically of great importance, freshwater crayfish have stimulated much economic activity and are the subject of many books and thousands of research articles.

Freshwater crayfish have served as model organisms for over 125 years in scientific research, from areas such as neurobiology and vision research to conservation biology and evolution. Recently, evolutionary histories in the form of phylogenies have served as a critical foundation for testing hypotheses in diverse research areas (e.g., Crandall 2006). Molecular methods have been applied widely to the phylogenetics and systematics of crayfish so that the status of perhaps the majority of species has been established with some confidence, though the phylogenetic relationships, particularly of the North American radiation, require further elucidation (Crandall and Buhay 2008). According to Burnham and Dawkins (2013), "freshwater species in general (and crayfish specifically) often have limited ranges with high species endemism within, and species turnover between, catchments. Freshwater species also face ever-increasing threats, and genetic diversity (both at and below the species level) is being lost as a result of these threats". They further state that "molecular taxonomy provides a tool by which this diversity can be rapidly (and relatively cheaply) uncovered before it is lost. Identifying previously unrealised diversity within crayfish via molecular techniques can act as a stimulus to further taxonomic investigations and conservation efforts". They also gave specific examples from the Australian crayfish fauna, where molecular data were used to highlight significant genetic diversity, which may correspond to previously overlook morphological variation. Their examples and results can be used to promote the undertaking of wide-scale molecular revisions of as many crayfish taxa as possible, looking for any previously unrecognised lineages within currently described species (akin to evolutionary significant units—ESUs) that may then warrant further revision.

Crayfish and threats

Two centers of crayfish diversity have been described, the first in the southern Appalachian Mountains of the southeastern United States (Northern Hemisphere center) and the second in southeastern Australia (Southern Hemisphere center) (Crandall and Buhay 2008); hot spots of diversity have also been identified for single families or genera (e.g., in Italy by Fratini et al. 2005). In recent times, however, their original distribution has been altered due to the massive human-mediated introduction of species outside of their native range, and the subsequent spread of some of these crayfish beyond the original area of introduction (Lodge et al. 2000, Holdich et al. 2009). Consequently, native crayfish diversity is in serious decline due to increased impacts due to habitat loss and degradation, often acting in synergy with the detrimental effects of invasions by alien species, over-harvesting, and chemical pollution. Roughly 50% of the species in the United States are imperiled (Taylor et al. 2007). In 2010, the International Union for Conservation of Nature (IUCN) wanted to comprehensively assess the status of the world's freshwater biodiversity in order to rapidly expand the taxonomic and geographic coverage of the IUCN Red List of Threatened Species (www.iucnredlist.org) in order to inform conservation strategies and management decisions. The priority taxa being assessed were freshwater fishes, molluscs, dragonflies and damselflies and crayfish worldwide. The global assessment was completed through a combination of regional assessments with a current major focus on Africa, Asia and Europe. According to Cumberlidge (2010), when examined at the level of individual zoogeographic regions, the accumulation of taxonomic knowledge is particularly contingent upon the productivity of a few regional experts. Although for some taxa, the accumulation curves flatten out (for instance Palearctic crayfish) thus demonstrating near completeness of the inventory, the majority of accumulation curves in other zoogeographical regions, as well as for individual taxa, demonstrate that we are nowhere near completing a full biodiversity inventory of the world's freshwater Decapoda.

Crayfish taxonomy and conservation

Studies utilising genetic data to examine the systematics of freshwater crayfish with morphology that is ambiguous or difficult to interpret suggest that morphologically based taxonomic studies of freshwater crayfish need to be interpreted with caution (e.g., Horwitz et al. 1990, Zeidler and Adams 1990, Campbell et al. 1994, Austin and Knott 1996). As noted by Austin and Knott (1996) the need for caution is because taxonomic characters may be more variable than realised, morphological and habitat differences may not equate with specific distinctions, and genetically distinct species need not be morphologically distinct.

Morphological plasticity has been demonstrated in decapod crustaceans. Examples from freshwater crayfish include those provided by Austin (1996) and Austin and Knott (1996), which suggest that the genus *Cherax* may display morphological plasticity in relation to environmental factors. *Cherax crassimanus* Riek, *Cherax quinquecarinatus* (Gray) and *Cherax preissii* Erichson each utilize an extremely wide range of freshwater habitats, ranging from deeper, permanent rivers to semi-permanent swamps. They found a direct correspondence between habitat variation and a large component of the morphological variation observed both within and between species. The morphological variation was found to correspond to habitat variation and was made up of a diverse range of traits, including several that have been considered previously (Riek 1967b, 1969) to be of taxonomic importance. A similar correlation of attributes to those reported by Austin and Knott (1996) was noted by Hobbs, Jr. (1975) among North American freshwater crayfish species. However, Austin and Knott (1996) were the first to show such a relationship within species.

The implication of these insights is that the conventional approach to the taxonomy of freshwater crayfish, where small anatomical differences are assumed to be reliable guides to specific distinctions, both in the Southern Hemisphere (e.g., Clark 1936, Riek 1951, 1956, 1967a,b, 1969, 1972, Sumner 1978, Swain et al. 1982, Morgan 1986, 1988, Hobbs, Jr. 1987) and in the Northern Hemisphere (Hobbs, Jr. 1989) may be flawed, and thus may extend to the existing taxonomic classifications of freshwater crayfish. Furthermore, the presence of potential morphological plasticity within freshwater crayfish suggest that, where habitat characteristics have been used as supporting information for the delineation of freshwater crayfish (based on an assumption that crayfish species tend to occupy narrow and distinct habitats), these errors may have been compounded (Austin and Knott 1996). Clearly the use of such convergent characteristics interpreted as the result of descent from a common ancestor will result in the construction of erroneous taxonomies and phylogenies (Fetzner and Crandall 2002). Addressing taxonomic and phylogenetic questions via the utilization of non-morphological characters (e.g., serology and genetics) has a long history in astacological research (e.g., Clark and Burnet 1942, Patak and Baldwin 1984, Patak et al. 1989, Austin 1996, Austin and Knott 1996) and more recently molecular data has been acknowledged in playing an important role in conservation biology through ensuring accurate definitions of species boundaries, facilitating detection of cryptic species, and providing boundaries for management units within species (Fratini et al. 2005). It has been noted that we are currently facing a global biodiversity crisis with a rapid loss of diversity occurring in all environments and at all levels, from ecosystems to genes (Browning et al. 2001), with population declines and species' extinctions occurring at an unprecedented rate (Dirzo and Raven 2003, DeSalle and Amato 2004, May 2010). It is evident that the scale of biodiversity loss globally makes the conservation of all threatened species virtually impossible; therefore certain units (whether ESUs, species, regions, etc...), must be made priorities.

According to Burnham and Dawkins (2013) crayfish taxonomy has often been in a state of flux, with different understandings of morphological and habitat variation within freshwater crayfish being common. An example of how examining additional data, and adding multiple data types (e.g., morphology, ecology/habitat, molecular), can affect our best estimate of taxonomy comes from the (now defunct) Tasmanian endemic genus *Parastacoides*. In 1936, Clark erected the monotypic genus, *Parastacoides*, with *Astacus tasmanicus* Erichson designated the type specimen; however, in 1939 Clark added another two species,

Parastacoides inermis Clark and *Parastacoides insignis* Clark, Riek (1951) described an additional two species *Parastacoides setosimerus* Riek and *Parastacoides leptomerus* Riek, but sixteen years later synonymised *P. setosimerus* and *Parastacoides tasmanicus* (Erichson) whilst adding two more species: *Parastacoides sternalis* Riek and *Parastacoides pulcher* Riek (Riek 1967a). Based on a numerical phenetic study, Sumner (1978) reviewed the genus and identified three groups, to which he gave sub-specific rank: *Parastacoides tasmanicus tasmanicus* (Erichson) (*P. tasmanicus, P. pulcher, P. leptomerus, P. setosimerus*), *Parastacoides tasmanicus inermis* (Clark) (*P. inermis, P. sternalis* Riek), *Parastacoides tasmanicus insignis* (Clark) (*P. insignis*): thus reducing the number of species back to one. Ecological work by Richardson and Swain (1980), however, suggested that habitat and morphological variation was more complicated than previously realised and was inconsistent with the recognition of only a single species of *Parastacoides* divided into three subspecies. Most recently, using a combination of molecular and morphological analyses, Hansen and Richardson (2006) divided *Parastacoides* into fourteen species within two newly erected genera, *Ombrastacoides* Hansen and Richardson and *Spinastacoides* Hansen and Richardson.

Final conclusion

Resolving taxonomy is a prerequisite for conserving and managing indigenous crayfish species

We have described how major geological and climatic changes have affected the present biogeographical spread of crayfish, resulting in evolutionary diversity. With more detailed research on populations, their overall taxonomy becomes less clear-cut and is clearly in a state of flux with serious legal implications for crayfish conservation. While this situation may fascinate researchers, a confused taxonomy means that conservation and management of threatened crayfish becomes weaker and more problematic. If a species is accepted as being under threat and is then protected under national or international legislation but is later shown to be a species complex or recommended to be split into a number of sub-species or sibling species, the legal status of its protection becomes unclear. Do we need to accept all populations with a degree of genetic segregation in order to define management and conservation units? It takes time and resources to get relevant legislation rewritten and passed, and the outcome may be unpredictable. If a recognisable 'deme' within a species is no longer seen as threatened, does this weaken the case for restricting trade and movement of non-indigenous crayfish species within its area of distribution?

To conclude, the worldwide crayfish distribution reveals a great variety of available information for different native crayfish species, ranging from the well-studied high diversity of the United States and Australia and the few species of Europe, to the still incomplete knowledge of the crayfish fauna of Mexico and South America.

The way forward is still to conduct complete molecular studies for each genus and then to link those results to a reliable morphological framework for each species.

Acknowledgements

Matt Longshaw and Paul Stebbing are warmly acknowledged for their invitation to participate to this book. We thank also Mauricio Pereira Almerão, Erich H. Rudolph and Keith A. Crandall who provided an update of the state of knowledge of South American crayfish species (taxonomy, phylogeny and distribution). Lucian Pârvalescu was of great help in providing figures for illustrating key to identify crayfish in European countries.

References

Adams, S.B. and G. Henderson. 2009. Mississippi Crayfishes Database.v. 3.1. USDA Forest Service, Oxford, MS. Available online at http://maps.fs.fed.us/crayfish/.

Adams, S.B., C.A. Taylor and C. Lukhaup. 2010. Crayfish fauna of the Tennessee River drainage in Mississippi, including new state species records. Southeastern Naturalist 9: 521–528.

Ahyong, S.T. and D. O'Meally. 2004. Phylogeny of the Decapoda Reptantia: Resolution using three molecular loci and morphology. Raffles Bull. Zool. 52: 673–93.

Alekhnovich, A. 2006. Production of a narrow-clawed crayfish population of *Astacus leptodactylus* (Eschscholtz, 1823) in Belorussian waterbodies. Crayfish News: IAA Newsletter 28: 5–7.

Allegrucci, G., F. Baldari, D. Cesaroni, R.S. Thorpe and V. Sbordoni. 1992. Morphometric analysis of interspecific and microgeographic variation of crayfish from a Mexican cave. Biol. J. Linn. Soc. 47: 455–468.

Almerão, M.P., E. Rudolph, C. Souty-Grosset, K. Crandall, L. Buckup, J. Amouret, A. Verdi, S. Santos and P.B. Araujo. 2015. The native South American crayfishes (Crustacea, Parastacidae): state of knowledge and conservation status. Aquatic Conservation: Marine and Freshwater Ecosystems 25:10.1002/aqc.v25.2, 288–301.

Alvarez, F., M. López-Mejía and J.L. Villalobos. 2007. A new species of crayfish (Crustacea: Decapoda: Cambaridae) from a salt marsh in Quintana Roo, Mexico. Proc. Biol. Soc. Wash. 120: 311–319.

Alves, N., J. Coughran, J. Furse and S. Lawler. 2010. *Euastacus armatus*. *In*: IUCN 2012. IUCN Red List of Threatened Species. Version 2012.2. <www.iucnredlist.org>.

Arens, A. and T. Taugbøl. 2005. Status of crayfish in Latvia. Bull. Fr. Pêche Piscic. 376-377: 519–28.

Austin, C.M. 1996. An electrophoretic and morphological taxonomic study of the freshwater crayfish genus *Cherax* (Decapoda: Parastacidae) in Northern and Eastern Australia. Aust. J. Zool. 44: 259–96.

Austin, C.M. and B. Knott. 1996. Systematics of the freshwater crayfish genus *Cherax* Erichson (Decapoda: Parastacidae) in south-western Australia: Electrophoretic, morphological and habitat variation. Aust. J. Zool. 44: 223–258.

Austin, C.M. and J. Bunn. 2010. *Cherax cainii*. *In*: IUCN 2012. IUCN Red List of Threatened Species. Version 2012.2. <www.iucnredlist.org>.

Beatty, S. 2005. Translocations of freshwater crayfish: contributions from life histories, trophic relations and diseases of three species in Western Australia. Ph.D. thesis, Murdoch University, Perth, Australia.

Bertocchi, S., S. Brusconi, F. Gherardi, F. Grandjean and C. Souty-Grosset. 2008. Genetic variability in the threatened crayfish *Austropotamobius italicus* in Tuscany: implications for its management. Fundamental and Applied Limnology Archiv für Hydrobiologie 173/2: 153–64.

Bouchard, R.W. and J.W. Bouchard. 1995. Two new species and subgenera (*Cambarus* and *Orconectes*) of crayfishes (Decapoda: Cambaridae) from the eastern United States. Notulae Naturae 471: 1–21.

Boyko, C.B., O.R. Ravoahangimalala, D. Randriamasimanana and T.H. Razafindrazaka. 2005. *Astacoides hobbsi*, a new crayfish (Crustacea: Decapoda: Parastacidae) from Madagascar. Zootaxa 1091: 41–51.

Braband, A., T. Kawai and G. Scholtz. 2006. The phylogenetic position of the East Asian freshwater crayfish *Cambaroides* within the Northern Hemisphere Astacoidea (Crustacea, Decapoda, Astacidea) based on molecular data. J. Zool. Syst. Evol. Res. 44: 17–24.

Bracken, H.D., A. Toon, D.L. Felder, J.W. Martin, M. Finley, J. Rasmussen, F. Palero and K.A. Crandall. 2009. The decapod tree of life: Compiling the data and moving toward a consensus of decapod evolution. Athropod Systematics and Phylogeny 67: 99–116.

Breinholt, J., M. Perez-Losada and K.A. Crandall. 2009. The timing of the diversification of the freshwater crayfishes. pp. 335–43. *In*: J.W. Martin, K.C. Crandall and D.L. Felder (eds.). Decapod Crustacean Phylogenetics. CRC Press Taylor and Francis Group, New York, USA.

Breinholt, J.W., M. Porter and K.A. Crandall. 2012. Testing phylogenetic hypotheses of the subgenera of the freshwater crayfish genus *Cambarus* (Decapoda: Cambaridae). PLoS ONE 7: e46105.

Browning, T.L., D.A. Taggart, C. Rummery, R.L. Close and M.D.B. Eldridge. 2001. Multifaceted genetic analysis of the "Critically Endangered" brush-tailed rock-wallaby *Petrogale penicillata* in Victoria, Australia: Implications for management. Conserv. Gen. 2: 145–156.

Buckup, L. 1998. Malacostraca - Eucarida. Astacidea. *In*: P.S. Young (ed.). Catalogue of Crustacea of Brazil. Rio de Janeiro. Museu Nacional. 373-375. (Série Livros n. 6).

Buckup, L. 1999. Família Parastacidae. *In*: L. Buckup and G. Bond-Buckup (eds.). Os Crustáceos do Rio Grande do Sul. Editora do Rio Grande do Sul, Porto Alegre, 319–327.

Buckup, L. and A. Rossi. 1980. O gênero *Parastacus* no Brasil (Crustacea, Decapoda, Parastacidae) Rev. Brasil. Biol. 40: 663–681.

Buhay, J.E. and K.A. Crandall. 2008. Taxonomic revision of cave crayfishes in the genus *Orconectes*, subgenus *Orconectes* (Decapoda: Cambaridae) along the Cumberland Plateau, including a description of a new species, *Orconectes barri*. J. Crust. Biol. 28: 57–67.

Buhay, J.E. and K.A. Crandall. 2009. Taxonomic revision of cave crayfish in the genus *Cambarus*, subgenus *Aviticambarus* (Decapoda: Cambaridae) with descriptions of two new species, *C. speleocoopi* and *C. laconensis*, endemic to Alabama, U.S.A. J. Crust. Biol. 29: 121–134.

Burnham, Q. and K.L. Dawkins. 2013. The role of molecular taxonomy in uncovering variation within crayfish and the implications for conservation. Freshw. Crayfish 19: 29–37.

Campbell, N.J.H., M.C. Geddes and M. Adams. 1994. Genetic variation in Yabbies, *Cherax destructor* and *C. albidus* (Crustacea: Decapoda: Parastacidae), indicates the presence of a single, highly sub-structured species. Aust. J. Zool. 42: 745–760.

Cherkashina, N.Y. 1999a. *Caspiastacus pachypus* (Rathke, 1837), its biology and distribution. Freshw. Crayfish 12: 846–853.

Cherkashina, N.Ya. 1999b. The state of populations of *Pontastacus cubanicus* (Birnstein and Winogradow, 1934) in the waterbodies of the lower Don area (Russia). Freshw. Crayfish 12: 643–654.

Clark, E. 1936. The freshwater and land crayfishes of Australia. Mem. Nat. Mus. Victoria 10: 5–58.

Clark, E. 1939. Tasmanian Parastacidae. Pap. Proc. R. Soc. Tasm. (1938): 117–128.

Clark, E. and E.M. Burnet. 1942. The application of serological methods to the study of the Crustacea. Aust. J. Exp. Biol. Med. 20: 89–95.

Collins, P.A., F. Giri and V. Williner. 2011. Biogeography of the freshwater decapods in the La Plata Basin, South America. J. Crust. Biol. 31: 179–191.

Cooper, J.E. and J.E. Price. 2010. A new spinose crayfish of the genus *Cambarus*, subgenus *Puncticambarus* (Decapoda: Cambaridae), from South Carolina. Proc. Biol. Soc. Wash. 123: 335–344.

Coughran, J. 2006. Biology of the Freshwater Crayfishes of Northeastern New South Wales, Australia. Unpublished Ph.D. Thesis. School of Environmental Science & Management, Southern Cross University.

Coughran, J. and R.B. McCormack. 2011. *Euastacus morgani* sp. n., a new spiny crayfish (Crustacea, Decapoda, Parastacidae) from the highland rainforests of eastern New South Wales, Australia. ZooKeys 85: 17–26.

Crandall, K.A. 2006. Applications of phylogenetics to issues in freshwater crayfish biology. Bull. Fr. Pêche Piscic. 380-381: 953–964.

Crandall, K.A. and J.F. Fitzpatrick, Jr. 1996. Crayfish molecular systematics: using a combination of procedures to estimate phylogeny. Systematic Biology 45: 1–26.

Crandall, K.A. and J.E. Buhay. 2008. Global diversity of crayfish (Astacidae, Cambaridae, and Parastacidae-Decapoda) in freshwater. Hydrobiologia 595: 295–301.

Crandall, K.A., J.W. Fetzner, C.G. Jara and L. Buckup. 2000. On the phylogenetic positioning of the South American freshwater crayfish genera (Decapoda: Parastacidae). J. Crust. Biol. 20: 530–540.

Cukerzis, J.M. 1988. *Astacus astacus* in Europe. Freshwater Crayfish 7: 309–40.

Cumberlidge, N. 2010. Freshwater decapod conservation: recent progress and future challenges. Proceedings of the 21st International Senckenberg-conference: Biology of freshwater decapods. Frankfurt. Germany, pp. 2.

De Grave, S., N.D. Pentcheff, S.T. Ahyong, T.-Y. Chan, K.A. Crandall, P.C. Dworschak, D.L. Felder, R.M. Feldmann, C.H.J.M. Fransen, L.Y.D. Goulding, R. Lemaitre, M.E.Y. Low, J.W. Martin, P.K.L. Ng, C.E. Schweitzer, S.H. Tan, D. Tshudy and R. Wetzer. 2009. A classification of living and fossil genera of decapod crustaceans. Raffles Bull. Zool. Suppl. No. 21: 1–109.

Dercourt, J., L.P. Zonenshain, L.E. Ricou, V.G. Kazmin, X. Le Pichon, A.L. Knipper, C. Grandjacquet, I.M. Sbortshikov, J. Geyssant, C. Lepvrier, D.H. Pechersky, J. Boulin, J.C. Sibuet, L.A. Savostin, O. Sorokhtin, M. Westphal, M.L. Bazhenov, J.P. Lauer and B. Biju-Duval. 1986. Geological evolution of the Tethys belt from the Atlantic to the Pamirs since the Lias. Tectonophysics 123: 241–315.

DeSalle, R. and G. Amato. 2004. The expansion of conservation genetics. Nature Reviews Genetics 5: 702–712.

Dirzo, R. and P.H. Raven. 2003. Global state of biodiversity and loss. Ann. Rev. Environ. Res. 28: 137–167.

Dixon, C.J., S. Ahyong and F.R. Schram. 2003. A new hypothesis of decapod phylogeny. Crustaceana 76: 935–975.

Edsman, L., L. Füreder, F. Gherardi and C. Souty-Grosset. 2010. *Astacus astacus*. *In*: IUCN 2012. IUCN Red List of Threatened Species. Version 2012.2. <www.iucnredlist.org>.

Environment Agency. 2007. News release. October 2007. RP/PR240/07/E.

Faxon, W. 1898. Observations on the astacidae in the United States National Museum and in the Museum of comparative zoology, with descriptions of new species. Proc. U.S. Nat. Mus. 20: 643–694.

Faxon, W. 1914. Notes on the crayfishes in the United States National Museum and the Museum of comparative zoology, with descriptions of new species and subspecies. Mem. Mus. Comp. Zool. Harvard Coll. 40: 351–427.

Feldmann, R.M., C.E. Schweitzer and J. Leahy. 2011. New Eocene crayfish from the McAbee beds in British Columbia: First record of Parastacoidea in the northern hemisphere. J. Crust. Biol. 31: 320–331.

Fetzner, J.W., Jr. 1996. Biochemical systematics and evolution of the crayfish genus *Orconectes* (Decapoda: Cambaridae). J. Crust. Biol. 16: 111–141.

Fetzner, J.W., Jr. 2005. The crayfish and lobster taxonomy browser: A global taxonomic resource for freshwater crayfish and their closest relatives. Accessed: 28 February 2013. http://iz.carnegiemnh.org/crayfish/NewAstacidea/.

Fetzner, J.W., Jr. and K.A. Crandall. 2002. Genetic variation. pp. 291–326. *In*: D.M. Holdich (ed.). Biology of Freshwater Crayfish. Blackwell Science, Oxford.

Fetzner, J.W. and K.A. Crandall. 2003. Linear habitats and the nested clade analysis: an empirical evaluation of geographic versus river distances using an Ozark crayfish (Decapoda: Cambaridae). Evolution 57: 2101–2118.

Filipová, L. 2012. Genetic variation in North American crayfish species introduced to Europe and the prevalence of the crayfish plague pathogen in their populations. M.S. Thesis, University of Praga and Poitiers, 132 pp.

Filipová, L., D.M. Holdich, F. Grandjean and A. Petrusek. 2009. Cryptic diversity within the invasive virile crayfish *Orconectes virilis* (Hagen, 1870) species complex: new lineages recorded in both native and introduced ranges. Biol. Inv. 12: 983–989.

Filipova, L., F. Grandjean, C. Chucholl, M. Soes and A. Petrusek. 2011. Identification of exotic North American crayfish in Europe by DNA barcoding. Knowl. Manage. Aquat. Ecosyst. 401: art. no. 11.

Fitzpatrick, J.F. 1983. How to Know the Freshwater Crustacea. Wm. C. Brown Company Publishers, Dubuque, Iowa. x + 227 pages, 216 figures.

Fratini, S., S. Zaccara, S. Barbaresi, F. Grandjean, C. Souty-Grosset, G. Crosa and F. Gherardi. 2005. Phylogeography of the threatened crayfish (genus *Austropotamobius*) in Italy: implications for its taxonomy and conservation. Heredity 94: 108–118.

Füreder, L., L. Edsman, D.M. Holdich, P. Kozak, Y. Machino, M. Pöckl, B. Renai, J. Reynolds, H. Schulz, R. Schulz, D. Sint, T. Taugbol and M.C. Trouilhé. 2006. Indigenous crayfish habitat and threats. pp. 26–47. *In*: C. Souty-Grosset,

D.M. Holdich, P.Y. Noel, J.D. Reynolds and P. Haffner (eds.). Atlas of Crayfish in Europe. Muséum National d'Histoire naturelle, Paris (Patrimoines naturels, 64).

Füreder, L., F. Gherardi and C. Souty-Grosset. 2010a. *Austropotamobius torrentium. In*: IUCN 2012. IUCN Red List of Threatened Species. Version 2012.2. <www.iucnredlist.org>.

Füreder, L., F. Gherardi, D. Holdich, J. Reynolds, P. Sibley and C. Souty-Grosset. 2010b. *Austropotamobius pallipes. In*: IUCN 2012. IUCN Red List of Threatened Species. Version 2012.2. <www.iucnredlist.org>.

Furse, J.M. 2010. Ecosystem Engineering by *Euastacus sulcatus* (Decapoda: Parastacidae) in the Hinterland of the Gold Coast, Queensland, Australia. Griffith University. Griffith School of Environment. Thesis (Ph.D.), 2010.

Furse, J.M., K.L. Dawkins and J. Coughran. 2013. Two new species of *Euastacus* (Decapoda: Parastacidae) from the Gondwana Rainforests of central eastern Australia. Freshwater Crayfish 19: 103–112.

Gherardi, F. and C. Souty-Grosset. 2010a. *Astacus leptodactylus. In*: IUCN 2012. IUCN Red List of Threatened Species. Version 2012.2. <www.iucnredlist.org>.

Gherardi, F. and C. Souty-Grosset. 2010b. *Astacus pachypus. In*: IUCN 2012. IUCN Red List of Threatened Species. Version 2012.2. <www.iucnredlist.org>.

Gherardi, F., C. Souty-Grosset, G. Vogt, J. Dieguez–Uribeondo and K.A. Crandall. 2010. Infraorder Astacidea Latreille, 1802 P.P.: The Freshwater Crayfish, Chapter 67: 269–423. In Treatise on Zoology – Decapoda, vol. 9A. Eds F.R. Schram and J.C. Vaupel Klein Brill, Leiden.

Grandjean, F. and C. Souty-Grosset. 1997. Preliminary results on the genetic variability of mitochondrial DNA in the signal crayfish, *Pacifastacus leniusculus* Dana. C. R. Acad. Sci. Life Sci. 320: 551–556.

Grandjean, F., D.J. Harris, C. Souty-Grosset and K.A. Crandall. 2000. Systematics of the European endangered crayfish species *Austropotamobius pallipes* (Decapoda, Astacidae). J. Crust. Biol. 20: 522–529.

Grandjean, F., D. Bouchon and C. Souty-Grosset. 2002a. Systematics of the European endangered crayfish species *Austropotamobius pallipes* (Decapoda: Astacidae) with a re-examination of the status of *Austropotamobius berndhauseri*. J. Crust. Biol. 22: 677–681.

Grandjean, F., M. Frelon-Raimond and C. Souty-Grosset. 2002b. Compilation of molecular data for the phylogeny of the genus *Austropotamobius*: one species or several? Bull. Fr. Pêche Piscic. 367: 671–680.

Hamr, P. 1998. Conservation status of Canadian freshwater crayfishes. World Wildlife Fund Canada and the Canadian Nature Federation, Toronto, 87 pp.

Hamr, P. 2002. *Orconectes*. pp. 585–608. *In*: D.M. Holdich (ed.). Biology of Freshwater Crayfish. Blackwell Science Ltd., Oxford.

Hansen, B. and A.M.M. Richardson. 2006. A revision of the Tasmanian endemic freshwater crayfish genus *Parastacoides* (Crustacea: Decapoda: Parastacidae). Invert. Syst. 20: 713–69.

Harlioğlu, M.M. 2007. A new record of recently discovered crayfish, *Austropotamobius torrentium* (Shrank, 1803), in Turkey. Bull. Fr. Pêche Piscic. 387: 1–5.

Harlioğlu, M.M. 2008. The harvest of the freshwater crayfish *Astacus leptodactylus* Eschscholtz in Turkey: harvest history, impact of crayfish plague, and present distribution of harvested populations. Aquacult. Int. 16: 351–360.

Harlioğlu, M.M. and D.M. Holdich. 2001. Meat yields in the introduced crayfish, *Pacifastacus leniusculus* and *Astacus leptodactylus*, from British water. Aquac. Res. 32: 411–417.

Hasiotis, S.T. 1999. The origin and evolution of freshwater crayfish based on crayfish body and trace fossils. Freshwater Crayfish 12: 49–70.

Hefti, D. and P. Stucki. 2006. Crayfish management for Swiss waters. Bull. Fr. Pêche Piscic. 380-381: 937–950.

Hobbs, H.H., Jr. 1943. Two new crayfishes of the genus *Procambarus* from Mexico (Decapoda, Astacidae). Lloydia 6: 198–206.

Hobbs, H.H., Jr. 1969. On the distribution and phylogeny of the crayfish genus *Cambarus*. pp. 93–178. *In*: Perry C. Holt (ed.). The Distributional History of the Biota of the Southern Appalachians, Part I: Invertebrates. Research Division Monograph 1, Virginia Polytechnic Institute, Blacksburg, Virginia.

Hobbs, H.H., Jr. 1973. Three new troglobitic decapod crustaceas from Oaxaca, Mexico. Association for Mexican Cave Studies, Bulletin 5: 25–38, figures 1–8.

Hobbs, H.H., Jr. 1972. Biota of Freshwater Ecosystems: Identification Manual No. 9. Crayfishes (Astacidae) of North and Middle America. For the Environmental Protection Agency, Project # 18050 ELD.

Hobbs, H.H., Jr. 1975. Adaptations and convergence in North American crayfish. Freshwater Crayfish 2: 541–551.

Hobbs, H.H. III. 1974. Observations on the cave-dwelling crayfishes of Indiana. Freshwater Crayfish 2: 405–414.

Hobbs, H.H., Jr. 1987. A review of the crayfish genus *Astacoides* (Decapoda: Parastacidae). Smiths. Contrib. Zool. 443: 1–50.

Hobbs, H.H., Jr. 1988. Crayfish distribution, adaptive radiation and evolution. pp. 52–82. *In*: D.M. Holdich and R.S. Lowery (eds.). Freshwater Crayfish, Biology, Management and Evolution. Croom Helm, London.

Hobbs, H.H., Jr. 1989. An illustrated checklist of the American crayfishes (Decapoda: Astacidae, Cambaridae, and Parastacidae). Smiths. Contrib. Zool. 480: 1–236.

Hobbs, H.H., Jr. and A.G. Grubbs. 1982. Description of a new troglobitic crayfish from Mexico and a list of Mexican crayfishes reported since the publication of the Villalobos monograph (1955) (Decapoda, Cambaridae). Assoc. Mex. Cave Stud. Bull. 8: 45–50.

Hobbs, H.H., Jr. and H.H. Hobbs III. 1990. A new crayfish (Decapoda: Cambaridae) from southeastern Texas. Proc. Biol. Soc. Wash. 103: 608–613.

Holdich, D.M. 2002. Distribution of crayfish in Europe and some adjoining countries. Bull. Fr. Pêche Piscic. 367: 611–650.

Holdich, D.M., J.D. Reynolds, C. Souty-Grosset and P. Sibley. 2009. A review of the ever increasing threat to European crayfish from non-indigenous crayfish species. Knowledge Management of Aquatic Ecosystems 394-395: 11p1–11p46.

Horwitz, P. and M. Adams. 2000. The systematics, biogeography and conservation status of species in the freshwater crayfish genus *Engaewa* Riek (Decapoda: Parastacidae) from south-western Australia. Invert. Taxon. 14: 655–680.

Horwitz, P., M. Adams and P. Baverstock. 1990. Electrophoretic contributions to the systematics of the freshwater crayfish genus *Engaeus* Erichson (Decapoda: Parastacidae). Invert. Taxon. 4: 615–641.

Huner, J.V. 2002. *Procambarus*. pp. 541–584. *In*: D.M. Holdich (ed.). Biology of Freshwater Crayfish. Blackwell Science, Oxford.

Huner, J.V. and J.E. Barr. 1991. Red Swamp Crawfish: Biology and Exploitation. 3rd edn. Louisiana State University, Baton Rouge, Louisiana.

Huxley, T.H. 1880. The Crayfish: An Introduction to the Study of Zoology. Kegan Paul, London.

Johnson, D.P. 2008. Descriptions of two new crayfishes of the genus *Fallicambarus* from southeast Texas with notes on the distribution of *F. (F.) macneesei*. Zootaxa 1717: 1–23.

Johnson, D.P. 2010. Four new crayfishes (Decapoda: Cambaridae) of the genus *Orconectes* from Texas. ZooTaxa 2626: 1–45.

Johnson, D.P. 2011a. *Fallicambarus (F.) wallsi* (Decapoda: Cambaridae), a new burrowing crayfish from eastern Texas. Zootaxa 2939: 59–68.

Johnson, D.P. 2011b. *Procambarus (Ortmannicus) luxus* (Decapoda: Cambaridae), a new crayfish from southern Texas. Zootaxa 2972: 59–68.

Jones, J.P.G., F.B. Andriahajaina, N.J. Hockley, K.A. Crandall and O.R. Ravoahangimalala. 2007. The ecology and conservation status of Madagascar's endemic freshwater crayfish (Parastacidae; *Astacoides*). Freshwater Biol. 52: 1820–33.

Karaman, M.S. 1962. Ein Beitrag zur Systematic der Astacidae (Decapoda). Crustaceana 3: 173–191.

Karaman, M.S. 1963. Studie der Astacidae (Crustacea, Decapoda) II. Teil. Hydrobiologia 22: 111–132.

Kawai, T. 1996. Distribution of the Japanese crayfish, *Cambaroides japonicus* in Hokkaido, Japan, and its loss of habitat in eastern Hokkaido. Mem. Kushiro City Mus. 20: 5–12.

Kawai, T. 2012. Morphology of the mandible and gill of the Asian freshwater crayfish *Cambaroides* (Decapoda: Cambaridae) with implications for their phylogeny. J. Crust. Biol. 32: 15–23.

Kawai, T., Y. Machino and H.S. Ko. 2003. Reassessment of *Cambaroides dauricus* and *C. schrenckii* (Crustacea: Decapoda: Cambaridae). Kor. J. Biol. Sci. 7: 191–196.

Köksal, F. 1988. *Astacus leptodactylus* in Europe. pp. 365–400. *In*: D.M. Holdich and R.S. Lowery (eds.). Freshwater Crayfish, Biology, Management and Exploitation. Croom Helm and Timber Press, London.

Koutrakis, E., C. Perdikaris, Y. Machino, G. Savvidis and N. Margaris. 2007. Distribution, recent mortalities and conservation measures of crayfish in Hellenic fresh waters. Bull. Fr. Pêche Piscic. 385: 25–44.

Larson, E.R., C.L. Abbott, N. Usio, N. Azuma, K.A. Wood, L.M. Herborg and J.D. Olden. 2012. The signal crayfish is not a single species: cryptic diversity and invasions in the Pacific Northwest range of *Pacifastacus leniusculus*. Freshwater Biol. 57: 1823–1838.

Latreille, P.A. 1802. Histoire Naturelle, Générale et Particulière des Crustacés et des Insectes. 14 volumes. Paris: F. Dufart.

Lodge, D.M., C.A. Taylor, D.M. Holdich and J. Skurdal. 2000. Nonindigenous crayfishes threaten North American freshwater biodiversity. Fisheries 25: 7–20.

Loughman, Z.J., T.P. Simon and S.A. Welsh. 2011. *Cambarus (Puncticambarus) smilax*, a new species of crayfish (Crustacea: Decapoda: Cambaridae) from the Greenbrier River basin of West Virginia. Proc. Biol. Soc. Wash. 124: 99–111.

Loughman, Z.J., D.A. Foltz, N.L. Garrison and S.A. Welsh. 2013a. *Cambarus (P.) theepiensis*, a new species of crayfish (Decapoda: Cambaridae) from the coalfields region of Eastern Kentucky and Southwestern West Virginia, USA. Zootaxa 3641: 63–73.

Loughman, Z.J., R.A. Fagundo, E. Lau, S.A. Welsh and R.F. Thoma. 2013b. *Cambarus (C.) hatfieldi*, a new species of crayfish (Decapoda: Cambaridae) from the Tug Fork River Basin of Kentucky, Virginia and West Virginia, USA. Zootaxa 3750: 223–236.

Loughman, Z.J., R.F. Thoma, J.W. Fetzner Jr. and G.W. Stocker. 2015. *Cambarus (Jugicambarus) pauleyi*, a new species of crayfish (Decapoda: Cambaridae) endemic to southcentral West Virginia, USA, with a re-description of *Cambarus (J.) dubius*. Zootaxa 3980(4): 526–546.

Lukhaup, C. 2015. *Cherax (Astaconephrops) pulcher*, a new species of freshwater crayfish (Crustacea, Decapoda, Parastacidae) from the Kepala Burung (Vogelkop) Peninsula, Irian Jaya (West Paupa), Indonesia. ZooKeys 502: 1–10.

Lukhaup, C. and B. Herbert. 2008. A new species of crayfish (Crustacea: Decapoda: Parastacidae) from the Fly River Drainage, Western Province, Papua New Guinea. Mem. Queensl. Mus. 52: 213–219.

Machino, Y., L. Füreder, P.J. Laurent and J. Petutschnig. 2004. Introduction of the white-clawed crayfish *Austropotamobius pallipes* in Europe. Ber. Nat.-med. Verein Innsbruck 91: 187–212.

Machino, Y. and D.M. Holdich. 2006. Distribution of crayfish in Europe and adjacent countries: updates and comments. Freshwater Crayfish 15: 292–323.

Martin, A.J., T.H. Rich, G.C.B. Poore, M.B. Schultz, C.M. Austin, L. Kool and P. Vickers-Rich. 2008. Fossil evidence in Australia for oldest known freshwater crayfish of Gondwana. Gondwana Res. 14: 287–296.

Mathews, L.M. and A.H. Warren. 2008. A new crayfish of the genus *Orconectes* Cope, 1872 from southern New England (Crustacea: Decapoda: Cambariidae). Proc. Biol. Soc. Wash. 121: 374–381.

Mathews, L.M., L. Adams, E. Anderson, M. Basile, E. Gottardi and M.A. Buckholt. 2008. Genetic and morphological evidence for substantial hidden biodiversity in a freshwater crayfish species complex. Mol. Phylog. Evol. 48: 126–135.

May, R.M. 2010. Ecological science and tomorrow's world. Phil. Trans. Roy. Soc. B: Biol. Sci. 365: 41–47.

McCormack, R.B. and J. Coughran. 2008. *Euastacus maccai*, a new freshwater crayfish from New South Wales. Fishes of Sahul 22(4): 471–476.

Mejía-Ortíz, L.M., R.G. Hartnoll, J.A. Viccon-Pale and A. José. 2003. A new stygobitic crayfish from Mexico, *Procambarus cavernicola* (Decapoda: Cambaridae), with a review of cave-dwelling crayfishes in Mexico. J. Crust. Biol. 23: 391–401.

Miller, G.C. 1960. The taxonomy and certain biological aspects of the crayfish of Oregon and Washington. Master's thesis, Oregon State College, Corvalis, pp. 216.

Miller, G.L. and S.R. Ash. 1988. The oldest freshwater decapod crustacean, from the Triassic of Arizona. Paleontology 31: 273–279.

Moore, M.J., R.J. DiStefano and E.R. Larson. 2013. An assessment of life-history studies for USA and Canadian crayfishes: Identifying biases and knowledge gaps to improve conservation and management. Freshwater Science 32(4): 1276–1287.

Morgan, G.J. 1986. Freshwater crayfish of the genus *Euastacus* Clark (Decapoda: Parastacidae) from Victoria. Mem. Mus. Vict. 47: 1–57.

Morgan, G.J. 1988. Freshwater crayfish of the Genus *Euastacus* Clark (Decapoda: Parastacidae) from Queensland. Mem. Mus. Vict. 49: 1–49.

Morgan, G.J. 1997. Freshwater crayfish of the genus *Euastacus* Clark (Decapoda: Parastacidae) from New South Wales, with a key to all species of the genus. Records of the Australian Museum, Supplement 23: 1–110.

Munasinghe, D.H.N., C.P. Burridge and C.M. Austin. 2004. The systematics of freshwater crayfish of the genus *Cherax* Erichson (Decapoda: Parastacidae) in eastern Australia re-examined using nucleotide sequences from 12S rRNA and 16S rRNA genes. Invert. Syst. 18: 215–225.

Nguyen, T.T.T., C.M. Austin, M.M. Meewan, M.B. Schultz and D.R. Jerry. 2004. Phylogeography of the freshwater crayfish *Cherax destructor* Clark (Parastacidae) in inland Australia: historical fragmentation and recent range expansion. Biol. J. Linn. Soc. 83: 539–350.

Olsen, P.E. 1977. Stop 11, Triangle Brick Quarry. pp. 59–60. *In*: G.L. Bain and B.W. Harvey (eds.). Field Guide to the Geology of the Durham Basin. Carolina Geological Survey Fortieth Anniversary Meeting, October, 1977.

Ortmann, A.E. 1902. The geographical distribution of freshwater decapods and its bearing upon ancient geography. Proc. Am. Phil. Soc. 41: 267–400.

Pârvulescu, L. 2010. Crayfish field guide of Romania. Editura Bioflux, Cluj-Napoca: 28 p. (ISBN 978-606-8191-08-9).

Patak, A. and J. Baldwin. 1984. Electrophoretic and immunochemical comparisons of haemocyanins from Australian freshwater crayfish (Family Parastacidae): phylogenetic implications. J. Crust. Biol. 4: 528–535.

Patak, A., J. Baldwin and P.S. Lake. 1989. Immunochemical comparisons of haemocyanins of Australasian freshwater crayfish: phylogenetic implications. Biochem. Syst. Ecol. 17: 249–252.

Patoka, J., M. Bláha and A. Kouba. 2015. *Cherax* (*Astaconephrops*) *gherardii*, a new crayfish (Decapoda: Parastaciade) from West Paupa, Indonesia. Zootaxa 3964(5): 526–536.

Pedraza-Lara, C. and I. Doadrio. 2015. A new species of dwarf crayfish (Decapoda: Cambaridae) from central México, as supported by morphological and genetic evidence. Zootaxa 3963(4): 583–594.

Pedraza-Olvera, C., A. Lopez-Romero and P.J. Gutierrez-Yurrita. 2004. Preliminary studies concerning phenotype and molecular differences among freshwater crayfish from the sub-genus *Procambarus* (*Ortmannicus*) in Sierra Gorda Biosphere Reserve, México. Freshwater Crayfish 14: 129–139.

Perdikaris, C., E. Koutrakis, V. Saraglidou and N. Margaris. 2007. Confirmation of occurrence of narrow-clawed crayfish *Astacus leptodactylus* Eschscholtz, 1823 in River Evros in Greece. Bull. Fr. Pêche Piscic. 385: 45–52.

Perry, W.L., J.L. Feder and D.M. Lodge. 2001a. Implications of hybridization between introduced and resident *Orconectes* crayfishes. Conserv. Biol. 15: 1656–1666.

Perry, W.L., J.L. Feder, G. Dwyer and D.M. Lodge. 2001b. Hybrid zone dynamics and species replacement between *Orconectes* crayfishes in a northern Wisconsin lake. Evolution 55: 1153–1166.

Porter, M.L., M. Perez-Losada and K.A. Crandall. 2005. Model-based multi-locus estimation of decapod phylogeny and divergence times. Mol. Phylog. Evol. 37: 355–69.

Reddell, J.R. 1981. A review of the cavernicole fauna of Mexico, Guatemala, and Belize. Texas Mem. Mus. Bull. 27: 1–327.

Richardson, A.M.M. and R. Swain. 1980. Habitat requirements and distribution of *Engaeus cisternarius* and three subspecies of *Parastacoides tasmanicus* (Decapoda: Parastacidae), burrowing crayfish from an area of south-western Tasmania. Mar. Freshw. Res. 31: 475–484.

Richardson, A.M.M., N. Doran and B. Hansen. 2006. The geographic ranges of Tasmanian crayfish: extent and pattern. Freshwater Crayfish 15: 347–64.

Riek, E.F. 1951. The freshwater crayfish (Family Parastacidae) of Queensland, with an appendix describing other Australian species. Rec. Aust. Mus. 22: 368–388.

Riek, E.F. 1956. Additions to the Australian freshwater crayfish. Rec. Aust. Mus. 24: 1–6.

Riek, E.F. 1967a. The Tasmanian freshwater crayfish genus *Parastacoidea* (Decapoda: Parastacidae). Aust. J. Zool. 15: 999–1006.

Riek, E.F. 1967b. The freshwater crayfish of Western Australia (Decapoda: Parastacidae). Aust. J. Zool. 15: 103–121.

Riek, E.F. 1969. The Australian freshwater crayfish (Crustacea: Decapoda: Parastacidae), with descriptions of new species. Aust. J. Zool. 17: 855–918.

Riek, E.F. 1972. The phylogeny of the Parastacidae (Crustacea: Astacoidea), and description of a new genus of Australian freshwater crayfishes. Aust. J. Zool. 20: 369–389.

Rojas, Y., F. Alvarez and J.L. Villalobos. 1999. A new species of crayfish of the genus *Procambarus* (Crustacea: Decapoda: Cambaridae) from Veracruz, México. Proc. Biol. Soc. Wash. 112: 396–404.

Rojas, Y., F. Alvarez and J.L. Villalobos. 2000. A new species of crayfish (Crustacea: Decapoda: Cambaridae) from Lake Catemaco, Mexico. Proc. Biol. Soc. Wash. 113: 792–798.

Rudolph, E.H. 2010. Sobre la distribución geográfica de las especies chilenas de Parastacidae (Crustacea: Decapoda: Astacidea). Boletín de Biodiversidad de Chile 3: 32–46.

Rudolph, E.H. and K.A. Crandall. 2005. A new species of burrowing crayfish, *Virilastacus rucapihuelensis* (Crustacea: Decapoda: Parastacidae), from southern Chile. Proc. Biol. Soc. Wash. 118: 765–776.

Rudolph, E.H. and K.A. Crandall. 2007. A new species of burrowing crayfish *Virilastacus retamali* (Decapoda: Parastacidae) from the southern Chilean peatland. J. Crust. Biol. 27: 502–512.

Rudolph, E.H. and K.A. Crandall. 2012. A new species of burrowing crayfish, *Virilastacus jarai* (Crustacea, Decapoda, Parastacidae) from central-southern Chile. Proc. Biol. Soc. Wash. 125: 258–275.

Scholtz, G. 2002. Phylogeny and evolution. pp. 30–52. *In*: D.M. Holdich (ed.). Biology of Freshwater Crayfish. Blackwell Scientific, Oxford.

Scholtz, G. and S. Richter. 1995. Phylogenetic systematics of the reptantian Decapoda (Crustacea, Malacostraca). Zool. J. Linn. Soc. 113: 289–328.

Schuster, G.A. 2008. *Orconectes* (*Trisellescens*) *taylori*, a new species of crayfish from western Tennessee (Decapoda: Cambaridae). Proc. Biol. Soc. Wash. 121: 62–71.

Schuster, G.A., C.A. Taylor and J. Johansen. 2008. An annotated checklist and preliminary designation of drainage distributions of the crayfishes of Alabama. Southeast. Nat. 7: 493–504.

Simić, V., A. Petrović, M. Rajković and M. Paunović. 2008. Crayfish of Serbia and Montenegro—the population status and the level of endangerment. Crustaceana 81: 1153–1176.

Simon, T.P. and P.D. McMurray, Jr. 2014. *Orconectes* (*Crockerinus*) *alluvius* (Decapoda: Cambaridae), a new crayfish species from the Crawford Upland and Mitchell Plain in southwestern Indiana. Proceedings of the Biological Society of Washington 127(2): 353–366.

Simon, T.P. and C.C. Morris. 2014. *Cambarus* (*Lacunicambarus*) *erythrodactylus*, a new species of crayfish (Decapoda: Cambaridae) of the *Cambarus diogenes* complex from Alabama and Mississippi, U.S.A. Proceedings of the Biological Society of Washington 127(4): 572–584.

Sinclair, E.A., F.W.J. Fetzner, J. Buhay and K.A. Crandall. 2004. Proposal to complete a phylogenetic taxonomy and systematic revision for freshwater crayfish (Astacidea). Freshwater Crayfish 14: 21–29.

Skurdal, J. and T. Taugbøl. 2002. *Astacus*. pp. 467–510. *In*: D.M. Holdich (ed.). Biology of Freshwater Crayfish. Blackwell Science, Oxford.

Śmietana, P., H.K. Schulz, S. Keszka and R. Schulz. 2006. A proposal for accepting *Pontastacus* as a genus of European crayfish within the family Astacidae based on a revision of the west and east taxonomic literature. Bull. Fr. Pêche Piscic. 380-381: 1041–1052.

Sokolsky, A., V. Ushivtsev, A.S. Mikouiza and E. Kalmikov. 1999. Influence of sea level fluctuations on wild crayfish populations in the Caspian Sea. Freshwater Crayfish 12: 655–664.

Sonntag, M.M. 2006. Taxonomic standing of the three subspecies of *Pacifastacus leniusculus* and their phylogeographic patterns in the Klamath Basin area. Brigham Young University, pp. 71.

Souty-Grosset, C., D.M. Holdich, P. Noël, J.D. Reynolds and P. Haffner. 2006. Atlas of crayfish in Europe. Collection Patrimoine Naturel Publications Scientifiques du Muséum National d'Histoire Naturelle, Paris, 187 pp.

Starobogatov, Ya.I. 1995. Taxonomy and geographical distribution of crayfishes of Asia and East Europe (Crustacea Decapoda Astacoidei). Arthropoda Selecta 4: 3–25.

Stephanou, D. 1987. Cyprus Country Report for 1986–1987, 3 p. http://www.fao.org/docrep/005/s7360b/S7360B02.htm.

Sumner, C.E. 1978. A revision of the genus *Parastacoides* Clark (Crustacea: Decapoda: Parastacidae). Aust. J. Zool. 26: 809–821.

Swahn, J.-Ö. 2004. The cultural history of crayfish. Bull. Fr. Pêche Piscic. 372-373: 243–251.

Swain, R., A.M.M. Richardson and M. Hortle. 1982. Revision of the Tasmanian genus of freshwater crayfish *Astacopsis* Huxley (Decapoda: Parastacidae). Mar. Freshwat. Res. 33: 699–709.

Taugbøl, T., A. Arens and A. Mitans. 2004. Freshwater crayfish in Latvia: Status and recommendations for conservation and sustainable use. NINA Project Report 29, 23 pp.

Taylor, C.A. 2002. Taxonomy and conservation of native crayfish stocks. pp. 236–257. *In*: D.M. Holdich (ed.). Biology of Freshwater Crayfish. Blackwell Science, Oxford.

Taylor, C.A. and G.A. Schuster. 2004. The Crayfishes of Kentucky. Illinois Natural History Survey Special Publication No. 28. viii + 219 pp.

Taylor, C.A. and G.A. Schuster. 2011. Monotypic no more, a description of a new crayfish of the genus *Barbicambarus* Hobbs, 1969 (Decapoda: Cambaridae) from the Tennessee River drainage using morphology and molecules. Proc. Biol. Soc. Wash. 123: 324–334.

Taylor, C.A., M.L. Warren, Jr., J.F. Fitzpatrick, Jr., H.H. III Hobbs, R.F. Jezerinac, W.L. Pflieger and H.W. Robison. 1996. Conservation status of crayfishes of the United States and Canada. Fisheries 21: 25–38.

Taylor, C.A., G.A. Schuster, J.E. Cooper, R.J. DiStefano, A.G. Eversole, P. Hamr, H.H. Hobbs, III, H.W. Robinson, C.E. Skelton and R.F. Thoma. 2007. A reassessment of the conservation status of crayfishes of the United States and Canada after 10+ years of increased awareness. Fisheries 22: 372–389.

Taylor, R.S., F.R. Schram and Y.B. Shen. 1999. Crayfish from the Upper Jurassic of Liaoning Province, China. Proceedings and Abstracts, 4th International Crustaceans Congress, Amsterdam, p. 76.

Tertyshny, A.S. and M.A. Panchishny. 2009. Use of effective microorganisms against parasites and disease of narrow-clawed crayfish. *In*: P. Kozák and A. Kouba (eds.). Abstract Book, Future of Native Crayfish in Europe, Regional European Crayfish Workshop: 7th–10th September 2009, Písek, Czech Republic, 42.

Thoma, R.F. and G.W. Stocker. 2009. *Orconectes* (*Procericambarus*) *raymondi* (Decapoda: Cambaridae), a new species of crayfish from southern Ohio. Proc. Biol. Soc. Wash. 122: 405–413.

Thoma, R.F., Z.J. Loughman and J.W. Fetzner, Jr. 2014. *Cambarus* (*Puncticambarus*) *callainus*, a new species of crayfish (Decapoda: Cambaridae) from the Big Sandy River basin in Kentucky, Virginia, and West Virginia, USA. Zootaxa 3900(4): 541–554.

Thoma, R.F. and J.W. Fetzner, Jr. 2015. *Cambarus* (*Jugicambarus*) *magerae*, a new species of crayfish (Decapoda: Cambaridae) from Virginia. Proceedings of the Biological Society of Washington 128(1): 11–21.

Toon, A., M. Pérez-Losada, C.E. Schweitzer, R.M. Feldmann, M. Carlson and K.A. Crandall. 2010. Gondwanan radiation of the Southern Hemisphere crayfishes (Decapoda: Parastacidae): evidence from fossils and molecules. J. Biogeog. 37: 2275–2290.

Villalobos, J.L., A. Cantu and E. Lira. 1993. Los crustáceos de agua dulce de México.Revista de la Sociedad Mexicana de Historia Natural, Volumen Especial (XLIV): 267–290.

von Martens, E. 1869. Südbrasilische Süss-und Brackwasser Crustaceen nach den Sammlungen des Dr. Reinh. Hensel. Archiv für Naturgeschicthe 35: 1–37.

Warren, M.L., Jr. and B.M. Burr. 1994. Status of freshwater fishes of the United States: Overview of an imperiled fauna. Fisheries 19: 6–18.

Whiting, A.S., S.H. Lawler, P. Horwitz and K.A. Crandall. 2000. Biogeographic regionalization of Australia: assigning conservation priorities based on endemic freshwater crayfish phylogenetics. Animal Conservation 3: 155–163.

Wildlife and Freshwater Fisheries Division, Alabama Department of Conservation and Natural Resources. 2005. Conserving Alabama's wildlife: a comprehensive strategy. Alabama Department of Conservation and Natural Resources, Montgomery, Alabama, 322 pp.

Williams, J.D., M.L. Jr. Warren, K.S. Cummings, J.L. Harris and R.J. Neves. 1993. Conservation status of freshwater mussels of the United States and Canada. Fisheries 18: 6–22.

Zaccara, S., F. Stefani, P. Galli, P.A. Nardi and G. Crosa. 2004. Taxonomic implications in conservation management of white-clawed crayfish (*Austropotamobius pallipes*) (Decapoda, Astacidae) in Northern Italy. Biological Conservation 120: 1–10.

Zeidler, W. and M. Adams. 1990. Revision of the Australian crustacean genus of freshwater crayfish *Gramastacus* Riek (Decapoda: Parastacidae). Invert. Syst. 3: 913–924.

Population Genetics of Crayfish: Endangered and Invasive Species

Catherine Souty-Grosset

"Conservation genetics encompasses genetic management of small populations to maximize retention of genetic diversity and minimize inbreeding, resolution of taxonomic uncertainties and delineation of management units, and the use of molecular genetic analyses in forensics and to understand species' biology".

(Frankham et al. 2009)

"The genetics and evolution of invasive species have received far less attention than their ecology. Invasive species may evolve both during their initial establishment and during subsequent range expansion, especially in response to selection pressures generated by the novel environment. Hybridization, either interspecific or between previously isolated populations of the same species, may be one important stimulus for the evolution of invasiveness".

(Sakai et al. 2001)

Introduction

Indigenous crayfish species (ICS)

Indigenous crayfish species are under pressure because of pollution, habitat loss, overfishing and overexploitation. Initially, conservation management plans consisted of restoring the habitat and also of population translocations but very often without knowledge of their taxonomic status. Even if ecological managers wanted to be informed, the "traditional taxonomy", based on morphological characters, was discouraging because it could indicate several types of classification. Sound knowledge of systematic relationships in a given taxon, especially within the problematic range from geographically separated populations to closely related species is essential for restocking operations, for promoting gene flow

Université de Poitiers, Laboratoire Ecologie & Biologie des Interactions - UMR CNRS 7267, Equipe Ecologie Evolution Symbiose - Bât B8, 5 rue Albert Turpin, TSA 51106, 86073, Poitiers Cedex 9, France.
Email: Catherine.souty@univ-poitiers.fr

by translocation of animals and for assessing priorities in the preservation of particular populations. Consequently, as explained in the previous chapter, priorities are to provide some contributions to taxonomic clarifications. For example the classical taxonomy of *Austropotamobius pallipes* (Lereboullet 1858) is a good example of this problem: the number of species, subspecies, or varieties of *A. pallipes* may vary depending on the philosophical stance of authors: the taxonomy obtained by studies of morphological characters revealed a complex species. Bott (1950) considered three subspecies: *A. pallipes pallipes* in France, England and Ireland; *A. pallipes lusitanicus* in Spain; *A. pallipes italicus* in Italy and Balkans. However, Karaman (1962) defined two species with three subspecies within *italicus* and Brodsky (1983) redefined two subspecies within *pallipes* and two within *italicus*. Thus the reliable taxonomy of the *Austropotamobius* complex was the first problem to be resolved before any conservation effort, as an inadequate taxonomy could have dramatic consequences in the management scheme. With the advent of molecular biology, resolution of the status of European Astacidae species was therefore undertaken (Grandjean et al. 2000, 2002a,b). Grandjean et al. (2002a) have analyzed mitochondrial 16S DNA sequences in several samples from Ireland, France and Corsica, Spain, Italy, Austria and Slovenia. In accordance with morphological data extracted from recent papers, a new classification, based on the presence of three subspecies (*italicus, carinthiacus* and *carsicus*) within *italicus*, was proposed. At present, *A. pallipes* is well recognised as a species complex and the identity of the taxon *A. pallipes* is clear across its northern and western range (particularly in France, Great Britain and Ireland). The situation is more complex with *A. italicus*, which comprises three subspecies in Spain, Italy, Austria and Balkans. Effectively Fratini et al. (2005) confirmed the presence of both *A. pallipes* and *A. italicus* in the Italian peninsula and the existence within the latter species of a strong intraspecific genetic variation, due to the occurrence of four subspecies with a well-defined geographic distribution. From a conservation viewpoint, Italy, with its high haplotype variability, may be considered a 'hot spot' for the genetic diversity of the European native crayfish *Austropotamobius*. It is why these authors suggested that re-introduction programs should be conducted with extreme caution in Italy, since not only the two *Austropotamobius* species but also the four *A. italicus* subspecies are genetically and taxonomically separate units and require independent conservation plans. This fact shows that classical taxonomical methods are sometimes not powerful enough to differentiate groups along phylogenetic lines or to provide a precise delimitation of closely related species or intraspecific taxa and this was exactly the case for *Austropotamobius pallipes*, native in much of south-western Europe (Souty-Grosset et al. 2006, Holdich et al. 2009).

In order to maintain both the genetic specificity of populations and the genetic diversity within and between populations, there is a need not only for increased taxonomic clarification but also descriptions of natural population distribution and biogeographical history. In this context, biodiversity may be measured not only at the level of ecosystem diversity, relating to the variety of habitats, biotic communities and ecological processes, as well as the tremendous diversity present within ecosystems in terms of habitat differences and of the variety of ecological processes but also at two other levels: (1) genetic diversity, occurring within and between populations of crayfish species as well as between species; (2) species diversity, quantified as the variety of living species in an ecological unit. Today, conservation genetics aims to maintain, on one hand, the genetic specificity of populations (genetic integrity principle) and, on the other hand, the genetic diversity within and between populations (biodiversity principle), these basic principles being considered both at the level of protection measures and management measures (Souty-Grosset et al. 2003). According to Weiss (2005) the main problem linking genetics to conservation is that small populations in the wild tend to suffer loss of genetic variation leading to reducing the possibility of adapting to environmental changes. For example as an endangered species (IUCN: Füreder et al. 2010), *A. pallipes* is subject to a loss of genetic diversity, a result of deterioration of water quality responsible for habitat fragmentation, with populations being confined to headwaters of the catchments. Consequently a certain degree of genetic variability must be absolutely maintained within the species because it governs the adaptation potential; the populations must be capable of responding to new environmental conditions.

If a recovery programme is to be initiated, then it is important to know how genetic variation is partitioned between remaining listed populations of endangered crayfish species. The conservation of genetic diversity is an important step in conservation strategies because the highest level of genetic variation is the rule for the long-term survival of a species. Up to 1996, little was known about population genetics in crayfish because only enzymatic electrophoretic analysis had been performed which did not provide useful markers for crayfish stock identification (Nemeth and Tracey 1979, Brown 1980, Albrecht and Von Hägen 1981, Attard and Vianet 1985, Busack 1988, 1989, Agerberg 1990, Fevolden and Hessen 1989). According to Fetzner and Crandall (2002), up to the year of their publication, the number of studies surveying population-level genetic variability was still very low (less than 15). After allozymes, another method of investigating genetic relationships has been developed from the analysis of mitochondrial DNA (mtDNA) using analysis of Restriction Fragment Length Polymorphism (RFLP). Owing to its maternal mode of inheritance and absence of recombination (Avise et al. 1987, Wilson et al. 1985), mtDNA was a favoured genetic system for analysis of population structure. Generally, mtDNA offers two important advantages over nuclear genetic markers such as isozymes: the phylogenetic relationships of mtDNA patterns reflect the history of maternal lineages within a population or species; the scarcity of papers about mtDNA variation analyses in crustaceans, indeed scarcely any on crayfish, was probably related to the difficulty of extracting total mtDNA for RFLP analysis. During the last decade, with the development of several PCR-based techniques, studies from several countries were conducted to first describe the distribution of the present natural populations and secondly, by studying sequences of mitochondrial DNA from the mitochondrial large (16S) subunit and mainly from the cytochrome oxidase subunit COI, to clarify the taxonomy (number and identification of the present species and subspecies by phylogenetic inferences) and to assess the biogeographical history. For maintaining the evolutionary heritage of populations, evolutionarily significant units (ESUs) must be identified. According to Moritz (1994), ESUs should be "reciprocally monophyletic for mtDNA and show significant divergence of allele frequencies at nuclear loci". Based on this definition, many ESUs have been identified within previously described crayfish species, including *Cherax tenuimanus* (Smith) from Western Australia (Nguyen et al. 2002b), and *Austropotamobius pallipes* (Lereboullet 1858) from France (Gouin et al. 2006), from Italy (Fratini et al. 2005) and from the Iberian Peninsula (Dieguez-Uribeondo et al. 2008). These preliminary steps are fundamental before defining conservation units and before working at the catchment level, using highly polymorphic nuclear markers as microsatellites. Effectively using both genetic markers (mtDNA and microsatellites) is powerful for defining management units (MUs according to Moritz 1994) within species. These recent approaches have provided a valuable framework for research leading to more frequent dialogues between geneticists and managers.

Non indigenous crayfish (NICS)

Non indigenous crayfish are now considered as the most important threats to indigenous European crayfish through competition for food or shelter, through direct aggression or cross-mating, and importantly also, through transmission of disease as the crayfish plague with the continuous spread of non-indigenous signal, red swamp and spiny-cheek crayfish (Holdich 2003). These species were introduced either intentionally, for harvesting for food, or unintentionally, as unused bait or unwanted aquarium pets (Holdich 1999, Taugbøl and Skurdal 1999, Lodge et al. 2000, Gherardi 2006, Taylor et al. 2007) with the consequence of illegal exploitation and trading of crayfish. In general, the application of molecular techniques to the study of crayfish has tended to be driven by concern for the conservation of declining native species, rather than exploring the invasion process of non-native species. But as NICS represent a major threat, the understanding of their population genetics is also one of the most important goals in conservation biology. Recently, it has been proposed that methods using molecular tools could help define efficient eradication strategies and should be a preliminary step in the management process. The genetics and evolution of Non Indigenous Species (NIS) has received far less attention than their ecology (Lee 2002), since genetic and evolutionary processes may be the key features in determining whether invasive species establish and spread (Sakai et al. 2001). Of great theoretical and practical importance is the ability to identify the location of origin of NIS and their route of invasion (Wilson et al. 1999, Kreiser et al. 2000, Cox 2004). Theoretical models of genetic organization and population structure following a founding effect can be described by two different scenarios. The first model predicts subpopulations to show strong genetic structuring and clinal variation, while the second involves extinction and recolonization that enhance gene flow and reduce interpopulation differentiation (Alvarez-Buylla and Garay 1994). Migration may be critical, not only as a source of continuing propagule pressure, but also as an important source of genetic variation to the colonizing population, if multiple invasions provide the genetic variation necessary for adaptive evolution. Multiple introductions can create invasive populations that are much more genetically diverse than any single source population when the invasive species is highly structured in its native range. Different colonizing populations of the same species are likely to be genetically divergent with different levels of genetic variation and therefore have different capacities to promote invasiveness; characteristics that promote invasiveness might evolve in some populations but not in others. Gene flow between populations could result in the spread of invasive genotypes. Alternatively, gene flow between populations that swamps out locally beneficial alleles could prevent evolution of invasiveness (Sakai et al. 2001). From this knowledge, useful information can be obtained about the vectors and the number of introductions and this may assist attempts to halt or to slow down the invasion process. Effective biological control agents of harmful NIS can be also found and it is also possible to understand to what degree the "enemy release" hypothesis (e.g., Keane and Crawley 2002) can explain invasions (Kreiser et al. 2000, Patti and Gambi 2001, Allendorf and Lundquist 2003). Molecular genetics techniques today offer a very powerful set of tools for characterizing populations of NIS and for relating them to the populations of their native and colonized geographical areas (Cox 2004). These molecular markers can provide an indication of the amount of genetic variation lost during a colonization bottleneck and furnish evidence for multiple population sources. They have been successfully used to pinpoint the source areas and the routes of dispersal followed by a number of freshwater non-indigenous crustaceans, including *Cercopagis pengoi* Ostroumov and the freshwater Cladocera (Cristescu et al. 2001, Hebert and Cristescu 2002). Barbaresi et al. (2007) undertook a pilot study on the red swamp crayfish *Procambarus clarkii* Girard in which they used molecular markers with the aim of understanding the dynamics of introductions of such a commercial invasive species that most often follow illegal paths. When historical human records are incomplete, inaccurate, or simply non-existent, molecular genetic studies often offer a powerful tool for the identification of relationships between introduced populations. Recently, Blanchet (2012) reviewed the potential uses of molecular tools to address issues in invasion biology on freshwater ecosystems, including the early detection of novel and cryptic non-native species, the identification of introduction pathways and vectors, the understanding of the drivers in successful invasions and the assessment of effective population sizes during the establishment of new populations. The usefulness of molecular tools to assess the ecological and evolutionary consequences of biological invasions is assessed with the opportunity of the latest techniques in molecular ecology (e.g., multiplex high-throughput sequencing and DNA barcoding).

This chapter develops the state of knowledge about conservation genetics of indigenous crayfish and provides the first data of novel investigations of the genetics of non indigenous, invasive crayfish species. Since Fetzner and Crandall (2002), the panel of molecular markers has been extended in order to assess divergence patterns and gene flow between populations of indigenous crayfish species and to evaluate possible hybridization events particularly between ICS and NICS.

Conservation genetics of indigenous crayfish species: case studies

Both habitat destruction and reduction of the size of the crayfish populations are responsible for the loss of indigenous crayfish populations. Populations are fragmented because of geographical isolation in headwaters. Indeed such decreasing populations are susceptible to stochastic events, i.e., unpredictable events as environmental changes or mutations, leading to genetic drift and finally the loss of genetic diversity.

According to Frankham et al. (2009), conservation genetics is defined as "the application of genetics to preserve species as dynamic entities capable of coping with environmental change. It encompasses genetic management of small populations, resolution of taxonomic uncertainties, defining management units (MUs) within species and the use of molecular genetics analyses in forensics and understanding species' biology." Information of the genetic diversity in crayfish is now a prerequisite for scientists and managers in forming strategies to preserve and protect indigenous and endangered species; the suitability of the target habitat, the stocking material and the stocking procedure itself are paramount during any reintroduction measure. Additionally, general water quality and structural parameters of a suitable habitat, as well as genetics of the stocking material must be considered (Souty-Grosset and Reynolds 2010).

Europe: the white-clawed crayfish *Austropotamobius pallipes* Lereboullet, 1858

According to Souty-Grosset et al. (1997), the white-clawed crayfish, *Austropotamobius pallipes pallipes*, still has a widespread distribution in France, but since the last century, populations have declined because of habitat alteration (due to human disturbance) and have also been eliminated by crayfish plague, for which introduced exotic species are a vector. Action plans for the conservation of *A. pallipes* are urgent and if recovery programmes are to be initiated in France and elsewhere, then it is important to estimate how much genetic variation is partitioned between remaining populations as the species is being currently threatened in all its European distribution. Souty-Grosset et al. (1997) were the first to utilize new molecular markers in the study of white-clawed crayfish populations. Mitochondrial DNA (mtDNA) variation in natural populations was examined by RFLP analysis in samples taken from fifteen French populations and six other European populations representative of three subspecies observed in *A. pallipes* in order to examine the extent of differentiation between populations. The study revealed a low level of genetic variation among English, Welsh and most French populations, corresponding to a genetic stock uniformity among *A. pallipes pallipes*. The only two French populations exhibiting a high level of intrapopulational genetic variation are in fact mixed samples: the comparison with results obtained in European populations revealed that the first population was composed of the two subspecies *A. pallipes pallipes* and *A. pallipes italicus* and the second of *A. pallipes italicus* and *A. pallipes lusitanicus*. Results showed that some repopulations, performed in the past from *A. pallipes italicus* and supposedly having failed, had been successful and as a result, the French stock did not correspond to the only subspecies *A. pallipes pallipes*. A first analysis of genetic variability observed on a regional scale revealed that there was no genetic structure according to the catchments and this could reflect human-mediated movement of crayfish stocks between these catchments. Consequently, mtDNA is indeed a relevant marker to measure genetic diversity between crayfish populations, to map how the subspecies are partitioned in France and what the importance of each is before any planning crayfish conservation strategies of this native crayfish. Later, Grandjean et al. (2002a) reviewed the phylogenetic relationships of the genus *Austropotamobius* in Europe by the compilation of two recent genetic studies based on the partial nucleotide sequence of the mitochondrial RNA 16S gene. The results showed a well-resolved phylogeny revealing four distinct clades: A1, A2, A3 and B, supported by high bootstrap values. The clades A (including A1, A2, A3) and B are separated by a high genetic divergence (5%). Based on morphological and nuclear data, two species could be defined: *A. italicus* and *A. pallipes*, respectively. The average of genetic divergence within the major group A and B was 2.1% ± 1.2 and 0.9% ± 0.6 respectively. The three clades A1, A2 and A3 correspond mainly to crayfish sampled from Austria-Switzerland, South of Balkans and Italy-Spain respectively. On the basis of morphological, genetic and distribution data, a new systematics-based on two species *A. pallipes* and *A. italicus* with 3 subspecies *A. i. carinthiacus*, *A. i. carsicus* and *A. i. italicus*—were proposed for the white-clawed crayfish species complex. Grandjean et al. (2002a) rejected the specific status of *A. berndhauseri* given for the endemic crayfish from southern Switzerland (Bott 1972) and redefined the species as *A. italicus carsicus*; moreover in the light of molecular data (Santucci et al. 1997, Grandjean et al. 2000) the status of *A. i. lusitanicus* concerning Spanish crayfish was also rejected.

In France, the analysis of mtDNA by RFLP on a greater number of populations (Grandjean and Souty-Grosset 2000) showed the existence of two genetically differentiated entities of the white clawed crayfish *A. pallipes*, corresponding to northern and southern populations. The same dichotomy was revealed using microsatellite markers (Gouin et al. 2006). Based on these results, the authors proposed the designation of two evolutionarily significant units for *A. pallipes* in France. Their data also support the maintenance of separate demographic management strategies (northern and southern populations) for crayfish inhabiting different river systems. However, in their study, the discovery of mixed populations poses some challenges for future management strategies.

Northern and southern populations could have been isolated for several thousand generations, which would place the separation between these two lineages near the last glaciation period of the Pleistocene. According to the present distribution of *A. pallipes*, the hypothesis of two refuges during the last glacial, one located at the Atlantic coast and one located at the Mediterranean coast, from which post-Pleistocene dispersion would have occurred, seems very likely and is in agreement with hypotheses already proposed for other freshwater species distributed in Western Europe (Gouin et al. 2001). Human-mediated translocation events are also significant in determining present indigenous crayfish distribution patterns. For example, translocation of white-clawed crayfish appears to have been a common practice throughout Western Europe (Souty-Grosset et al. 1997, Largiadèr et al. 2000, Grandjean et al. 2001, Gouin et al. 2003, Fratini et al. 2005, Trontelj et al. 2005), and particularly in France (Machino et al. 2004). It is therefore plausible that crayfish could have been translocated from east to west, leading to the establishment of new populations or to genetic admixture between Atlantic and Mediterranean stocks. The presence of crayfish on European islands is also of interest. Grandjean et al. (1997a) have shown through molecular genetics that English crayfish were probably introduced from France. In Ireland, the white-clawed crayfish (defined as *A. pallipes*) is widespread throughout the central lowlands. Ireland's biogeography has long been of interest—cut off soon after the last glaciation, a number of endemic species or subspecies have been identified, but there are also affinities with the fauna and flora of distant areas: Nordic, Celtic, Continental and of particular interest, Lusitanian. One hypothesis is that Irish crayfish may be a relic of Lusitanian stocks, but there was also a suspicion that crayfish were most likely to have been introduced from Great Britain (Lucey 1999). When Irish stocks were investigated with mtDNA RFLP, only one haplotype was found across all Irish populations which corresponds to the less frequent of the two haplotypes found in the French region Poitou-Charentes (Gouin et al. 2003). Moreover, the haplotype is quite different from the Spanish haplotypes (Dieguez-Uribeondo et al. 2008). Consequently, Irish crayfish may not have been introduced from England nor from Spain but from France. There were monastic orders from France in Ireland as early as the 12th century and perhaps introductions from western France were made by them. Moreover, translocations from the south to the north of Ireland were ascertained by genetic investigations. The colonization history may have happened step by step among geographically close locations. In Spain, genetic findings are also in accordance with the theoretical genetic consequence of translocation from a limited number of individuals. Only one haplotype was found which was also shared with Italian ones (Grandjean et al. 2001). This confirms that human introductions of crayfish from Italy were the origin of most, if not all, Spanish populations. Moreover, the results suggest a drastic bottleneck (severe reduction in population size) during the history of Spanish populations. This relative lack of genetic variation in the Spanish *A. pallipes* stock could be the result of different but not exclusive events; not only human-mediated introductions but also selection (for example, impact of the crayfish plague involving a selection and restriction of the genetic diversity) and recent historical events (acute population fragmentation in the last few decades) (Dieguez-Uribeondo et al. 2008). More recently, Pedraza-Lara et al. (2010) found that the genetic variation observed in Iberian populations of *A. i. italicus* could be linked to demographic responses to the retreat and recovery of the ice during and after the last glacial maxima (LGM). Considering the hypothesis proposed by Karaman (1962) and supported by Dieguez-Uribeondo et al. (2008), it is possible that the sub-species *A. i. italicus* had a unique distribution area ranging from the Iberian Peninsula eastwards until at least the Central Apennines. Matallanas et al. (2011) used a 1184 bp-length sequence of mtDNA COI gene and found a sensitive tool for assessing the genetic diversity and structure of the Spanish populations of *A. italicus*, because they found eight haplotypes, i.e., corresponding to the highest diversity reported in Spanish crayfish populations and a substantial genetic differentiation among populations with a clear geographic pattern. Thus, given the risk status of the species across its range, this variability in certain populations offers some hope for the species from a management point of view. Matallanas et al. (2012) analyzed three Spanish populations with nuclear (microsatellites) and mitochondrial markers (COI and 16S rDNA) and found four haplotypes at mitochondrial level and polymorphism for four microsatellite loci. Despite this genetic variability, bottlenecks were detected in the two natural Spanish populations tested. In addition, the distribution of the mitochondrial haplotypes and Simple Sequence Repeats (SSR) alleles show a similar geographic pattern and the genetic differentiation between these samples is mainly due to genetic drift.

In contrast to Spain, Italy appears to be a hot spot of genetic variability for the genus *Austropotamobius* with the presence of both *A. pallipes*, confined to North-Western Italy, and *A. italicus*, distributed across the peninsula (Fratini et al. 2005: Fig. 1 and Fig. 2).

Although both species overlap in the Ligurian Apennine (Santucci et al. 1997), there is no evidence of hybridization events. Additionally, within *A. italicus*, a strong intra-specific genetic variation was found, due to the occurrence of four subspecies with a well-defined geographic distribution, *A. i. italicus,*

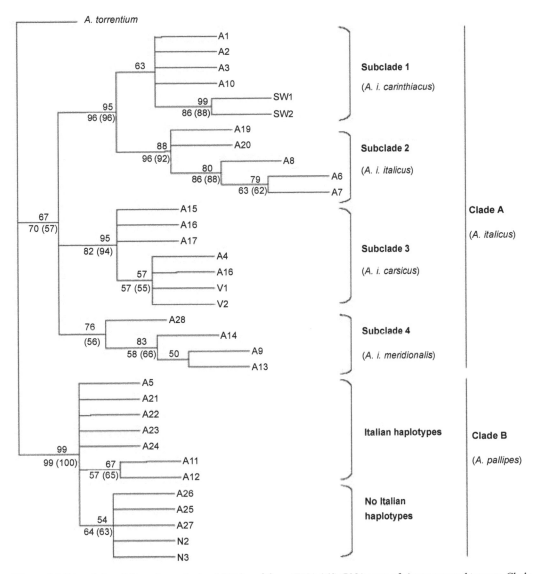

Figure 1. NJ tree inferred from the analysis of 486 bp of the mtDNA 16S rRNA gene of *Austropotamobius* spp. Clade A corresponds to *A. italicus* and clade B to *A. pallipes*. The subclades designed 1–4 indicate the *A. italicus* subspecies (according to Fratini et al. 2005).

A. i. carsicus, A. i. carinthiacus, and *A. i. meridionalis*, this last subspecies being described for the first time (Fratini et al. 2005). As a result, Bertocchi et al. (2008) investigated the population differentiation of *A. italicus* in Tuscany, based on sequences of a fragment of 16S mtDNA analysis in *A. italicus* populations across 5 catchments: River Magra, R. Serchio, R. Bisenzio, R. Sieve and R. Arno (from west to east). The use of markers with higher resolution, such as microsatellites, are essential in order to understand their genetic structure and then to define concrete conservation measures (Gouin et al. 2000). Following recommendations that a combination of different molecular methods are used (Haig 1998, Souty-Grosset et al. 1999, 2003), Bertocchi et al. (2008) characterized the genetic structure by analysing microsatellites for identifying, within the *A. italicus italicus* lineage, demographically and genetically independent populations that may be designated as separate management units (MU *sensu* Moritz 1994) and for defining conservation priorities. Management efforts and economic resources should be concentrated on populations with the highest genetic variability, universally known to be crucial for determining its

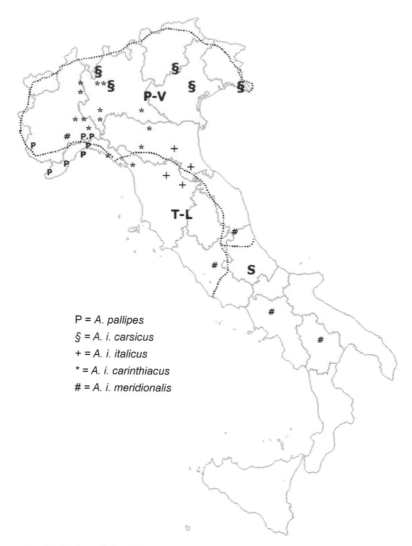

P = *A. pallipes*
§ = *A. i. carsicus*
+ = *A. i. italicus*
* = *A. i. carinthiacus*
# = *A. i. meridionalis*

Figure 2. Geographic distribution of *A. pallipes* and *A. Italicus* in Italy. PV: Padan-Venetian ichthyogeographic district; T: Tuscan-Latium district; S: Southern Italy district (according to Fratini et al. 2005).

evolutionary potential to adapt to changing environmental conditions (Soulé and Mills 1992, Primack 2000). Their study aimed at (i) defining the local phylogeography of *A. italicus* in Tuscany; (ii) describing the extent of the genetic variability within and between populations; (iii) interpreting the distribution of the genetic variability according to historical events and the phylogeography of the populations and (iv) integrating genetic knowledge with the ecological data collected in the same water courses (Renai et al. 2006). Bertocchi et al. (2008) concluded that Tuscany, where two *Austropotamobius italicus* subspecies are present, is the depository of a quite elevated haplotype genetic variability and strong genetic distances. On the other hand, through microsatellites analysis, low levels of genetic diversity, significant differentiation among the different river drainages and populations highly structured within rivers were described. Any conservation program and re-introduction plan should keep in mind the geographic distribution of both the taxonomic units to ensure the preservation of independent genetic pools. Moreover, in the light of the recorded possible human effect on the distribution of the different evolutionary units, it is desirable that any population would be genetically screened before any management action. Moreover particular attention is needed in the Italian overlapping areas between subspecies where genetically distinct units live together. Whatever the country (for example France: Gouin et al. 2006; Italy: Bertocchi et al. 2008), the translocation of indigenous crayfish populations is frequent and is leading to mixed populations.

Recent advances shows again the frequency of events of translocations such as in islands: Amouret et al. (2015) published the first record of the white-clawed crayfish in Sardinia Island (Italy). Using a fragment of the mitochondrial DNA 16S rRNA gene, the haplotypes corresponded to the *A. italicus meridionalis* subclade. Results improve the existing knowledge about the phylogeography of the taxon across Italy, confirming its complex pattern of distribution. In addition to the non-native status of the Sardinian *A. i. meridionalis* crayfish, the most proximal Mediterranean population of white-clawed crayfish existing in Corsica belongs to *A. pallipes* from Southern France. The authors concluded that, due to its "permit of residence", it is important to encourage the protection and conservation of the local habitat of the nominate specimen to face any habitat degradation, increased human fishing as the spread of the North American crayfish, *Procambarus clarkii*. Continental translocation events were also discovered both in Maritim alps (Stefani et al. 2011) and the record of a population of the southern lineage of *A. italicus*, i.e., a non native taxon, in northern Alps (Chucholl et al. 2015).

Europe: the stone crayfish *Austropotamobius torrentium* Schrank, 1803

The stone crayfish *A. torrentium* (Schrank, 1803) is widely distributed in Southeastern and Central Europe. *A. torrentium* shows a distribution with almost no overlap with the western European *A. pallipes* complex, except in the Ausserfern Region (Tyrol, Austria) according to Sint et al. (2006). The species ranges from eastern France and Luxembourg, throughout southern Germany, northern Switzerland, Austria and northeastern Italy, far southeast from Slovenia and Croatia into the Balkans all the way into northern Greece, Bulgaria and into Black Sea Turkey (Souty-Grosset et al. 2006). It is confined to headwaters and small water systems, adapted to water with turbulent flow and a rocky environment (Huber and Schubart 2005). Trontelj et al. (2005) investigated the phylogenetic and phylogeographic relationships in the crayfish *Austropotamobius torrentium* inferred from mitochondrial COI gene sequences. They sampled stone crayfish from 32 localities in central Europe, i.e., from many parts of the Danubian and Adriatic drainage in Slovenia, and the Southeastern Alps, along with populations from the upper Rhine drainage, which all share the same, monophyletic origin. The numerous, genetically very similar haplotypes are no older than the second half of the Pleistocene. The high haplotype diversity in the small area on the southeastern edge of the Alps suggests that many populations survived the final cycles of glaciation in nearby microrefugia. Schubart and Huber (2006) sampled stone crayfish from 18 localities throughout southern Germany and used two mitochondrial genetic markers (16S rRNA and COI gene). Results revealed that German populations from the Danube and Rhine drainage and all shared identical haplotypes both from 16S rRNA and COI gene. Rare haplotypes from COI gene were occasionally encountered and restricted to southwestern Bavaria. Only three variable sites were found in 45 German, Swiss and Austrian stone crayfish resulting in five different haplotypes, with the prevalence of one most common haplotype. The authors showed that in German populations rare haplotypes are not randomly distributed, but found in higher frequencies in the Bavarian Alps of the Allgäu and in adjacent Tyrol. On the other hand, stone crayfish population from the Bavarian Forest and the Rhine tributaries appear genetically impoverished, so far only showing the most common haplotype. Consequently, there are significant differences between the Allgäu populations and the rest of the German populations in haplotype frequencies, resulting in a relatively high F_{ST} value. This finding is of importance for future conservation efforts of stone crayfish populations in Germany and Austria. More recently Iorgu et al. (2011) described the first microsatellite loci for the stone-crayfish by cross-species amplification. These microsatellite markers will be useful for population genetic studies of the stone crayfish at the microgeographical level.

Concerning the situation in Balkans, by studying Mitochondrial 16S rRNA and COI genes, Klobučar et al. (2013) considered phylogeography and phylogeny of the stone crayfish in order to elucidate the role of the Dinaric Karst geology in shaping the evolutionary history and genetic diversity of aquatic fauna in the western Balkans. The main finding is the existence of geographically isolated and deeply divergent cryptic monophyletic phylogroups within the species in the northern-central Dinaric region following the gradual north–south expansion of stone crayfish during the pre-Pleistocene.

Europe: the noble crayfish *Astacus astacus* Linnaeus, 1758

According to Holdich et al. (2009) the noble crayfish—considered as the most highly valued freshwater crayfish in Europe—is currently found in 39 territories across most of northern Europe, although its range was greater before the introduction of crayfish plague (*Aphanomyces astaci*). The most abundant populations exist in Nordic and Baltic countries, e.g., it is the dominant crayfish in Latvian waters, having been found in 220 out of 258 crayfish localities (Taugbøl 2004, Arens and Taugbøl 2005). *Astacus astacus* is indigenous to Scandinavia and was present in Finland, where its distribution has been reduced dramatically during the past century, due to environmental changes and above all due to crayfish plague outbreaks and stocking of alien signal crayfish (*Pacifastacus leniusculus*). For conservation purposes it is essential to know if the populations are autochthonous or if they are a mixture of several populations. Investigation of the *A. astacus* GA-repeat in the ITS1 region near the 5′ end was conducted in a population genetics study by Edsman et al. (2002). These authors showed that ITS fragments have the potential to be used as markers in population investigations of *A. astacus*. Following Edsman et al. (2002), Alaranta et al. (2006) used the same method for undertaking genetic studies of the Finnish noble crayfish populations and to investigate population diversity among selected noble crayfish populations in Finland, Sweden and Estonia. Based on the ITS1 fragment variation, some Finnish noble crayfish populations were most likely original populations or originated from one source population. They differed from the other populations according to the population divergence test. However, there were no differences between some of the Finnish populations and this may be a consequence of multiple stockings. Five of the Finnish populations differed from the Swedish and the Estonian populations. One population, Lake Saimaa, did not differ from one Estonian and two Swedish populations. Furthermore, a population from northern Finland was not different from a population in northern Sweden. The Estonian populations had a larger number of fragments present in their genotypes compared to the Finnish and the Swedish populations. The authors concluded that it would be important to know more about the copy number of the rDNA repeats and their location in the chromosomes as well as the inheritance mechanism and that consequently it would also be interesting to compare results using different techniques, such as random amplified polymorphic DNA polymerase chain reaction (RAPD-PCR) and inter-simple sequence repeat polymerase chain reaction (ISSR-PCR) and microsatellites with several loci developed for *A. astacus*. Recently, Schrimpf et al. (2011) provided the first large-scale study of haplotype diversity of *A. astacus* covering a large portion of its distribution range, including river catchments of the North and Baltic Seas in central Europe and the Black Sea in southeastern Europe. They analyzed a partial sequence of the mitochondrial gene cytochrome oxidase subunit I (COI) from a large number of crayfish stocks (92) sampled in three European river basins (Black Sea, North Sea and Baltic Sea). Twenty-two haplotypes were identified, with one common haplotype found across the whole study area. They detected differences in the genetic diversity between major river catchments. The highest haplotype diversity was found in southeastern Europe and a high number of unique haplotypes suggested a glacial refuge in the Balkan area. However they found very low haplotype diversity in central Europe that could be a result of human translocation and/or founder effects due to postglacial re-colonization. Nevertheless, the high frequency of unique haplotypes in all major catchment areas indicates a differentiation of noble crayfish populations throughout Europe despite the extensive human translocation of crayfish. The results of this study suggest a glacial refuge in the Balkan area and a postglacial re-colonization of central Europe. Despite human translocations, which were revealed by the unexpected distribution of some haplotypes, a differentiation of noble crayfish populations in all major catchment areas supports the establishment of distinct ESUs allowing the establishment of conservation management plans for this vulnerable species. In a second step Schrimpf (2013) focused on the postglacial recolonization of central Europe by *Astacus astacus* by using nuclear microsatellites. Results allowed the authors to calculate the first calibrated tree of noble crayfish and to correlate the split of lineages with climatic events. A very distinct lineage was discovered in the Western Balkans that may have served as an isolated glacial refugium during the last glacial maximum. A second independent refugium was suspected in the eastern Black Sea basin from where the species re-colonized central Europe through the Danube and through a second migratory route in Eastern Europe (Schrimpf et al. 2014, Fig. 3).

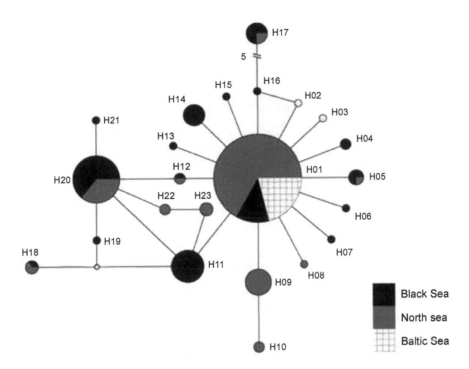

Figure 3. Median joining network of COI haplotypes (350 bp) from 416 individuals of *Astacus astacus*. The size of the circles is proportional to the frequency of the haplotypes. Median vectors are indicated as white dots. The number of base pair (bp) changes are given; no number = 1 bp change (Courtesy from Anne Schrimpf).

Gross et al. (2013) conducted a larger scale population genetic survey based on 10 newly developed microsatellite markers. They investigated crayfish sampled from Baltic Sea areas (Estonia, Finland and Sweden) where the largest proportion of the remaining populations exists and from the Black Sea catchment (the Danube drainage). Two highly differentiated population groups were identified corresponding to the Baltic Sea and the Black Sea catchments, respectively. The Baltic Sea catchment populations had significantly lower genetic variation and unique allele numbers than the Black Sea catchment populations. Within the Baltic Sea area, a clear genetic structure was revealed with population samples corresponding well to their geographic origin, suggesting little impact of long-distance translocations. Both cladogram and STRUCTURE analyses revealed a similar and clear structuring of the studied populations into two highly differentiated groups according to their catchment of origin: the Baltic Sea (Swedish, Finnish and Estonian populations) and the Black Sea (the Czech and southern German populations of the Danube drainage). These results are in agreement with those obtained on mtDNA COI gene haplotype frequencies (Schrimpf et al. 2011). Finally, Gross et al. (2013) suggested multiple glacial refugia as it is the case for *Austropotamobius pallipes* (Santucci et al. 1997, Grandjean and Souty-Grosset 2000, Gouin et al. 2001, 2006) and concluded that the clear genetic structure strongly suggests that the choice of stocking material for re-introductions and supplemental releases needs to be based on empirical genetic knowledge. Finally Bláha et al. (2015) found a significant genetic structure was found among populations that originated from Central compared to Southern Bohemia populations and match the source and translocated populations too. They conclude that a particular population is suitable for conservation management purposes and makes it reasonable to treat populations of noble crayfish as a single genetic unit.

USA: the golden crayfish *Orconectes luteus* Creaser, 1933

Fetzner and Crandall (2003) analyzed the genetic variation of the golden crayfish (*Orconectes luteus*) populations from the Ozarks region of Missouri. They found high levels of divergence among populations corresponding to watershed fragmentation created by Pleistocene glacial events. They also found general patterns of within watershed haplotype uniformity, where each watershed appeared to contain a unique haplotype or set of haplotypes. Consequently populations are isolated and the transfer of individuals between watersheds is a rare event.

USA: the case of cave dwelling *Orconectes* species

Buhay and Crandall (2005) compared three obligate cave-dwelling *Orconectes* species—*Orconectes incomptus* (Hobbs and Barr 1972), *Orconectes australis* (Rhoades 1944) and *Orconectes sheltae* (Cooper and Cooper 1997)—to two common surface stream-dwelling *Orconectes* species *Orconectes luteus* (wide-ranging in Missouri) and *Orconectes juvenilis* Hagen, 1870 (restricted range in the Upper Cumberland River and Kentucky River basins). When compared to the cave crayfish, it was found that the surface crayfish studied were showing a decline in genetic variability. Results show recent drastic declines in genetic variability in both species and particularly *O. luteus* suggesting conservation management.

USA: the Williams crayfish *Orconectes williamsi* Fitzpatrick, 1966

Fetzner and DiStefano (2008) studied the population genetics of an imperiled crayfish, *Orconectes williamsi* Fitzpatrick, from the upper White River Drainage of Missouri and northern Arkansas which had originally been described as the Williams crayfish, *Orconectes (Procericambarus) williamsi*. In this study, 24 sampling localities for *O. williamsi* were examined for levels of genetic variation within a 659 base pair region of the mitochondrial COI gene. *Orconectes williamsi* was found to be quite variable across its range, with a total of 53 distinct haplotypes being detected among the 326 sampled individuals. A nested clade phylogeographic analysis (NCPA) of *O. williamsi* populations resulted in four distinct haplotype networks that were quite divergent from one another. The majority of populations from the eastern portion of the Missouri range grouped into a single large network that was further divided into three distinct subgroups. The populations associated with the different networks detected by the NCPA were quite different from each other and should be considered evolutionary significant units (ESUs), as they are reciprocally monophyletic (i.e., show fixed haplotype differences) for their respective mtDNA profiles. In addition, the three divergent population subgroups that make up the main network should each be considered an individual conservation management unit (MU).

Australia: the yabby *Cherax destructor*, Clark, 1936

Population genetic studies on the Australian freshwater crayfish, *Cherax destructor*, revealed that there is population differentiation between watersheds, particularly between Northern and Southern watersheds (Hughes and Hillyer 2003, Nguyen et al. 2005). Campbell et al. (1994) found a high degree of morphological variability exhibited by *Cherax destructor* and studied their population structure. Genetic divergence between morphotypes was relatively low compared with known interspecific levels and levels of divergence between populations within the morphotypes. Nguyen et al. (2005) found that gene flow was constricted in multiple contiguous watersheds, which resulted in the differentiation of populations. They attributed this to behavioural or life history features restricting dispersal.

Australia: the swamp crayfish *Tenuibranchiurus glypticus*, Riek, 1951

Dawkins et al. (2010) studied the distribution and population genetics of the threatened freshwater crayfish genus *Tenuibranchiurus glypticus* (the world's second smallest freshwater crayfish) inhabiting coastal swamps in central-eastern Australia. Although only one species is described in the genus *Tenuibranchiurus*, it was expected that populations isolated through habitat fragmentation would be highly divergent. Their study aimed to determine if the populations of *Tenuibranchiurus* are genetically distinct, and if ancient divergence, as indicated in other species in the region, was evident. Analysis of two mitochondrial DNA gene regions revealed two highly divergent clades, with numerous additional subclades. Both clades and subclades were strongly congruent with geographical location, and were estimated to have diverged from each other during the Miocene or Pliocene era. Little sharing of haplotypes between subpopulations was evident, indicating negligible gene flow, and genetic differentiation between subclades possibly suggested distinct species. The coastal distribution of *Tenuibranchiurus*, severe habitat fragmentation and clear differences between subclades suggested also that they should be recognized as evolutionarily significant units, and be treated as such if conservation and management initiatives are to be undertaken.

Australia: the slender yabby *Cherax dispar* Riek, 1951

Bentley et al. (2010) showed that the freshwater crayfish *Cherax dispar* (Decapoda: Parastacidae) exhibits intraspecific genetic diversity in a biodiversity hotspot, i.e., in coastal regions and islands of South East Queensland, Australia. They used two mitochondrial genes (cytochrome oxidase subunit I and 16S ribosomal DNA) and one nuclear gene (ITS region). Deep genetic divergences were found within *C. dispar*, including four highly divergent (up to 20%) clades. The geographic distribution of each clade revealed strong latitudinal structuring along the coast rather than structuring among the islands. A restricted distribution was observed for the most divergent clade, which was discovered only on two of the sand islands (North Stradbroke Island and Moreton Island). Furthermore, strong phylogeographic structuring was observed within this clade on North Stradbroke Island, where no haplotypes were shared between samples from opposite sides of the island. The authors concluded that this low connectivity within the island supports the idea that *C. dispar* rarely disperse terrestrially (i.e., across watersheds).

Australia: the gilgie *Cherax quinquecarinatus* Gray, 1845

Gouws et al. (2010) investigated the phylogeographic structure of the gilgie (*Cherax quinquecarinatus*), the more widespread species among the six native south-western Australian freshwater crayfish species. The cytochrome c oxidase subunit I mitochondrial DNA gene was amplified. Three geographically-restricted lineages were identified: from the northwestern, southern coastal and intermediate/south-western regions. The extent of genetic differentiation among lineages was comparable to that observed in the koonac *Cherax preissii* (Gouws et al. 2006) suggesting temporal congruence of the historical events responsible for the observed structure.

Australia: *Geocharax gracilis* Clark, 1936

Sherman et al. (2012) studied the crayfish *Geocharax gracilis* found both in natural and agricultural drainage systems from south-eastern Australia. To investigate population structure, genetic diversity and patterns of connectivity in natural and human-altered ecosystems, they isolated 24 microsatellite loci using next generation sequencing. They detected high to moderate levels of genetic variation across most loci with a mean allelic richness of 8.42 and observed heterozygosity of 0.629. They showed significant deviations from Hardy-Weinberg expectations. They concluded that these 24 variable markers will provide an important tool for future population genetic assessments in natural and human altered environments.

Australia: the marron *Cherax tenuimanus* Smith, 1912

Nguyen et al. (2002a) studied the genetic variation of *Cherax tenuimanus*, one of the world's largest freshwater crayfish species which is endemic in Western Australia and recently classified by IUCN as Critically Endangered (Austin and Bunn 2010). They investigated the mtDNA gene region 16S between populations from Western Australia as well as translocated populations located in Southern Australia and Victoria. Two distinct genetic groups were identified (the first in Southern Western Australia and the second in all other populations); consequently they defined two ESUs. The high conservation value of the Margaret River population was recognized. The authors also identified introgression and outbreeding depression, which can lead to decreased population fitness of *Cherax tenuimanus*. Following this study, Bunn et al. (2008) described extensive translocations and hybridization and highlighted how the genetic integrity of this species has been raised as a serious conservation concern. They claimed that how maintaining genetic diversity is a management priority for both commercial and wild populations.

Australia: the giant freshwater crayfish *Astacopsis gouldi* Clark, 1936

Sinclair et al. (2011) studied the giant Tasmanian freshwater crayfish *Astacopsis gouldi*, the world's largest freshwater invertebrate. This species is known only from river drainages in northern Tasmania. A narrow distribution, pollution of habitat and over-harvesting has led to the rapid decline of populations and subsequent loss from a number of drainages. They collected mtDNA sequences to assess population structure and genetic diversity from throughout the species' distribution. They found a lineage from north-eastern Tasmania, which was genetically divergent compared with the remaining distribution in north-western Tasmania. Populations from the remaining distribution, including haplotypes found across a noted faunal barrier (Tamar River), were genetically homogeneous. This finding is concordant with the hypothesis of more interconnected drainages associated with lower sea levels in the past. The new cryptic lineage from north-east Tasmania requires further investigation and may be of extremely high conservation value. Conservation efforts for *A. gouldi*, combining habitat restoration with *in situ* management of wild populations and some population augmentation into once occupied rivers, will also have a positive impact for conservation of freshwater ecosystems in northern Tasmania (Richardson et al. 2006). The species was assessed as Endangered (IUCN: Walsh and Doran 2010).

New Zealand: *Paranephrops planifrons* White, 1842

This species has a wide distribution over the North Island and the West Coast district of the South Island. Smith and Smith (2009) found a small-scale population-genetic differentiation in the crayfish *Paranephrops planifrons* endemic to this country. A portion of the cytochrome c oxidase I gene in *Paranephrops planifrons* was used to evaluate population genetic differentiation and define conservation units at small spatial scales among neighbouring catchments in the central-west North Island of New Zealand. They found 23 haplotypes. Haplotype diversity was high in most catchments. *Paranephrops planifrons* had a great proportion of the total genetic diversity distributed between catchments (c. 72%) and much less within streams (c. 18%); most catchments had unique haplotypes, with only one shared among neighbouring catchments. *P. planifrons* has no inter-catchment dispersal, except where there are downstream freshwater connections. Translocation of populations may be necessary to restore areas and should preferably be from within catchments.

Japan: *Cambaroides japonicus* De Haan, 1841

Koizumi et al. (2012) studied the endangered Japanese crayfish *Cambaroides japonicus.* Intra-specific genetic diversity is important not only because it influences population persistence and evolutionary potential, but also because it contains past geological, climatic and environmental information. They showed unusually clear genetic structure of *Cambaroides japonicus*, a sedentary species in northern Japan. Over the native range, most populations consisted of unique 16S mtDNA haplotypes, resulting in significant genetic divergence (overall $F_{ST} = 0.96$). Owing to the simple and clear structure, a new graphic approach showed a detailed evolutionary history; regional crayfish populations were comprised of two distinct lineages that had experienced contrasting demographic processes (i.e., rapid expansion vs. slow stepwise range expansion) following differential drainage topologies and past climate events. Nuclear DNA sequences also showed deep separation between the lineages. Current ocean barriers to dispersal did not significantly affect the genetic structure of the freshwater crayfish, indicating the formation of relatively recent land bridges. Koizumi et al. (2012) illustrated one of the best examples of how phylogeographic analysis can unravel a detailed evolutionary history of a species and how this history contributes to the understanding of the past environment in the region. Ongoing local extinctions of the crayfish lead not only to loss of biodiversity but also to the loss of a significant information regarding past geological and climatic events. In the same year, Dawkins and Furse (2012) pointed out the need to examine the genetic diversity within *Cambaroides japonicus* in order to identify a number of potential conservation options, including the identification of ESUs. Such studies will also substantially add to the current information available on this sole native Japanese species of freshwater crayfish, by providing effective populations sizes, sex ratios, and possibly measures of inbreeding or migration.

Korea: *Cambaroides similis* Koelbel, 1892

Ahn et al. (2011) studied the Korean freshwater crayfish, *Cambaroides similis*, a native species threatened because of range reduction and habitat degradation caused by environmental changes and water pollution. For the conservation and restoration of this species, it is necessary to understand the current population structures of Korean *C. similis* using estimation of their genetic variation. Eight microsatellite loci were developed. The observed heterozygosities and expected heterozygosities ranged from 0.000 to 0.833 and from 0.125 to 0.943, respectively, and the former values were significantly lower than the latter ones expected under the Hardy-Weinberg equilibrium. No significant linkage disequilibrium was revealed between any of the locus pairs. The genetic distances of populations were significantly correlated with geographic distances. This result may show the regional differentiation caused by restricted gene flow between northern and southern populations within Seoul. The microsatellite markers have well the potential for analyses of the genetic diversity and population structure of *C. similis* species, with implications for its conservation and management plans.

Population genetics of invasive non-indigenous crayfish species: case studies

In order to understand the consequences of introduction and invasion of crayfish species, it is necessary to first understand important aspects of phylogeography, population genetics and molecular ecology. Human commercial activities, specifically aquaculture, legal or illegal stocking, live food trade, aquarium and pond trade (Lodge et al. 2000), have led to the deliberate introduction of, e.g., several crayfish species from North America to Europe (Gherardi and Holdich 1999). Today at least eight NICS are established in European ecosystems. Holdich et al. (2009) named 'Old NICS' the North American crayfish (the red swamp crayfish *Procambarus clarkii* Girard 1852; the signal crayfish *Pacifastacus leniusculus* Dana 1852 and the spiny-cheek crayfish *Orconectes limosus* Rafinesque 1817) mostly introduced before 1975 'New NICS', being more recently introduced such as the North American species (*Orconectes immunis*, *Orconectes juvenilis*, *Orconectes virilis*, *Procambarus* sp. and *Procambarus acutus*) and the Australian species (*Cherax destructor* and *Cherax quadricarinatus*) all of which have much narrower ranges in Europe.

The three old NICS present in Europe

The three 'old NICS' *Procambarus clarkii*, *Pacifastacus leniusculus*, *Orconectes limosus* are the most widespread invasive American crayfishes in Europe but differ in their colonization history. Whereas the signal crayfish (since the 1960s) and the red swamp crayfish (since 1973) were introduced in Europe several times and in large numbers; the spiny-cheek crayfish was brought in 1890 with only one introduction of less than hundred animals (Souty-Grosset et al. 2006). Genetic investigations were performed above all on old NICS.

The red swamp crayfish *Procambarus clarkii* Girard, 1852

In Europe, the red swamp crayfish was first introduced into Spain in 1973 using specimens from Louisiana (Habsburgo-Lorena 1986). Reasons for its introduction were varied and included aquaculture, live food trade, bait, and the pet aquarium trade (Huner 1977, Huner and Avault 1979, Hobbs 1989). From Spanish waters, the species spread to southern Portugal (Henttonen and Huner 1999). In less than 20 years from this first introduction, new populations of *P. clarkii* were reported in several countries of Europe, including

Portugal, Cyprus, England, France, Germany, Italy, Mallorca, Netherlands, and Switzerland (Gherardi and Holdich 1999). Populations found in Italy were described for northern and central Italy by Barbaresi and Gherardi (2000). In northern Italy, *P. clarkii* is undergoing a great expansion in both the Po and the Reno drainage basins in Piedmont and in Emilia-Romagna, respectively. In central Italy, the species is widespread in Tuscany, especially in Massaciuccoli Lake after the establishment of a farm in 1990. It is hypothesized that all the *P. clarkii* populations appearing in Tuscany after 1990 probably originated from man-made translocations from this lake. Since its first introduction, the red swamp crayfish has now also been found in Liguria. As of 2005, *P. clarkii* had invaded 13 countries and is considered a major freshwater pest (Souty-Grosset et al. 2006). Holdich et al. (2009) recorded its presence in 15 territories, including a number of islands, i.e., São Miguel (Azores), Cyprus, Majorca, Sardinia, Sicily and Tenerife. Different mechanisms play a role in the displacement of the crayfish outside their native ranges including (1) natural, such as active dispersal, (2) accidental, such as escape from holding facilities, or (3) deliberate through human activities (Barbaresi and Gherardi 2000).

There are only anecdotal reports of the geographic source of most introduced *P. clarkii* populations. Consequently the population genetic approach is of great help in understanding the contribution of each mechanism to its actual distribution. In order to outline the history of the invasion process throughout Europe, the objectives of Barbaresi et al. (2003, 2007) were to address the role of single versus multiple dispersal events through the comparison of the genetic structure of different European *P. clarkii* populations. The first introduction of *P. clarkii* into Spain from Louisiana is well documented. On the other hand, the events leading to subsequent expansion of the species are only partially known (Laurent 1997, Gherardi and Holdich 1999). A study was undertaken using a population genetic approach that aimed at analyzing the invasion process of this introduced crayfish (Fetzner 1996, Fetzner and Crandall 2002). Because allozyme variability is very low in crayfish (reviewed in Fetzner and Crandall 2002), a preliminary study using RAPD markers was initially performed by Barbaresi et al. (2003). It revealed high levels of genetic variability in five European populations, suggesting multiple introductions of individuals coming from different source locations (Table 1). This finding is in agreement with recent studies conducted in the Australian redclaw crayfish *Cherax quadricarinatus* von Martens (Macaranas et al. 1995) and in the native European white-clawed crayfish *Austropotamobius pallipes* Lereboullet, 1858 (Souty-Grosset et al. 1999, Gouin et al. 2001,

Table 1. *P. clarkii* populations: sampling location, site code, number of sampled individuals for microsatellite and mtDNA (in parenthesis) analysis, and status of the population (I = introduced, N = native) are indicated. For introduced populations, the Table shows the year of introduction and the source population derived from either anecdotic or published (*) data (Habsburgo-Lorena 1986). [+] indicates populations previously studied using RAPDs (Barbaresi et al. 2003).

Collection location	Site code	Sample size	Status	Year	Source
Veta La Palma, Doñana, Sevilla, Spain	VPA	10 (5)	I	1974	Louisiana *
Le Tatre, Givrezac Charente, France	TAT	10 (5)	I	1976	Spain
Évora, Alentejo, Portugal [+]	EVO	10 (5)	I	Around 1985	Spain
Carmagnola, Torino, Piedmont, Italy	CAR	10 (5)	I	1989	Unknown
Fucecchio, Tuscany, Italy [+]	FUC	10 (5)	I	Around 1995	Massaciuccoli
Firenze, Tuscany, Italy [+]	FIR	10 (4)	I	Around 1995	Massaciuccoli
Malalbergo, Bologna, Emilia Romagna, Italy [+]	MAL	10 (5)	I	Around 1990	Unknown
Mellingen, Aargau, Switzerland	MEL	10 (4)	I	Around 1990	Unknown
Massaciuccoli, Tuscany, Italy [+]	MAS	10 (3)	I	1992	Spain (Doñana)
Sarzana, La Spezia, Liguria, Italy	SAR	10 (3)	I	1998	Unknown
New Orleans, Louisiana, USA [+]	NOR	10 (5)	N	---	---
Chihuahua, Mexico	CHI	9 (4)	N	---	---

2006). Investigating mtDNA, Barbaresi et al. (2007)—as they previously investigated 16S mtDNA and found only one haplotype (Barbaresi et al. 2003)—used CO1 mtDNA to find more variability.

However, all haplotypes were closely related, differing by very few mutational steps (Fig. 4). This low level of sequence divergence is expected showing that the colonisation of *P. clarkii* is recent (35 years); a similar situation has been also described in the Chinese mitten crab *Eriocheir sinensis* Milne-Edwards, which colonized Europe less than 100 years ago (Hänfling and Kollmann 2002). In *P. clarkii*, the numbers of haplotypes and within-population gene diversity were higher in the source population NOR (New Orleans) than in introduced populations. However some variation in gene diversity within Europe was also found. All the populations shared at least one haplotype except FIR (Firenze, Italy), which is the population harbouring two haplotypes (4 and 6) deriving from the most frequent haplotype 2. This result sharply contrasts with the anecdotal information stating this population is derived from MAS (Massaciuccoli, Italy). The detection of unique haplotypes in this population suggests a different origin of source individuals, possibly a source not sampled in this study. These haplotypes could be the result of a different introduction possibly from China following the immigration of a Chinese community to Florence. As the first introduction of *P. clarkii* from Louisiana to Asia was done in 1918 in Japan and in a second step, a translocation was made from Japan to China (1948) (Laurent 1997), the introduced individuals in FIR could have evolved differently because of successive translocations and more than 80 years of divergence. This hypothesis should be verified by investigation of Japanese and Chinese populations.

The study was also the first using microsatellite loci characterized by Belfiore and May (2000) in *P. clarkii*. A high genetic variation within sampled populations emerged from the use of microsatellite markers. In addition, microsatellites revealed a high inter-population differentiation with an F_{ST} up to 0.461. For comparison, Gouin et al. (2006) found also an F_{ST} of 0.461 in southern French populations of the indigenous white-clawed crayfish *A. pallipes*, indicating an absence of gene flow due to the fragmentation of the populations. Considering the degree of variability of *P. clarkii* in its native range, contradictory results emerged. In fact, while in the sample from Louisiana a high heterozygosity value was found similar to the major part of the European populations, heterozygosity was low in the Mexican population sampled

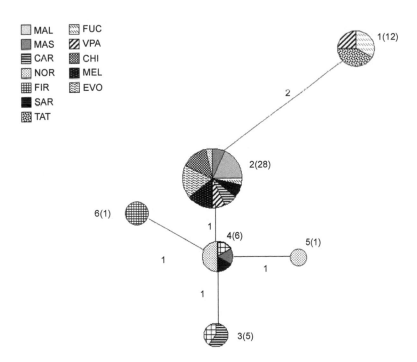

Figure 4. Haplotype network of the mtDNA sequence data. Numbers along branches denote the number of nucleotide substitutions between haplotypes. The haplotype number (in bold) follows Table 1. The frequency of each haplotype is indicated in parenthesis. Sample names follow Table 1 above.

in the field and naturally dispersed. The level of diversity observed in the Louisiana population could be explained as the result of the intense commercial exploitation of the species in the Southern United States (Louisiana accounts for most of the national aquaculture production). In this area, exchanges of crayfish from natural habitats to culture ponds can be common. In addition, stock translocation is reported to be a common practice (Busack 1988). Heterozygosity found in the European populations was high and could be explained by the fact that the initial genetic structure of a successful invasive population depends on several factors, including the effective population size of the introduction event(s), the genetic diversity of the source population(s), and the number of founding sources. Since this is the first study using microsatellites in this species, no comparison can be made with the results from other authors; in particular, the degree of variability in native populations is unknown. The high genetic diversity revealed by microsatellite markers in some introduced populations of *P. clarkii* could be the result of different types of introduction events, i.e., (1) multiple introduction events with individuals from different sources, (2) a single introduction of a large number of individuals from a genetically diverse source population, and (3) a combination of these events. Regarding the population from Doñana, historical data report that this population originated from a stock coming from Louisiana and introduced for aquaculture purpose (Gherardi and Holdich 1999). The high genetic variability of the Spanish population could thus be explained by a single introduction event from a very heterogeneous pool coming from that area. The degree of heterozygosity found in this study would not have been detected in the case of a severe bottleneck. The lack of heterozygote deficiencies is predicted for bioinvasions and commonly encountered as, for example, in freshwater mussels (Holland 2001). An exception is represented by the population from Piedmont, where a bottleneck effect was revealed (according to the Bottleneck's test). For this population, the origin from a small number of introduced individuals can be hypothesised. As shown by pairwise F_{ST}, no differentiation was found between samples from Sarzana (Liguria) and Massaciuccoli (Tuscany). In accordance with data obtained with mt-DNA, this result seems to confirm the hypothesis of a single origin of the population sampled from Liguria, possibly by active dispersal (through marshes) and/or human translocation of crayfish coming from near Massaciuccoli. From a management point of view, this result points out how dispersal capability could be favoured, at least at a microgeographical scale. The usefulness of microsatellites as evidence of bottlenecks and gene flow has been outlined by Colautti et al. (2005) for the Eurasian spiny waterflea, whose spread was found to depend on long-distance jump dispersal (Suarez et al. 2001). The use of six microsatellites in the quagga mussel, *Dreissena rostriformis bugensis* Andrusov, gave evidence of considerable gene flow (multiple invasions) among populations: its genetic diversity was consistent with the existence of a large metapopulation that has not experienced bottlenecks or founder effects (Therriault et al. 2005). *Procambarus clarkii*'s spread may involve both long-distance jump dispersal and natural dispersal at a microgeographical scale. The high level of genetic diversity in introduced population of this species and its corresponding success of establishment support the hypothesis that high genetic variability is an important characteristic of successful invasive populations (e.g., Ehrlich 1986, Holland 2000). However, this result cannot be generalized. For example, Tsutsui et al. (2000) used microsatellite markers in the invasive ant *Linepithema humile* and showed that the loss in genetic diversity of the introduced populations of this species was associated with the reduced intraspecific aggression among spatially separate colonies with the formation of interspecifically dominant supercolonies. In that case, genetic bottlenecks have led to widespread ecological success. Results on *P. clarkii* confirm the model suggested by Barbaresi et al. (2003), in which the colonization of Europe by this species derives from subsequent introductions of individuals coming from different source populations. This model is consistent with both the high genetic diversity observed (introduction of different sets of individuals) and the genetic differentiation of populations resulting generally from the casual bias of introductions.

As the species is the most distributed alien species in the world (present now in Belize, Brazil, Chile, China; Colombia, Costa Rica, Cyprus, Dominican Republic, Ecuador, Egypt, Georgia, Israel, Japan, Kenya, Mexico, Philippines, Puerto Rico, South Africa, Sudan, Taiwan, Province of China, Uganda, United States, Venezuela, Zambia)—following the study of Barbaresi et al. (2007)—other studies were performed in other invaded countries. Zhu and Yue (2008) isolated eleven microsatellites from the red swamp crayfish. All 11 microsatellites were polymorphic and unlinked. These markers are being used to study the invasion routine, genetic diversity and population structure of the species *P. clarkii*. Evidence of founder effects were found in all populations studied and no support based on geographical distance was demonstrated; it is why the authors suggested translocation by humans (Yue et al. 2010). In China also, Li et al. (2012) investigated the population genetic structure and post-establishment dispersal patterns of the red swamp crayfish following its introduction in the early 20th century. It has been spread to almost all forms of fresh water bodies including lakes, rivers and even paddy fields in most provinces of China. To clarify issues such as the initial entry point(s), dispersal pattern, genetic diversity and genetic structure of *P. clarkii* in China, the genetic structure and diversity of *P. clarkii* populations at 37 sampling sites (35 from China, one from the USA and one from Japan) were analyzed using both mitochondrial gene sequences (COI and 16S rRNA) and 12 nuclear microsatellites. Multiple tests including phylogenetic analyses, Bayesian assignment and analysis of isolation by distance showed that (1) the population from Japan and those collected from China, particularly from Nanjing and its some neighboring sites have similar genetic composition, (2) relatively high genetic diversity was detected in Chinese populations, (3) *P. clarkii* populations in China did not experience significant population expansions. Taken together, Nanjing, Jiangsu province is the presumed initial entry point, and human-mediated dispersal and adaptive variation are likely responsible for the observed genetic pattern of *P. clarkii* in China. Finally, recent findings from Zhu et al. (2013) are of interest for aquaculture in China (see related paragraph). In Southern America, Torres and Álvarez (2012) compared the genetic variation in the mitochondrial gene COI of *P. clarkii* among samples from the native range in Mexico and the United States and from introduced populations in western and southern Mexico and Costa Rica. Three populations (Illinois and Louisiana, United States and northern Coahuila, Mexico) represented the native range and six populations came from areas where the species has been introduced (central Coahuila, southern Nuevo León, Durango, Chihuahua and Chiapas, Mexico, and Cartago, Costa Rica). A 689 bp fragment was amplified from 37 samples. Uncorrected genetic distances among sequences were p = 0 to 0.02031 and 12 haplotypes were found. A phylogenetic reconstruction shows that the three populations from the native range remain very similar to each other and some introduced populations can be directly associated to one of them. The populations from Nuevo León, central Coahuila and Costa Rica were the most divergent ones. Overall the genetic variation found in *P. clarkii* in both native and introduced populations is low.

The signal crayfish *Pacifastacus leniusculus* Dana, 1852

At the level of its native range, according to Larson and Olden (2011), the signal crayfish *Pacifastacus leniusculus* is the most widely distributed and best known of the crayfishes native to the Pacific Northwest in USA. The native range of *P. leniusculus* is in Washington, Oregon, Montana, Idaho, Nevada, Utah, and also in British Columbia, Canada (Souty-Grosset ct al. 2006). Three subspecies of *Pacifastacus leniusculus* are currently recognized: *P. l. leniusculus* (Dana, 1852), *P. l. klamathensis* (Stimpson, 1857) and *P. l. trowbridgii* (Stimpson, 1857). The question of recognizing three subspecies was still debated until the recent paper from Larson et al. (2012): first they discovered cryptic diversity within *P. leniusculus*, previously unrecognized by both morphology (Miller 1960) and in past molecular investigations of this species (Agerberg and Jansson 1995, Sonntag 2006). Secondly they confirmed that the range of morphological variability characterizing three historical *P. leniusculus* subspecies persists. However the morphology of some subspecies (*P. l. klamathensis*, *P. l. trowbridgii*) spans both *P. leniusculus* and cryptic groups, while subspecies morphology (*P. l. leniusculus*) occurs predominantly within *P. leniusculus*. At present the signal crayfish is becoming invasive in California and is among the most problematic invaders in freshwater systems of Europe (Henttonen and Huner 1999) and Japan (Usio et al. 2007).

In Europe the species is the most widespread of the NICS and is recorded in 27 territories (Holdich et al. 2009). Most populations have been derived from implants made into Sweden in 1959 and subsequently. These were mainly derived from Lake Tahoe in California. As at least two subspecies are present in this Lake Tahoe (Miller 1960, Riegel 1959), more than one signal crayfish lineage could be expected in Europe. Hayes (2012) did the first population level genetic characterization of an invasive crayfish species in United Kingdom and found overall high levels of genetic diversity in populations. He used a combination of two markers, as mtDNA markers confer relatively more ancestral information pertaining to the relationships between individuals and populations, whilst microsatellites allow for a finer, landscape scale analysis to be made and might therefore elucidate patterns between populations within catchments. He concluded that the diversity found within some populations surpassed expectations and displayed markedly high levels, suggesting that non-native species might in some circumstances see increased diversity in populations within their introduced ranges. The same year, Filipová (2012) analyzed cytochrome c oxidase subunit I (COI) gene fragment of signal crayfish from 17 European countries and compared the results obtained with data from the Klamath River basin in North America (Sonntag 2006) which included COI sequences of individuals representing three known subspecies of *P. leniusculus*. Samples from six North American populations were added. With this sampling, Filipová (2012) found only haplotypes related to *Pacifastacus leniusculus leniusculus*, suggesting that this is the major subspecies introduced in Europe. Similar to populations from the United Kingdom, high genetic variation was also found in European populations which were not found in another widespread invader in Europe (see below the case the spiny-cheek crayfish *Orconectes limosus* (Filipová et al. 2011a).

   The status of signal crayfish introduced to Japan was also investigated. Between 1926 and 1930, signal crayfish were brought to Japan from the Columbia River basin, northwestern USA (Usio et al. 2007), and Hobbs (1989) noted that both *Pacifastacus l. leniusculus* and *P. l. trowbridgii* were introduced there. However, based on morphological examination, the subspecific status of signal crayfish in Japan could not be assessed (Kawai et al. 2003). To provide an effective tool to analyze population genetics of this alien species, Azuma et al. (2011) developed five polymorphic microsatellite DNA markers from the genome of this species. The number of alleles and expected heterozygosity in each locus ranged between 4–33 and 0.304–0.941, respectively, indicating the utility of the markers in population analysis. They analyzed 212 individuals of *P. leniusculus* from seven locations in Hokkaido and Honshu, Japan using these markers. When they used a Bayesian clustering analysis, the three clusters, Akashina, Tankai and Hokkaido, were discriminated from each other with different allele frequencies. In addition, a recently found population in Tone River in central Honshu appeared to have originated from Hokkaido, consistent with a previous study using ectosymbiont worms. Thus, the new microsatellite markers are useful in identifying the population structure and genetic connectivity of *P. leniusculus* in Japan. They discussed potential applications of microsatellite analysis in tracking dispersal pathways and defining eradication units for this invasive crayfish.

The spiny-cheek crayfish *Orconectes limosus* Rafinesque, 1817

The North American spiny-cheek crayfish, *Orconectes limosus* (Rafinesque, 1817), a widespread invader in Europe, seems to have been introduced there successfully only once. According to available literature, 90 individuals of unclear origin were released in Poland in 1890. Despite this apparent bottleneck, the species has successfully colonized various aquatic habitats and has displaced native crayfish species in many places. Filipová et al. (2011a) test whether different European populations were likely to have come from a single source and to identify their possible origin, we analyzed the diversity of the mitochondrial gene for cytochrome c oxidase subunit I (COI) of *O. limosus* individuals from Europe and from its original range in North America, including the presumed source region of European populations, the Delaware River watershed (eastern USA). Two haplotypes were found in European populations. One haplotype was widespread; the other was present in a single population. In contrast, 18 haplotypes were detected in North America. This result supports the hypothesis of a single overseas introduction of *O. limosus* and suggests that the high invasion success of this species was not limited by an introduction bottleneck. Two divergent clades were detected in North American *O. limosus* populations. One, which includes the dominant haplotype in Europe, was found in a large part of the species' present range. The 2nd (diverging by 0.1%) was mostly restricted to a limited area in southeastern Pennsylvania. *Orconectes limosus* populations in the northern part of the species' North American range, at least some of which are non-indigenous themselves, may share the source area with European *O. limosus*. The endangered status of *O. limosus* populations in southeastern Pennsylvania and northeastern Maryland, where much of the species' genetic diversity resides, should be considered in conservation management.

   At the national level, in Czech Republic, Filipová et al. (2009) used allozymes and hypothesized a low genetic variability resulting from a bottleneck effect during introduction in European spiny-cheek crayfish populations. On the other hand, the fast spread of *O. limosus* in Europe and colonisation of various habitats suggest that this species does not suffer from inbreeding depression due to an introduction bottleneck. They analysed 14 *O. limosus* populations from the using allozyme electrophoresis to evaluate the level of intra- and inter-population genetic variation. Out of eight well-scoring allozyme loci chosen for detailed analysis, six were variable in studied populations, suggesting that sufficient variability was maintained during the introduction. Genetic differentiation of Czech populations of the spiny cheek crayfish was relatively low and did not show any clear geographic pattern, probably due to long-range translocations by humans.

NICS studied outside Europe

If the three most widely-spread NICS are the North American species: *Pacifastacus leniusculus*, *Orconectes limosus* and *Procambarus clarkii* being considered as "Old NICS" (because introduced before 1975) are well studied in Europe (Holdich et al. 2009), the "New NICS", introduced after 1980, are less studied in genetics because of their narrower ranges in Europe. Here we report some case studies about these species investigated outside Europe.

The rusty crayfish *Orconectes rusticus* Girard, 1852

This species has invaded much of Minnesota, Wisconsin, Michigan and Illinois in United States. Rusty crayfish were also found to hybridize with *Orconectes limosus* (Smith 1981). While *Orconectes limosus* numbers declined four years later, no conclusions regarding the cause of the decline were discussed (Perry et al. 2001a,b); consequently the rusty crayfish ecologically displaces and, through hybridization, genetically assimilates and morphologically extirpates the northern clearwater crayfish populations.

The virile crayfish *Orconectes virilis* Hagen, 1870

Ballast water and the use of crayfish as fish bait have led to anthropogenic introductions of *Orconectes virilis* in freshwater systems (Lodge et al. 2000) and the species is established outside of its native range, leading to introgression with the endemic gene pool (Perry et al. 2001a,b, 2002). The species has a widespread native range in Canada from Saskatchewan to Ontario and in the United States from Montana to New York; it is considered an invasive species over the rest of the United States (Hobbs 1974). Studies have revealed cryptic lineages. Mathews et al. (2008) investigated genetically and morphologically the *O. virilis* species complex over parts of its geographic range. Mitochondrial and nuclear genetic data was used and found that many cryptic lineages exist, although the context of this hidden biodiversity is not completely understood; McKniff (2012) used AFLP (Amplified Fragment-length polymorphism) markers to obtain insights into the population genetics of an invasive species of crayfish, *Orconectes virilis*. Studying a set of populations of freshwater crayfish in New England because freshwater organisms are subject to biodiversity changes, McKniff (2012) found a cryptic lineage studying populations from Blackstone River Valley (New England).

Application in aquaculture

The human-mediated movement of crayfish around the world follows a multiplicity of pathways that result in either accidental, in ballast or via canals, or deliberate introductions of NICS for aquaculture, stocking, live food commerce, aquarium and pond trade, live bait, biological supply, etc. (Holdich and Pöckl 2007). For centuries at least, large crayfish have been a valued addition to the human diet and in pursuit of their exploitation, some crayfish stocks have been overfished and others moved around for aquaculture, both processes causing damage to habitats and communities in which indigenous crayfish species live (Reynolds and Souty-Grosset 2012). The three American species *Pacifastacus leniusculus*, *Procambarus clarkii* and *Orconectes limosus* and *Astacus leptodactylus* from Eastern Europe are listed among the top 27 animal alien species introduced in Europe for aquaculture and related activities (with twenty freshwater or anadromous fishes, two marine/estuarine bivalves and one marine peneid shrimp) (Savini et al. 2010). These authors demonstrated that *Procambarus clarkii* and *Pacifastacus leniusculus* are responsible for the largest range of impacts (i.e., crayfish plague dissemination, bioaccumulation of pollutants, community dominance, competition and predation on native species, habitat modifications, food web impairment, herbivory and macrophytes removal). In fact, in Europe, there is no (or few and isolated) genetic studies performed on the three species coming from farms in Europe but rather genetic studies of wild populations (see paragraph on population genetics of NICS). Here we report genetic studies performed on species according to the country where there are the most important aquaculture.

The red swamp crayfish *Procambarus clarkii* Girard, 1852

Concerning *Procambarus clarkii*, this crayfish dominates both culture and capture fisheries in its native state of Louisiana (USA), as well as where it has been introduced, e.g., China and Spain. Production levels in Europe are very small compared to the 50000 and 70000 tonnes produced per annum by the USA and China respectively (Souty-Grosset et al. 2006). Effectively Zhu et al. (2013) stated that the red swamp crayfish has become one of the most important freshwater aquaculture species in China. As little was known about its population genetics and geographic distribution in China, they studied the genetic diversity among 6 crayfish populations from 4 lakes using AFLPs. They obtained 3 major clusters by principal coordinate analysis and cluster analysis. The estimated average GST value across all loci was 0.4186, suggesting (very) low gene flow among the different localities. Indeed there is high genetic differentiation among crayfish in the middle and lower reaches of the Yangtze River. Results will be useful for the development of artificial propagation and genetic improvement programs for crayfish, which give populations the ability to adapt to environmental changes and stresses. Crayfish is an extremely important economic resource in the middle and lower reaches of the Yangtze River. Furthermore, the supply of offspring from hatcheries cannot meet the demand of the market. Zhu et al. (2013) found that one of the populations has considerable genetic variety and concluded therefore that parental crayfish for artificial propagation should be collected from this population to ensure the conservation of wild resources and genetic diversity and consider that this information will help in the selection of high quality individuals for artificial reproduction.

The yabby *Cherax destructor*, Clark, 1936

In Australia < 500 tonnes are produced from fisheries and < 50 tonnes from culture per annum. Although *C. destructor*, an ecologically and commercially important species that is widespread throughout the freshwater systems of central Australia, are produced from ponds on purpose-built farms, most come from trapping wild individuals (Souty-Grosset et al. 2006). First Austin et al. (2003) solved the taxonomy and phylogeny of the *Cherax destructor* complex (Decapoda: Parastacidae) using mitochondrial 16S sequences. Phylogenetic analysis found three distinct clades that correspond to the species *C. rotundus*, *C. setosus* and *C. destructor*. *C. rotundus* is largely confined to Victoria, and *C. setosus* is restricted to coastal areas north of Newcastle in New South Wales. After this study Nguyen and Austin (2004) were the first to demonstrate inheritance of molecular markers and sex in the Australian freshwater crayfish *Cherax destructor* Clark. Three kinds of molecular genetic markers (mtDNA, random-amplified polymorphic DNAs -RAPDs- and allozymes) and sex were investigated in crossbreeding experiments between three populations of *Cherax destructor*. The finding that the inheritance of allozyme and RAPD markers conforms to Mendelian expectation gives greater confidence in the use of these markers for taxonomic, population genetic and aquaculture-related studies with particularly the allozyme and RAPD markers. These markers can be of potential value for quantitative trait loci (QTLs) detection, marker-assisted selection (MAS), communal rearing studies and genome mapping. An increasing number of studies that use sophisticated molecular techniques for genetic improvement in livestock are being applied to aquatic species of significance to aquaculture. In this context, this study demonstrates that both allozyme and RAPD markers have potential for these kinds of applications. Finally, Nguyen et al. (2005) examined allozymes and RAPD variation in *Cherax destructor*. At the intra-population level, allozymes revealed a similar level of variation to that found in other freshwater crayfish; RAPDs showed less diversity than allozymes, which was unexpected. At the inter-population level, both techniques revealed significant population structure, both within and between drainages. RAPD results were consistent with phylogeographic patterns previously identified using mtDNA. The findings have implications for aquaculture. In selective breeding of *C. destructor*, the fixed allelic differences they observed among several populations are potentially useful for communal rearing experiments, for monitoring the genetic effects of selection during breeding programs, broodstock management and for developing markers for QTL selection. The large range of intra-population variation suggests that some populations with high variation may be more effectively used as base-line stocks for selective breeding than others. Given the high level of interest in the aquaculture of *C. destructor* and the existence of three distinct clades, the strategy for the management of this species would be to restrict translocations for aquaculture to within the geographical boundaries of each of these clades, as has been recommended for other species of freshwater crayfish (Busack 1988, Fetzner et al. 1997).

The redclaw *Cherax quadricarinatus* von Martens, 1868

The redclaw crayfish *Cherax quadricarinatus* is a freshwater crayfish species endemic to northern Australia and Papua New Guinea that is the focus of a growing culture industry in number of regions around the world as this species is considered as one of the best crayfish for aquaculture. Effectively it can achieve a weight of 500 g and this is a robust species, with a simple life cycle requiring simple foods. The production technology is straightforward; it is economic to produce, has a good meat taste and yield and fetches a good price. Many countries in Asia, North and South America, Africa and parts of Europe have obtained stock during the 1990s. This species accounts for almost the entire commercial production of freshwater crayfish in Australia. Its culture potential has also been recognized in other countries and consequently it was introduced to China in the early 1990s for commercial culture (Souty-Grosset et al. 2006). Macaranas et al. (1995) were the first to find genetic variability in this species. Analyses of allozymes and RAPD markers in redclaw populations from the Northern Territory and North Queensland showed that if allozymes revealed a low level of genetic variation (as already found by Austin 1996), RAPD could distinguish each river population and also grouped them according to geographic proximity. RAPD analyses also revealed significant genetic variability both within the species and within individual populations, which could be used to improve their culture. In the same species, Xie et al. (2010) isolated and characterized 15 microsatellite loci. Thirteen of 15 microsatellite loci were polymorphic. Number of alleles per locus ranged from two to seven while observed and expected heterozygosities ranged from 0.172 to 0.985 and from 0.373 to 0.778, respectively. Eleven loci conformed to Hardy-Weinberg equilibrium in the sampled population. These microsatellite loci developed provided an important resource for studying genetic diversity and population structure in redclaw crayfish. Since this study, He et al. (2013) developed and optimized 15 polymorphic microsatellites isolated from *C. quadricarinatus*. The variability of these microsatellites was investigated in unrelated individuals cultured in China. All microsatellite loci were polymorphic and indeed provide the opportunity for studying genetic diversity and population structure in redclaw crayfish and for monitoring relative levels of genetic diversity in sampled populations.

A further species *Cherax tenuimanus*, endemic in Western Australia, is an important crayfish in aquaculture (Imgrund et al. 1997). Results of the genetic population structure of the species were described by Nguyen et al. (2002a,b).

The particular case of the marbled crayfish *Procambarus fallax* Hagen, 1870 and pet trade

The major pathway for the introduction of the marbled crayfish is the deliberate release of aquarium specimens (Souty-Grosset et al. 2006). Scholtz et al. (2003) were the first to describe this crayfish of marbled appearance and of uncertain geographical origin that was introduced into the German aquarium trade in the mid-1990s, being capable of unisexual reproduction (parthenogenesis). They demonstrated that this crayfish-named '*Marmorkrebs*'—is indeed parthenogenetic under laboratory conditions and used morphological and molecular analysis to show that it belongs to the American Cambaridae family. Because of this parthenogenetic reproduction, the marbled crayfish is also a potential ecological threat by outcompeting native species in the absence of sexual reproduction. Martin et al. (2007), testing its exclusively parthenogenetic reproduction mode, used molecular markers and clearly proved the genetic uniformity of the offspring. This was the first molecular study showing that this crayfish produces genetically uniform clones. They tested various generations of a Marmorkrebs population by microsatellite markers and found that all specimens were identical in their allelic composition. Moreover the analysis of the two mitochondrial genes convincingly supports the assumption of a close relationship between Marmorkrebs and *P. fallax* (Martin et al. 2010). To conclude, parthenogenesis leads to a high reproductive potential, and Marmorkrebs can overpopulate an aquarium very quickly; consequently aquarium hobbyists will readily sell ornamental individuals (often by internet) (Souty-Grosset et al. 2006). Chucholl et al. (2012) highlighted that the number of Marmorkrebs established populations in Europe will further increase over time because the species is still widespread in the European pet trade, including in Ireland, a European location without NICS (Faulkes 2015).

Interestingly, the situation in Europe since 2003 led to investigations in North America; Faulkes (2010) surveyed online North American pet owners with the aims of trying to track when Marmorkrebs entered the North American pet trade; he found dates since at least 2004, with the number of people increasing every year. The increasing spread of Marmorkrebs through the pet trade -through online sources- increases the probability that Marmorkrebs will be released into North American ecosystems.

Conclusions

In 2002, Fetzner and Crandall estimated the number of studies surveying population-level genetic variability to be still very low (less than 15) and considered the exploration of levels of genetic variation in crayfish as an open field of study. They underlined the urgency to use crayfish genetic diversity due to the conservation pressures that affect a large percentage of crayfish species throughout the world. Since this publication, the number of papers has increased significantly because of the urgency to save native crayfishes and to fight against invasive crayfishes. If restocking programs continue to translocate individuals with no regard for their population's genetic structure, the natural genetic make-up will further dissolve (Souty-Grosset et al. 2003), which is accompanied by a reduction in intraspecific diversity. Conservation strategies, therefore, need to manage crayfish populations as distinct ESUs and give the highest priority to the populations with high genetic diversity and unique haplotypes. Mitochondrial DNA is a widely used marker to reconstruct the phylogeographic history of species. Over the last decades, the number of stocking events that disregard the genetic structure within and between populations (Souty-Grosset and Reynolds 2010) and cross-basin translocations in response to rapidly declining stocks have led to a contamination of local stocks (Largiadèr et al. 2000). This has led to repeated calls for modern conservation programs that consider the genetic origin of the stocking material (e.g., Schulz et al. 2004, Bertocchi et al. 2008, Souty-Grosset and Reynolds 2010, Strayer and Dudgeon 2010). Genetic screening is essential, including determining what alleles are present. In recent years, a number of population genetic studies on the European white clawed crayfish, *Austropotamobius pallipes sensu lato*, have received a lot of attention. The studies by Grandjean et al. (2000, 2002a, 2002b), Fratini et al. (2005) and Trontelj et al. (2005) revealed that this species complex probably has its centre of radiation (and found its Pleistocene refugia) in Istra (Croatia) and Italy, where five distinct genetic forms can be recognized, which are in part referred to as separate species or subspecies: *A. pallipes* (Lereboullet, 1858) and *A. italicus* (Faxon, 1914), the second possibly consisting of the subspecies *italicus, carsicus, carinthiacus*, and *meridionalis* (Grandjean et al. 2002b, Fratini et al. 2005, discussed by Trontelj et al. 2005). On the other hand, the genetic diversity of this crayfish from the British Isles, Ireland and Spain is comparatively impoverished and the latter two faunas probably represent more or less recent human-mediated introductions from Italy into Spain (Albrecht 1982, Grandjean et al. 2000), and from France into Ireland (Gouin et al. 2001). These results are of crucial importance for conservation biology, because they allow discerning the areas with the highest genetic richness within this species complex and determining which populations should be given highest priority in terms of conservation measurements. With *A. italicus*, this is also a good example of the importance of genetic information as given by Bertocchi et al. (2008): populations sampled in one basin showed no heterozygotes and a high level of inbreeding. Thus knowledge of the genetic structure of studied populations, combined with information on their ethology, ecology, and demography, is an essential prerequisite for any action aimed at reintroducing this threatened species. In *Astacus astacus*, Gross et al. (2013) highlighted how, from a conservation point of view, identification of priority populations for conservation (e.g., by means of genetic methods as described herein) should be the first step, followed by avoiding the introduction of alien crayfish species into those areas. Based on the data from this study, the genetically most distinct and variable populations within each of the identified genetic clusters should be considered for conservation to retain a maximum of the species genetic and evolutionary potential.

Consequences of invasion by non native crayfishes are not only the elimination of native crayfish by competition, deterioration of the habitat and dissemination of the new and novel diseases but also could led to hybridization (Perry et al. 2001a,b, 2002); recently, Zuber et al. (2012) estimate that hybridization between native and invasive species is still little-studied (few case studies) whereas it is an important factor during crayfish invasions. They report genetic evidence of hybridization between invasive *O. rusticus* (Girard, 1852) and native *O. sanbornii* (Girard, 1852) in the Huron River in north-central Ohio by studying several molecular markers as nuclear DNA, mitochondrial DNA and allozymes and were able to confirm the presence of individuals of hybrid ancestry. This study is the second to detect hybridization between *O. rusticus* and a congener, which suggests that this may be an important mechanism of invasion by *O. rusticus* when closely related species are present and it is necessary to assess the implications of hybridization for the native species. Ibrasheva (2011), knowing the invasive nature of several species of freshwater crayfish of the genus *Procambarus*, predicted that some of the watersheds within Massachusetts area could be affected by the invasive species. They examined 14 organisms of *Procambarus* genus, but undetermined species sampled from 5 sites in Massachusetts. All of the organisms collected either belonged to one species—*Procambarus acutus* that is not considered to be an invasive species, or are hybrids of both invasive *Procambarus clarkii* and *Procambarus acutus*. Phylogenetic analysis grouped all of the organisms of undetermined species with only one crayfish, *Procambarus acutus* a crayfish originated from the Cape Fear River, Randolph County, North Carolina. It was predicted that the specimen collected within Massachusetts could have possibly been related to the specimen from North Carolina.

This chapter of course is not exhaustive and only some case studies were chosen. Moreover genetics investigations are still lacking within the antipodean Parastacidae as highlighted by Almerão et al. (2015) for South American Parastacidae. Conservation management-based on conservation genetics and resolving the genetic paradox in invasive species (Frankham 2005)—are to be prioritized following the assessment of the world's 590 crayfish species; Richman et al. (2015) evaluated their extinction risk using the IUCN Categories and Criteria and found they were one of the most threatened taxonomic groups assessed to date. The level of extinction risk was different between families, with more threatened species in the Parastacidae and Astacidae than the Cambaridae.

In the world, crayfish are under pressure because of pollution, habitat loss, overfishing and overexploitation as well introduction of alien species. In order to maintain both the genetic specificity of populations and the genetic diversity within and between populations, there is a need for increased taxonomic clarifications, descriptions of natural population distribution and biogeographical history.

Acknowledgements

Matt Longshaw and Paul Stebbing are warmly acknowledged for their invitation to participate in this book. I also wish to thank Anne Schrimpf who was of great help in providing figures for illustrating genetics studies in *Astacus astacus*. Thanks are due to Julian Reynolds for giving advices and improving the manuscript.

References

Agerberg, A. 1990. Genetic-variation in three species of freshwater crayfish *Astacus astacus* L., *Astacus leptodactylus* Esch., and *Pacifastacus leniusculus* (Dana), revealed by isozyme electrophoresis. Hereditas 113: 101–108.

Agerberg, A. and H. Jansson. 1995. Allozymic comparisons between three subspecies of the freshwater crayfish *Pacifastacus leniusculus* (Dana), and between populations introduced to Sweden. Hereditas 122: 33–39.

Ahn, D.-H., M.H. Park, J.H. Jung, M.J. Oh, S. Kim, J. Jung and G.S. Min. 2011. Isolation and characterization of microsatellite loci in the Korean crayfish, *Cambaroides similis* and application to natural population analysis. Animal Cells and Systems 15: 37–43.

Alaranta, A., P. Henttonen, J. Jussila, H. Kokko, T. Prestegaard, L. Edsman and M. Halmekyto. 2006. Genetic differences among noble crayfish (*Astacus astacus*) stocks in Finland. Sweden and Estonia based on ITS1 – Region. Bull. Fr. Pêche Piscic. 380-381: 965–976.

Albrecht, H. 1982. Das System der europäischen Flußkrebse (Decapoda, Astacidae): Vorschlag und Begründung. Mitteilungen aus dem Hamburgischer Zoologischen Museum und Institut 79: 187–210.

Albrecht, H. and H.O. Von Hägen. 1981. Differential weighting of electrophoretic data in crayfish and fiddler crabs (Decapoda: Astacidae and Ocypodidae). Comp. Biochem. Physiol. B 70: 393–399.

Allendorf, F.W. and L.L. Lundquist. 2003. Introduction: population biology, evolution, and control of invasive species. Conservation Biology 17: 24–30.

Almerão, M.P., E. Rudolph, C. Souty-Grosset, K. Crandall, L. Buckup, J. Amouret, A. Verdi, S. Santos and P.B.D. Araujo. 2015. The native South American crayfishes (Crustacea, Parastacidae): state of knowledge and conservation status. Aquatic Conserv. Mar. Freshw. Ecosyst. 25: 288–301. doi: 10.1002/aqc.2488.

Alvarez-Buylla, E.R. and A.A. Garay. 1994. Population genetic structure of *Cecropiaobtusifolia*, a tropical pionneer tree species. Evolution 48: 437–453.

Amouret, J., S. Bertocchi, S. Brusconi, M. Fondi, F. Gherardi, F. Grandjean, L. Chessa, E. Tricarico and C. Souty-Grosset. 2015. The first record of translocated white-clawed crayfish from the *Austropotamobius pallipes* complex in Sardinia (Italy) J. Limnol. 74: 491–500.

Attard, J. and R. Vianet. 1985. Variabilité génétique et morphométrique de cinq populations de l'écrevisse européenne *Austropotamobius pallipes* (Lereboullet 1858) (Crustacea, Decapoda). Can. J. Zool. 63: 2933–2939.

Arens, A. and T. Taugbøl. 2005. Status of crayfish in Latvia. Bull. Fr. Pêche Piscic. 376-377: 519–528.

Avise, J.C., J. Arnold, R.M. Ball, E. Bermingham, T. Lamb, J.E. Neigel, C.A. Reeb and N.C. Saunders. 1987. Intraspecific phylogeography: the mitochondrial DNA bridge between population genetics and systematics. Ann. Rev. Ecol. Syst. 18: 489–522.

Austin, C.M. 1996. Electrophoretic and Morphological Systematics Studies of the Genus *Cherax* (Decapoda: Parastacidae) in Australia. Ph.D. thesis, University of Western Australia, Perth.

Austin, C.M. and J. Bunn. 2010. *Cherax tenuimanus*. *In*: IUCN 2013. IUCN Red List of Threatened Species. Version 2013.1. <www.iucnredlist.org>.

Austin, C.M., T.T.T. Nguyen, M. Meewan and D.R. Jerry. 2003. The taxonomy and evolution of the *Cherax destructor* complex (Decapoda: Parastacidae) re-examined using mitochondrial 16S sequences. Aust. J. Zool. 51: 99–110.

Azuma, N., N. Usio, T. Korenaga, I. Koizumi and N. Takamura. 2011. Genetic population structure of the invasive signal crayfish *Pacifastacus leniusculus* in Japan inferred from newly developed microsatellite markers. Plankton Benthos Res. 6: 187–194.

Barbaresi, S. and F. Gherardi. 2000. The invasion of the alien crayfish *Procambarus clarkii* in Europe, with particular reference to Italy. Biol. Invasions 2: 259–264.

Barbaresi, S., R. Fani, F. Gherardi, A. Mengoni and C. Souty-Grosset. 2003. Genetic variability in European populations of an invasive American crayfish: preliminary results. Biol. Invasions 5: 269–274.

Barbaresi, S., F. Gherardi, A. Mengoni and C. Souty-Grosset. 2007. Genetics and invasion biology in fresh waters: A pilot study of *Procambarus clarkii* in Europe. pp. 381–400. *In*: F. Gherardi (ed.). Biological Invaders in Inland Waters: Profiles, Distribution, and Threats. Invading Nature. Springer Series in Invasion Ecology. Springer, Dordrecht, The Netherlands.

Belfiore, N.M. and B. May. 2000. Variable microsatellite loci in red swamp crayfish, *Procambarus clarkii*, and their characterization in other crayfish taxa. Mol. Ecol. 9: 2155–2234.

Bentley, A.I., D.J. Schmidt and J.M. Hughes. 2010. Extensive intraspecific genetic diversity of a freshwater crayfish in a biodiversity hotspot. Freshw. Biol. 55: 1861–1873.

Bertocchi, S., S. Brusconi, F. Gherardi, F. Grandjean and C. Souty-Grosset. 2008. Genetic variability in the threatened crayfish *Austropotamobius italicus* in Tuscany: implications for its management. Fundamental and Applied Limnology Archiv für Hydrobiologi 173: 153–64.

Bláha, M., M. Žurovcová, A. Kouba, T. Policar and P. Kozák. 2015. Founder event and its effect on genetic variation in translocated populations of noble crayfish (*Astacus astacus*). J. Appl. Gen. DOI 10.1007/s13353-015-0296-3.

Blanchet, S. 2012. The use of molecular tools in invasion biology: an emphasis on freshwater ecosystems. Fish. Managt. Ecol. 19: 120–132.

Bott, R. 1950. Die Flußkrebse Europas (Decapoda, Astacidae). Abhandlungen Senckenberg naturforsh. Gesell. (Frankfurt) 483: 1–36.

Bott, R. 1972. Beseidlungsgeschischte und Systematik der Astaciden W-Europas unter besonderer Berücksichtigung der Schweiz. Revue Suisse de Zoologie 79: 387–408.

Brodsky, S.Y. 1983. On the systematics of palaearctic crayfishes (Crustacea, Astacidae). Freshwater Crayfish, Charles R. Goldman editor, V. Davis, California, USA, 1981. Westport, Connectitut: AVI Publishing Co. 5: 464–470.

Brown, K. 1980. Low genetic variability and high similarities in the crayfish genera *Cambarus* and *Procambarus*. Am. Midl. Nat. 105: 225–232.

Buhay, J.E. and K.A. Crandall. 2005. Subterranean phylogeography of freshwater crayfishes shows extensive gene flow and surprisingly large population sizes. Mol. Ecol. 14: 4259–4273.

Bunn, J., A. Koenders, C.M. Austin and P. Horwitz. 2008. Identification of hairy, smooth and hybrid marron (Decapoda: Parastacidae) in the Margaret River: Morphology and allozymes. Freshw. Crayfish 16: 113–121.

Busack, C.A. 1988. Electrophoretic variation in the red swamp (*Procambarus clarkii*) and white river crayfish (*P. acutus*) (Decapoda: Cambaridae). Aquaculture 69: 211–226.

Busack, C.A. 1989. Biochemical systematics of crayfishes of the genus *Procambarus*, subgenus *Scapulicambarus* (Decapoda: Cambaridae). J. N. Am. Benthol. Soc. 8: 180–186.

Campbell, N.J.H., M.C. Geddes and M. Adams. 1994. Genetic variation in yabbies (*Cherax destructor* Clark and *C. albidus* Clark (Crustacea: Decapoda: Parastacidae)) indicates the presence of a single, highly sub-structured, species. Aust. J. Zool. 42: 745–760.

Chucholl, C., K. Morawetz and H. Groß. 2012. The clones are coming – strong increase in Marmorkrebs [*Procambarus fallax* (Hagen, 1870) f. *virginalis*] records from Europe. Aquat. Invasions 7: 511–519.

Chucholl, C., A. Mrugała and A. Petrusek. 2015. First record of an introduced population of the southern lineage of white-clawed crayfish (*Austropotamobius "italicus"*) north of the Alps. Knowledge and Management of Aquatic Ecosystems 10. doi: 10.1051/kmae/2015006.

Colautti, R.I., M. Mance, M. Viljanen, H.A.M. Ketelaars, H. Bürgi, H.J. MacIsaac and D. Heath. 2005. Invasion genetics of the Eurasian spini waterflea: evidence for bottleneck and gene flow using microsatellites. Mol. Ecol. 14: 1869–1879.

Cox, G.W. 2004. Alien Species and Evolution. Island Press, Washington, Covelo, London.

Cristescu, M.E.A., P.D.N. Hebert, J.D.S. Witt, H.J. MacIsaac and I.A. Grigorovich. 2001. An invasion history for *Cercopagis pengoi* based on mitochondrial gene sequences. Limnol. Oceanogr. 46: 224–229.

Dawkins, K. and J. Furse. 2012. Conservation genetics as a tool for conservation and management of the native Japanese freshwater crayfish *Cambaroides japonicus* (De Haan). Crustacean Research, Special Number 7: 35–43.

Dawkins, K., J. Furse, C. Wild and J. Hughes. 2010. Distribution and population genetics of the freshwater crayfish genus *Tenuibranchiurus* (Decapoda: Parastacidae). Mar. Freshw. Res. 61: 1048–1055.

Dieguez-Uribeondo, J., F. Royo, C. Souty-Grosset, A. Ropiquet and F. Grandjean. 2008. Low genetic variability of the white-clawed crayfish in the Iberian Peninsula: its origin and management implications. Aquat. Conserv. Mar. Freshwat. Ecosyst. 18: 19–31.

Edsman, L., J.S. Farris, M. Källersjö and T. Prestegaard. 2002. Genetic differentiation between noble crayfish, *Astacus astacus* (L.) populations detected by microsatellite length variation in the rDNA ITS1 region. Bull. Fr. Pêche Piscic. 367: 691–706.

Ehrlich, P.R. 1986. Which animals will invade? pp. 79–95. *In*: J.A. Drake and H.A. Mooney (eds.). Ecology of Invasions of North American and Hawaii. Springer-Verlag, New York, NY.

Faulkes, Z. 2010. The spread of the parthenogenetic marbled crayfish, Marmorkrebs (*Procambarus* sp.) in the North American pet trade. Aquat. Invasions 5: 447–450.

Faulkes, Z. 2015. A bomb set to drop: parthenogenctic Marmorkrebs for sale in Ireland, a European location without non-indigenous crayfish. Management of Biological Invasions 6: 111–114.

Fetzner, J.W. 1996. Biochemical systematics and evolution of the crayfish genus *Orconectes* (Decapoda: Cambaridae). J. Crust. Biol. 16: 111–141.

Fetzner, J.W. and K.A. Crandall. 2002. Genetic variation. pp. 291–326. *In*: D.M. Holdich (ed.). Biology of Freshwater Crayfish. Blackwell Science, Oxford, U.K.

Fetzner, J.W. and K.A. Crandall. 2003. Linear habitats and the nested clade Analysis: An empirical evaluation of geographic vs. river distances using an Ozark crayfish (Decapoda: Cambaridae). Evolution 57: 2101–2118.

Fetzner, J.W. and R.J. DiStefano. 2008. Population genetics of an imperiled crayfish from the White River drainage of Missouri, USA. Freshw. Crayfish 16: 131–146.

Fetzner, J.W.J., R.J. Sheehan and L.W. Sheeb. 1997. Genetic implications of broodstock selection for crayfish aquaculture in the Midwestern United States. Aquaculture 154: 39–55.

Fevolden, S.E. and D.O. Hessen. 1989. Morphological and genetic differences among recently founded populations of noble crayfish (*Astacus astacus*). Hereditas 110: 149–158.

Filipová, L. 2012. Genetic variation in North American crayfish species introduced to Europe and the prevalence of the crayfish plague pathogen in their populations. M.S. Thesis, University of Praga and Poitiers, 132 pp.

Filipová, L., E. Kozubíková and A. Petrusek. 2009. Allozyme variation in Czech populations of the invasive spiny-cheek crayfish *Orconectes limosus* (Cambaridae). Knowledge and Management of Aquatic Ecosystems 394-395: art. no. 10.

Filipová, L., D.A. Lieb, F. Grandjean and A. Petrusek. 2011. Haplotype variation in the spiny-cheek crayfish *Orconectes limosus*: colonization of Europe and genetic diversity of native stocks. J. N. Am. Benthol. Soc. 30: 871–881.

Frankham, R. 2005. Invasion biology—resolving the genetic paradox in invasive species. Heredity 94: 385.

Frankham, R., J.D. Ballou and D.A. Briscoe. 2009. Introduction to Conservation Genetics. 2nd edition. Cambridge University Press, 618 pp.

Fratini, S., S. Zaccara, S. Barbaresi, F. Grandjean, C. Souty-Grosset, G. Crosa and F. Gherardi. 2005. Phylogeography of the threatened crayfish (genus *Austropotamobius*) in Italy: implications for its taxonomy and conservation. Heredity 94: 108–18.

Füreder, L., F. Gherardi, D. Holdich, J. Reynolds, P. Sibley and Souty-Grosset. 2010. *Austropotamobius pallipes*. *In*: IUCN 2010. IUCN Red List of Threatened Species. Version 2010.4. <www.iucnredlist.org>.

Gherardi, F. 2006. Crayfish invading Europe: the case study of *Procambarus clarkii*. Invited review paper. Mar. Freshw. Behav. Physiol. 39: 175–191.

Gherardi, F. and D.M. Holdich (eds.). 1999. Crayfish in Europe as Alien Species–How to Make the Best of a Bad Situation? (Crustacean Issues, 11). A.A. Balkema, Rotterdam, 299 p.

Gouin, N., F. Grandjean and C. Souty-Grosset. 2000. Characterization of microsatellite loci in the endangered freshwater crayfish *Austropotamobius pallipes* (Astacidae) and their potential use in other decapods. Molecular Ecology 9: 636–638.

Gouin, N., F. Grandjean, D. Bouchon, J. Reynolds and C. Souty-Grosset. 2001. Genetic structure of the endangered freshwater crayfish *Austropotamobius pallipes*, assessed using RAPD markers. Heredity 87: 80–87.

Gouin, N., F. Grandjean, S. Pain, C. Souty-Grosset and J.D. Reynolds. 2003. Origin and colonization history of the white-clawed crayfish, *Austropotamobius pallipes*, in Ireland. Heredity 81: 70–77.

Gouin, N., F. Grandjean and C. Souty-Grosset. 2006. Population genetic structure of the endangered crayfish *Austropotamobius pallipes* in France based on microsatellite variation: biogeographical inferences and conservation implications. Freshw. Biol. 51: 1369–1387.

Gouws, G., B.A. Stewart and S.R. Daniels. 2006. Phylogeographic structure of a freshwater crayfish (Decapoda: Parastacidae: *Cherax preissii*) in south-western Australia. Mar. Fresh. Res. 57: 837–848.

Gouws, G., B.A. Stewart and S.R. Daniels. 2010. Phylogeographic structure in the gilgie (Decapoda: Parastacidae: *Cherax quinquecarinatus*): a south-western Australian freshwater crayfish. Biol. J. Linn. Soc. 101: 385–402.

Grandjean, F. and C. Souty-Grosset. 2000. Mitochondrial DNA variation and population genetic structure of the white-clawed crayfish *Austropotamobius pallipes pallipes*. Conserv. Gen. 1: 309–319.

Grandjean, F., C. Souty-Grosset and D.M. Holdich. 1997. Mitochondrial DNA variation in four British populations of the white-clawed crayfish *Austropotamobius pallipes*: implications for management. Aquat. Liv. Resources 10: 121–126.

Grandjean, F., D.J. Harris, C. Souty-Grosset and K.A. Crandall. 2000. Systematics of the European endangered crayfish species *Austropotamobius pallipes* (Decapoda, Astacidae). J. Crust. Biol. 20: 522–529.

Grandjean, F., N. Gouin, C. Souty-Grosset and J. Diéguez-Uribeondo. 2001. Drastic bottlenecks in the endangered crayfish species *Austropotamobius pallipes* in Spain and implications for its colonization history. Heredity 86: 431–438.

Grandjean, F., D. Bouchon and C. Souty-Grosset. 2002a. Systematics of the European endangered crayfish species *Austropotamobius pallipes* (Decapoda: Astacidae) with a re-examination of the status of *Austropotamobius berndhauseri*. J. Crust. Biol. 22: 677–681.

Grandjean, F., M. Frelon-Raimond and C. Souty-Grosset. 2002b. Compilation of molecular data for the phylogeny of the genus *Austropotamobius*: one species or several? Bull. Fr. Pêche Piscic. 367: 671–680.

Gross, R., S. Palm, K. Koiv, T. Prestegaard, J. Jussila, T. Paaver, J. Geist, H. Kokko, A. Karjalainen and L. Edsman. 2013. Microsatellite markers reveal clear geographic structuring among endangered noble crayfish (*Astacus astacus*) populations in Northern and Central Europe. Conserv. Genet. 14: 809–821.

Habsburgo-Lorena, A.S. 1986. The status of the *Procambarus clarkii* population in Spain. Freshw. Crayfish 6: 131–136.

Haig, S.M. 1998. Molecular contributions to conservation. Ecology 79: 413–425.

Hänfling, B. and J. Kollmann. 2002. An evolutionary perspective of biological invasions. Trend. Ecol. Evol. 17: 545–546.

Hayes, R.B. 2012. Consequences for lotic ecosystems of invasion by signal crayfish. School of Biological & Chemical Sciences. Queen Mary University of London, 272 pp.

He, L., J. Xie, Q. Li, Y. Zhao, Y. Wang and Q. Wang. 2013. Isolation and characterization of polymorphic microsatellite loci in the redclaw crayfish, *Cherax quadricarinatus*. J. Aquac. Res. Development 4: 162.

Hebert, P.D.N. and M.E.A. Cristescu. 2002. Genetic perspectives on invasions: the case of the Cladocera. Can. J. Fish. Aquat. Sci. 59: 1229–1234.

Henttonen, P. and J.V. Huner. 1999. The introduction of alien species of crayfish in Europe: a historical introduction. pp. 13–22. *In*: F. Gherardi and D.M. Holdich (eds.). Crayfish in Europe as Alien Species. How to Make the Best of a Bad Situation? Rotterdam, A.A. Balkema.

Hobbs, H.H. III 1974. Observations on the cave-dwelling crayfishes of Indiana. Freshw. Crayfish 2: 405–14.

Hobbs, H.H. 1989. An illustrated checklist of the American crayfishes (Decapoda: Astacidae, Cambaridae, and Parastacidae). Smithson. Contrib. Zool. 480.

Holdich, D.M. 1999. The negative effect of established crayfish introductions. Crustacean issues, A.A. Balkema Publishers, Rotterdam, Netherlands 11: 31–47.

Holdich, D.M. 2003. Crayfish in Europe—an overview of taxonomy, legislation, distribution, and crayfish plague outbreaks. pp. 15–34. *In*: D.M. Holdich and P.J. Sibley (eds.). Management & Conservation of Crayfish, Proceedings of a Conference Held in Nottingham on 7th November, 2002. Environment Agency, Bristol.

Holdich, D.M. and M. Pöckl. 2007. Invasive crustaceans in European inland waters. pp. 29–75. *In*: F. Gherardi (ed.). Biological Invaders in Inland Waters: Profiles, Distribution and Threats. Springer, The Netherlands.

Holdich, D.M., J.D. Reynolds, C. Souty-Grosset and P. Sibley. 2009. A review of the ever increasing threat to European crayfish from non-indigenous crayfish species. Knowledge of Management of Aquatic Ecosystems 394–395: 11, http://dx.doi.org/10.1051/kmae/2009025.

Holland, B.S. 2000. Genetics of marine bioinvasions. Hydrobiologia 420: 63–71.

Holland, B.S. 2001. Invasion without a bottleneck: microsatellite variation in natural and invasive populations of the brown mussel, *Perna perna* (L.). Mar. Biotech. 3: 407–415.

Huber, M.G.J. and C.D. Schubart. 2005. Distribution and reproductive biology of *Austropotamobius torrentium* in Bavaria and documentation of a contact zone with the alien crayfish *Pacifastacus leniusculus*. Bull. Fr. Pêche Piscic. 376-377: 759–776.

Hughes, J.M. and M. Hillyer. 2003. Patterns of connectivity among populations of the freshwater crayfish *Cherax destructor* (Decapoda: Parastacidae) in western Queensland, Australia. Mar. Freshw. Res. 54: 587–596.

Huner, J.V. 1977. Introductions of the Louisiana red swamp crayfish, *Procambarus clarkii* (Girard): an update. Freshw. Crayfish 3: 193–202.

Huner, J.V. and J.W. Avault, Jr. 1979. Introductions of *Procambarus* spp. Freshw. Crayfish 4: 191–194.

Ibrasheva, D. 2011. Phylogenetic analysis of freshwater crayfish of Massachusetts: The genus *Procambarus*. Dissertation Bachelor, 37 pp.

Iorgu, E., I.O.P. Popa, A.M. Petrescu and L.O. Popa. 2011. Cross-amplification of microsatellite loci in the endangered stone-crayfish *Austropotamobius torrentium* (Crustacea: Decapoda). Knowledge and Management of Aquatic Ecosystems 401: 08, DOI: 10.1051/kmae/2011021.

Imgrund, J., D. Groth and J. Wetherall. 1997. Genetic analysis of the freshwater crayfish *Cherax tenuimanus*. Electrophoresis 18: 1660–1665.

Karaman, M.S. 1962. Ein Beitrag zur Systematik der Astacidae (Decapoda). Crustaceana 3: 173–191.

Kawai, T., T. Mitamura and A. Othaka. 2003. The taxonomic status of the introduced North American signal crayfish, *Pacifastacus leniusculus* (Dana, 1852) in Japan, and the source of specimens in the newly reported population in Fukushima prefecture. Crustaceana 77: 861–870.

Keane, R.M. and M.J. Crawley. 2002. Exotic plant invasions and the enemy release hypothesis. Trend. Ecol. Evol. 17: 164–170.

Klobučar, G., M. Podnar, M. Jelić, D. Franjević, M. Faller, A. Štambuk, S. Gottstein, V. Simić and I. Maguire. 2013. Role of the Dinaric Karst (western Balkans) in shaping the phylogeographic structure of the threatened crayfish *Austropotamobius torrentium*. Freshwater Biology 58: 1089–1105.

Koizumi, I., N. Usio, T. Kawai, N. Azuma and R. Masuda 2012. Loss of genetic diversity means loss of geological information: the endangered Japanese crayfish exhibits remarkable historical footprints. PLoS ONE 7: e33986.

Kreiser, B.R., J.B. Mitton and J.D. Woodling. 2000. Single versus multiple sources of introduced populations identified with molecular markers: a case study of a freshwater fish. Biol. Invasions 2: 295–304.

Largiadèr, C.R., F. Herger, M. Lörtscher and A. Choll. 2000. Assessment of natural and artificial propagation of the white-clawed crayfish (*Austropotamobius pallipes* species complex) in the Alpine region with nuclear and mitochondrial markers. Mol. Ecol. 9: 25–37.

Larson, E.R. and J.D. Olden. 2011. The State of Crayfish in the Pacific Northwest. Fisheries 36: 60–73.

Larson, E.R., C.L. Abbott, N. Usio, N. Azuma, K.A. Wood, L.M. Herborg and J.D. Olden. 2012. The signal crayfish is not a single species: cryptic diversity and invasions in the Pacific Northwest range of *Pacifastacus leniusculus*. Freshw. Biol. 57: 1823–1838.

Laurent, P. 1997. Introductions d'écrevisses en France et dans le monde, historique et conséquences. Bull. Fr. Pêche Piscic. 344-345: 345–356.

Lee, C.E. 2002. Evolutionary genetics of invasive species. Trend Ecol. Evol. 17: 386–391.

Li, Y., X. Guo, X. Cao, W. Deng, W. Luo and W. Wang. 2012. Population genetic structure and post-establishment dispersal patterns of the red swamp crayfish *Procambarus clarkii* in China. PLoS ONE 7(7): e40652.

Lodge, D.M., C.A. Taylor, D.M. Holdich and J. Skurdal. 2000. Nonindigenous crayfishes threaten North American freshwater biodiversity: Lessons from Europe. Fisheries 25: 7–20.

Lucey, J. 1999. A chronological account of the crayfish *Austropotamobius pallipes* (Lereboullet) in Ireland. Bull. Irish Biogeogr. Soc. 23: 143–161.

Macaranas, J.M., P.B. Mather, P. Hoeben and M.F. Capra. 1995. Assessment of genetic variation in wild populations of the redclaw crayfish (*Cherax quadricarinatus*, von Martens 1868) by means of allozyme and RAPD-PCR markers. Mar. Freshw. Res. 46: 1217–1228.

Machino, Y., L. Füreder, P.J. Laurent and J. Petutschnig. 2004. Introduction of the white-clawed crayfish *Austropotamobius pallipes* in Europe. Berichte des naturwissenschaftlich-medizinischen Vereins in Innsbruck 85: 223–229.

Martin, P., K. Kohlmann and G. Scholtz. 2007. The parthenogenetic Marmorkrebs (marbled crayfish) produces genetically uniform offspring. Naturwissenschafte 94: 843–846.

Martin, P., N.J. Dorn, T. Kawai, C. van der Heiden and G. Scholtz. 2010. The enigmatic Marmorkrebs (marbled crayfish) is the parthenogenetic form of *Procambarus fallax* (Hagen, 1870). Contrib. Zool. 79: 107–118.

Matallanas, B., M.D. Ochando, A. Vivero, B. Beroiz, F. Alonso and C. Callejas. 2011. Mitochondrial DNA variability in Spanish populations of *A. italicus* inferred from the analysis of a *COI* region. Knowledge and Management of Aquatic Ecosystems 401: article 30.

Matallanas, B., C. Callejas and M.D. Ochando. 2012. A genetic approach to spanish populations of the threatened *Austropotamobius italicus* located at three different scenarios. The Scientific World Journal Article ID 975930: 9 pages.

Mathews, L., L. Adams, E. Adnerson, M. Basile, E. Gottardi and M. Buckholt. 2008. Genetic and morphological evidence for substantial hidden biodiversity in a freshwater crayfish species complex. Unpublished report. Worcester, MA: W.P.I.

McKniff, J. 2012. Investigation of the population genetics of crayfish (*Orconectes virilis*) using AFLP markers. Project Report Worcester Polytechnic Institute, 44 pp.

Miller, G.C. 1960. The taxonomy and certain biological aspects of the crayfish of Oregon and Washington. Master's thesis, Oregon State College, Corvalis, pp. 216.

Moritz, C. 1994. Defining 'evolutionarily significant units' for conservation. Trend. Ecol. Evol. 9: 373–375.

Nemeth, S.T. and M.L. Tracey. 1979. Allozyme variability and relatedness in six crayfish species. J. Hered. 70: 37–43.

Nguyen, T.T.T. and C. Austin. 2004. Inheritance of molecular markers and sex in the Australian freshwater crayfish, *Cherax destructor* Clark. Aquacult. Res. 35: 1328–1338.

Nguyen, T.T.T., M. Meewan, S. Ryan and C.M. Austin. 2002a. Genetic diversity and translocation in the marron, *Cherax tenuimanus* (Smith): implications for management and conservation. Fish. Managt. Ecol. 9: 163–173.

Nguyen, T.T.T., N.P. Murphy and C.M. Austin. 2002b. Amplification of multiple copies of mitochondrial cytochrome *b* gene fragments in the Australian freshwater crayfish, *Cherax destructor* Clark (Parastacidae: Decapoda). Animal Genetics 33: 304–308.

Nguyen, T.T.T., C.P. Burridge and C.M. Austin. 2005. Population genetic studies on the Australian freshwater crayfish, *Cherax destructor* (Parastacidae: Decapoda) using allozyme and RAPD markers. Aquatic Living Resources 18: 55–64.

Patti, F.P. and M.C. Gambi. 2001. Phylogeography of the invasive polychaete *Sabella spallanzanii* (Sabellidae) based on the nucleotide sequence of internal transcribed spacer 2 (ITS2) of nuclear rDNA. Mar. Ecol. Prog. Ser. 215: 169–177.

Pedraza-Lara, C., F. Alda, S. Carranza and I. Doadrio. 2010. Mitochondrial DNA structure of the Iberian populations of the white-clawed crayfish, *Austropotamobius italicus italicus* (Faxon, 1914). Mol. Phylog. Evol. 57: 327–342.

Perry, W.L., J.L. Feder, G. Dwyer and D.M. Lodge. 2001a. Hybrid zone dynamics and species replacement between *Orconectes* crayfishes in a northern Wisconsin lake. Evolution 55: 1153–1166.

Perry, W.L., J.L. Feder and D.M. Lodge. 2001b. Implications of hybridization between introduced and resident *Orconectes* crayfishes. Conservation Biology 15: 1656–1666.

Perry, W.L., D.M. Lodge and J.L. Feder. 2002. Importance of hybridization between indigenous and non indigenous freshwater species: an overlooked threat to North American biodiversity. Sysm. Biol. 51: 255–275.

Primack, R.B. 2000. Conservation at the population and species level. pp. 121–182. *In*: R.B. Primack (ed.). A Primer of Conservation Biology 2nd edn. Sinauer Associates, Sunderland, MA.

Renai, B., S. Bertocchi, S. Brusconi, F. Gherardi, F. Grandjean, M. Lebboroni, B. Parinet, C. Souty Grosset and M.C. Trouilhé. 2006. Ecological characterisation of streams in Tuscany (Italy) for the management of the threatened crayfish *Austropotamobius italicus*. Bull. Fr. Pêche Piscic. 380-381: 1095–1114.

Reynolds, J. and C. Souty-Grosset. 2012. Management of Freshwater Biodiversity: Crayfish as Bioindicators. Cambridge University Press, 384 pp.

Richardson, A., N. Doran and B. Hansen. 2006. The geographic ranges of Tasmanian crayfish: extent and pattern. Freshwater crayfish 15: 1–17.

Richman, N.I., M. Böhm, S.B. Adams, F. Alvarez, E.A. Bergey, J.J.S. Bunn, Q. Burnham, J. Cordeiro, J. Coughran, K.A. Crandall, K.L. Dawkins, R.J. DiStefano, N.E. Doran, L. Edsman, A.G. Eversole, L. Füreder, J.M. Furse, F. Gherardi, P. Hamr, D.M. Holdich, P. Horwitz, K. Johnston, C.M. Jones, J.P.G. Jones, R.L. Jones, T.G. Jones, T. Kawai, S. Lawler, M. López-Mejía, R.M. Miller, C. Pedraza-Lara, J.D. Reynolds, A.M.M. Richardson, M.B. Schultz, G.A. Schuster, P.J. Sibley, C. Souty-Grosset, C.A. Taylor, R.F. Thoma, J. Walls, T.S. Walsh and B. Collen. 2015. Multiple drivers of decline in the global status of freshwater crayfish (Decapoda: Astacidea). Phil. Trans. R. Soc. Lond. B: Biol. Sci. 370: 20140060.

Riegel, J.A. 1959. The systematics and distribution of crayfishes in California. Calif. Fish Game 45: 29–50.

Sakai, A.K., F.W. Allendorf, J.S. Holt, D.M. Lodge, J. Molofsky, K.A. With, S. Baughman, R.J. Cabin, J.C. Cohen, N.C. Ellstrand, D.E. McCauley, P. O'Neil, I.M. Parker, J.N. Thompson and S.G. Weller. 2001. The population biology of invasive species. Ann. Rev. Ecol. Syst. 32: 305–332.

Santucci, F., M. Iaconelli, P. Andreani, R. Cianchi, G. Nascetti and L. Bullini. 1997. Allozyme diversity of European freshwater crayfish of the genus *Austropotamobius*. Bull. Fr. Pêche Piscic. 347: 663–676.

Savini, D., A. Occhipinti–Ambrogi, A. Marchin, E. Tricarico, F. Gherardi, S. Olenin and S. Gollasch. 2010. The top 27 animal alien species introduced into Europe for aquaculture and related activities. J. Appl. Ichthyol. 26: 1–7.

Scholtz, G., A. Braband, L. Tolley, A. Reimann, B. Mittmann, C. Lukhaup, F. Steuerwald and G. Vogt. 2003. Parthenogenesis in an outsider crayfish. Nature 421: 806–806.

Schrimpf, A. 2013. DNA-based methods for freshwater biodiversity conservation—Phylogeographic analysis of noble crayfish (*Astacus astacus*) and new insights into the distribution of crayfish plague. Dissertation Universität Koblenz-Landau, 182 pp.

Schrimpf, A., H. Schulz, K. Theissinger, L. Pârvulescu and R. Schulz. 2011. First large-scale genetic analysis of the vulnerable noble crayfish *Astacus astacus* reveals low haplotype diversity of Central European populations. Knowledge and Management of Aquatic Ecosystems 401: 35.1111.

Schrimpf, A., K. Theissinger, J. Dahlem, I. Maguire, L. Pârvulescu, H.K. Schulz and R. Schulz. 2014. Phylogeography of noble crayfish (*Astacus astacus*) reveals multiple refugia. Freshwater Biology 59: 761–776.

Schubart, C.D. and M.G.J. Huber. 2006. Genetic comparisons of German populations of the stone crayfish, *Austropotamobius torrentium* (Crustacea: Astacidae). Bull. Fr. Pêche Piscic. 380-381: 1019–1028.

Schulz, H.K., P. Smietana and R. Schulz. 2004. Assessment of DNA variations of the noble crayfish (*Astacus astacus* L.) in Germany and Poland using Inter-Simple Sequence Repeats (ISSRs). Bull. Fr. Pêche Piscic. 372-373: 387–399.

Sherman, C.D.H., D. Ierodiaconou, A.M. Stanley, K. Weston, M.G. Gardner and M.B. Schultz. 2012. Development of twenty-four novel microsatellite markers for the freshwater crayfish, *Geocharax gracilis* using next generation sequencing. Conserv. Genet. Res. 4: 555–558.

Sinclair, E.A., A. Madsen, T. Walsh, J. Nelson and K.A. Crandall. 2011. Extensive gene flow across independent river drainages in the giant Tasmanian freshwater crayfish, *Astacopsis gouldi* (Decapoda: Parastacidae); implications for conservation. Animal Conservation 14: 87–97.

Sint, D., J. Dalla Via and L. Füreder. 2006. The genus *Austropotamobius* in the Ausserfern Region (Tyrol. Austria) with an overlap in the distribution of *A. torrentium* and *A. pallipes* populations. Bull. Fr. Pêche Piscic. 380-381: 1029–1040.

Smith, P.J. and B.J. Smith. 2009. Small-scale population-genetic differentiation in the New Zealand caddisfly *Orthopsyche fimbriata* and the crayfish *Paranephrops planifrons*. N.Z. J. Mar. Freshwat. Res. 43: 723–734.

Sonntag, M.M. 2006. Taxonomic standing of the three subspecies of *Pacifastacus leniusculus*, and their phylogeographic patterns in the Klamath Basin area. Brigham Young University, pp. 71.

Soulé, Me and L.S. Mills. 1992. Conservation genetics and conservation biology: a troubled marriage. pp. 55–69. *In*: O.T. Saundlund, K. Hindar and A.H.D. Brown (eds.). Conservation of Biodiversity for Sustainable Development. Scandinavian University Press, Oslo, Scandinavia.

Souty-Grosset, C. and J.D. Reynolds. 2010. Current ideas on methodological approaches in European crayfish conservation and restocking procedure. Knowledge & Management of Aquatic Systems 394-395: 01.

Souty-Grosset, C., F. Grandjean, R. Raimond, M. Frelon, C. Debenest and M. Bramard. 1997. Conservation genetics of the white-clawed crayfish *Austropotamobius pallipes*: the usefulness of the mitochondrial DNA marker. Bulletin Français de la Pêche et de la Protection des Milieux Aquatiques 70: 677–692.

Souty-Grosset, C., F. Grandjean and N. Gouin. 1999. Molecular genetic contributions to conservation biology of the European native crayfish *Austropotamobius pallipes*. Freshwater Crayfish 12: 371–386.

Souty-Grosset, C., F. Grandjean and N. Gouin. 2003. Involvement of genetics in knowledge, stock management and conservation of *Austropotamobius pallipes* in Europe. Bulletin Français de la Pêche et de la Pisciculture 370-371: 167–179.

Souty-Grosset, C., D. Holdich, P. Noël, J. Reynolds and P. Haffner (eds.). 2006. Atlas of Crayfish in Europe. Coll. Patrimoine Naturel. Publ. Sci. du Museum National d'Histoire Naturelle, Paris vol. 64 (187 pp.).

Stefani, F., S. Zaccara, B. Giovanni, G.B. Delmastro and M. Buscarino. 2011. The endangered white-clawed crayfish *Austropotamobius pallipes* (Decapoda Astacidae) east and west of the Maritim Alps, a result of human translocation? Conservation Genetics 12: 51–60.

Strayer, D.L. and D. Dudgeon. 2010. Freshwater biodiversity conservation: recent progress and future challenges J. N. Am. Benthol. Soc. 29: 344–358.

Suarez, A.V., D.A. Holway and T.J. Case. 2001. Patterns of spread in biological invasions dominated by long-distance jump dispersal: Insights from Argentine ants. Proc. Natl. Acad. Sci. U.S.A. 98: 1095–1100.

Taugbøl, T. 2004. Reintroduction of noble crayfish *Astacus astacus* after crayfish plague in Norway. Bull. Fr. Pêche Piscic. 372-372: 83–96.

Taugbøl, T. and J. Skurdal. 1999. The future of native crayfish in Europe: How to make the best of a bad situation? pp. 271–279. *In*: F. Gherardi and D.M. Holdich (eds.). Crayfish in Europe as Alien Species – How to Make the Best of a Bad Situation? Crustacean Issues 11. A.A. Balkema, Rotterdam.

Taylor, C.A., G.A. Schuster, J.E. Cooper, R.J. DiStefano, A.G. Eversole, P. Hamr, H.H. Hobbs III, H.W. Robison, C.E. Skelton and R.F. Thomas. 2007. A reassessment of the conservation status of crayfishes of the United States and Canada: the effects of 10+ years of increased awareness. Fisheries 32: 372–389.

Therriault, T.W., M.I. Orlova, M.F. Docker, J.J. MacIsaac and D.D. Heath. 2005. Invasion genetics of a freshwater mussel (*Dreissena rostriformis bugensis*) in Eastern Europe: high gene flow and multiple introductions. Heredity 95: 16–23.

Torres, E. and F. Alvarez. 2012. Genetic variation in native and introduced populations of the red swamp crayfish *Procambarus clarkii* (Girard, 1852) (Crustacea, Decapoda, Cambaridae) in Mexico and Costa Rica. Aquat. Invasions 7: 235–241.

Trontelj, P., Y. Machino and B. Sket. 2005. Phylogenetic and phylogeographic relationships in the crayfish genus *Austropotamobius* inferred from mitochondrial COI gene sequences. Mol. Phylog. Evol. 34: 212–226.

Tsutsui, N.D., A.V. Suarez, D.A. Holway and T.J. Case. 2000. Reduced genetic variation and the success of invasive species. Proc. Natl. Acad. Sci. U.S.A. 97: 5948–5953.

Usio, N., K. Nakata, T. Kawai and S. Kitano. 2007. Distribution and control status of the invasive signal crayfish (*Pacifastacus leniusculus*) in Japan. Jap. J. Limnol. 68: 471–482.

Walsh, T. and N. Doran. 2010. *Astacopsis gouldi. In*: IUCN 2013. IUCN Red List of Threatened Species. Version 2013.1. <www.iucnredlist.org>.

Weiss, S. 2005. Conservation genetics of freshwater organisms. Bull. Fr. Pêche Piscic. 376-377: 571–583.

Wilson, A.B., K.A. Naish and E.G. Boulding. 1999. Multiple dispersal strategies of the invasive quagga mussel (*Dreissena bugensis*) as revealed by microsatellite analysis. Can. J. Fish. Aquat. Sci. 56: 2248–2261.

Wilson, A.C., R.L. Cann and S.M. Carr. 1985. Mitochondrial DNA and two perspectives on evolutionary genetics. Biol. J. Linn. Soc. 26: 375–400.

Xie, Y., L. He, J. Sun, L. Chen, Y. Zhao, Y. Wang and Q. Wang. 2010. Isolation and characterization of fifteen microsatellite loci from the redclaw crayfish, *Cherax quadricarinatus*. Aquat. Liv. Res. 23: 231–234.

Yue, G.H., J. Li, Z. Bai, C.M. Wang and F. Feng. 2010. Genetic diversity and population structure of the invasive alien red swamp crayfish. Biol. Invasions 12: 2697–2706.

Zhu, B.H., Y. Huang, Y.G. Dai, C.W. Bi and C.Y. Hu. 2013. Genetic diversity among red swamp crayfish (*Procambarus clarkii*) populations in the middle and lower reaches of the Yangtze River based on AFLP markers. Genet. Mol. Res. 12: 791–800.

Zhu, Z.Y. and G.H. Yue. 2008. Eleven polymorphic microsatellites isolated from red swamp crayfish, *Procambarus clarkii*. Mol. Ecol. Res. 8: 796–798.

Zuber, S.T., K. Muller, R.H. Laushman and A.J. Roles. 2012. Hybridization between an invasive and a native species of the crayfish genus *Orconectes* in north-central Ohio. J. Crust. Biol. 32: 962–971.

Crayfish Growth and Reproduction

Colin L. McLay[1,*] and *Anneke M. van den Brink*[2]

Introduction

Growth and reproduction are competitive processes with the first being an essential prerequisite for the second. In crayfish both of these processes continue until the animal dies although exactly how these are coordinated and scheduled remains poorly understood. They can be viewed as two cyclic intermittent processes, sometimes alternating sometimes not. Often in decapods growth and mating (the first step in reproduction) are closely linked, but this is not the case in crayfish. Growth requires intermittent moulting, which in mature females affects sperm storage, a result that has consequences for both sexes because all stored sperm are lost and females are restored to a virginal state. Males in particular have to do something about this if they are to see any of their genes passed on to the next generation. Growth really only concerns the individual whereas mating is the concern of two individuals, male and female, whose interests may be different. So if mating and moulting are not linked there has to be an alternative means of mate attraction to ensure that mating occurs. Significant recent progress has been made into understanding the basis of mate attraction in crayfish and how urine-based signals are used for communication (Breithaupt 2011). An important consequence of the de-linkage of moulting and mating is that females can have multiple partners and thus there is the possibility of sperm competition. We have organized and summarized the data in this chapter so that it leads towards an understanding of the evolution of growth and mating behaviour in crayfish and how these are integrated. Such an evolutionary discussion needs to have in mind a sister group that helps to inform us about ancestral character states and derived characters that can be attributed to the colonization of freshwater. As will become evident we assume that the marine sister group for crayfish is the clawed nephropoid lobsters.

Like all arthropods freshwater crayfish have a multi-layered exoskeleton hardened by calcium salts except around the joints where the integument is soft and flexible. The body consists of 20 segments, which are divided into three regions: head with the eyes, sensory appendages, mandible and maxillae; thorax with maxillipeds and five pairs of leg-like appendages; and abdomen equipped with pleopods used for swimming and egg-carrying in females. The head and the thorax form one unit, the cephalothorax. All

[1] Biological Sciences, Canterbury University, Christchurch, NZ, 8004.
[2] HZ University of Applied Sciences, Edisonweg 4, 4382 NW Vlissingen, The Netherlands.
  Email: anneke.brink@gmail.com
* Corresponding author: colin.mclay@canterbury.ac.nz

of these specialized, jointed appendages are covered by an integument that is replaced when the crayfish moults. A heavy exoskeleton ensures protection of the internal organs, but it limits growth to short periods when it is soft enough to expand. However, moulting provides an opportunity to repair damage and to replace lost appendages, something that is likely to occur in an animal that can live for decades (Vogt 2012). In all malacostracans moulting consists of withdrawing the soft body, and all of its appendages, through a dorsal split in the exoskeleton between the cephalothorax and the abdomen. [Abbreviations used: CL = caparace length; TL = total length.]

Growth

The crayfish lifecycle

Unlike their marine cousins, freshwater crayfish have no larval stages, and all the development typical of free-living decapod larvae occurs inside the egg, so what emerges from the egg is a juvenile crayfish that already possesses most of the normal adult appendages. Juveniles remain attached to the female pleopods by the unique telson thread. At this stage they are lecithotrophic, relying on the remaining egg yolk. Juveniles are attached to the mother for the first two stages with the moulted exoskeleton disintegrating, thereby preserving the integrity of the telson thread and not jeopardizing parental protection by detachment (Vogt 2008a). When they become free-living, moulting involves casting an intact shell allowing the soft body the opportunity to expand in size. Periodic moulting results in a stair-case like pattern in size increase over time: short periods of rapid size increase followed by much longer intermoult periods when size remains the same. Sexual maturity is marked by a pubertal moult in both sexes after which some body features depart from isometric growth. Freshwater crayfish have indeterminate growth with moulting continuing after they reach sexual maturity and until such time as they die. A key requirement for freshwater crayfish is shelter and protection from predators and if there is none available then they can easily excavate their own burrows. This essential resource makes them relatively easy to capture by sinking an old opened perforated paint tin with cord attached into likely habitat. No bait is necessary to attract them!

Hormonal control of maturation and moulting

The crayfish exoskeleton is multi-layered, thickened and robust and consists of calcified chitin and protein material (Fig. 1). The outermost layer, the epicuticle, is underlain by the thicker procuticle, both of which are calcified to varying degrees. This may be partitioned into an exocuticle, which can be pigmented with melanin, and an endocuticle overlying the non-calcified membranous layer. An alternative way of labelling these layers uses the time of formation as a basis: pre-ecdysial and post-ecdysial can be used for all layers below the eipcuticle. These inert layers are all underlain by the epidermis (or hypodermis), live cuboidal cells responsible for secreting the protective cuticle that consists of alternating layers of chitin and protein, which make the exoskeleton tough, yet pliable. Flexibility is greatest at the arthrodial joints, which lack the calcium mineralized epicuticle that gives the rest of the exoskeleton its strength.

As in all of the Arthropoda moulting is caused by hormones released by endocrine organs. It consists of a series of physiological steps each controlled by different hormones released by the neuroendocrine X-organ and the epithelial Y-organ. Ecdysteroids such as ecdysone (20-hydroxyecdysone) released by the thoracic Y-organ promote preparation for moulting, while neuropeptides produced by cells in the eyestalks X-organ (sinus gland) inhibit moulting by suppressing secretions from the Y-organ. The interaction between these two organs can be demonstrated by ablating the eyes which can precipitate moulting. The levels of ecdysone fluctuate during the moult cycle: immediately post-moult levels are very low and this continues during the intermoult period, but during the pre-moult stage levels begin to increase (Smith and Chang 2007). Peptides involved in moult regulation are members of a family of compounds, which may also be involved in regulating reproduction and metabolism.

The longest part of the moult cycle is the intermoult, stage C4, when the cuticle layers are complete and it has hardened and calcified. Crayfish resume normal feeding patterns and begin to accumulate reserves

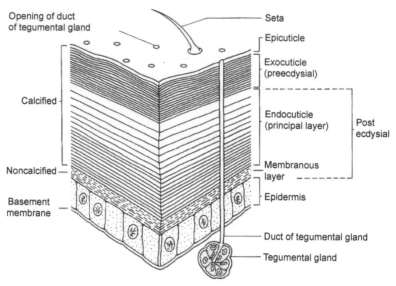

Figure 1. Section through the crayfish body exoskeleton (from Reynolds 2002). Strength and rigidity are provided by the combination of inner layers of protein and chitin with calcified outer layers. Note that the integument making up the arthrodial membrane on limbs is less calcified and therefore more flexible.

in the hepatopancreas and/or gonads if they are to reproduce. Preparations for the next moult commence in stage D0 with separation of the epidermis from the overlying cuticle, resorption of old cuticle, formation of the epicuticle and secretion of the endocuticle (D2-4) (Table 1). The crayfish becomes less active and calcium is withdrawn from the exoskeleton and stored as gastroliths (Greenaway 1985).

As all crayfish owners know their pets readily moult in captivity. Now that many people have internet access there are many recordings of crayfish moulting, uploaded by pet owners who have video cameras. The following account is based on more or less complete moulting sequences of the following species: *Pacifastacus leniusculus* (Dana, 1852), *Procambarus clarkii* (Girard, 1852), *Pr. spiculifer* (Le Conte, 1856), *Pr. fallax* (Hagen, 1870) (marmorkrebs), *Cherax quinquecarinatus* (Gray, 1865) and *Cherax quadricarinatus* (von Martens, 1868) obtained by searching YOUTUBE using the terms "crayfish molting". When it is ready to moult the crayfish becomes less active and they may seek shelter of some kind where protection is available, but such shelter is not necessarily a prerequisite for moulting. The animal usually lies on one side with its appendages extended out from the body. These are often quite active, although not performing any particular action such as feeding or walking, and the activity is probably a reflection of what is happening beneath the exoskeleton. Here the soft new exoskeleton needs to be separated from the old so that it can be withdrawn, sometimes through narrowed sections of the exit path such as the joints which characterize arthropods. The movement of the segmental appendages is akin to quivering and is most evident with the pereopods and pleopods. The first sign that the new crayfish is going to emerge is seen when the carapace (sometimes called the "saddle" by aquarists) separates from the first abdominal terga in the mid-dorsal line as a result of the animal partially folding the abdomen anteriorly thereby creating the split. At this stage the abdomen remains in place, but the thorax begins to swell pushing the crayfish outwards as the old carapace separates from the new one. The crayfish continues to withdraw its thorax from the old one and the first appendages to emerge are the sensory antennae and the eyes at the anterior end. It seems necessary to assume that the animal does this by straightening its cephalothorax, using leverage provided by its pereopods which are still in place and the "elasticity" remaining in the cephalothorax sternum. Otherwise the long antennae, for example, have no way of being withdrawn from their old habitus. The same leverage that began at the anterior end continues posteriorly with the mouthparts and stomach lining being extracted. This seems to be easily done so there must have been some loosening activity beforehand which facilitates the withdrawal of these intimate structures. Once the feeding appendages have been extracted moulting continues with the chelae and walking legs (pereopods). Once the appendages of the

Table 1. Moult stages of *Orconectes virilis* (Cambaridae) and *Astacus leptodactylus* (from Reynolds 2002). Currently recognized stages in the moult cycle are based on five stages, A–E, recognized by Drach (1939).

Orconectes virilis	*Astacus leptodactylus*

Stage A
   Soft integument
   Reduction of epidermal cells
   Epi- and exocuticle formed
	Stage A
	   Soft integument
	A1 Soft pereopods
	A2 More rigid pereopods

Stage B
   Parchment-like integument
   Calcification of exocuticle
	Stage B
	   Parchment-like integument
	B1 Propodus and merus flexible
	B2 Propodus and merus brittle
	   Start of endocuticle secretion

Stage C
   Hardening of cuticle
   Epidermis and cuticle in contact
	Stage C

C1	C1 Flexible carapace
C2	C2 Rigid carapace

C3 Endocutcile completed
   Membranous layer formation
	C3 Carapace calcified
	   Membranous layer formation

C4 Calcification of whole cuticle	C4 Intermoult
Normal functional period of crayfish life cycle—intermoult	

Stage D
   Preparation for moult
	Stage D
	   Preparation for moult

D0 Separation of epidermis	D0 Apolysis of cuticle Formation of epicuticle
D1 Epidermal cell elongation	D1 Achievement of epiculture

D2 Resorption of old cuticle
   Formation of epicuticle
   Secretion of endocuticle
	D2 Flexible edge to branchiostegites
	   Secretion of exocuticle

D3 Resorption of old cuticle
   Thickening of epicuticle
   Exocuticle completed
	D3 Soft edge to edge to branchiostegites

D4 As for D3	D4 Split between thorax and abdomen

Stage E
   Ecdysis or moult
	Stage E
	   Ecdysis or moult

cephalothorax have been withdrawn moulting is normally completed rapidly by a flick of the abdomen. This can take less than a second with the new abdomen suddenly parting company with the rest the skeleton. For this to happen there must have been a longer preparatory period wherein the pleopods had been readied for removal. The short duration of this part of moulting is no doubt attributable to the fact that all the abdominal segments are separate and the joints between them are flexible. Thus moulting is completed when the crayfish swims away with a single flick of the abdomen leaving the intact old shell behind along with anything that might be attached, such as ectoparasites (Evans and Edgerton 2002) or spermatophores in the case of mated females.

The now soft crayfish rests on the bottom although not able to support itself by using its limbs. Limb movements are gentle and in some cases the animal lies on its side extending its pereopods, in particular, as far forward as possible as though it was undertaking stretching exercises. During the whole moulting sequence it is presumably taking in as much water as possible in order to stretch and inflate the soft exoskeleton before it hardens. Meanwhile the soft crayfish remains sheltered away from possible attacks by others. The first part of the exoskeleton to become quinone hardened and calcified is the cuticle. In some circumstances the crayfish may consume parts of the old exoskeleton, but not before mineralization of the chelae and stomach ossicles has begun. Huner and Lindquist (1985) compared remineralisation in *Astacus astacus* (Linnaeus, 1758), *Pacifastacus leniusculus* (Astacidae), *Procambarus clarkii* and *P. acutus* (Girard, 1852) (Cambaridae). The warm water astacids have a greater level of mineralisation than the cold water cambarids. In *Astacus astacus* the hardening process is largely complete after 2–4 days (Taugbol et al. 1997).

Vogt (2012) makes the interesting point that by regular moulting, Crustacea in general and Decapoda in particular, can avoid the effects of mechanical aging and senescence. By renewing their stomach ossicles they do not need to go to the dentist, by casting their exoskeleton they have no need for new hip joints, and by discarding all the setae they have no need for a hair transplant! With such advantages resulting from indeterminate growth it is difficult to explain why some cease moulting after their pubertal moult, but at least the Astacidea are not amongst these and have not given up the chance to repair and regenerate damaged or lost limbs. It has been shown that stem cell activity persists in *Procambarus fallax* (marmorkrebs) (Vogt 2010) and in some prawns and homarid lobsters (Vogt 2012). Decapods with determinate growth may have entered a state of somatic senescence, but not reproductive senescence as they expend more energy on propagating their genes rather than on somatic house-keeping.

Components of crustacean growth

Crayfish growth is discontinuous and can be resolved into two components: moult increment and intermoult interval. Together these components can be used to generate size vs. age relationships if we have an adequate model that explains these components. Such models could include pre-moult size (normally carapace length, CL), temperature and density, but apart from size, data such as these are not always available. However, there is abundant evidence that water temperature is a primary factor in determining the rate of growth with about 10°C being the lower growth limit for temperate crayfish (Momot 1984, Lowery 1988, Merrick 1993, Lodge and Hill 1994, Parkyn et al. 2002). Moulting may therefore be restricted to summer months for some populations. A simple model using premoult size as the independent variable for both components and assuming a linear relationship (negative slope for moult increment and positive slope for moult interval) would predict a staircase of decreasing tread height and increasing tread length for size vs. age resulting in a decreasing growth rate. Size would appear to be asymptotic to a maximum, but only depends upon survival rather than a theoretical maximum. Some crayfish grow rapidly and live only a short time while others may only moult annually, or at greater intervals, and live for several decades. Many cambarid crayfish have rapid growth and maturation while many astacids and parastacids grow much more slowly not reaching maturity for several years (Honan and Mitchell 1995). A feature of crustacean growth that limits the use of size as an accurate estimate of age is the fact that growth increments at each moult get smaller and smaller, so that it becomes increasingly difficult to separate the members of each cohort. In addition the cumulative effects of variation in both increment and interval (i.e., moult frequency) also mean that predicting age from size becomes increasingly difficult as size increases. Independent ways of estimating age are difficult when all the hard parts are shed at each moult (typically at least 11 moults, Holdich 2002). However, use of tags that can survive moulting enables us to follow the growth trajectories by taking repeated measurements of the same crayfish. There is no evidence that crayfish lose the ability to moult with increasing size or age, so they all have an indeterminate growth format.

For example, growth of *Paranephrops planifrons* White, 1842 living in a North Island, New Zealand pasture stream, based on mark-recapture data, showed that moult increment increased with increasing pre-moult size up to 20 mm CL, but percentage moult increment declined from around 16% at 5 mm CL to 8% at 30 mm CL. Moult frequency also declined with size. Mean CL when leaving the female was 3.5 mm, after one year it was 11.4–11.8 mm (nine moults), after two years 18.2–19.3 mm (three moults), after three years 22.1–23.8 mm (two moults), and after four years 26.7–29.6 mm (one moult) (Hopkins 1967b). In a more detailed study Parkyn et al. (2002) found that growth of juveniles in Waikato pasture streams (modified habitat) was rapid with females reaching reproductive maturity at around 20 mm CL in their first year. However, in the native forest habitat growth was much slower due to lower moult frequency and smaller moult increments caused by lower water temperatures. Females from forest streams took close to two years to reach maturity. Changes in land use can have significant impact on freshwater crayfish life cycles particularly if they result in changes to water temperature. *Paranephrops zealandicus* (White, 1847) from a South Island, New Zealand headwater stream, grew much more slowly because water temperatures only exceeded 10°C (the lower limit for moulting) for 60 days per year (Whitmore and Huryn 1999). Females took six or more years to mature and eggs and juveniles remained attached to the parent for at least 15 months. Crayfish over 16 years old were common and the largest animals were estimated to be over 28 years.

Kawai et al. (1997) analysed the population structure of *Cambaroides japonicus* (de Haan, 1842) in a small stream, near Atsuta, Hokkaido, Japan, measuring CL and then returning crayfish to the stream. Some individuals were kept captive under near-natural conditions and their growth recorded. Crayfish smaller than 10 mm CL could not be sexed so were divided equally between males and females. Polymodal analysis separated out size groups (assumed to conform to a normal distribution) in monthly field samples and growth data from captive animals was used to estimate the number of months since hatching for each size group (Fig. 2). Size vs. age was estimated by the von Bertalanffy growth model (Fig. 3). Reproduction and recruitment occurs in summer. They were able to estimate survivorship and thus generate a static life table for *C. japonicus*: male longevity was 11 yrs and for females 10 yrs. Both sexes became sexually mature after 5–6 yrs. Scalici et al. (2008) present a similar analysis for *Austropotamobius italicus* (Faxon, 1914), including data from marked-recaptured individuals, indicating a similar longevity.

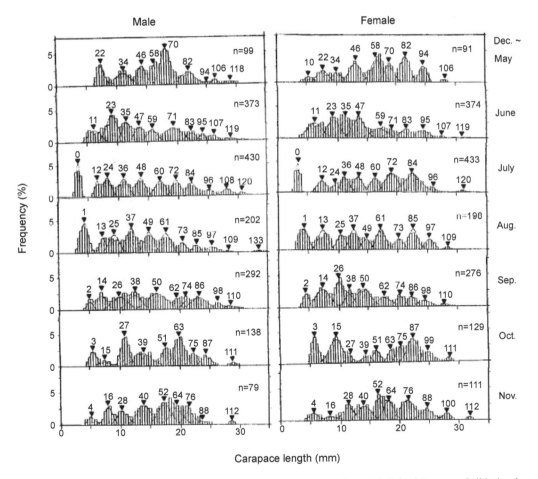

Figure 2. Population size structure of male and female *Cambaroides japonicus* in a small Hokkaido stream. Solid triangles are the mean CL of the normal distributions described by the dashed curves for each age cohort. The numbers above the triangles indicate the number of months since hatching. Sample sizes (n) are given for each month (from Kawai et al. 1997).

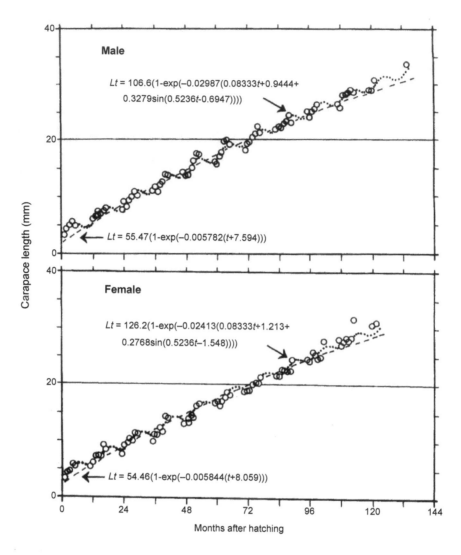

Figure 3. The fit of von Bertalanffy growth equations to the mean CL data of male and female *Cambaroides japonicus* collected in a small stream in Hokkaido. The upper half of each panel gives the sine-wave von Bertalanffy equation (Pauly and Gashutz 1979) that describes seasonal variation in mean size (represented by ....), while the lower half gives the ordinary von Bertalanffy equation that describes the size specific annual growth trend (represented by --------). L_t is the CL at age t (in months) after recruitment (from Kawai et al. 1997).

There have been some suggestions that obligate troglobitic crayfish may be long-lived compared to surface dwellers. *Orconectes australis* (Rhoades, 1941) from the eastern United States has been widely cited as a species which may live to almost 180 years and take 29–105 yr to reach a size when females produced their first brood (Cooper 1975, Culver 1982). However, a detailed five year mark-recapture study based on more than 3800 crayfish in three isolated cave systems provided better estimates of longevity (Venarsky et al. 2012). For example, they measured growth rates of *O. australis* in Hering Cave, Alabama, by a mark-recapture method and estimated the annual size-specific CL increments (Fig. 4). They then applied these increments to a juvenile crayfish CL = 3 mm and iteratively generated growth curves (Fig. 4). Since size-specific growth rates declined with increasing CL the slope of growth curves gradually decreased as crayfish became larger. Females took only five or six yr to produce their first broods and less than 5% exceed 22 yr longevity. These milestones are comparable to other cave-dwelling and surface crayfish for

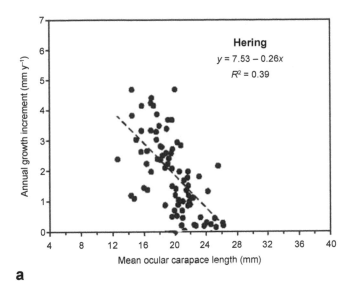

a

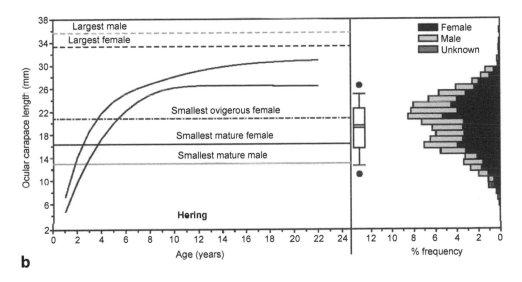

b

Figure 4. (a) Relationship between annual growth increments and mean CL (mm) for *Orconectes australis* in Hering Cave, Alabamba. Dashed line is the result of plotting the least-squared regression fit of the equation shown. (b) Growth model for *O. australis*: solid lines are upper and lower 95% confidence interval; the pooled size-frequency distribution for the population is plotted to the right of the growth model; to the left of the size-frequency distribution is a box and whisker plot with the box indicating the 25th and 75th percentile containing the mean (dashed line) and median (solid line) and the whisker is the error bar (dots are the 5th and 95th percentiles) (from Venarsky et al. 2012).

which reliable data exist and show that caves are not in fact a 'Shangrila' habitat where 'immortality' might be possible.

Reliable methods of estimating age of arthropods are hard to come by, but one that has shown some potential is measurement of lipofuscin granules in neurological tissues (Vogt 2012). Lipofuscin is an index rather than direct measurement of chronological time (Belchier et al. 1998, Reynolds 2002). In the Australian red-claw crayfish, *Cherax quadricarinatus*, lipofuscin was a better estimate of age than body size because within a cohort individual variation of lipofuscin was less than that for size. Furthermore, it continues to accumulate over time at a higher rate than size increments (Sheehy 1989). In Swedish

Pacifastacus leniusculus lipofuscin concentration was linearly related to age and was a better predictor of age than CL (Belchier et al. 1998). In the case of this species the estimated longevity was 16 yr, a decade longer than estimates from a field study of cohorts (Guan and Wiles 1999). However, for practical use any surrogate of age must be easily measured for the large sample sizes that are necessary when making population estimates of demographic parameters. For this reason modal progression of cohorts based on CL continues to be used, despite the accuracy decreasing with age. There are two contributors to this increasing uncertainty: firstly the cumulative effects of variation in both moult increment and intermoult interval affecting CL and secondly the cumulative effects of mortality that decrease the number of old animals. If large animals make a substantial contribution to recruitment, then it could be important to have better estimates of their age, and make lipofuscin worth the extra effort. The technique could be used to validate age estimates based on CL.

Among decapods, the Astacidea tend to have quite long life spans (Momot 1984, Vogt 2012). Table 2 lists 32 species whose life spans range from 1.5 to about 60 yr. Across all the species the mean maximum life span is 12.7 yr. The Astacidae (8.3 yr) and Cambaridae (7.0 yr) have smaller mean maxima, but the Parastacidae (25.6 yr) live considerably longer because growth is much slower in cooler habitats and many live in burrows where survivorship may be better. Actual mean longevity of cohorts would be considerably less than the maximum values listed, although the absence of any larval stages means that astacids live much longer, on average, than many marine decapods. Prolonged post-hatching maternal brood care also improves the chances of survival.

Modelling individual crayfish growth (as opposed to aquacultural pond production) has scarcely begun. For the most part growth is dealt with as though it was continuous, as found in fish, and accepted as an approximation to a process that we know is discontinuous. The only example of individual modelling that we are aware of is *Cambaroides japonicus* for which the relationship between post-moult size and pre-moult size of individuals was modelled by a non-linear equation (Easton and Misra 1988). Dealing with a population of different-sized individuals rather than a few cohorts of different ages introduces a whole new level of complexity, but one which should be well within the capability of modern desk-top computers. The two basic quantities, growth increment and inter-moult interval, represent size and age respectively. When the two quantities are accumulated according to some model we generate a size vs. age relationship. With growth being indeterminate there is no implied or assumed maximum or asymptotic size as is part of some continuous growth models (e.g., von Bertalanffy model). The starting point is the initial size and age at recruitment, which are then propagated cumulatively by the model and the variables that influence size increment and moulting probability. The use of a more biologically meaningful growth model allows one to make predictions of the effects of the environmental variables, such as temperature, and stress on growth of crayfish. However, measuring the effects of temperature, for example, on moult increment and probability under controlled conditions may be challenging.

Chang et al. (2012) recently reviewed models of crustacean growth which used different ways of describing the relationships between moult increment and inter-moult interval (dependent variables) and pre-moult size (independent variable). They used the recently developed information theory approach that employs a multi-model inference procedure to select the best model to predict the relationship of increment and interval to pre-moult CL. They did not use data for freshwater crayfish, but for the marine American lobster (*Homarus americanus* H. Milne Edwards, 1837) a linear model of moult increment over all sizes of males was best, but for females a "bent-line" model was best. This difference is a reflection of the fact that females tend to have a similar or even decreasing increment once they begin to produce eggs. For moult interval of the American lobster there was again a difference between the sexes: for females the relationship was best assumed to be linear, but for males a non-linear increasing model was better than the linear model. The results of modelling *H. americanus* may well be indicative of what will be found for crayfish. Models incorporating the effects of temperature are likely to predict faster growth, although there is likely to be some trade-off between increment and interval (Hartnoll 2001). Increased temperature decreases inter-moult interval, but can also decrease moult increment because less time is available for feeding and gathering energy for growth. However, the net effect is still to increase growth rate (Hartnoll 1982). Low temperatures have the opposite effect and for both marine lobsters and freshwater crayfish (~10°C) is a limit below which moulting ceases.

Table 2. Life spans of freshwater crayfish. Species are listed according to estimated maximum longevity from shortest to longest. Note: life spans tend to reflect environmental (especially water temperature and habitat) rather than species' characteristics. The same species in a different habitat may have a different life span.

Species	Family	Life Span (yr)	Reference
Procambarus clarkii	Cambaridae	1–1.5	Reynolds et al. (1992)
Cambarellus shufeldtii (Faxon, 1884)	Cambaridae	1.5	Walls (2009)
Cambarus halli Hobbs, 1968	Cambaridae	2+	Dennard et al. (2009)
Fallicambarus gordoni	Cambaridae	2–3	Johnston and Figiel (1997)
Cherax quadricarinatus	Parastacidae	3	Sheehy (1992)
Cambarus hubbsi	Cambaridae	3	Larson and Magoulick (2011)
Cambarus longulus	Cambaridae	3	Smart (1962)
Procambarus suttkusi	Cambaridae	3	Baker et al. (2008)
Orconectes williamsi Fitzpatrick, 1966	Cambaridae	3–4	DiStefano et al. (2013)
Paranephrops planifrons White, 1842	Parastacidae	4–7	Hopkins (1967b), Parkyn et al. (2002)
Procambarus fallax (Hagen, 1870) f. *virginalis*	Cambaridae	4.4	Vogt (2010, 2011) "marmorkrebs"
Cambarus elkensis	Cambaridae	5	Jones and Eversole (2011)
Fallicambarus fodiens	Cambaridae	6	Norrocky (1991)
Pacifastacus leniusculus	Astacidae	~6	Guan and Wiles (1999)
Cambarus dubius	Cambaridae	7	Loughman (2010)
Astacus leptodactylus	Astacidae	7.4	Deval et al. (2007)
Orconectes inermis	Cambaridae	9–10	Weingartner (1977)
Parastacoides tasmanicus	Parastacidae	10	Hamr and Richardson (1994)
Astacus astacus	Astacidae	> 10	Skurdal and Taugbol (2002)
Cambaroides japonicus	Cambaridae	11	Kawai et al. (1997)
Cambarus bartonii	Cambaridae	13	Huryn and Wallace (1987)
Procambarus erythrops Relyea and Sutton, 1975	Cambaridae	16	Streever (1996)
Austropotamobius pallipes	Astacidae	6–14	Reynolds et al. (1992)
Austropotamobius italicus	Astacidae	10–13	Grandjean et al. (2000), Scalici et al. (2008)
Paranephrops zealandicus	Parastacidae	16–28	Whitmore and Huryn (1999)
Parastacus pugnax	Parastacidae	19	Ibarra and Arana (2012)
Astacoides betsileoensis Petit, 1923	Parastacidae	20	Jones et al. (2007)
Orconectes australis	Cambaridae	22	Venarsky et al. (2012)
Astacoides granulimanus	Parastacidae	25+	Jones and Coulson (2006)
Paranephrops zealandicus	Parastacidae	29	Whitmore and Huryn (1999)
Euastacus armatus (von Martens, 1866)	Parastacidae	50	Alves et al. (2010)
Astacopsis gouldi	Parastacidae	60	Lukhaup and Pekny (2008)

Relative growth

Many previous studies of crayfish secondary sexual characters are inadequate: they resort to the use of ratios whereby the character is divided by the reference dimension and then plotted against the same reference dimension. The slopes of resulting lines cannot be easily compared across species because proportionate growth of the secondary sexual character cannot be estimated from fitting simple linear models to such data. Frequently the actual data are omitted from graphs and only the best fit lines are plotted making growth patterns difficult to interpret. One cannot easily see whether there is any spread in the sizes of maturing animals making it impossible to accurately estimate the size (CL) at maturity. ANOVA tables summarizing linear regressions are a poor substitute for the data themselves. Relative growth is best studied by fitting the 'power equation' to the variable in question and the reference dimension, which involves fitting log–log data by linear regression. The slope of such lines is an estimate of the growth constant that describes proportionate increase. A slope of 1.0 indicates isometry, so growth in size of the character is simply what you would expect from increased body size while any slope less than or greater than 1.0 indicates disproportionate or allometric growth.

This disproportionate growth requires separate explanation. Changes in the intercept of the growth equation can also indicate an abrupt change even if the slope remains the same. When mature crayfish have different breeding and non-breeding dimensions (e.g., change in form) the intercept can be used to compare the two growth forms. For all of these reasons, use of the "power model" is more productive when interpreting size data.

When a dimension is allometric and differs between males and females we refer to it as a being a secondary sexual character, as opposed to such primary sexual characters as the position of the gonopore or shape of abdominal pleopods. Growth of secondary sexual characters is usually isometric in juveniles, sometimes different between the sexes, but changes on their proportionality can be used to estimate the size at onset of sexual maturity. For decapods the two most useful characters are cheliped size (length and width) and abdominal width. The latter is more straightforward being only a single measurement, but more complex with the cheliped since it can be decomposed into several parts consisting of sizes of the merus, carpus, propodus and dactyl articles. Usually the size of the propodus (or palm) carries the most information about relative growth and is easiest to measure. The size of the abdomen is a direct measure of the egg-carrying capacity of the female while the size of the cheliped can be important in male contests for mates. The net result is that males end up with more muscle tissue in their chelae compared to females while females end up with more muscle in their abdomens.

The most detailed study of relative growth in crayfish is by Stein et al. (1977) who examined *Orconectes propinquus* (Girard, 1852) from Wisconsin using log-transformed data (Fig. 5). This account of relative growth is made a little more complex because, like all cambarids, *O. propinquus* undergoes form changes (see below) after reaching maturity. The mean minimum size at maturity for *O. propinquus* in Trout Lake was 18.5 mm CL for both sexes. Immature and mature females could be separated by comparing their annulus ventralis: immature females had a flat seventh sternite while in mature females the surface was sunken to provide a narrow groove with two pronounced swellings on either side. Growth of *Orconectes propinquus* secondary sexual characters was different for males and females. Male chelae growth was tri-phasic with relative growth continuity broken into three phases. The first break was in the juvenile stage at 13–15 mm CL and the second was associated with sexual maturity around 20 mm CL. In juveniles the break is produced by males moulting to form II during the summer when they are around 13–15 mm CL. Subsequently males moult back to form I when they reach maturity in the fall. Consequently the interaction between sexual maturation and form change means that the best model describing male chelae growth is tri-phasic. Female chelae growth is di-phasic with a discontinuity at 13 mm CL. Growth of the abdomen was tri-phasic in females but only di-phasic for males. Both sexes had isometric growth during the juvenile period, up to 13 mm CL which corresponds to the end of the first summer. Then the abdomen in females began to grow faster than in males which remained isometric. At maturity, 20 mm CL, females showed evidence of a single step pubertal moult after which growth continued to be isometric, but at an elevated size. Independent evidence of the pubertal moult was the development of the annulus ventralis where spermatophores could be stored. In *O. propinquus* both sexes have a tri-phasic pattern of relative

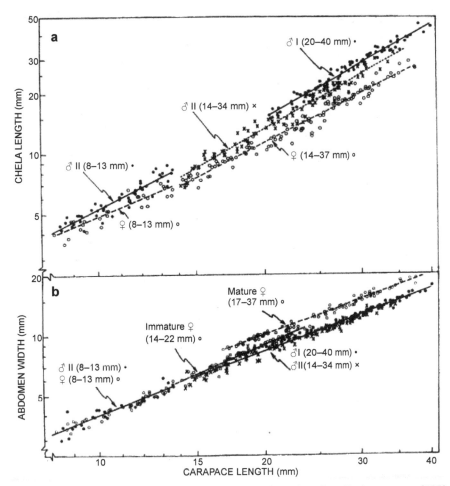

Figure 5. Relative growth in *Orconectes propinquus* from Trout Lake, Wisconsin, collected in the summers of 1973–74. Chela length (a) and abdomen width (b) are plotted against carapace length (mm) on log-log scales. Sex and maturity of crayfish were determined by examining the gonopores and annulus ventralis of females. For both sexes three size groups (CL) are plotted: females are divided into Immature/Mature while males are divided into Form I and II. Data for each group are shown using symbols as labelled (from Stein et al. 1977).

growth of chelae in males and the abdomen in females. Although the chosen CL range for each phase was somewhat arbitrary, there does not seem to be great overlap between crayfish in different phases, suggesting that the same number of moults maybe required for them to reach each phase. More detailed analysis of the data around the discontinuities could be used to estimate the break points more precisely, but in the case of *O. propinquus* this is not really necessary.

Rhodes and Holdich (1979) investigated the secondary sexual characters of British *Austropotamobius pallipes* (Lereboullet, 1858) and found that growth of cheliped length in males changed suddenly around 30 mm CL while females continued to grow in linear fashion. Similarly with abdomen width which began to grow more rapidly in females around 25 mm CL, while males continued to grow at the same rate. The change from immature to mature seems to occur at around the same CL so there is little size overlap between the two stages, although the actual data points were not plotted so it is difficult to be sure of this. If there is no overlap in CL between immature and mature crayfish this would suggest that the same number of moults were required to reach that size (cf. Stein et al. 1977 for *O. propinquus*). Also data were not log-transformed so it is impossible to estimate proportionate growth. Similar results using untransformed data were obtained by Streissel and Hodl (2002) for *A. torrentium* in Austria, with male size at maturity at around 25 mm CL and for females it was in the range 29–32 mm CL. Male *A. italicus* from

Florence, Italy, show strong positive allometric cheliped growth whereas female chelae grow isometrically (Gherardi et al. 2000).

Kato and Miyashita (2003) compared relative growth of functional vs. non-functional pleopods in both males and females of two crayfish: *Pacifastacus trowbridgii* (Stimpson, 1857) and *Procambarus clarkii*. In males the first pleopod is specialized and used to transfer sperm whereas the third pleopod (the 'control') is not modified. In females the pleopods are used to carry eggs and juveniles when they hatch. Pleopods of both sexes showed evidence of the effects of selection: the male first pleopod had negative allometry and less variation around the line, suggesting stabilizing selection, whereas female pleopods had strong positive allometry suggesting directed selection. The male gonopod is able to copulate with a wider size range of females by limiting growth, while females are able to carry more eggs and juveniles if they have relatively larger pleopods.

The ontogenetic growth patterns of crayfish seem to follow the pattern of a pubertal moult in both sexes whereby male chelae and female abdominal width become positively allometric. The growth pattern in these freshwater crayfish is probably similar to what is found in related marine clawed lobsters such as *Nephrops norvegicus* (Linnaeus, 1758) (Farmer 1974c) at least for the Astacidae and Parastacidae. However, further accurate analyses of data from a wider range of species would help to establish this tentative generalization.

Cyclic dimorphism

In most decapods (including astacid and parastacid crayfish) changes in relative growth of secondary sexual characters like chelae size are irreversible, but some cambarids exhibit a cyclic alternation of isometric and allometric form that is related to the seasonal breeding. In summer males capable of breeding (with larger form I chelae) moult to a non-breeding stage (with smaller form II chelae) and then moult back again 8–10 weeks later. Form I males have hooks on the ischia of the last two pairs of walking legs for holding females during copulation.

During the northern summer *Orconectes propinquus* males moult twice, firstly from form I to form II in mid-June, and then back to form I again in August. During the summer break they are reproductively inactive. In form I males (breeding) had longer and wider chelae than females and form II males of the same CL. By contrast form II males (non-breeding) had chelae of intermediate size (Stein et al. 1977). When form I males moult in summer they grow in CL but not in chelae size (see Fig. 5). These size differences influence male mating success: males use their chelae in male-male interactions and also to grasp and hold females during copulation. Those with larger chelae gained more mating opportunities than males with smaller chelae (Stein 1976). Chelae in form I males can comprise about 40% of total dry weight. In *Orconectes rusticus* (Girard, 1852) larger cheliped size in form I males also results in them winning more encounters (Rutherford et al. 1995). The impact of form differences on behavioural outcomes depends upon the extent to which moulting is synchronized: a high degree of synchrony results in little overlap in occurrence of the two forms and thus their interaction, so there is little advantage in being one form or the other.

In *Orconectes limosus* (Rafinesque, 1817) form I (breeding) gonopods are longer, wider, more robust, more rigid and more sharply pointed than form II (non-breeding) (Fig. 6). When the male moults into the non-breeding form the terminal elements of the gonopods become shorter, sclerotized, calcified and rounded, similar to juvenile gonopods (Buřič et al. 2010). All known species of *Orconectes* show form changes (Hobbs 1974). It has been assumed that *Orconectes* males moult twice a year in the summer and fall, but Buřič et al. (2010) found that in *O. limosus* from the Czech Republic initially form I, around 85% moulted twice in a year (form I to form II and then back to form I), but about 9% of the males moulted only once (without form change) or did not moult at all (6%).

Dimorphism has been documented in males of *Orconectes luteus* (Creaser, 1933) (see Muck et al. 2002), *Procambarus digueti* Bouvier, 1897 and *P. bouvieri* (Ortmann, 1909) (see Gutiérrez-Yurrita and Latournerie-Cervera 1999). Wetzel (2002) also found that males of *Orconectes illinoiensis* Brown, 1956, *O. indianensis* (Hay, 1896), *O. kentuckiensis* Rhoades, 1944 [*O. cf. propinquus*], and *O. virilis* (Hagen, 1870) showed chelae form changes, and was also able to document parallel changes in females. Form

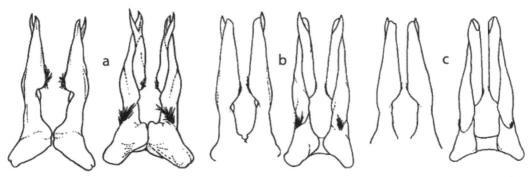

Figure 6. Lateral and mesial views (left and right respectively) of the first gonopod (G1) of captive *Orconectes limosus*, from an invasive population in the Czech Republic, showing form change: (a) form I, (b) form II, (c) juvenile. The sex ratio in tanks was 2 females per male and crayfish were tagged (from Buřič et al. 2010).

I females (sexually active) show one of the following: swollen white glair glands, dependent offspring (embryos to 3rd stage juveniles), or remnants of eggs still attached to pleopods. These females have wider abdomens than same-sized form II females and only form I females were observed mating with form I males. Form II females are most common during the summer growing season while form I females first appear in the fall mating season and are most common through to the end of the spring spawning season. Female *O. illinoiensis* kept captive under natural conditions moulted from form I to form II and vice versa, in both cases the moult resulted in an increase in CL and a size decrease/increase of the abdomen width (Wetzel 2002) (Fig. 7). In populations of *Cambarus robustus* Girard, 1852 from Ontario, Canada, both forms coexist during the year and in size-matched encounters form I males were almost always dominant (Guiasu and Dunham 1998). Similar form changes have been reported for both male and female *Cambarus elkensis* Jezerinac and Stocker, 1993 from West Virginia (Jones and Eversole 2011).

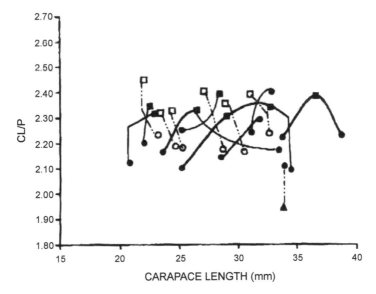

Figure 7. Form change in captive female *Orconectes illinoiensis* abdomen under natural temperature and photo-period. Relative width (P)/carapace length (CL) (mm) of the second abdominal segmentis plotted against CL. Each point represents an instar and lines connect moults of the same crayfish. Plot symbols indicate: circles are form I, squares are form II. Solid symbols and lines represent crayfish collected as form I; empty symbols and dashed lines represent females collected as form II (from Wetzel 2002).

Cambaroides japonicus males do not show form alteration typical of cambarids (Kawai and Saito 2001). The absence of form changes and presence of a simpler annulus ventralis suggests that *C. japonicus* belongs to the putative sister group of the American cambarids (Scholtz 2002). Furthermore some argue, on the basis of the male second gonopod and molecular differences in the COI, 12S rRNA and 16S rRNA, that all 6 species of Asian cambarids should be placed in a separate family, Cambaroididae (Kawai et al. 2013).

This cyclic pattern of moulting in cambarids is very curious: why would a mature female moult into a non-reproductive instar rather than simply grow larger and remain ready to breed? One moult may be necessary to clean the pleopods of egg shells, and to repair damage to the body such as limb-loss, but moulting is always a risky venture, because newly-moulted crayfish are vulnerable, so one wonders what the rewards of this moulting are? In the case of males if they are going to moult twice, then why not use the opportunity to grow the chelae even larger, rather than only once? Why would a male crayfish moult from form I to form II and become an almost certain loser in size matched encounters between the two forms? The strategy would only seem to deliver benefits (whatever they may be) as long as all crayfish did the same thing at the same time. It seems reasonable to predict that form I crayfish which chose to moult and grow larger, without changing form, would always be winners. The precise benefits of form alternation in cambarids remain an open question.

Maturation

The size at which crayfish become sexually mature is an important element in understanding population dynamics and in managing captive populations for aquaculture. A non-invasive way of detecting size at maturity is to analyse changes in secondary sexual features such as chelipeds and the abdomen. On a log-log plot changes in slope or intercept of the fitted line can be used to estimate the size at sexual maturity (see above). In *Cherax quadricarinatus* androgenic gland hormones (AGH) control masculinisation, especially development of male secondary sexual characters and sexual behaviour (Barki et al. 2003). In females, prominent cement glands on the abdomen are evidence of maturity and in some species may be an external sign of ovary maturation. *Orconectes illinoiensis* form II females moulting to form I developed glair glands a few days later (Wetzel 2002). Development of adominal glair glands, opening into pores on the sterna, pleura, pleopods and uropods, seems to be a reliable way to judge female maturity. Changes in male behaviour towards females can be a sign of approaching sexual maturity. In crabs male guarding behaviour can be a reliable indicator of gonad maturation (Brockerhoff and McLay 2005), but crayfish do not seem to use this method. Alternatively in grapsid crabs examination of the female gonopore operculum can also be used: females with mobile opercula are receptive to mating whereas females with rigid opercula are not. However crayfish do not have equivalent structures that could signal receptivity/attractiveness.

In female cambarid crayfish there is an external sperm storage site, the annulus ventralis, formed by a depression on the sternal sclerite of the penultimate thoracic segment. Andrews (1906a,b) described the ontogeny of the annulus ventralis of *Orconectes limosus*: when juveniles hatch from the egg the annulus is absent, but after two moults there is a slight transverse depression as well as the first indication of where the gonopores will be in the coxae of the sixth pereopods. After four moults an asymmetrical ridge and groove are apparent and after six moults the pair of rounded elevations along the anterior annulus margin is apparent. At sexual maturity the transversely elongated depression, behind the rounded elevations, is crossed by the S-shaped sperm tube. Post-maturity moults continue to increase the size of the annulus ventralis.

Reproduction

By way of introduction, we use *Paranephrops planifrons*, a slow-growing NZ parastacid, to illustrate the reproductive schedule of events in a freshwater crayfish and compare these in two contrasting habitats (Parkyn et al. 2002). There is a difference in timing of recruitment and how long it takes this crayfish to reach sexual maturity in pastures forest streams (Fig. 8). Recruitment of young into pasture stream populations (mean temp = 14–16°C) occurs two months earlier (in summer) than in forest streams (mean temp = 10–12°C) giving them a growth advantage in their first year. This means that they reach reproductive size after only ~18 mon whereas those in forest streams need ~36 mon. However both of these populations

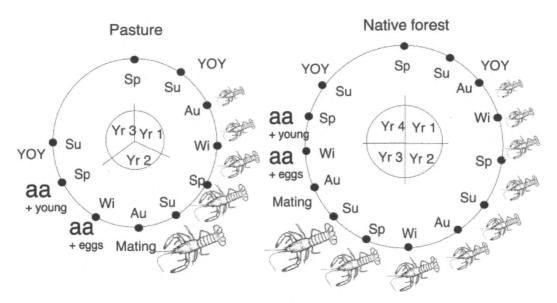

Figure 8. Reproductive cycle of *Paranephrops planifrons* in native forest and pasture streams in Waikato region, New Zealand. Abbreviations used: aa = females; YOY = young of year, i.e., recruits; SP, SUM, AU, WI = annual seasons, spring, summer, autumn and winter respectively; Yr = year 1–4 (from Parkyn et al. 2002).

mature much faster than the other NZ species, *P. zealandicus*, which lives in southern NZ where it is colder and require 6–7 yr (Whitmore and Huryn 1999). Growth of *P. planifrons* occurs year-round, but is much slower in winter when fewer than 50% moulted. Higher moult frequency in pasture streams lead to faster growth rates. After mating in autumn (March–June) embryos are carried by the female during the winter for ~6 mon and after they hatch juveniles remain with the mother until they are released during the warmer months (later in forest streams than in pasture streams). Since females in forest streams grow to a larger size, they probably produce more recruits than females in pasture streams, leading to higher density, although they may not reproduce every year (Parkyn et al. 2002). It appears that *P. planifrons* in these habitats may only produce 1–3 broods per female lifetime which lasts 3–5 yr.

Reproductive system morphology

A valuable overview and comparison of crustacean reproductive systems can be found in López Greco (2012). Crayfish are for the most part gonochoristic. For both sexes the gonads are in the thorax, beneath the pericardial sinus and above the hind-gut and hepatopancreas. Their size depends upon the age and reproductive activity of the crayfish. During the breeding season they enlarge considerably with testis taking on a milky-white colour due to the presence of sperm and the ovaries become distended with yellow-brownish eggs (Vogt 2002). Some hermaphrodites have been recorded in the Parastacidae (Rudolph 1995a,b), but it is not clear whether these protandric individuals are truly functional, and capable of passing on gametes, or whether there are other explanations for their different sexual system.

The male reproductive system of both cambarids and astacids consists of a tri-lobed organ (Y-shaped) with paired anterior lobes and a single posterior lobe joined in the middle where the vas deferen ducts arise and emerge in the coxae of the fifth pereopods. However in parastacids the testis is more H-shaped, the posterior part consists of paired lobes while the anterior lobes are much reduced, while the vas deferens arise separately more from the separate lobes, rather than from the connection between the three lobes as seen in the Y-shaped testis (Hobbs et al. 2007) (Fig. 9). The vas deferens is highly convoluted and is the place where the sperm are packaged into spermatophores ultimately to be delivered to the male gonopore from where they are delivered to the female. In all the crayfish the distal muscular section, the ejaculatory duct, plays an important part in copulation, more so in parastacid males which lack gonopods. In these

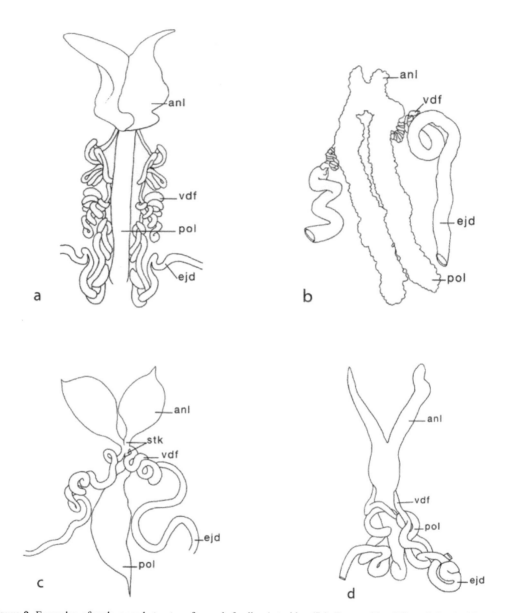

Figure 9. Examples of male gonad structure for each family: Astacidae (7a), Parastacidae (7b) and Cambaridae (7c-d). (a) *Pacifastacus trowbridgii*; (b) *Parastacoides tasmanicus*; (c) *Cambarus bartonii cavatus* (Hay, 1902); (d) *Cambaroides japonicus*. Abbreviations used: anl = anterior lobule; ejd = ejaculatory duct; pol = posterior lobule; stk = stalk; vdf = proximal vas deferens (from Hobbs et al. 2007).

crayfish the spermatophores are applied directly to the female sternum by the extended vas deferens or 'penis' while in astacids and cambarids the gonopods form a vital link in transferring spermatophores to the female sternum or annulus ventralis respectively. South American parastacids, such as *Parastacus defossus* Faxon, 1898, lack the convoluted (coiled) section of the vas deferens that is found in the Australian parastacids, such as *Cherax* spp. (Hobbs et al. 2007). In the latter the spermatophores are composed of a double layered coating while in the former only a single layer protects the spermatophores (Noro et al. 2007). The morphology of the male reproductive system and spermatophore formation in *Cherax quadricarinatus* has been described by López Greco et al. (2007).

Crayfish sperm develop in acini which bud off on the wall of the collecting tubules, increase in size, fill with sperm that are discharged into the collecting tubule, and then they degenerate and shrink. In

Cambarus acuminatus Faxon, 1884, for example, the next generation of acini begin on a new section of the wall (Fig. 10). Sperm are gathered together in the proximal vas deferens where they are packaged into spermatophores by secretions from the glandular epithelial wall of the duct (Hobbs et al. 2007). In *Cherax albidus* Clarke, 1936 extruded spermatophores are around 10 mm long (Beach and Talbot 1987). The beginning of the ejaculatory part of the duct in *Cambarus acuminatus* is marked by the thickening of the muscular layer, both longitudinal and circular muscle fibres. It seems likely that the anchored longitudinal fibres are involved in propelling the terminal part of the tube out through the gonopore, while the circular muscles also contract and squeeze out the spermatophores into the base of the gonopod. The terminus of the vas deferens is probably eversible. The terminal part of the vas deferens is lined with cuticle and so will be shed, along with any content such as spermatophores, when the crayfish moults (Hobbs et al. 2007).

The female system in *Orconectes limosus* and *Procambarus clarkii* consists of three lobes, two anterior and one posterior, plus the oviducts which connect to the middle part of the ovary and open in the coxae of the third pereopods (Ando and Makioka 1998, Unis and Erkan 2012). The Y-shaped gonads of these female cambarids are similar to those of males. Female and male *Astacus astacus* also have Y-shaped gonads (Huxley 1881). Beatty et al. (2003, 2005b) described the growth and maturation of the ovaries in

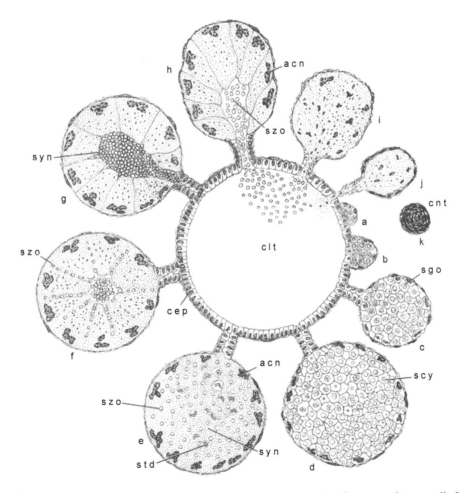

Figure 10. Diagram of the life cycle of acinus in testes of *Cambarus acuminatus*: (a) rudimentary acinus on wall of collecting tubule; (b) youngacinus; (c) with spermatogonia; (d) with spermatocytes; (e) with spermatids and spermatozoa in massive syncytium; (f) with spermatozoa being concentrated centrally; (g) with secondary syncytia enveloping spermatozoa; (h) with spermatozoa being expelled; (i, j) degeneration of acinus; (k) completely degenerated with connective tissue remaining as remnant. (Explanation of labels used: acn – accessory nucleus; cnt – connective tissue; clt – collecting tubule; sgo – spermatogonia; scy – spermatocyte; syn – syncytium; std – spermatid; szo – spermatozoa.) (from Hobbs et al. 2007).

Cherax cainii Austin, 2002 and *Ch. quinquecarinatus*, dividing it into seven stages (see Table 3). Oocyte maximum diameter increases as yolk granules are added and the ovary becomes swollen with dark grey eggs. After these are discharged a few oocytes of mixed size remain in a creamy ovarian matrix.

Table 3. Stages of ovarian development for female *Cherax quinquecarinatus* (after Beatty et al. 2005b).

Ovarian stage	Macroscopic description	Maximum oocyte diameter (µm)	Histological description
I/II Immature/recovering	Ovaries very thin, string-like; some very pale orange oocytes discernable in an otherwise creamy ovarian matrix.	600	Oogonia, chromatin nucleolar and perinucleolar oocytes dominate. Post-spent ovaries also contain atretic oocytes and post-ovulatory follicles.
III Developing (yolk vesicle)	Ovaries slightly thickened, with bright orange oocytes easily discernable.	1100	Perinucleolar oocytes that have undergone primary vitellogenesis dominate ovary. Oogonia oocytes still present.
IV Developed (late yolk vesicle)	Ovaries thickened with an obvious increase in size of oocytes, which are grey-green.	1800	Oocytes have distinct cytoplasmic yolk vesicle region and yold granules present indicating secondary vitellogenesis. Perinucleolar oocytes present.
V Mature or gravid (yolk vesicle)	Ovaries slightly swollen, with oocytes becoming dark grey.	2200	Yolk granules dominate the cytoplasm indicating further vitellogenesis. Ovarian epithelium with follicle cells surrounds oocytes.
VI Ripe/spawning	Ovaries very swollen, containing very dark grey oocytes.	2500	Cytoplasm of oocytes dominated by yolk vesicles. Perinucleolar oocytes still present.
VII Spent	Ovaries thickened compared to virgins; orange oocytes of mixed sizes discernable in a predominantly creamy ovarian matrix.	1600	Post-ovulatory follicles present along with large unextruded ova and perinucleolar oocytes.

In *Procambarus fallax* (Hagen, 1870), the parthenogenic marmorkrebs, the ovary (Fig. 11) first becomes evident at TL ~12 mm as a short tube with small eggs, but without oviducts. The ovary and oviducts are well developed in adolescents of TL ~19 mm. The ovary consists of numerous oogenetic pouches each composed of follicle epithelium enclosing an oocyte and a central lumen lined with epithelium that includes several germaria. This lumen is continuous with the lumina of the oviducts which are simple tubes connecting with the gonopores. The oviducts arise from the medial portion of the ovary, which connects the three lobes, and they are lined with a folded single layer of epithelium and a peripheral layer of connective tissue and musculature. During vitellogenesis large amounts of yolk and energy reserves are deposited in the oocytes to be used by the embryo and early post-hatching stages. In *P. fallax*, mature eggs ready for spawning are 1–1.2 mm diameter (Vogt et al. 2004).

Gonad maturation

In the parastacid *Cherax quinquecarinatus* changes in the gonosomatic index (GSI) for females in Bull Creek, Western Australia shows that this species undergoes prolonged spawning from August (spring) to February (autumn). Maximum water temperature was 24.2°C in January and minimum water temperature was 17.8°C in September. The pattern suggests that there were three spawning peaks: August, October and

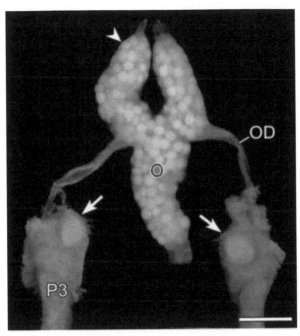

Figure 11. Complete female (5.5 mm TL) reproductive system of *Procambarus fallax* (marmorkrebs). Labels: o – ovary; od – oviduct; P3 – third pereopod. Arrows – gonopores; arrowhead – maturing oocyte in ovary; scale bar = 2 mm (from Vogt et al. 2004).

December-January. Stage V ovaries were present in most months. During the colder months (February–July) the GSI of females with stage III–VI ovaries remained low and similar to the GSI of females with stage I/II ovaries (Fig. 12). Male GSI followed a similar pattern (Beatty et al. 2005b).

Gonad maturation in male *Cherax quadricarinatus*, native range Northern Territory and Queensland, Australia, has been investigated by López Greco et al. (2007) in Argentina. In mature males ready to mate the distal section of the vas deferens (VD) is greatly enlarged by a white mass of stored soft spermatophores. The annual cycle shows testes weight increasing in winter, as males produce sperm and prepare for the forthcoming breeding season, and then in the summer, sperm count and weight of the VD rise as mating commences. In autumn VD weight decreases as sperm is used up and not replaced (Bugnot and López Greco 2009). Soon after deposition on the female sternum the spermatophores hydrate and begin to harden and after 48 hr they can begin dehiscing (López Greco et al. 2008). Breakdown is complete after 102 hr suggesting that the spermatophores only have a short shelf-life before releasing the sperm.

Ovarian maturation is affected by seasonal changes in day length and water temperature in many species. Aiken (1969) found that photoperiod could be used to manipulate scheduling of ovary maturity and spawning in *Orconectes virilis*. Increased water temperature in spring induces egg laying provided that the animals have been kept in constant darkness at low temperature during winter. Under constant darkness ovaries matured much more rapidly than under the ambient photoperiod (Stephens 1952). Mating and spawning by *Astacus leptodactylus* were accelerated over winter months by keeping them under darkness (Harlioğlu and Duran 2010). Similar results were obtained for *Procambarus llamasi* Villalobos, 1954 (Carmona-Osalde et al. 2002). For species in which females normally lay their eggs while secluded in a burrow, reproduction in the laboratory may be enhanced by replicating conditions closer to nature. However, it is clear that day-length has a major effect on ovarian maturation in crayfish although it may affect different species in different ways. For example, *Astacus leptodactylus* (Eschscholtz, 1823) begins to mate and spawn when day-length and temperature are decreasing, but in *Procambarus llamasi* and *Orconectes virilis*, the same responses result when day-length and temperature are increasing.

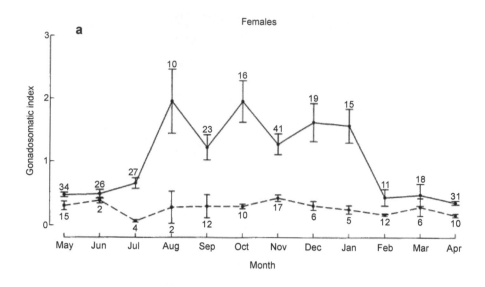

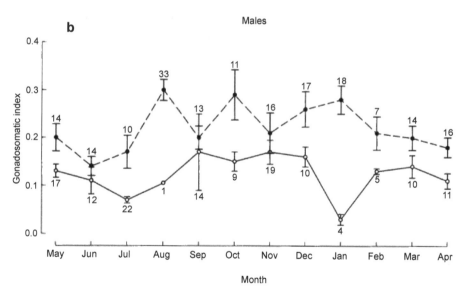

Figure 12. *Cherax quinquecarinatus* mean gonadosomatic indices (+/– 1 S.E.) for females (a) and males (b) with immature (i.e., gonad stages I/II, lower line) and mature/maturing (i.e., gonad stages III–VI, upper line) (from Beatty et al. 2005b).

Intersexuality and parthenogenesis

As far as it is known, the Astacidea have a gonochoristic sexual system: separate sexes and usually no sex change of individuals during their life history. Their reproductive systems are bilaterally symmetrical with pairs of gonads and ducts that open in the coxae of the third (female) or fifth (male) pereopods. However, there have been numerous records of intersexuality wherein some individuals show simultaneous occurrence of both male and female primary sexual characters (gonopores): this can include either or both external characters and gonad differentiation (Martinez and Rudolph 2011). Intersex crayfish have been recorded in species of several parastacid genera: *Cherax, Engaeus, Engaewa* and *Euastacus* from Australia as well as *Parastacus, Samastacus* and *Virilastacus* from South America. Amongst other crayfish intersexuality has been recorded in *Astacus, Pacifastacus* (Astacidae), and in *Cambarus, Procambarus, Orconectes* and *Cambarellus* (Cambaridae) (Rudolph 1995a,b).

In their study of *Euastacus bipsinosus* Clark, 1936, Honan and Mitchell (1995) compared the occurrence of crayfish with aberrant gonopores at three river sites. In total 21 crayfish had aberrant gonopores: all had 1 or 2 male gonopores on the P5 coxae, identical to normal males, and 19 also had female gonopores, which were small and lacked setae. These gonopores were on coaxe of either P3 or P4 and superficially similar to immature females, recessed and with a membranous cover rather than non-recessed and calcified as in normal females. At one site no abnormal crayfish were found and at another site only 1% was found, but at the third site abnormal crayfish were much more common. In Jerusalem Creek, South Australia, of 109 crayfish captured 63% were normal females and 21% normal males, but 17% were aberrant all between 50–65 mm CL. Similar levels of abnormal patterns have been found in other species of *Euastacus* and other parastacid crayfish (Honan and Mitchell 1995).

In a captive Israeli population of *Cherax quadricarinatus*, Parnes et al. (2003) recorded seven different combinations of male and female gonopore characters in different individuals. Through breeding experiments they investigated the genetic basis for these combinations by crossing intersex males with females (WW). They tested a simple chromosomal model of sex determination, which assumed males to be the homogametic sex (ZZ), and found that intersex individuals, which are functionally males, are in fact genetically female (WZ). In *Ch. quadricarinatus* females are the heterogametic sex. Barki et al. (2006) compared the agonistic and mating behaviour of intersex *Cherax quadricarinatus* with normal crayfish. They found that intersex individuals, despite being genetically female, generally behaved like males: they engaged in fighting and mated with receptive females, but duration of fights was intermediate between normal males and females and copulations were remarkably short.

In some species of burrow-living *Parastacus* a small percentage (1–2%) show evidence of protandric sequential hermaphroditism while males and females are approximately equally common amongst the rest (Noro et al. 2008). Two sexual systems can be distinguished: gonochorism with permanent intersexuality and partial protandric hermaphroditism. More intensive study by Martinez and Rudolph (2011) of *Parastacus* spp. suggests that sexuality may not be fixed for each species, but could be an adaptation to the requirements of different habitats.

Parthenogenesis has recently been discovered in the Astacidea and may be an important factor in alien colonization of new areas enabled by human transport. Scholtz et al. (2003) found the first parthenogenetic crayfish, called "marmorkrebs" because of their marbled colour pattern, being sold as aquarium pets in Germany and established that these females probably belong to a species of *Procambarus*. Further work verified that these crayfish are a parthenogenetic form of *Procambarus fallax* (Martin et al. 2010) and that there are also intersex individuals (with first pleopods that are male-like) that are still functionally female (Martin and Scholtz 2012). Vogt (2008b, 2011) has argued that this species could become a valuable model organism for all kinds of crustacean studies, in the same way that the fruit-fly *Drosophila* has become irreplaceable to biologists in general. Vogt et al. (2008) have shown that marbled crayfish, raised under identical conditions, can produce a range of phenotypes among clone-mates, despite the fact that they are all genetically identical. Parthenogenesis is not necessarily a bad thing, resulting in reduced fitness and inability to adapt to environmental change. Parthenogenetic *Procambarus clarkii* have recently been discovered in China (Yue et al. 2008). More recently Buřič et al. (2011) have found evidence that exotic *Orconectes limosus* from a Czech Republic stream are capable of facultative parthenogenesis.

Mating and moulting link

For decapod crustaceans the most important elements for understanding their reproduction are: (1) whether there is a link between mating and moulting or not; (2) whether females have indeterminate or determinate growth; and (3) how sperm are stored and subsequently used to fertilize the eggs (McLay and López Greco 2011). If sperm are stored by the female then the details of the storage organ and how it is connected to the oviduct is also of vital importance. The ancestral condition for decapods is that mating occurs when females are soft-shelled, immediately after moulting (for example caridean shrimps and portunid and cancrid crabs). Hard-shell or inter-moult mating occurs in palinurid lobsters, some hermit crabs (coenobitids and diogenids) and in some brachyuran crabs (leucosiids, xanthids, majids, grapsoid and ocypodoid crabs) (Asakura 2009). As shown earlier crayfish growth is indeterminate so does not cease once they reach a certain size or age. The consequence of this is that, with external sperm storage, all sperm are lost each time the female moults so they have to mate again in order to continue breeding. If moulting causes sperm loss then we would expect that mating would be linked to moulting so that breeding can continue. If mating is not linked to moulting then the question arises as to how females attract a mate. In this section we are concerned with mate attraction, not with mate choice and mating behaviour (see Behaviour chapter herein).

Somewhat surprisingly in most species of freshwater crayfish mating is not linked to moulting. In the Astacidae *Pacifastacus trowbridgii, P. leniusculus, Austropotamobius pallipes, A. italicus* (Faxon, 1914), *A. torrentium* (Schrank, 1803), *Astacus astacus,* and *A. leptodacylus* a female moult does not precede mating. Similarly, in the Cambaridae there are many examples including *Orconectes nais* (Faxon, 1885), *O. immunis* (Hagen, 1870), *O. limosus, O. rusticus, O. propinquus, O. virilis, O. inermis* Cope, 1872, *O. pellucidus* (Tellkampf, 1844), *Cambaroides japonicus, Faxonella clypeata* (Hay, 1899), *Procambarus alleni* (Faxon, 1884), *Procambarus blandingi* (Harlan, 1830), *P. clarkii,* and *P. hayi* (Faxon, 1884). This means that females cannot benefit from male protection when they are in a soft-shell state and therefore most vulnerable to attack by predators (see references to studies of all these species in Asakura 2009). In the Parastacidae *Cherax quadricarinatus* is the only one, out of approximately 140 species, where mating is known, but many of the species in this family are difficult to study because they live underground in burrows. *Cherax quadricarinatus* is a popular species in the aquarium and trade and so is better known. Details of the mating behaviour of this species are readily available by searching YOUTUBE videos with the terms "crayfish mating". In *Ch. quadricarinatus* females do not undergo any external changes beforehand that might indicate receptivity, but it seems that they initiate mating by approaching the male, giving the impression that it was a "spur-of-the-moment" decision. From the non-aggressive front-to-front position both partners raised themselves and aided by the female, the male inserted his abdomen under the female and at the same time rolled over on to his dorsum. After copulation it was the female who broke off the liaison by tail-flipping away. There was no sign of post-copulatory guarding by the male and both partners were in a hard-shell condition (Barki and Karplus 1999). Given that there was no advertising and courtship behaviour was not observed, one wonders what exactly the female had to offer the male and vice versa.

In *Parastacoides tasmanicus* (Erichson, 1846) males and females live alone for most of the year, but during the breeding season (February–April) they are found paired in burrows with most females just about to moult or recently moulted. Mating and spawning occurred within one month of pairing (Hamr and Richardson 1994). It appears that males are attracted to female burrows and are in attendance when she moults with spawning occurring soon thereafter. This kind of seasonal cohabitation is likely to be more common amongst burrow-living parastacids than in the other families, but at present knowledge is sparse. By contrast burrowing *Distocambarus crockeri* Hobbs and Carlson, 1983 Form I males were recorded in the same burrows as females in most months, except for December and January (winter), but they were most common in March, April and May (springtime) (Eversole and Welch 2013). Ovigerous females were found in April–June (peak in May) 2005–2006 suggesting only a single generation per year.

In non-burrowing species males cannot use indications that moulting by a female might be imminent to decide whether or not to remain close at hand, so how do females advertise and males recognize receptivity? Chemical communication amongst crayfish and its history has recently been reviewed by Breithaupt (2011) so here we will concentrate on recent detailed studies of the use of pheromones in representative species. An overview of earlier research into crayfish pheromones can be found in Bechler (1995). When mating is not linked to female moulting, and occurs in the intermoult, the female does not need male protection prior to copulating.

Male crayfish do not show any courtship behaviour that might mark the beginning of a reproductive interaction. Stebbing et al. (2003) used *Pacifastacus leniusculus*, mature females CL > 30 mm, to condition water that was then tested by bioassaying the behaviour of males CL > 35 mm. The males clearly responded differently to this water than they did to immature female or male conditioned water. Males were stimulated to court and try and mate with the source of the female odour (an air-bubble stone!). Copulation lasted 32 min. Female maturity was assessed on the presence or absence of well developed glair glands (Ingle and Thomas 1974). Berry and Breithaupt (2008) tested urine taken from females during the breeding season and found that males responded by significantly increasing mounting behaviour, but not all of the sexual behaviour normally seen. Berry and Breithaupt (2010) injected fluroescein to label urine produced by both males and females and found that females produced a signal that elicited male mating behaviour. Urine-blocking prevented any male courtship behaviour. Visualization using fluorescent dye showed that female urine production coincided with aggressive behaviours, but not with female submissive behaviours in reproductive interactions. In these interactions urine was predominantly produced by females during the precopulatory phase. They conclude that the coincidence of chemical signalling and aggressive behaviour suggests that release of urine has evolved as an aggressive signal in both sexes. Whether the coupling of a signal, indicating receptivity, with an aggressive signal should be interpreted as a test of male strength (perhaps 'genetic quality') remains an open question. Males who responded positively to the invitation to copulate might have assumed that the female was about to spawn, but unfortunately the state of the ovary was not assessed. To use his sperm supply efficiently it would be prudent to not waste time or sperm on females unless they were on the verge of egg-laying. If the shelf-life of sperm is short then males need to make informed decisions about the imminence of egg laying. This is especially important in crayfish which lack a protected sperm storage site. The field study of *P. leniusculus* in the Great Ouse River by Guan and Wiles (1999) revealed a schedule of moulting, mating and egg-laying that suggests that females were 'advertising' the fact that their ovaries were reaching maturity and that males who mated would likely fertilize eggs, but the exact timing of mating prospects remains unknown.

In the cambarid crayfish *Orconectes rusticus* reproductive pairs both males and females in chela contact released urine and generated currents at a higher rate than non-reproductive pairs (Simon and Moore 2007). Each crayfish may have been assessing their partner's readiness for mating. Attempts to copulate only occurred when males were Form I and females showed glair development. It appears that they were communicating the state of their gonads, although whether mate assessment was reciprocal is not clear. Perhaps it was only the male assessing the imminence of egg laying and thus the prospects for fertilization and paternity. In another cambarid *Procambarus clarkii* males only require a chemical signal, but females require both chemical and visual information (Aquiloni and Gherardi 2008b, Aquiloni et al. 2009). Studies of urine signalling have not been done on any parastacids so, while we are reluctant to extend knowledge of astacid and cambarid mating systems to parastacids which are their sister group, it seems likely that similar chemical communication occurs.

The best male strategy would be to minimize the delay between copulation and fertilization. For the three species discussed above there is wide range in the length of the delay: in *Pacifastacus leniusculus* egg laying occurs in the week after copulation (Guan and Wiles 1999, Nakata et al. 2004) so it is possible that females release a signal indicating ripe ovaries. However, for the other two species, *Procambarus clarkii* and *Orconectes rusticus*, the delay could be as short as a few weeks or as long as eight months, respectively (Ameyaw-Akumfi 1981, Berrill and Arsenault 1982, 1984, Snedden 1990). This suggests that the ovary was not ready to release eggs so it is possible that these cambarid crayfish use a different mate attraction signal.

Sperm transfer and storage in crayfish

Mating in crayfish follows a similar pattern in all three families. The male and female place their bodies "face-to-face" with the male on top, using his chelae to hold the female's chelae (see Fig. 13 A- *Austropotamobius pallipes* (Astacidae); B- *Procambarus alleni* (Cambaridae); C- *Cherax destructor* (Parastacidae). *Cambaroides japonicas* is an exception because the male lies beneath the female, not grasping her using his chelae, as in other crayfish, but using his walking legs instead (Kawai and Saito 2001).

In all the Astacidea the last thoracic sternite is mobile and this feature is a very important part of the mechanics of reproduction. It facilitates the transfer and placement of spermatophores on the female. The Astacidae males have the first two pairs of pleopods modified for sperm transfer: they consist of a stout sub-tubular first pleopod (lacking terminal ornamentation) with the second pleopod inserted into the tube. In *Austropotamobius pallipes* (Astacidae) the fifth pereopods are used as part of the sperm transfer apparatus: either the right or the left pereopod is positioned transversely across the sternum in front of the gonopods and the proximal part of the coupled pleopods are brought into close contact with the genital papillae (extensions of the vas deferens sometimes referred to as "penises") extending from the male gonopores. The second pleopod completes the tubular structure of the first by forming the inner wall.

Figure 13. Copulation in crayfish captured from YouTube videos, showing male above female. (A) *Austropotamobius pallipes* Astacidae (courtesy of Underwater-Ireland.com); (B) *Procambarus alleni* Cambaridae (courtesy of Pavel Angelov); (C) *Cherax destructor* Parastacidae (courtesy of Natcrayfish).

The male presses the tips of the interlocked gonopods on to the female sternal area, between P4 and P5, and spermatophores are forced along the tube by rapid lengthwise plunging movements of the second pleopod. Spermatophores are attached to the female as white tubular deposits (Ingle and Thomas 1974), rather than having it plastered over the sternal area and on proximal articles of the pereopods. Sperm transfer in *Pacifastacus trowbridgii* is similar to *A. pallipes* except that the fifth pereopods are not always used to support the gonopods (Mason 1970c). Copulation in *Pacifastacus leniusculus* results in many spermatophores being transferred to the female (Dudenhausen and Talbot 1983).

In the family Parastacidae males (and females) lack first pleopods on the abdomen and pleopods are not modified for sperm transfer, which is achieved by bringing an extension from male gonopore (a modest "penis"), on mesial surface of P5 coxa, close to the female sternal area (Fig. 14 A male "penises" of *Cherax quadricarinatus*). The distal end of the vas deferens does not show any particular modifications for transferring sperm and making it different from what is found in astacids or cambarids (Laura López-Greco, pers. com.). In *Cherax quadricarinatus* only a single sperm mass is transferred at copulation where its sticky surface attaches it to the female sternum and the exterior coating begins to harden (López Greco and Lo Nostro 2008) (Fig. 14B spermatophores attached to sternum of *C. quadricarinatus* female).

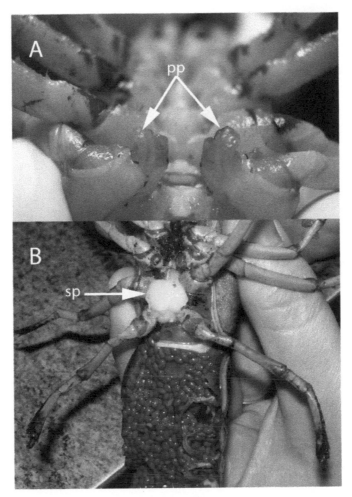

Figure 14. (A) Ventro-posterior view of "penises" (pp) of male *Cherax quadricarinatus*; (B) Ventro-posterior view of spermatophores (sp) attached to sternum of an ovigerous female *Cherax quadricarinatus* (photos by Laura López Greco).

Sperm transfer is probably similar to what occurs in *Macrobrachium rosenbergii* De Man, 1879 where the behaviour has been studied in more detail (Smith and Ritar 2008).

The Cambaridae males also have well developed ornamented, gonopods for sperm transfer (Fig. 15). Since a fairly high degree of precision is needed when inserting gonopods into the aperture of the annulus ventralis, males have hooks on the ischia of P3 which they use to grasp the female P4 and firmly couple with and keep the gonopods in place during the long process of sperm transfer. Andrews (1904, 1910a,b, 1911) made the first detailed study of sperm transfer in the American cambarid crayfish, *Orconectes limosus*. The behaviour of this crayfish is remarkably similar to that of *A. pallipes* and showed the importance of crossing by the fifth pereopod: it ensures that the gonopods are correctly positioned to receive sperm from the genital papillae, which emerges from the coxae of the same limb. Only one fifth pereopod is used at a time, it does not seem to matter which. The fifth pereopods are also used by another cambarid, *Procambarus clarkii* to support the gonopods during copulation (Ameyaw-Akumfi 1981). Sperm transfer in *Orconectes limosus* can be a prolonged affair: Andrews (1910b) observed that undisturbed pairs could remain in copula for as long as nine hr, although it is not clear whether sperm was being transferred continuously. Some of this time might be better considered as mate guarding: preventing the female from mating with other males.

The cambarid sperm storage organ (Fig. 16) is a complex invagination of the seventh sternite. Hagen (1870) gave the name "annulus ventralis" to the organ found on the sternum of female crayfish, guessing correctly that it had some function in reproduction. The first and almost only work on the nature and

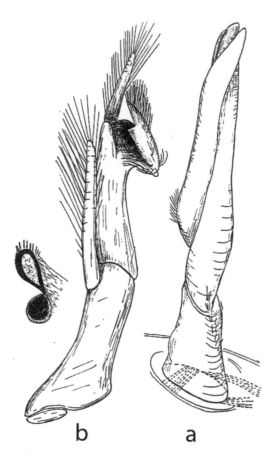

Figure 15. *Orconectes limosus* male gonopods. (a) posterior face of left first gonopod (G1); (b) anterior face of left G2, and beside is an enlarged section near its tip (from Andrews 1911).

structure of the sperm storage organ in cambarid crayfish has been by Hagen (1870) and Andrews (1904, 1906a,b, 1908a,b). Others have used the shape of this organ in taxonomy, but without any of the internal details (Hobbs 1988, Taylor 2002). Andrews (1906a) describes the annulus ventralis of *Orconectes limosus* as follows: "the exoskeleton covering the annulus furnishes a thick-walled case about a long bent tube, which opens to the exterior at the anterior end and enlarges as a two-horned pouch at its posterior end, while along its entire length it communicates with the surface by curved slits leading to the zigzag suture. Essentially the annulus is a bent pocket lined by exoskeleton. When the crayfish casts off its exoskeleton the above described shell of the annulus is thrown off as part of the entire exoskeleton". He recognized that this meant a post-moult female crayfish would need to mate again before she could lay fertile eggs. Andrews (1906a) kept post-copulatory, unmoulted females isolated from males for 5 months and found that they still successfully laid fertilized eggs after that time. To eliminate the possibility that the crayfish was parthenogenic, he carefully removed the annulus ventralis full of sperm from females about to lay their eggs and found that, while egg laying was not affected, none of the eggs developed. Performing the same excision after the eggs had been laid had no effect on their development.

Sperm is forced by mechanical movements into the inner-most recesses of the annulus ventralis, presumably filling them first, so that the sperm transferred last would be in the vestibule. In the absence of any mixing, last in would be first out! In *Orconectes limosus* the male may leave behind a sperm plug protruding from the aperture as the gonopods were withdrawn. Andrews (1906a) termed the annulus ventralis an "*indirect sperm transfer*" organ in recognition of its separation from the gonopore, where the

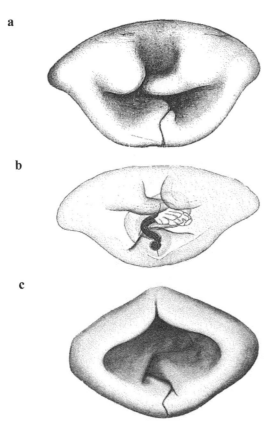

Figure 16. Some examples of the female cambarid annulus ventralis: (a) *Orconectes limosus* ventral view of right-handed annulus; (b) *Orconectes limosus* ventral view of left-handed annulus showing a sperm plug in the orifice with the tube and recess full of sperm stained a darker colour; (c) *Orconectes virilis* ventral view of empty annulus (from Andrews 1906b).

ova emerge. This important insight indicated that he understood that copulation and fertilization were not linked, but happened at some later time and furthermore that fertilization must be external because the sperm were not in a place where they could meet the ova prior to them encountering the water as happens in eubrachyuran crabs for example. The shape of the annulus ventralis is not necessarily the same in every female of a species. It is an asymmetrical curved pocket the shape of which can be dextral (right-handed) or sinistral (left-handed). Perhaps this variation in the shape of the annulus ventralis is the reason why males need similar generalized gonopod shape. In the Asian cambarids the annulus ventralis is less complex. Kawai and Scholtz (2002) show details of the open annulus ventralis of *Cambaroides japonicus*. Sperm are placed directly in the annulus ventralis, not spread over the rest of the surrounding sternum as can happen in astacids and parastacids.

Finally, it seems necessary to assume that spermatophores deposited by the three families of crayfish must be different: spermatophores transferred by astacids and parastacids must be adhesive, because they are deposited on the female sternum without the benefit of protection. Sticky spermatophores would seem to be essential especially for parastacids given the female-over-male copulatory position. However, those transferred by cambarid males must be non-adhesive because they are deposited in the annulus ventralis, and when fertilization occurs, they are not accessible to mechanical disruption to liberate the sperm. The way that crayfish store spermatophores means that any energy needed by the sperm must accompany them at the time of copulation. Therefore we might expect that cambarid males would provision their spermatophores with more energy than either astacids or parastacids because they may need a longer shelf life.

Promiscuity and sperm plugs

Of particular relevance here is the fact that neither pre-copulatory nor post-copulatory guarding by males has been reported for any species of freshwater crayfish. Given that male crayfish do not overtly display any form of mate guarding we might expect females to mate with several males. From a female point of view being able to mate with as many males as possible may be an advantage provided there are no injuries. The number of partners that female crayfish mate with in the wild is difficult to estimate, so most data comes from captive animals in controlled conditions. Ingle and Thomas (1974) reported multiple mating in *Austropotamobius pallipes*: one female mated with five males and some larger males mated at least six times. Males seemed to restrict deposition of spermatophores to the female sternum between the last two pairs of legs. If a female had already mated then males appeared to show preference for unoccupied areas where adhesion may have been better, although this may be just coincidental. When a female is carrying spermatophores from more than one male, multiple paternities is possible, but not necessarily assured. Male *Austropotamobius italicus* mated with virgin females on successive days managed to successfully copulate with up to four females although most of them (42.5%) only mated with one female. Ejaculate size decreased considerably with each successive mating (Rubolini et al. 2007). Accurate measurements of the sperm re-charge rate are not available for any male crayfish.

Given the absence of mate guarding the level of promiscuity in cambarids may be affected by males depositing sperm plugs in the entrance to the female annulus ventralis (Fig. 16). Andrews (1904, 1908b) described sperm transfer to the annulus ventralis of female *Orconectes limosus* followed by deposition of a sperm plug. Berrill and Arsenault (1982, 1984) found sperm plugs in *Orconectes rusticus* and Andrews (1908a) found *Cambarellus montezumae* (Sassure 1857) females with plugs. Plugs were also found in *Procambarus verrucosus* (Hobbs 1952). In *Procambarus suttkusi* (Hobbs, 1953) in the Choctawatchee River Form I males were found in May to September and females with sperm plugs were found in October peaking in the following April and May (Baker et al. 2008). It seems likely that use of sperm plugs is widespread amongst the Cambaridae, but because they do not entirely close off the cavity, the extent of their impact on promiscuity and fertilization is yet to be established. They may not be much of a substitute for mate guarding. Females may delay egg-laying for several weeks, thus providing the opportunity to mate with other males.

The only evidence of female multiple mating in the wild is for two cambarid crayfish. Walker et al. (2002) collected mated *Orconectes placidus* (Hagen, 1870) females from a Tennessee river and micro-satellite DNA was extracted from parents, eggs and attached juveniles. Forty percent of broods were sired by a single male, but 60% were sired by up to four males. Broods had two fathers on average. Within mixed broods the percentage sired by the males was highly skewed, ranging from > 85% to around 50% for the primary father. *O. placidus* females store sperm in their annulus ventralis whose entrance can sometimes be blocked by a male sperm plug. It is unclear whether any of the *O. placidus* had sperm plugs, but if they did, the plug was not effective in preventing multiple paternities. Also the data do not reveal whether the sperm of the last or any other male contributed more than the others. All of the juveniles tested were progeny of the female which was carrying them.

In the other case, *Procambarus clarkii* females carrying juveniles, collected from 3 locations in China, were genotyped using four microsatellites (Yue et al. 2010). Mothers were the exclusive maternal parent of their offspring, but 29 of 30 mothers (96.7%) had mated multiple males (2–4, mean of 2.7) who had sired different numbers of young in each brood. Male parentage was skewed with the male contributions to the brood (in rank order) being: 69.7% (Father1), 21.8% (F2), 8.5% (F3), 5.8% (F4) (cf. *O. placidus*). The reasons why some males fertilized more eggs than others cannot be determined, but may be the result of mating order.

Unfortunately, no parentage analysis has been done on species from the other two families (lacking an annulus ventralis) that would allow comparison of the effects of open vs. enclosed sperm storage on paternity. Galeotti et al. (2007) suggest that *Austropotamobius italicus* females should produce offspring sired by the last male to mate because in this species males can remove sperm deposited by earlier males before depositing their own. It will be important to analyse parentage to see whether this is the case.

Fertilization

The difference in sperm storage in the three families means that there are likely to be differences in the mechanism of fertilization. Release of the eggs is quite straight forward, but the most interesting and least understood part is how sperm are liberated at the critical moment. The other important factor is the fact that crayfish sperm is aflagellate and therefore immobile. Thus we have to try and explain how two cells, lacking any means of locomotion, manage to rendezvous and combine to initiate development. We have detailed studies of fertilization in the Astacidae and Cambaridae, but nothing is known for sure about the Parastacidae. Here we look in detail at *Austropotamobius pallipes* (Astacidae, exposed sperm storage) and *Orconectes limosus* (Cambaridae, closed sperm storage) and compare them. In both examples a chamber is formed along the margins of the curled abdomen by the release of mucous-like glair from the cement glands on the pleopods thereby linking the abdomen to the sternum. Niksirat et al. (2014) showed that *Astacus leptodactylus* (Astacidae) compensate for the lack of motility of the sperm by having a mechanism to facilitate egg-sperm binding: before the eggs are released the female glair secretion encounters the spermatophores and dissolves the wall releasing the sperm into the fertilization chamber. The process of fertilization is enhanced by mixing of the gametes by the female pleopods.

In *Austropotamobius pallipes* fertilization can be broken down into five stages (Ingle and Thomas 1974). The first stage begins with preening, about 72 hr after mating, with the female elevating her body, using the full extension of her pereopods and telson, and then employing the dactylus comb of the fifth pereopods to preen the setae of the telson, uropods, pleopods and abdominal segments. This continued for about 36 hr. The second stage, lasting around 10 min, is marked by abdominal contractions when the telson is held tightly against the sternum and slid forward and backwards. The third stage begins with the female extending her chelipeds and, using her pereopods, turns on to her back with the abdomen still held close taking about 30–40 min (Fig. 17a). Then she alternately raises and lowers the right and left chelipeds thereby causing her to roll from side to side, interrupted by resting (Fig. 17b). After about 30 min of rolling and resting the abdomen is relaxed revealing the eggs immersed in almost transparent glair (Fig. 17c). Given an average number of eggs (about 80) the fertilization rate would only be around three per min. (This is almost an order of magnitude slower than what is found in the slowest brachyuran crab (see McLay and López-Greco 2011)). In the fourth stage the female regains her normal stance by using the fourth and fifth pereopods and chelae to lift her body off the bottom and roll forwards, end over end, with the abdomen still folded, all of which takes around 15–20 min (Fig. 17d–f). During this time the glair-encapsulated fertilization chamber remains closed. The final stage consists of the female resting for about 30 min and then rolling her body from side to side using the pereopods on alternate sides continuing for about 12 hr (Fig. 17g). This seems to be the time needed to attach all the eggs to the pleopods, the whole process dependent on gravity. After that the abdomen is extended, with the eggs now firmly attached to pleopod setae, and frequent preening of the egg mass by the fifth pereopods begins (and possibly also the chelate second to fourth pereopods). This preening also removes the residual glair thereby destroying the fertilization chamber and allowing free circulation of water through the embryos.

Austropotamobius italicus males can allocate sperm by varying the ejaculate size according to female size (Rubolini et al. 2006). Since larger females are likely to produce larger broods this has the effect of minimizing sperm wastage or shortage that might result from delivering the same ejaculate size to all females regardless of their size. They found that copulation duration and number of ejaculations accurately predicted the amount of sperm transferred to the female. There was also the possibility that males delivered less sperm as they grew larger (= older) suggesting that they were more able than small males to economize on sperm use or that they may become senescent.

Mason (1970b) has also carefully recorded the spawning behaviour of another astacid crayfish, *Pacifastacus trowbridgii*. The sequence of behaviours involved in spawning is remarkably similar to that described by Ingle and Thomas (1974) for *A. pallipes*. He provides photos of females before and after spawning with a female whose sternum was mostly covered by spermatophores beforehand and another recently spawned ovigerous female whose sternum did not carry any spermatophores at all. Although the time elapsed since spawning is not given, it appears as though the entire complement of spermatophores was used to fertilize the brood of around 150 eggs.

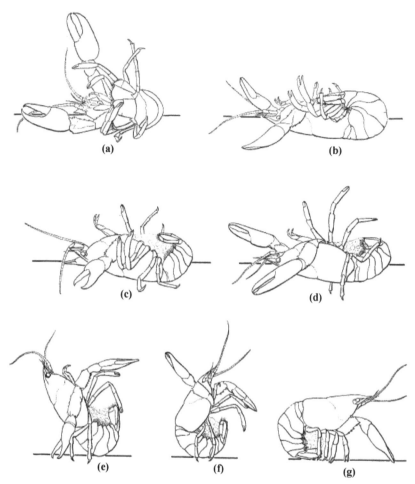

Figure 17. Spawning and fertilization in *Austropotamobius pallipes*. (a) female turns over on her back; (b) left and right chelae are used to roll from side to side during which glair is secreted from the pleopods; (c) eggs are released into the abdominal chamber and fertilized as they pass over the spermatophores; (d–f) female regains an upright stance, with abdomen still curled and glair intact, using her fourth and fifth pereopods and chelae; (g) side to side rolling ensures eggs are distributed and attached to all the pleopods (from Ingle and Thomas 1974).

Andrews (1904, 1906c) made the first detailed observations of egg-laying by *Orconectes limosus*. He divided the whole process into four stages: preparatory cleansing, glairing, egg extrusion, and rhythmic body turning (Fig. 18a–d). The first stage is equivalent to the "preening" stage seen in *A. pallipes*, involving the same body elevation, but employing not only the fifth pereopods with the dactyl comb, but also the second and third chelate limbs to thoroughly clean the abdomen and pleopods. Glairing occurs with the abdomen folded and in *O. limosus* cement secretion lasts about 30 min and may begin before the female rolls over on to her back. Nocturnal egg extrusion by six females took an average of 16.6 min (range 10–30 min) to lay 200–600 eggs (Fig. 18b). Eggs generally emerged from both oviducts at a rate of 12–60 per min from each gonopore sometimes continuously other times in groups of three. Fertilization rate would be an order of magnitude faster than *A. pallipes* (on average 12–36 per min compared to 3 per min see above). Gravitation is believed to carry the eggs across the annulus ventralis, where they are fertilized, towards the abdominal pleopods. Andrews (1906c) observed eggs emerging from the oviducts and being channelled posteriorly along the sternum aided by the slope created by the female elevating the anterior end of the body using her pereopods. Finally, the female returns to her normal orientation standing on her walking legs (Fig. 18c-d) and eggs are attached by threads (part of the outer case) during rhythmic

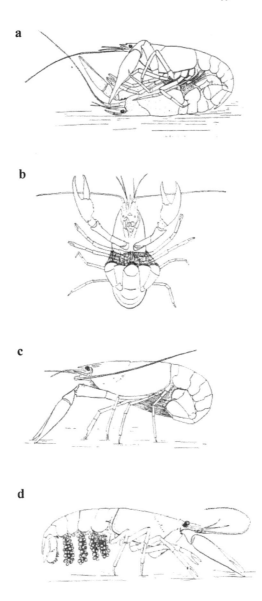

Figure 18. Mating, fertilization and egg laying in *Orconectes limosus*. (a) The male holding female and with fifth leg supporting gonopods that are about to transfer sperm to the annulus of the female; (b) Female lying on back with legs held rigid, abdomen folded and glair connected to thorax, thereby sealing the fertilization chamber as the eggs are being laid; (c) Female standing up after laying: glair and bent abdomen still as in b; (d) Glair has now been removed by chelate pereopods and female is aerating eggs by raising and straightening the abdomen and waving pleopods back and forth (from Andrews 1904).

side to side body rolling as seen in *A. pallipes*. During egg laying the females enter a trance-like state with limbs locked in position and can be handled for closer observation of fertilization and attachment of eggs to the pleopods. This is similar to what happens during conjugation when the mating couple appear oblivious to their surroundings and can be lifted out of the water together, without interruption (Andrews 1910b). In the parthenogenetic marmorkrebs, *Procambarus fallax*, the female still goes through the same pre-spawning, spawning and post-spawning behaviours even though no fertilization occurs (Vogt and Tolley 2004). Curiously *Cherax quadricarinatus* females raised at 30°C without males also spawned and attached their eggs without mating although it is not clear whether these eggs developed further (Tropea et al. 2010). We refer to the use of glair to form a "fertilization chamber", but we need to recognize that egg attachment occupies a much longer time and so it can equally be described as a "spawning chamber" wherein the fertilized eggs become attached.

The comparison of spawning behaviour of *Austropotamobius* and *Orconectes* reveals a lot of similarity although Ingle and Thomas (1974) recognized more stages than did Andrews (1904, 1906c), and because the stages were defined differently, the duration of these stages are not directly comparable. Both crayfish preened the abdomen, pleopods, and telson after mating, then used glair, generated from the pleopods, to form a fertilization chamber between the telson and the sternum and then used body-rolling to help distribute and attach the embryos. The delay between mating and commencement of preening was considerably shorter in *Austropotamobius* than in *Orconectes* (see next section and Table 4). The major difference between these representatives of two crayfish families, which store sperm differently, is that the rate of egg-laying is much higher in the cambarid than in the astacid. The possible consequences of this are discussed below. In order to facilitate future studies of spawning behaviour it would be useful to adopt a more detailed and standardized series of stages that encompassed all the important behaviours that affect fertilization. We suggest the following stages: (1) Abdominal preening and the duration of time

Table 4. Delay between mating and fertilization in freshwater crayfish.

Species	Timing of mating	Fertilization timing	Delay	Reference
Astacidae				
Astacus astacus (Europe)	October-November	November-December	Several d – a few wk	Cukerzis (1988), Taugbol and Skurdal (1990)
Astacus leptodactylus (Europe)	October-November	November-December	4–6 wk	Koksal (1988)
Pacifastacus leniusculus (Japan)	Mid-October (exotic) (native to NW of US)	Mid-October	2–3 d	Nakata et al. (2004)
Pacifastacus leniusculus (U.K.)	September–November (exotic)	September–November	< 1 wk	Guan and Wiles (1999)
Austropotamobius italicus (Italy)	October-November	October-November	Several d/wk	Galeotti et al. (2007)
Austropotamobius pallipes (Spain)	September	October	Min 2 d – max 14 d	Carral et al. (1994)
Austropotamobius pallipes (U.K.)	Late October	October-November	Several d - 2 wk	Ingle and Thomas (1974), Brewis and Bowler (1985)
Austropotamobius torrentium (Europe)	October	October-November	A few wk	Hubenova et al. (2010)
Cambaridae				
Cambarus robustus (Ontario)	June-July	August	1–2 mon	Corey (1990)
Cambaroides japonicus (Japan)	September-October	May	6 mon	Kawai and Saito (2001)
Orconectes limosus (U.K.)	Spring (April)	Spring (May)	~1 mon	Holdich and Black (2007)
Orconectes limosus (Quebec, Canada)	September-October and March-April (2 periods of mating)	May	6 to ~3 mon	Hamr (2002)
Orconectes limosus (Baltimore and Maryland)	October–April	March-April	~5 to 1 mon	Andrews (1904)

Table 4. contd....

Table 4. contd.

Species	Timing of mating	Fertilization timing	Delay	Reference
Orconectes limosus (Switzerland)	Late August–April (over winter)	Mid April	7 to ~1 mon	Stucki and Staub (1999)
Orconectes rusticus (Ontario, Canada)	Autumn - Mid-March (2 periods of mating)	April	8 to 2	Berrill and Arsenault (1982, 1984)
Orconectes virilis (Ontario)	July–September	May	9	Weagle and Ozburn (1972)
Orconectes williamsi (Missouri)	October	March	5–6	DiStefano et al. (2013)
Procambarus clarkii (Michigan)	March-April	?May–July	Several to	Ameyaw-Akumfi (1981), Huner (1988)
Procambarus hayi (Mississippi)	May–August	Late-August to Mid-November	?4	Payne (1972)
Parastacidae				
Cherax destructor	Begins Spring-early summer	December–February peak	Spawn 2–3 times per yr	Beatty et al. (2005a); reproduction can be seasonal to almost continuous depending upon conditions
Cherax quadricarinatus (Argentina)	?Frequency of mating	Spawning independent of mating	Spawn 3–5 times per yr	Tropea et al. (2010)
Cherax quadricarinatus (Israel)	3–5 matings per yr	3–5 spawns per yr	15–19 on average between broods	Barki et al. (1997)
Cherax quadricarinatus (Australia)	Captive in culture. 3–5 spawnings per yr	Captive in culture. Breeding season is spring-summer 6 months	12–24 hr post-mating	Jones (1995), Beatty et al. (2005b)
Euastacus bispinosus (Victoria, South Australia)	?April-May	April-May	?a few	Honan and Mitchell (1995)
Parastacoides tasmanicus (Tasmania, Australia)	February–April	February–April	?a few	Hamr and Richardson (1994)

Footnote: In *Astacus astacus* the delay depends upon water temperature: if mating occurs early, when the temperature is higher, the delay can be several weeks, but if it occurs closer to winter, when the temperature is lower, the delay is only a few days.

since the last mating occurred; (2) Abdominal contractions and folding beneath the thorax accompanied by the release of glair (= cement of some authors) to form the fertilization chamber; (3) female turns on to her dorsum, with abdomen held closely to the sternum, then rolls from side to side while the eggs are released from the gonopores and fertilized, noting especially duration of this stage and pleopod activity; (4) female regains her normal orientation standing on legs; (5) duration of egg attachment to pleopods by side to side rolling; (6) breaking open of glair chamber by extention of abdomen and preening to remove glair remanents. Where appropriate it would be valuable to know what role sperm plugs might have played and especially whether females were able to remove them so as to aid the release of sperm. All these behaviours are important if we are to understand the role that females may have in affecting the outcome of fertilization.

Andrews (1906c) proposes a mechanical explanation of how sperm get out of the annulus ventralis, as a result of movements of the last thoracic somite. He examined an *O. limosus* female before fertilization occurred and found that the annulus ventralis was full, but 24 hr after egg laying the annulus ventralis of

15 females was mostly empty. He estimated that in one female annulus ventralis there were of the order of 50–60,000 sperm which were used to fertilize only a few hundred eggs, so on an individual basis sperm were clearly not in short supply. Observations of a female during the second stage when glair is released suggested that the female forced the spermatophores out of the annulus ventralis by pressing the sternite between the fifth pereopods (the 'post-annular sclerite' of Hobbs 1989) against the posterior margin of the annulus ventralis. Using females, which had mated, but not laid eggs, he was able to apply pressure to the fifth sternite and make spermatophores ooze out of the annulus ventralis aperture. The presence of such a mechanism in this cambarid crayfish might explain why the rate of egg-laying in *O. limosus* is an order of magnitude faster than in *Astropotamobius*. If the female has some degree of direct control over the release of sperm from the annulus ventralis then coordination of egg and sperm release might facilitate faster and more efficient fertilization. Whereas as in *Austropotamobius* egg laying is slower because the eggs must pass more slowly over the spermatophores attached to the sternum and give them time to burst open and fertilize the eggs. This hypothesis implies that the eggs themselves carry a chemical signal that causes spermatophores to burst.

How do the sperm get out of the spermatophores and fertilize the eggs? What makes the spermatophore burst open at the right moment? It would seem that the most efficient way to achieve this would be for the eggs to carry a chemical stimulus that would cause the release of sperm into the fertilization chamber (as suggested above for *Austropotamobius pallipes*). It has been suggested that the glair substance produced by *Austropotamobius italicus* females contains a stimulus that causes spermatophores to burst open (Galeotti et al. 2012). To our knowledge there have not been any studies of spermatophore bursting and what the nature of a chemical stimulus might be and where exactly it might come from. While the shelf-life of sperm in spermatophores may be weeks or months, once the spermatophore has burst the liberated sperm may only survive for a few hr: *Cherax quadricarinatus* sperm liberated into saline solution lasted at least 2 h, but the maximum limit has not been measured (López Greco and Lo Nostro 2008).

When a female crayfish's eggs are fertilized by more than one male, the spatial pattern of embryo attachment on the pleopods might provide additional information about the way in which sperm were used in fertilization. The only investigation of this was done by Walker et al. (2002) who found that while the average number of sires of *Orconectes placidus* eggs was two males, whose contributions were often skewed, neither attached eggs nor juveniles showed any non-random distribution amongst pleopods that might reflect biased fertilization. Admittedly the attachment process might have randomized the embryos after fertilization so the question remains open.

Some recent research by Aiken et al. (2004) on fertilization in the American lobster, *Homarus americanus*, may answer some of the above questions about freshwater crayfish. These are part of the group of clawed lobsters that includes *Nephrops* and the crayfish and so, by out-group comparison, they provide insights as to the ancestral reproductive character states. To establish external fertilization, Aiken et al. (2004) occluded the entrance to the sperm storage organ using epoxy resin, but found that this did not prevent egg fertilization in *H. americanus*. However, if both the entrance and the posterolateral grooves on sternite 7 were blocked then no fertilization occurred. The female controls the release of sperm by contraction of muscles attached to the wall of the storage organ. The entrance to the organ may be the route by which males deposit sperm, but it is not the route followed by sperm when they exit to meet the eggs. This experiment supports the hypothesis of Andrews (1906b) that female cambarid crayfish use muscular contractions that move the mobile last thoracic segment and apply pressure to the annulus ventralis to expel the sperm. However the structure of the cambarid annulus ventralis is different: in homarids the external features of the sperm storage organ are a Y-shaped gap with the base of the Y being the entrance for sperm and the two posteriorly directed arms (posterolateral grooves) the exit routes, but in cambarids there is a single variously-shaped groove crossing sternite seven. Hence a similar occlusal experiment on a cambarid would be instructive to not only establish the routes followed by the spermatophores, but also to test for the effectiveness of any sperm plugs that might be deposited. It is not clear whether the cambarid sperm plug needs to be removed to allow fertilization or whether the sperm can circumvent that blockage by emerging from the annulus on either side.

The role of the sperm plug may be to interfere with mating attempts of further males, but the plug in *H. americanus* does not interfere with fertilization. Andrews (1906b) assumed that the eggs passed over the annulus ventralis as a result of gravity, but this may not entirely explain what happens: in *H. americanus* the pleopods have a role in creating a current which draws the eggs posteriorly into the abdominal chamber. The first pair of pleopods is held erect, creating a barrier that retains eggs on the sternal plate where they accumulate near the sperm release site momentarily, and then the pleopods are folded back, allowing the current to draw the now-fertilized eggs in amongst the other pleopods for attachment. Thus in lobsters eggs are released in pulses: egg laying and sperm release continues until the entire brood has been laid (Aiken et al. 2004). The same process may also be found in crayfish although whether it is the same in cambarids and the other crayfish without sperm storage remains to be established. Female control of the fertilization process, by timing both the release of eggs and the release of sperm, means that cambarid eggs do not need to carry a chemical signal that causes spermatophore dehiscence. This signal seems to be a necessary element in astacid and parastacid crayfish that have external sperm storage, but it has yet to be identified.

Delay between mating and fertilization

In all pleocyemate decapods males transfer spermatophores to females who then use them to fertilize eggs and carry the embryos on their abdomen. External fertilization is the norm, except in the Eubrachyura where it is internal. In all crayfish there is a delay between the transfer of sperm and its use to fertilize eggs. The longer the delay, the greater the chance that females can be promiscuous and accumulate sperm from several different partners. There are few very precise data (measured in hr or d) about the delay between mating and fertilization, but there are several examples (measured in months) which give an approximate estimate of the schedule (see Table 4).

Amongst the Astacidae the delay ranges from 2–3 d to several weeks in *Astacus*, *Pacifastacus* and *Austropotamobius*. There can be differences that may be related to where the species is living, as in the case of *Pacifastacus leniusculus*, perhaps reflecting differences in water temperature. Cambarids have by far the longest delay, especially species of *Orconectes* and *Procambarus* which mate in the autumn, but do not lay eggs until the following spring. *Cambaroides japonicus* can also delay fertilization for six months (Kawai and Saito 2001). In one of the few attemps to measure sperm longevity, Andrews (1906a) found that mated *Orconectes limosus* females isolated from males could still produce fertilized eggs after 5 mon and perhaps longer. Similarly, captive *O. virilis* females kept isolated in a cool room for 5–9 mon were still able to produce viable eggs (Rogowski et al. 2013). However, in some cambarids the shelf life of sperm may be limited: Berrill and Arsenault (1982) found that *O. rusticus* females which did not mate in the spring produced mostly infertile eggs emphasizing the need to have fresh supplies on hand. Parastacid crayfish are similar to astacids in laying eggs a few days after mating. The species of *Cherax* used in aquaculture can spawn several times a year, but it is not clear whether females must mate again before each brood. Therefore, it is not surprising that female cambarids, who store sperm for long periods, provide protection in the form of the annulus ventralis, whereas females of the other two families simply carry the spermatophores on their exposed sternum unprotected. If the female provides minimal protection to the spermatophores obtained by mating then one might expect the delay between mating and fertilization to be shorter than if they were stored in a more secure site and generally speaking this trend is confirmed by the data in Table 4. A further consequence of the shorter delay could be a decrease in the number of males that a female could mate with. Thus we might anticipate a lower level of sperm competition in astacid and parastacid crayfish.

Sperm competition

In their recent review of crayfish reproduction, Gherardi and Aquiloni (2011) were the first to address the question about the role of sexual selection in moulding sexual behaviour wherein they examined some of the elements of sperm competition. They were primarily interested in focusing on the ways in which males could maximize the success of their own sperm or at least minimize multiple paternities. The essential prerequisites for sperm competition in Decapoda are: females having multiple male partners, sperm storage, and a delay between mating and fertilization (Diesel 1991). Crayfish exhibit all of these characteristics although in different ways. In preceding sections of this chapter we have presented evidence about growth format and reproductive behaviours which are relevant. To summarize for crayfish these are: indeterminate growth, mating not normally linked to moulting, multiple mating, sperm storage in open (sternum) or closed sites (annulus ventralis), and lack of male mate guarding. However, a number of open questions remain about how females attract males, when there is no immediate prospect of fertilizing eggs, what the shelf-life of stored sperm might be, and given that there is no mate guarding, is mating order important? The absence of mate guarding means that females can have more male partners and increase the level of sperm competition.

Wild spring mating explosions of *Orconectes rusticus* reported east of the Rockies, in Ontario, in which large males copulate with whomever they meet in Thompson's Creek (Berrill and Arsenault 1984) deserve more detailed attention. These mating episodes during the hours of darkness were attributed to temperatures rising above 4°C (Berrill and Arsenault 1982). Males fought with each other and even interrupted copulating pairs. Such waste of limited sperm supplies seems irresponsible on the part of any males who might want to prolong their blood-line, and is the most extreme example of a prevailing view that male crayfish will mate with anything that moves, provided that the weather is right on the day! The conclusion that it was all brought to an end by females who retired to their burrows to brood seems to ignore the best interests of both male and female crayfish behaviour. However, the evidence did show that for males, size does matter. Almost all previous studies of crayfish mating have focussed on the environmental conditions that might have been correlated with, and therefore 'explained' male mating behaviour: photoperiod, water temperature and the past history (i.e., cumulative exposure) to these variables. It is implied that such behaviour did not evolve, but was brought on by the weather. Apparently it was not considered necessary to ask any questions that warranted study of the females and their readiness to mate or not. Why males might be attracted to females and why they in turn may or may not want to mate has until now been deemed as largely irrelevant.

What might an evolutionary explanation of crayfish mating look like and what are the important variables that would need to be included? When fertilization is internal, as in eubrachyuran crabs, eggs are delivered directly to the seminal receptacle where they encounter sperm, made up of multiple ejaculates, which may be layered or mixed, in a three-dimensional space (McLay and López Greco 2011). In general sperm closest to the oviduct has the best chance to fertilize the eggs. However, when fertilization is external, sperm is often encountered in two dimensional spaces above the surface of the sternum (assuming that there is only a single layer of spermatophores). The prime position for immotile sperm should still be closest to the gonopores in the coxae of the third pereopods in astacids and parastacids, and closest to the entrance of the annulus ventralis in cambarids.

If there are highly skewed numbers of embryos sired by one male, it suggests that when there are multiple copulations, one male has an advantage over the others. Snedden's (1990) experiment using irradiated (sterilized) *O. rusticus* males suggested that when two males are mated with a female, the last male fertilized 92% of the eggs. Similar highly skewed paternity was found in *O. placidus* and *Procambarus clarkii* (Walker et al. 2002, Yue et al. 2010) although male mating order in those cases was unknown. Certainly the data do not support the idea of well-mixed sperm. These are all cambarid crayfish which store sperm in an annulus ventralis so it is possible that it is a case of 'last in first out'. The lack of mate guarding, especially in cambarids, may be explained by the long delay between mating and fertilization. If the schedule of ovarian development does not coincide with the mating schedule, then the best male strategy would be to fill up as many annulus ventrali as possible rather than wait around for a particular female to spawn. Buřič et al. (2009) found that radio-tagged male *Orconectes limosus* in a stream moved more in search of mates during the mating season, covering a maximum of about 100 m per day. In astacids and parastacids it seems that there might be a closer correspondence between the ovarian and mating schedules, so if there was to be any mate guarding, these are the families in which we might find it. However there are other alternatives to guarding.

Austropotamobius italicus (Astacidae) males take radical steps during mating to modify fertilization outcomes by removing and eating many (average 77.2%) of the spermatophores that they find. A third of the following males removed all of the first male's sperm before depositing their own (Villanelli and Gherardi 1998, Galeotti et al. 2007). The ability of *A. italicus* males to vary the amount of sperm transferred according to female size (Rubolini et al. 2006) may also be a valuable trait when faced with the prospect of sperm competition, since small ejaculates given to large females would be largely a waste of time (and sperm). Female *Cherax quadricarinatus* (Parastacidae) are said to use their fifth pereopods to manipulate the spermatophores, stuck on to her sternum by the male, and initiate or modify the fertilization process as spawning occurs 12–24 hr post-mating. Barki and Karplus (1999) developed a more detailed behavioural assay for female receptivity in *Cherax quadricarinatus* and also report "sternum rubbing" by the female, using her last pair of pereopods, when not disturbed by a male. This curious behaviour deserves further study in order to establish what the effects of the rubbing is on the spermatophores and how many of them might be removed or broken open. It may be a female equivalent of the male *Austropotamobius italicus* habit of removing spermatophores deposited by previous partners (see Galeotti et al. 2007).

Parental care

Parental care in crayfish is provided by the females of all three families with some variation in details and has been examined to various degrees in all three families: Astacidae [*Astacus astacus* by Huxley (1881); *Austropotamobius pallipes* by Holdich and Reeve (1988); *Pacifastacus leniusculus* by Andrews (1907); *Pacifastacus trowbridgii* by Mason (1970a)]; Cambaridae [*Cambarus longulus* Girard, 1852 by Smart (1962); *Cambaroides japonicus* by Scholtz and Kawai (2002); *Orconectes limosus* by Andrews (1907) and Mathews (2011); *Orconectes luteus* by Muck et al. (2002); *Orconectes neglectus* (Faxon, 1885) by Price and Payne (1984); *Procambarus clarkii* by Figler et al. (1997) and Aquiloni and Gherardi (2008a); *Procambarus fallax* by Vogt and Tolley (2004)]; Parastacidae [*Astacopsis gouldi* Clark, 1936 by Hamr (1992); *Astacopsis franklinii* (Gray, 1845) by Hamr (1992); *Cherax cainii* by Burton et al. (2007); *Engaeus cisternarius* Suter, 1977 by Suter (1977); *Paranephrops planifrons* by Hopkins (1967a); *Parastacoides tasmanicus* by Hamr (1992); *Parastacus pilimanus* (von Marten, 1869) by Dalosto et al. (2012); *Virilastacus araucanius* (Faxon, 1914) by Rudolph and Rojas (2003)]. The total duration of parental care can be considerable: in the species listed above eggs are carried by the female for several weeks prior to caring for the juveniles, but in *Paranephrops zealandicus* eggs are carried by the female for more than a year and it is at least 15 mon until the first free-living crayfish enter the population (Whitmore and Huryn 1999).

Crayfish have direct development so that hatchlings emerge from the large eggs as miniature non-feeding adults (see Fig. 19). Hatching in *Pacifastacus leniusculus* is facilitated by the embryo as the inner lining is eroded by secretion of an enzyme, causing the egg shell to lose almost two thirds of its strength by the time of eclosion (Pawlos et al. 2010). The juveniles are initially attached to the female by telson threads (formed from the embryonic cuticle and the egg shell) so that their first habitat is the surface of the mother's abdomen. The presence of the telson thread is a unique feature of the freshwater crayfish and is one of the main characters used to argue for their monophyletic invasion of freshwater from the sea (Scholtz 2002, Scholtz and Kawai 2002). Juveniles remain with the mother because the sense organs, necessary for an independent existence, are not fully developed until the third stage, i.e., after two moults (Vogt and Tolley 2004). Brood care for crayfish is indispensable in the freshwater environment (both rivers

Figure 19. Female of the marbled crayfish, *Procambarus fallax*, carrying stage 2 juveniles (arrowed) attached to pleopods beneath her abdomen (from Vogt 2013).

and lakes) and even occurs in burrowing crayfish where the mother and her offspring are confined (Hamr 1992). The mother provides a refuge from predators that may include other crayfish. Juveniles remain with the mother for periods ranging from a few weeks to several months, leaving her for short periods and then returning.

Mason (1970a) found that ovigerous female *Pacifastacus trowbridgii* approaching hatching time are more aggressive. Most broods hatch over a two-day period. Juveniles remain with the parent for 21–25 d during which they pass through three stages (moults) before taking up an independent life. Maternal female *Pacifastacus* lead a solitary existence.

Given the chance, some non-ovigerous female crayfish and males will eat juveniles. However, once females become ovigerous they do not attack juveniles (Gherardi 2002). After juvenile *Orconectes sanbornii* (Faxon, 1884), *Orconectes virilis* and *Procambarus clarkii* hatch they are strongly attracted to the female parent and remain close to her during the second and third instar (first instar larvae are physically attached). The effectiveness of a presumed attractive chemical signal declines over the following 7–21 d, depending on species, and females become predatory again (Little 1975, 1976). The chemical nature of the brooding attractant remains unknown (Vogt and Tolley 2004), although its function seems to be protection of defenceless juveniles against predation by keeping them close to the female. Females provide a safe refuge, but perhaps not close surveillance that might provide warnings of imminent danger.

Juveniles are attracted to conspecific maternal females rather than their particular parent (Gherardi 2002). The maternal pheromone seems to be specific to the species, but not to a particular mother. This lack of juvenile discrimination by the mother between their own and those of other females in *Procambarus clarkii* was demonstrated by Aquiloni and Gherardi (2008a). Fostering will even continue after the maternal female moults (Figler et al. 1997). Maternal *P. clarkii* also show heightened levels of aggression towards other crayfish. More recently, Mathews (2011) discovered evidence of kin recognition by recently independent *Orconectes limosus* juveniles and their mothers, suggesting that parental care may continue longer than previously believed. The South American burrowing crayfish, *Parastacus pilimanus* show a high degree of tolerance of juveniles by the mother. Juveniles left their mother after just nine days. No cannibalism was observed over the six months after hatching, although the juveniles did not always return to the female's pleopods (Dalosto et al. 2012). This contrasts markedly with non-burrowing epigean species which will attack juveniles only a few days after they leave the maternal female (Gherardi 2002). Low parental aggression is essential if generations are to overlap and coexist in their chosen refuge. In *Distocambarus crockeri* burrows multiple occupancy by more than two adults was rare (~1%) and young-of-the-year remained with their presumed parents around a year, but their mortality rate was quite high with only a third alive after that time (Eversole and Welch 2013).

Male crayfish do not seem to have a direct active role in parental care of the juveniles. All the experimental work on parental care has only involved females, perhaps because it was assumed that males were more likely to eat the juveniles rather than care for them and of course there are no abdominal attachment sites available. However, they may have an indirect role if the female with whom the male mated uses his burrow to nurture the offspring. Horwitz and Richardson (1986) cite examples of male and females cohabiting in burrows. Admittedly it is difficult to verify population structure of burrowing animals because of the sampling problems. The study of *Fallicambarus fodiens* (Cottle, 1863) by Norrocky (1991), which employed pipe-traps at burrow entrances, is an exception. However, there do not seem to be many attempts to link male cohabitation to female reproductive state. Cohabitation of *Parastacoides tasmanicus* couples seems to be restricted to just prior to the female moulting (biennial) and mating so females are alone when the eggs hatch. In general male parental care would not be expected if the male could not be sure of paternity. By the same token if females do not recognize their own offspring, and accept any nearby juveniles, then they could end up fostering progeny of other females. However, this may only be a theoretical possibility, rather than a practical one, because females are often living in their own burrow or refuge without other females. For crayfish, 'parental care' would be more accurately described as 'maternal care' (more informative than 'brood care') (cf. Thiel 2000). Prolonged parental care is an essential element in the success of crayfish colonization of freshwater environments (Vogt and Tolley 2004) because these crustaceans do not have a dispersive larval stage. Furthermore each offspring has been provisioned by the mother with enough yolk reserves to become juveniles so there are fewer of them, making each one a high-value item. The value of parental care can be evaluated by comparing the fecundity of crayfish (see Table 5) with that of a similar-sized Norway lobster: the average brood size of crayfish is only ~2% of the brood size of comparable *N. norvegicus* (Tuck et al. 2000).

Theories about evolution of brood care in crayfish provide important insights into the phylogeny of the Astacida (Scholtz and Richter 1995). It is hypothesized that crayfish originated from a marine crayfish

Table 5. Reproductive characteristics of freshwater crayfish. (Summarized from data in Honan and Mitchell 1995, with the addition of post-1995 published data.)

	Max CL (mm)	Female CL at first maturity (mm)	Egg numbers (over female size range)	Max egg diameter (mm)	Number of breeding seasons	Reference
Astacidae						
Astacus astacus	67.5	32.8	10–240	2.3–3.5	4–5 Incubation period 240 d	Skurdal and Taugbol (2002), Reynolds et al. (1992)
Astacus leptodactylus	63.9	35.1	18–465	2.5	5–7	Berber and Mazlum (2009)
Austropotamobius italicus	55	~25	1 brood per yr	~2.7	10 Mature at 3 yrs	Galeotti et al. (2007)
Austropotamobius pallipes	60	26	20–165	2.3–3.3	3–11 Incubation period 260 d	Honan and Mitchell (1995), Reynolds (2002)
Pacifastacus leniusculus	75	30	50–320; 100–470 Nakata et al. 2004 Mean = 158 Guan and Wiles (1999)	2.56; 2.4–2.9 Nakata	2–4 Incubation period 240 d	Honan and Mitchell (1995) Guan and Wiles (1999)
Cambaridae						
Cambarus bartoni	35	24	20–85	2.6	2	Honan and Mitchell (1995)
Cambarus longulus	28	18–22	Only ovarian # known 30–120	2.4	2	Honan and Mitchell (1995)
Cambarus robustus	57	35	10–230	2.7	1	Honan and Mitchell (1995)
Cambarus elkensis	> 50	29	100–216	~2.5	2	Jones and Eversole (2011)
Cambaroides japonicus	32	15.2	22–75	2.1–2.5	5–6	Kawai et al. (1997), Nakata and Goshima (2004)
Orconectes eupunctus	~22	14	13–97	~2.0	?	Larson and Magoulick (2008)
Orconectes kentuckiensis	36.2	15	50–250	> 2.1	1–2	Honan and Mitchell (1995)
Orconectes neglectus	~25	14	21–149	~2.0	?	Larson and Magoulick (2008)
Orconectes palmeri	41	18	?	2.0	1	Honan and Mitchell (1995)
Orconectes propinquus	26.1	13	21–249	0.96	1	Corey (1987)
Orconectes rusticus	51	17	50–575	1.2	1	Hamr (2002)

Table 5. contd....

Table 5. contd.

	Max CL (mm)	Female CL at first maturity (mm)	Egg numbers (over female size range)	Max egg diameter (mm)	Number of breeding seasons	Reference
Orconectes virilis	30	14	20–310	1.0	2 Incubation period 60 d	Corey (1987)
Orconectes williamsi	26	11.9–15.1	20–140	1.9–2.7	3–4	DiStefano et al. (2013)
Orconectes lutens	29	11–12	20–260	2.0	1–2	Muck et al. (2002)
Procambarus clarkii	70	45	140–595	2.0	?4–5 20 d incubation period per brood	Honan and Mitchell (1995), Gutiérrez-Yurrita and Montes (1999)
Procambarus hayi	70	28	~250	2.2	2–3	Honan and Mitchell (1995)
Procambarus fallax ("Marmorkrebs")	38	17.3	50–500	1.5	7	Vogt (2008b)
Parastacidae						
Astacoides betsileoensis	78	55.5	100–390	3.8	~25	Honan and Mitchell (1995), Jones et al. (2007)
Astacoides caldwelli	70	50	50–240	3.8	?	Jones et al. (2007), Hobbs (1987)
Astacoides crosnieri	49.2	32	21–50	3.8	~20	Honan and Mitchell (1995), Jones et al. (2007)
Astacoides granulimanus	71	40.6	20–290	4–4.6	~25	Honan and Mitchell (1995), Jones et al. (2007)
Astacopsis franklinii	53	33	35–118	3.2–3.9	?	Honan and Mitchell (1995)
Astacopsis gouldi	178	99	244–1300	4.7–5.7	?	Honan and Mitchell (1995)
Cherax cainii	90	30	71–707	3.0	1/yr	Beatty et al. (2003)
Cherax destructor	150	33.5	124–1000	1.8–2.0	2–3/yr	Honan and Mitchell (1995), Beatty et al. (2005a)
Cherax quadricarinatus	?100	42	200–1000	?2.5	3–4/yr	Sammy (1988), Honan and Mitchell (1995)
Cherax quinquecarinatus	~41	18.8	40–147	2.6	3/yr	Beatty et al. (2005b)
Cherax tenuimanus	103	48	20–760+	2.4–2.8	?	Honan and Mitchell (1995)
Engaeus cisternarius	31.8	25.6	45–75	2.45	> 1	Honan and Mitchell (1995)
Engaeus laevis	21.5	13	29–184	1.8–2.0	?	Honan and Mitchell (1995)

Table 5. contd....

Table 5. contd.

	Max CL (mm)	Female CL at first maturity (mm)	Egg numbers (over female size range)	Max egg diameter (mm)	Number of breeding seasons	Reference
Engaeus leptorhynchus	33.5	33.5	108	1.7	~4	Honan and Mitchell (1995)
Engaeus orientalis	26.9	14.2	9–19	?	?	Honan and Mitchell (1995)
Engaeus tuberculatus	34.6	14.9	22–63	1.5–1.7	?	Honan and Mitchell (1995)
Euastacus armatus	146.2	40–100	44–155 ?500–1000	?4	?1	Honan and Mitchell (1995), Morgan (1997)
Euastacus australasiensis	59.4	30–40	44–155	?	?	Honan and Mitchell (1995)
Euastacus bispinosus	13.3	86	300–812	3.9–4.1	> 3/fm	Honan and Mitchell (1995)
Euastacus spinifer	105	75	260–780	3.2–3.9	> 4/fm	Honan and Mitchell (1995)
Paranephrops plantfrons	32	24	20–170	1.31	~4/fm	Hopkins (1967a)
Parastacoides tasmanicus	32	21	23–85	2.4–3.0	~6/fm	Honan and Mitchell (1995), Hamr and Richardson (1994)
Parastacus pugnax	55	30	5–46	?	16 annual broods	Ibarra and Arana (2012)

ancestor, whose females carried eggs until development was advanced. When they colonized freshwater it was advantageous to have direct development and maternal brood care, including the ability to attach to the mother post-hatching. The parental care characters of the northern hemisphere Cambaridae and southern hemisphere Parastacidae are considered to be convergent: different solutions to the same problems resulting from the fact that all freshwater eventually flows downhill to the sea. To prevent down-stream displacement juveniles find it advantageous to remain with their mother. The evolutionary patterns of post-embryonic development in crayfish represent a step-wise extension of maternal care through refined adaptations to the freshwater environment (Scholtz 1995, Scholtz and Kawai 2002).

Fecundity and egg size

Comparisons of egg numbers produced by female crayfish are difficult because they can be estimated at several different stages of the reproductive cycle: ovarian numbers at different stages of gonad maturity usually using histological methods (sometimes called the 'ovarian fecundity') and direct counts of numbers of unhatched eggs or hatched juveniles attached to the abdomen (sometimes called the 'pleopodal fecundity'). Comparing fecundity between species needs to use females of roughly the same size (CL). The habit of many crayfish to seek shelter in burrows, and be refractory towards traps and nets, makes gathering enough ovigerous females a significant challenge for some species. Published studies sometimes include both ovarian and pleopodal fecundities, but only rarely. Pleopodal egg numbers are probably the easiest to count, although one must take into account another variable which is egg mortality. For example Celada et al. (2005) found that female *Pacifastacus leniusculus* in culture had lost almost 50% of pleopodal eggs by the time hatching was near. Recording the stage of egg development would ideally help reduce the variance in pleopodal fecundity estimates. Huner and Lindqvist (1991) recommend provision of ovarian fecundities as a matter of course. Corey (1991) suggests that as a rule of thumb pleopodal fecundity will be about 50% of ovarian fecundity, although there can be significant differences between species and individuals depending upon their circumstances. Assuming that there is sufficient sperm available, pleopodal fecundity is ultimately limited by the size of the female abdomen, which is an indicator of the number of attachment sites for fertilized eggs. The same problems arise when trying to estimate egg-size which is a measure of the level of female investment per offspring. It is worth noting that whereas egg numbers are estimated on a linear scale, egg investment has to be based on diameter[3] thereby making accurate measurement and standardization of the stage of development even more important.

Given that the marine ancestors of astacid crayfish are likely to have had much smaller eggs and larger brood sizes, because they had planktonic larval stages, albeit limited in duration, the initial steps in colonization of freshwater must have involved production of large eggs. The provision of enough energy for each offspring to complete their development, without a feeding larval stage, requires packaging of the reproductive investment into larger but fewer eggs. Given fewer eggs each one had greater value and concomitant evolution of female parental care became worthwhile. The more abbreviated the development the larger are the eggs.

Astacid crayfish have much larger eggs than most decapods. Egg size changes during embryonic development: egg volume in *Pacifastacus leniusculus* increases by almost 40% while being carried by the female (Pawlos et al. 2010). Table 5 lists selected reproductive data about more than 40 species across all three families. Mean egg diameter is 2.8 mm (2.6–3.0) for the Astacidae; 1.9 mm (0.96–2.7) for the Cambaridae; 3.0 mm (1.3–5.2) for the Parastacidae. Mean brood size is 158 eggs (74–300) for the Astacidae; 138 eggs (40–275) for the Cambaridae; 217 eggs (15–772) for the Parastacidae. Brood size usually increases linearly with female CL (Momot 1984, Corey 1991). For the species listed in Table 5 the data confirm the general pattern that crayfish mature at one-third to one half their maximum sizes (Black 1966, Huner and Romaire 1979, Huner and Barr 1984).

Reproductive strategy is made up of egg size, egg number and reproductive effort (the product of these two) and should maximize the number of recruits per female. For the present purposes we assume that the annual brood number is the same for each species as we are not attempting to estimate lifetime effort. Comparison of the mean reproductive effort (effectively the brood volume in mm[3]) of the three crayfish families shows the Astacidae 1816 (range 1454–2234); the Cambaridae 437 (64–1425); and the Parastacidae 3138 (250–16,006) (data in Table 5). Clearly on this scale, the Cambaridae have the lowest reproductive effort while the Parastacidae have by far the highest, with the Astacidae in between. Most of these differences are the result of differences in egg size rather than differences in brood size. Many environmental factors contribute to differences in crayfish reproductive strategy (Huner and Lindqvist 1991) and the main message of the reproductive data is that these crustaceans are very adaptable in adjusting to whatever freshwater habitat they colonize. Production of offspring is dependent on female size which can be strongly influenced by habitat, especially water temperature. Some cambarids are short lived and have short incubation periods meaning that they can produce multiple broods each year, whereas the longer lived astacids and parastacids only produce an annual brood after attaining maturity. One major piece of information that is lacking for most crayfish, but especially short lived species, is the relationship between the breeding cycle and the moult cycle. Where there is a long incubation period we can make a reasonable guess that there is only one brood per moult cycle, but when the brood cycle is only a matter of a few weeks it is unknown whether inter-brood moulting occurs. Moulting produces a larger female and therefore larger broods, but it would require females to re-mate in order to obtain sperm. In cambarids we need to know how many broods can be fertilized by a female with a replete annulus ventralis.

Overview synthesis of growth and reproduction

The pattern of crayfish life history is mostly the result of their 'adopted' habitat, namely freshwater. In looking for comparisons with other species it does not make any sense to compare them with insect inhabitants of freshwater, for example, because they have a different evolutionary history. Four main groups of decapods have independently invaded freshwater since the Triassic: carideans, astacids, anomuran aeglids and brachyurans (Vogt 2013). Note that we do not include decapods that live on land, but migrate to the sea to release their larvae, for, e.g., the coconut crab, *Birgus latro* (Linnaeus, 1767). Within those groups there have been multiple colonisations, resulting in different families, e.g., within the Brachyura, "freshwater crabs" are a polyphyletic group. Besides the crayfish notable examples are the Aeglidae, an endemic freshwater family of crab-like anomurans from South America (Tudge 2003) and the remarkable Jamiacan sesarmids (Diesel et al. 2000). Each of these groups has found its own adaptive solution to living in freshwater (with its lower salinity) or amongst land plants, which involves reduction in egg number, production of large eggs, direct development and extension of parental care beyond the egg stage (Vogt 2013). As far as we know these evolutionary pathways, embarked upon independently by these decapods, have all involved one-way traffic because none have gone back to the sea.

Like their ancestors, crayfish are iteroparous, producing multiple broods over a time range: several broods in one year; single broods every year (or in alternate years) for several years ranging up to several decades (see Table 2 for examples of life spans).

We have presented an outline of crayfish growth and reproductive features, but which of these features are apomorphic and which are plesiomorphic? To do this we need to specify a marine out-group of clawed lobsters which inform us about the likely character states in the ancestor. It seems that the natural group to compare crayfish with is the Nephropidae, for example *Nephrops norvegicus*. The nephropid lobsters are grouped with other clawed lobsters in the Homarida, which are their sister group (Scholtz and Richter 1995). In summary, Farmer (1974a) found that in *Nephrops* growth is indeterminate; mating is linked to moulting so males are instantly attracted to recently moulted females or water from their tank suggesting a pheromone; during mating the male holds the female down by grasping her chelipeds; females have a closed thelycum in the seventh sternite for external sperm storage; male transfers sperm to the thelycum using the first two pairs of pleopods, both pleopods remain obviously biramous like the following pleopods, but at least the first one is grooved; females have to remate after moulting; only one moult/brood are produced per year; gonads are H-shaped and maturation in females follows an annual cycle coordinated with moulting, but in males spermatogenesis seems to occur throughout the year; for egg-laying and fertilization the *N. norvegicus* female lies on her back, releases eggs from the gonopore which are fertilized in a glair chamber as they pass over the thelycum towards the abdomen where they attach; egg size is around 1 mm diameter at laying and swells to 2 mm during development; egg numbers range over 250–1500 depending on female CL; there are three planktonic larval stages. Moulting/mating in *N. norvegicus* occurs May to August and egg laying occurs August to September (Farmer 1974a) so there is a delay of approximately two months from May to August for fertilization to occur during which females could mate with other males. Microsatellite studies by Streiff et al. (2004) found evidence of multiple paternity in six out of 11 broods analysed. In these broods two to three sires were implicated, but their contributions were fairly even in each brood, suggesting that stored sperm from multiple copulations may well have been well-mixed or at least none had priority.

In the other group of clawed lobsters, homarids, mating and moulting are linked and males will guard females both before and after mating (Aiken et al. 2004). However, in *Homarus americanus* intermoult mating also occurs, especially in small females. Some females may have multiple partners but the processes of mate attraction and guarding, which involves the use of shelters, may limit the number of males that a female can mate with (Waddy and Aiken 1991, Waddy et al. 2013). Detailed discussion of homarid mating is beyond the scope of the present discussion, but does give an indication of variation in the reproductive life styles amongst the ancestral stock of clawed lobsters.

So when we compare freshwater crayfish with *N. norvegicus*, what differences do we find? How many of the crayfish characters were likely derived from the ancestor (plesiomorphic) and how many evolved after colonization of freshwater (apomorphic)? The following characters seem to be plesiomorphic: growth is indeterminate, male grasps female by her chelipeds when mating, females have to remate after moulting, single brood per moult cycle, male pleopods modified for sperm transfer, fertilization in a glair chamber, comparatively small number of large eggs laid and there is a fertilization delay after mating. The only apomorphic characters are: cyclic dimorphism in cambarids, mating not linked to moulting, sperm storage is mostly external except for cambarids (see further discussion below about this character), no larval stages, but maternal care instead. It is clear that most of the growth and reproductive characters of crayfish are in fact derived from the ancestral condition.

The method of sperm storage deserves further discussion. Farmer (1974b) described the ontogeny of the thelycum in *Nephrops norvegicus* females on the sternum of the penultimate thoracic segment. Growth at the margins leaves an inverted-Y shaped entrance to a cavity which stores sperm transferred by the male. Eggs emerge from the female gonopores in the coxa of the sixth pereopods must pass over the thelycum in order to reach the abdomen where they become attached to pleopods. The annulus ventralis is the homologous structure found in female cambarid crayfish and it varies in shape according to the species. The enigmatic Asian genus *Cambaroides* has a much less specialized sperm storage structure on the same sternite as cambarids, but whether it should be called an "annulus ventralis" depends in which family *Cambaroides* is allocated. Although Scholtz (2002) argued for its inclusion in the Cambaridae, more recent work based on both extant and fossil species suggests that *Cambaroides* should be included in the Astacidae (Crandall et al. 2000, Rode and Babcock 2003) recognizing that *Cambaroides*, like all the Astacidae, does not have protected sperm storage and alternation of reproductive forms I and II. These same studies also show that the Astacoidea is a monophyletic group so how do we account for the absence of sperm storage in two families? The sister group is the clawed nephropoid lobsters (Crandall et al. 2000) which have sperm storage organs, so we have to conclude that sperm storage has been lost in the astacids and parastacids and furthermore that sperm storage in the cambarids is a plesiomorphic character. The loss of sperm storage in astacoid crayfish seems counter-intuitive given that they colonized an environment consisting of small bodies of water and consequently small population units where opportunities to mate may have been limited. Thus sperm storage and its origins in crayfish remains an open question for the moment.

Research agenda for the future

There are still many unanswered questions about growth and reproduction and about how these interact to facilitate the adaptation of crayfish to their environment. Some of their reproductive behaviour seems counter-intuitive, or at least enigmatic, and so demands an answer.

Perhaps the best progress can be made by studying individuals that are marked/tagged in some way or kept captive in small groups where the progress of the same animals can be followed over time, preferably over the entire lifespan. It seems to us that it would be better to study a small number of crayfish intensively over time, rather than trying to take representative samples from large populations and then try and infer what the "average" individual is doing. There is uncertainty associated with both approaches, but at least there is a high level of accuracy when individuals are followed over their lifespan.

One of the strangest aspects of crayfish growth patterns is the form changes undertaken by cambarids (except for *Cambaroides* spp.) where reproductive forms alternate with non-reproductive forms. Apparently not all *Orconectes limosus* moult and change form because Buřič et al. (2010) found that 9% moulted without form change, while 6% did not moult at all. We need to know what the advantages are of changing form and whether in other cambarid crayfish part of the population does not change form.

What patterns of variation are there in the coincidence of the moult cycle with the brood cycle in different species of crayfish? It seems that the two cycles are mutually exclusive. In mature crayfish one or more moults must occur between each brood which is thus produced during an intermoult. When moulting and mating are not linked, how do females attract a male? If their ovaries are not mature what exactly are they advertising because they cannot be offering the opportunity to fertilize a brood of eggs?

What is the spatial pattern of sperm storage in cambarids (closed) and in astacids and parastacids (open)? In the case of cambarids does the last male to mate have priority in fertilization? Is there an order of priority in astacids and parastacids based on where on the sternum the spermatophores are placed? With closed sperm storage in females do the males produce non-adhesive spermatophores while in the other two families, where females have open storage, do males produce adhesive spermatophores? We probably can already guess that if there is a delay in fertilization then astacid and parastacid males must produce adhesive spermatophores.

One of the most intriguing questions about crayfish is how long is the shelf-life of sperm? This character is very important to both females and males: if females solicit males when their ovaries are not ready to spawn, then the sperm must last until the eggs are laid; and for males who are competing with other males to fertilize the eggs, they need to provide viable sperm to be in the race. There may be a higher provisioning cost for a male to produce sperm with a longer shelf-life, but will they fertilize more eggs or perhaps shorter-lived sperm might do better? What about shelf life of spermatophores in species where mating and fertilization are separated by a much longer period than a few hours/days? Given that cambarids have internal sperm storage does their sperm have a longer shelf life than sperm of the other two families?

In traditional terms we would probably describe fertilization in crayfish as being external, as opposed to it being internal in most crabs. But calling it external is not quite accurate because for each brood produced, a female creates a temporary fertilization chamber by cupping the abdomen and secreting glair which has the effect of restricting the movement of gametes and increases the probability of eggs and sperm meeting. That part is clear, but what is not clear is how the simultaneous release of eggs and sperm is achieved. There is some evidence that in cambarids females may be able to achieve this by movements of the mobile last thoracic segment, expelling sperm from the annulus ventralis, but be that as it may, how is the rendezvous arranged in astacids and parastacids? Do their eggs produce a dehiscent signal to the spermatophores? Furthermore does this imply that the probability of paternity by multiple male partners is different for these three families of crayfish? Could it be that the lack of male mate guarding in crayfish is a consequence of sperm competition resulting from uncertainty about paternity? Spermatophores are not necessarily cheap to produce so with limits on sperm supply what is the best way for males to invest their gametes?

Research on homarid lobsters suggests that females can control the whole fertilization process by timing the release of ova and sperm although this cannot be construed as some kind of female choice because it is just a mechanical process for releasing the gametes rather than influencing which gametes might enjoy an advantage. Similar experiments involving occlusion of the annulus ventralis in cambarids could prove very interesting.

When comparing the gametic strategies of marine decapod Crustacea, Sainte-Marie (2007, p.204) came to a conclusion that is well-worth quoting here: "With very few exceptions, biological research and management practices for decapods continue to reflect the archaic assumption that males, once sexually mature, have virtually unlimited resources for inseminating females ... future investigations should put as much emphasis on deciphering the gametic strategies of males as for females." This advice is even more salient for the freshwater crayfish because of smaller brood sizes, and perhaps more limited opportunities to breed, which mean that lifetime fecundity for both sexes may be much reduced compared to their marine ancestors.

Future research resulting in answers to these and further questions will fill in the many gaps in current knowledge and provide a more complete understanding of growth and reproduction in freshwater crayfish, and how they might be affected by environmental factors. In particular we would like to remind researchers not to forget that crayfish, like all living things, have an evolutionary history: what we see today is for the most part explained by what happened yesterday!

Acknowledgements

We are grateful to Laura López Greco for her valuable advice and permission to reproduce the photographs of *Cherax quadricarinatus* male and female, Fig. 14.

Crayfish Names Used in Chapter 3

Astacoidea

Astacidae (7 spp.)
Astacus astacus (Linnaeus, 1758)
Astacus leptodactylus (Eschsholtz, 1823)
Austropotamobius pallipes (Lereboullet, 1858)
Austropotamobius italicus (Faxon, 1914)
Austropotamobius torrentium (Schrank, 1803)
Pacifastacus leniusculus (Dana, 1852)
Pacifastacus trowbridgii (Stimpson, 1857)

Cambaridae (46 spp.)
Cambarellus montezumae (Sassure, 1857)
Cambarellus shufeldtii (Faxon, 1884)
Cambaroides japonicus (de Haan, 1842)
Cambarus acuminatus (Faxon, 1884)
Cambarus bartonii cavatus (Hay, 1902)
Cambarus dubius (Faxon, 1884)
Cambarus elkensis (Jezerinac and Stocker, 1993)
Cambarus hubbsi (Creaser, 1931)
Cambarus longulus (Girard, 1852)
Cambarus robustus (Girard, 1852)
Distocambarus crockeri (Hobbs and Carlson, 1983)
Fallicambarus fodiens (Cottle, 1863)
Fallicambarus gordoni (Fitzpatrick, 1987)
Faxonella clypeata (Hay, 1899)
Orconectes australis (Rhoades, 1941)
Orconectes eupunctus (Williams, 1952)
Orconectes illinoiensis (Brown, 1956)
Orconectes immunis (Hagen, 1870)
Orconectes indianensis (Hay, 1896)
Orconectes inermis (Cope, 1872)
Orconectes kentuckiensis (Rhoades, 1944)
Orconectes limosus (Rafinesque, 1817)
Orconectes luteus (Creaser, 1933)
Orconectes nais (Faxon, 1885)
Orconectes neglectus (Faxon, 1885)
Orconectes palmeri (Faxon, 1884)
Orconectes pellucidus (Tellkampf, 1844)
Orconectes placidus (Hagen, 1870)
Orconectes propinquus (Girard, 1852)
Orconectes rusticus (Girard, 1852)
Orconectes sanbornii (Faxon, 1884)
Orconectes virilis (Hagen, 1870)
Orconectes williamsi (Fitzpatrick, 1966)
Procambarus acutus (Girard, 1852)
Procambarus alleni (Faxon, 1884)
Procambarus blandingi (Harlan, 1830)
Procambarus bouvieri (Ortmann, 1909)
Procambarus clarkii (Girard, 1852)
Procambarus digueti (Bouvier, 1897)
Procambarus erythrops (Relyea and Sutton, 1975)
Procambarus fallax (Hagen, 1870)
Procambarus hayi (Faxon, 1884)
Procambarus llamasi (Villalobos, 1954)
Procambarus spiculifer (Le Conte, 1856)
Procambarus suttkusi (Hobbs, 1953)
Procambarus verrucosus (Hobbs, 1952)

Parastacidae (28 spp.)
Astacoides betsileoensis (Petit, 1923)
Astacoides caldwelli (Bate, 1865)
Astacoides crosnieri (Hobbs, 1987)
Astacoides granulimanus (Monod and Petit, 1929)

Astacopsis franklinii (Gray, 1845)
Astacopsis gouldi (Clark, 1936)
Cherax albidus (Clarke, 1936)
Cherax cainii (Austin, 2002)
Cherax destructor (Clark, 1936)
Cherax quadricarinatus (von Martens, 1868)
Cherax quinquecarinatus (Gray, 1865)
Cherax tenuimanus (Smith, 1912)
Engaeus cisternarius (Suter, 1977)
Engaeus laevis (Clark, 1941)
Engaeus leptorhynchus (Clark, 1936)
Engaeus orientalis (Clark, 1941)
Engaeus tuberculatus (Clark, 1936)
Euastacus armatus (von Martens, 1866)
Euastacus australasiensis (H. Milne Edwards, 1837)
Euastacus bispinosus (Clark, 1936)
Euastacus spinifer (Heller, 1865)
Paranephrops planifrons (White, 1842)
Paranephrops zealandicus (White, 1847)
Parastacoides tasmanicus (Erichson, 1846)
Parastacus defossus (Faxon, 1898)
Parastacus pugnax (Poeppig, 1835)
Parastacus pilimanus (von Marten, 1869)
Virilastacus araucanius (Faxon, 1914)

Species used as outgroup comparisons
Birgus latro (Linnaeus, 1767)
Homarus americanus (H. Milne Edwards, 1837)
Macrobrachium rosenbergii (De Man, 1879)
Nephrops norvegicus (Linnaeus, 1758)

References

Aiken, D.E. 1969. Ovarian maturation, egg laying in the crayfish *Orconectes virilis*: influence of temperature and photoperiod. Can. J. Zool. 47: 931–935.

Aiken, D.E., S.L. Waddy and S.M. Mercer. 2004. Confirmation of external fertilization in the American lobster, *Homarus americanus*. J. Crust. Biol. 24: 474–480.

Alves, N., J.R. Merrick, J. Kohen and D. Gilligan. 2010. Juvenile recruitment of the Murray crayfish, *Euastacus armatus* (Decapoda: Parastacidae), from southeastern Australia. Freshwater Crayfish 17: 167–175.

Ameyaw-Akumfi, C. 1981. Courtship in the crayfish *Procambarus clarkii* (Girard) (Decapoda, Astacidea). Crustaceana 40: 57–64.

Ando, H. and T. Makioka. 1998. Structure of the ovary and mode of oogenesis in a freshwater crayfish, *Procambarus clarkii* (Girard). Zoological Science 15: 893–901.

Andrews, E.A. 1904. Breeding habits of crayfish. Am. Nat. 38: 165–206.

Andrews, E.A. 1906a. Ontogeny of the annulus ventralis. Biol. Bull. 10: 122–137.

Andrews, E.A. 1906b. The annulus ventralis. Proc. Boston Soc. Nat. Hist. 32: 427–479.

Andrews, E.A. 1906c. Egg-laying of crayfish. Am. Nat. 40: 343–356.

Andrews, E.A. 1907. The young of the crayfishes *Astacus* and *Cambarus*. Smithson. Contrib. Knowledge 35: 1–79.

Andrews, E.A. 1908a. The annulus of a Mexican crayfish. Biol. Bull. 14: 121–133.

Andrews, E.A. 1908b. The sperm receptacle in the crayfishes, *Cambarus cubensis* and *C. paradoxus*. J. Wash. Acad. Sci. 10: 167–185.

Andrews, E.A. 1910a. The anatomy of the stylets of *Cambarus* and of *Astacus*. Biol. Bull. 18: 79–97.

Andrews, E.A. 1910b. Conjugation in the crayfish, *Cambarus affinis*. J. Exp. Zool. 9: 235–264.

Andrews, E.A. 1911. Male organs for sperm-transfer in the crayfish, *Cambarus affinis*: their structure and use. J. Morphol. 22: 239–293.

Aquiloni, L. and F. Gherardi. 2008a. Extended mother–offspring relationships in crayfish: the return behaviour of juvenile *Procambarus clarkii*. Ethology 114: 946–954.

Aquiloni, L. and F. Gherardi. 2008b. Assessing mate size in the red swamp crayfish *Procambarus clarkii*: effects of visual versus chemical stimuli. Freshwat. Biol. 53: 461–469.

Aquiloni, L., A. Massolo and F. Gherardi. 2009. Sex identification in female crayfish is bimodal. Naturwissen. 96: 103–110.

Asakura, A. 2009. The evolution of mating systems in decapod crustaceans. *In*: J.W. Martin, K.A. Crandall and D.L. Felder (eds.). Decapod Crustacean Phylogenetics. CRC Press, Taylor and Francis Group, Boca Raton, London, New York 18: 121–182.

Baker, A.M., P.M. Stewart and T.P. Simon. 2008. Life history study of *Procambarus suttkusi* in Southeastern Alabama. J. Crust. Biol. 28: 451–460.

Barki, A. and I. Karplus. 1999. Mating behavior and a behavioral assay for female receptivity in the red-claw crayfish *Cherax quadricarinatus*. J. Crust. Biol. 19: 493–497.

Barki, A., T. Levi, G. Hulata and I. Karplus. 1997. Annual cycle of spawning and molting in the red-claw crayfish, *Cherax quadricarinatus*, under laboratory conditions. Aquaculture 157: 239–249.

Barki, A., I. Karplus, I. Khalaila, R. Manor and A. Sagi. 2003. Male-like behavioral patterns and physiological alterations induced by androgenic gland implantation in female crayfish. J. Exp. Biol. 206: 1791–1797.

Barki, A., I. Karplus, R. Manor and A. Sagi. 2006. Intersexuality and behavior in crayfish: the de-masculinization effects of androgenic gland ablation. Horm. Behav. 50: 322–331.

Beach, D. and P. Talbot. 1987. Ultrastructural comparison of sperm from the crayfishes *Cherax tenuimanus* and *C. albidus* Journal of Crustacean Biology 7: 205–218.

Beatty, S.J., D.L. Morgan and H. Gill. 2003. Reproductive biology of the large freshwater crayfish *Cherax cainii* in south-western Australia. Mar. Freshwat. Res. 54: 597–608.

Beatty, S.J., D.L. Morgan and H. Gill. 2005a. Role of the life history strategy in the colonisation of Western Australian aquatic systems by the introduced crayfish *Cherax destructor* Clark, 1936. Hydrobiologia 549: 219–237.

Beatty, S.J., D.L. Morgan and H. Gill. 2005b. Life history and reproductive biology of the gilgie, *Cherax quinquecarinatus*, a freshwater crayfish endemic to southwestern Australia. J. Crust. Biol. 25: 251–262.

Bechler, D.L. 1995. A review and prospectus of sexual and interspecific pheromonal communication in crayfish. Freshw. Cray. 10: 657–667.

Belchier, M., L. Edsman, M.R.J. Sheehy and P.M.J. Shelton. 1998. Estimating age and growth in long-lived temperate crayfish using lipofucsin. J. Freshwat. Biol. 39: 439–446.

Berber, S. and Y. Mazlum. 2009. Reproductive efficiency of the narrow-clawed crayfish, *Astacus leptodactylus*, in several populations in Turkey. Crustaceana 82: 531–542.

Berrill, M. and M. Arsenault. 1982. Spring breeding of a northern temperate crayfish, *Orconectes rusticus*. Can. J. Zool. 60: 2641–2645.

Berrill, M. and M. Arsenault. 1984. The breeding behaviour of a northern temperate orconectid crayfish, *Orconectes rusticus*. Anim. Behav. 32: 333–339.

Berry, F.C. and T. Breithaupt. 2008. Development of behavioural and physiological assays to assess discrimination of male and female odours in crayfish, *Pacifastacus leniusculus*. Behaviour 145: 1427–1446.

Berry, F.C. and T. Breithaupt. 2010. To signal or not to signal? Chemical communication by urine-borne signals mirrors sexual conflict in crayfish. BMC Biology 8: 1–11.

Black, J.B. 1966. Cyclic male reproductive activities in the dwarf crawfishes *Cambarellus shufeldii* (Faxon) and *Cambarellus puer* Hobbs. Trans. Am. Microsc. Soc. 85: 214–232.

Breithaupt, T. 2011. Chemical communication in crayfish. pp. 257–276. *In*: T. Breithaupt and M. Thiel (eds.). Chemical Communication in Crustacea. Springer, Berlin.

Brewis, J.M. and K. Bowler. 1985. A study of reproductive females of the freshwater crayfish *Austropotamobius pallipes*. Hydrobiologia 121: 145–149.

Brockerhoff, A.M. and C.L. McLay. 2005. Comparative analysis of the mating strategies in grapsid crabs with special reference to two common intertidal crabs, *Cyclograpsus lavauxi* and *Helice crassa* (Decapoda: Grapsidae), from New Zealand. J. Crust. Biol. 25: 507–520.

Bugnot, A.B. and L.S. López Greco. 2009. Sperm production in the red claw crayfish *Cherax quadricarinatus* (Decapoda, Parastacidae). Aquaculture 295: 292–299.

Buřič, M., A. Kouba and P. Kozák. 2009. Spring mating period in *Orconectes limosus*: the reason for movement. Aqua. Sci. 71: 473–477.

Buřič, M., A. Kouba and P. Kozák. 2010. Molting and growth in relation to form alternations in the male spiny-cheek crayfish *Orconectes limosus*. Zool. Stud. 49: 28–38.

Buřič, M., M. Hulka, A. Kouba, A. Petrusek and P. Kozák. 2011. A successful crayfish invader is capable of facultative parthenogenesis: a novel reproductive mode in decapod crustaceans. PLoS ONE 6(5 e20281): 1–5.

Burton, T., B. Knott, D. Judge and P. Vercoe. 2007. Embryonic and juvenile attachment structures in *Cherax cainii* (Decapoda: Parastacidae): Implications for maternal care. Am. Midl. Nat. 157: 127–136.

Carmona-Osalde, C., M. Rodriguez-Serna and A.M. Olvera Novoa. 2002. The influence of the absence of light on the onset of first maturity and egg laying in the crayfish *Procambarus (Austrocambarus) llamasai* (Villalobos, 1955). Aquaculture 212: 289–298.

Carral, J.M., J.D. Celada, J. González, M. Sáes-Royuela and V.R. Gaudioso. 1994. Mating and spawning of freshwater crayfish, *Austropotamobius pallipes* Lereboullet, under laboratory conditions. Aquacult. Fish. Manag. 25: 721–727.

Chang, Y.-J., C.-L. Sun, Y. Che and S.-Z. Yeh. 2012. Modelling the growth of crustacean species. Rev. Fish Biol. Fisheries 22: 157–187.

Celada, J.D., J.I. Antolín and R. Rodriguez. 2005. Successful sex ratio of 1M:4F in the astacid crayfish *Pacifastacus leniusculus* Dana under captive breeding conditions. Aquaculture 244: 89–95.

Cooper, J.E. 1975. Ecological and behavioral studies in Shelta cave, Alabama, with emphasis on decapod crustaceans. Ph.D. Thesis, University of Kentucky, Lexington, KY, U.S.A.

Corey, S. 1987. Comparative fecundity of four species of crayfish in Southwestern Ontario, Canada (Decapoda, Astacidea). Crustaceana 52: 276–286.

Corey, S. 1990. Life history of *Cambarus robustus* Girard in the Eramosa-Speed River system of southwestern Ontario, Canada (Decapoda, Astacidea). Crustaceana 59: 225–230.

Corey, S. 1991. Comparative potential reproduction and actual production in several species of North American crayfish. *In*: A.M. Wenner and A.M. Kuris (eds.). Crustacean Egg Production. Rotterdam 7: 69–76.

Crandall, K.A., D.J. Harris and J.W. Fetzner Jr. 2000. The monophyletic origin of freshwater crayfish estimated from nuclear and mitochondrial DNA sequences. Proceedings of the Royal Society of London 267: 1679–1686.

Cukerzis, J.M. 1988. *Astacus astacus* in Europe. pp. 309–340. *In*: D.M. Holdich and R.S. Lowery (eds.). Freshwater Crayfish Biology, Management and Exploitation. Croom Helm, London.

Culver, D.C. 1982. Cave Life: Evolution and Ecology. Harvard University Press, Cambridge, MA, U.S.A.

Dalosto, M.M., A.V. Palaoro and S. Santos. 2012. Mother-offspring relationship in the neotropical burrowing crayfish *Parastacus pilimanus* (von Martens, 1869) (Decapoda, Parastacidae). Crustaceana 85: 1305–1315.

Dennard, S., J.T. Peterson and E.S. Hawthorne. 2009. Life history and ecology of *Cambarus halli* (Hobbs). Southeast. Nat. 8: 479–494.

Deval, M.C., T. Bök, C. Ateş and Z. Tosunoğlu. 2007. Length based estimates of growth parameters, mortality rates, and recruitment of *Astacus leptodactylus* (Eschscholtz, 1823) (Decapoda, Astacidae) in unexploited inland waters of the northern Marmara region, European Turkey. Crustaceana 80: 655–665.

Diesel, R. 1991. Sperm competition and the evolution of mating behavior in Brachyura, with special reference to spider crabs (Decapoda, Majidae). pp. 145–163. *In*: R.T. Bauer and J.W. Martin (eds.). Crustacean Sexual Biology. Columbia University Press, New York.

Diesel, R., C.D. Schubart and M. Schuh. 2000. A reconstruction of the invasion of land by Jamaican crabs (Grapsidae: Sesarminae). J. Zool. 250: 141–160.

DiStefano, R.J., T.R. Black, S. Herleth-King, Y. Kanno and H.T. Mattingly. 2013. Life histories of two populations of the imperiled crayfish *Orconectes williamsi* (Decapoda: Cambaridae) in southwestern Missouri, U.S.A. J. Crust. Biol. 33: 15–24.

Drach, P. 1939. Mue et cycle d'intermue chez les Crustaces Decapodes. Ann. Inst. Oceanogr. Paris (N. S.) 19: 103–391.

Dudenhausen, E.E. and P. Talbot. 1983. An ultrastructural comparison of soft and hardened spermatophores from the crayfish *Pacifastacus leniusculus* Dana. Canadian J. Zool. 61: 182–194.

Easton, M.D.L and R.K. Misra. 1988. Mathematical representation of crustacean growth. ICES Journal of Marine Science 45: 61–72.

Evans, L.H. and B.F. Edgerton. 2002. Pathogens, parasites and commensals. pp. 377–438. *In*: D.M. Holdich (ed.). Biology of Freshwater Crayfish. Blackwell Science Ltd., London.

Eversole, A.G. and M.E. Welch. 2013. Ecology of the primary burrowing crayfish *Distocambaris crockeri*. J. Crust. Biol. 33: 660–666.

Farmer, A.S.D. 1974a. Reproduction in *Nephrops norvegicus* (Decapoda: Nephropidae). J. Zool. 174: 161–183.

Farmer, A.S.D. 1974b. The development of the external sexual characters of *Nephrops norvegicus* (L.) (Decapoda: Nephropidae). J. Nat. Hist. 8: 241–255.

Farmer, A.S.D. 1974c. Relative growth in *Nephrops norvegicus* (L.) (Decapoda: Nephropidae). J. Nat. Hist. 8: 605–620.

Figler, M.H., G.S. Blank and H.V.S. Peeke. 1997. Maternal aggression and post-hatch care in red swamp crayfish, *Procambarus clarkii* (Girard): the influences of presence of offspring, fostering, and maternal molting. Mar. Freshw. Behav. Phy. 30: 173–194.

Galeotti, P., F. Pupin, D. Rubolini, R. Sacchi, P. Nardi and M. Fasola. 2007. Effects of female mating status on copulation behaviour and sperm expenditure in the freshwater crayfish *Austropotamobius italicus*. Behav. Ecol. Sociobiol. 61: 711–718.

Galeotti, P., G. Bernini, L. Locatello, R. Sacchi, M. Fasola and D. Rubolini. 2012. Sperm traits negatively covary with size and asymmetry of a secondary sexual trait in a freshwater crayfish. PLoS ONE 7(8): 1–8.

Gherardi, F. 2002. Behaviour. pp. 258–290. *In*: D.M. Holdich (ed.). Biology of Freshwater Crayfish. Blackwell Science Ltd., London.

Gherardi, F. and L. Aquiloni. 2011. Sexual selection in crayfish: a review. Crust. Monogr. 15: 213–223.

Gherardi, F., P. Acquistapace and S. Barbaresi. 2000. The significance of chelae in the agonistic behaviour of the white-clawed crayfish, *Austropotamobius pallipes*. Mar. Freshw. Behav. Phy. 33: 187–200.

Grandjean, F., D. Bouchon and C. Souty-Grosset. 2000. Systematics of the European endangered crayfish species *Austropotamobius pallipes* (Decapoda: Astacidae). J. Crust. Biol. 20: 522–529.

Greenaway, P. 1985. Calcium balance and moulting in the Crustacea. Biol. Rev. 60: 425–454.

Guan, R.-Z. and P.R. Wiles. 1999. Growth and reproduction of the introduced crayfish *Pacifastacus leniusculus* in a British lowland river. Fish. Res. 42: 245–259.

Guiasu, R.C. and D.W. Dunham. 1998. Inter-form agonistic contests in male crayfishes, *Cambarus robustus* (Decapoda, Cambaridae). Invertebr. Biol. 117: 144–154.

Gutiérrez-Yurrita, P.J. and J.R. Latournerie-Cervera. 1999. Ecological features of Procambarus digueti and *Procambarus bouvieri* (Cambaridae), two endemic crayfish species of Mexico. Freshwater Crayfish 12: 605–619.

Gutiérrez-Yurrita, P.J. and C. Montes. 1999. Bioenergetics and phenology of reproduction of the introduced red swamp crayfish, *Procambarus clarkii*, in Doñana National Park, Spain, and implications for species management. Freshw. Biol. 42: 561–574.

Hagen, H.A. 1870. Illustrated catalogue of the Museum of comparative zoology at Harvard College. No. 3. Monograph of the North American Astacidae. Mem. Mus. Comp. Zool. 2: viii + 109 pp., 11 pls.

Hamr, P. 1992. Embryonic and postembryonic development in the Tasmanian freshwater crayfishes *Astacopsis gouldi, Astacopsis franklinii* and *Parastacoides tasmanicus tasmanicus* (Decapoda: Parastacidae). Aust. Mar. Freshwater Res. 43: 861–878.

Hamr, P. 2002. *Orconectes.* pp. 585–608. *In*: D.M. Holdich (ed.). Biology of Freshwater Crayfish. Blackwell Science, London.

Hamr, P. and A.M.M. Richardson. 1994. Life history of *Parastacoides tasmanicus tasmanicus* Clark, a burrowing freshwater crayfish from South-western Tasmania. Aust. Mar. Freshwater Res. 45: 455–470.

Harlioğlu, M.M. and T.Ç. Duran. 2010. The effect of darkness on mating and pleopodal egg production time in a freshwater crayfish, *Astacus leptodactylus* Eschscholtz. Aquac. Int. 18: 843–849.

Hartnoll, R.G. 1982. Growth. *In*: The biology of Crustacea. L. G. Abele ed, New York, Academic Press, Inc. 2: 111–196.

Hartnoll, R.G. 2001. Growth in Crustacea - twenty years on. Hydrobiologia 449: 111–122.

Hobbs, H.H., Jr. 1974. Synopsis of the families and genera of crayfishes (Crustacea: Decapoda). Smithson. Contr. Zool. 164: 1–32.

Hobbs, H.H., Jr. 1987. A review of the crayfish genus *Astacoides* (Decapoda: Parastacidae). Smithson. Contr. Zool. 443: 1–50.

Hobbs, H.H., Jr. 1988. Crayfish distribution, adaptive radiation and evolution. pp. 52–82. *In*: D.M. Holdich and R.S. Lowery (eds.). Freshwater Crayfish Biology, Management and Exploitation. Croom Helm, London.

Hobbs, H.H., Jr. 1989. An illustrated checklist of the American crayfishes (Decapoda: Astacidae, Cambaridae, and Parastacidae). Smithson. Contr. Zool. 480: 1–236.

Hobbs, H.H., Jr., M.C. Harvey and H.H. Hobbs III. 2007. A comparative study of functional morphology of the male reproductive systems in the Astacidea (Crustacea: Decapoda) with emphasis on the freshwater crayfishes. Smithson. Contr. Zool. 624: 1–76.

Holdich, D.M. 2002. Background and functional morphology. pp. 3–29. *In*: D.M. Holdich (ed.). Biology of Freshwater Crayfish. Blackwell Science, London.

Holdich, D.M. and I.D. Reeve. 1988. Functional morphology and anatomy. pp. 11–51. *In*: D.M. Holdich and R.S. Lowery (eds.). Freshwater Crayfish Biology, Management and Exploitation. Croom Helm, London.

Holdich, D.M. and J. Black. 2007. The spiny-cheek crayfish, *Orconectes limosus* (Rafinesque, 1817) [Crustacea: Decapoda: Cambaridae], digs into the UK. Aq. Inv. 2: 1–15.

Honan, J.A. and B.D. Mitchell. 1995. Reproduction of *Euastacus bispinosus* Clark (Decapoda: Parastacidae), and trends in the reproductive characteristics of freshwater crayfish. Mar. Freshw. Res. 46: 485–499.

Hopkins, C.L. 1967a. Breeding in the freshwater crayfish *Paranephrops planifrons* White. N.Z. J. Mar. Freshwat. Res. 1: 51–58.

Hopkins, C.L. 1967b. Growth rate in a population of the freshwater crayfish, *Paranephrops planifrons* White. N.Z. J. Mar. Freshwat. Res. 1: 464–474.

Horwitz, P.H.J. and A.M.M. Richardson. 1986. An ecological classification of the burrows of Australian freshwater crayfish. Aust. J. Mar. Freshwat. Res. 37: 237–242.

Hubenova, T., P. Vasileva and A. Zaikov. 2010. Fecundity of stone crayfish *Austropotamobius torrentium* from two different populations in Bulgaria. Bulg. J. Agric. Sci. 16: 387–393.

Huner, J.V. 1988. *Procambarus* in North America and elsewhere. pp. 239–261. *In*: D.M. Holdich and R.S. Lowery (eds.). Freshwater Crayfish Biology, Management and Exploitation. Croom Helm, London.

Huner, J.V. and R.P. Romaire. 1979. Size at maturity as a means of comparing populations of *Procambarus clarkii* (Girard) (Crustacea, Decapoda, Cambaridae) from different habitats. Freshw. Cray. 4: 53–64.

Huner, J.V. and J.E. Barr. 1984. Red Swamp Crayfish: Biology and Exploitation. Louisiana Sea Grant Program, Louisiana State University, Baton Rouge, Louisiana. (Revised edition).

Huner, J.V. and O.V. Lindquist. 1985. Exoskeleton mineralization in astacid and cambarid crayfishes (Decapoda, Crustacea). Comp. Biochem. Physiol. 80A: 515–521.

Huner, J.V. and O.V. Lindqvist. 1991. Special problems in freshwater crayfish egg production. *In*: A.M. Wenner and A.M. Kuris (eds.). Crustacean Egg Production. A. A. Balkema, Rotterdam, Crustacean Issues 7: 235–246.

Huryn, A.D. and J.B. Wallace. 1987. Production and litter processing by crayfish in an Appalachian mountain stream. J. Freshwat. Biol. 18: 277–286.

Huxley, T.H. 1881. The Crayfish: An Introduction to the Study of Zoology. C. Kegan Paul and Co., London.

Ibarra, M.A. and P.M. Arana. 2012. Biological parameters of the burrowing crayfish, *Parastacus pugnax* (Poeppig, 1835), in Tiuquilemu, Bio-Bio Region, Chile. Lat. Am. J. Aquat. Res. 40: 418–427.

Ingle, R.W. and W. Thomas. 1974. Mating and spawning of the crayfish *Austropotamobius pallipes* (Crustacea: Astacidae). J. Zool. 173: 525–538.

Johnston, C.E. and C. Figiel. 1997. Microhabitat parameters and life-history characteristics of *Fallicambarus gordoni* Fitzpatrick, a crayfish associated with pitcher-plant bogs in Southern Mississippi. J. Crust. Biol. 17: 687–691.

Jones, C.M. 1995. Production of juvenile redclaw crayfish, *Cherax quadricarinatus* (von Martens) (Decapoda, Parastacidae) I. Development of hatchery and nursery procedures. Aquaculture 138: 221–238.

Jones, D.R. and A.G. Eversole. 2011. Life history characteristics of the Elk River crayfish. J. Crust. Biol. 31: 647–652.

Jones, J.P.G. and T. Coulson. 2006. Population regulation and demography in a harvested freshwater crayfish from Madagascar. Oikos 112: 602–611.

Jones, J.P.G., F.B. Andriahajaina, N.J. Hockley, K.A. Crandall and O.R. Ravoahangimalala. 2007. The ecology and conservation status of Madagascar's endemic freshwater crayfish (Parastacidae; *Astacoides*). Freshwater Biol. 52: 1820–1833.

Kato, N. and T. Miyashita. 2003. Sexual difference in modes of selection on the pleopods of crayfish (Decapoda: Astacoidea) revealed by the allometry of developmentally homologous traits. Can. J. Zool. 81: 971–978.

Kawai, K. and G. Scholtz. 2002. Behavior of juveniles of the Japanese endemic species *Cambaroides japonicus* (Decapoda: Astacidea: Cambaridae), with observations on the position of spermatophore attachment on adult females. J. Crust. Biol. 22: 532–537.

Kawai, T. and K. Saito. 2001. Observations on the mating behavior and season, with no form alternation, of the Japanese crayfish, *Cambaroides japonicus* (Decapoda, Cambaridae), in Lake Komadome, Japan. J. Crust. Biol. 21: 885–890.

Kawai, T., T. Hamano and S. Matsuura. 1997. Survival and growth of the Japanese crayfish *Cambaroides japonicus* in a small stream in Hokkaido. Bul. Mar. Sci. 61: 147–157.

Kawai, T., V.S. Labay and L. Filipova. 2013. Taxonomic re-examination of *Cambaroides* (Decapoda: Cambaridae) with a re-description of *C. schenckii* from Sakhalin Island Russia and phylogenetic discussion of the Asian cambarids based on morphological characteristics. J. Crust. Biol. 33: 702–717.

Koksal, G. 1988. *Astacus leptodactylus* in Europe. pp. 365–400. *In*: D.M. Holdich and R.S. Lowery (eds.). Freshwater Crayfish Biology, Management and Exploitation. Croom Helm, London.

Larson, E.R. and D.D. Magoulick. 2008. Comparative life history of native (*Orconectes eupunctus*) and introduced (*Orconectes neglectus*) crayfishes in the spring river drainage of Arkansas and Missouri. Am. Midl. Nat. 160: 323–341.

Larson, E.R. and D.D. Magoulick. 2011. Life-history notes on *Cambarus hubbsi* Creaser (Hubbs Crayfish) from the South Fork Spring River, Arkansas. Southeast. Nat. 9: 121–132.

Little, E.E. 1975. Chemical communication in maternal crayfish behaviour of crayfish. Nature 255: 400–401.

Little, E.E. 1976. Ontogeny of maternal behavior and brood pheromone in crayfish. J. Comp. Physiol. A112: 133–142.

Lodge, D.M. and A.H. Hill. 1994. Factors governing species composition, population size, and productivity of cool water crayfishes. Nord. J. Freshw. Res. 69: 111–36.

López Greco, L.S. 2012. Functional anatomy of the reproductive system. pp. 413–450. *In*: L. Watling and M. Thiel (eds.). Functional Morphology and Diversity. Oxford University Press, New York.

López Greco, L.S. and F.L. Lo Nostro. 2008. Structural changes in the spermatophore of the freshwater 'red claw' crayfish *Cherax quadricarinatus* (Von Martens, 1898) (Decapoda, Parastacidae). Acta Zool. 89: 149–155.

López Greco, L.S., F. Vazquez and E.M. Rodriguez. 2007. Morphology of the male reproductive system and spermatophore formation in the freshwater 'red claw' crayfish *Cherax quadricarinatus* (Von Martens, 1898) (Decapoda, Parastacidae). Acta Zoolog. 88: 223–229.

Loughman, Z.J. 2010. Ecology of *Cambarus dubius* in north-central West Virginia. Southeast. Nat. 9(Spec. Iss. 3): 217–230.

Lowery, R.S. 1988. Growth, moulting and reproduction. pp. 83–113. *In*: D.M. Holdich and R.S. Lowery (eds.). Freshwater Crayfish Biology, Management and Exploitation. Croom Helm, London.

Lukhaup, C. and R. Pekny. 2008. Süßwasserkrebse aus aller Welt, 2. Auflage. Dähne Verlag, Ettlingen.

Martin, P. and G. Scholtz. 2012. A case of intersexuality in the parthenogenetic Marmorkrebs (Decapoda: Astacida: Cambaridae). J. Crust. Biol. 32: 345–350.

Martin, P., N.J. Dorn, T. Kawai, C. van der Heiden and G. Scholtz. 2010. The enigmatic Marmorkrebs (marbled crayfish) is the parthenogenetic form of *Procambarus fallax* (Hagen, 1870). Contrib. Zool. 79: 107–118.

Martinez, A.W. and E.H. Rudolph. 2011. Records of intersexuality in the burrowing crayfish, *Parastacus pugnax* (Poeppig, 1835) (Decapoda, Parastacidae) with comments on the sexuality of *Parastacus*. Crustaceana 84: 221–241.

Mason, J.C. 1970a. Maternal-offspring behavior of the crayfish, *Pacifastacus trowbridgi* (Stimpson). Am. Midl. Nat. 84: 463–473.

Mason, J.C. 1970b. Egg-laying in the western North American crayfish, *Pacifastacus trowbridgii* (Stimpson) (Decapoda, Astacidae). Crustaceana 19: 37–44.

Mason, J.C. 1970c. Copulatory behavior of the crayfish, *Pacifastacus trowbridgii* (Stimpson). Can. J. Zool. 48: 969–976.

Mathews, L.M. 2011. Mother–offspring recognition and kin-preferential behaviour in the crayfish *Orconectes limosus*. Behaviour 148: 71–87.

McLay, C.L. and L.S. López-Greco. 2011. A hypothesis about the origin of sperm storage in the Eubrachyura, the effects of seminal receptacle structure on mating strategies and the evolution of crab diversity: how did a race to be first become a race to be last? Zoolog. Anz. 250: 378–406.

Merrick, J.R. 1993. Freshwater Crayfishes of New South Wales. Linnean Society of New South Wales, Milson's Point, N.S.W., 127 pp.

Momot, W.T. 1984. Crayfish production: a reflection of community energetics. J. Crust. Biol. 4: 35–54.

Morgan, G.J. 1997. Freshwater crayfish of the genus *Euastacus* Clark (Decapoda: Parastacidae) from New South Wales, with a key to all species of the genus. Rec. Aust. Mus. 23: 1–110.

Muck, J.A., C.F. Rabeni and R.J. Distefano. 2002. Reproductive biology of the crayfish *Orconectes luteus* (Creaser) in a Missouri stream. Am. Midl. Nat. 147: 338–351.

Nakata, K. and S. Goshima. 2004. Fecundity of the Japanese crayfish, *Cambaroides japonicus*: ovary formation, egg number and egg size. Aquaculture 242: 335–343.

Nakata, K., A. Tanaka and S. Goshima. 2004. Reproduction of the alien crayfish species *Pacifasticus leniusculus* in Lake Shikaribetsu, Hokkaido, Japan. J. Crust. Biol. 24: 496–501.

Niksirat, H., A. Kouba and P. Kozak. 2014. Post-mating morphological changes in the spermatozoon and spermatophore wall of the crayfish *Astacus leptodactylus*: insight into a non-motile spermatozoon Anim. Reprod. Sci. 149: 325–334.

Noro, C.K., D. da Silva-Castiglioni, L. López-Greco, L. Buckup and G. Bond-Buckup. 2007. Morphology of the vasa deferentia of *Parastacus defossus* and *P. varicosus* and comparison within the Parastacidae. Nauplius 15: 43–48.

Noro, C.K., L.S. López-Greco and L. Buckup. 2008. Gonad morphology and type of sexuality in *Parastacus defossus* Faxon 1898, a burrowing, intersexed crayfish from southern Brazil (Decapoda: Parastacidae). Acta Zool. 89: 59–67.

Norrocky, M.J. 1991. Observations on the ecology, reproduction and growth of the burrowing crayfish *Fallicambarus (Creaserinus) fodiens* (Decapoda: Cambaridae) in North-central Ohio. Am. Midl. Nat. 125: 75–86.

Parkyn, S.M., K.J. Collier and B.J. Hicks. 2002. Growth and population dynamics of crayfish *Paranephrops planifrons* in streams within native forest and pastoral land uses. N.Z. J. Mar. Freshwat. Res. 36: 847–861.

Parnes, S., I. Khalia, G. Hulata and A. Sagi. 2003. Sex determination in crayfish: are intersex *Cherax quadricarinatus* (Decapoda, Parastacidae) genetically females? Genet. Res. 82: 107–116.

Pauly, D. and G. Gashutz. 1979. A simple method for fitting oscillating length growth data, with aprogram for pocket calculators. I.C.E.S.CM. 1979/G: 24 Demersal Fish Committee, 26p.

Pawlos, D., K. Formicki, A. Korzelecka-Orkisz and A. Winnicki. 2010. Hatching process in the signal crayfish, *Pacifastacus leniusculus* (Dana, 1852) (Decapoda, Astacidae). Crustaceana 83: 1167–1180.

Payne, J.F. 1972. The life history of *Procambarus hayi*. Am. Midl. Nat. 87: 25–35.

Price, J.O. and J.F. Payne. 1984. Postembryonic growth and development in the crayfish *Orconectes neglectus chaenodactylus* Williams, 1952 (Decapoda, Astacidea). Crustaceana 46: 176–194.

Reynolds, J.D. 2002. Growth and reproduction. pp. 152–191. *In*: D.M. Holdich (ed.). Biology of Freshwater Crayfish. Blackwell Science Ltd., London.

Reynolds, J.D., J.D. Celada, J.M. Carral and M.A. Matthews. 1992. Reproduction of astacid crayfish in captivity-current developments and implications for culture, with special reference to Ireland and Spain. Invertebr. Reprod. Dev. 22: 253–266.

Rhodes, C.P. and D.M. Holdich. 1979. On size and sexual dimorphism in *Austropotamobius pallipes* (Lereboullet): a step in assessing the commercial exploitation potential of the native British freshwater crayfish. Aquaculture 17: 345–358.

Rode, A.L. and L.E. Babcock. 2003. Phylogeny of fossil and extant freshwater crayfish and some closely related nephropid lobsters. Journal of Crustacean Biology 23: 418–435.

Rogowski, D.L., S. Sitko and S.A. Bonar. 2013. Optimising control of invasive crayfish using life-history information. Freshwat. Biol. 58: 1279–1291.

Rubolini, D., P. Galeotti, G. Ferrari, M. Spairani, F. Bernini and M. Fasola. 2006. Sperm allocation in relation to male traits, female size, and copulation behaviour in a freshwater crayfish species. Beh. Ecol. and Sociobiol. 60: 212–219.

Rubolini, D., P. Galeotti, F. Pupin, R. Sacchi, P.A. Nardi and M. Fasola. 2007. Repeated matings and sperm depletion in the freshwater crayfish *Austropotamobius italicus*. J. Freshwat. Biol. 52: 1898–1906.

Rudolph, E.H. 1995a. Partial protandric hermaphroditism in the burrowing crayfish *Parastacus nicoleti* (Philippi, 1882) (Decapoda: Parastacidae). J. Crust. Biol. 15: 720–732.

Rudolph, E.H. 1995b. A case of gynandromorphism in the freshwater crayfish *Samastacus spinifrons* (Philippi, 1882) (Decapoda, Parastacidae). Crustaceana 68: 705–711.

Rudolph, E.H. and C.S. Rojas. 2003. Embryonic and early postembryonic development of the burrowing crayfish, *Virilastacus araucanius* (Faxon, 1914) (Decapoda, Parastacidae) under laboratory conditions. Crustaceana 76: 835–850.

Rutherford, P.L., D.W. Dunham and V. Allison. 1995. Winning agonistic encounters by male crayfish *Orconectes rusticus* (Girard) (Decapoda, Cambaridae): chela size matters but chela symmetry does not. Crustaceana 68: 526–529.

Sainte-Marie, B. 2007. Sperm demand and allocation in decapod crustaceans. pp. 191–210. *In*: J.E. Duffy and M. Thiel (eds.). Evolutionary Ecology of Social and Sexual Systems. Oxford University Press, New York.

Sammy, N. 1988. Breeding biology of *Cherax quadricarinatus* in the Northern Territory. Proceedings of the 1st Australian Shellfish Conference, Curtin University of Technology, Curtin University of Technology: Perth, WA, Curtin University of Technology.

Scalici, M., A. Belluscio and G. Gibertini. 2008. Understanding population structure and dynamics in threatened crayfish. J. Zool. 275: 160–171.

Scholtz, G. 2002. Phylogeny and evolution. pp. 10–16. *In*: D.M. Holdich (ed.). Biology of Freshwater Crayfish. Blackwell Science, London.

Scholtz, G. and T. Kawai. 2002. Aspects of embryonic and postembryonic development of the Japanese freshwater crayfish *Cambaroides japonicus* (Crustacea, Decapoda) including a hypothesis on the evolution of maternal care in the Astacida. Acta Zoologica 83: 203–212.

Scholtz, G., A. Braband, L. Tolley, A. Reimann, B. Mittmann, C. Lukhaup, F. Steuerwald and G. Vogt. 2003. Parthenogenesis in an outsider crayfish. Nature 421: 806.

Scholtz, G. and S. Richter. 1995. Phylogenetic systematics of the reptantian Decapoda (Crustacea, Malacostraca). Zoological Journal of the Linnean Society 113: 289–328.

Sheehy, M.R.J. 1989. Crustacean brain lipofuscin: an examination of the morphological pigment in the fresh-water crayfish *Cherax cuspidatus* (Parastacidae). J. Crust. Biol. 9: 387–391.

Sheehy, M.R.J. 1992. Lipofuscin age-pigment accumulation in the brains of ageing field- and laboratory-reared crayfish *Cherax quadricarinatus* (von Martens) (Decapoda: Parastacidae). J. Exp. Mar. Biol. Ecol. 161: 79–89.

Simon, J.L. and P.A. Moore. 2007. Male–Female communication in the crayfish *Orconectes rusticus*: the use of urinary signals in reproductive and non-reproductive pairings. Ethology 113: 740–754.

Skurdal, J. and T. Taugbol. 2002. *Astacus*. pp. 467–510. *In*: D.M. Holdich (ed.). Biology of Freshwater Crayfish. Blackwell Science, Oxford.

Smart, G.C. 1962. The life history of the crayfish *Cambarus longulus longulus*. Am. Midl. Nat. 68: 83–94.

Smith, G.G. and A.J. Ritar. 2008. Reproduction and growth of decapod crustaceans in relation to aquaculture. pp. 457–490. *In*: E. Mente (ed.). Reproductive Biology of Crustaceans: Case Studies of Decapod Crustaceans. Science Publishers, Plymouth.

Smith, S.G. and E.S. Chang. 2007. Molting and growth. pp. 197–254. *In*: V.S. Kennedy and L.E. Cronin (eds.). The Blue Crab *Callinectes sapidus*. Maryland Sea Grant College, Maryland.

Snedden, W.A. 1990. Determinants of male mating success in the temperate crayfish *Orconectes rusticus*: chela size and sperm competition. Behaviour 115: 100–113.

Stebbing, P.D., M.G. Bentley and G.J. Watson. 2003. Mating behaviour and evidence for a female released courtship pheromone in the signal crayfish *Pacifastacus leniusculus*. J. Chem. Ecol. 29: 465–475.

Stein, R.A. 1976. Sexual dimorphism in crayfish chelae: functional significance linked to reproductive activities. Can. J. Zool. 54: 220–227.

Stein, R.A., M.L. Murphy and J.J. Magnusen. 1977. External morphological changes associated with sexual maturity in the crayfish (*Orconectes propinquus*). Am. Midl. Nat. 97: 495–501.

Stephens, G.J. 1952. Mechanisms regulating the reproductive cycle in the crayfish *Cambarus*. I. The female cycle. Phys. Zool. 25: 70–84.

Streever, W.J. 1996. Energy economy hypothesis and the troglobitic crayfish *Procambarus erythrops* in Sim's Sink Cave, Florida. Am. Midl. Nat. 135: 357–366.

Streiff, R., S. Mira, M. Castro and M.L. Cancela. 2004. Multiple paternity in Norway Lobster (*Nephrops norvegicus* L.) assessed with microsatellite markers. Mar. Biotech. 6: 60–66.

Streissel, F. and W. Hödl. 2002. Growth, morphometrics, size at maturity, sexual dimorphism and condition index of *Austropotamobius torrentium* Schrank. Hydrobiologia 477: 201–208.

Stucki, T. and E. Staub. 1999. Distribution of crayfish species and legislation concerning crayfish in Switzerland. Crayfish in Europe as alien species. How to make the best of a bad situation. Crustacean Issues 11: 141–147.

Suter, P.J. 1977. The biology of two species of *Engaeus* (Decapoda: Parastacidae) in Tasmania II. Life history and larval development, with particular reference to *E. cisternarius*. Aust. J. Mar. Freshwat. Res. 28: 85–93.

Taugbol, T. and J. Skurdal. 1990. Reproduction, molting and mortality of female noble crayfish, *Astacus astacus* (L. 1758), from five Norwegian populations subjected to indoor culture conditions (Decapoda, Astacoidea). Crustaceana 58: 113–123.

Taugbol, T., J. Skurdal, A. Burba, C. Munoz and M. Saez-Royuela. 1997. A test of crayfish predatory and nonpredatory fish species as bait in crayfish traps. Fish. Manage. Ecol. 4: 127–134.

Taylor, C.A. 2002. Taxonomy and conservation of native crayfish stocks. pp. 236–257. *In*: D.M. Holdich (ed.). Biology of Freshwater Crayfish. Blackwell Science, London.

Thiel, M. 2000. Extended parental care behavior in crustaceans—a comparative overview. Crustacean Issues 12: 211–226.

Tropea, C., Y. Piazza and L.S. López Greco. 2010. Effect of long-term exposure to high temperature on survival, growth and reproductive parameters of the 'redclaw' crayfish *Cherax quadricarinatus*. Aquaculture 302: 49–56.

Tuck, I.D., R.J.A. Atkinson and C.J. Chapman. 2000. Population biology of the Norway lobster, *Nephrops norvegicus* (L.) in the Firth of Clyde, Scotland II: fecundity and size at onset of sexual maturity. ICES J. Mar. Sci. 57: 1227–1239.

Tudge, C.C. 2003. Endemic and enigmatic: the reproductive biology of *Aegla* (Crustacea: Anomura: Aeglidae) with observations on sperm structure. Mem. Mus. Vict. 60: 63–70.

Unis, C. and M.B. Erkan. 2012. Morphology and development of the female reproductive system of *Astacus leptodactylus* (Eschscholtz, 1823) (Decapoda, Astacidae). Turk. J. Zool. 36: 775–784.

Venarsky, M.P., A.D. Huryn and J.P. Benstead. 2012. Re-examining extreme longevity of the cave crayfish *Orconectes australis* using new mark–recapture data: a lesson on the limitations of iterative size-at-age models. J. Freshwat. Biol. 57: 1471–1481.

Villanelli, F. and F. Gherardi. 1998. Breeding in the crayfish, *Austropotamobius pallipes*: mating patterns, mate choice and intermale competition. J. Freshwat. Biol. 40: 305–315.

Vogt, G. 2002. Functional anatomy. pp. 53–151. *In*: D.M. Holdich (ed.). Biology of Freshwater Crayfish. Blackwell Science Ltd., London.

Vogt, G. 2008a. Investigation of hatching and early post-embryonic life of freshwater crayfish by *in vitro* culture, behavioral analysis, and light and electron microscopy. J. Morphol. 269: 790–811.

Vogt, 2008b. The marbled crayfish: a new model organism for research on development, epigenetics and evolutionary biology. J. Zool. Lond. 276: 1–13.

Vogt, G. 2010. Suitability of the clonal marbled crayfish for biogerontological research: a review and perspective, with remarks on some further crustaceans. Biogerontology 11: 643–669.

Vogt, G. 2011. Marmorkrebs: Natural crayfish clone as emerging model for various biological disciplines. J. Bioscience 36: 377–382.

Vogt, G. 2012. Ageing and longevity in the Decapoda (Crustacea): a review. Zool. Anz. 251: 1–25.

Vogt, G. 2013. Abbreviation of larval development and extension of brood care as key features of the evolution of freshwater Decapoda. Biol. Rev. 88: 81–116.

Vogt, G. and L. Tolley. 2004. Brood care in freshwater crayfish and relationship with the offspring's sensory deficiencies. J. Morphol. 262: 566–582.

Vogt, G., L. Tolley and G. Scholtz. 2004. Life stages and reproductive components of the marmorkrebs (marbled crayfish), the first parthenogenetic decapod. J. Morph. 261: 286–311.

Vogt, G., M. Huber, M. Thiermann, G. van den Boogaart, O.J. Schmitz and C.D. Schubart. 2008. Production of different phenotypes from the same genotype in the same environment by developmental variation. J. Exp. Biol. 211: 510–523.

Waddy, S.L. and D.E. Aiken. 1991. Mating and insemination in the American Lobster, *Homarus americanus*. pp. 126–144. *In*: R.T. Bauer and J.W. Martin (eds.). Crustacean Sexual Biology. Columbia University Press, New York.

Waddy, S.L., N. Hamilton-Gibson and D.E. Aiken. 2013. Female American lobsters (*Homarus americanus*) do not delay their molt if they cannot find a mate. Inver. Reprod. Deve. 57: 101–104.

Walker, D., B.A. Porter and J. Avise. 2002. Genetic parentage assessment in the crayfish *Orconectes placidus*, a high-fecundity invertebrate with extended maternal brood care. Mol. Ecol. 11: 2115–2122.

Walls, J.G. 2009. Crawfishes of Louisiana. Florida State University Press, Baton Rouge, LA.

Weagle, K.V. and G.W. Ozburn. 1972. Observations on aspects of the life history of the crayfish, *Orconectes virilis* (Hagen), in northwestern Ontario. Can. J. Zool. 50: 366–370.

Weingartner, D.L. 1977. Production and trophic ecology of two crayfish species cohabiting an Indiana Cave, Ph.D. Thesis. Michigan State University, East Lansing, MI, U.S.A.

Wetzel, J.E. 2002. Form alternation of adult female crayfishes of the genus *Orconectes* (Decapoda: Cambaridae). Am. Midl. Nat. 147: 326–337.

Whitmore, N. and A.D. Huryn. 1999. Life history and production of *Paranephrops zealandicus* in a forest stream, with comments about the sustainable harvest of a freshwater crayfish. Freshwat. Biol. 42: 467–478.

Yue, G.H., G.L. Wang, B.Q. Zhu, C.M. Wang, Z.Y. Zhu and L.C. Lo. 2008. Discovery of four natural clones in a crayfish species *Procambarus clarkii*. Int. J. Biol. Sci. 4: 279–282.

Yue, G.H., J.L. Li, C.M. Wang, J.H. Xia, G.L. Wang and J.B. Feng. 2010. High prevalence of multiple paternity in the invasive crayfish species, *Procambarus clarkii*. Int. J. Biol. Sci. 6: 107–115.

Behavior of Crayfish

Ana M. Jurcak, Sara E. Lahman, Sarah J. Wofford and *Paul A. Moore**

Introduction

Freshwater crayfish have been used for ethological and behavioral studies for decades. They are ubiquitous polytrophic consumers within freshwater ecosystems that show high levels of aggression inter-and intraspecifically. Thus, they are well suited for field and laboratory studies as well as studies focused on social behavior, predator–prey interactions, and foraging behaviors. Their sensory systems have been well documented as has their use of chemical signals in a number of different behavioral situations. In addition, many different species are highly mobile (both from natural and anthropogenic means) and are considered an invasive species in a large number of aquatic basins.

Predator and prey interactions

Crayfish as prey

Crayfish act as prey to both aquatic and terrestrial predators (Englund and Krupa 2000). Aquatic predators include different fish species such as small mouth bass, *Micropterus dolomieui* (Stein and Magnuson 1976, Stein 1977), and brown trout, *Salmo trutta*, long-finned eel, *Anguilla dieffenbachii* (Shave et al. 1994), and snapping turtles, *Chrysemys picta* (Hazlett and Schoolmaster 1998). Terrestrial predators include mammals, such as raccoons, *Procyo lotor*, and wading birds, such as herons, *Butorides* and *Ardea* (Englund and Krupa 2000, Dekar and Magoulick 2013). Both types of predators have an effect on the behavior of crayfish.

When predators are present, crayfish will adjust their behavior, typically reducing their movement (Stein and Magnuson 1976, Stein 1977, Hazlett and Schoolmaster 1998, Hazlett 1999) or moving away from the predator (Englund and Krupa 2000). *Orconectes virilis* and *Orconectes rusticus* reduce their feeding and cleaning movements when exposed to the predator odor of snapping turtles, *Cheldrya serpentina* (Hazlett and Schoolmaster 1998). Both pond and stream *O. virilis* responded very similarly to the predator odor of snapping turtles (Hazlett 1999). *Paranephrops zealandicus* reduced both stationary and walking behavior as well as sought cover more when exposed to predator odors of long-finned eel and brown trout (Shave et al. 1994). Like many other species of crayfish, *Procambarus clarkii* will exhibit the anti-predator escape behavior of tail flipping away from predators (Herberholz et al. 2004).

Laboratory for Sensory Ecology, 217 Life Sciences Building, Bowling Green State University, Bowling Green, Ohio 43403.
* Corresponding author: pmoore@bgsu.edu

Size of crayfish will also play a role in the types of behaviors exhibited when a predator is present. Small or juvenile crayfish will respond differently than adult or large crayfish under predation pressure (Stein and Magnuson 1976, Englund and Krupa 2000). When small *Orconectes propinquus* were exposed to predator odors from smallmouth bass, their movements such as walking, climbing, feeding and grooming were drastically reduced compared to larger crayfish. While larger crayfish did show a reduction in their movements when a predator cue was present, crayfish increased behaviors where they would use their chela for defense (Stein and Magnuson 1976). Smaller *Cambarus bartonii* and *Orconectes putnami* would shift their habitat location from deep water to shallow water to avoid predation by green sunfish, *Lepomis cyanellus*, and creek chub, *Semotilus atromaculatus* (Englund and Krupa 2000).

Crayfish as predators

Crayfish serve a complex role in aquatic ecosystems and serve as prey items to various species as well as terrestrial organisms (Minkley and Craddock 1961, Hanson et al. 1990) but crayfish are also polytrophic omnivores and predators to some aquatic species (Guan and Wiles 1997). Crayfish, especially the more aggressive invasive species, have been known to affect a habitat by overexploitation. Predation by both male and female *Orconectes virilis* has been shown to significantly impact the abundance of other macroinvertebrates, notably snails (*Stagnicola elodes* and *Physa gyrina*) (Hanson et al. 1990). Likewise, *Orconectes rusticus* significantly reduce the amount of macrophytes and snails in experimental enclosures (Lodge et al. 1994). In addition to macroinvertebrates, crayfish also affect fish populations through consumption of small individuals and eggs. For example, *O. virilis* reduce egg abundance and affect reproductive success in two substrate-nesting sunfish (*Lepomis gibbosus* and *L. macrochirus*) (Dorn and Mittelbach 2004). Guan and Wiles (1997) found that an increase in the abundance of the crayfish species *Pacifastacus leniusculus* significantly increased mortality of the benthic fish species *Cottus gobio* and *Noemacheilus barbatulus* (Guan and Wiles 1997). However, crayfish also have non-consumptive predatory effects on other members of an aquatic ecosystem. For instance, crayfish (*O. rusticus*) evict johnny darters (*Etheostoma nigrum*) from shelters, increasing the darters' activity levels and subjecting them to a higher probability of predation by small-mouth bass (*Micropterus dolomieui*). Conversely, activity of the bass forced the darters into shelters which were occupied by the crayfish, subjecting the darters to potential predation by the crayfish (Rahel and Stein 1988). The presence of a crayfish predator was also shown to alter the behavior of the freshwater snail, *Physella gyrina*. When exposed to a predatory fish odor, snails sought shelter near the benthos in order to escape a fish predator. However, when snails were exposed to a crayfish predator odor, snails avoided benthic cover and escaped towards the surface of the water (Turner et al. 1999).

Shelters and burrows

Shelter use

Shelters are an important resource for crayfish as some species build burrows and others use natural substrates for protection. Shelters are used for protection against conspecifics (Gherardi 2002, Gherardi and Daniels 2004), extreme conditions, environmental changes (Gherardi 2002, Martin III and Moore 2007), and predators (Gherardi 2002, Gherardi and Daniels 2004, Martin III and Moore 2007), during vulnerable stages such as moulting (Jones and Ruscoe 2001, Gherardi 2002), as well as attracting potential mates (Bergman and Moore 2003, Gherardi and Daniels 2004). Non-burrowing crayfish use vegetation, gravel and large rocks for shelters (Jones and Ruscoe 2001). More fossorial species of crayfish will construct burrows for shelter use (Gherardi 2002, Dalosto et al. 2013, Palaoro et al. 2013).

Crayfish will compete with conspecifics and other species for the use of shelters. Crayfish species often engage in agonistic bouts with conspecifics over shelter usage (Bergman and Moore 2003, Gherardi and Daniels 2004, Martin III and Moore 2007, 2008). Small benthic fish species such as bullheads, *Cottus gobio*, and Johnny darters *Etheostoma nigrum*, will compete and are often excluded from shelters by crayfish (Griffiths et al. 2004). Rusty crayfish, *Orconectes rusticus*, were found to evict johnny darters, *Etheostoma nigrum*, using shelters for protection against small mouth bass, *Micropterus dolomieui* (Rahel and Stein 1988).

Burrowing crayfish

The type of burrows constructed and used by crayfish range from simple holes to complex burrows with multiple openings and many side chambers (Gherardi 2002, Palaoro et al. 2013, Helms et al. 2013, Barbaresi et al. 2004, Stoeckel et al. 2011). Two super families (Astacoidea and Parastacoidea) contain many different burrowing species of crayfish (Crandall and Buhay 2007, Noro and Buckup 2010). Parastacoidean crayfish are found in South America, Madagascar, Tasmania, New Zealand, New Guinea, and Australia in varying types of habitats. These habitats include lakes, streams, peat bogs, areas with clay soils, and sandy soils (Noro and Buckup 2010). Species in this super family include: *Parastacus brasilienis* and *Parastacus pilimanus* from Brazil (Dalosto et al. 2013), and *Virilastacus araucanius* from Chile, Southern Brazil, and Uruguay (Noro and Buckup 2010). Astacoidea are found in the Northern Hemisphere, such as the United States (Crandall and Buhay 2007). Species in Astacoidea include *Cambarus straitus* (Stoeckel et al. 2011) and *Cambarus harti*, Peidmont blue burrower (Helms et al. 2013), from the United States. Many resources, such as food, protection against predators, and protection from environmental extremes, are provided by these burrows (Gherardi 2002, Barbaresi et al. 2004, Dalosto et al. 2013, Palaoro et al. 2013). Species from the Parastacoidea family have been assumed to use chambers which include roots from vegetation as "feeding chambers" (Noro and Buckup 2010). Despite many studies on individual species of burrowing crayfish, much is still unknown about the biology and ecology of burrowing crayfish compared to more open water crayfish species (Helms et al. 2013).

Categorization of burrowing crayfish

Crayfish can be categorized into three different types of burrowers: primary, secondary, and tertiary burrowers (Welch and Eversole 2006, Stoeckel et al. 2011, Helms et al. 2013). Primary burrowers spend almost their entire lives in or near constructed burrows and seldom enter into open water (Stoeckel et al. 2011). These crayfish burrow in flood plains of agricultural, industrial, or residential areas that were previously forested areas (Loughman et al. 2013). Many primary burrowers have complex burrows including many tunnels, chambers, and surface openings (Welch et al. 2008, Stoeckel et al. 2011). The Camp Shelby burrowing crayfish, *Fallicambarus gordoni*, from the United States, constructs multiple branch burrows 1 meter below the surface with a vertical tunnel and a small terminal chamber at the end. These branches include backfilled tunnels which may be temporary based on the season. Some horizontal tunnels included roots which were clipped by the crayfish (Welch et al. 2008).

Secondary burrowers spend a majority of their lives inside and around their burrows but can be found in open water (Welch and Eversole 2006, Stoeckel et al. 2011). Typically, they inhabit lotic or lentic waters during the wet seasons. The burrows of secondary burrowers have less tunnels and a smaller number of chambers than the burrows of primary burrower species (Noro and Buckup 2010). Secondary burrows usually consist of a single sub vertical passageway (Gherardi 2002). *Procambarus clarkii*, the red swamp crayfish, is a secondary burrower which constructs burrows of simple design without multiple branches or multiple chambers (Gherardi 2002, Barbaresi et al. 2004).

Tertiary burrowers use burrows the least compared to primary and secondary burrowing crayfish. These crayfish inhabit open water and construct relatively simple burrows with fewer chambers and branches than those of primary and secondary burrowers (Welch and Eversole 2006, Stoeckel et al. 2011). Burrows are used seasonally, generally only for reproduction, as well as avoiding extreme temperatures which can cause freezing or desiccation (Gherardi 2002, Stoeckel et al. 2011). *Cherax destructor*, a tertiary burrower from Australia, burrows when conditions become unfavorable, such as a drop in water level and water temperature (Gherardi 2002). While burrowing crayfish in general are understudied, tertiary burrowers are more often studied compared to both primary and secondary burrowers due to sampling difficulties with the latter burrowing crayfish types (Loughman et al. 2013).

Burrow construction

Crayfish burrows are located in variety of habitats with a range of substrates including permeable and dense soils made of sand, mud, or clay (Grow 1981, Noro and Buckup 2010, Stoeckel et al. 2011, Helms et al. 2013). Burrows consist of a chimney-like structure made from mud pellets at the openings of the burrows (Grow 1981, Hobbs, Jr. and Whiteman 1991, Helms et al. 2013). The chimney leads to a vertical tunnel which can vary in depth. The amount of chambers or branches of a burrow will depend on the intricacy of the burrow design, with more complex burrows having many tunnels and chambers (Grow 1981, Welch et al. 2008, Noro and Buckup 2010, Stoeckel et al. 2011, Helms et al. 2013). The more intricate burrows may contain multiple surface openings, chambers containing water to prevent desiccation, tunnels with roots that may be used for food, or chambers housing juveniles (Horwitz and Richardson 1986). Seasonal water levels may have an effect on the design and construction of crayfish burrows (Stoeckel et al. 2011, Helms et al. 2013).

Crayfish use different motor patterns for constructing burrows (Grow 1981). Crayfish perform "pushing" actions using their chelae, third maxillipeds, and second pair of pereiopods to form a V-shaped wedge and push soil (Grow 1981). Carrying behavior occurs when soil is lifted by the chelae of the crayfish and transferred to its third maxillipeds. These appendages are used to carry the soil to the intended destination (Grow 1981). Finally, fanning, which is used in concert with pushing, occurs when crayfish use beating pleopods to move soil posteriorly behind the uropods and telson (Grow 1981).

Behavior differences between open water and burrowing crayfish

Burrowing crayfish and non-burrowing crayfish exhibit differences in behavioral patterns which are thought to be tied to a concept of territoriality. Non-burrowing crayfish have been shown to be more aggressive and live in a solitary fashion as compared to primary burrowers (Palaoro et al. 2013, Bergman and Moore 2003). In addition, aggressive differences were found within the different types of burrowing crayfish. For example, Dalosto et al. (2013) found that *P. brasiliensis* (primary burrower) was more aggressive than *P. pilimanus* (secondary burrower). Compared to true open water species, both burrowers were less aggressive. Open water crayfish showed more clearly escalated agonistic behaviors than both species in the study. Some explanation of the lower aggression level in burrowing crayfish may be attributed to the reduced need for acquiring resources for which open water crayfish need to compete (Dalosto et al. 2013).

Burrowing crayfish overlap in generations and coexist through burrow sharing with conspecifics (Palaoro et al. 2013). Many species of burrowing crayfish will share burrows such as *Fallicambarus fodiens* of north central Ohio, *Procambarus gracilis* of southeastern Wisconsin, and the genus *Engaeus* of Australia. *F. fodiens* will share burrows with males in both reproductive (Form I) and non-reproductive (Form II) forms, adult females, as well as the species *Cambarus diogenes.* An adult male, an adult female and many generations of their offspring of the Australian crayfish genus *Engaeus* will all inhabit the same burrow (Punzalan et al. 2001).

Agonism

Fighting behavior is found throughout the animal kingdom and is necessary to obtain resources vital to increase fitness through survivability and fecundity (Parker 1974, Smith 1974). Several taxa of decapod crustaceans, including various crayfish species, have been established as models for agonistic behavior. Consequently, crayfish have been utilized in experiments on eavesdropping (Aquiloni et al. 2008, Zulandt et al. 2008, Aquiloni and Gherardi 2010), winner and loser effects (Daws et al. 2002, Bergman et al. 2003, Hock and Huber 2006, Seebacher and Wilson 2007), and assessment strategies utilized during fighting behavior (Huber et al. 1997, Schroeder and Huber 2001). Bovbjerg (1953) called attention to a lack of invertebrate studies of dominance hierarchy establishment by collecting observational data on "tension contacts" and hierarchy establishment of *Orconectes virilis*. Consequently, these observations, along with other studies, brought the utility of crustaceans as models for agonism to the forefront (Bovbjerg 1956, 1959, 1970). The possession of weaponry in many species, the occupation of defendable home sites, and relatively small size and propensity for constant activity make crayfish an ideal candidate for answering questions concerning the mechanics and evolution of agonistic behavior (Dingle 1983).

How do crayfish fight?

Crayfish are multimodal in their abilities to gather information from the surrounding environment and potential opponents, but like other aquatic species, crayfish rely more heavily on chemosensory systems (Bergman and Moore 2005b, Horner et al. 2007, Callaghan et al. 2012). As discussed in other chapters (3, 5 and 7), crayfish utilize chemical stimuli to locate food sources (Moore and Grills 1999, Steele et al. 1999) and mates (Aquiloni et al. 2008, Durgin et al. 2008). However, they also use information in the form of chemical stimuli to locate conspecifics (Adams and Moore 2003, Schneider et al. 2008), and these stimuli have been shown to dictate fight dynamics (Breithaupt and Eger 2002, Acquistapace et al. 2003, Horner et al. 2007). These all-important chemical cues and signals are hypothesized to be utilized in individual recognition (Bergman et al. 2003, Schneider et al. 2008) for the purposes of establishing dominance hierarchies, eavesdropping on other fighting pairs (Zulandt et al. 2008) to gain information about potential future opponents, and to signal dominance or intent to retreat during a fight (Rubenstein and Hazlett 1974, Bergman et al. 2003). Crayfish with impaired or damaged chemosensory appendages are unable to elucidate status recognition with opponents. This lack of recognition leads to longer fights with higher probabilities of injuries (Bergman et al. 2003). While the exact mode of action that chemical signals are used within a fighting context are not yet well understood, studies have exhibited that the absence or manipulation of these chemical stimuli or chemical sensing organs significantly affect fight dynamics and outcome (Bergman and Moore 2005a, Cook and Moore 2008, Schneider et al. 2008, Callaghan et al. 2012). Horner et al. (2007) showed that selective ablation of chemosensory specific receptors (aesthetasc sensilla) significantly altered fight dynamics, causing longer and less productive fights in ablated crayfish (Horner et al. 2007). Blocking the release of urine (chemical cue) in dyadic interactions of *Orconectes rusticus* yielded similar results. Fights were longer and more intense when urine cues were absent versus when they were present (Schneider et al. 2008).

The most commonly studied morphological traits used as Resource Holding Potential (RHP) variables are weight, body size (Rabeni 1985, Edsman and Jonsson 1996) and chelae length (Garvey and Stein 1993, Gherardi et al. 2000). Typically crayfish 10% larger than a conspecific have an increased probability of winning contests (Figler et al. 1995a, Klar and Crowley 2012). Likewise, if body length or carapace length is similar, an individual with larger weaponry (chelae) will likely be victorious (Schroeder and Huber 2001). Chelae are heavily utilized in agonistic encounters for less intense behaviors such as closed chelae boxing or pushing as well as for more intense behaviors like grasping and grabbing at the opponents appendages (Bergman and Moore 2003). Consequently, chelae are frequently lost or damaged in agonistic behaviors. Studies of *Austropotamobius pallipes* showed that one-clawed individuals exhibited similar agonistic patterns as their two chelae counterparts. However, individuals missing a claw showed a lower motivation to fight (Gherardi et al. 2000), likely due to a strong RHP asymmetry (Smith and Parker 1976).

Well established fight ethograms have been utilized for several years to judge the intensity of agonistic behaviors (Bruski and Dunham 1987, Bergman et al. 2003). These behaviors range from a simple display known as the meral spread, in which the aggressor approaches the opponent with outspread chelae, to unrestrained fighting in which one or both individuals attempt to cause physical damage to the opponent using the chelae (Stocker and Huber 2001). However, even during more intense fighting bouts, individuals will periodically pause the fighting behavior and remain stationary. Rubenstein and Hazlett (1974) documented that during these pauses, *Orconectes virilis* exhibit motion of the maxillipeds, which generates a water current between the dyad, and the antennules, which potentially signals a retreat or is used for sampling (Rubenstein and Hazlett 1974). Fights are concluded when one of the opponents adopts a submissive or subordinate posture or behavior. In *Procambarus clarkii*, this submissive posture involves the subordinate individual holding its body flat against the substrate with chelae held forward (Huner and Barr 1984). Another submissive behavior is known as a tailflip and is characterized by the submissive individual propelling itself backwards over the substrate (Rubenstein and Hazlett 1974). These submissive behaviors are preceded or accompanied by the release of urine which is hypothesized to carry information about the internal state of the individual or the opponent and signal social status (Zulandt-Schneider and Moore 2000, Moore and Bergman 2005).

Why do crayfish fight?

Like other taxa, crayfish fight to obtain resources. Due to the reproductive cycling of both male and female crayfish (Martin III and Moore 2010, Stein 1976), fights for and with conspecifics for the purpose of mating is only prevalent throughout about half of the year (Villanelli and Gherardi 1998). Based on field observations and laboratory experiments crayfish seem to fight more often for shelter than food resources. A field study of both *O. rusticus* and *O. virilis* found that fights in the presence of shelters were longer and more intense than fights over food resources (Bergman and Moore 2003). However, certain preferred food resources may garner more intense fighting than others. For instance, a study of *Procambarus clarkii* stomach contents showed a higher preference towards animal prey items rather than detritus or macrophytes (Alcorlo et al. 2004). Consequently, macroinvertebrate prey or fish eggs may illicit increased occurrence of high intensity fights than a detrital mass.

Crayfish populations in the field are very dense and consequently individuals of that population are in constant contact (Huryn and Wallace 1987, Parkyn et al. 2002), allowing for increased probability of fighting behavior. Crayfish utilize fighting behavior and individual recognition to establish and maintain dominance hierarchies (Gherardi and Daniels 2003, Schneider et al. 2008). These established hierarchies likely reduce the amount of time needed to dedicate to fighting so that more time and energy can be used to obtain resources such as food, shelter and mates (Goessmann et al. 2000, Sato and Nagayama 2012). Shelters play a role in social dominance of crayfish and agonistic encounters with conspecifics (Bergman and Moore 2003, Gherardi and Daniels 2004, Fero et al. 2007, Martin III and Moore 2007). Fero et al. (2007) found that shelter usage correlates to dominance hierarchies in *O. rusticus*. When shelters are present, the intensity of fights between crayfish increases, more so than in the presence of food (Bergman and Moore 2003). Ownership of shelters also plays a role in the level of aggression in agonistic encounters with crayfish (Tricarico and Gherardi 2010).

Establishment and maintenance of dominance hierarchies

Dyadic interactions in crayfish result in a dominant and subordinate individual and the establishment of these relationships result in dominance hierarchies (Herberholz et al. 2001). These organized relationships, coupled with a capacity for individual recognition, allow for decreased fighting time so that more time can be spent on foraging, mating, or avoiding predators (Gherardi and Daniels 2003, Fero and Moore 2008, Sato and Nagayama 2012). While the benefits of holding the dominant position in a hierarchy can be beneficial, constant vigilance and aggressive behaviors are necessary to maintain this position (Gherardi and Daniels 2003, Fero et al. 2006). A study of *O. rusticus* hierarchy establishment in the presence of a shelter resource showed that dominant ranked crayfish actually spent significantly less time in the shelter than lower ranked individuals. Since the top-ranked individuals participated in the most agonistic interactions, the authors hypothesize that this decreased shelter usage was likely due to the motivation of the dominant ranked individual to reinforce status (Fero et al. 2006).

Intersexual fighting behavior

As mentioned previously, both male and female individuals go through reproductive cycling. Both male and female reproductive crayfish tend to be more aggressive than their non-reproductive counterparts due to physiological shifts in hormonal states (Martin III and Moore 2010) which facilitate higher levels of aggression. They found that both male and female *O. rusticus* were more likely to win a one-on-one fight than non-reproductive individuals (Martin III and Moore 2010). Moreover, in some crayfish species the males must fight and overcome the female for copulation to take place. Observations of *Pacifastacus trowbridgii* showed that a pre-copulatory chelae contact phase occurred (Mason 1970). Unfortunately, studies on female and intersexual fights are limited and have only been explored in a few crayfish species and behavioral contexts (Figler et al. 1995a, Martin III and Moore 2010), some of these studies only use female fights in the context of maternal aggression (Figler et al. 1995b). However, recent studies have shown that there is cause to think that males and females assess and decide differently in agonistic contests (Wofford et al. 2015).

Interspecific competition

Due to the propensity of crayfish species to become invasive outside of their native range, there are several studies on the interactions of various crayfish species (Gherardi and Daniels 2004, Chucholl et al. 2008, Gherardi and Acquistapace 2007, Hanshew and Garcia 2012). While the rules of engagement and the hypothesized mechanisms of fighting do not seem to differ between crayfish species, there are notable differences in aggression levels for certain species. Studies of a common invader, *O. rusticus*, have shown that these crayfish consistently outcompete *Orconectes limosus*, *O. virilis*, and *O. propinquus* in agonistic contests and in resource acquisition (Hill and Lodge 1999, Klocker and Strayer 2004). Heightened levels of aggression coupled with morphologies which favor larger claws make species such as *O. rusticus* and *P. clarkii* formidable invaders in several areas throughout the American Midwest and Italy (Garvey and Stein 1993, Figler et al. 1995a, Klocker and Strayer 2004, Gherardi and Acquistapace 2007, Schroeder and Huber 2001, Stocker and Huber 2001).

In addition to competition amongst other crayfish species, crayfish also compete with taxa found in the same trophic level or feeding guild. These competitors commonly include other benthic invertebrates (Charlebois and Lamberti 1996, Wilson et al. 2004) as well as some fish species (Carpenter 2005, Ilhéu et al. 2007). Interactions with these other taxa include instances of both interference and exploitative competition.

Assessment

While crayfish benefit from frequent fighting behavior, knowing when to stop or retreat from fights is equally important. Assessment strategies can be broadly categorized as self-assessment (i.e., energy reserves, fight capability, size), cumulative assessment (i.e., components of self-assessment in addition to the effects of opponent-inflicted injury), or mutual assessment (i.e., comparative energy reserve, size differential: Arnott and Elwood 2009). The assessment of an opponent or oneself to gain information and make decisions about how long to let a fight persist is a blossoming field (Hsu et al. 2008, Arnott and Elwood 2008, 2009). However, not many studies specifically focusing on assessment and decision making in contests have yet been performed for crayfish (Pavey and Fielder 1996, Schroeder and Huber 2001, Stocker and Huber 2001). Current hypotheses state that crayfish assessment is likely a type of mutual assessment which is based on differences in chelae and carapace size (Schroeder and Huber 2001) or internal state gleaned from urine cues (Schneider et al. 2008). Agonism and assessment alike are currently dominated by male centric behavioral studies in the crayfish model system (Rubenstein and Hazlett 1974, Panksepp and Huber 2001, Daws et al. 2002, Bergman et al. 2003, Schneider et al. 2008, Callaghan et al. 2012), and more female and mixed sex fight studies are needed to advance our knowledge.

Foraging behavior

Feeding strategies and composition of diet

Crayfish species have been described as opportunistic and generalist feeders, allowing these crustaceans to occupy polytrophic roles within an ecosystem (Momot et al. 1978, Creed 1994, Momot 1995, Dorn and Wojdak 2004). Given the freshwater habitats in which they reside, crayfish will eat a variety of organic matter as food items including plant and animal detritus (Morshiri and Goldman 1969, Mason 1975, Avault et al. 1983, Huner et al. 1988, Wiernicki 1984, Avault and Brunson 1990), snails and snail egg masses (Hofkin et al. 1991, 1992), benthic macroinvertebrates with reduced escape responses (Odonata, Ephemeroptera larvae and planorbid snails (Rickett 1974, Matthews et al. 1993, Ilhéu and Bernardo 1993a)), macrophytes, small invertebrates, fish, and fish eggs (Prins 1968, Dean 1969, Rickett 1974, Mason 1975,

Covich 1977, Abrahamsson 1996). Gut analysis and subsequent behavioral studies have confirmed that crayfish are primarily omnivorous detrivores and prefer herbivory over carnivory, but will readily switch strategies based on availability of food (Schoener 1971, Ilhéu and Bernardo 1993b, Smart et al. 2002). Adult crayfish are mainly herbivorous (Mason 1975, Olsen et al. 1991, Ilhéu and Bernardo 1993a) while juveniles tend to be carnivorous or omnivorous. Additionally, cannibalism is common among most species of crayfish (Lorman and Magnuson 1978, Ilhéu and Bernardo 1993a).

Analysis of the stomach contents of *Procambarus clarkii* in Egypt revealed that the dietary composition of individual animals varied depending on the size of the crayfish (Habashy et al. 2011). Crayfish in the early stages of life (0.8 to 1.7 cm carapace lengths) primarily consumed phytoplankton and zooplankton, diatoms and molluscs, fish scales, insect parts, and fishes. Crayfish of medium size (carapace lengths measuring 1.8 to 5.2 cm) preferred to feed upon diatoms, fish scales, and molluscs along with smaller percentages of zooplankton and fish. Large sized crayfish (carapace lengths measuring between 5.3 and 7.3 cm) were shown to primarily consume macrophytes, diatoms, and molluscs. In the larger size group, fishes and plankton were shown to be moderately consumed, while insect parts were very rare. This particular study confirmed that *P. clarkii* are able to consume a variety of food items and that dietary preference shifts at different stages of life.

Response to food

The omnivorous nature of crayfish feeding patterns is illustrated by the response of sensory organs to various amino acids and carbohydrates commonly found in plants, other crayfish species, insect hemolymph and larvae, and other macroinvertebrates. Feeding behavior can be elicited by single compounds and specific mixtures of compounds that are present in prey organisms that may function as signals for food detection (McLeese 1970, Shelton and Mackie 1971, Mackie 1973, Carr et al. 1984, Carr and Derby 1986). Tierney and Atema (1988) demonstrated that *O. virilis* demonstrated feeding movements when stimulated with L-isoleucine, glycine, hydroxy-L-proline, L-glutamate, L-valine, and B-alanine, while *O. rusticus* responded to cellobiose, sucrose, glycine, maltose, glycogen, nicotinic acid methyl ester, putrescine, and L-glutamate. Hazlett (1994) demonstrated that three different species of crayfish (*O. rusticus, O. virilis,* and *Cambarus robustus*) required previous experience with a food odor in order to exhibit a feeding response. Stimulation of the chemosensory system in response to food odors results in a typical pattern of behavioral movement. For example, in lobsters, the highest frequency of antennular flicking is exhibited at the onset of chemical stimulation (Schmitt and Ache 1979, Reeder and Ache 1980, Devine and Atema 1982). Once a food source is detected, decapod crustaceans commence walking upstream, sampling the temporal and spatial distribution through antennular flicking. The fluctuating distribution of an odor signal has been hypothesized to dictate the magnitude and direction of turning and heading angles as well as the velocity of the moving crustacean (Moore and Atema 1988, Moore and Grills 1999, Keller et al. 2001, Grasso and Basil 2002). However, the exact characteristics of the odor signal that drive changes in orientation to food sources remain unknown.

Alterations to feeding responses

Much research has demonstrated that changes to surrounding water chemistry will impact the feeding behaviors of crayfish. Allison et al. (1992) and Uiska et al. (1994) demonstrated that under acidic conditions *C. bartoni* exhibited low response to food odors through slower locomotory behavior toward food and decreased antennular flicking. Elevated levels of copper increased latency of food response and overall success in locating food in *C. bartoni* (Sherba et al. 2000).

In addition to changes to water chemistry due to anthropogenic chemicals, the presence of alarm cues of conspecifics and odors from a predator can alter the foraging behavior of crayfish. Rusty crayfish *O. rusticus* were shown to have a decreased behavioral response to food odors in the presence of an injured conspecific odor (Pecor and Hazlett 2006). Both *O. virilis* and *O. rusticus* performed less active feeding behaviors, including scraping the substrate with sensory appendages, locomotion toward an odor source, and a raised body position, when exposed to both a food odor and predator odor (Acquistapace et al. 2003).

Stronger responses to food odors by *O. virilis* were shown to be influenced by differences in hydrodynamics (flow), however, crayfish were less responsive to food odors as compared to responses to alarm odors in flowing environments (Hazlett et al. 2006). Research has recently demonstrated that infections by the trematode parasite *Microphallus* also reduced foraging behavior in *O. rusticus* (Sargent et al. 2014).

Feeding patterns

Foraging excursions of a threatened species of crayfish *Austropotamobius pallipes* typically do not exceed one hour during the non-reproductive phase in the summer (Schoener 1971). During these excursions, feeding crayfish have efficient pathways and do not spend extraneous amounts of time and energy on other activities while feeding. However, *A. pallipes* moved at a decreased speed and did not cover as broad an area as other decapods have been shown to cover (Gherardi et al. 1989).

Higher frequency of feeding activity of *O. rusticus* was observed when overall crayfish activity was higher (i.e., when competing with of the more active *O. propinquus*). Pintor and Sih (2009) suggested that *O. rusticus* increases foraging when in a group with a higher number of active individuals, thereby proportionally reducing an individual's risk of predation (i.e., safety in numbers). An increase in active foraging has strong implications for the composition and structure of littoral zones. Results of a study completed by Chambers et al. (1990) showed that the addition of the crayfish *O. virilis* significantly affected the biomass, density, and/or shoot morphology of four macrophyte species (*Potamogeton richardsonii, Myriophyllum exalbescens, Nuphar variegatum* and *Sparganium eurycarpum*), indicating that even relatively low densities of crayfish could impact the growth of submersed aquatic plants. High densities of foraging *O. virilis* will also significantly affect the abundance and distribution of macroinvertebrates (Hanson et al. 1990). The increased presence of *Pacifastacus leniusculus* in lowland British rivers was shown to be significantly correlated with decreased abundance of benthic fishes due to predation (Guan and Wiles 1997).

Reproductive behavior

Mate choice

Until recently, the prevailing thought within the crayfish community presumed that mate choice was mainly performed by the females per classical differences in gamete production (Trivers 1972). Most studies to date have found some aspect of female selection in regard to mate choice (Villanelli and Gherardi 1998, Gherardi et al. 2006, Aquiloni and Gherardi 2008), but selection appears to be based on male size for females in the species *Austropotamobius italicus* (Woodlock and Reynolds 1988, Gherardi et al. 2006). Number of chelae is also a factor in female mate selection (Villanelli and Gherardi 1998). Prior to mate choice and copulation, mating interactions take the form of agonistic encounters, but switch from an aggressive interaction to a mating interaction presumably through the use of chemical signals (Acquistapace et al. 2002, Martin III and Moore 2010). Recently, work has shown that male selection also plays a role in mate choice (Aquiloni and Gherardi 2008). Females have also demonstrated the ability for multiple matings within a single mating season.

Male *Procambarus clarkii* select unmated females over mated females and most likely detect the reproductive state of females through the use of chemical signals (Aquiloni and Gherardi 2008). There is evidence of sperm competition in crayfish and the detection of the mating status of females could allow male crayfish an increased chance of reproductive success by mating with virgin females or by mating first with females (Sneddon 1990). After mating, males insert a mating plug to reduce the future success of mating encounters with other males (Crocker and Barr 1968). Yet, *O. rusticus* males may use the copulatory stylets to remove the mating plug deposited by a previous male (Berrill and Arsenault 1984). Male crayfish can also modulate the size of the sperm packet as well as the number of sperm within that packet that is transferred to the female crayfish. Larger sperm packets are transferred to larger females which should also increase the percentage of offspring fertilized by the copulating male (Rubolini et al. 2006).

The use of multimodal communication during mating interactions is well documented, although chemical signals appear to play a predominate role in sex and reproductive status recognition (Acquistapace et al. 2003, Martin III and Moore 2010). Although a sex pheromone has not been isolated to date, Ameyaw-Akumfi and Hazlett (1976) has hinted at a pheromonal based recognition system for sex and reproductive status. Given the role of chemical signals in other socially motivated behavior, it is highly likely that some form of a chemically-based recognition exists within crayfish.

Acknowledgements

The authors would like to acknowledge the important role that Dr. Francesca Gherardi has played in elucidating the many different behaviors of the crayfish. Dr. Gherardi has been a leader in the field for decades and her multidisciplinary work has served as a guiding light to the field. Her contributions will be greatly missed.

References

Abrahamsson, S.A.A. 1996. Dynamics of an isolated population of the crayfish *Astacus astacus*. Oikos 17: 96–107.

Acquistapace, P., L. Aquiloni, B.A. Hazlett and F. Gherardi. 2002. Multimodal communication in crayfish: sex recognition during mate search by male *Austropotamobius pallipes*. Can. J. Zool. 80: 2041–2045.

Acquistapace, P., B.A. Hazlett and F. Gherardi. 2003. Unsuccessful predation and learning of predator cues by crayfish. J. Crust. Biol. 23: 364–370.

Adams, J.A. and P.A. Moore. 2003. Discrimination of conspecific male molt odor signals by male crayfish, *Orconectes rusticus*. J. Crust. Biol. 23: 7–14.

Alcorlo, P., W. Geiger and M. Otero. 2004. Feeding preferences and food selection of the red swamp crayfish, *Procambarus clarkii*, in habitats differing in food item diversity. Crustaceana 77: 435–453.

Allison, V., D.W. Dunham and H.H. Harvey. 1992. Low pH alters response to food in the crayfish *Cambarus bartoni*. Can. J. Zool. 70: 2416–2420.

Ameyaw-Akumfi, C. and B.A. Hazlett. 1976. Sex recognition in the crayfish, *Procambarus clarkii*. Sci. 190: 1225–1226.

Aquiloni, L. and F. Gherardi. 2008. Mutual mate choice in crayfish: large body size is selected by both sexes, virginity by males only. J. Zool. 274: 171–179.

Aquiloni, L. and F. Gherardi. 2010. Crayfish females eavesdrop on fighting males and use smell and sight to recognize the identity of the winner. Anim. Behav. 79: 265–269.

Aquiloni, L., M. Buric and F. Gherardi. 2008. Crayfish females eavesdrop on fighting males before choosing the dominant mate. Curr. Biol. 18: R462–R463.

Arnott, G. and R.W. Elwood. 2008. Information gathering and decision making about resource value in animal contests. Anim. Behav. 76: 529–542.

Arnott, G. and R.W. Elwood. 2009. Assessment of fighting ability in animal contests. Anim. Behav. 77: 991–1004.

Avault, J.W. and M.W. Brunson. 1990. Crawfish forage and feeding systems. Rev. Aquat. Sci. 3: 1–10.

Avault, J.W., R.P. Romaire and M.R. Miltner. 1983. Feeds and forages for red swamp crawfish, *Procambarus clarkii*: 15 years research at Louisiana State University reviewed. Freshwater Crayfish 3: 62–369.

Barbaresi, S., E. Tricarico and F. Gherardi. 2004. Factors inducing the intense burrowing activity of the red-swamp crayfish, *Procambarus clarkii*, an invasive species. Naturwissenschaften 91. 342–345.

Bergman, D.A. and P.A. Moore. 2003. Field observations of intraspecific agonistic behavior of two crayfish species, *Orconectes rusticus* and *Orconectes virilis*, in different habitats. Biol. Bull. 205: 26–35.

Bergman, D.A. and P.A. Moore. 2005a. Prolonged exposure to social odours alters subsequent social interactions in crayfish (*Orconectes rusticus*). Anim. Behav. 70: 311–318.

Bergman, D.A. and P.A. Moore. 2005b. The role of chemical signals in the social behavior of crayfish. Chem. Senses 30: 305–306.

Bergman, D.A., C.P. Kozlowski, J.C. McIntyre, R. Huber, A.G. Daws and P.A. Moore. 2003. Temporal dynamics and communication of winner-effects in the crayfish, *Orconectes rusticus*. Behaviour 140: 805–825.

Berrill, M. and M. Arsenault. 1984. The breeding behaviour of a northern temperate orconectid crayfish, *Orconectes rusticus*. Anim. Behav. 32: 333–339.

Bovbjerg, R. 1953. Dominance order in the crayfish *Orconectes rusticus* (Hagen). Physiol. Zool. 26: 173–178.

Bovbjerg, R. 1956. Some factors affecting aggressive behavior in crayfish. Physiol. Zool. 29: 127–136.

Bovbjerg, R. 1959. Density and dispersal in laboratory crayfish populations. Ecology 40: 504–506.

Bovbjerg, R. 1970. Ecological isolation and competitive exclusion in two crayfish (*Orconectes virilis* and *Orconectes immunis*). Ecology 51: 225–236.

Breithaupt, T. and P. Eger. 2002. Urine makes the difference: chemical communication in fighting crayfish made visible. J. Exp. Biol. 205: 1221–1231.

Bruski, C.A. and D.W. Dunham. 1987. The importance of vision in agonistic communication of the crayfish *Orconectes rusticus*. Behaviour 103: 83–107.

Callaghan, D., C.D. Weisbord, W.A. Dew and G.G. Pyle. 2012. The role of various sensory inputs in establishing social hierarchies in crayfish. Behaviour 149(13-14): 1443–1458.

Carpenter, J. 2005. Competition for food between an introduced crayfish and two fishes endemic to the Colorado River basin. Env. Biol. Fish. 72: 335–342.

Carr, W.E.S. and C.D. Derby. 1986. Behavioral chemoattractants for the shrimp, *Palaemontes pugio*: identification of active components in food extracts and evidence of synergistic mixture interactions. Chem. Senses 11: 49–64.

Carr, W.E.S., J.C. Netherton III and M.L. Milstead. 1984. Chemoattractants of the shrimp, *Palaemonetes pugio*: variability in responsiveness and the stimulatory capacity of mixtures containing amino acids, quaternary ammonium compounds, purines and other substances. Comp. Biochem. Physiol. 77A: 469–474.

Chambers, P.A., J. Hanson, J.M. Burke and E.E. Prepas. 1990. The impact of the crayfish *Orconectes virilis* on aquatic macrophytes. Freshwater Biol. 24: 81–91.

Charlebois, P.M. and G.A. Lamberti. 1996. Invading crayfish in a Michigan stream: direct and indirect effects on periphyton and macroinvertebrates. J. N. Am. Benthol. Soc. 15: 551–563.

Chucholl, C., H.B. Stich and G. Maier. 2008. Aggressive interactions and competition for shelter between a recently introduced and an established invasive crayfish: *Orconectes immunis* vs. *O. limosus*. Fund. Appl. Limnol. 172: 27–36.

Cook, M.E. and P.A. Moore. 2008. The effects of the herbicide metolachlor on agonistic behavior in the crayfish, *Orconectes rusticus*. Arch. Environ. Con. Tox. 55: 94–102.

Covich, A.P. 1977. How do crayfish respond to plants and Mollusca as alternate food resources? Freshwater Crayfish 3: 165–179.

Crandall, K.A. and J.E. Buhay. 2007. Global diversity of crayfish (Astacidae, Cambaridae, and Parastacidae-Decapoda) in freshwater. Hydrobiologia 595: 295–301.

Creed, R.P. 1994. Direct and indirect effects of crayfish grazing in a stream community. Ecology 75: 2091–2103.

Crocker, D.W. and D.W. Barr. 1968. Handbook of the Crayfishes of Ontario. University of Toronto Press, Toronto.

Dalosto, M.M., A.V. Palaoro, J.R. Costa and S. Santo. 2013. Aggressiveness and life underground: the case of burrowing crayfish. Behavior 150: 3–22.

Daws, A.G., J. Grills, K. Konzen and P.A. Moore. 2002. Previous experiences alter the outcome of aggressive interactions between males in the crayfish, *Procambarus clarkii*. Mar. Freshw. Behav. Phys. 35: 139–148.

Dean, J.L. 1969. Biology of the crayfish *Orconectes causeyi* and its use for control of aquatic weeds in trout lakes. Technical Papers, United States Bureau of Sport Fisheries and Wildlife 24: 1–15.

Dekar, M.P. and D.D. Magoulick. 2013. Effect of predators on fish and crayfish survival in intermittent streams. Southeast. Nat. 12: 197–208.

Devine, D.V. and J. Atema. 1982. Function of chemoreceptor organs in spatial orientation of the lobster, *Homarus americanus*: differences and overlap. Biol. Bull. 163: 144–153.

Dingle, H. 1983. Strategies of agonistic behavior in crustacea. pp. 85–111. *In*: S. Rebach and D.W. Dunham (eds.). Studies in Adaptation: The Behavior of Higher Crustacea. John Wiley and Sons, New York.

Dorn, N.J. and G.G. Mittelbach. 2004. Effects of a native crayfish (*Orconectes virilis*) on the reproductive success and nesting behavior of sunfish (*Lepomis* spp.). Can. J. Fish. Aquat. Sci. 61: 2135–2143.

Dorn, N.J. and J.M. Wojdak. 2004. The role of omnivorous crayfish in littoral communities. Oecologia 140: 150–159.

Durgin, W.S., K.E. Martin, H.R. Watkins and L.M. Mathews. 2008. Distance communication of sexual status in the crayfish *Orconectes quinebaugensis*: female sexual history mediates male and female behavior. J. Chem. Ecol. 34: 702–707.

Edsman, L. and A. Jonsson. 1996. The effect of size, antennal injury, ownership, and ownership duration on fighting success in male signal crayfish, *Pacifastacus leniusculus* (Dana). Nord. J. Freshw. Res. 72: 80–87.

Englund, G. and J.J. Krupa. 2000. Habitat use by crayfish in stream pools: influence of predators, depth, and body size. Freshwater Biol. 43: 75–83.

Fero, K. and P.A. Moore. 2008. Social spacing of crayfish in natural habitats: what role does dominance play? Behav. Ecol. Sociobiol. 62: 1119–1125.

Fero, K., J.L. Simon, V. Jourdie and P.A. Moore. 2006. Consequences of social dominance on crayfish resource use. Behaviour 144: 61–82.

Fero, K., J.L. Simon and P.A. Moore. 2007. Consequences of social dominance on crayfish use. Behaviour 144: 61–82.

Figler, M.H., M. Twum, J.E. Finkelstein and H.V.S. Peeke. 1995a. Intruding male red swamp crayfish, *Procambarus clarkii*, immediately dominate members of established communities of smaller, mixed-sex conspecifics. Aggressive Behav. 21: 225–236.

Figler, M.H., M. Twum, J.E. Finkelstein and H.V.S. Peeke. 1995b. Maternal aggression in red swamp crayfish (*Procambarus clarkii*, Girard): The relation between reproductive status and outcome of aggressive encounters with male and female conspecifics. Behaviour 132: 107–125.

Garvey, J.E. and R.A. Stein. 1993. Evaluating how chela size influences the invasion potential of an introduced crayfish (*Orconectes rusticus*). Am. Midl. Nat. 129: 172–181.

Gherardi, F. 2002. Behaviour. pp. 258–290. *In*: D.M. Holdich (ed.). Biology of Freshwater Crayfish. Blackwell Science Ltd., Oxford.

Gherardi, F. and W.H. Daniels. 2003. Dominance hierarchies and status recognition in the crayfish *Procambarus acutus acutus*. Can. J. Zool. 81: 1269–1281.

Gherardi, F. and W.H. Daniels. 2004. Agonism and shelter competition between invasive and indigenous crayfish species. Can. J. Zool. 82: 1923–1932.

Gherardi, F. and P. Acquistapace. 2007. Invasive crayfish in Europe: the impact of *Procambarus clarkii* on the littoral community of a Mediterranean lake. Freshwater Biol. 52: 1249–1259.

Gherardi, F., F. Tarducci and F. Micheli. 1989. Energy maximization and foraging strategies in *Potamon fluviatile* (Decapoda, Brachyura). Freshwater Biol. 22: 233–245.

Gherardi, F., P. Acquistapace and S. Barbaresi. 2000. The significance of chelae in the agonistic behaviour of the white-clawed crayfish, *Austropotamobius pallipes*. Mar. Freshw. Behav. Phy. 33: 187–200.

Gherardi, F., B. Renai, P. Galeotti and D. Rubolini. 2006. Nonrandom mating, mate choice, and male-male competition in the crayfish *Austropotamobius italicus*, a threatened species. Arch. Hydrobiol. 165: 557–576.

Goessmann, C., C. Hemelrijk and R. Huber. 2000. The formation and maintenance of crayfish hierarchies: behavioral and self-structuring properties. Behav. Ecol. Sociobiol. 48: 418–428.

Grasso, F.W. and J.A. Basil. 2002. How lobsters, crayfishes, and crabs locate sources of odor: current perspectives and future directions. Curr. Opin. Neurobiol. 12: 721–727.

Griffiths, S.W., P. Collen and J.D. Armstrong. 2004. Competition for shelter among over-wintering signal crayfish and juvenile Atlantic salmon. J. Fish Biol. 65: 436–447.

Grow, L. 1981. Burrowing behaviour in the crayfish, *Cambarus diogenes diogenes* (Girard). Anim. Behav. 29: 351–356.

Guan, R.Z. and P.R. Wiles. 1997. Ecological impact of introduced crayfish on benthic fishes in a British lowland river. Conserv. Biol. 11: 641–647.

Habashy, M.M., M.A. El-Kasheif and S.A. Ibrahim. 2011. Studies on the feeding behavior of the exotic freshwater crayfish *Procambarus clarkii*, with references on their distribution in Egypt. pp. 173–187. *In*: Proceedings of the 4th Global Fisheries and Aquaculture Research Conference, the Egyptian International Center for Agriculture, Giza, Egypt, 3–5 October 2011. Massive Conferences and Trade Fairs.

Hanshew, B.A. and T.S. Garcia. 2012. Invasion of the shelter snatchers: behavioural plasticity in invasive red swamp crayfish, *Procambarus clarkii*. Freshwater Biol. 57: 2285–2296.

Hanson, J., P.A. Chambers and E.E. Prepas. 1990. Selective foraging by the crayfish *Orconectes virilis* and its impact on macroinvertebrates. Freshwater Biol. 24: 69–80.

Hazlett, B.A. 1994. Crayfish feeding responses to zebra mussels depend on microorganisms and learning. J. Chem. Ecol. 20: 2623–2630.

Hazlett, B.A. 1999. Responses to multiple chemical cues by the crayfish *Orconectes virilis*. Behaviour 136: 161–177.

Hazlett, B.A. and D.R. Schoolmaster. 1998. Responses of cambarid crayfish to predator odor. J. Chem. Ecol. 24: 1757–1769.

Hazlett, B.A., P. Acquistapace and F. Gherardi. 2006. Responses of the crayfish *Orconectes virilis* to chemical cues depend upon flow conditions. J. Crust. Biol. 26: 94–98.

Helms, B.A., C. Figiel, J. Rivera, J. Stoeckel, G. Stanton and T.A. Keller. 2013. Life-history observations, environmental associations, and soil preferences of the Piedmont Blue Burrower (*Cambarus [Depressicambarus] harti*) Hobbs. Southeast. Nat. 12: 134–160.

Herberholz, J., F.A. Issa and D.H. Edwards. 2001. Patterns of neural circuit activation and behavior during dominance hierarchy formation in freely behaving crayfish. J. Neurosci. 21: 2759–67.

Herberholz, J., M.M. Sen and D.H. Edwards. 2004. Escape behavior and escape circuit activation in juvenile crayfish during prey-predator interactions. J. Exp. Biol. 207: 1855–1863.

Hill, A.M. and D.M. Lodge. 1999. Replacement of resident crayfishes by an exotic crayfish: The roles of competition and predation. Ecol. Appl. 9: 678–690.

Hobbs, H.H., Jr. and M. Whiteman. 1991. Notes on the burrows, behavior, and color of the crayfish *Fallicambarus (F.). devastator* (Decapoda: Cambaridae). Southwest. Nat. 36: 127–135.

Hock, K. and R. Huber. 2006. Modeling the acquisition of social rank in crayfish: winner and loser effects and self-structuring propertics. Behaviour 143: 325–346.

Hofkin, B.V., G.M. Mkoji, D.K. Koech and E.S. Loker. 1991. Control of schistosome-transmitting snails in Kenya by the North American crayfish *Procambarus clarkii*. Am. J. Trop. Med. Hyg. 45: 339–344.

Hofkin, B.V., D.M. Hofinger, D.K. Koech and E.S. Loker. 1992. Predation of *Biomphalaria* and non-target molluscs by the crayfish *Procambarus clarkii*: implications for the biological control of schistosomiasis. Ann. Top. Med. Parasit. 86: 663–670.

Horner, A.J., M. Schmidt, D.H. Edwards and C.D. Derby. 2007. Role of the olfactory pathway in agonistic behavior of crayfish, *Procambarus clarkii*. Invertebr. Neurosci. 8: 11–8.

Horwitz, P.H.J. and A.M.M. Richardson. 1986. An ecological classification of burrows of Australian freshwater crayfish. Aust. J. Mar. Fresh. Res. 37: 237–242.

Hsu, Y., S. Lee, M.H. Chen, S.Y. Yang and K. Cheng. 2008. Switching assessment strategy during a contest: fighting in killifish *Kryptolebias marmoratus*. Anim. Behav. 75: 1641–1649.

Huber, R., K. Smith, A. Delago, K. Isaksson and E.A. Kravitz. 1997. Serotonin and aggressive motivation in crustaceans: altering the decision to retreat. Proc. Natl. Acad. Sci. USA 94: 5939–5942.

Huner, J.V. and J.E. Barr. 1984. Red Swamp Crawfish: Biology and Exploitation. Louisiana State University, Baton Rouge, LA.

Huner, J.V., D.M. Holdich and R.S. Lowery. 1988. *Procambarus* in North America and elsewhere. pp. 239–261. *In*: D.M. Holdich and R.S. Lowery (eds.). Freshwater Crayfish: Biology, management and Exploitation. Croom Helm (Chapman & Hall), London.

Huryn, A.D. and J.B. Wallace. 1987. Production and litter processing by crayfish in an Appalachian mountain stream. Freshwater Biol. 18: 277–286.

Ilhéu, M. and J.M. Bernardo. 1993a. Aspects of trophic ecology of red swamp crayfish (*Procambarus clarkii* Girard) in Alentejo, South of Portugal. pp. 417–423. *In*: Actas VI Congreso Español de Limnología, Granada.

Ilhéu, M. and J.M. Bernardo. 1993b. Experimental evaluation of food preference of red swamp crayfish, *Procambarus clarkii*: vegetal versus animal. Freshwater Crayfish 9: 359–364.

Ilhéu, M., J.M. Bernardo and S. Fernandes. 2007. Predation of invasive crayfish on aquatic vertebrates: the effect of *Procambarus clarkii* on fish assemblages in Mediterranean temporary streams. pp. 543–558. *In*: F. Gherardi (ed.). Biological Invaders in Inland Waters: Profiles, Distribution, and Threats. Springer, Rotterdam.

Jones, C.M. and I.M. Ruscoe. 2001. Assessment of five shelter types in the production of redclaw crayfish *Cherax quadricacrinatus* (Decapoda: Parastacidae) under earthen pond conditions. J. World Aquacult. Soc. 32: 41–52.

Keller, T.A., A.M. Tomba and P.A. Moore. 2001. Orientation in complex chemical landscapes: Spatial arrangement of chemical sources influences food-finding efficiency in artificial streams. Limnol. Oceanogr. 46: 238–247.

Klar, N.M. and P.H. Crowley. 2012. Shelter availability, occupancy, and residency in size-asymmetric contests between rusty crayfish, *Orconectes rusticus*. Ethology 118: 118–126.

Klocker, C.A. and D.L. Strayer. 2004. Interactions among an invasive crayfish (*Orconectes rusticus*), a native crayfish (*Orconectes limosus*), and native bivalves (Sphaeriidae and Unionidae). Northeast. Nat. 11: 167–178.

Lodge, D.M., M.W. Kershner, J.E. Aloi and A.P. Covich. 1994. Effects of an omnivorous crayfish (*Orconectes rusticus*) on a freshwater littoral food web. Ecology 75: 1265–1281.

Lorman, J. and J.J. Magnuson. 1978. The role of crayfish in aquatic ecosystems. Fisheries 3: 8–10.

Loughman, Z.J.D.A. Fotlz II and S.A. Welsh. 2013. Baited lines: An active nondestructive collection method of burrowing crayfish. Southeast. Nat. 12: 809–815.

Mackie, A. 1973. The chemical basis of food detection in the lobster, *Homarus gammarus*. Mar. Biol. 21: 103–108.

Martin III, A.L. and P.A. Moore. 2007. Field observations of agonism in the crayfish, *Orconectes rusticus*: Shelter use in a natural environment. Ethology 113: 1192–1201.

Martin III, A.L. and P.A. Moore. 2008. The influence of dominance on shelter preference and eviction rates in crayfish, *Orconectes rusticus*. Ethology 114: 351–360.

Martin III, A.L. and P.A. Moore. 2010. The influence of reproductive state on the agonistic interactions between male and female crayfish (*Orconectes rusticus*). Behaviour 147: 1309–1325.

Mason, J.C. 1970. Copulatory behavior of the crayfish, *Pacifastacus trowbridgii* (Stimpson). Can. J. Zool. 48: 969–976.

Mason, J.C. 1975. Crayfish production in a small woodland stream. Freshwater Crayfish 2: 449–479.

Matthews, M.A., J.D. Reynolds and M.J. Keatinge. 1993. Macrophyte reduction and benthic community alteration by the crayfish *Austropotamobius pallipes* (Lereboullet). Freshwater Crayfish 9: 289–299.

McLeese, D. 1970. Detection of dissolved substances by the American lobster (*Homarus americanus*) and olfactory attraction between lobsters. J. Fish. Res. Board Can. 27: 1371–1378.

Minkley, W.L. and J.E. Craddock. 1961. Active predation of crayfish on fishes. Prog. Fish Cult. 23: 120–123.

Momot, W.T. 1995. Redefining the role of crayfish in aquatic ecosystems. Rev. Fish. Sci. 3: 33–63.

Momot, W.T., H. Gowing and P.D. Jones. 1978. The dynamics of crayfish and their role in ecosystems. Am. Midl. Nat. 99: 10–35.

Moore, P.A. and J. Atema. 1988. A model of a temporal filter in chemoreception to extract directional information from a turbulent odor plume. Biol. Bull. 174: 355–363.

Moore, P.A. and J.L. Grills. 1999. Chemical orientation to food by the crayfish *Orconectes rusticus*: influence of hydrodynamics. Anim. Behav. 58: 953–963.

Moore, P.A. and D.A. Bergman. 2005. The smell of success and failure: the role of intrinsic and extrinsic chemical signals on the social behavior of crayfish. Integr. Comp. Biol. 45: 650–657.

Morshiri, G.A. and C.K. Goldman. 1969. Estimation of assimilation efficiency in crayfish *Pacifastacus leniusculus*. Arch. Hydrobiol. 68: 298–306.

Noro, C.K. and L. Buckup. 2010. The burrows of *Parastacus defossus* (Decapoda: Parastacidae), a fossorial freshwater crayfish from Southern Brazil. Zoologia 27: 341–346.

Olsen, T.M., D.M. Lodge, G.M. Capelli and R.J. Houlihan. 1991. Mechanisms of impact of an introduced crayfish (*Orconectes rusticus*) on littoral congeners, snails, and macrophytes. Can. J. Fish. Aquat. Sci. 48: 1853–1861.

Palaoro, A.V., M.M. Dalosto, C. Coutinho and S. Santos. 2013. Assessing the importance of burrows through behavioral observations of *Parastacus brasiliensis*, a Neotropical burrowing crayfish (Crustacea), in laboratory conditions. Zool. Stud. 52: 2–9.

Panksepp, J.B. and R. Huber. 2001. Chronic alterations in serotonin function: Dynamic neurochemical properties in agonistic behavior of the crayfish, *Orconectes rusticus*. J. Neurobiol. 50: 276–290.

Parker, G. 1974. Assessment strategy and the evolution of fighting behaviour. J. Theor. Biol. 47: 223–243.

Parkyn, S., K.J. Collier and B.J. Hicks. 2002. Growth and population dynamics of crayfish *Paranephrops planifrons* in streams within native forest and pastoral land uses. New Zeal. J. Mar. Fresh. 36: 847–861.

Pavey, C.R. and D.R. Fielder. 1996. The influence of size differential on agonistic behaviour in the freshwater crayfish *Cherax cuspidatus* (Decapoda: Parastacidae). J. Zool. 238: 445–457.

Pecor, K.W. and B.A. Hazlett. 2006. A test of temporal variation in risk and food stimuli on behavioral tradeoffs in the rusty crayfish, *Orconectes rusticus*: risk allocation and stimulus degradation. Ethology 112: 230–237.

Pintor, L.M. and A. Sih. 2009. Differences in growth and foraging behavior of native and introduced populations of an invasive crayfish. Biol. Invasions 11: 1895–1902.

Prins, R. 1968. Comparative ecology of the crayfish *Orconectes rusticus* and *Cambarus tenebrosus* in Doe Run, Meade County, Kentucky. Int. Rev. Ges. Hydrobio. Hydrogr. 53: 667–714.

Punzalan, D., R.C. Guiaşu, D. Belchior and D.W. Dunham. 2001. Discrimination of conspecific-built chimneys from human-built ones by the burrowing crayfish, *Fallicambarus fodiens* (Decapoda, Cambaridae). Invertebr. Biol. 120: 58–66.

Rabeni, C.F. 1985. Resource partitioning by stream-dwelling crayfish: The influence of body size. Am. Midl. Nat. 113: 20–29.

Rahel, F.J. and R.A. Stein. 1988. Complex predator-prey interactions and predator intimidation among crayfish, piscivorous fish, and small benthic fish. Oecologia 75: 94–98.

Reeder, P.B. and B.W. Ache. 1980. Chemotaxis in the Florida spiny lobster, *Panulirus argus*. Anim. Behav. 28: 831–839.

Rickett, J.D. 1974. Trophic relationship involving crayfish of the genus *Orconectes* in experimental ponds. Prog. Fish Cult. 36: 207–211.

Rubenstein, D.I. and B.A. Hazlett. 1974. Examination of the agonistic behaviour of the crayfish *Orconectes virilis* by character analysis. Behaviour 50: 193–216.

Rubolini, D., P. Galeotti, G. Ferrari, M. Spairani, F. Bernini and M. Fasola. 2006. Sperm allocation in relation to male traits, female size, and copulation behaviour in a freshwater crayfish species. Behav. Ecol. Sociobiol. 60: 212–219.

Sargent, L.W., A.K. Baldridge, M. Vega-Ross, K.M. Towle and D.M. Lodge. 2014. A trematode parasite alters growth, feeding behavior, and demographic success of invasive rusty crayfish (*Orconectes rusticus*). Oecologia 175: 947–958.

Sato, D. and T. Nagayama. 2012. Development of agonistic encounters in dominance hierarchy formation in juvenile crayfish. J. Exp. Biol. 215: 1210–1217.

Schmitt, B.C. and B.W. Ache. 1979. Olfaction: responses of a decapod crustacean are enhanced by flicking. Science 205: 204–206.

Schneider, R.A.Z., R. Huber and P.A. Moore. 2008. Individual and status recognition in the crayfish, *Orconectes rusticus*: the effects of urine release on fight dynamics. Behaviour 138: 137–153.

Schoener, T.W. 1971. Theory of feeding strategies. Annu. Rev. Ecol. Syst. 2: 369–403.

Schroeder, L. and R. Huber. 2001. Fight strategies differ with size and allometric growth of claws in crayfish, *Orconectes rusticus*. Behaviour 138: 1437–1449.

Seebacher, F. and R.S. Wilson. 2007. Individual recognition in crayfish (*Cherax dispar*): the roles of strength and experience in deciding aggressive encounters. Biol. Letters 3: 471–4.

Shave, C.R., C.R. Townsend and T.A. Crowl. 1994. Anti-predator behaviours of a freshwater crayfish (*Paranephrops zealandicus*) to a native and an introduced predator. New Zeal. J. Ecol. 18: 1–10.

Shelton, R. and A. Mackie. 1971. Studies on the chemical preferences of the shore crab, *Carcinus maenas* (L.). J. Exp. Mar. Biol. Ecol. 7: 41–49.

Sherba, M., D.W. Dunham and H.H. Harvey. 2000. Sublethal copper toxicity and food response in the freshwater crayfish *Cambarus bartonii* (Cambaridae, Decapoda, Crustacea). Ecotox. Environ. Safe. 46: 329–333.

Smart, A.C., D.M. Harper, F. Malaisse, S. Schmitz, S. Coley and A.C.G. de Beauregard. 2002. Feeding of the exotic Louisiana red swamp crayfish, *Procambarus clarkii* (Crustacea, Decapoda), in an African tropical lake: Lake Naivasha, Kenya. Hydrobiologia 488: 129–142.

Smith, J.M. 1974. The theory of games and the evolution of animal conflicts. J. Theor. Biol. 47: 209–221.

Smith, J.M. and G.A. Parker. 1976. The logic of asymmetric contests. Anim. Behav. 24: 159–175.

Sneddon, W.A. 1990. Determinants of male mating success in the temperate crayfish, *Orconectes rusticus*: Chelae size and sperm competition. Behav. 115: 101–113.

Steele, C., C. Skinner, P. Alberstadt and C. Mathewson. 1999. Organization of chemically activated food search behavior in *Procambarus clarkii* (Girard) and *Orconectes rusticus* (Girard) crayfishes. Biol. Bull. 196: 295–302.

Stein, R.A. 1976. Sexual dimorphism in crayfish chelae: functional significance linked to reproductive activities. Can. J. Zool. 54: 220–227.

Stein, R.A. 1977. Predation, optimal foraging, and the predator-prey interaction between fish and crayfish. Ecology 58: 1237–1253.

Stein, R.A. and J.J. Magnuson. 1976. Behavioral response of crayfish to a fish predator. Ecology 57: 751–761.

Stocker, A.M. and R. Huber. 2001. Fighting strategies in crayfish *Orconectes rusticus* (Decapoda, Cambaridae) differ with hunger state and presence of food cues. Ethology 107: 727–736.

Stoeckel, J.A., B.S. Helms and E. Cash. 2011. Evaluation of a crayfish burrowing chamber design with simulated groundwater flow. J. Crust. Biol. 31: 50–58.

Tierney, A.J. and J. Atema. 1988. Behavioral responses of crayfish (*Orconectes virilis* and *Orconectes rusticus*) to chemical feeding stimulants. J. Chem. Ecol. 14: 123–133.

Tricarico, E. and F. Gherardi. 2010. Past ownership makes crayfish more aggressive. Behav. Ecol. Sociobiol. 64: 575–581.

Trivers, R.L. 1972. Parental investment and sexual selection. pp. 136–179. *In*: B. Campbell (ed.). Sexual Selection and the Descent of Man. Heinemann, London, UK.

Turner, A.M., S.A. Fetterolf and R.J. Bernot. 1999. Predator identity and consumer behavior: differential effects of fish and crayfish on the habitat use of a freshwater snail. Oecologia 118: 242–247.

Uiska, E., D.W. Dunham and H.H. Harvey. 1994. Cumulative pattern in pH change alters response to food in the crayfish *Cambarus bartonii*. Can. J. Zool. 72: 187–190.

Villanelli, F. and F. Gherardi. 1998. Breeding in the crayfish, *Austropotamobius pallipes*: mating patterns, mate choice and intermale competition. Freshwater Biol. 40: 305–315.

Welch, S.M. and A.G. Eversole. 2006. The occurrence of primary burrowing crayfish in terrestrial habitat. Biol. Conserv. 130: 458–464.

Welch, S.M., J.L. Waldron, A.G. Eversole and J.C. Simoes. 2008. Seasonal variation and ecological effects of Camp Shelby burrowing crayfish (*Fallicambarus gordoni*) burrows. Am. Midl. Nat. 159: 378–384.

Wiernicki, C. 1984. Assimilation efficiency by *Procambarus clarkii* fed Elodea (*Egera densa*) and its products of decomposition. Aquaculture 36: 203–215.

Wilson, K.A., J.J. Magnuson, D.M. Lodge, A.M. Hill, T.K. Kratz, W.L. Perry and T.V. Willis. 2004. A long-term rusty crayfish (*Orconectes rusticus*) invasion: dispersal patterns and community change in a north temperate lake. Can. J. Fish. Aquat. Sci. 61: 2255–2266.

Wofford, S.J., R.L. Earley and P.A. Moore. 2015. Evidence for assessment disappears in mixed-sex contests of the crayfish, *Orconectes virilis*. Behaviour 152: 995–1018.

Woodlock, B. and J.D. Reynolds. 1988. Laboratory breeding studies of freshwater crayfish, *Austropotamobius pallipes* (Lereboullet). Fresh. Biol. 19: 71–78.

Zulandt, T., R.A. Zulandt-Schneider and P.A. Moore. 2008. Observing agonistic interactions alters subsequent fighting dynamics in the crayfish, *Orconectes rusticus*. Anim. Behav. 75: 13–20.

Zulandt-Schneider, R.A. and P.A. Moore. 2000. Urine as a source of conspecific disturbance signals in the crayfish *Procambarus clarkii*. J. Exp. Biol. 203: 765–771.

5

Chemical Ecology of Crayfish

Thomas Breithaupt,[1,*] *Francesca Gherardi,*[#] *Laura Aquiloni*[2,a] and
Elena Tricarico[2,b]

Introduction

Aquatic animals such as crayfish live in a soup of organic molecules, which mainly originate from other aquatic organisms. These chemicals are inevitable by-products of life's processes including feeding, digestion, excretion, predation or decay. They contribute to the nutrient flow between organisms within an ecological network. In addition, the molecules may carry information and influence important decisions and behaviours of other organisms. Numerous studies have shown that most aquatic organisms respond to minute concentrations of chemical substances released by other organisms (Brönmark and Hansson 2012). Interactions between organisms based on such infochemicals (Dicke and Sabelis 1988)—also referred to as semiochemicals (Wyatt 2011)—are subject to the field of chemical ecology, a very active and important subdiscipline of ecology. Based on the nature of interaction, different types of infochemicals are recognized as (1) *pheromones*, they mediate interactions between organisms of the same species (e.g., sex pheromones or dominance pheromones) and the information transfer is beneficial to both sender and receiver; (2) *kairomones*, they mediate interactions between individuals of different species where the information transfer is beneficial for the receiver but not the sender of the chemical (e.g., the detection of chemicals originating from a potential prey organisms or a predator is beneficial to the receiver but may lead to death or missed feeding opportunity of the source organism); and (3) *alarm substances* (alarm cue, "Schreckstoff"), they are chemicals released by an injured prey that evoke defensive or escape responses in conspecific receivers (see also Brönmark and Hansson 2012). The detection of infochemicals is crucial for the survival and reproductive success of crayfish. Chemicals released by potential prey organism can inform a crayfish about the presence and location of food (section 'Foraging behaviour'). Excretory products of a predator or body fluids set free during a predation event alarm the crayfish and may trigger fight or flight responses (section 'Predator detection'). Pheromones originating from conspecifics carry

[1] School of Biological, Biomedical and Environmental Sciences, University of Hull, Hull, HU6 7RX, United Kingdom.
[2] Department of Biology, University of Florence, Via Romana 17, 50125 Florence (Italy).
[a] Email: laura.aquiloni@unifi.it
[b] Email: elena.tricarico@unifi.it
[*] Corresponding author: t.breithaupt@hull.ac.uk

[#] Francesca Gherardi had the initial idea for this chapter but sadly passed away on 14th February 2013 at the age of 57. She was an unusually creative scientist, a dear colleague, mentor, teacher and friend, who has greatly influenced the research of crayfish ecology as well as many other areas of crustacean biology.

information about aggressive state or vulnerability of a competitor (dominance pheromones; section 'Fighting in Social behaviour'). They may also inform about the sexual receptivity or health status of a potential mating partner (sex pheromones; section 'Reproduction in Social Behaviour') or may attract freshly hatched juveniles to the protective tail of their mother (brood pheromones; section 'Parent-Offspring interactions in Social Behaviour'). In view of the potential advantages and opportunities offered by the detection and recognition of organic molecules, it comes as no surprise that crayfish have evolved a set of excellent chemoreceptive sensors (section 'Chemoreception'). Chemoreception is not only important for detection of food and predators. It also forms the basis of a well-developed intraspecific chemical communication system. In this chapter, we will review the current knowledge of the chemical ecology of crayfish. Due to its availability, suitable size, easiness of maintenance in the lab and robustness, crayfish have served as important model systems in animal behaviour, neuroethology and ecology. Within these disciplines, many studies that include chemical signals and kairomones have been published, but knowledge of crayfish chemical ecology has not been reviewed in one chapter. Following an overview of crayfish chemoreceptors, our review focuses on three major aspects of the crayfish chemical ecology: (i) how are infochemicals used to find food and avoid predation, (ii) how do they mediate communication between conspecifics during aggressive interaction, sexual interactions and parent-offspring interactions, and (iii) how can this knowledge be used for predicting the effect of pollutants on crayfish behaviour, for the management of crayfish species in aquaculture and for the control of invasive crayfish species.

Chemoreception

Chemoreception is arguably the most important sensory modality of crayfish. It is complemented by a variety of other sensory organs that detect environmental stimuli. These include well-developed compound eyes (Vogt 1980) as well as extra-ocular photoreceptors (Edwards 1984). However, the nocturnal lifestyle of crayfish combined with reduced performance of the superposition eyes at night (Warrant and McIntyre 1990), and the high turbidity of many crayfish habitats limits the behavioural utility of vision. Crayfish possess external mechanoreceptors, such as the long second antennae and a multitude of small sensory hairs, which are speckled over the entire body surface (Breithaupt and Tautz 1990). Crayfish seem not able to hear (even if they can generate acoustic signals: Buscaino et al. 2012), but they can sense even the weakest water currents and use it to perceive the main direction of ambient flow as well as vibrations produced by other animals (Breithaupt 2002). Detection of the main flow direction helps in orientation towards chemical sources (see below). The chemical senses are most important for their survival and successful reproduction. Chemoreceptors are involved in the detection of food, predators and conspecifics and provide more specific information about the source of the stimulus than the visual and mechanoreceptive senses.

In terrestrial animals, we discriminate olfaction and taste (gustation) based on the medium they operate in, the behavioural functions and the region in the brain where the stimuli are processed (Caprio and Derby 2008). Olfaction is used for the detection of volatile molecules in air and taste for the detection of water-soluble molecules. For aquatic animals such as crayfish, this distinction is not possible as all chemical stimuli are water-soluble. Taste mediates simple, reflective behaviours such as grabbing, biting and swallowing, whereas olfaction mediates more complex behaviours such as courtship, agonistic behaviour, finding or avoiding distant odour sources (e.g., food, mate, and predator). This distinction may apply both in air and in water. Finally, all olfactory stimuli are processed in a specific region of the brain (olfactory lobe in vertebrates, antennal lobe in insects, and olfactory lobe in crustaceans) that is organised into glomeruli (Caprio and Derby 2008). Glomeruli process specific chemical categories of odours mediated by specific receptor neurons.

In decapod crustaceans, we discriminate two general types of chemosensory organs: unimodal olfactory sensilla and bimodal non-olfactory sensilla (Hallberg and Skog 2011); the latter are also called taste or gustatory sensilla following the criteria outlined above. The crustacean olfactory sensilla are called aesthetascs and are located on the upper (lateral) branch of the first antenna ("antennule"; Fig. 1). Non-olfactory bimodal sensilla are found on almost all appendages including the mouthparts, legs and antenna (Fig. 1). These sensilla contain both chemo- and mechanoreceptors (Hatt 1986).

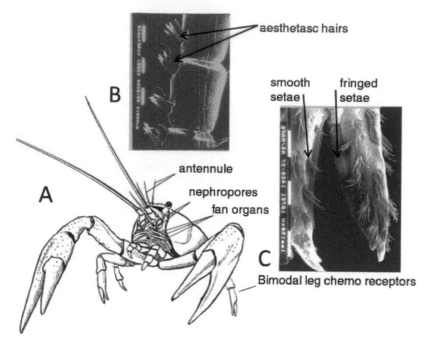

Figure 1. Structures involved in chemoreception and chemical communication in crayfish. (A) location of antennules (site of olfaction), chelae of 2nd walking leg (site of bimodal chemoreceptors), nephropores (release of chemical signals) and maxilliped mouthparts (site of flow generating fan organs). (B) SEM photograph of groups of aesthetasc hairs on the flagellum of the lateral (upper) branch of the antennule of *Astacus leptodactylus*. (C) two types of bimodal receptors on the propodit and dactylopodit of the distal segments of the 2nd walking leg of *A. leptodactylus*. White bars show 1 mm scale. Inset imageA reprinted from Breithaupt (2001), Inset B modified after Breithaupt and Hardege (2012).

Olfactory sensilla

Aesthetasc sensilla have been investigated in three different species of crayfish: *Orconectes propinquus* (Tierney et al. 1986), *Procambarus clarkii* (Mellon et al. 1989, Mellon 2012) and *Pacifastacus leniusculus* (Hallberg et al. 1997). Up to 150 aesthetascs are found on the ventral side of the lateral antennular flagellum and are arranged on each segment in a proximal and a distal group. Each group contains from two to eight sensilla depending on species (Tierney et al. 1986, Mellon et al. 1989) (Fig. 1). The sensilla are 100–200 µm long and 10–20 µm in diameter with variations between species. The cuticle in the distal 2/3 of the aesthetasc is no more than 1 µm thick and permeable to dissolved odourants (Tierney et al. 1986). Each sensillum typically contains 100–300 olfactory receptor neurons that all project into the olfactory lobe in the brain (Tierney et al. 1986, Mellon et al. 1989, Hallberg et al. 1997).

Olfactory stimulus acquisition by flicking and fanning

When closely observing a crayfish it is noticeable that the antennules often undergo individual twitches or trains of twitches. This so-called "flicking" behaviour consists of quick downward and slower upward movements of the aesthetasc bearing flagella of the antennules and is an important component of olfaction. Flicking has been shown to temporarily enhance olfactory stimulus uptake and is comparable to sniffing in mammals (Schmitt and Ache 1979). In crayfish, flicks occur spontaneously in irregular intervals of up to 4 min in the absence of external stimuli (Mellon 1997). Flicking rate is strongly increased to one or two flicks per second upon odour stimulation or mechanical stimulation by transient water movements (Moore and Kraus-Epley 2013). Individual flicks have average speeds of 1.8 cm s^{-1} for the downstroke and 1.1 cm s^{-1} for the return stroke (Mellon 1997). During the fast downstroke, water penetrates the chemosensory array of sensilla and carries odour molecules to the receptor surface (Koehl et al. 2001). The odour is retained near the aesthetasc during the slow return stroke (see Fig. 2; Pravin et al. 2012). During the return stroke, molecular diffusion moves the molecules across the membrane into the sensillum. It is only at the next downstroke that the previously captured odour molecules are flushed and up to 97.6% of these odourants are replaced by new ones (Pravin et al. 2012). Therefore, flicking allows discrete sampling in time and space of concentration profiles within the variable odour plumes encountered in the natural environment (Koehl et al. 2001).

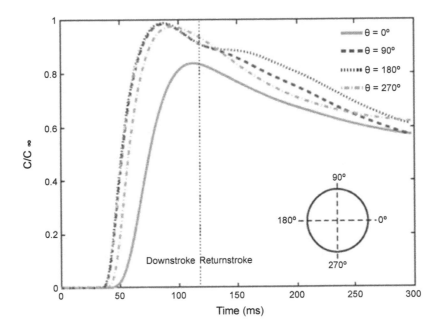

Figure 2. Time course of odorant concentration at four locations on the surface of an aesthetasc sensillum during a complete course of flicking. In this geometry, $\vartheta = 180°$ represent the location that faces the ambient flow, while $\vartheta = 270°$ represents the location that faces the neighbouring aesthetasc. Modified after (Pravin et al. 2012), with permission of Springer.

Another active mechanism used by decapod crustaceans to enhance olfactory perception is to create water currents that draw odour molecules to the antennules. Crustaceans can generate water currents by pumping and fanning appendages (Atema 1985). Self-generated water currents play important roles in gill ventilation, locomotion, suspension feeding, chemical signalling and chemoreception (Atema 1995). Chemoreception is facilitated by the so called "fan organs" (Atema 1995, Breithaupt 2001). The fan organs consist of the outer flagella ("exopodites") of the three-paired maxillipeds (mouthparts located under the mouth opening) and generate water currents that draw odour molecules towards the antennules (Breithaupt 2001, Burrows 2009). Similar to antennular flicking, fanning behaviour can be elicited by any disturbances of visual, chemical or mechanical nature. Each fan consists of a multi-segmental flattened stem, which is distally feathered by laterally emerging setae (Fig. 3; Breithaupt 2001, Burrows 2009). During the powerstroke, the setae are extended and the fan acts like a paddle moving water in a specific direction. During the recovery stroke, the setae fold in and the flagella is flexed to provide minimal drag when returning to the starting position. The exopodites of one side beat sequentially (Burrows 2009) and create a continuous flow of water (Breithaupt 2001). Crayfish can create one or two outward jets using the fan organs of one or both sides, respectively (Denissenko et al. 2007). These jets induce an inflow that draws odour towards the antennules (see Fig. 3; Denissenko et al. 2007). The inward flow is relatively slow and slows down even more with increasing distance from the fans; its use in odour acquisitions does not extend beyond one body length even in stagnant water (Denissenko et al. 2007).

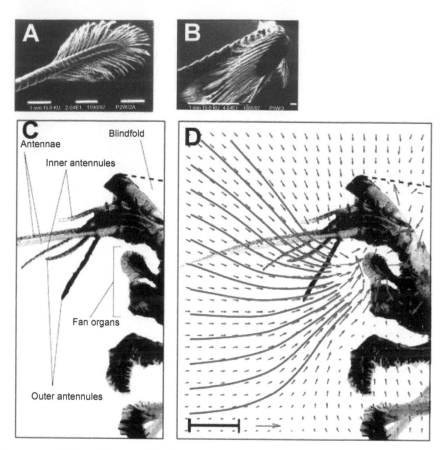

Figure 3. Fan organs and the flow fields created by fanning. During the power stroke (SEM picture of *Procambarus clarkii* fan) the setae on the distal flagellum are extended (A). During the recovery stroke they are folded (B). (C) shows in side view the position of the fan organs below the antennae and antennules. The flow field (D) generated by the fan organs (see photograph on the left) measured at the vertical plane and averaged over 30 instantaneous measurements (30 seconds). The fluid converges towards the fan organs in the plane of measurements and escapes in the form of jets in a perpendicular direction (not shown). A reference segment at the bottom is 2 cm, a reference vector is 1 cm/s. Insets A and B are reprinted from Breithaupt (2001). Inset C is modified after Denissenko et al. (2007), with permission of The Journal of Experimental Biology.

Bimodal sensilla

Non-olfactory bimodal sensilla are found on almost all appendages. These sensilla contain both chemo- and mechanoreceptors. The chemoreceptors do not project into the olfactory lobe but to other centres in the brain and into the ventral nerve chord (Schmidt and Mellon 2011). Typical bimodal sensilla that are found in decapod crustaceans including crayfish are the hedgehog hairs (also called stout setae), the smooth and squamous setae and the serrate setae (Derby 1982; for a recent review of types of setae see also Garm and Watling 2013). The hedgehog hairs (Derby 1982), named "fringed setae" in Altner et al. (1983), are stout conical structures organized in a row lining the inner cutting edge of the chelae of the second and third walking legs (Derby 1982, Altner et al. 1983, see Fig. 1C). The smooth setae (named "large setae" in Hatt and Bauer 1980) are arranged in small groups ("tufts") rising at intervals out of small depressions along the propodite and dactylopodite of the walking legs (Fig. 1C). Using electrophysiology, Hatt (1986), Hatt and Bauer (1980) and Derby (1982) confirmed that hedgehog hairs and smooth setae respond both to mechanical and chemical stimuli.

Foraging behaviour

The role of chemoreceptors in foraging behaviour

Some of the most essential functions of chemoreception are related to foraging behaviour. Chemoreceptors are used to detect, recognize, find food and control ingestion of food items (Derby and Atema 1982). The diminished visual conditions at night (the preferred activity period of decapods crustaceans) and in turbid waters, and the absence of hearing organs makes the chemical senses even more important in controlling foraging activities. In contrast to the other senses, chemoreception provides detailed information about the quality of the food item and will guide the animal's decisions in the process of searching for food and feeding. Crayfish are omnivorous, feeding on macrophytes (Nyström et al. 1996), live macro-invertebrates (Hanson et al. 1990), fish carrion (Willman et al. 1994), periphyton and detritus (Hogger 1988). They also predate on freshly moulted conspecifics, small live fish and tadpoles when opportunities arise (Stein 1977, Breithaupt et al. 1995, Guan and Wiles 1997, Cruz et al. 2006). Crayfish respond to a wide variety of chemicals including carbohydrates, amino acids, nucleotides and derivates, and amines. This reflects their omnivorous foraging habits. Sensitivity to individual chemicals or mixtures of chemicals have been tested by electrophysiological (Hodgson 1958, Hatt and Bauer 1980, Altner et al. 1983, Hatt 1986, Corotto and O'Brien 2002) and behavioural studies (Tierney and Atema 1988, Corotto et al. 2007). The carbohydrates cellobiose and sucrose are the most stimulatory compounds in *Orconectes rusticus* and *O. virilis* (Tierney and Atema 1988). They are indicative of a macrophyte diet since cellobiose is produced by hydrolysis of cellulose, and sucrose occurs abundantly in plants (Tierney and Atema 1988). Sensitivity to a wide variety of amino acids reflects a carnivorous diet. For example, *Cambarus bartoni* shows a high antennular sensitivity to the amino acids glycine and glutamate (Hodgson 1958). These amino acids that naturally occur in the hemolymph of insects and in the cuticle of crayfish (Tierney and Atema 1988) stimulate walking (Corotto et al. 2007) and feeding movements in crayfish (Tierney and Atema 1988). The leg chemoreceptors of *Austropotamobius torrentium* are sensitive to many amino acids, the most potent being serine, alanine, histidine, β-alanine, ornithine, and proline (Hatt and Bauer 1980, Hatt 1984, 1986). Differences in sensitivity to different chemicals were found between different species (Hatt 1984, Tierney and Atema 1988, Corotto et al. 2007). It should be interesting to see if these differences reflect adaptations to different diets.

Foraging behaviour in crayfish is organised into several successive stages starting with detection followed by far-field (distant) food search, and leading up to near-field (local) search (Steele et al. 1999). Far-field food search is elicited by low concentrations of food odour (e.g., diluted feeding stimulants), and mediated by the antennules (Derby and Atema 1982). High concentration stimuli are experienced as the crayfish approaches the food source and will elicit near-field search including leg probing behaviour (Moore et al. 1991, Steele et al. 1999). Food items are picked up by the chelae of the walking legs and forwarded to the mouthparts where manipulation and ingestion occurs (Derby and Atema 1982, Dunham et al. 1997).

Chemo-orientation

Odour plumes

Crayfish are able to find distant sources of odour (food or mates) in a variety of flow situations (Moore and Grills 1999, Keller et al. 2001). This is surprising for two reasons. Firstly, chemical stimuli have no inherent directionality. Secondly, water currents in riverine environments are generally turbulent making the odour distribution less predictable than in laminar (unidirectional) flow conditions. Odour molecules either leak from an organism or they are actively released into water currents such as the ventilation currents as in molluscs and crustaceans generated by the source organism. As the molecules are carried away from the source by the ambient flow, they develop into a plume (Weissburg 2011, Atema 2012). The particular structure of the plume depends on the hydrodynamic conditions (ranging from laminar to fully turbulent) at the source and those encountered on the journey away from the source (Webster and Weissburg 2001). A rough surface structure of the riverbed (e.g., gravel or cobble) and high flow velocities create much more turbulent conditions than a smooth substrate (e.g., sand) and low flow velocities (Moore et al. 2000). Eddies and smaller turbulence initially causes disruption of the odour plume into patches of different height and steepness (Moore et al. 2000). To the crayfish, this causes signal intermittency. Odour peaks are detected as pulses in time (Atema 2012). Both peak height and steepness decrease with distance downstream and across stream from the odour source (Fig. 4), creating an odour landscape. Odour landscapes differ between environments with the odour peak and steepness distribution as well as the plume width being typical for a particular flow condition (Moore et al. 2000). Over time, turbulent mixing homogenizes the plume and widens it (Webster and Weissburg 2001, Weissburg 2011).

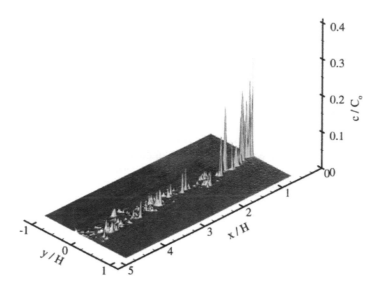

Figure 4. Instantaneous concentration field of dye released from an upstream nozzle into a flume (100 cm width, 2400 cm length, 0.2 cm water depth) at a flow speed of 5 cm/s. Measurements were taken in a plane 2.5 cm above the bed parallel to the ground using laer-induced fluorescence (LIF) technique. x is the downstream flow coordinate, y is horizontal cross-stream coordinate, H is water depth. Concentration is normalized by source concentration C_0. From Webster et al. (2001), with permission of John Wiley & Sons.

How do crayfish find an odour source?

The exact search strategy used by crayfish (or by any other decapod crustacean) to find an odour source in a complex flow environment is not yet known. Theoretically, they could use one of three different strategies to track odour to its source: (i) eddy chemotaxis (or eddy chemo-rheotaxis), (ii) odour-gated rheotaxis, or (iii) plume edge tracking (Grasso and Basil 2002, Atema 2012). In the first strategy, crayfish may use features of the odour landscape that change as a function of the distance from the source to navigate through the flume (Atema 1996). They would find the source by comparing the steepness of concentration gradients (or peak heights) in the different odour patches (see Fig. 4) they encounter and by following a route that leads towards increasing odour patch steepness (or peak height), a strategy named "eddy chemotaxis" (Atema 1996). To use this strategy, animals need to detect the gradient by comparing the input to bilateral appendages (e.g., left and right antennules) or by comparing receptors along the axis of one sensory appendage. There is evidence in some decapod crustaceans that bilateral comparison is used for chemo-orientation. Behavioural studies using selective removal of chemosensory appendages in spiny lobster (*Panulirus argus*) and American lobsters (*Homarus americanus*) showed that the aesthetasc bearing lateral filaments of the antennules play important roles in guiding the chemical search (Reeder and Ache 1980, Devine and Atema 1982). Lobsters lacking a lateral antennule lose the ability to find the odour source, whereas animals lacking either a medial antennule or the chemoreceptors on the walking legs are not affected in their orientation (Devine and Atema 1982). It is likely that during their orientation decapods not only rely on chemical but also use directional information from the fluid flow (Keller et al. 2001). Atema (2012) suggested that aquatic animals may use the spatial and temporal coincidence of odour and flow information to detect flavoured eddies and use this to steer their orientation in an odour landscape. Recordings of neural activity of the crayfish brain in response to bimodal stimulation support this hypothesis (Mellon 2005). Simultaneous onset of flow and odour stimuli leads to a two-fold response amplification in Type-I interneurons compared to odour stimuli that are preceded by flow onset (Mellon 2005, Atema 2012). On the other hand, Weissburg (2011) argues that the temporal resolution of the olfactory receptors may not be good enough to resolve the peaks in the odour landscape. Lobsters can resolve 5 Hz of chemical stimulus pulses (Gomez and Atema 1996). This may be too slow to detect the short odour pulses (< 100 ms) that occur even in relatively slow flow speeds (5 cm/s) (Webster and Weissburg 2001, Weissburg 2011). (ii) Odour-gated rheotaxis is a much simpler strategy that does not depend on evaluating each individual eddy or odour patch. Animals using this strategy walk upstream as long as they are stimulated by odour. They use the time-averaged flow direction as a directional cue. When they lose contact with the plume, they move cross-stream until they either re-contact the plume or abandon the search (Grasso and Basil 2002). Cross-stream casting is not commonly seen in plume tracking decapod crustaceans. In addition, crayfish do not head straight upstream when tracking a plume (Moore and Grills 1999). Hence, their search strategy may be more complex. (iii) "plume edge tracking" refers to a strategy whereby animals move along the edge of the plume in order to reach the odour source (Grasso and Basil 2002). Animals can determine the edge by keeping one sensor (of a bilateral sensor pair like the antennules) inside the plume and the other sensor outside the plume. A combination of this strategy and rheotaxis has been suggested as a search strategy for blue crabs (*Callinectes sapidus*) (Weissburg 2011). Moore and Grills (1999) suggest that crayfish (*Orconectes rusticus*) stay within the plume boundaries and do not follow the edge of the plume. However, the lack of plume structure data makes interpretation of their results difficult (Weissburg 2011). Crayfish appear to perform particularly well in turbulent odour plumes. Moore and Grills (1999) showed that *O. rusticus* finds an odour source faster in a river with a rough surface (cobble bed) generating strong turbulence than on a smooth sandy surface with little turbulence (Fig. 5). Marine blue crabs, in contrast, show decreased performance in more turbulent flows (Weissburg and Zimmerfaust 1993). In summary, while we have a good idea about what directional cues crayfish could use for plume tracking, the exact search strategy is still unknown.

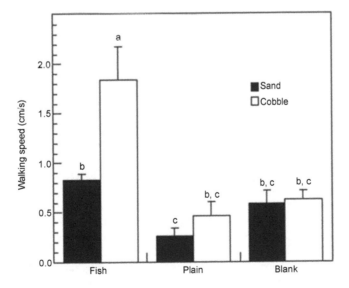

Figure 5. Mean +– walking speed for crayfish orienting in an artificial flume on different surface substrates to three stimulus treatments (Fish = Fish odour in gelatine; Plain = gelatine; Blank = no odour). Bars with the same letter do not differ significantly from each other. Modified after Moore et al. (1999), with permission of Elsevier.

Many species of crayfish inhabit lakes with little or no flow. We can speculate that under these conditions crayfish use a random walks or similar strategies to find their food (Benhamou 2007). In the near field of the odour source, fanning and leg probing are important to find the source (Moore and Grills 1999, Breithaupt 2001, Denissenko et al. 2007).

Predator detection

Alarm/predator cues and crayfish response

Avoiding predation is costly as it interferes with fitness-related activities, such as feeding or mating; prey are thus expected to modulate their responses to the different predator species in function of the level of risk that they individually pose (Ferrari et al. 2009).

In freshwater ecosystems, where visual and auditory senses are often ineffective, prey species commonly rely on chemical cues to assess predation risks. Chemicals indicating danger to a receiver may emanate from disturbed or injured conspecifics (disturbance or alarm pheromones), from injured heterospecifics that are typically preyed by the same predators as the receiver (alarm kairomones) (Chivers and Smith 1998) and/or from the predators themselves (predator kairomones) (e.g., Mathis and Smith 1993, Hazlett 1994, Schoeppner and Relyea 2009). Alarm pheromones and alarm kairomones (hereafter collectively referred to as "alarm odours") provide more immediate, but obviously less reliable, information about predation risks (reviewed by Kats and Dill 1998) than predator kairomones (hereafter referred to as "predator odour"). In fact, by perceiving predator odours, prey can theoretically recognize each single predator species and can thus exhibit defensive behaviours that are appropriate to both the riskiness of the closest predator and its hunting strategy (Turner et al. 1999, Relyea 2003).

An abundant literature shows that several crayfish species are capable of detecting chemical stimuli released by disturbed (Hazlett 1985a, 1989, 1990, Zulandt Schneider and Moore 2000) or injured conspecifics (Hazlett 1994, Mitchell and Hazlett 1996). These substances function as indicators of predation risk and may cause appropriate changes of behaviour in the alerted individuals by inducing, e.g., avoidance of areas of potential danger, freezing or reduced activity, and increased use of cover and watchful posture (Hazlett 1994). The numerous studies conducted so far on crayfish have been mostly centred on their responses to the alarm odours released by injured individuals (exceptions in Hazlett 1985a, Zulandt Schneider and Moore 2000, Hazlett et al. 2007, who investigated behaviour in the presence of stress chemicals released by crayfish urine). These studies evidenced (a) the high number of species capable of reacting to alarm substances (see Hazlett 2011); (b) the ability of some species, particularly alien species, to respond to heterospecific alarm cues (e.g., Hazlett 2000, Gherardi et al. 2002, Hazlett et al. 2003): the use of a greater variety of danger cues is crucial for invasive species in a new environments where they are exposed to unknown predators; (c) the integration of these chemical cues with other sources of stimuli (e.g., Bowma and Hazlett 2001, Tomba et al. 2001), (d) the plasticity of the response (e.g., crayfish from aquaculture ponds, where predation risks are reduced, displayed no fright responses when exposed to conspecific alarm odour; Acquistapace et al. 2004), and (e) the potential role of learning (Acquistapace et al. 2004).

Upon the detection of predator odour crayfish usually switch to a lowered posture while decreasing non-locomotory movements (e.g., *Orconectes virilis* in Hazlett and Schoolmaster 1998), even if predator odour does not always elicit this response, as in the crayfish *O. propinquus* (Hazlett 1994). *Orconectes propinquus* and *O. virilis* are congeneric, and ecologically similar, coexisting in the same area and being thus subject to the same predation risks. However, *O. virilis*, as many other crayfish species, is primarily nocturnal (Hazlett and Schoolmaster 1998), while *O. propinquus* is more diurnal (Hazlett 1994). Nocturnal animals may rely primarily on non-visual cues for predator avoidance, such as chemical and tactile inputs, while in diurnal animals vision may be a reliable method for locating approaching predators.

Alarm odours are usually more powerful than predator odours: Gherardi et al. (2011a) found that intensity of the behavioural response of crayfish from different populations is higher when they are exposed to conspecific alarm odours rather than to fish odours, pinpointing how conspecific alarm odours seems to provide a more reliable indication of possible predation events (Fig. 6).

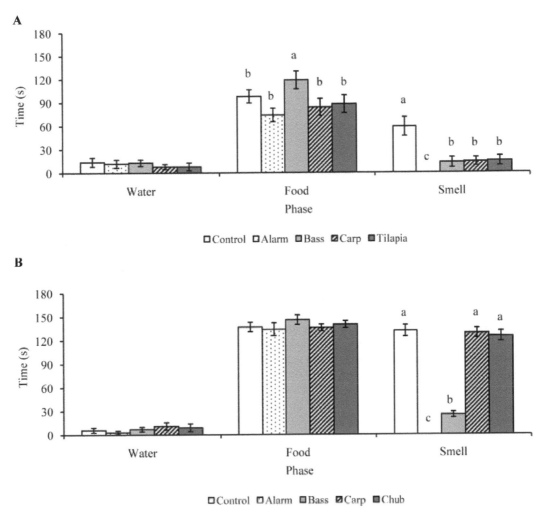

Figure 6. Comparisons among treatments in each of the three phases (water, food and smell) for the time spent feeding by 15 crayfish from each of the two populations tested, i.e., from (A) the Malewa River, Kenya and (B) Lake Trasimeno, Italy. Treatments differed for the odours tested during the smell phase: conspecific alarm odour, fish odours (i.e., odours of bass, carp and tilapia/chub) and no odour (plain water). Letters over bars denote the hierarchy among pairs after Tukey's HSD test for post hoc comparisons. Bars are means (± SE). Modified after Gherardi et al. (2011a), with permission of John Wiley and Sons.

Moreover, it seems that crayfish are able to discriminate by odour the highest-risk predator (bass; Fig. 6) from the other fish species but that their fear is not always "innate", despite the evolutionary history they can share. Without a previous direct experience with the test fish species, crayfish might perceive a general fish odour, which alerts it to risk whether or not the fish shares the same native range (Height and Whisson 2006, Gherardi et al. 2011a). Indeed, innate recognition of historical predators could be a successful strategy in ecosystems where the predictability of attack from a predator is high, and the diversity of predators is low. In contrast, prey exposed to a variety of predator species that are unpredictable in their probability of attack are expected to display plasticity in their antipredator response (Ferrari et al. 2009, Gherardi et al. 2011a).

In their natural habitats, crayfish can detect multiple stimuli in different sensory modalities and this can alter the responses to alarm/predator cues: for example, in *Pacifastacus leniusculus* both chemical and visual cues associated with predation risk elicited a higher response (decrease of locomotion and increase in shelter use) than either stimuli presented alone (Blake and Hart 1993). Crayfish response depends on hunger level, strength of odour cue and recent experiences with odours. When the odour of crushed conspecifics is presented, it almost eliminates responses to food odours in *O. virilis* and *O. rusticus* (Hazlett 2003a), increasing the time taken to find food (Tomba et al. 2001). However, ten days of starvation significantly reduce this effect in *O. virilis* (Hazlett 2003a), and the same species varied its response to differing combinations of food and predator cues, reducing (but not eliminating) feeding when predator odours were less than full strength (Hazlett 1999).

Chemical nature

Alarm substances released by injured conspecifics are contained in the hemolymph (Hazlett 2011): individuals of *P. clarkii*, when presented with solutions at different concentrations of hemolymph combined with food odour, responded in a similar fashion as after the insertion of solutions made from damaged conspecifics (Acquistapace et al. 2005). The exact chemical nature of alarm substances is still unknown but they are peptides: Acquistapace et al. (2005) found that frozen hemolymph had no bioactivity when tested 24 h after its extraction. Freezing may alter the tertiary structure of proteins, thus reducing or eliminating their bioactivity: this lability of alarm chemicals suggests their peptidic nature.

On the contrary, cues of predators such as fish may be released by the excreta or, more likely, by skin secretion. Following the hypothesis provided by Boriss et al. (1999), trimethylamine (TMA), a recurrent component of the cocktail of substances that produce fish odour and responsible for the odour emanated from decaying fish, could be the chemical component of the predator cue. It is also ubiquitous in live fish and in some other taxa (including zooplankton species: de Angelis and Lee 1994). TMA is the result of the reduction by bacteria of the mucus on fish skin, which contains trimethylamine-N-oxide (TMAO), an important cell volume regulator and protein stabilizer in both marine and freshwater fish (Boriss et al. 1999). Substances such as TMA should be thus tested as chemical stimuli in further experiments involving detection of predator odour to assess their efficacy.

Learning

Prey animals often need to learn to recognize cues from potential predators by associating those cues with direct evidence of predation such as haemolymph released from conspecifics during predation. The formation of a learned association has also been demonstrated in crayfish (Hazlett and Schoolmaster 1998): individuals of *O. virilis* with no previous experience with snapping turtles (*Chrysemys picta*) did not respond to the introduction of turtle odour alone. Following simultaneous exposure to turtle and crushed conspecific odours, they start to response to turtle odour (Fig. 7).

Crayfish can even learn to show predation-avoidance behaviours to non-predators such as goldfish, *Carassius auratu*s or gyrinid beetles or snails, when goldfish and invertebrate cues are paired with alarm odours (Hazlett et al. 2002, Hazlett 2007). Since predators of crayfish often have generalist and opportunistic feeding habits (Hobbs 1993), it seems advantageous for a species to use a broad range of information about predation risks, for instance by associating to the enemy the alarm substances emitted by heterospecifics that are members of the same 'prey guild' (Hazlett 1994, 2000). This allows individuals of that species to cope with new types of predators and is particularly advantageous when they occupy a novel environment. The ability to form associations may be especially well developed in invasive species because the introduction into new habitats may expose individuals to unknown predators. The faster they can learn (Hazlett et al. 2002), the more efficient their predator avoidance behaviours will be in decreasing the risk of predation (Mathis and Smith 1993). Invasive species such as *O. rusticus* and *P. clarkii* remember such learned associations longer than native species that are not expanding ranges (Hazlett et al. 2002). *Procambarus clarkii* is able to assess an unknown fish species as risky based on a single pairing of conspecific alarm odour and fish odour, and remembers this association without reinforcement for up to three weeks (Hazlett et al. 2002).

Latent inhibition (several repeated exposure to the predator odour alone prior the combined presentation of predator and crushed conspecific odours) and learned irrelevance (random exposure to the odours prior to their combined presentation) can impede the learned association (Acquistapace et al. 2003, Hazlett

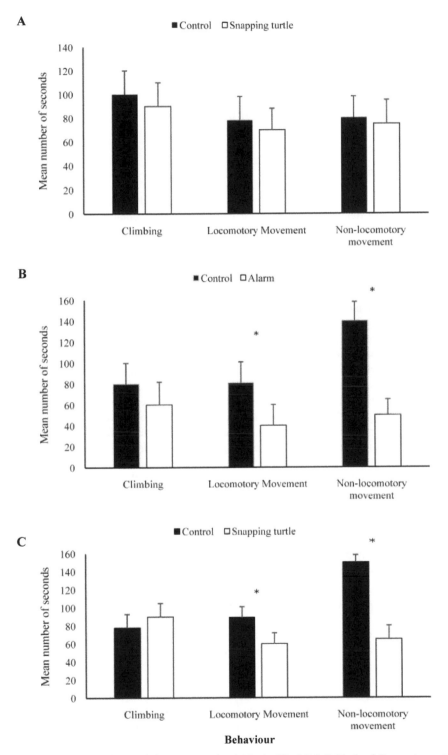

Figure 7. Mean (+ SE) number of seconds in postures and activities by 20 adult individuals of *Orconectes rusticus* during 5-min observation periods following introduction of: (A) snapping turtle odour to naive crayfish, (B) conspecific alarm odour to naive crayfish, and (C) snapping turtle odour to crayfish that had experienced simultaneous introduction of alarm and snapping turtle odours. Asterisks indicate a significant difference between test and control values. Modified after Hazlett and Schoolmaster (1998), with permission of Springer.

2003b). Crayfish can also show a second-order conditioning (Hazlett 2007): individuals of *O. rusticus* consider as a danger cue an odour (formerly neutral) when it is paired with another odour that has been previously combined with the odour of crushed conspecific (and thus has been considered as danger cue). As also reaffirmed by Hazlett (2007), this is due to a cost-benefit balance: it is better to spent less time in feeding than ignoring a predation cue.

Social behaviour

Fighting

Dominance hierarchies

Since Bovbjerg's studies (1953, 1956), crayfish have been considered as good model organisms to understand several important aspects of agonistic behaviour in invertebrates. They can indeed form and maintain, at least in confined environments, stable linear dominance hierarchies (in *O. virilis*: Bovbjerg 1953; *Cambarellus shufeldtii*: Lowe 1956; *P. clarkii*: Copp 1986; *Procambarus acutus acutus*: Gherardi and Daniels 2003; *A. italicus*: Tricarico et al. 2005; reviewed in Zulandt et al. 2008). In crayfish, the formation of dominance hierarchies is developed through dyadic social interactions, whereby winners obtain access to food, shelter and mates (reviewed in Zulandt et al. 2008).

Three possible mechanisms are responsible of the changes in dominant and subordinate behaviour in crayfish as well as in other aquatic invertebrates (Gherardi and Daniels 2003). In the first, the so-called "winner and loser effects" (Dugatkin 1997), an animal behaves in accordance to its own experience independently of its rival: a prior winning experience increases, and a prior losing experience decreases, the probability of victories. In the second, an animal can recognize an opponent's status by a pheromone, a posture or a behaviour, without any previous direct experience with it, as hypothesized for example in some crayfish species (*P. clarkii*: Copp 1986, *O. rusticus*: Zulandt Schneider et al. 2001). The third mechanism consists of recognizing the previously encountered opponents from chemical or visual cues exclusive to them ("true individual recognition") or proper of one of two categories ("binary individual recognition"; Gherardi et al. 2012). As suggested by Moore and Bergman (2005), extrinsic (e.g., previous history, sensory communication) and intrinsic chemical processes (e.g., the neurochemical state) are determinants not only for the formation of dominance relationships in crayfish but also for their maintenance. Several studies (Copp 1986, Zulandt Schneider et al. 2001, Breithaupt and Eger 2002, Bergman et al. 2003) support the role that status recognition plays in maintaining hierarchies ("assessment hierarchies": Barnard and Burk 1979), but others (Rubenstein and Hazlett 1974, Daws et al. 2002) underline the influence of past social experience, in the form of "winner and loser effects" ("confidence hierarchies": Goessmann et al. 2000). The two mechanisms may however coexist, as suggested by Gherardi and Daniels (2003) for *P. acutus acutus*, by Bergman and Moore (2003) for *Orconectes* sp., and by Tricarico et al. (2005) for *A. italicus*.

Chemical signals in dominance hierarchies

There is a heated debate around the mechanisms maintaining dominance hierarchies in crayfish and the substances involved. Crayfish can recognize higher and lower-ranked conspecifics, and the putative badges of status of several crustacean decapod species, particularly crayfish, are chemicals. The perception of those substances may induce responses typical of a "winner" (or of a "loser") in the opponent (Bergman et al. 2003). When tested in a flow-through Y-maze, *P. clarkii* individuals, irrespective of sex and previous experience, increased rates of locomotion in the presence of conspecific odour. Particularly, both naïve and experienced males also responded more aggressively to water from dominant animals than those from subordinate ones, supporting the hypothesis of dominance recognition (Zulandt Schneider et al. 1999) (Fig. 8).

The main hypothesis is that crayfish signal their status through the emission of chemical substances in their urine (Zulandt Schneider et al. 2001, Breithaupt and Eger 2002, Bergman et al. 2003, Berry and Breithaupt 2010). In dyads composed by crayfish experimentally deprived of the ability to detect chemical

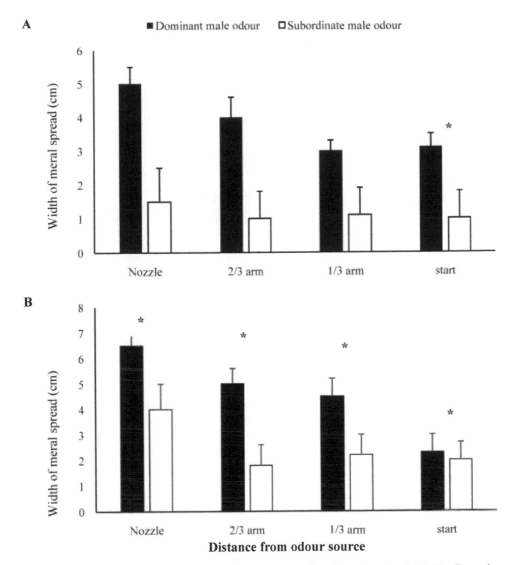

Figure 8. The effect of male conspecific odor on naive male (A, N = 9) and experienced male (B, N = 7) meral spreads to dominant (black) and subordinate (white) odors. Measurements were made at the nozzle, 2/3 the way up the arm, 1/3 the way up the arm, and at the start of each arm. Bars are mean meral spread widths (+ SEM). Asterisks mark significant differences between odor treatments at P < 0.05 using a nested ANOVA design. Modified after Zulandt-Schneider et al. (1999), with permission of Springer.

cues, by, e.g., obstructing their chemoreceptors or preventing urine release through the block of nephropores or removing the olfactory aesthetasc sensilla on the antennules, the intensity of aggression increases and fights become longer (Zulandt Schneider et al. 2001, Bergman et al. 2003, Horner et al. 2008). Urine can also influence an opponent's internal state and behaviour: in the absence of other sensory contact with a sender, exposure to social odours for five consecutive days alters the subsequent agonistic behaviour of a receiver with individuals of *O. rusticus* exposed to the odour of losers (or winners) tending to behave as winners (or losers) (Bergman and Moore 2005). The importance of urine-borne chemicals is also demonstrated by the link between aggressive interactions and urine release (Breithaupt and Eger 2002). Visualizing the urine in the blindfolded individuals of *Astacus leptodactylus* (Fig. 9), the authors found that urine is more likely to be released during social interactions than other activities, more by the future winners, and increases in quantity as aggression level increases. Moreover, it seems that *A. leptodactylus*

Figure 9. Urine release of two fighting male *Astacus leptodactylus*. Urine was visualised by injection of fluorescein dye into the circulatory system (see Breithaupt and Eger 2001). Crayfish were blindfolded reversible by opaque plastic film so they did not respond to the dye cloud. Modified after Breithaupt (2011), with permission of Springer.

uses forward-directed gill currents for chemical signalling during fights, while urine released spontaneously is carried laterally by fanning the exopodites of the mouthparts (Breithaupt and Eger 2002).

A plausible hypothesis is that the chemical substances released in the urine are metabolites of the biogenic amines serotonin and/or octopamine (Moore and Bergman 2005), the concentration of which in the hemolymph changes in function of the different social states in decapods (reviewed in Kravitz 2000). Similarly to other decapods (reviewed in Tricarico and Gherardi 2007), crayfish with serotonin injected into their hemolymph increase their aggressive motivation (*P. clarkii*, Livingstone et al. 1980; *Astacus astacus*, Huber et al. 1997, Huber and Delago 1998) and assume a dominant posture: their chelipeds are spread out and raised ("meral spread") and the abdomen is semi-flexed. Subordinates treated with serotonin show a reduced tendency to retreat and fight for longer and more strongly against dominant crayfish (Huber et al. 1997, Huber and Delago 1998). Octopamine has the opposite effects on decapod aggression. In crayfish, the injection of octopamine leads to a decrease in aggression (*P. clarkii*, Livingstone et al. 1980) and to the acquisition of a posture typical of subordinate individuals, i.e., tonic extension of the extremities with the chelipeds lowered. Tricarico and Gherardi (2007) studied the influence of biogenic amines on the agonistic behaviour of *P. clarkii*, investigating whether the hierarchical rank of fighting individuals might be altered by injecting solutions of either serotonin or octopamine into their hemolymph. The authors assessed the effect and duration of the bioamines on the behaviour, posture, and chelar force of 60 adult males paired for size. They also examined the potential of bioamines to modify dominance hierarchies by observing, for two hours after the treatment, the behaviour of three categories of familiar size-matched pairs: (1) 20 "control pairs" (both individuals injected with a physiological solution), (2) 20 "reinforced pairs" (the dominant individual, alpha, injected with serotonin, and the subordinate individual, beta, with octopamine), and (3) 20 "inverted pairs" (alpha injected with octopamine, and beta with serotonin). The authors found that the two bioamines were able to alter the posture and aggressiveness of the treated individuals in opposite directions, without however affecting their chelar force. However, the large majority of the "inverted pairs" retained their former position in the hierarchy (Fig. 10), suggesting that (1) the role that intrinsic characteristics (such as body size, weight, and chelae dimensions) and prior social experience play in maintaining dominance hierarchies in crayfish can be more relevant or that (2) other hormonal substances can influence aggression in crustaceans (Panksepp and Huber 2002).

To investigate this latest issue, Aquiloni et al. (2012) conducted similar experiments to Tricarico and Gherardi (2007) on the red swamp crayfish *P. clarkii*, using the crustacean Hyperglycemic Hormone (cHH), a multifunctional member of the eyestalk neuropeptide family and a phosphate saline solution as

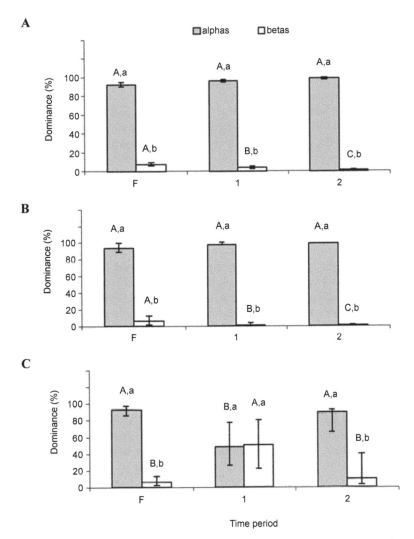

Figure 10. Means (and SE) of percentage of dominance recorded in (A) control pairs (N = 20), (B) reinforced pairs (N = 20) and (C) inverted pairs (N = 20) during the familiarization phase (F, the individuals were matched and let fight; no injections were performed) and in the first (1) and in the second (2) hour of the experimental phase (individuals were observed fighting during the first and second hour after the injection of bioamines or physiological solution). Capital letters over bars denote the hierarchy among time periods within each rank after the Student Newman Keuls' multiple comparisons, small letters, the hierarchy between alphas and betas within each time period after a paired samples Student's t test. Modified after Tricarico and Gherardi (2007), with permission of Elsevier.

control. They found that, independently of the crayfish's prior social experience, cHH injections induced in *P. clarkii* the expression of dominance that differs in relation to the original rank: in comparison with control individuals, fights of treated alphas were longer and reached a higher intensity then in treated betas. These behavioural changes after cHH injections lead to a temporary reversal of the hierarchy associated with an increased glycemia in the crayfish hemolymph, but also with the reduced time spent motionless. These results demonstrate, for the first time, that serotonin-cHH-glycemia physiological axis explains both the mechanisms through which cHH controls agonism and the expression and timing of dominant behaviours triggered by either cHH or serotonin injections. As reaffirmed by Tricarico and Gherardi (2007), other substances can be involved in crayfish agonistic behaviour, such as dopamine or 5-carboxamidotryptamine maleate. Tierney and Mangiamele (2001) found indeed that the level of aggression in *P. clarkii* was reduced by serotonin but was however enhanced by a serotonin analogue, 5-carboxamidotryptamine maleate.

Specimens of *O. rusticus* implanted with 5-HT were indistinguishable from the controls in terms of fighting behaviour when the rate of the substance release was slow, but were more aggressive when it was fast (Panksepp and Huber 2002). The moulting hormone ecdysone is another candidate, since crayfish fighting and escape behaviour changes dramatically over the moult cycle. However, the chemical identity of these pheromones remains elusive, and no bioassay has been established that could serve to guide fractionation of the urinary compounds (Breithaupt 2011).

As underlined by Zulandt Schneider et al. (2001), chemical signals play a central role in determining dynamics and outcome of social interactions in the laboratory, but it is unclear whether the hydrodynamics of natural habitats allows for the successful use of chemical signals during social interactions in nature (Bergman et al. 2006). A form of multimodality (odour combined with sight) can be used by *P. clarkii* females to recognize and choose the dominant male after having eavesdropped on two fighting males (Aquiloni and Gherardi 2010). *Cherax destructor*, too, recognizes familiar conspecifics, using either chemical or visual cues (Crook et al. 2004), even if a subsequent series of experiments on this species showed that crayfish prevalently use vision in a form of recognition that shows several properties of true individual recognition (Van der Velden et al. 2008). Also, Zulandt et al. (2008) found in *O. rusticus* that bystander crayfish lost significantly more to a tester crayfish if they have previously only seen two crayfish fighting (the eavesdropping effect). The authors hypothesized that (1) observing fights could alter the stress level of the bystander crayfish, which could decrease the functioning of serotonin, leading to a reduction of aggression and subsequent losses in aggressive interactions, or that (2) the bystander crayfish release chemical signals during the observational stage of the experiment, not having enough reserves of chemical signals to use during subsequent encounters. The lack of ability to use appropriate chemical signals during a subsequent interaction would lead to a decrease in aggression, slowed escalation and negative fight outcomes (Zulandt Schneider et al. 2001).

However, the importance of chemical signals in fighting behaviour of crayfish is undoubtable: Bergman et al. (2005) and Simon and Moore (2007) have shown that urine release during social interactions is often carried out in conjunction with specific behaviours, and it is often quite limited in duration. Breithaupt and Eger (2002) showed that aggressive fight elements are only effective in repelling the opponent if they are accompanied by urine signals. Crayfish can thus use chemical signals alone or coupled with visual ones depending on the situation. However, the substances involved as well as the information transferred still await clarification.

Reproduction

Searching for mates and assessing the quality of mates requires an efficient system of information exchange between individuals. Such system relies on composite stimuli perceived through several sensory channels.

To improve the reliability of communication and the memorability of the information, multiple sensory channels are often engaged simultaneously (i.e., multimodality, Rowe and Guilford 1999) also in relation to the ecology of the habitat occupied by the species, the timing of communication and the information conveyed (Bradbury and Vehrencamp 2011). In the aquatic environment, olfaction plays the main role in the communication, because chemicals convey long-distance signals at most conditions while visual signals are restricted to a limited period of the day, and to low turbidity conditions. In some contexts, however, other sensory channels participate in gathering of the information: turbulent water quickly disperses the chemicals, whereas stagnant water disperses them extremely slowly, so additional sensory channels could be necessary to correctly locate the source of a signal (Vickers 2000). All these general considerations apply also for crayfish. The literature is full of examples on crayfish showing the role of olfaction during the different phases of reproduction, as well as in other social behaviours (Gherardi 2002, Breithaupt 2011). Olfaction acts alone or together with vision depending on the species (Acquistapace et al. 2002), sex of the receiver (Aquiloni et al. 2009) and the conveyed message (Aquiloni and Gherardi 2010), whereas the use of other sensory channels is less known (e.g., acoustic signals, Favaro et al. 2011) or still anecdotal (e.g., contact signals, as in *Paleomonetes pugio*: Caskey and Bauer 2005). In general, chemical signals involved in communication among conspecific are known as pheromones (Wyatt 2014) and their role has been demonstrated in many decapod crustaceans (*Callinectes sapidus*: Gleeson 1980;

Carcinus maenas: Hardege et al. 2002; *Homarus americanus*: McLeese 1970, Dunham 1979, Cowan 1991), crayfish included (e.g., *Orconectes virilis*: Hazlett, 1985b; *Pacifastacus leniusculus*: Stebbing et al. 2003a, Berry and Breithaupt 2010; *Procambarus clarkii*: Ameyaw-Akumfi and Hazlett 1975, Bechler et al. 1988, Dunham and Oh 1992, 1996). Through pheromones, crayfish recognize the species and the sex of individuals, they select potential mates based on some preferred traits and, in combination with visual stimuli, they are capable to recognize individuals, as described below.

Recognition of species

To identify the correct species, crayfish typically rely on pheromones as confirmed by studies investigating recognition between sympatric species, e.g., *O. virilis* and *Orconectes propinquus* (Tierney and Dunham 1982, 1984). When this mechanism fails, mating may occur between males and females of two different species resulting in either reproductive interference (between *Austropotamobius pallipes* and *Astacus leptodactylus* or *P. leniusculus* in England; Holdich et al. 1995) or hybridization with the eventual genetic assimilation of one species, as in the case of the invasive *O. rusticus* replacing *O. propinquus* in Michigan (Perry et al. 2001), or hybridization with loss of genetic biodiversity, as in the case of the two endangered species *A. pallipes* and *A. italicus* in Italy (Ghia et al. 2011).

Recognition of sex

Pheromones are also involved in sex recognition. Receptive females of several species (*Cambarus robustus*, *O. propinquus*, *O. virilis*, and *P. clarkii*) are known to emit urine-borne sex pheromones (reviewed in Gherardi 2002). The release by females of chemicals that stimulate mating behaviour in males was first shown by Stebbing et al. (2003a) in a laboratory experiment. The exposure of males *via* air-stones to the water conditioned by mature females induced in them the "handling" of air-stones; handling included the behavioural patterns typical of mating, i.e., seizure, mounting, and spermatophore deposition. Berry and Breithaupt (2010) used a behavioural bioassay involving both a male and a receptive female. The nephropores of the female were blocked and, when urine of a receptive female was artificially introduced, the males were more likely to display mating attempts towards the female than upon introduction of water. The correlation between urine-borne chemicals and mating was well illustrated by Simon and Moore (2007) using urine visualization: within reproductive pairs, both male and female crayfish generated currents and released urine at higher rates than the pairs in other treatment groups, thus suggesting the ability by crayfish to adapt the use of hydrodynamic and chemosensory communication as a function of their reproductive state (Fig. 11). These results also support the hypothesis that urine contains a putative chemical signal used for identification of reproductive *status* and information of the mate quality (Aquiloni and Gherardi 2008a).

Crayfish may also use vision for sex identification (Dunham and Oh 1996), particularly in some species, such as *P. clarkii*, with a more diurnal timing of activity. More recent studies pinpointed the bimodal nature of sex recognition in this species: crayfish rely on both smell and sight but the relative importance of chemical and visual cues varies between sexes. Chemical recognition seems to be a male prerogative (Aquiloni et al. 2009, Fig. 12a,c), whereas in other species that inhabit clearer waters, such as *A. pallipes*, males use both olfaction and vision (Acquistapace et al. 2002). Contrary to males, females seem to make more extensive use of vision, although visual stimuli are not sufficient to identify the sex of a conspecific but they should be combined with the male odour to suppress female aggressiveness (Aquiloni et al. 2009, Fig. 12b,d). From a theoretical point of view, this means that in the females of *P. clarkii* one signal component (odour) modulates the "message" of another (vision) in a non-redundant bimodal system of communication, rarely described in invertebrates.

Mate choice

In many animal species, because the cost of reproduction for males is generally lower than for females, males are less selective in their mate choice than females (Trivers 1972). In crayfish, on the other hand, the great investment for the production of spermatophores (Dewsbury 1982) together with other factors,

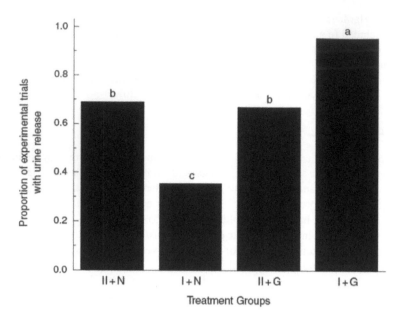

Figure 11. Test of urine current visualization in pairs at different reproductive stages in the crayfish *Orconectes rusticus*. Proportion of experimental trials that resulted in urine release from either the male or the female crayfish. Titles along the x-axis represent the reproductive state of the male (first letters) and the female (second letters). Initial letter in the treatment groups refers to the reproductive status of males (I: reproductive; II: non-reproductive), while the second letter refers to the reproductive status of females.

(N: non-reproductive; G: reproductive). Bars with different letters above them are significantly different from each other using a chisquared analysis for proportions followed by a multiple comparisons test for proportions (p < 0.05). Reprinted from Simon and More (2007), with permission of John Wiley and Sons.

such as an unfavourable operational sex-ratio, a restricted mating period and a huge difference in female quality (reviewed in Gherardi 2002), contribute to an increase of the reproductive costs in males, leading to the occurrence of a mutual mate choice.

Mate choice by females

Several male traits have been suggested to serve as choice criteria by females. As a first, females of a wide range of species were found to select mates with large body size: *A. astacus* (Furrer 2004), *A. pallipes* (Villanelli and Gherardi 1998, Gherardi et al. 2006), *O. rusticus* (Berrill and Arsenault 1984), and *P. clarkii* (Aquiloni and Gherardi 2008b). This preference might have evolved because large males are relatively more fertile with respect to smaller individuals (but in *A. pallipes* the extent of ejaculates decreases with the increased male size as the effect of senescence; Rubolini et al. 2006, 2007), and they offer high-quality vital resources to females, such as breeding burrows, because they are dominant in intra-sexual competition (but in *O. rusticus* females are known to extrude and brood their eggs in isolation; Berrill and Arsenault 1982). How do females recognize and select the larger male? In *P. clarkii*, females used a combination of visual and chemical stimuli to choose the larger male (Aquiloni and Gherardi 2008b), whereas the sight alone or the odour alone of an individual, independently of its size or sex, elicited aggression (Aquiloni et al. 2009, Fig. 2d). That is, the sight of a mate of a larger size is not *per se* an index of the "best" partner but it must be confirmed by chemical stimuli that, in turn, provides information about the species, the sex, and the reproductive condition of the potential mate.

Recognition of the dominant mate by the female appears to be a more complex process. Although size is a reliable predictor of fighting outcome (Bovbjerg 1953), a variety of other intrinsic and extrinsic factors might affect the dominance status (e.g., dietary effects: Vye et al. 1997; moult stage: Tamm and Cobb 1978; the experience of previous agonistic encounters: Rubenstein and Hazlett 1974). In addition,

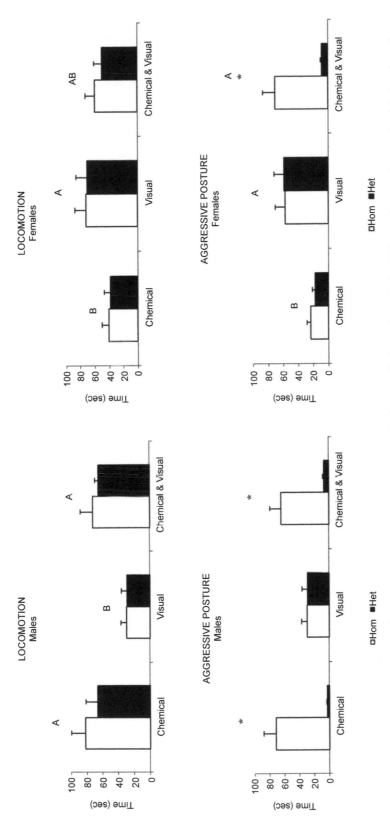

Figure 12. Behavioural test to assess the role played by chemical and visual cues for sex discrimination in *Procambarus clarkii*. Differences between the control and the test phases (mean ± SE) in the time spent (in sec) by males and females in locomotion (a, b) and in the aggressive posture (c, d): comparisons among treatments and sex of the source crayfish (Hom: the same sex as the test crayfish; Het: the other sex) per sex of the test crayfish. Letters over bars denote the hierarchy among treatments (C, V, and CV); asterisk denotes significant differences at p < 0.001 after a Wilcoxon signed ranks test between Hom and Het. N = 20 per treatment. Reprinted from Aquiloni et al. (2009), with permission of Springer.

a slight alteration of social contexts induces a quick switch of the order in a hierarchy, e.g., when a group is repeatedly reconstituted (Dugatkin et al. 1994) or when the order followed to add the same individuals to the reconstituting group is reverted (Landau 1965). Thus, although it has been shown that crayfish recognize the social status of conspecifics from their odour (Zulandt Schneider et al. 1999, 2001), making a choice among potential mates in a crowded context could be quite complex. Aquiloni et al. (2008) found that *P. clarkii* females are able to recognize the dominant between to equally sized males only after having eavesdropped on them fighting. By social eaves dropping, females seem to make low-cost, direct comparisons between the two potential mates, obtain information about the quality of the signallers, and then use this information to guide their future decisions. Again, female choice seems to rely on a combination of visual and chemical stimuli (but note that in *P. leniusculus*, the males do not use chemical signals during reproductive interactions; Berry and Breithaupt 2010).

Other targets of female choice are the size, the symmetry and the quality of chelae as ornament. Large chelae serve as powerful weapons during intraspecific fights, being obvious determinants of wins (in: *C. robustus*, Guiasu and Dunham 1998; *O. propinquus*, Stein 1976; *O. rusticus*, Schroeder and Huber 2001; *P. clarkii*, Gherardi et al. 1999). Males with large chelae are in general more successful than those with small chelae in copulating with females, but large chelae may entail some costs: in *C. dispar*, the larger chelae of males were associated with decreased escape performance. It is thus expected that they are subject to intersexual selection, being evolved because of females' preference (Villanelli and Gherardi 1998). This statement has been questioned in *P. clarkii*: females do not show any preference when simultaneously offered with similarly-sized males with large and small chelae (Aquiloni and Gherardi 2008b). Similarly, *A. pallipes* females adopt two co-occurring strategies of maternal allocation depending on male traits, laying fewer but larger eggs for relatively small-sized and large-clawed males and the reverse for relatively large-sized and small-clawed males (Galeotti et al. 2006). Chelar asymmetry, due to the loss of one of the two chelae, has no apparent effect on mate choice, at least in *P. clarkii* (Aquiloni and Gherardi 2008b), whereas in *A. pallipes*, owning asymmetric claws, decreases the ability of males to win fights, to secure females for copulation (Rubolini et al. 2006), and to remove the spermatophores of the previous mate when they are the second mates (Galeotti et al. 2008). In some species, chelae ornaments are objects of female choice. In particular, the location of a vulnerable red patch membrane on the cheliped propodi of *C. quadricarinatus* males renders this structure a handicap for the bearer: males that can manage in spite of a handicap send a message of their proven quality. The red patch might be an honest signal of male quality as a reflection of his health status because its colour derives from carotenoids that crayfish cannot synthesize but obtain from the diet (Karplus et al. 2003).

Mate choice by males

Due to their polygynous habit and the long time needed to produce sperm, males may be limited in their sperm supply: in *A. pallipes*, at the end of the mating season, *vasa deferentia* weight of laboratory-mated males was up to 55% lower than in the pre-mating season (Woodlock and Reynolds 1988). This was also found in a laboratory experiment: when *A. pallipes* males were offered different receptive females in sequence, independently of their size and of the size of the mates, their ejaculate size decreased with consecutive matings (Rubolini et al. 2007). Male sperm limitation might induce the females to choose non-sperm-depleted males on one hand (but no evidence has been collected so far; Aquiloni and Gherardi 2008c) and for the males to evolve forms of mate selection on the other. Male choosiness might also be determined by the long copulation time of some species, large investment in sperm production, and restricted mating periods (less than one month in *A. pallipes*; Villanelli and Gherardi 1998). As a confirmation of this, a recent study showed that the selection criterion followed by *P. clarkii* males is female size, with the larger and more fecund females being preferred (Aquiloni and Gherardi 2008b). It is in fact a general rule in crayfish that pleopodal egg number increases with female body size (e.g., in: *A. astacus*, Cukerzis 1988; *A. leptodactylus*, Köksal 1988; *A. pallipes*, Rubolini et al. 2006; *C. japonicus*, Nakata and Goshima 2004; and *P. clarkii*, Nobblitt et al. 1995). So, *A. pallipes* males were found to adjust the volume of their ejaculate to the size of the females (Rubolini et al. 2006): males individually paired in the laboratory with receptive females of different size allocated more sperm (assessed by the area of the female pleon covered by the deposited spermatophores) to larger females, i.e., the mates that provide the greatest fertilization returns. In Cambaridae, on the contrary, spermatophores are inserted into the *annulus ventralis*, which makes sperm inaccessible for the subsequent males. In these instances, sperm competition may be avoided by adjusting the length of copulation as a function of the female mating status (suggested for *O. rusticus*; Snedden 1990), depositing a mating plug in the opening of the receptacle (suggested for *O. rusticus*; Crocker and Barr 1968), removing the plug of a previous male with copulatory stylets (Berrill and Arsenault 1984), or selecting virgin females identified through some pheromones (in *P. clarkii*, Aquiloni and Gherardi 2008b; in *Orconectes quinebaugensis*, Durgin et al. 2008), and defending them. As shown in *P. clarkii*, sight and smell work differently in male and female, possibly due to the diverse role played in crayfish by the two sexes during mating (Aquiloni and Gherardi 2008a): in male, contrary to female that rely on the co-occurrence of sight and smell, the choice of high quality mate (= larger mate) depends only on smell, while visual and chemical cues together render them willing to mate.

Recognition of individuals

Individual recognition in crayfish, as well as other invertebrates, has been controversial because of the apparently complex neuronal machinery involved and the unanswered question about the adaptive significance of such a refined form of social recognition in some "asocial" species (Gherardi et al. 2012). However, some "asocial" contexts certainly favour the evolution of individual recognition because this ability may reduce the costs in case inflicted and, at the same time, brings considerable benefits to both the signaller and the receiver (Tibbetts and Dale 2007). In the solitary *C. destructor*, Crook et al. (2004) have shown that individuals discriminate familiar from unfamiliar crayfish using a combination of visual and chemical cues and adopt a strategy, known as the "dear enemy" (*sensu* Fisher 1954), of approaching and spending more time with familiar individuals in order to reduce the energetic cost and physical damage from high intensity fights that occur between unfamiliar crayfish. The same species was found to recognize familiar conspecifics using "facial" features learned while fighting with it and to retain memory of it for at least 24 h (Van der Velden et al. 2008, Fig. 13). The width and colour of the crayfish "face"

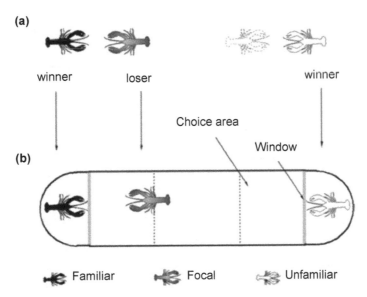

Figure 13. Test paradigm for visual recognition in the crayfish *Cherax destructor*. (a) Fights between size-matched crayfish to familiarize opponents. These fights occurred, one pair at a time, in the central area of the tank shown. The tank was cleaned between encounters. (b) Winners and losers were transferred to the test arena with a pen and choice area at each end. The focal losing crayfish could spend time in any of the three areas, two of which would indicate preference for proximity to a specific animal. The figure shows a focal crayfish visiting the familiar animal from the previous encounter. Features, for example facial width and colour, were varied between the stimulus animals in the pens. The window prevents chemical and mechanical cues from passing, but does not interfere with vision. Reprinted from Van der Velden et al. (2008), Fig. 1.

(i.e., the region anterior to the cephalic groove) were uncorrelated between each other and presented a relatively high degree of variability within the population, thus suggesting that both may be favoured for recognition in this species (Van der Velden et al. 2008). More refined ability was found in *P. clarkii* females, which recognize and choose the dominant as a mate after having eavesdropped on two fighting males (Aquiloni et al. 2008). To do this, females do not use any characteristic proper of a dominant male, including his *status* pheromone, since they became incapable of distinguishing the dominant in unknown pairs. *P. clarkii* females are able to recognize the winners as individuals—and not as generic dominants-using the co-occurrence of visual and chemical stimuli emitted by the eavesdropped male (Aquiloni and Gherardi 2010). This finding supports the hypothesis of true individual recognition in crayfish, opening new intriguing question about the cognitive ability of this taxon.

Parent-offspring interactions

Parent-offspring associations have been reported from a variety of aquatic and terrestrial crustacean species (Thiel 1999), but information of social interactions of family groups is available for only 40 of the 136 known crustacean species that lives in parent-offspring group, corresponding just to 29% of the total (Thiel 2007). Among these reports, only 10 are from crustacean decapods, including five from crayfish (fourin Brachyura and one in Caridea: Thiel 2007) due to the narrow breadth of the species studied so far in this taxon. This drawback is particularly evident when important questions about evolutionary mechanisms through which parental behaviour develops are waiting for a response. According to Clutton-Brock (1991), all the activities towards the offspring that represents a cost for the parent qualify as parental care (Clutton-Brock 1991). The return for these expenditures by parents is increased offspring growth and survival (e.g., Thiel 2003), which improves the inclusive fitness of parents. If parents continue to care potentially self-sufficient juveniles, this behaviour is termed extended parental care (hereafter: XPC; Thiel 1999). There are many reports of crustaceans with XPC from terrestrial and freshwater environments, even though the large majority of crustacean species are marine (Thiel 2003). This imbalance suggests that the evolution of XPC may have been favoured in the stressful conditions encountered in terrestrial and freshwater environments (Hazlett 1983); freshwater crayfish are among the few decapods with XPC (Hazlett 1983). Unlike most of their marine relatives, freshwater crayfish inhabit environments characterized by harsh and unsettled condition of temperature, oxygen and, in some cases, water availability. They often live in dense populations in which they interact aggressively to establish dominance hierarchies, a social structure expected under conditions of limited resources (Wilson 1975), and in which juveniles and weak specimens are frequently cannibalized (Nyström 2002). Such conditions have exerted an evolutionary pressure both towards a direct development of offspring, resembling the adults just after hatching, and complex XPC on hatched juveniles (Hazlett 1983).

   Certainly due to the poor number of species studied in this taxon, little is known about the interaction between parents in family groups as well as about the role played by males in XPC. However, in contrast to the most other crustaceans where males and females perform very brief mating associations, in some crayfish males and females may cohabit for prolonged period (Thiel 2007). This is well illustrated in the case of *O. virilis*. Males of this species build a burrow where, following the copulation, they continue to cohabit with the female through the winter (Ameyaw-Akumfi 1976), leaving it before offspring release (Hazlett 1983). Prolonged male-female cohabitation was also reported for *P. clarkii* and *P. acutus* (Thiel 2007), but the male's contribution towards the offspring is not clear. To our knowledge, the mother may be considered the exclusive caregiver of the offspring (Gherardi 2002). Mother-offspring interactions in this taxon may involve a wide array of different behaviours of increased complexity ranging from a continuous ventilation of eggs and hatchlings (Reynolds 2002), grooming or feeding of developing offspring (Pandian 1994), adjustment of pleopod beating in response to changed microclimatic conditions (Ameyaw-Akumfi 1976, Bechler 1981), selective removal of non-viable eggs (Tack 1941) or dead juveniles (Ameyaw-Akumfi 1976, Hazlett 1983), to aggressive defence of juveniles against attacks or predation events (Figler et al. 2001), and support of juveniles in learning processes and in dangerous situations (Ameyaw-Akumfi 1976). In some species, females assist in the hatching process by lifting their abdomens and vigorously waving their pleopods (Hazlett 1983). However, despite the appeal of this topic, the knowledge of the parental behaviours in crayfish is mostly anecdotal due to brief occasional observations and a general lack of specific studies. The literature is also scant on chemical signals involved in such interactions. Most studies on parent offspring communication in crayfish were conducted in the 1970's and 1980's and, as noted by Levi et al. (1999), these interactions have been generally described without analysing variation of specific acts of the female during the developmental stages. In this section, we propose a synthetic view of the available literature on the behavioural interactions between a female crayfish and her offspring during the several stages of their development, highlighting the role of chemical cues in such a complex relationship.

First and second stages of juveniles

From hatching until the third stage of development, offsprings remain attached to their mother's abdomen by transient structures (the telson thread and pereopodal hooks; Vogt and Tolley 2004) with some differences among families, as reported by Levi et al. (1999): southern-hemisphere crayfish (Parastacidea) differ from the northern-hemisphere ones (Astacidae and Cambaridae) in that they cling to their mother in an upside down position, using small curved hooks on pereiopods 4 and 5 (Suter 1977, Sokol 1988, Merrick and Lambert 1991). Afterwards, they freely crawl onto her body for a period ranging between few weeks in most species and three-four months or more in *P. clarkii* (Huner and Barr 1991, Huner 1994) and *Paranephrops zealandicus* (Whitmore and Huryn 1999).

During this period, the mother shows a limited number of locomotion bouts and a low speed of movement (Mason 1970, Hazlett 1983, Hamr and Richardson 1994). In some species, she remains secluded in shelter or burrows until the young have reached independence (see Thiel 2007) and, contrary to non-maternal adults, executes few cleaning and feeding acts just towards non-viable eggs or death juveniles (Little 1976, Lundberg 2004, Aquiloni and Gherardi 2008d). This behaviour avoids excess contamination of the water in the burrow (Little 1976), but the mechanism that the mothers have to discriminate between viable or non-viable offspring is still unknown. In several species (*Pacifastacus trowbridgi*, Mason 1970; *Astacus astacus*, *O. virilis* and *P. clarkii*, Burba 1983; *P. clarkii*, Figler et al. 1995, 1997, 2001; *Cherax quadricarinatus*, Levi 1997), while aggression by mothers toward juveniles appears to be suppressed, aggression toward other adults, males and non-reproductive females is strongly enhanced probably as a mechanism for protecting their young. These findings suggest that behavioural interactions among conspecifics are under hormonal control, even though present knowledge on this topic remains inconclusive (Thiel 2007). Otherwise, the maintenance of the maternal behaviour seems to be the result of a complex interaction between the hormonal *milieu* of the brooding female and the feedback she receives from her brood (Levi et al. 1999). Studies by Little (1976) suggested that mothers do not need visual input from offspring, since they maintain the same feeding patterns and the same responsiveness to the water with odour of juveniles even if blinded. However, it is also interesting to note that the odour alone is not enough to maintain the maternal behaviours: without any physical contact with the juveniles for more than 12 days, the female starts to cannibalize them. Cannibalism also appears when the number of young attached to the female body becomes less than 10 (Little 1976), indicating that mechanoreceptors on the female pleopods are involved (Figler et al. 1997, 2004). More recently, another study highlights the possible role played by chemio-mechanoreceptors (serrate setae) on the fifth walking legs: reproductive females have a higher number of such receptors than non-reproductive females or males (Belanger and Moore 2013).

Third stage of juveniles

The third-stage juveniles, once sense organs have been fully developed (Vogt and Tolley 2004), occasionally leave their mothers to explore the environment and feed for short time periods. At the end of their excursions or if disturbed, they return under their mother's abdomen or attach on other parts of her body for shelter (Hazlett 1983, Ameyaw-Akumfi 1976, Figler et al. 1997, Gherardi 2002). A similar "return" behaviour has been also described in a number of peracarids (*Neohaustorius schmitzi*, Croker 1968; *Gammarus palustris*, Borowsky 1980; *Parallorchestes ochotensis*, Kobayashi et al. 2002) but in these species the mothers simply allow juveniles to associate with them (Dick et al. 1998). In crayfish, on the contrary, mothers seems to facilitate the juveniles' return to the female's abdomen behaving as follows: they show reduced locomotion, execute few cleaning and feeding acts, do not show any attempts to grab the approaching juveniles, and also assume a characteristic "spoon-like" telson posture with the abdomen extended and the caudal fan opened and slightly folded (Aquiloni and Gherardi 2008d).

At this stage, the interactions between mothers and offspring appear to be mediated by chemical signals. Little (1975, 1976) and Ameyaw-Akumfi (1976) showed that third-stage juveniles of at least five crayfish species orient towards the water in which a brooding female (either the biological or the non-biological mother) has been held, but not towards the water conditioned by the odour of a male or of a non-brooding female, thus suggesting the release by the former of a maternal pheromone. Levi et al. (1999) reported evidence on the role of contact pheromones in mother–offspring relationships: dependent juveniles of a moulted female of *C. quadricarinatus* clustered to the mother's exuviae rather than to their mother likely due to chemical cues deposited on it, but absent in the newly moulted mother.

While, on one hand, this shows the existence of a 'brood' pheromone in crayfish that allows dependent juveniles to discriminate between maternal and non-maternal crayfish, on the other hand it is still unclear whether juveniles are able to discriminate their own mothers from other maternal crayfish. Figler et al. (1997) showed that both brooding females and juveniles of *P. clarkii* are receptive to cross-fostering, indicating that neither dependent juveniles nor maternal females of that species discriminate between related and unrelated individuals. However, this issue has not been completely solved. In another laboratory experiment carried out by Aquiloni and Gherardi (2008d), *P. clarkii* juveniles accepted foster mothers but not as fast as biological mothers, leading the authors to suggest that juveniles can discriminate their biological mother from other brooding females but are ready to accept also the latter after having experienced their maternal behaviour (Fig. 14). The decision of whether to escape from or to attach to the adult may be more likely

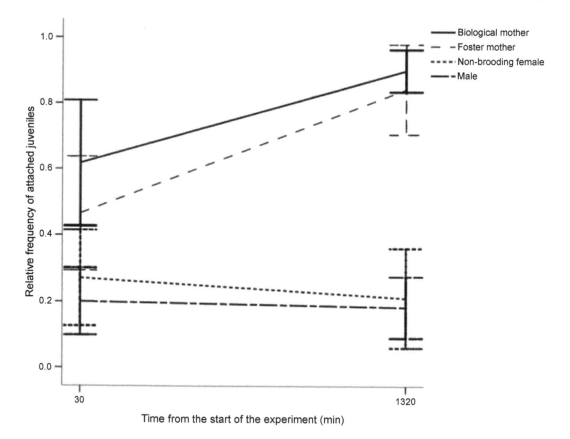

Figure 14. Relative mean frequency of the juveniles (n = 13) that were counted on their putative mother's body 30 and 1320 min after the start of the experiment compared among treatments. Error bars show standard error. Data from Aquiloni and Gherardi 2008d.

taken by combining chemical information with visual stimuli associated with maternal behaviour (Hazlett 1983). On the contrary, juveniles of the cambarid *O. limosus* are able to distinguish between their own and unfamiliar mothers based entirely on dispersed chemical cues (Mathews 2011). Thus, the specificity of recognition between mothers and juveniles and the mechanisms involved in such recognition remain unclear.

Towards independence

After the fourth stage, all sense organs necessary for an independent life, such as eyes, olfactory aesthetascs, gustatory fringed setae, hydrodynamic receptor hairs are now well developed and operating (Vogt and Tolley 2004), thus reducing progressively brood care till the weaning that consists in the young definitively leaving their mother. At this stage, siblings become progressively independent with some differences among individuals: while the larger ones are already completely autonomous, medium and smaller ones remain longer attached to the mother's pleopods (Ameyaw-Akumfi 1976). These differences in size and behaviour determine aggressive interactions among cohabiting juveniles, during which someone could be cannibalized (Mason 1970).

While juveniles are reaching the independence, the mothers become a menace to their offspring, as during the latest phases of XPC they start to feed on the approaching juveniles (Mason 1970, Hazlett 1983). These evidences, as reported above, confirm the role that offsprings play in the maintenance of maternal behaviour: the lack of appropriate mechanical, and likely chemical, stimuli due to progressive parting of juveniles from the mother's body induces the disappearance of maternal hormonal status (Little 1976, Figler et al. 1997, 2004).

Little is available in literature about interactions between mothers and offspring or among siblings at this stage, and even fewer studies on the recognition system and the possible cues involved. In any case, leaving their mother, juveniles have two possibilities: they disperse or they remain associated, more or less strictly, with relatives. Consequently, the modalities for the recognition of related conspecifics should be more or less sophisticated. Most crayfish species with independent juveniles appear incapable of recognizing their relatives (Thiel 2007). In such species, the likelihood of an individual encountering the own relatives should be extremely low, thus a refined mechanism to recognize them has no adaptive value. Conversely, some crayfish species are known to maintain a long association with relatives. Among these, there are species of the genus *Engaeus*, in which family groups (Healy and Yaldwyn 1970) were found to inhabit communal burrows (Clark 1936, Suter and Richardson 1977). These species are expected to display complex social behaviours similar to those described in terrestrial isopod *Hemilepistus reamuri* (Linsenmair and Linsenmair 1971), but no studies have ever been carried out to test this hypothesis. Moreover, the likelihood of interactions with relatives is also high when juveniles dig their burrow in the neighbourhood of that of the mother (e.g., some *Orconectes* species, Fitzpatrick 1987) or inhabit the same pond (e.g., in *Procambarus layi*, Payne 1972). In these species, mothers are likely to encounter their own juveniles regularly after they have become independent and consequently a refined mechanism to recognize the own offspring might have an adaptive value. Under some ecological conditions, in fact, the ability to discriminate between kin and non-kin may facilitate behavioural interactions that are mutually beneficial: the reduction of antagonistic interactions (sea trout, Höjesjö et al. 1998), or the aggression towards conspecifics (lobsters, Karavanich and Atema 1998), or even cannibalism, as described both in vertebrates (Loekle et al. 1982, Pfennig et al. 1993, 1994, Green et al. 2008) and invertebrates (Nummelin 1989, Bilde and Lubin 2001, Schausberger and Croft 2001).

An experiment carried out by Mathews (2011) on *O. limosus* suggests the existence of kin recognition in crayfish. *Orconectes limosus* juveniles become increasingly independent from their mother but remain associated with one another for long, so that a reciprocal recognition could have clear fitness benefits. During the experiment, mothers showed a rapid decline in "feeding inhibition" once they were separated from their newly independent juveniles, and they accepted food regularly within a few days after separation. However, mothers were significantly less likely to cannibalise their own young than young of other females for at least 10 days later (Fig. 15). In the same period, juveniles detected their own mothers based on chemical cues even after they had apparently become independent of maternal care (Fig. 16). As the author suggests, the mother–offspring recognition could extend substantially beyond the period of direct

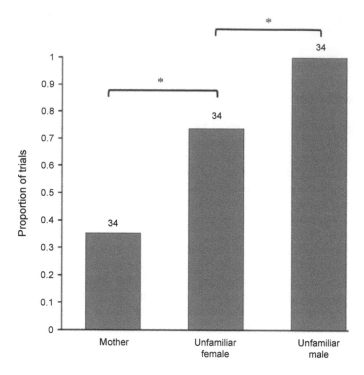

Figure 15. Proportions of trials in which juvenile crayfish were consumed by adults who were either their mothers, unfamiliar females, or unfamiliar males. Numbers above bars indicate sample sizes; asterisks indicate statistical significance at $\alpha = 0.05$. Data from Mathews 2011.

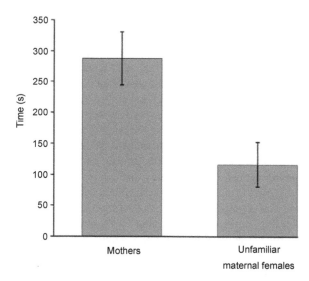

Figure 16. Mean time spent by juveniles in arms of the Y-maze containing water conditioned by exposure either to their mothers or to unfamiliar maternal females. Error bars show standard error. Data from Mathews 2011.

maternal care in cases of long-lasting associations between mothers and offspring, and perhaps between siblings. This finding has the potential to open avenues for the future research on the still understudied field of mother-offspring relationship in crayfish.

Applied aspects

Effect of pollutants on chemosensory behaviour

Anthropogenic activities have led to discharges of chemicals into freshwater environments that are detrimental to the health of aquatic organisms including crayfish. While traditionally the impact of the chemicals on animals was measured as dose dependent mortality (e.g., LD_{50}), it is now recognized that sublethal effects can exert strong ecological effects (Lürling and Scheffer 2007, Lürling 2012). Pollutants have been shown to disrupt endocrine processes in crustaceans such as those underlying moulting and reproduction (Rodriguez et al. 2007) and can even disrupt vital information transfer between organisms. This has been called "info-disruption" (Lürling and Scheffer 2007, Lürling 2012). A variety of contaminants has been shown to interfere with chemical information transfer in crayfish. These include pesticides, metals and pH changes (Table 1). Info-disruptors can interfere with foraging behaviour by impairing chemical orientation and reducing the likelihood of finding the food (Lahman et al. 2015). They can also affect intraspecific agonistic interactions (Cook and Moore 2008), and anti-predator responses in crayfish (Wiggington et al. 2010). The mechanism of chemical info-disruption is still unclear. Heavy metals such as copper do not appear to destroy the chemoreceptor cells since the behavioural impairment is reversible when copper is removed (Lahman et al. 2015). Metals and pesticides could act as competitive inhibitors, binding to the olfactory receptor site preventing odour molecules from doing so (Olsén 2011). pH changes could affect either the odorant or the chemoreceptor by changing the shape of the molecule (Leduc et al. 2013). Hydrocarbons, surfactants, humic acid and nutrients were all shown to impair chemosensory behaviour in aquatic organisms (Lürling 2012) but their effect on crayfish behaviour has not been tested yet. The list of pollutants (Table 1) is indeed incomplete, and the exact mechanisms of info-disruption is unknown for any of them. More research is thus needed to understand how chemoreception is altered by info-disrupting chemicals.

Table 1. Examples of pollutant induced info-disruption in crayfish.

Pollutant	Disruption	Concentration/pH	
Pesticides			
Metolachlor	Crayfish *Orconectes rusticus* unable to located food, altered response to alarm pheromones	25–75 µg l⁻¹ 25–75 ppb	Wolf and Moore 2002
Metalochlor	*Orconectes rusticus* were less likely to initiate and win encounters against naive conspecifics	80 ppb	Cook and Moore 2008
pH changes			
pH reduction	Increased time to find food, decreased antennular flicking in *Cambarus bartoni*	pH 4.5	Allison et al. 1992
pH reduction	Feeding movements and antennular movements in response to feeding stimulants (amino acids) were progressively reduced when lowering pH in *Orconectes virilis* and *Procambarus acutus*	pH 4.5 pH 3.5	Tierney and Atema 1986
Metals			
Various metals	Responses of crayfish *P. clarkii, O. rusticus, C. bartoni* to feeding stimulants were suppressed by a mixture of copper, chromium, arsenic and selenium	Not measured	Steele et al. 1992
Copper	Crayfish unable to locate food	0.02, 0.2 mg l⁻¹	Sherba et al. 2000
Copper	Impairs chemical orientation in *O. rusticus* at the lowest concentration tested	0.005, 0.045, 0.45 mg l⁻¹	Lahman et al. 2015
Cadmium	Anti-predator behaviour (tail-flips, claw-raising) decreased in *Orconectes placidus, O. virilis, Procambarus acutus, P. alleni* and *P. clarkii*	0.3 mg l⁻¹	Wigginton et al. 2010
Cadmium	Anti-predator behaviour (claw raising response) increased in *P. clarkii*	3.5 mg l⁻¹	Wigginton et al. 2010

Use of chemical cues/pheromones for the management of invasive crayfish species

Crayfish are the largest and amongst the longest lived invertebrate organisms in temperate freshwater environments and they represent good candidates for invading aquatic systems (Moyle and Light 1996). Once arrived in a new ecosystem, non-indigenous crayfish species (NICS) may lead to dramatic direct and indirect effects on the ecosystem at all levels of ecological organization (e.g., Lodge et al. 1998, Nyström 1999). The modes of resource acquisition by crayfish and their capacity to develop new trophic relationships, coupled with their action as bio-turbator, have the potential to impose "considerable environmental stress" and, in several instances, they may induce "irreparable shifts in species diversity" (Hobbs 1989). In the last decade, several attempts have been made to contain the spread of NICS, but none has been definitive (reviewed in Freeman et al. 2010, Gherardi et al. 2011b). The traditional methods, including manual removal (Peay and Hiley 2001), trapping (Frutiger et al. 1999), fish predators (Blake and Hart 1993, Frutiger and Müller 2002), use of natural pesticides (Peay et al. 2006), or also a combination of the previous ones (i.e., trapping and fish predators, Hein et al. 2007) provide only temporary results such that repeated efforts need to be implemented in crayfish management.

Ideal control methods have a maximum impact on the target species with a minimum impact on indigenous species, environment and health, while also being economically viable. The use of semiochemicals, including pheromones, largely fulfils the necessary environmental and economic criteria of an ideal control method and represents a promising challenge for managing invasive crayfish. Sex pheromones, in particular, have been well documented in insects, where a concentrated research effort has gone into unravelling the molecular structures of their chemical signal, now successfully applied in integrated pest management in agriculture (El-Sayed et al. 2006). It is commonly known that crustacean decapods, similarly to insects, use sex pheromones to recognize and locate potential mates (for a review see the above section on reproduction) with some differences among species and between sexes. As reported in the aforementioned section, whereas chemical or visual cues presented alone were sufficient to attract potential mates in *P. clarkii* (Dunham and Oh 1996) or *O. virilis* (Hazlett 1985b), *A. pallipes* required the co-occurrence of the visual stimuli independently of the sex of the transmitter (Acquistapace et al. 2002). In addition, *P. clarkii* females require both chemical and visual stimuli from a male to recognize a potential mate, whereas the odour alone of a receptive female triggers mate search in males (Aquiloni et al. 2009) and also convey information about female quality (i.e., the size, Aquiloni and Gherardi 2008a). Therefore, in those crayfish species (including invasive crayfish species) that may locate mates by means of odour cues alone, sex pheromones could be potentially applicable for control, at least during the mating season. There are three possible applications: (1) to increase trapping success of crayfish, (2) to attract them into areas where trapping is more effective (Rogers and Holdich 1998), and (3) to disturb the mate searching activities through the release of large quantities of sex pheromones in an area. Once males are removed from the population or are prevented from findings mates, less mating might take place and a quick reduction in the size of the population could be achieved.

The major drawback to face is that, up to now, the molecular structure of pheromones has only been identified in one species of decapod crustaceans, the shore crab *Carcinus maenas* (Hardege et al. 2011) and is unknown in other species, if we exclude the controversial case of *Erimacrus isenbeckii* (Asai et al. 2000). Consequently, field activities may be only conducted using both the water conditioned by receptive females (Stebbing et al. 2004 for the signal crayfish, *P. leniusculus* in UK) or live crayfish (Aquiloni and Gherardi 2010 for the red swamp crayfish, *P. clarkii* in Italy).

Stebbing et al. (2003b) conducted experiments on the use of pheromones as a method of controlling signal crayfish. This research focused on four categories of pheromones: sex, stress, alarm and avoidance pheromones. The sex pheromone investigated is a female-released chemical that attracts and stimulates mating behaviour in males during the breeding season (Stebbing et al. 2003a). Stress, alarm and avoidance pheromones are all repellents, in extreme cases stimulating an escape response; the difference between the categories is their source of release. Stress pheromones are released from stressed but undamaged conspecifics; alarm pheromones are released from a damaged conspecific, while avoidance pheromones are released directly from a repellent stimulus, i.e., a predatory fish (Zulandt Schneider and Moore 2000).

The conditioned water from these sources was freeze-dried and embedded into a gel matrix to use as bait in a standard Swedish traps, which were left for 24 hours (Stebbing et al. 2004). Trapping took place year round (except for sex pheromones that were only tested during the mating season) at two field sites. The results show that, although the sex pheromone baited traps did not appear to be any more effective than food, sex-pheromone-baited traps were attractive to males (Stebbing et al. 2004). The failure of the stress and alarm pheromones to repel individuals may have more to do with the design of the experiment than the tested chemicals. It is possible that the food placed into the traps was a greater attractant than the repellents were a deterrent. This idea is supported by the fact that there was no significant difference between the numbers of crayfish found in the stress and alarm baited traps and the food-baited traps. Additionally, Stebbing et al. (2004, 2005) have shown that the sex pheromones are a highly selective (probably species-specific) bait particularly useful when indigenous crayfish species are present in the target area: sex-pheromone of signal crayfish are in fact no attractive for the native *A. pallipes*, minimizing the risk of capturing them.

Another field study was conducted in an Italian wetland invaded by the red swamp crayfish *P. clarkii* (Aquiloni and Gherardi 2010). In this case, standard traps had been baited with live sexually receptive individuals, either males or females, and the number, sex and size of the obtained catches were compared with empty traps and with traps baited with food. Similarly to the previous study described, the traps containing receptive females attracted more males than females. This confirms that in *P. clarkii* the females—and not the males—release sex pheromones and orient the males to the female location, as previously shown in laboratory studies (Aquiloni et al. 2009). A second interesting result was that the crayfish attracted by receptive individuals had a smaller body size than those captured using food as bait. Since in this species body size is related to age (Huner 2002), the ability to attract young individuals with more reproductive seasons ahead might be an advantage. However, the efficacy of the method is low since pheromones attract relatively fewer crayfish than food. Indeed, confinement in traps might cause stress on the senders with the consequent reduced emission of pheromones (Hazlett 1999). Purification and concentration of the molecules involved in sexual communication might improve the efficacy of the method but the chemical nature of sex pheromones is still unknown and research in the field is long and expensive (Holdich et al. 1999).

Notwithstanding the strong limitation of the method at the mating season only, we are confident that sex pheromones, if purified and concentrated, might be adopted at least as a means of early detection of low density populations (e.g., early stages of colonization of invasive populations) in relatively small and confined areas (Gherardi et al. 2011b). On the contrary, since the structure and function of the pheromone has yet to be identified, the adoption of this method for the control of established populations is far away and, as suggested by Bills and Marking (1988), it should be complemented by the simultaneous use of other methods (trapping, predators, and the Sterile Male Release Technique, reviewed in Gherardi et al. 2011b) in order to hit the different targets of the same population all the year round.

Conclusions and outlook

Even though many achievements have been accomplished in our understanding of chemical communication in crustaceans, this aspect is still in its infancy compared to insects and vertebrates. Notwithstanding our good knowledge about crayfish phylogeny (Gherardi et al. 2010) and the large number of extant species described (≈640), over 75% of studies on their biology have been focused on 10 species only, often alien species for Europe, with studies on *P. clarkii* largely prevailing. However, crayfish represent an excellent model for the study of chemical ecology as the structures of olfaction involved in emission/reception are well known and the recognition that chemicals cues and signals are essential—alone or in combination with other sensory channels—in mediating most of crayfish's behaviours. Multimodality is indeed emerging as one of the main issues to investigate, particularly for the social behaviours (fighting and mating). It is not always clear whether the natural habitats allow for the successful use of chemical signals during social interactions (Bergman et al. 2006). In some contexts, olfaction cannot be sufficient to convey the signals, and other sensory channels (e.g., vision) participate in gathering of the information, possibly in a non-redundant system of communication, as evidenced for the choice of males in *P. clarkii* females (Aquiloni et al. 2008). However, almost all the studies have been performed under laboratory conditions, with some exceptions involving the management of invasive crayfish species: conducting field studies would be ideal to observe the behaviour (and response) of animals exposed to multiple chemical –and visual and/or acoustic but also tactile-signals in a natural environment, but their feasibility seems for the moment difficult.

The use of chemical stimuli is undoubtedly crucial for foraging and anti-predator behaviours, even if several questions still need to be addressed, e.g., whether the observed differences among species in sensitivity to different food odours reflect adaptations to different diets or are due to the components of alarm and predator odours. Indeed, as for dominance and sexual pheromones, one of the future challenges is to identify and purify the substances responsible for eliciting these behaviours. To this end, it is necessary to develop a reliable and efficacious bioassay, similar to those already developed for assessing anti-predator behaviours, and established only in part for sexual interactions, but not for fighting ones (see Breithaupt 2011). The identification of sex pheromones would be of great importance for their use in the integrated management protocols for invasive species, such as *P. clarkii* and *P. leniusculus*.

Another intriguing and unexplored aspect of chemical ecology is in further understanding of the mother-offspring relationship. This aspect, despite being known in crayfish since the 1980's, and the presence of kin recognition mediated by chemical signals are really understudied, and represent a stimulating field to explore for chemical ecology and for crustacean behaviour in general.

In conclusion, our understanding of the crayfish chemical ecology has improved markedly in recent years but a great deal of work is still waiting for us: these new (and old) achievements can be considered only the "tip of the iceberg" because they are continuously leading to new intriguing questions and hypotheses that need to be faced in the future ahead.

References

Acquistapace, P., L. Aquiloni, B.A. Hazlett and F. Gherardi. 2002. Multimodal communication in crayfish: sex recognition during mate search by male *Austropotamobius pallipes*. Can. J. Zool. 80: 2041–2045.

Acquistapace, P., B.A. Hazlett and F. Gherardi. 2003. Unsuccessful predation and learning of predator cues by crayfish. J. Crust. Biol. 23: 364–370.

Acquistapace, P., W.H. Daniels and F. Gherardi. 2004. Behavioral responses to "alarm odors" in potentially invasive and noninvasive crayfish species from aquaculture ponds. Behaviour 141: 691–702.

Acquistapace, P., L. Calamai, B.A. Hazlett and F. Gherardi. 2005. Source of alarm substances in crayfish and their preliminary chemical characterization. Can. J. Zool. 83: 1624–1630.

Allison, V., D.W. Dunham and H.H. Harvey. 1992. Low pH alters response to food in the crayfish, *Cambarus bartoni*. Can. J. Zool. 70: 2416–2420.

Altner, I., H. Hatt and H. Altner. 1983. Structural properties of bimodal chemosensitive and mechanosensitive setae on the pereiopod chelae of the crayfish, *Austropotamobius torrentium*. Cell. Tissue Res. 228: 357–3374.

Ameyaw-Akumfi, C. 1976. Some aspects of the breeding biology of crayfish. Ph.D. Thesis, University of Michigan, USA.

Ameyaw-Akumfi, C. and B.A. Hazlett. 1975. Sex recognition in the crayfish, *Procambarus clarkii*. Science 190: 1225–1226.

Aquiloni, L. and F. Gherardi. 2008a. Assessing mate size in the red swamp crayfish *Procambarus clarkii*: effects of visual versus chemical stimuli. Freshw. Biol. 53: 461–469.

Aquiloni, L. and F. Gherardi. 2008b. Mutual mate choice in crayfish: large body size is selected by both sexes, virginity by males only. J. Zool. 274: 171–179.

Aquiloni, L. and F. Gherardi. 2008c. Evidence of cryptic mate choice in crayfish. Biol. Lett. 4: 163–165.

Aquiloni, L. and F. Gherardi. 2008d. Extended mother-offspring relationships in crayfish: the return behaviour of *Procambarus clarkii* juveniles. Ethology 114: 946–954.

Aquiloni, L. and F. Gherardi. 2010. Crayfish females eavesdrop on fighting males and use smell and sight to recognize the identity of the winner. Anim. Behav. 79: 265–269.

Aquiloni, L., M. Buřič and F. Gherardi. 2008. Crayfish females eavesdrop on fighting males before choosing the dominant mate. Curr. Biol. 18: 462–463.

Aquiloni, L., A. Massolo and F. Gherardi. 2009. Sex identification in female crayfish is bimodal. Naturwissenschaften 96: 103–110.

Aquiloni, L., P.G. Giulianini, A. Mosco, C. Guarnaccia, E. Ferrero and F. Gherardi. 2012. Crustacean Hyperglycemic Hormone (cHH) as a modulator of aggression in crustacean decapods. PLoS ONE 7: e50047.

Asai, N., N. Fusetani, S. Matsugana and J. Sasaki. 2000. Sex pheromones of the hair crab *Erimacrus isenbeckii*. Part 1: isolation and structures of novel ceramides. Tetrahedron 56: 9895–9899.

Atema, J. 1985. Chemoreception in the sea: adaptations of chemoreceptors and behaviour to aquatic stimulus conditions. Symp. Soc. Exp. Biol. 39: 386–423.

Atema, J. 1995. Chemical signals in the marine-environment—dispersal, detection, and temporal signal analysis. Proc. Natl. Acad. Sci. USA 92: 62–66.

Atema, J. 1996. Eddy chemotaxis and odor landscapes: exploration of nature with animal sensors. Biol. Bull. 191: 129–138.

Atema, J. 2012. Aquatic odor dispersal fields: opportunities and limits of detection, communication and navigation. pp. 1–18. *In*: C. Broenmark and L.A. Hansson (eds.). Chemical Ecology in Aquatic Systems. Oxford University Press, Oxford.

Barnard, C.J. and T. Burk. 1979. Dominance hierarchies and the evolution of "individual recognition". J. Theor. Biol. 81: 65–73.

Bechler, D.L. 1981. Copulatory and maternal offspring behavior in the hypogean crayfish *Orconectes inermis inermis* Cope and *Orconectes pellucidus* (Tell Kampf) (Decapoda, Astacidea). Crustaceana 40: 136–143.

Bechler, D.L., X. Deng and B. McDonald. 1988. Interspecific communication between sympatric crayfish of the genus *Procambarus* (Decapoda, Astacidae). Crustaceana 54: 153–162.

Belanger, R.M. and P.A. Moore. 2013. A comparative analysis of setae on the pereiopods of reproductive male and female *Orconectes rusticus* (Decapoda: Astacidae). J. Crust. Biol. 33: 309–316.

Benhamou, S. 2007. How many animals really do the Levy walk? Ecology 88: 1962–1969.

Bergman, D.A. and P.A. Moore. 2003. Field observations of intraspecific agonistic behaviour of two crayfish species, *Orconectes rusticus* and *Orconectes virilis*, in different habitats. Biol. Bull. 205: 26–35.

Bergman, D.A. and P.A. Moore. 2005. Prolonged exposure to social odours alters subsequent social interactions in crayfish (*Orconectes rusticus*). Anim. Behav. 70: 311–318.

Bergman, D.A., C.P. Kozlowsky, J.C. McIntyre, R. Huber, A.G. Daws and P.A. Moore. 2003. Temporal dynamics and communication of winner-effects in the crayfish, *Orconectes rusticus*. Behaviour 140: 805–82.

Bergman, D.A., A.L. Martin and P.A. Moore. 2005. The control of information flow by the manipulation of mechanical and chemical signals during agonistic encounters by crayfish, *Orconectes rusticus*. Anim. Behav. 70: 485–496.

Bergman, D.A., C.N. Redman, K.C. Fero, J.L. Simon and P.A. Moore. 2006. The impacts of flow on chemical communication strategies and fight dynamics of crayfish. Mar. Freshw. Behav. Physiol. 39: 245–258.

Berrill, M. and M. Arsenault. 1982. Spring breeding of a northern temperate crayfish *Orconectes rusticus*. Can. J. Zool. 60: 2641–2645.

Berrill, M. and M. Arsenault. 1984. The breeding behaviour of a northern temperate orconectid crayfish, *Orconectes rusticus*. Anim. Behav. 32: 333–339.

Berry, F.C. and T. Breithaupt. 2010. To signal or not to signal? Chemical communication by urine-borne signals mirrors sexual conflict in crayfish. BMC Biol. 8: 25.

Bilde, T. and Y. Lubin. 2001. Kin recognition and cannibalism in a subsocial spider. J. Evol. Biol. 14: 959–966.

Bills, T.D. and L. Marking. 1988. Control of nuisance populations of crayfish with traps and toxicants. Prog. Fish-Culturist 50: 103–106.

Blake, M.A. and P.J.B. Hart. 1993. The behavioural response of juvenile signal crayfish *Pacifastacus leniusculus* to stimuli from perch and eels. Freshw. Biol. 29: 89–97.

Boriss, H., M. Boersma and K.H. Wiltshire. 1999. Trimethylamine induces migration of waterfleas. Nature 398: 382.

Borowsky, B. 1980. Factors that affect juvenile emergence in *Gammarus palustris* (Bousfield, 1969). J. Exp. Mar. Biol. Ecol. 42: 213–223.

Bovbjerg, R.V. 1953. Dominance order in the crayfish *Orconectes virilis* (Hagen). Physiol. Zool. 26: 173–178.

Bovbjerg, R.V. 1956. Some factors affecting aggressive behavior in crayfish. Physiol. Zool. 29: 127–136.

Bowma, P. and B.A. Hazlett. 2001. Integration of multiple predator cues by the crayfish *Orconectes propinquus*. Anim. Behav. 61: 771–776.

Bradbury, J.W. and S.L. Vehrencamp. 2011. Principles of animal communication. Sinauer Associates, Incorporated.

Breithaupt, T. 2001. Fan organs of crayfish enhance chemical information flow. Biol. Bull. 200: 150–154.

Breithaupt, T. 2002. Sound perception in aquatic crustaceans. pp. 548–559. *In*: K. Wiese (ed.). The Crustacean Nervous System. Springer, Berlin.

Breithaupt, T. 2011. Chemical communication in crayfish. pp. 257–276. *In*: T. Breithaupt and M. Thiel (eds.). Chemical Communication in Crustaceans. Springer, New York.

Breithaupt, T. and J. Tautz. 1990. The sensitivity of crayfish mechanoreceptors to hydrodynamic and acoustic stimuli. pp. 114–120. *In*: K. Wiese, W.-D. Krenz, J. Tautz, H. Reichert and B. Mulloney (eds.). Frontiers in Crustacean Neurobiology. Advances in Life Sciences. Birkhäuser, Basel.

Breithaupt, T. and P. Eger. 2002. Urine makes the difference: chemical communication in fighting crayfish made visible. J. Exp. Biol. 205: 1221–1232.

Breithaupt, T. and J.D. Hardege. 2012. Pheromones mediating sex and dominance in aquatic animals. pp. 39–56. *In*: C. Bronmark and L.-A. Hansson (eds.). Chemical Ecology in Aquatic Systems. Oxford University Press, Oxford.

Breithaupt, T., B. Schmitz and J. Tautz. 1995. Hydrodynamic orientation of crayfish (*Procambarus clarkii*) to swimming fish prey. J. Comp. Physiol. A 177: 481–491.

Brönmark, C. and L.-A. Hansson. 2012. Chemical ecology in aquatic systems. Oxford University Press, Oxford.

Burba, A. 1983. Chemical regulation in crayfish behaviour during postembryonic development. Freshw. Crayfish 5: 451–458.

Burrows, M. 2009. A single muscle moves a crustacean limb joint rhythmically by acting against a spring containing resilin. BMC Biol. 7: 27.

Buscaino, G., G. Alonge, F. Filiciotto, V. Maccarrone, G. Buffa, V. Di Stefano, B. Patti, C. Buscaino, S. D'Angelo and S. Mazzola. 2012. The underwater acoustic activities of the red swamp crayfish *Procambarus clarkii*. J. Acoust. Soc. Am. 132: 1792–1798.

Caprio, J. and C.D. Derby. 2008. Aquatic animal models in the study of chemoreception. pp. 97–134. *In*: S. Firestein and G.K. Beauchamp (eds.). Olfaction & Taste, Vol. 4. Elsevier, San Diego.

Caskey, J.L. and R.T. Bauer. 2005. Behavioral tests for a possible contact sex pheromone in the caridean shrimp *Palaemonetes pugio*. J. Crust. Biol. 25: 35–40.

Chivers, D.P. and R.J.F. Smith. 1998. Chemical alarm signalling in aquatic predator-prey systems: a review and prospectus. Ecoscience 5: 338–352.

Clark, E. 1936. Notes on the habits of land crayfishes. Victorian Nat. 53: 65–8.

Clutton-Brock, T.H. 1991. The evolution of parental care. Princeton University Press, Princeton, New Jersey.

Cook, M.E. and P.A. Moore. 2008. The effects of the herbicide metolachlor on agonistic behavior in the crayfish, *Orconectes rusticus*. Arch. Environ. Toxicol. 55: 94–102.

Copp, N. 1986. Dominance hierarchies in the crayfish *Procambarus clarkii* and the question of learned individual recognition. Crustaceana 51: 9–24.

Corotto, F.S. and M.R. O'Brien. 2002. Chemosensory stimuli for the walking legs of the crayfish *Procambarus clarkii*. J. Chem. Ecol. 28: 1117–1130.

Corotto, F.S., M.J. McKelvey, E.A. Parvin, J.L. Rogers and J.M. Williams. 2007. Behavioral responses of the crayfish *Procambarus clarkii* to single chemosensory stimuli. J. Crust. Biol. 27: 24–29.

Cowan, D.F. 1991. The role of olfaction in courtship behaviour of the American lobster, *Homarus americanus*. Biol. Bull. 181: 307–402.

Crocker, D.W. and D.W. Barr. 1968. Handbook of the Crayfish of Ontario. University of Toronto Press, Toronto.

Croker, R.A. 1968. Return of juveniles to the marsupium in the amphipod *Neohaustorius schmitzi*. Crustaceana 14: 215.

Crook, R., B.W. Patullo and D.L. Macmillan. 2004. Multimodal individual recognition in the crayfish *Cherax destructor*. Mar. Fresh. Behav. Physiol. 37: 271–285.

Cruz, M.J., R. Rebelo and E.G. Crespo. 2006. Effects of an introduced crayfish, *Procambarus clarkii*, on the distribution of south-western Iberian amphibians in their breeding habitats. Ecography 29: 329–338.

Cukerzis, J.M. 1988. *Astacus astacus* in Europe. pp. 309–340. *In*: D.M. Holdich and R.S. Lowery (eds.). Freshwater Crayfish: Biology, Management & Exploitation. Cambridge University Press, Cambridge.

Daws, A.G., J. Grills, K. Konzen and P.A. Moore. 2002. Previous experiences alter the outcome of aggressive interactions between males in the crayfish, *Procambarus clarkii*. Mar. Freshw. Behav. Physiol. 35: 139–148.

de Angelis, M.A. and C. Lee. 1994. Methane production during zooplankton grazing on marine phytoplankton. Limnol. Oceanogr. 39: 1298–1308.

Denissenko, P., S. Lukaschuk and T. Breithaupt. 2007. The flow generated by an active olfactory system of the red swamp crayfish. J. Exp. Biol. 210: 4083–4091.

Derby, C.D. 1982. Structure and function of cuticular sensilla of the lobster *Homarus americanus*. J. Crust. Biol. 2: 1–21.

Derby, C.D. and J. Atema. 1982. The function of chemo- and mechanoreceptors in lobster (*Homarus americanus*) feeding behaviour. J. Exp. Biol. 98: 317–328.

Devine, D.V. and J. Atema. 1982. Function of chemoreceptor organs in spatial orientation of the lobster, *Homarus americanus*— differences and overlap. Biol. Bull. 163: 144–153.

Dewsbury, D.A. 1982. Ejaculate cost and male choice. Am. Nat. 119: 601–610.

Dick, J.T.A., S.E. Faloon and R.W. Elwood. 1998. Active brood care in an amphipod: influences of embryonic development, temperature and oxygen. Anim. Behav. 56: 663–672.

Dicke, M. and M.W. Sabelis. 1988. Infochemical terminology: based in cost-benefit analysis rather than origin of compounds? Funct. Ecol. 2: 131–139.

Dugatkin, L.A. 1997. Winner and loser effects and structure of dominance hierarchies. Behav. Ecol. 8: 583–587.

Dugatkin, L.A., M.S. Alfieri and A.J. Moore. 1994. Can dominance hierarchies be replicated? Form–re-form experiments using the cockroach (*Nauphoeta cinerea*). Ethology 97: 94–102.

Dunham, D.W. and J.W. Oh. 1992. Chemical sex discrimination in the crayfish *Procambarus clarkii*: role of antennules. J. Chem. Ecol. 18: 2363–2372.

Dunham, D.W. and J.W. Oh. 1996. Sex discrimination by female *Procambarus clarkii* (Girard, 1852) (Decapoda, Cambaridae): use of chemical and visual stimuli. Crustaceana 69: 534–542.

Dunham, D.W., K.A. Ciruna and H.H. Harvey. 1997. Chemosensory role of antennules in the behavioral integration of feeding by the crayfish *Cambarus bartonii*. J. Crust. Biol. 17: 27–32.

Dunham, P.J. 1979. Mating in the American lobster: stage of molt cycle and sex pheromone. J. Mar. Behav. Physiol. 6: 1–11.

Durgin, W.S., K.E. Martin, H.R. Watkins and L.M. Mathews. 2008. Distance communication of sexual status in the crayfish *Orconectes quinebaugensis*: female sexual history mediates male and female behavior. J. Chem. Ecol. 34: 702–707.

Edwards, D.H. 1984. Crayfish extraretinal photoreception. 1. Behavioral and motoneuronal responses to abdominal illumination. J. Exp. Biol. 109: 291–306.

El-Sayed, A.M., D.M. Suckling, C.H. Wearing and J.A. Byers. 2006. Potential of mass trapping for long-term pest management and eradication of invasive species. J. Econ. Entomol. 99: 1550–1564.

Favaro, L., T. Tirelli, M. Gamba and D. Pessani. 2011. Sound production in the red swamp crayfish *Procambarus clarkii* (Decapoda: Cambaridae). Zool. Anz. 250: 143–150.

Ferrari, M.C.O., G.E. Brown, F. Messier and D.P. Chivers. 2009. Threat-sensitive generalization of predator recognition by larval amphibians. Behav. Ecol. Sociobiol. 63: 1369–1375.

Figler, M.H., M. Twum, J.E. Finkelstein and H.V.S. Peeke. 1995. Maternal aggression in red swamp crayfish (*Procambarus clarkii*, Girard): the relation between reproductive status and outcome of aggressive encounters with male and female conspecifics. Behaviour 132: 107–125.

Figler, M.H., G.S. Blank and H.V.S. Peeke. 1997. Maternal aggression and post-hatch care in red swamp crayfish, *Procambarus clarkii* (Girard): the influences of presence of offspring, fostering and maternal molting. Mar. Freshwater Behav. Physiol. 30: 173–194.

Figler, M.H., G.S. Blank and H.V.S. Peeke. 2001. Maternal territoriality as an offspring defense strategy in red swamp crayfish (*Procambarus clarkii*, Girard). Aggress. Behav. 27: 391–403.

Figler, M.H., H.V.S. Peeke, M.J. Snyder and E.S. Chang. 2004. Effects of egg removal on maternal aggression, biogenic amines, and stress indicators in ovigerous lobsters (*Homarus americanus*). Mar. Freshwater Behav. Physiol. 37: 43–54.

Fisher, J. 1954. Evolution and bird sociality. pp. 71–83. *In*: J. Huxley, A.C. Hardy and E.B. Ford (eds.). Evolution as a Process. Allen & Unwin, London.

Fitzpatrick, J.F., Jr. 1987. The subgenera of the crawfish genus *Orconectes* (Decapoda: Cambaridae). Proc. Biol. Soc. Wash. 100: 44–74.

Freeman, M.A., J.F. Turnbull, J.F. Yeomans and C.W. Bean. 2010. Prospects for management strategies of invasive crayfish populations with an emphasis on biological control. Aquat. Conserv. Mar. Freshw. Ecosyst. 20: 211–223.

Frutiger, A. and R. Müller. 2002. Controlling unwanted *Procambarus clarkii* populations by fish predation. Freshw. Crayfish 13: 309–315.

Frutiger, A., S. Borner, T. Büsser, R. Eggen, R. Müller, S. Müller and H.R. Wasmer. 1999. How to control unwanted populations of *Procambarus clarkii* in Central Europe? Freshw. Crayfish 12: 714–726.

Furrer, S.C. 2004. Untersuchungen des Partnerwahlverhaltens beim Edel- und Galizierkrebs sowie der Life History beim Steinkrebs. Ph.D. Thesis Dissertation, University of Zürich, Switzerland.

Galeotti, P., D. Rubolini, G. Fea, D. Ghia, P.A. Nardi, F. Gherardi and M. Fasola. 2006. Female freshwater crayfish adjust egg and clutch size in relation to multiple male traits Proc. R. Soc. B 273: 1105–1110.

Galeotti, P., D. Rubolini, F. Pupin, R. Sacchi and M. Fasola. 2008. Sperm removal and ejaculate size correlate with chelae asymmetry in a freshwater crayfish species. Behav. Ecol. Sociobiol. 62: 1739–1745.

Garm, A. and L. Watling. 2013. The crustacean integument: setae, setules, and other ornamentation. pp. 167–198. *In*: L. Watling and M. Thiel (eds.). Functional Morphology and Diversity. Oxford University Press, Oxford.

Gherardi, F. 2002. Behaviour. pp. 258–290. *In*: D.M. Holdich (ed.). Biology of Freshwater Crayfish. Blackwell Science, Oxford.

Gherardi, F. and W.H. Daniels. 2003. Dominance hierarchies and status recognition in the crayfish, *Procambarus acutus acutus*. Can. J. Zool. 81: 1269–1281.

Gherardi, F., S. Barbaresi and A. Raddi. 1999. The agonistic behaviour in the red swamp crayfish, *Procambarus clarkii*: functions of the chelae. Freshw. Crayfish 12: 233–243.

Gherardi, F., P. Acquistapace, B.A. Hazlett and G. Whisson. 2002. Behavioural responses to alarm odours in indigenous and non-indigenous crayfish species: a case study from Western Australia. Mar. Freshw. Res. 53: 93–98.

Gherardi, F., B. Renai, P. Galeotti and D. Rubolini. 2006. Nonrandom mating, mate choice, and male-male competition in the crayfish *Austropotamobius italicus*, a threatened species. Arch. Hydrobiol. 165: 557–576.

Gherardi, F., C. Souty-Grosset, G. Vogt, J. Diéguez-Uribeondo and K.A. Crandall. 2010. Infraorder Astacidea Latreille, 1802. The freshwater crayfish. pp. 269–423. *In*: F.R. Schram and J.C. von Vaupel Klein (eds.). Treatise on Zoology - Anatomy, Taxonomy, Biology - The Crustacea, Decapoda, Volume 9 Part A -Eucarida: Euphausiacea, Amphionidacea, and Decapoda (partim). Brill, Leiden, The Netherlands.

Gherardi, F., K.M. Mavuti, N. Pacini, E. Tricarico and D.M. Harper. 2011a. The smell of danger: chemical recognition of fish predators by the invasive crayfish *Procambarus clarkii*. Freshw. Biol. 56: 1567–1578.

Gherardi, F., L. Aquiloni, J. Diéguez-Uribeondo and E. Tricarico. 2011b. Managing invasive crayfish: is there a hope? Aquat. Sci. 73: 185–200.

Gherardi, F., L. Aquiloni and E. Tricarico. 2012. Revisiting social recognition systems in invertebrates. Anim. Cogn. 15: 745–762.

Ghia, D., G. Fea, F. Bernini and P.A. Nardi. 2011. Reproduction experiment on *Austropotamobius pallipes* complex under controlled conditions: can hybrids be hatched? Knowl. Managt. Aquatic Ecosyst. 401: 11.

Gleeson, R.A. 1980. Pheromone communication in the reproductive behaviour of the blue crab, *Callinectes sapidus*. Mar. Ecol. Prog. Ser. 244: 179–189.

Goessmann, C., C. Hemerlrijk and R. Huber. 2000. The formation and maintenance of crayfish hierarchies: behavioral and self-structuring properties. Behav. Ecol. Sociobiol. 48: 418–428.

Gomez, G. and J. Atema. 1996. Temporal resolution in olfaction: stimulus integration time of lobster chemoreceptor cells. J. Exp. Biol. 199: 1771–1779.

Grasso, F.W. and J.A. Basil. 2002. How lobsters, crayfishes, and crabs locate sources of odor: current perspectives and future directions. Curr. Opin. Neurobiol. 12: 721–727.

Green, W.W., R.S. Mirza and G.G. Pyle. 2008. Kin recognition and cannibalistic behaviours by adult male fathead minnows (*Pimephales promelas*). Naturwissenschaften 95: 269–272.

Guan, R.Z. and P.R. Wiles. 1997. Ecological impact of introduced crayfish on benthic fishes in a British lowland river. Conserv. Biol. 11: 641–647.

Guiasu, R.C. and D.W. Dunham. 1998. Inter-form agonistic contests in male crayfishes, *Cambarus robustus* (Decapoda, Cambaridae). Invert. Biol. 117: 144–154.

Hallberg, E. and M. Skog. 2011. Chemosensory sensilla in crustaceans. pp. 103–121. *In*: T. Breithaupt and M. Thiel (eds.). Chemical Communication in Crustaceans. Springer, New York.

Hallberg, E., K.U.I. Johansson and R. Wallen. 1997. Olfactory sensilla in crustaceans: morphology, sexual dimorphism, and distribution patterns. Int. J. Insect Morphol. Embryol. 26: 173–180.

Hamr, P. and A.M.M. Richardson. 1994. Life history of *Parastacoides tasmanicus tasmanicus* Clark, a burrowing freshwater crayfish from south-western Tasmania. Aust. J. Mar. Freshwat. Res. 45: 455–470.

Hanson, J.M., P.A. Chambers and E.E. Prepas. 1990. Selective foraging by the crayfish *Orconectes virilis* and its impact on macroinvertebrates. Freshw. Biol. 24: 69–80.

Hardege, J.D., A. Jennings, D. Hayden, C.T. Müller, D. Pascoe, M.G. Bentley and A.S. Clare. 2002. Novel behavioural assay and partial purification of a female derived sex pheromone in *Carcinus maenas*. Mar. Ecol. Progr. Ser. 244: 179–189.

Hardege, J.D., H.D. Bartels-Hardege, N. Fletcher, J.A. Terschak, M. Harley, M.A. Smith, L. Davidson, D. Hayden, C.T. Müller, M. Lorch, K. Welham, T. Walther and R. Bublitz. 2011. Identification of a female sex pheromone in *Carcinus maenas*. Mar. Ecol. Progr. Ser. 436: 177–189.

Hatt, H. 1984. Structural requirements of amino-acids and related compounds for stimulation of receptors in crayfish walking leg. J. Comp. Physiol. 155: 219–231.

Hatt, H. 1986. Responses of a bimodal neuron (chemo- and vibration-sensitive) on the walking legs of the crayfish. J. Comp. Physiol. A 159: 611–617.

Hatt, H. and U. Bauer. 1980. Single unit analysis of mechanosensitive and chemosensitive neurons in the crayfish claw. Neurosci. Lett. 17: 203–207.

Hazlett, B.A. 1983. Parental behavior in decapod crustaceans. pp. 171–193. *In*: S. Rebach and D.W. Dunham (eds.). Studies in Adaptation—The Behavior of Higher Crustacea. John Wiley & Sons Inc., New York, USA.

Hazlett, B.A. 1985a. Disturbance pheromones in the crayfish *Orconectes virilis*. J. Chem. Ecol. 11: 1695–1711.

Hazlett, B.A. 1985b. Chemical detection of sex and condition in the crayfish *Orconectes virilis*. J. Chem. Ecol. 11: 181–189.

Hazlett, B.A. 1989. Additional sources of disturbance pheromone affecting the crayfish *Orconectes virilis*. J. Chem. Ecol. 15: 381–385.

Hazlett, B.A. 1990. Source and nature of disturbance-chemical system in crayfish. J. Chem. Ecol. 16: 2263–2275.

Hazlett, B.A. 1994. Alarm responses in the crayfish *Orconectes virilis* and *Orconectes propinquus*. J. Chem. Ecol. 20: 1525–1535.

Hazlett, B.A. 1999. Responses to multiple chemical cues by the crayfish *Orconectes virilis*. Behaviour 136: 161–177.

Hazlett, B.A. 2000. Information use by an invading species: do invaders respond more to alarm odors than native species? Biol. Inv. 2: 289–294.

Hazlett, B.A. 2003a. The effects of starvation on crayfish responses to alarm odor. Ethology 109: 587–592.

Hazlett, B.A. 2003b. Predator recognition and learned irrelevance in the crayfish *Orconectes virilis*. Ethology 109: 765–780.

Hazlett, B.A. 2007. Conditioned reinforcement in the crayfish *Orconectes rusticus*. Behaviour 144: 847–859.

Hazlett, B.A. 2011. Chemical cues and reducing the risk of predation. pp. 355–370. *In*: T. Breithaupt and M. Thiel (eds.). Chemical Communication in Crustaceans. Springer, New York.

Hazlett, B.A. and D.R. Schoolmaster. 1998. Responses of cambarid crayfish to predator odor. J. Chem. Ecol. 24: 1757–1770.

Hazlett, B.A., P. Acquistapace and F. Gherardi. 2002. Differences in memory capabilities in invasive and native crayfish. J. Crust. Biol. 22: 439–448.

Hazlett, B.A., A. Burba, F. Gherardi and P. Acquistapace. 2003. Invasive species use a broader range of predation-risk cues than native species. Biol. Inv. 5: 223–228.

Hazlett, B.A., S. Lawler and G. Edney. 2007. Agonistic behaviour of the crayfish *Euastacus armatus* and *Cherax destructor*. Mar. Freshw. Behav. Physiol. 40: 257–266.

Healy, A. and J. Yaldwyn. 1970. Australian crustaceans in colour. pp. 1–112. Tuttle, Rutland, Vermont.

Height, S.G. and G.J. Whisson. 2006. Behavioural responses of Australian freshwater crayfish (*Cherax cainii* and *Cherax albidus*) to exotic fish odour. Aust. J. Zool. 54: 399–407.

Hein, C.L., M.J. Vander Zanden and J.J. Magnuson. 2007. Intensive trapping and increased fish predation cause massive population decline of an invasive crayfish. Freshw. Biol. 52: 1134–1146.

Hobbs, H.H. 1989. An illustrated checklist of the American crayfishes (Decapoda: Astacidae, Cambaridae and Parastacidae). Smith. Contr. Zool. 480: 1–236.

Hobbs, H.H. III. 1993. Trophic relationships of North American freshwater crayfishes and shrimps. Contr. Milwaukee Public Mus. 85: 1–110.

Hodgson, E.S. 1958. Electrophysiological studies of Arthropods chemoreception. iii. Chemoreceptors of terrestrial and freshwater Arthropods. Biol. Bull. 115: 114–125.

Höjesjö, J., J.I. Johnsson, E. Petersson and T. Järvi. 1998. The importance of being familiar: individual recognition and social behavior in sea trout (*Salmo trutta*). Behav. Ecol. 9: 445–451.

Hogger, J.B. 1988. Ecology, population biology and behaviour. pp. 114–144. *In*: D.M. Holdich and R.S. Lowery (eds.). Freshwater Crayfish: Biology, Management and Exploitation. Croom Helm, Timber Press, London, Portland.

Holdich, D.M., J.P. Reader, W.D. Rogers and M. Harlioğlu. 1995. Interactions between three species of crayfish (*Austropotamobius pallipes*, *Astacus leptodactylus* and *Pacifastacus leniusculus*). Freshw. Crayfish 10: 46–56.

Holdich, D.M., R. Gydemo and W.D. Rogers. 1999. A review of possible methods for controlling nuisance populations of alien crayfish. pp. 245–270. *In*: F. Gherardi and D.M. Holdich (eds.). Crayfish in Europe as Alien Species—How to Make the Best of a Bad Situation? Crustacean Issues 11. Balkema, Rotterdam.

Horner, A.J., M. Schmidt, D.H. Edwards and C.D. Derby. 2008. Role of the olfactory pathway in agonistic behavior of crayfish, *Procambarus clarkii*. Invert. Neurosci. 8: 11–18.

Huber, R. and A. Delago. 1998. Serotonin alters decisions to withdraw in fighting crayfish, *Astacus astacus*: the motivational concept revisited. J. Comp. Physiol. A 182: 573–583.

Huber, R., K. Smith, A. Delago, K. Isaksson and E.A. Kravitz. 1997. Serotonin and aggressive motivation in crustaceans: altering the decision to retreat. Proc. Natl. Acad. Sci. 94: 5939–5942.

Huner, J.V. 1994. Freshwater crayfish culture. pp. 5–89. *In*: J.V. Huner (ed.). Freshwater Crayfish Aquaculture in North America, Europe, and Australia. Food Products Press, Binghamton, New York.

Huner, J.V. 2002. *Procambarus*. pp. 541–584. *In*: D.M. Holdich (ed.). Biology of Freshwater Crayfish. Blackwell, Oxford.

Huner, J.V. and J.E. Barr. 1991. Red swamp crawfish: biology and exploitation. 3rd edn. Sea Grant College Program, Baton Rouge, Louisiana.

Karavanich, C. and J. Atema. 1998. Individual recognition and memory in lobster dominance. Anim. Behav. 56: 1553–1560.

Karplus, I., A. Sagi, I. Khalaila and A. Barki. 2003. The soft red patch of the Australian freshwater crayfish *Cherax quadricarinatus* (von Martens): a review and prospects for future research. J. Zool. 259: 375–379.

Kats, L.B. and L.M. Dill. 1998. The scent of death: chemosensory assessment of predation risk by animals. Ecoscience 5: 361–394.

Keller, T.A., A.M. Tomba and P.A. Moore. 2001. Orientation in complex chemical landscapes: spatial arrangement of chemical sources influences crayfish food-finding efficiency in artificial streams. Limnol. Oceanogr. 46: 238–247.

Kobayashi, T., S. Wada and H. Mukai. 2002. Extended maternal care observed in *Parallorchestes ochotensis* (Amphipoda, Gammaridea, Talitroidea, Hyalidae). J. Crust. Biol. 22: 135–142.

Koehl, M.A.R., J.R. Koseff, J.P. Crimaldi, M.G. McCay, T. Cooper, M.B. Wiley and P.A. Moore. 2001. Lobster sniffing: antennule design and hydrodynamic filtering of information in an odor plume. Science 294: 1948–1951.

Köksal, G. 1988. *Astacus leptodactylus* in Europe. pp. 365–400. *In*: D.M. Holdich and R.S. Lowery (eds.). Freshwater Crayfish: Biology, Management and Exploitation. Croom Helm, Timber Press, London, Portland.

Kravitz, E. 2000. Serotonin and aggression: insights gained from a lobster model system and speculations on the role of amine neurons in a complex behavior. J. Comp. Physiol. A 186: 221–238.

Lahman, S.E., K.R. Trent and P.A. Moore. 2015. Sublethal copper toxicity impairs chemical orientation in the crayfish *Orconectes rusticus*. Ecotoxicol. Environ. Saf. 113: 369–377.

Landau, H.G. 1965. Development of structure in a society with a dominance relation when new members are added successively. Bull. Math. Biophys. 27: 151–160.

Leduc, A.O.H.C., P.L. Munday, G.E. Brown and M.C.O. Ferrari. 2013. Effects of acidification on olfactory-mediated behaviour in freshwater and marine ecosystems: a synthesis. Phil. Trans. Roy. Soc. London. B, 368.

Levi, T. 1997. Reproductive cycle, dynamic relations between mother and offspring, and the variation in aggressive behaviour in female redclaw crayfish, *Cherax quadricarinatus*, of different reproductive state. M.Sc. thesis, The Hebrew University of Jerusalem, Faculty of Agricultural Food and Environmental Quality.

Levi, T., A. Barki, G. Culata and I. Karplus. 1999. Mother-offspring relationships in the red-claw crayfish *Cherax quadricarinatus*. J. Crust. Biol. 19: 477–484.

Linsenmair, K.E. and C. Linsenmair. 1971. Paarbildung und Paarzusammenhalt bei der monogamen Wüstenassel *Hemilepistus reaumuri* (Crustacea, Isopoda, Oniscoidea). Z. Tierpsychol. 29: 134–155.

Little, E.E. 1975. Chemical communication in maternal behavior of crayfish. Nature 255: 400–401.

Little, E.E. 1976. Ontogeny of maternal behaviour and brood pheromone in crayfish. J. Comp. Physiol. A 112: 133–142.

Livingstone, M.S., R. Harris-Warrick and E.A. Kravitz. 1980. Serotonin and octopamine produce opposite postures in lobsters. Science 208: 76–79.

Lodge, D.M., R.A. Stein, K.M. Brown, A.P. Covich, C. Brönmark, J.E. Garvey and S.P. Klosiewski. 1998. Predicting impact of freshwater exotic species on native biodiversity: challenges in spatial scaling. Aust. J. Ecol. 23: 53–67.

Loekle, D.M., D.M. Madison and J.J. Christian. 1982. Time dependency and kin recognition of cannibalistic behavior among poeciliid fishes. Behav. Neur. Biol. 35: 315–318.

Lowe, M.E. 1956. Dominance-subordinate relationships in the crawfish *Cambarellus shufeldtii*. Tulane Stud. Zool. 4: 139–170.

Lürling, M. 2012. Infodisruption: pollutants interfering with the natural chemical information conveyance in aquatic systems. pp. 250–271. *In*: C. Bronmark and L.-A. Hansson (eds.). Chemical Ecology in Aquatic Systems. Oxford University Press, Oxford.

Lürling, M. and M. Scheffer. 2007. Info-disruption: pollution and the transfer of chemical information between organisms. Trends Ecol. Evol. 22: 374–379.

Lundberg, U. 2004. Behavioural elements of the noble crayfish, *Astacus astacus* (Linnaeus, 1758). Crustaceana 77: 137–162.

Mason, J.C. 1970. Maternal-offspring behavior of the crayfish, *Pacifastacus trowbridgi* (Stimpson). Am. Midl. Nat. 84: 463–473.

Mathews, L.M. 2011. Mother–offspring recognition and kin-preferential behaviour in the crayfish *Orconectes limosus*. Behaviour 148: 71–87.

Mathis, A. and R.J.F. Smith. 1993. Chemical alarm signals increase the survival time of fathead minnows (*Pimephales promelas*) during encounters with northern pike (*Esox lucius*). Behav. Ecol. 4: 260–265.

McLeese, D.W. 1970. Detection of dissolved substances by the American lobster (*Homarus americanus*) and olfactory attraction between lobsters. J. Fish. Res. Board Can. 27: 1371–1378.

Mellon, D., Jr. 1997. Physiological characterization of antennular flicking reflexes in the crayfish. J. Comp. Physiol. A 180: 553–565.

Mellon, D., Jr. 2005. Integration of hydrodynamic and odorant inputs by local interneurons of the crayfish deutocerebrum. J. Exp. Biol. 208: 3711–3720.

Mellon, D., Jr. 2012. Smelling, feeling, tasting and touching: behavioral and neural integration of antennular chemosensory and mechanosensory inputs in the crayfish. J. Exp. Biol. 215: 2163–2172.

Mellon, D., Jr., H.R. Tuten and J. Redick. 1989. Distribution of radioactive leucine following uptake by olfactory sensory neurons in normal and heteromorphic crayfish antennules. J. Comp. Neurol. 280: 645–662.

Merrick, J.R. and C.N. Lambert. 1991. The yabby, marron and red-claw, production and marketing. Merrick Publications, Artamon, New South Wales, Australia 2064.

Mitchell, B. and B.A. Hazlett. 1996. Predator avoidance strategies of the crayfish *Orconectes virilis*. Crustaceana 69: 400–412.

Moore, P.A. and J.L. Grills. 1999. Chemical orientation to food by the crayfish *Orconectes rusticus*: influence of hydrodynamics. Anim. Behav. 58: 953–963.

Moore, P.A. and D.A. Bergman. 2005. The smell of success and failure: the role of intrinsic and extrinsic chemical signals on the social behavior of crayfish. Integr. Comp. Biol. 45: 650–657.

Moore, P.A. and K.E. Kraus-Epley. 2013. The impact of odor and ambient flow speed on the kinematics of the crayfish antennular flick: implications for sampling turbulent odor plumes. J. Crust. Biol. 33(6): 772–783.

Moore, P.A., N. Scholz and J. Atema. 1991. Chemical orientation of lobsters, *Homarus americanus*, in turbulent odor plumes. J. Chem. Ecol. 17: 1293–1307.

Moore, P.A., J.L. Grills and R.W.S. Schneider. 2000. Habitat-specific signal structure for olfaction: an example from artificial streams. J. Chem. Ecol. 26: 565–584.

Moyle, P.B. and T. Light. 1996. Biological invasions of fresh water: empirical rules and assembly theory. Biol. Conserv. 78: 149–161.

Nakata, K. and S. Goshima. 2004. Fecundity of the Japanese crayfish, *Cambaroides japonicus*: ovary formation, egg number and egg size. Aquaculture 242: 335–343.

Nobblitt, S.B., J.F. Payne and M. Delong. 1995. A comparative study of selected physical aspects of the eggs of the crayfish *Procambarus clarkii* (Girard, 1852) and *P. zonangulus* (Hobbs and Hobbs, 1990) (Decapoda, Cambaridae). Crustaceana 68: 575–582.

Nummelin, M. 1989. Cannibalism in water striders (Heteroptera: Gerridae): is there kin recognition? Oikos 56: 87–90.

Nyström, P. 1999. Ecological impact of introduced and native crayfish on freshwater communities: European perspectives. pp. 63–84. *In*: F. Gherardi and D.M. Holdich (eds.). Crayfish in Europe as Alien Species—How to Make the Best of a Bad Situation? Crustacean Issues 11. Balkema, Rotterdam.

Nyström, P. 2002. Ecology. pp. 192–235. *In*: D.M. Holdich (ed.). Biology of Freshwater Crayfish. Blackwell Science, Oxford.

Nyström, P., C. Bronmark and W. Graneli. 1996. Patterns in benthic food webs: a role for omnivorous crayfish? Freshwater Biol. 36: 631–646.

Olsén, K.H. 2011. Effects of pollutants on olfactory mediated behaviors in fish and crustaceans. pp. 507–529. *In*: T. Breithaupt and M. Thiel (eds.). Chemical Communication in Crustaceans. Springer, New York.

Pandian, T.J. 1994. Arthropoda-Crustacea. pp. 39–166. *In*: K.G. Adiyodi and R.G. Adiyodi (eds.). Reproductive Biology of Invertebrates: Asexual Propagation and Reproductive Strategies. John Wiley & Sons Inc., Chichester, United Kingdom.

Panksepp, J.B. and R. Huber. 2002. Chronic alterations in serotonin function: dynamic neurochemical properties in agonistic behavior of the crayfish *Orconectes rusticus*. J. Neurobiol. 50: 276–290.

Payne, J.F. 1972. The life history of *Procambarus hayi*. Am. Mid. Nat. 87: 25–35.

Peay, S. and P.D. Hiley. 2001. Eradication of alien crayfish. Phase II. Environment Agency Technical Report W1-037/TR1, Environment Agency, Bristol.

Peay, S., P.D. Hiley, P. Collen and I. Martin. 2006. Biocide treatment of ponds in Scotland to eradicate signal crayfish. Bull. Fr. Peche Piscic. 380-381: 1363–1379.

Perry, W.L., J.E. Feder and D.M. Lodge. 2001. Implications of hybridization between introduced and resident *Orconectes* crayfish. Conserv. Biol. 15: 1656–1666.

Pfennig, D.W., H.K. Reeve and P.W. Sherman. 1993. Kin recognition and cannibalism in spadefoot toad tadpoles. Anim. Behav. 46: 87–94.

Pfennig, D.W., P.W. Sherman and J.P. Collins. 1994. Kin recognition and cannibalism in polyphenic salamanders. Behav. Ecol. 5: 225–232.

Pravin, S., D. Mellon, Jr. and M.A. Reidenbach. 2012. Micro-scale fluid and odorant transport to antennules of the crayfish, *Procambarus clarkii*. J. Comp. Physiol. A 198: 669–681.

Reeder, P.B. and B.W. Ache. 1980. Chemotaxis in the florida spiny lobster *Panulirus argus*. Anim. Behav. 28: 831–839.

Relyea, R.A. 2003. How prey respond to combined predators: a review and an empirical test. Ecology 84: 1827–1839.

Reynolds, J.D. 2002. Growth and reproduction. pp. 152–191. *In*: D.M. Holdich (ed.). Biology of Freshwater Crayfish. Blackwell Science, Oxford.

Rodriguez, E.M., D.A. Medesani and M. Fingerman. 2007. Endocrine disruption in crustaceans due to pollutants: a review. Comp. Biochem. Physiol. A 146: 661–671.

Rogers, D. and D. Holdich. 1998. Eradication of alien crayfish populations. Environment Agency, Bristol, R and D Technical Report W169.

Rowe, C. and T. Guilford. 1999. The evolution of multimodal warning displays. Evol. Ecol. 13: 655–671.

Rubenstein, D.L. and B.A. Hazlett. 1974. Examination of the agonistic behaviour of the crayfish *Orconectes virilis* by character analysis. Behaviour 50: 193–216.

Rubolini, D., P. Galeotti, G. Ferrari, M. Spairani, F. Bernini and M. Fasola. 2006. Sperm allocation in relation to male traits, female size, and copulation behaviour in a freshwater crayfish species. Behav. Ecol. Sociobiol. 60: 212–219.

Rubolini, D., P. Galeotti, F. Pupin, R. Sacchi, A.P. Nardi and M. Fasola. 2007. Repeated matings and sperm depletion in the freshwater crayfish *Austropotamobius italicus*. Freshw. Biol. 52: 1898–1906.

Schausberger, P. and B.A. Croft. 2001. Kin recognition and larval cannibalism by adult females in specialist predaceous mites. Anim. Behav. 61: 459–464.

Schmidt, M. and D. Mellon, Jr. 2011. Neuronal processing of chemical information in crustaceans. pp. 123–147. *In*: T. Breithaupt and M. Thiel (eds.). Chemical Communication in Crustaceans. Springer, New York.

Schmitt, M. and B.W. Ache. 1979. Olfaction: responses of a decapod crustacean are enhanced by flicking. Science 205: 204–206.

Schoeppner, N.M. and R.A. Relyea. 2009. Interpreting the smells of predation: how alarm cues and kairomones induce different prey defences. Funct. Ecol. 23: 1114–1121.

Schroeder, L. and R. Huber. 2001. Fighting strategies in small and large individuals of the crayfish, *Orconectes rusticus*. Behaviour 138: 1437–1449.

Sherba, M., D.W. Dunham and H.H. Harvey. 2000. Sublethal copper toxicity and food response in the freshwater crayfish *Cambarus bartonii* (Cambaridae, Decapoda, Crustacea). Ecotoxicol. Environ. Safety 46: 329–333.

Simon, J.L. and P.A. Moore. 2007. Male-female communication in the crayfish *Orconectes rusticus*: the use of urinary signals in reproductive and non-reproductive pairings. Ethology 113: 740–754.

Snedden, W.A. 1990. Determinants of male mating success in the temperate crayfish *Orconectes rusticus*: chela size and sperm competition. Behaviour 115: 100–113.

Sokol, A. 1988. The Australian yabby. pp. 401–425. *In*: D.M. Holdich and R.S. Lowery (eds.). Freshwater Crayfish: Biology, Management and Exploitation. Croom Helm, London, England.

Stebbing, P.D., M.G. Bentley and G.J. Watson. 2003a. Mating behaviour and evidence for a female released courtship pheromone in the signal crayfish *Pacifastacus leniusculus*. J. Chem. Ecol. 29: 465–475.

Stebbing, P.D., G.J. Watson, M.G. Bentley, D. Fraser, R. Jennings, S.P. Rushton and P.J. Sibley. 2003b. Reducing the threat: the potential use of pheromones to control invasive signal crayfish. Bull. Fr. Peche Piscic. 370-371: 219–224.

Stebbing, P.D., G.J. Watson, M.G. Bentley, D. Fraser, R. Jennings, S.P. Rushton and P.J. Sibley. 2004. Evaluation of the capacity of pheromones for control of invasive non-native crayfish: part 1. English Nature Research Reports No. 578, English Nature, Peterborough, UK.

Stebbing, P.D., G.J. Watson, M.G. Bentley, D. Fraser, R. Jennings and P.J. Sibley. 2005. Evaluation of the capacity of pheromones for control of invasive non-native crayfish: part 2. English Nature Research Reports No. 633, English Nature, Peterborough, UK.

Steele, C.W., S. Strickler-Shaw and D.H. Taylor. 1992. Attraction of crayfishes *Procambarus clarkii*, *Orconectes rusticus* and *Cambarus bartoni* to a feeding stimulant and its suppression by a blend of metals. Environ. Toxicol. Chem. 11: 1323–1329.

Steele, C., C. Skinner, C. Steele, P. Alberstadt and C. Mathewson. 1999. Organization of chemically activated food search behavior in *Procambarus clarkii* Girard and *Orconectes rusticus* Girard crayfishes. Biol. Bull. 196: 295–302.

Stein, R.A. 1976. Sexual dimorphism in crayfish chelae: functional significance linked to reproductive activities. Can. J. Zool. 54: 220–227.

Stein, R.A. 1977. Selective predation, optimal foraging, and predator-prey interaction between fish and crayfish. Ecology 58: 1237–1253.

Suter, P.J. and A.M.M. Richardson. 1977. The biology of two species of *Engaeus* (Decapoda: Parastacidae) in Tasmania. III. Habitat, food, associated fauna and distribution. Aust. J. Mar. Freshw. Res. 28: 95–103.

Tack, P.I. 1941. Reproductive biology of an epibenthic amphipod (*Dyopedos monacanthus*) with extended parental care. J. Mar. Biol. Assoc. UK 77: 1059–1072.

Tamm, G.R. and S.J. Cobb. 1978. Behavior and the crustacean molt cycle: changes in aggression of *Homarus americanus*. Science 200: 79–81.

Thiel, M. 1999. Parental care behaviour in crustaceans—a comparative overview. Crustac. Issues 12: 211–226.

Thiel, M. 2003. Extended parental care in crustaceans—an update. Rev. Chil. Hist. Nat. 76: 205–218.

Thiel, M. 2007. Social behaviour of parent-offspring groups in crustaceans. pp. 294–318. *In*: J.E. Duffy and M. Thiel (eds.). Evolutionary Ecology of Social and Sexual Systems: Crustaceans as Model Organisms. Oxford University Press, Oxford.

Tibbetts, E.A. and J. Dale. 2007. Individual recognition: it is good to be different. Trends Ecol. Evol. 22: 529–537.

Tierney, A.J. and D.W. Dunham. 1982. Chemical communication in the reproductive isolation of the crayfishes *Orconectes propinquus* and *Orconectes virilis* (Decapoda, Cambaridae). J. Crust. Biol. 2: 544–548.

Tierney, A.J. and D.W. Dunham. 1984. Behavioral mechanisms of reproductive isolation in the crayfishes of the genus *Orconectes*. Am. Midl. Nat. 111: 304–310.

Tierney, A.J. and J. Atema. 1986. Effects of acidification on the behavioral responses of crayfishes (*Orconectes virilis* and *Procambarus acutus*) to chemical feeding stimuli. Aquat. Toxicol. 9: 1–11.

Tierney, A.J. and J. Atema. 1988. Behavioral responses of crayfish (*Orconectes virilis* and *Orconectes rusticus*) to chemical feeding stimulants. J. Chem. Ecol. 14: 123–133.

Tierney, A.J. and L.A. Mangiamele. 2001. Effects of serotonin and serotonin analogs on posture and agonistic behavior in crayfish. J. Comp. Physiol. A 187: 757–767.

Tierney, A.J., C.S. Thompson and D.W. Dunham. 1986. Fine structure of aesthetasc chemoreceptors in the crayfish *Orconectes Propinquus*. Can. J. Zool. 64: 392–399.

Tomba, A.M., T.A. Keller and P.A. Moore. 2001. Foraging in complex odor landscapes: chemical orientation strategies during stimulation by conflicting chemical cues. J. N. Am. Benthol. Soc. 20: 211–222.

Tricarico, E. and F. Gherardi. 2007. Biogenic amines influence aggressiveness in crayfish but not their force or hierarchical rank. Anim. Behav. 74: 1715–1724.

Tricarico, E., B. Renai and F. Gherardi. 2005. Dominance hierarchies and status recognition in the threatened crayfish, *Austropotamobius italicus*. Bull. Fr. Peche Piscic. 376–377: 655–664.

Trivers, R.L. 1972. Parental investment and sexual selection. pp. 136–179. *In*: B. Champbell (ed.). Sexual Selection and the Descent of Man. Aldine, Chicago.

Turner, A.M., S.A. Fetterolf and R.J. Bernot. 1999. Predator identity and consumer behavior: differential effects of fish and crayfish on the habitat use of a freshwater snail. Oecologia 118: 242–247.

Van der Velden, J., Y. Zheng, B.W. Patullo and D.L. Macmillan. 2008. Crayfish recognize the faces of fight opponents. PLoS ONE 3: e1695.

Vickers, N.J. 2000. Mechanisms of animal navigation in odor plumes. Biol. Bull. 198: 203–212.

Villanelli, F. and F. Gherardi. 1998. Breeding in the crayfish, *Austropotamobius pallipes*: mating patterns, mate choice and intermale competition. Freshw. Biol. 40: 305–315.

Vogt, G. and L. Tolley. 2004. Brood care in freshwater crayfish and relationship with the offspring's sensory deficiencies. J. Morph. 262: 566–582.

Vogt, K. 1980. Optical system of the crayfish eye. J. Comp. Physiol. A 135: 1–19.

Vye, C., J.S. Cobb, T. Bradley, J. Gabbay, A. Genizi and I. Karplus. 1997. Predicting the winning or losing of symmetrical contests in the American lobster *Homarus americanus* (Milne-Edwards). J. Exp. Mar. Biol. Ecol. 217: 19–29.

Warrant, E.J. and P.D. McIntyre. 1990. Limitations to resolution in superposition eyes. J. Comp. Physiol. A 167: 785–803.

Webster, D.R. and M.J. Weissburg. 2001. Chemosensory guidance cues in a turbulent chemical odor plume. Limnol. Oceanogr. 46: 1034–1047.

Webster, D.R., S. Rahman and L.P. Dasi. 2001. On the usefulness of bilateral comparison to tracking turbulent chemical odor plumes. Limnol. Oceanogr. 46(5): 1048–1053.

Weissburg, M.J. 2011. Waterborne chemical communication stimulus dispersal dynamics and orientation strategies in crustaceans. pp. 63–83. *In*: T. Breithaupt and M. Thiel (eds.). Chemical Communication in Crustaceans. Springer, New York.

Weissburg, M.J. and R.K. Zimmerfaust. 1993. Life and death in moving fluids - hydrodynamic effects on chemosensory-mediated predation. Ecology 74: 1428–1443.

Whitmore, N. and A.D. Huryn. 1999. Life history and production of *Paranephrops zealandicus* in a forest stream, with comments about the sustainable harvest in a freshwater crayfish. Freshw. Biol. 42: 467–478.

Wigginton, A.J., R.L. Cooper, E.M. Fryman-Gripshover and W.J. Birge. 2010. Effects of cadmium and body mass on two anti-predator behaviors of five species of crayfish. Int. J. Zool. Res. 6: 39–51.

Willman, E.J., A.M. Hill and D.M. Lodge. 1994. Response of 3 crayfish congeners (*Orconectes* spp.) to odors of fish carrion and live predatory fish. Am. Midl. Nat. 132: 44–51.

Wilson, E.O. 1975. Sociobiology. Harvard University Press, Cambridge, MA, 697 pp.

Wolf, M.C. and P.A. Moore. 2002. The effects of the herbicide metolachlor on the perception of chemical stimuli by *Orconectes rusticus*. J. N. Am. Benthol. Soc. 21: 457–467.

Woodlock, B. and J.D. Reynolds. 1988. Laboratory breeding studies of freshwater crayfish, *Austropotamobius pallipes* (Lereboullet). Freshw. Biol. 19: 71–78.

Wyatt, T.D. 2011. Pheromones and behavior. pp. 23–38. *In*: T. Breithaupt and M. Thiel (eds.). Chemical Communication in Crustaceans. Springer, New York.

Wyatt, T.D. 2014. Pheromones and animal behaviour: chemical signals and signatures. Cambridge University Press, Cambridge, 419 pp.

Zulandt, T., R.A. Zulandt Schneider and P.A. Moore. 2008. Observing agonistic interactions alters subsequent fighting dynamics in the crayfish, *Orconectes rusticus*. Anim. Behav. 75: 13–20.

Zulandt Schneider, R.A. and P.A. Moore. 2000. Urine as source of conspecific disturbance signals in the crayfish *Procambarus clarkii*. J. Exp. Biol. 203: 765–771.

Zulandt Schneider, R.A., R.W.S. Schneider and P.A. Moore. 1999. Recognition of dominance status by chemoreception in the red-swamp crayfish, *Procambarus clarkii*. J. Chem. Ecol. 25: 781–794.

Zulandt Schneider, R.A., R. Huber and P.A. Moore. 2001. Individual and status recognition in the crayfish, *Orconectes rusticus*: the effects of urine release on fight dynamics. Behaviour 138: 137–153.

6

Parasites, Commensals, Pathogens and Diseases of Crayfish

Matt Longshaw

Introduction

Crayfish are hosts to a wide range of different commensals, parasites and pathogens including viruses, bacteria, fungi, protistans, trematodes, cestodes, acanthocephalans, nematodes, branchiobdellids, temnocephalids and arthropods. The anthropogenic movement of crayfish either for aquaculture purposes or the aquarium trade has resulted in the translocation of a number of disease conditions of concern, some of which have led to great reductions in native crayfish populations. One of the most studied disease conditions of crayfish is the so-called crayfish plaque, caused by *Aphanomyces astaci*. Whilst this fungal-like organism has no doubt caused complete or near extinction of some native crayfish populations, its infamous reputation has meant that some unexplained mortalities might have been incorrectly attributed to this disease despite clear diagnostic evidence. This view was ably summarised by Edgerton et al. (2004) who stated that such an "extreme emphasis on *A. astaci* has created inertia in European astacology, which has curtailed researchers, state fish-disease diagnosticians, and resource managers from fully assessing and considering the existence of other serious pathogens of freshwater crayfish and the ensuing consequences. As a result, basic skills in crayfish pathology have been lost or underdeveloped. Moreover, some management schemes aimed at conserving native European freshwater crayfish are less likely to be effective, and might actually be harmful, in achieving their goal because of a lack of appreciation of the presence or significance of certain pathogens". Another issue is that few studies of crayfish diseases consider the pathology associated with the infection meaning that assessment of the individual impact is often lacking; the bulk of studies on crayfish diseases appear to have been focused on descriptive and lifecycle studies. Whilst advances in diagnostic methods in recent years have ensured that, in part, correct diagnosis of an aetiological agent has been possible, this has been negatively balanced with a lack of crayfish pathology specialists worldwide. Of those individuals actively involved in this work, most are based in Australia, Europe and the USA. This geographical bias has meant that not all species of crayfish have been examined for pathogens and lead to the exciting possibilities in the future of new and novel disease conditions being described in crayfish across their distribution.

Benchmark Animal Health Ltd., Bush House, Edinburgh Technopole, Milton Bridge, Edinburgh, Scotland, EH26 0BB.
  Email: matt.longshaw@bmkanimalhealth.com

Extensive systematic reviews of crayfish pathogens have been completed by Alderman and Polglase (1988), Edgerton et al. (2002a) and Longshaw (2011). This current chapter builds on these reviews by providing an overview of the main taxonomic features of the different disease agents including lifecycle information and host-pathogen lists, and important references for each species reported; the chapter ends with a focussed list of parasites and disease agents of economically and ecologically important crayfish species and considers future research areas in crayfish pathology.

Host responses

The carapace of crayfish acts as the main barrier against mechanical damage and limits the invasion of a number of disease agents. It follows that the breaching of this barrier will allow the ingress of parasites and diseases into the host. In addition, these external surfaces can also act as a suitable substrate for parasites, which may increase external fouling and possibly reduce mobility. However, through the natural process of moulting, numbers of ectoparasites may be reduced.

Crayfish, like most invertebrates, have an open circulatory system with haemolymph transported around the body via arteries into the main organs and returned via venous channels to the gills and heart. The combination of opsonins and haemocytes within the haemolymph act to recognise non-self organisms, and following their migration to the site of injury or attack, there is subsequent phagocytosis or encapsulation of these foreign bodies. Recognition of non-self bodies is initiated by pattern recognition proteins such as β-glucan-binding protein (βGBP) and lipopolysaccharide- and glucan-binding protein (LGBP) which bind to β-1,3-glucans, masquerade-like proteins and serine proteinase homologues (SPH's) and lectins; binding of these proteins to β-1,3-glucan and/or lipopolysaccharides triggers the prophenoloxidase (proPO) cascade (Duvic and Söderhäll 1990, Kopáček et al. 1993a, Middleton et al. 1996, Huang et al. 2000, Lee and Söderhäll 2001, Lanz et al. 2009, Fang et al. 2013).

One of the main features of the immune response in crayfish is the proPO cascade, which enhances phagocytosis, initiates nodule or capsule formation, mediates coagulation, produces fungistatic substances and terminates in melanisation of the non-self molecules (Jiravanichpaisal et al. 2006). Activation of the proPO cascade is initiated by the LGBP and βGBP and degranulation of the haemocytes and is regulated by serine proteinases and melanisation inhibition proteins (MIPs) (Aspán et al. 1990, Aspán and Söderhäll 1991, Johansson and Söderhäll 1985, Liu et al. 2013, Söderhäll et al. 2009).

Circulating haemocytes are mainly involved in the recognition, phagocytosis/encapsulation, melanisation and degradation of foreign bodies. In common with other crustaceans, crayfish have three classes of haemocytes, namely phagocytic hyaline cells (HC), semigranular cells (SGC) involved in early pathogen detection and granular cells (GC) which are responsible for activating the proPO system as well as containing antimicrobial peptides and cell adhesion proteins, including peroxinectin (Jiravanichpaisal et al. 2006, Johansson et al. 1995, Liu et al. 2009, Shi et al. 2005, Sricharoen et al. 2005, Taylor et al. 2009). Haematopoiesis is continuous in crayfish with haemocyte formation occurring within the haematopoietic tissue where five cell types involved in haematopoesis are recognised including two main proliferating cell types and three precursors of GC's and SGC's (Chaga et al. 1995). Differentiation of the hematopoetic cells into either GC's or SGC's occurs after release of these precursor cells into the haemolymph and is supported by the cytokines Astakine 1 and Astakine 2 (Lin and Söderhäll 2011, Lin et al. 2011, Söderhäll 2013) and by β-thymosins (Saelee et al. 2013). Clotting of the haemolymph is aided by a number of clotting proteins (Hall et al. 1999, Kopáček et al. 1993b, Vafopoulou 2009).

Crayfish also utilise antimicrobial/antibacterial peptides (AMP/ABP), including lysozymes, and other peptides like procambarin, astacidin 1, astacidin 2, *Pc*Ast, AMP-14 and AMP-16 (Jiravanichpaisal et al. 2007, Lee et al. 2003, Shi et al. 2014, Sricharoen et al. 2005, Zeng 2013, Zhang et al. 2010) as well as apoptosis, cytosolic manganese superoxide dismutase and prohibitins (Lan et al. 2013, Liu et al. 2009, 2013).

Taxonomic review of crayfish pathogens

This section covers the known parasites, pathogens and commensals reported or described from crayfish throughout their range. Despite my best efforts, it is possible that some references and a very small number of agents may have been overlooked. This is inevitable but I hope will not detract from the overall utility of the chapter. However, there is a major note of caution required when reading this chapter. Where I could, I updated the nomenclature of the crayfish hosts and their pathogens/parasites based on the best available up to date data. Notwithstanding, I am aware that some names of host and/or pathogen given in some of the early literature are patently wrong; where possible I have tried to correct these accepting that some of the records of parasite/commensal/disease agents will need to be re-evaluated using appropriate tools. Thus, any errors in the nomenclatural changes or in the interpretation of the many papers are entirely my fault—I therefore ask that, like any good scientist should, that you go back to the original manuscripts where possible and confirm for yourself that I got it right first time.

Viruses

In general, viruses from crayfish have not been fully characterised with most of the taxonomic characteristics derived from histopathological and ultrastructural studies. This is partly due to a global lack of expertise in crustacean pathology, partly due to many viral infections being rare or present at low prevalence in asymptomatic animals, partly due to lack of appropriate molecular primers, and partly to a lack of suitable cell lines to culture viruses, this being a potential area for research in the future. Most crayfish viruses appear to be relatively host and tissue specific and, within the crayfish, most have been reported from parastacids. For example, in a series of reports described below, Edgerton and colleagues describe a plethora of viral infections from mainly *Cherax* spp. It does not follow that *Cherax* spp. are any more susceptible to viral infections compared to their astacid counterparts; rather it reflects a concerted effort by antiopodean researchers to survey commercial species for viral infections. Equivalent efforts in Europe, the USA and elsewhere are limited with few reports of viruses in native species outside of Australia. It is highly probable that more viral infections remain to be discovered and described from astacid and cambarid crayfish with equivalent efforts. Viral infections have been reported from the hepatopancreas and gut, the gills and more rarely the haemocytes with few systemic infections being noted.

Hepatopancreatic viral infections

The vast majority of viral infections of crayfish have been described from the hepatopancreas. Pathology is broadly similar with most hepatopancreatic cell types involved in the infection; pathology ranges from mild hypertrophy of the affected nuclei, through to large hypertrophied nuclei with marginated or rarefied chromatin and formation of intranuclear septae. Necrosis and sloughing of affected cells can occur although mortalities are rare or negligible. Reported intranuclear bacilliform (double stranded DNA) viruses of the hepatopancreas include *Astacus astacus* bacilliform virus (*Aa*BV) (Edgerton et al. 1996b), *Austropotamobius pallipes* bacilliform virus (*Ap*BV) (Edgerton 2003, Edgerton et al. 2002b, Longshaw et al. 2012b), *Cherax destructor* bacilliform virus (*Cd*BV) (Edgerton 1996), *Pacifastacus leniusculus* bacilliform virus (*Pl*BV) (Hauck et al. 2001, Longshaw et al. 2012a) and *Cherax quadricarinatus* bacilliform virus (*Cq*BV) (=hepatopancreatic baculovirus of *Cherax quadricarinatus* = *Cherax* baculovirus (CBV) (Anderson and Prior 1992, Edgerton 1996, Edgerton and Owens 1999, Groff et al. 1993, Hauck et al. 2001, Romero and Jiménez 2002). Juvenile crayfish can become infected within two weeks of exposure to macerated, *Cq*BV-infected hepatopancreas (Edgerton and Owens 1997), although feeding material to naïve crayfish does not appear to instigate new infections (Claydon et al. 2004a). Experimental transmission of *Penaeus merguiensis* densovirus (*Pmerg*DNV) to *C. quadricarinatus* via intramuscular injection or orally in overcrowded and normal stocking densities lead to some mortalities in crayfish; necrotic, eosinophilic cytoplasmic lesions were noted in the hepatopancreas but intranuclear inclusions were absent (La Fauce and Owens 2007). The authors suggested that death was due to confounding factors including overcrowding and opportunistic infections rather than *Pmerg*DNV, that virus replication does not take place in the host and that *C. quadricarinatus* are short-term carriers of the virus. Two other intranuclear hepatopancreatic viruses are *Cherax quadricarinatus* Giardiavirus-like virus, (GCV - a double stranded RNA member of the Totiviridae) (Edgerton et al. 1994, Edgerton and Owens 1997, 1999) and a Picornaviridae or Circoviridae single stranded RNA virus of *Cherax albidus*; the virus is also noted in the labyrinth epithelium of the antennal gland (Jones and Lawrence 2001). Finally, Edgerton et al. (2000) described a double stranded RNA Reoviridae in the hepatopancreas of moribund *C. quadricarinatus*. The infection was characterised by cytoplasmic inclusions in hepatopancreatocytes, some closely associated with the nucleus (La Fauce and Owens 2007). Naïve crayfish were successfully infected following experimental transmission of infected material via intraperitoneal injection into the hepatopancreas or through feeding with infected animals appearing lethargic with a weakened tail-flip response (Hayakijkosol and Owens 2011).

Haemocytic viral infections

Halder and Ahne (1988a,b) were able to transmit infectious pancreatic necrosis virus (IPNV), a double stranded RNA member of the Birnaviridae normally found in teleost fish, to the haemocytes of *Astacus astacus* via exposure to infected water for 1 hour, by co-habitation with infected trout, through intraperitoneal injection and through feeding of infected fish tissues to crayfish. Positive isolation of virus from haemolymph of exposed crayfish was possible two days post-transmission. Excretion of viable virus from crayfish was demonstrated through successful re-infection of trout fry and eggs with IPNV. Recently, Soowannayan et al. (2015) were able to transmit yellow head virus (YHV) to red claw crayfish, which were asymptomatic for the disease. The authors were subsequently able to transmit the infection from *C. quadricarinatus* to black tiger shrimp (*Penaeus monodon*) by injection and cohabitation, and suggested that red claw crayfish were excellent carriers for this viral infection.

Viral infections of the gill

Viral infections of crayfish gills are apparently rare. Edgerton et al. (2000) described *Cherax quadricarinatus* parvo-like virus (*Cq*PlV); affected nuclei in the gills were hypertrophied with marginated chromatin and peripheral nucleoli. Initial electron microscopy studies revealed evidence of electron-dense, rounded virus-like particles scattered throughout the nucleus. Injection of affected crayfish with 7 µg/kg Ivermectin led to a 68% reduction in numbers of hypertrophied nuclei in crayfish (Nguyen et al. 2014). A similar, if not identical infection in the gills of *C. quadricarinatus* in Ecuador has also been noted (Romero and Jiménez 2002). Subsequently, Rusaini et al. (2013) applied suppression subtractive hybridisation (SSH) to crayfish showing typical lesions as reported by Edgerton et al. (2000). The method was used to compare genes of interest in infected and non-infected individuals and although a large number of sequences were isolated, none had homology to parvovirus or other viral genes. The lack of amplification of relevant genes led the authors to suggest an idiopathic aetiology for the lesions. Halder and Ahne (1988a) provided a single image of virus-like particles in the connective tissue of *A. astacus* gills. No further details were provided and it does not appear to have been reported since.

Systemic viral infections

Edgerton et al. (1997) described *Cherax destructor*–systemic parvo-like virus (*Cd*SPV), a single stranded DNA member of the Parvoviridae. Cowdry type A inclusions (CAIs), hypertrophied nuclei, rarefied chromatin and enlarged nucleoli were recorded in most organs but were most common in the gills. In the single animal examined, the authors also noted extensive necrotic foci and an opaque musculature. A systemic parvovirus, apparently distinct from that reported by Edgerton et al. (1997) called Decapod ambidensovirus, variant *Cherax quadricarinatus* densovirus (*Cq*DV), causes mortalities of up to 96% in farmed *Cherax quadricarinatus* (Bowater et al. 2002, Bochow et al. 2015). Clinical signs included lethargy, weakness, anorexia and red colouration of carapace. Intranuclear inclusion bodies ranging in size and appearance from small, eosinophilic inclusions to large prominent basophilic inclusion bodies occurred mainly in the gills, cuticular and gut epithelium, connective tissue and occasionally in the eye. Longevity of affected animals is marginally increased via injection of the hosts with 7 µg/kg Ivermectin (Nguyen et al. 2014).

   Cherax quadricarinatus are susceptible to the single stranded RNA Nodaviridae *Macrobranchium rosenbergii* nodavirus (*Mr*NV), the causative agent of white tail disease of prawns. Hayakijkosol et al. (2011) were able to experimentally transmit the infection by feeding and inoculation routes; affected crayfish had an inability to swim normally, showed a reduced appetite, pale exoskeleton and ultimately died. Whilst there was evidence of muscle necrosis and myositis, no viral inclusions were observed and as a result the authors suggest that limited replication occurs in crayfish. Prevention of replication by the virus and reductions in mortalities have been achieved through the use of gene-silencing technologies (RNAi) (Hayakijkosol and Owens 2012).

Cherax quadricarinatus are also susceptible to the prawn disease Spawner-isolated mortality virus (SMV). Typically, affected animals showed no external clinical signs and presented with normal tissues in histology. Application of DIG-labelled SMV probes resulted in positive signals in nuclei of several organs, including the hepatopancreas, midgut and reproductive organs (Owens and McElnea 2000).

One of the best-studied viral infections of decapod crustaceans is white spot disease (WSD). Concerns have been raised over its ability to infect all decapod crustaceans, albeit with differential pathogenicity and the risks associated with transnational transport of live crayfish (Edgerton 2002, Holdich et al. 2009, Longshaw et al. 2012a, Mrugała et al. 2015). Crayfish reported as susceptible to WSD include *Astacus astacus, A. leptodactylus, Austropotamobius pallipes, Cherax quadricarinatus, C. albidus, Orconectes (Faxonius) limosus, Orconectes (Procericambarus) punctimanus, Orconectes (Gremicambarus) virilis, Pacifastacus (Pacifastacus) leniusculus, Procambarus (Scapulicambarus) clarkii* and *Procambarus (Ortmannicus) zonangulus* (Bateman et al. 2012, Baumgartner et al. 2009, Corbel et al. 2001, Davidson et al. 2010, Edgerton 2004a,b, Gao et al. 2014, Heidarieh et al. 2013, Jiravanichpaisal et al. 2001, 2004, Longshaw et al. 2012a, Mrugała et al. 2015, Wu et al. 2012). All attempts to experimentally transmit the virus to naïve crayfish have successfully led to the development of the disease in those animals (Bateman et al. 2012, Edgerton 2004a,b, Huang et al. 2001, Jiravanichpaisal et al. 2001, Maeda et al. 2000, Shi et al. 2000, Soowannayan and Phanthura 2011, Yan et al. 2007) and it is thus highly likely that all crayfish species are susceptible to WSD. Attempts have been made to increase resistance to WSD, primarily in *P. (S.) clarkii*, through a number of routes, including exposure of crayfish to inactivated virus or to envelope proteins, use of gene silencing technology, use of prohibitin, manipulation of the host immune responses and through altering water temperatures (Du et al. 2006, 2008, 2013, Gao et al. 2014, Heidarieh et al. 2013, Jiravanichpaisal et al. 2004, Lan et al. 2013, Wu et al. 2012, Xu et al. 2006, Zeng 2013, Zhang et al. 2012, Zhu et al. 2009, Zuo et al. 2015). Methods for the successful detection of infections in crayfish and other decapods are well established (Claydon et al. 2004b, Du et al. 2007, Huang et al. 2001, Lo et al. 1999, Poulos et al. 2001, Stentiford et al. 2009, Xie et al. 2005) although caution has been expressed with their use in certain circumstances (Baumgartner et al. 2009).

Originally reported by Edgerton and Owens (1999), Edgerton (2000) described a number of idiopathic conditions in farmed *C. quadricarinatus*, some of which he suggested may have a viral aetiology. These included a haemocytic enteritis typified by necrosis of the midgut and rounded up hepatocreatocytes with pyknotic nuclei, and the association of cytoplasmic inclusions, suggestive of a viral infection, in the antennal gland, haemolymph vessels and the mandibular organ. The relationship of these pathologies to that described by Romero and Jiménez (2002) is unclear who described pyknosis, karyorrhexis and necrosis of the stomach hypodermis and of the antennal gland as well as Cowdry type A inclusions. They discounted the possibility that the pathologies were related to infectious hypodermal and haematopoietic necrosis virus (IHHNV) or white spot syndrome virus (WSSV). No further reports or characterisation of these tissue changes have been reported.

Bacteria

A number of bacteria have been isolated from crayfish. These have mainly been from asymptomatic animals, and more often than not, from haemolymph of these animals. Importantly, many of the isolates were identified using primary tests and Analytical Profile Index (API) strips. Whilst these have a value in signposting the worker to a potential identification of the bacteria, they can be equivocal and some of the results obtained by this method should be treated with caution. As an example, Topić Popović et al. (2014) compared results of bacteria isolated from apparently healthy *Astacus astacus* and *A. leptodactylus* using the phenotypic API20E test and matrix assisted laser induced desorption ionisation connected to the time of flight mass spectrometry (MALDI-TOF MS). There was an incongruence between the results obtained by the phenotypic test and that obtained from the MALDI-TOF test with the MALDI-TOF method apparently able to more accurately identify bacteria. Of the 23 bacterial isolates obtained, there was only agreement between both methods for one bacterial isolate (*Hafnia alvei*). However, few isolates were reliably identified to species using either API20E or MALDI-TOF. Thus, data provided below should be treated with caution and, as with many of the infections reported in crayfish, will need re-evaluation in light of newer techniques and methodologies.

Bacteria (Phylum Actinobacteria)

Some of the Gram-positive bacteria within this phylum isolated from crayfish are of medical significance. Several species have been isolated from the haemolymph of apparently healthy crayfish including *Arthrobacter* sp. from *Orconectes* (*Gremicambarus*) *virilis* and *Procambarus* (*Scapulicambarus*) *clarkii* (Davidson et al. 2010, Scott and Thune 1986b), *Corynebacterium* sp. from *Cherax albidus, Cherax quadricarinatus* and *P.* (*S.*) *clarkii* (Bowater et al. 2002, Scott and Thune 1986b, Wong et al. 1995), *Micrococcus* spp. from *C. albidus* and *C. quadricarinatus* (Wong et al. 1995) and *Mycobacterium chelonae* from *C. quadricarinatus* (Sewell and Cannon 1994). A putative *Nocardia* sp. identified from histological sections was reported from a single sluggish and unresponsive *Austropotamobius pallipes* in the UK (Alderman et al. 1986). It has not been recorded since. A serious case of Buruli ulcer in three members of the same family in Japan, was attributed to *Mycobacterium ulcerans* subsp. *shinshuense* (Ohtsuka et al. 2014). The authors were able to detect the same bacterium from samples of an unidentified crayfish in a water channel surrounding the house; direct transmission from the crayfish to humans was not considered possible.

Bacteria (Phylum Bacteroidetes)

Several of these Gram-negative bacteria have been isolated from the haemolymph of apparently healthy crayfish, including *Chryseobacterium* sp. in *Pacifastacus* (*Pacifastacus*) *leniusculus* (Jiravanichpaisal et al. 2009), *Flavobacterium dormitator* from *Procambarus* (*Scapulicambarus*) *clarkii* (Scott and Thune 1986b), *Flavobacterium* spp. from *Cherax albidus, C. quadricarinatus* and *P.* (*S.*) *clarkii* (Jones and Lawrence 2001, Scott and Thune 1986b, Wong et al. 1995), *Sphingobacterium multivorum* (=CDC group IIk-2) from *P.* (*S.*) *clarkii* (Scott and Thune 1986b), *Elizabethkingia meningosepticum* from *Astacus astacus* (Oidtmann and Hoffman 1999) and *Weeksella virosa* (=CDC Group IIf) from *Cambarellus* (*Cambarellus*) *patzcuarensis* (Longshaw et al. 2012a).

Bacteria (Phylum Firmicutes)

None of these Gram-positive bacteria isolated from crayfish have been associated with mortality. With the exception of *Bacillus mycoides* isolated from the intestine of healthy *Cherax cainii* (Ambas et al. 2015), they have all been isolated from haemolymph. Species found include *Bacillus* spp. that have been isolated from *Cherax albidus, C. quadricarinatus,* and *Procambarus* (*Scapulicambarus*) *clarkii* (Amborski et al. 1975, Jones and Lawrence 2001, Scott and Thune 1986b, Wong et al. 1995), *Listeria monocytogenes* from *Astacus leptodactylus* (Khamesipour et al. 2013, Li et al. 2015), *Kurthia* sp. and *Staphylococcus* spp. from *C. albidus* (Wong et al. 1995), *Staphylococcus cohnii* from *C. quadricarinatus* (Wong et al. 1995), *S. epidermidis* from *C. quadricarinatus* and *P.* (*S.*) *clarkii* (Scott and Thune 1986b, Wong et al. 1995) and *Streptococcus* sp. from *P.* (*S.*) *clarkii* (Scott and Thune 1986b).

Bacteria (Phylum Proteobacteria)

The bulk of the bacteria isolated from crayfish belong to the phylum Proteobacteria (Table 1). The genera isolated belong to a range of orders including Aeromonadales (*Aeromonas*), Alteromonadales (*Shewanella*), Burkholderiales (*Alcaligenes, Oligella*), Campylobacterales (*Campylobacter*), Caulobacterales (*Phenylobacterium*), Enterobacteriales (*Citrobacter, Cronobacter, Edwardsiella, Enterobacter, Erwinia,*

Table 1. List of species of bacteria in the Phylum Proteobacteria, in alphabetical order, reported from crayfish hosts.

Genus/species	Host(s)	Reference(s)
Acinetobacter antitratus	*Procambarus (Scapulicambarus) clarkii*	Scott and Thune (1986b)
A. calcoaceticus	*P. (S.) clarkii*	Scott and Thune (1986b)
A. lwoffi	*Cherax albidus, P. (S.) clarkii*	Scott and Thune (1986b), Wong et al. (1995)
Acinetobacter sp.	*Cherax quadricarinatus, Orconectes (Gremicambarus) virilis, Pacifastacus (Pacifastacus) leniusculus, P. (S.) clarkii*	Davidson et al. (2010), Jiravanichpaisal et al. (2009), Scott and Thune (1986b), Wong et al. (1995)
Aeromonas hydrophila	*Astacus astacus, Astacus leptodactylus, Austropotamobius pallipes, C. albidus, Cherax cainii, Cherax quadricarinatus, P. (P.) leniusculus, P. (S.) clarkii*	Avenant-Oldewage (1993), Edgerton et al. (1995), Jiravanichpaisal et al. (2009), Jones and Lawrence (2001), Longshaw et al. (2012a), Oidtmann and Hoffman (1999), Quaglio et al. (2006a), Raissy et al. (2014), SamCookiyaei et al. (2012), Scott and Thune (1986b)
A. liquefacieus	*P. (S.) clarkii*	Amborski et al. (1975)
A. sobria	*A. astacus, Cambarellus (Cambarellus) patzcuarensis, C. albidus, C. quadricarinatus, Orconectes (Crockerinus) propinquus, Procambarus (Ortmannicus) fallax*	Krugner-Higby et al. (2010), Longshaw et al. (2012a), Oidtmann and Hoffman (1999), Wong et al. (1995)
Aeromonas sp.	*C. albidus*	Jones and Lawrence (2001)
A. veroni	*C. albidus*	Jones and Lawrence (2001)
Alcaligenes sp.	*C. albidus, C. quadricarinatus*	Bowater et al. (2002), Wong et al. (1995)
Campylobacter spp.	*A. leptodactylus*	Raissy et al. (2014)
Citrobacter freundii	*A. astacus, A. pallipes, C. (C.) patzcuarensis, C. albidus, C. quadricarinatus, P. (O.) fallax, P. (S.) clarkii*	Amborski et al. (1975), Bowater et al. (2002), Oidtmann and Hoffman (1999), Quaglio et al. (2006a,b), Longshaw et al. (2012a), Wong et al. (1995)
C. gillenii	*P. (P.) leniusculus*	Jiravanichpaisal et al. (2009)
C. murliniae/freundii	*P. (P.) leniusculus*	Jiravanichpaisal et al. (2009)
Citrobacter sp.	*A. pallipes, C. quadricarinatus*	Romero and Jiménez (2002), Vey et al. (1975)
Coliform-like spp.	*C. albidus*	Wong et al. (1995)
Coxiella cheraxi	*C. quadricarinatus*	Cooper et al. (2007), Jiménez and Romero (1997), La Fauce and Owens (2007), Tan and Owens (2000)
Cronobacter sakazakii	*A. astacus*	Oidtmann and Hoffman (1999)
Edwardsiella tarda	*C. quadricarinatus*	Bowater et al. (2002)
Enterobacter aerogenes	*P. (S.) clarkii*	Scott and Thune (1986b)
E. agglomerans	*C. quadricarinatus*	Bowater et al. (2002)
E. cloacae	*P. (S.) clarkii*	Scott and Thune (1986b)
E. intermedium	*C. quadricarinatus*	Eaves and Ketterer (1994)
Erwinia sp.	*A. astacus, O. (G.) virilis*	Davidson et al. (2010), Oidtmann and Hoffman (1999)
Escherichia coli	*A. leptodactylus, C. albidus, C. quadricarinatus*	Eaves and Ketterer (1994), Jones and Lawrence (2001), Raissy et al. (2014)
Francisella tularensis biovar palaearctica	*P. (S.) clarkii*	Anda et al. (2001)

Table 1. contd....

Table 1. contd.

Genus/species	Host(s)	Reference(s)
Grimontia hollisae	*P. (O.) fallax*	Longshaw et al. (2012a)
Hafnia alvei	*Astacus astacus, A. pallipes, C. albidus, Cherax destructor, P. (P.) leniusculus*	Jones and Lawrence (2001), Quaglio et al. (2008), Longshaw et al. (2012a), Oidtmann and Hoffman (1999), Orozova et al. (2014)
Klebsiella pneumoniae	*C. quadricarinatus*	Edgerton et al. (1995)
Micrococcus luteus	*C. quadricarinatus, P. (S.) clarkii*	Scott and Thune (1986b), Wong et al. (1995)
M. roseus	*P. (S.) clarkii*	Scott and Thune (1986b)
Moraxella sp.	*C. quadricarinatus*	Bowater et al. (2002)
Oligella ureolytica (=CDC Group IVe)	*P. (S.) clarkii*	Scott and Thune (1986b)
Pasteurella multocida	*P. (O.) fallax*	Longshaw et al. (2012a)
Phenylobacterium sp.	*O. (G.) virilis*	Davidson et al. (2010)
Plesiomonas shigelloides	*C. albidus, C. quadricarinatus*	Edgerton et al. (1995), Wong et al. (1995)
Proteus morganii	*A. leptodactylus, A. pallipes, Orconectes (Faxonius) limosus*	Toumanoff (1965)
Proteus sp.	*C. albidus*	Jones and Lawrence (2001)
P. vulgaris	*A. leptodactylus, A. pallipes, O. (F.) limosus*	Toumanoff (1965)
Pseudomonas aeruginosa	*A. astacus, A. pallipes*	Vey (1981)
P. alcaligenes	*P. (S.) clarkii*	Amborski et al. (1975), Scott and Thune (1986b)
P. cepacia	*C. quadricarinatus*	Wong et al. (1995)
P. fluorescens	*A. pallipes*	Vey et al. (1975)
P. guinea/peli	*P. (P.) leniusculus*	Jiravanichpaisal et al. (2009)
P. libanensis/gessardii	*P. (P.) leniusculus*	Jiravanichpaisal et al. (2009)
P. luteola	*A. astacus*	Oidtmann and Hoffman (1999)
P. maltophila	*C. quadricarinatus*	Wong et al. (1995)
P. mendocina	*P. (S.) clarkii*	Scott and Thune (1986b)
P. putida	*A. astacus, A. pallipes*	Vey (1981)
P. putrefaciens	*P. (S.) clarkii*	Scott and Thune (1986b)
Pseudomonas sp.	*C. albidus, C. destructor, C. quadricarinatus, P. (P.) leniusculus, P. (S.) clarkii*	Amborski et al. (1975), Jiravanichpaisal et al. (2009), Jones and Lawrence (2001), Quaglio et al. (2006b), Sewell and Cannon (1994)
P. stutzeri	*P. (S.) clarkii*	Scott and Thune (1986b)
Rickettsia-like organisms	*C. quadricarinatus, P. (O.) fallax*	Edgerton and Prior (1999), Vogt et al. (2004)
Serratia sp.	*O. (G.) virilis*	Davidson et al. (2010)
Shewanella putrefaciens	*A. astacus, C. albidus, C. destructor, C. quadricarinatus*	Edgerton et al. (1995), Oidtmann and Hoffman (1999), Wong et al. (1995)
Shewanella sp.	*C. cainii*	Ambas et al. (2015)
Sphingomonas paucimobilis	*A. astacus*	Oidtmann and Hoffman (1999)

Table 1. contd....

Table 1. contd.

Genus/species	Host(s)	Reference(s)
Vibrio alginolyticus	*A. leptodactylus, P. (P.) leniusculus, P. (S.) clarkii*	Longshaw et al. (2012a), Raissy et al. (2014), Scott and Thune (1986b)
V. anguillarum	*C. albidus*	Jones and Lawrence (2001)
V. cholerae	*C. albidus, C. quadricarinatus, P. (S.) clarkii*	Thune et al. (1991), Wong et al. (1995)
V. harveyi	*A. leptodactylus*	Raissy et al. (2014)
V. mimicus	*C. albidus, C. quadricarinatus, A. leptodactylus, P. (S.) clarkii*	Eaves and Ketterer (1994), Raissy et al. (2014), Thune et al. (1991), Wong et al. (1995)
V. vulnificus	*A. leptodactylus*	Raissy et al. (2014)
Vibrio sp.	*P. (S.) clarkii*	Scott and Thune (1986b)

Escherichia, Hafnia, Klebsiella, Plesiomonas, Proteus, Serratia), Legionellales (*Coxiella*), Pasteurellales (*Pastuerella*), Pseudomonadales (*Acinetobacter, Moraxella, Pseudomonas, Sphingomonas*) Thiotrichales (*Francisella*), Vibrionales (*Grimontia, Vibrio*).

Most of these Gram-negative have not been associated with disease in crayfish and many have been isolated from the haemolymph of apparently healthy animals. However, mortalities of *Procambarus* (*Scapulicambarus*) *clarkii* have been associated with *Vibrio mimicus* and *V. cholerae* (Thune et al. 1991). Affected animals were lethargic and daily mortality rates were 5–25%, with death rates decreasing with increased aeration in the ponds. Mortalities associated with *V. mimicus* have also been reported for *Cherax albidus* and *Cherax quadricarinatus* (Eaves and Ketterer 1994, Wong et al. 1995); the potential for zoonotic transmission of *Vibrio* spp. via the consumption of infected crayfish has been promulgated (Bean et al. 1998).

A systemic infection of *C. quadricarinatus* due to *Coxiella cheraxi* has been shown to be responsible for high mortalities under aquaculture conditions (Tan and Owens 2000). Infected animals were lethargic with a focus of infection in the gills and hepatopancreas (La Fauce and Owens 2007, Tan and Owens 2000). A similarly virulent, systemic infection of the same host showing similar pathology and mortality patterns in Ecuador has been reported (Romero and Jiménez 2002); the relationship between the two isolates remains unknown. A second *Rickettsia*-like organism reported from the hepatopancreas of a single moribund *C. quadricarinatus* appears to be distinct (Edgerton and Prior 1999).

Bacteria (Phylum Tenericutes)

Several members of this phylum can cause disease in their hosts, including plants and in arthropods, where they can act as male killing organisms (Kageyama et al. 2007, Regassa and Gasparich 2006). *Spiroplasma eriocheiris*, causing tremor disease in mitten crabs, occurs systemically in *Procambarus* (*Scapulicambarus*) *clarkii* with infected animals typically show signs of weakness and muscle tremors (Wang et al. 2005, 2010, Ding et al. 2013). Early discrepancies in the ability to transmit this bacterium to crayfish and minor differences in strain data appear to have been resolved (Ding et al. 2015) and tools for its identification and distribution have been developed (Bi et al. 2008, Ding et al. 2007, 2012, 2013, 2015, Wang et al. 2003, 2005, 2009, 2010). A putative *Spiroplasma* sp. infecting the Sertoli cells of feral *Pacifastacus* (*Pacifastacus*) *leniusculus* populations in the UK has been described by Longshaw et al. (2012a). Affected tubules were normally devoid of sperm and degeneration of Sertoli and epithelial cells was noted in latter stages of the infection. Similar infections have not been reported in other crayfish populations. An uncharacterised Mollicute-like organism has been described from the degenerate cuticular epithelium of moribund *Cherax quadricarinatus* in Ecuador (Jiménez et al. 1998). The pathogen, like the *Spiroplasma* of Longshaw et al. (2012a), requires isolation, culture and molecular characterisation to confirm their identity.

Fungi and fungal-like organisms

Infections with fungi, and fungal–like organisms are relatively commonplace in crayfish. It is recognised that the taxonomy of the group is in a state of flux and no attempt has been made to impose any viewpoint regarding to validity of one or other taxonomic grouping. Instead, fungi and fungi-like organisms are arranged below at the level of phylum. Reports of mortality and/or pathology associated with some fungi are lacking. It is possible that some fungi are present on crayfish but do not present a risk to that species in that locality. Furthermore, pathogenesis of the fungi may be dependent on the interaction between the physiological state of the host, the fungal strain and its environment.

Fungi (Phylum Oomycota)

Oomycetes, or water moulds are not true fungi and are recognised as both opportunistic pathogens as well as saprophytes. Oomycetes in crayfish occur in three main orders, namely Lagenidiales, which includes the genus *Lagenidium*, the Order Peronosporales containing the genera *Phytophthora*, *Phytopythium* and *Pythium* and the Order Saprolegniales containing the genera *Achlya*, *Aphanomyces*, *Dictyuchus*, *Leptolegnia*, *Saprolegnia* and *Scoliolegnia* (Table 2).

Traditionally, oomycetes were identified using a range of morphological features. Application of molecular tools to various isolates have shown that several of those genera and species that were previously considered to be well characterised are made up of a complex of morphotypes, species and strains. For example, *Aphanomyces astaci*, the causative agent of the so-called "crayfish plague", comprises of a number of strains and *Aphanomyces repetans* consists of at least two strains (Diéguez-Uribeondo et al. 1995, Kozubíková et al. 2011a, Oidtmann et al. 2002, Royo et al. 2004, Viljamaa-Dirks et al. 2013). A range of approaches to reducing the impact, transmission and increasing resistance to infection have been developed or applied including disinfectants/chemotherapeutants (Alderman and Polglase 1985a, Cerenius et al. 1992, Jussila et al. 2011, Rantamaki et al. 1992, Söderhäll and Ajaxon 1982). The potential for biological control of *A. astaci* using a mycoplasma-like organism found in the hyphae of a lab strain of *A. astaci* unable to produce zoospores does not appear to have been realised (Heath and Unestam 1974). Diagnositic methods for the rapid and accurate identification of *Aphanomyces* infections in crayfish have been developed (Hochwimmer et al. 2009, Huang et al. 1994, Kozubíková et al. 2011a,b, Oidtmann et al. 2002, Royo et al. 2004, Strand et al. 2011, Tilmans et al. 2014, Vennerström et al. 1998, Viljamaa-Dirks et al. 2013) although concerns have been raised over potential lack of specificity with some tools (Ballesteros et al. 2009).

Fungi (Phylum Entomophthoromycota)

The phylum Entomophthoromycota was previously known as the Zygomycota and contains the two orders Mortierellales and Mucorales that contain representatives of fungi reported from crayfish. This group of fungi have no motile stages with passive transmission between hosts occurring. The majority of these fungi have been found in association with *Procambarus* (*Scapulicambarus*) *clarkii*, including *Absidia fusca*, *A. glauca*, *Mortierella* sp., *M. turficola*, *Mucor* sp., *M. hiemalis*, *M. plumbleus*, *Rhizopus* sp. and *R. stolonifer* (Dörr et al. 2012, Garzoli et al. 2014, Quaglio et al. 2006b). Garzoli et al. (2014) suggested that whilst many of the fungi that they isolated from the digestive system had the potential to be pathogenic to *P.* (*S.*) *clarkii*, it was probable that the hosts selected the fungi to assist with the breakdown of plant material in the gut, which may facilitate the invasive nature of the crayfish allowing it to select a wider range of food. The crayfish was considered to be a potential vector of plant diseases as some of the isolated fungi were known to be phytopathogenic. In addition, *Mucor* sp. and *Circinella muscae* have been isolated from *Pacifastacus* (*Pacifastacus*) *leniusculus* (Geasa 2014) while *Mucor hiemalis* and *M. racemosus* have been isolated from *Astacus astacus* (Makkonen et al. 2010).

Table 2. List of Oomycota, in alphabetical order, reported from crayfish hosts. Where *= transmission demonstrated through experimental studies only.

Genus/species	Host(s)	Reference(s)
Achlya sp.	*Astacus astacus, Cherax quadricarinatus*	Sewell and Cannon (1994), Vey (1981)
Aphanomyces astaci	*Astacopsis franklinii*, Astacopsis gouldi*, A. astacus, Astacus leptodactylus, Austropotamobius pallipes, Austropotamobius torrentium, Cambarellus (Cambarellus) patzcuarensis, Cambaroides japonicus, Cherax destructor*, Cherax papuanus*, C. quadricarinatus, Euastacus crassus*, Euastacus kershawi*, Euastacus spinifer*, Geocharax gracilis*, Orconectes (Faxonius) limosus, Orconectes (Gremicambarus) virilis, Orconectes (Trisellescens) immunis, Pacifastacus (Pacifastacus) leniusculus, Procambarus (Austrocambarus) cf. llamasi, Procambarus (Austrocambarus) vazquezae, Procambarus (Leconticambarus) alleni, Procambarus (Ortmannicus) enoplosternum, Procambarus (Scapulicambarus) clarkii, Procambarus (Ortmannicus) fallax*	Andersson and Cerenius (2002), Aquiloni et al. (2011), Diéguez-Uribeondo et al. (2009), Keller et al. (2014), Kozubíková et al. (2011a), Marino et al. (2014), Mrugała et al. (2015), Schrimpf et al. (2013), Tilmans et al. (2014), Unestam (1976)
A. frigidophilus	*A. astacus, A. pallipes*	Ballesteros et al. (2006), Vrålstad et al. (2009)
A. repetans	*A. pallipes, P. (S.) clarkii, P. (P.) leniusculus*	Cammà et al. (2010), Royo et al. (2004)
Aphanomyces sp.	*P. (P.) leniusculus*	Diéguez-Uribeondo et al. (2009)
Dictyuchus sp.	*P. (P.) leniusculus*	Vey (1977)
Lagenidium sp.	*C. quadricarinatus*	Sewell and Cannon (1994)
Leptolegnia sp.	*A. astacus*	Diéguez-Uribeondo et al. (2009)
Phytophthora inundata-P. humicola	*O. (F.) limosus*	Kozubíková-Balcarová et al. (2013)
Phytopythium sp.	*A. astacus, C. quadricarinatus*	Kozubíková-Balcarová et al. (2013), Sewell and Cannon (1994)
Pythium spp.	*A. astacus*	Kozubíková-Balcarová et al. (2013)
Saprolegnia australis	*A. astacus, A. pallipes, Orconectes (Crockerinus) propinquus, O. (F.) limosus, P. (P.) leniusculus*	Hirsch et al. (2008), Kozubíková-Balcarová et al. (2013), Krugner-Higby et al. (2010), Makkonen et al. (2010), Vrålstad et al. (2009)
S. diclina	*O. (F.) limosus*	Hirsch et al. (2008)
S. ferax	*A. astacus, A. pallipes, O. (F.) limosus*	Kozubíková-Balcarová et al. (2013)
S. hypogyna	*A. astacus*	Kozubíková-Balcarová et al. (2013)
S. littoralis	*A. astacus*	Diéguez-Uribeondo et al. (2007)
S. parasitica	*A. astacus, A. leptodactylus, A. pallipes, O. (F.) limosus, P. (P.) leniusculus, P. (S.) clarkii*	Diéguez-Uribeondo et al. (1994), Kozubíková-Balcarová et al. (2013), Smith and Söderhäll (1986)
Saprolegnia sp.	*A. leptodactylus, A. pallipes, Cherax cainii, C. destructor, C. quadricarinatus, P. (P.) leniusculus, P. (G.) simulans*	Fard et al. (2011), Geasa (2014), Herbert (1987), Lahser, Jr. (1975), Quaglio et al. (2006a), Sewell and Cannon (1994)
Saprolegniales I	*A. astacus*	Kozubíková-Balcarová et al. (2013)
Saprolegniales II	*A. astacus, O. (F.) limosus*	Hirsch et al. (2008), Kozubíková-Balcarová et al. (2013)
Saprolegniales III	*A. astacus, O. (F.) limosus*	Hirsch et al. (2008), Kozubíková-Balcarová et al. (2013)
Saprolegniales IV	*O. (F.) limosus*	Hirsch et al. (2008)
Scoliolegnia asterophora	*A. astacus*	Makkonen et al. (2010)

Fungi (Phylum Chytridiomycota)

The chytrid fungus *Batrachochytrium dendrobatidis* has been implicated in the decline of amphibian populations worldwide, with a number of non-amphibian hosts demonstrated to be involved in its transmission (Brannelly et al. 2015, McMahon et al. 2013). Crayfish, in particular *Orconectes* (*Gremicambarus*) *virilis*, *Procambarus* (*Leconticambarus*) *alleni* and *Procambarus* (*Scapulicambarus*) *clarkii*, have been reported as being able to carry the infection and transmit it to naïve tadpoles (McMahon et al. 2013). Mortalities were noted in crayfish exposed to *B. dendrobatidis* along with a reduction in growth rates in survivors. McMahon et al. (2013) suggested that the fungus might release chemicals that are in themselves toxic to crayfish, as mortalities occurred in crayfish exposed to filtered water that previously contained *B. dendrobatidis* zoospores. Clear seasonality in crayfish infections were noted by Brannelly et al. (2015) in both farmed and wild crayfish; the authors suggested that *P. (S.) clarkii* represented a serious risk to amphibian populations due to its worldwide trade and that the risk needed to be considered in biosecurity measures.

Fungi (Phylum Blastocladiomycota)

Members of this phylum are filamentous fungi that form uniflagellated zoospores (Walker et al. 2011). Sewell and Cannon (1994) listed *Cherax quadricarinatus* as a host for *Allomyces* sp. in a review paper. No further data was provided for the infection, which had been reported from an unpublished conference proceeding.

Fungi (Phylum Ascomycota)

In general, members of this phylum that occur on crayfish tend to be associated with melanised lesions on the cuticle and gills giving rise to the venacular names such as "burn spot disease" or "black gill disease" (Table 3). There is a need to reevaluate some of the early records of these agents due to the incongruence between morphology and molecular data (e.g., *Fusarium solani*, a morphospecies containing at least 60 species). In addition, the pathology of these infections and potential impact on host survival need reassessing; some studies have suggested that they retard host moulting or death (Alderman and Polglase 1985b).

Table 3. List of Ascomycota, in alphabetical order, reported from crayfish hosts.

Genus/species	Host(s)	Reference(s)
Acremonium chrysogenum	*Procambarus (Scapulicambarus) clarkii*	Dörr et al. (2012)
A. kiliense	*P. (S.) clarkii*	Dörr et al. (2012)
A. persicinum	*P. (S.) clarkii*	Garzoli et al. (2014)
Acremonium sp.	*Astacus leptodactylus, Austropotamobius pallipes, Pacifastacus (Pacifastacus) leniusculus, P. (S.) clarkii*	Diler and Bolat (2001), Dörr et al. (2012), Geasa (2014), Quaglio et al. (2006a)
A. (=Cephalosporium) leptodactyli	*A. leptodactylus*	Mann (1940)
Alternaria alternata	*P. (S.) clarkii*	Dörr et al. (2012)
A. cheiranthi	*P. (S.) clarkii*	Dörr et al. (2012)
A. chlamydospora	*P. (P.) leniusculus, P. (S.) clarkii*	Dörr et al. (2012), Geasa (2014)
Alternaria sp.	*A. leptodactylus, A. pallipes, P. (P.) leniusculus, Procambarus (Girardiella) simulans, Procambarus (Ortmannicus) acutus, Fallicambarus hedgpethi*	Fard et al. (2011), Geasa (2014), Lahser, Jr. (1975), Quaglio et al. (2006a)
Arthrinium sp.	*P. (S.) clarkii*	Garzoli et al. (2014)
A. phaeospermum	*P. (S.) clarkii*	Garzoli et al. (2014)

Table 3. contd....

Table 3. contd.

Genus/species	Host(s)	Reference(s)
Aspergillus album	*P. (S.) clarkii*	Quaglio et al. (2006b)
A. brasiliensis	*P. (S.) clarkii*	Garzoli et al. (2014)
A. clavatus	*P. (S.) clarkii*	Quaglio et al. (2006b)
A. flavus	*A. leptodactylus, P. (S.) clarkii*	Fard et al. (2011), Garzoli et al. (2014)
A. fumigatus	*P. (S.) clarkii*	Garzoli et al. (2014)
A. glaucus	*P. (S.) clarkii*	Garzoli et al. (2014)
A. niger	*P. (S.) clarkii*	Dörr et al. (2012)
Aspergillus sp.	*A. pallipes, P. (P.) leniusculus, P. (S.) clarkii*	Dörr et al. (2012), Geasa (2014), Quaglio et al. (2006a)
A. terreus	*P. (S.) clarkii*	Dörr et al. (2012)
A. versicolor	*P. (S.) clarkii*	Garzoli et al. (2014)
Aureobasidium pullulans var. *melanogenum*	*P. (S.) clarkii*	Garzoli et al. (2014)
A. p. var. *pullulans*	*P. (S.) clarkii*	Dörr et al. (2012)
Cephalotrichum microsporum	*P. (S.) clarkii*	Garzoli et al. (2014)
Chaetomella raphigera	*P. (S.) clarkii*	Dörr et al. (2012)
Chaetomium sp.	*P. (S.) clarkii*	Garzoli ct al. (2014)
Cladosporium chlorocephalum	*P. (S.) clarkii*	Dörr et al. (2012)
C. cladosporoides	*P. (S.) clarkii*	Dörr et al. (2012)
Cladosporium (=Hormodendrum) sp.	*P. (S.) clarkii, P. (G.) simulans*	Lahser, Jr. (1975), Quaglio et al. (2006b)
Clonostachys rosea	*P. (S.) clarkii*	Dörr et al. (2012)
Coniella sp.	*P. (S.) clarkii*	Garzoli et al. (2014)
Drechslera sp.	*P. (S.) clarkii*	Quaglio et al. (2006b)
Emericellopsis sp.	*P. (S.) clarkii*	Dörr et al. (2012)
Epicoccum nigrum	*Astacus astacus, P. (S.) clarkii*	Dörr et al. (2012), Makkonen et al. (2013)
Fusarium avenaceum	*A. astacus*	Makkonen et al. (2013)
F. dimerum	*P. (S.) clarkii*	Dörr et al. (2012)
F. graminearum	*P. (P.) leniusculus*	Edsman et al. (2015)
F. negundis	*P. (P.) leniusculus*	Edsman et al. (2015)
F. oxysporum	*A. leptodactylus, A. pallipes, P. (S.) clarkii*	Dörr et al. (2012), Maestracci and Vey (1987)
F. proliferatum	*P. (S.) clarkii*	Dörr et al. (2012)
"*F. solani*"	*A. astacus, A. leptodactylus, A. pallipes, P. (P.) leniusculus*	Chinain and Vey (1987), Smith and Söderhäll (1986)
Fusarium sp.	*A. pallipes, A. leptodactylus, P. (P.) leniusculus, P. (S.) clarkii, P. (G.) simulans*	Fard et al. (2011), Geasa (2014), Lahser, Jr. (1975), Quaglio et al. (2006a,b)
F. tricinctum	*P. (P.) leniusculus*	Edsman et al. (2015)
F. verticilloides	*P. (S.) clarkii*	Dörr et al. (2012)
Geotrichum spp.	*A. pallipes*	Quaglio et al. (2008)
Gliocladium sp.	*A. pallipes, P. (S.) clarkii*	Quaglio et al. (2006b, 2008)
Graphium sp.	*P. (S.) clarkii*	Dörr et al. (2012)
Hemicarpenteles ornatum	*P. (S.) clarkii*	Garzoli et al. (2014)

Table 3. contd....

Table 3. contd.

Genus/species	Host(s)	Reference(s)
Hormisum sp.	*P. (S.) clarkii*	Lahser, Jr. (1975)
Khuskia oryzae	*P. (S.) clarkii*	Garzoli et al. (2014)
Microdochium bolleyi	*P. (S.) clarkii*	Dörr et al. (2012)
Oidiodendron flavum	*P. (S.) clarkii*	Garzoli et al. (2014)
Paecilomyces farinosus	*P. (S.) clarkii*	Garzoli et al. (2014)
P. inflatus	*P. (S.) clarkii*	Garzoli et al. (2014)
P. lilacinum	*P. (S.) clarkii*	Dörr et al. (2012)
Paecilomyces sp.	*P. (S.) clarkii*	Quaglio et al. (2006b)
Pestalotiopsis guepinii	*P. (S.) clarkii*	Garzoli et al. (2014)
Penicillium expansum	*A. leptodactylus*	Fard et al. (2011)
Penicillium sp.	*A. pallipes, P. (P.) leniusculus, P. (S.) clarkii*	Dörr et al. (2012), Geasa (2014), Quaglio et al. (2008)
P. verrucosum	*P. (S.) clarkii*	Garzoli et al. (2014)
Phoma glomerata	*P. (S.) clarkii*	Dörr et al. (2011, 2012)
Phoma sp.	*P. (S.) clarkii*	Quaglio et al. (2006b)
Plectosporium (=Fusarium) tabacinum	*A. pallipes*	Alderman and Polglase (1985b), Palm et al. (1995), Smith and Söderhäll (1986)
Ramularia astaci	*A. astacus*	Mann and Pieplow (1938), Smith and Söderhäll (1986)
R. (=Didymaria) cambari	*Orconectes (Faxionus) limosus*	Mann and Pieplow (1938)
Scopulariopsis sp.	*P. (S.) clarkii*	Garzoli et al. (2014)
Sordaria fimicola	*P. (S.) clarkii*	Garzoli et al. (2014)
Talaromyces flavus	*P. (S.) clarkii*	Garzoli et al. (2014)
Trichoderma sp.	*A. pallipes, P. (P.) leniusculus, P. (S.) clarkii*	Dörr et al. (2012), Geasa (2014), Quaglio et al. (2006a)
T. viridae	*P. (S.) clarkii*	Quaglio et al. (2006b)
Ulocladium sp.	*P. (S.) clarkii*	Quaglio et al. (2006b)
Uncinula sp.	*P. (G.) simulans*	Lahser, Jr. (1975)

Fungi (Phylum Basidiomycota)

The phylum contains a wide range of species including mushrooms and toadstools as well as rust and smut fungi. Other than the report of Goodrich (1956) who considered that *Cryptococcus gammari* was likely to be lethal to its crayfish host, no mortailities have been reported in any crayfish species to date associated with the Basidiomycota. Only four species of Basidiomycota have been isolated from crayfish, including *Cryptococcus gammari* from *Austropotamobius pallipes* (Goodrich 1956), *C. laurentii* and *Rhodotorula* sp. from *Procambarus (Scapulicambarus) clarkii* (Quaglio et al. 2006b) and *Trichosporon beigelii (=cutaneum)* from *Astacus astacus* (Söderhäll et al. 1993).

Fungi (Phylum Microsporidia)

Microsporidia are obligate, intracellular pathogens with a number described from crayfish (Table 4). Traditionally considered as protistans, they are now recognised as basal fungi based on a number of molecular approaches (Gill and Fast 2006, Hirt et al. 1999, Lee et al. 2008, Walker et al. 2011). Lifecycles are normaly dimorphic within the crayfish host (Lom et al. 2001, Moodie et al. 2003a,b). A lack of understanding of this phenomenon may have led Goodrich (1956) to incorrectly name solitary *Thelohania contejeani* spores in *Austropotamobius pallipes* as *Nosema* sp.; Edgerton et al. (2002a) further confounded the issue by erroneously referring to these as *Ameson* sp. The musculature is the primary site for infections by Microsporidia in crayfish, with the exception of a *Bacillidium*-like microsporidian detected by PCR in the eggs of a single *P. (P.) leniusculus* collected in the United Kingdom (Dunn et al. 2009) and an undescribed *Thelohania* sp. in the connective tissue of *Cherax quadricarinatus* reported by Edgerton and Owens (1999). Clinical signs of muscle infections are limited and usually consist of sluggishness, a reduced tail flick response and opacity of the musculature (Herbert 1987, 1988, Langdon 1991, Moodie et al. 2003c) and, in some cases, mortality of the host ensues (Goodrich 1956, Sogandares-Bernal 1962a, 1965, Moodie et al. 2003c). Reports of *T. contejeani* in *Paranephrops* spp. from New Zealand and in *Orconectes (Gremicambarus) virilis* from the USA (Quilter 1976, Jones 1980, Graham and France 1986) most probably represent new, undescribed species. Additionally, there remains a large number of Microsporidia in crayfish described from light microscopy only that require re-examination and a formal description.

Table 4. List of Microsporidia, in alphabetical order, reported from crayfish species.

Genus/Species	Host(s)	Reference(s)
Bacillidium sp.	*Pacifastacus (Pacifastacus) leniusculus*	Dunn et al. (2009)
Cystosporogenes sp.	*P. (P.) leniusculus*	Imhoff et al. (2010)
Microsporidium sp.	*Cherax cainii, P. (P.) leniusculus*	Dunn et al. (2009), O'Donoghue et al. (1990)
Pleistophora sp.	*Cherax destructor*	O'Donoghue et al. (1990)
P. soganderesi	*Cambarellus (Pandicambarus) puer*	Sogandares-Bernal (1962a), Sprague (1966)
Thelohania sp.	*Orconectes (Crockerinus) propinquus, Orconectes (Gremicambarus) virilis, Paranephrops planifrons, Paranephrops zealandicus, Cambarellus (Dirigicambarus) shufeldtii, C. cainii, C. destructor, Cherax quadricarinatus, Cherax quinquecarinatus*	Edgerton and Owens (1999), Graham and France (1986), Herbert (1987, 1988), Jones (1980), Krugner-Higby et al. (2010), O'Donoghue and Adlard (2000), Quilter (1976), Sewell and Cannon (1994), Sogandares-Bernal (1965)
T. cambari	*Cambarus (Cambarus) bartonii bartonii*	Sprague (1950)
T. contejeani	*Astacus astacus, Astacus leptodactylus, Austropotamobius pallipes, Orconectes (Faxonius) limosus, P. (P.) leniusculus*	Dunn et al. (2009), Edgerton et al. (2002a), Lom et al. (2001), Longshaw et al. (2012b)
T. montirivulorum	*C. destructor*	Moodie et al. (2003a)
T. parastaci	*Cherax albidus, C. destructor, Cherax rotundus*	Moodie et al. (2003b)
Vairimorpha cheracis	*C. destructor*	Moodie et al. (2003c)
Vavraia parastacida	*C. albidus, C. cainii, C. quinquecarinatus, C. quadricarinatus*	Langdon (1991), Langdon and Thorne (1992)
Vittaforma sp.	*P. (P.) leniusculus*	Dunn et al. (2009)
Vittaforma corneae	*P. (P.) leniusculus*	Imhoff et al. (2010)

Mesomycetozoea

The Mesomycetozoea (Ichthyosporea or DRIPs clade) are a monophyletic group of fungus-like protists branching near to the animal/fungal divergence (Glockling et al. 2013, Ragan et al. 1996). One of the best known Mesomycetozoea of crayfish are *Psorospermium* spp. Original reports of these infections in European crayfish and wider have been referred to as *Psorospermium haeckli*. However, it is clear that many morphotypes exist and initial molecular data would suggest that at least two of these morphotypes can be discriminated (Bangyeekhun et al. 2001). In spite of evidence to the contrary, most records continue to classify infections in crayfish as being due to *P. haeckli*. It has been suggested that at least four morphotypes exist worldwide with two occurring in Europe (oval or elongate), one in North America (elongate) and one in Australia (oval or curved) (Bangyeekhun et al. 2001). However, this simplistic view ignores some of the subtleties noted by researchers with Boshko (1981) describing three morphotypes in *Astacus leptodactylus* (elongate, oval or roundish), whilst Henttonen et al. (1994) reported four morphotypes (long American pointed, long American round pointed, short American round pointed and short American curved) from north American *Procambarus, Orconectes* and *Pacifastacus* spp. Lucić et al. (2004) reports on three morphotypes occurring in mixed populations of *Astacus astacus* and *Astacus leptodactylus* whilst Longshaw et al. (2012a) provide evidence for at least two morphotypes in samples of *Cherax (Cherax) peknyi*. Scheer (1979) described *Psorospermium orconectis* from the lumen of *Orconectes (Faxonius) limosus* with an absence of cuticular plates being of taxonomic importance. It is highly likely that a species complex or indeed a number of distinct species exist within the genus.

The extent to which spore maturation and host affects morphology is unknown; the use of cross species transmission studies may help to determine any genotypic and phenotypic variation that may exist. It is therefore clear that a reappraisal of the morphological and molecular characteristics of the group need to be conducted in order to understand the relationships between the various types described. As such, and to avoid further confusion in the literature it is suggested that *Psorospermium* spp. should not be ascribed to a species unless a full description is provided to include molecular, morphological and host data. Hosts for *Psorospermium* spp. in Europe include *A. astacus, A. leptodactylus, Austropotamobius pallipes, Austropotamobius torrentium*, and *Pacifastacus (Pacifastacus) leniusculus* (Diéguez-Uribeondo et al. 1993, Gydemo 1996, Henttonen et al. 1997, Longshaw et al. 2012b, Lucić et al. 2004, Vogt et al. 1996); hosts in North America include *Cambarus (Lacunicambarus) diogenes, Orconectes (Gremicambarus) virilis, Orconectes (Crockerinus) propinquus, Orconectes (Procericambarus) rusticus, Orconectes (Trisellescens) immunis, Pacifastacus (Pacifastacus) leniusculus, Procambarus (Leconticambarus) alleni, Procambarus (Ortmannicus) fallax, Procambarus (Ortmannicus) zonangulus, Procambarus (Scapulicambarus) clarkii* (Henttonen et al. 1994, 1997, Klarberg et al. 2000, Krugner-Higby et al. 2010); *Parastacus pugnax* has been noted as a host in Chile (Rudolph et al. 2007); Australian hosts include *Cherax cainii, Cherax quadricarinatus, Cherax albidus* (Edgerton and Owens 1999, Herbert 1987, Henttonen et al. 1997, Jones and Lawrence 2001), whilst Longshaw et al. (2012a) reported two morphotypes from *Cherax (Cherax) peknyi* imported into the United Kingdom from Indonesia and Singapore.

The lifecycle of *Psorospermium* spp. has not been fully elucidated although Vogt and Rug (1999) suggest that a diphasic lifecycle is probable with a stage occurring outside the crayfish host. Alternate hosts have not been reported although *Psorospermium* spp. have been reported from *Asellus* sp. and from a cockroach (*Blatella* sp.) and the stools of a human (Bouckenooghe and Marino 2001, Gatta et al. 2009). Transmission of the parasite between crayfish hosts is also possible (Gydemo 1996, Vogt et al. 1996). Development within the host moves from an amoeboid form, through a series of size increases and changes in shape and formation of the cell wall to produce "mature" forms in various tissues (Henttonen et al. 1997). *Psorospermium* spp. can elicit a host response in crayfish including activation of the proPO system, haemocyte encapsulation and a melanisation reaction around the parasite (Cerenius et al. 1991, Thörnqvist and Söderhäll 1993, Vranckx and Durliat 1981).

Protista

The majority of protistans found on or in crayfish are ectoparasitic ciliates, occurring mainly on the carapace and occasionally on the gills; however they are rarely associated with mortalities (Brown et al. 1993, Ninni 1864). The genera are found in the following orders within the Ciliata: Apostomatida (*Hyalophysa*), Chlamydodontida (*Chilodonella*), Colpodida (*Colpoda*), Dysteriida (*Trochilia*), Endogenida (*Acineta, Tokophrya, Trichophrya*), Euplotida (*Euplotes*), Evaginogenida (*Anarma, Discophrya*), Exogenida (*Paracineta, Podophyra*), Heterotrichida (*Climacostomum, Stentor*), Hymenostomatida (*Tetrahymena*), Peniculida (*Paramecium*), Sessilida (*Epistylis, Carchesium, Cothurnia, Cyclodonta, Lagenophrys, Opercularia, Orbopercularia, Paralagenophrys, Platycola, Propyxidium, Pseudovorticella, Pyxicola, Setonophrys, Sincothurnia, Thuricola, Vaginicola, Vorticella* and *Zoothamnium*) and Sporadotrichida (*Stylonichia*). Representative examples are shown in Fig. 1. The other two genera of protistans reported

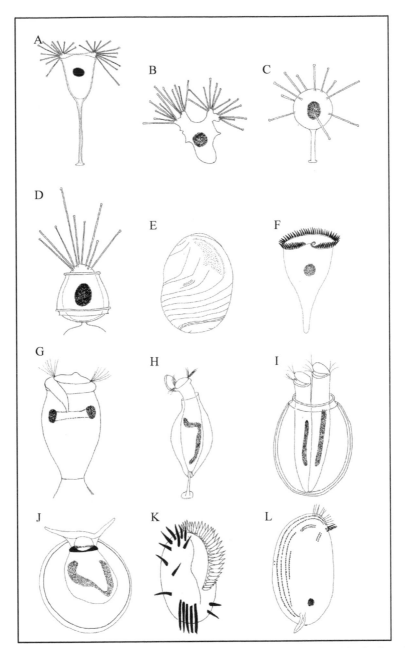

Figure 1. Representative examples of protistan parasites reported from crayfish. (A) *Acineta*. (B) *Trichophyra*. (C) *Podophyra*. (D) *Paracineta*. (E) *Hyalophysa*. (F) *Stentor*. (G) *Epistylis*. (H) *Pyxicola*. (I) *Platycola*. (J) *Setonophrys*. (K) *Euplotes*. (L) *Trochilia*. Figures A–D, F–I and L after Lee et al. 2000, Fig. E after Browning and Landers 2012, Fig. J after Clamp 1991. Not to scale.

from crayfish are the coccidian *Mantonella* (Phylum Apicomplexa) and the flagellate *Bodo* within the order Kinetoplastida. A list of the known protistans parasites from crayfish is provided in Table 5.

Table 5. List of Protista, in alphabetical order, reported from crayfish.

Genus/species	Host(s)	Reference(s)
Acineta fluviatilis	*Cherax cainii, Cherax destructor*	O'Donoghue et al. (1990)
A. laucastris	*Procambarus (Girardiella) simulans*	Lahser, Jr. (1975)
A. tuberosa	*Cambarellus (Cambarellus) patzcuarensis, C. cainii, C. destructor, Astacus leptodactylus*	Fernandez-Leborans and Tato-Porto (2000b), Mayén-Estrada and Aladro-Lubel (2001), O'Donoghue et al. (1990)
Acineta sp.	*C. (C.) patzcuarensis, Cherax quadricarinatus, Pacifastacus (Pacifastacus) leniusculus leniusculus, Procambarus (Scapulicambarus) clarkii*	Cuellar et al. (2002), Mayén-Estrada and Aladro-Lubel (2001), Romero and Jiménez (2002), Scott and Thune (1986a), Vogelbein and Thune (1988)
Anarma multiruga	*Cambarellus (Cambarellus) zempoalensis*	Lopez-Ochoterena and Gasea (1971)
Bodo sp.	*P. (P.) l. leniusculus, P. (G.) simulans*	Cuellar et al. (2002), Lahser, Jr. (1975)
Carchesium granulatum	*Cambarus* sp.	Sprague and Couch (1971)
C. polypinum	*C. (C.) patzcuarensis*	Mayén-Estrada and Aladro-Lubel (2002)
Chilodonella sp.	*A. leptodactylus*	Fard et al. (2011)
Climacostomum virens	*P. (G.) simulans*	Lahser, Jr. (1975)
Colpoda sp.	*P. (G.) simulans, P. (S.) clarkii*	Lahser, Jr. (1975)
Cothurnia (=Cothurniopsis) astaci	*Astacus astacus, A. leptodactylus, Orconectes (Faxonius) limosus*	Fernandez-Leborans and Tato-Porto (2000a), Sprague and Couch (1971), Warren and Paynter (1991)
C. bavarica	*A. leptodactylus, O. (F.) limosus*	Fernandez-Leborans and Tato-Porto (2000a), Warren and Paynter (1991)
C. curva (=gracilis)	*A. astacus, A. leptodactylus, O. (F.) limosus*	Fernandez-Leborans and Tato-Porto (2000a), Morado and Small (1995), Sprague and Couch (1971)
C. plachteri	*A. astacus, Austropotamobius torrentium*	Fernandez-Leborans and Tato-Porto (2000a), Warren and Paynter (1991)
C. sieboldii	*A. astacus, A. leptodactylus, Austropotamobius pallipes, A. torrentium*	Fard et al. (2011), Quaglio et al. (2006a), Sprague and Couch (1971), Warren and Paynter (1991)
Cothurnia sp.	*A. pallipes, C. cainii, C. destructor, C. quadricarinatus, P. (S.) clarkii*	Longshaw et al. (2012a,b), O'Donoghue et al. (1990), O'Donoghue and Adlard (2000), Quaglio et al. (2006b), Scott and Thune (1986a), Sewell and Cannon (1994), Vogelbein and Thune (1988)
C. (=Daurotheca) tespa	*P. (P.) l. leniusculus*	Fernandez-Leborans and Tato-Porto (2000a), Warren and Paynter (1991)
C. (=D.) transoceanica	*P. (P.) l. leniusculus*	Fernandez-Leborans and Tato-Porto (2000a), Warren and Paynter (1991)

Table 5. contd....

Table 5. contd.

Genus/species	Host(s)	Reference(s)
C. (=D.) ussurina	*Cambaroides dauricus, Cambaroides schrenckii, Cambarus* spp., *Orconectes* spp., *Pacifastacus* spp.	Fernandez-Leborans and Tato-Porto (2000a), Warren and Paynter (1991)
C. (=D.) variabilis (=marginata)	*A. astacus, A. leptodactylus, Cambarus (Cambarus) bartonii bartonii, C. (C.) patzcuarensis, Orconectes (Crockerinus) propinquus, O. (F.) limosus, Orconectes (Hespericambarus) difficilis, Pacifastacus (Hobbsastacus) gambelli, P. (G.) simulans, Procambarus (Ortmannicus) acutus, Fallicambarus (Creaserinus) fodiens*	Lahser, Jr. (1975), Mayén-Estrada and Aladro-Lubel (2002), Morado and Small (1995), Warren and Paynter (1991)
Cyclodonta bipartita (=Cothurnia affinis = trilobata = voigti) = Cothurniopsis rheotypica (=longipes)	*A. leptodactylus, O. (F.) limosus*	Fernandez-Leborans and Tato-Porto (2000a), Warren and Paynter (1991)
Discophrya (=Podophrya = Tokophrya) astaci (=inclinata)	*A. astacus, A. leptodactylus, A. torrentium, O. (F.) limosus*	Fernandez-Leborans and Tato-Porto (2000b), Morado and Small (1995)
D. lichtensteinii	*Astacus* sp.	Morado and Small (1995)
Epistylis astaci	*A. astacus, A. leptodactylus, A. torrentium*	Fernandez-Leborans and Tato-Porto (2000a), Morado and Small (1995)
E. bimarginata	*A. astacus, C. (C.) patzcuarensis*	Mayén-Estrada and Aladro-Lubel (2001), Sprague and Couch (1971)
E. branchiophila	*C. (C.) patzcuarensis*	Mayén-Estrada and Aladro-Lubel (2001)
E. chrysemidis	*A. leptodactylus*	Fard et al. (2011)
E. cambari	*Cambarus* sp., *A. leptodactylus, P. (G.) simulans*	Fernandez-Leborans and Tato-Porto (2000a), Lahser, Jr. (1975), Sprague and Couch (1971)
E. carinogammari	*C. (C.) patzcuarensis*	Mayén-Estrada and Aladro-Lubel (2001)
E. crassicollis	*A. astacus, A. leptodactylus*	Fernandez-Leborans and Tato-Porto (2000a)
E. gammari	*C. (C.) patzcuarensis*	Mayén-Estrada and Aladro-Lubel (2001)
E. lacustris	*C. (C.) patzcuarensis*	Mayén-Estrada and Aladro-Lubel (2001)
E. niagarae	*A. leptodactylus, A. torrentium, O. (F.) limosus, C. (C.) patzcuarensis, Cambarus* sp.	Fernandez-Leborans and Tato-Porto (2000a), Harlioğlu (1999), Mayén-Estrada and Aladro-Lubel (2001)
Epistylis sp.	*A. leptodactylus, A. pallipes, Cherax albidus, C. cainii, C. destructor, C. quadricarinatus, Orconectes (Gremicambarus) virilis, Orconectes (Procericambarus) rusticus, P. (P.) l. leniusculus, P. (S.) clarkii*	Brown et al. (1993), Cuellar et al. (2002), Herbert (1987), Hüseyin and Selcuk (2005), Jones and Lawrence (2001), O'Donoghue et al. (1990), Quaglio et al. (2006a,b), Romero and Jiménez (2002), Sewell and Cannon (1994), Vogelbein and Thune (1988)

Table 5. contd....

Table 5. contd.

Genus/species	Host(s)	Reference(s)
E. stammeri	*C. (C.) patzcuarensis*	Mayén-Estrada and Aladro-Lubel (2001)
E. variabilis	*C. (C.) patzcuarensis, C. cainii, C. destructor*	Mayén-Estrada and Aladro-Lubel (2001), O'Donoghue et al. (1990), O'Donoghue and Adlard (2000)
Euplotes sp.	*P. (G.) simulans, P. (S.) clarkii*	Lahser, Jr. (1975)
Hyalophysa bradburyi	*Cambarus (Lacunicambarus) diogenes, Procambarus (Pennides) spiculifer, Procambarus (Pennides) suttkusi, Procambarus (Pennides) versutus*	Browning and Landers (2012)
H. clampi	*Cambarus (Depressicambarus) latimanus, Cambarus (Depressicambarus) striatus, Procambarus (Ortmannicus) acutissimus, P. (P.) spiculifer, P. (P.) suttkusi, P. (P.) versutus*	Browning and Landers (2012)
H. lwoffi	*Cambarus* sp.	Grimes (1976)
Lagenophrys andos	*Parastacus pugnax*	Rudolph (2013)
L. antichos	*Parastacus defossus, Parastacus nicoleti, P. pugnax, Parastacus saffordi, Parastacus varicosus*	Clamp (1988)
L. darwini	*C. cainii, C. destructor, C. quadricarinatus*	Kane (1965), O'Donoghue et al. (1990)
L. dennisi	*C. (C.) patzcuarensis, C. (C.) b. bartonii, Cambarus (Hiaticambarus) chasmodactylus, Orconectes (Crockerinus) illinoiensis*	Clamp (1987), Mayén-Estrada and Aladro-Lubel (2000)
L. deserti	*C. cainii, Cherax quinquecarinatus*	Kane (1965)
L. (=Circolagenophrys) diogenes (=incompta)	*C. (L.) diogenes?, O. (C.) illinoiensis*	Clamp (1987)
L. dungogi	*Euastacus* sp.	Kane (1965)
L. engaei	*Engaeus* sp., *Engaeus hemicirratulus, Engaeus victoriensis*	Kane (1965)
L. (=C.) leniusculus (=oregonensis)	*Pacifastacus (Hobbsastacus) connectens, P. (P.) l. leniusculus, Pacifastacus (Pacifastacus) leniusculus trowbridgii*	Clamp (1987)
L. novazealandae	*Paranephrops zealandicus*	Clamp (1994)
L. petila	*Parastacoides tasmanicus* complex	Clamp (1994)
L. rugosa	*Geocharax falcata*	Kane (1965)
Lagenophrys sp.	*C. albidus, C. quadricarinatus, P. (S.) clarkii*	Herbert (1987), Jones and Lawrence (2001), Scott and Thune (1986a)
L. willisi	*C. albidus, C. cainii, C. destructor, Cherax setosus*	Kane (1965), O'Donoghue et al. (1990)
Lernaeophrya capitata	*A. leptodactylus*	Fernandez-Leborans and Tato-Porto (2000b)
Opercularia allensi (=ramosa)	*A. leptodactylus*	Fernandez-Leborans and Tato-Porto (2000a)
O. articularia	*A. leptodactylus*	Fard et al. (2011)

Table 5. contd....

Table 5. contd.

Genus/species	Host(s)	Reference(s)
O. crustaceorum	*A. astacus, A. torrentium*	Fernandez-Leborans and Tato-Porto (2000a)
O. nutans	*A. leptodactylus*	Fernandez-Leborans and Tato-Porto (2000a)
Operculigera asymmetrica	*P. pugnax, Samastacus spinifrons*	Clamp (1991)
O. insolita	*P. pugnax*	Clamp (1991)
O. madagascarensis	*Astacoides granulimanus*	Clamp (1992)
O. parastacis	*P. nicoleti, P. pugnax*	Fernandez-Leborans and Tato-Porto (2000a), Rudolph (2013)
O. seticola	*P. pugnax*	Clamp (1991)
O. striata	*P. pugnax*	Fernandez-Leborans and Tato-Porto (2000a), Rudolph (2013)
O. taura	*P. pugnax*	Clamp (1991)
Orbopercularia (=Opercularia) astacicola	*A. torrentium*	Fernandez-Leborans and Tato-Porto (2000a)
Paracineta fixa	*Cambarus* sp.	Morado and Small (1995)
Paralagenophrys singularis	*C. (C.) b. bartonii*	Morado and Small (1995)
Paramecium sp.	*P. (S.) clarkii*	Lahser, Jr. (1975)
Platycola decumbens	*C. (C.) patzcuarensis*	Mayén-Estrada and Aladro-Lubel (2002)
Podophrya fixa	*A. leptodactylus*	Fard et al. (2011)
P. astaci	*A. astacus*	Sprague and Couch (1971)
P. sandi	*C. (C.) patzcuarensis*	Mayén-Estrada and Aladro-Lubel (2001)
Propyxidium asymmetrica	*A. astacus*	Fernandez-Leborans and Tato-Porto (2000a)
Pseudovorticella quadrata	*C. (C.) patzcuarensis*	Mayén-Estrada and Aladro-Lubel (2002)
Pyxicola annulata	*A. leptodactylus*	Fard et al. (2011)
P. bicalceata	*C. destructor*	O'Donoghue and Adlard (2000), Sewell and Cannon (1994)
P. carteri	*C. cainii, C. destructor*	O'Donoghue et al. (1990)
P. jacobi	*C. destructor*	O'Donoghue et al. (1990)
P. pusilla	*C. cainii, C. destructor*	O'Donoghue et al. (1990)
Setonophrys (=Lagenophrys) bispinosa	*C. setosus*	Clamp (1991), Kane (1965)
S. (=L.) communis (=latispinosa = lawri)	*C. albidus, C. cainii, C. destructor, C. quadricarinatus, Cherax rotundus, Engaeus quadrimanus, Engaeus* sp., *Euastacus armatus, Euastacus crassus*	Clamp (1991), Kane (1965), O'Donoghue et al. (1990)
S. (=L.) lingulata	*C. albidus, C. cainii, Cherax depressus, C. destructor, C. rotundus*	Clamp (1991), Kane (1965), O'Donoghue et al. (1990)
S. (=L.) occlusa	*C. albidus, C. destructor, C. rotundus*	Clamp (1991), Kane (1965)
S. (=L.) seticola	*C. albidus, C. destructor, C. setosus, Engaeus fultoni, Engaeus* sp., *E. crassus, G. falcata*	Clamp (1991), Kane (1965)

Table 5. contd....

Table 5. contd.

Genus/species	Host(s)	Reference(s)
S. (=L.) spinosa	*C. cainii, C. destructor*	Clamp (1991), Kane (1965), O'Donoghue et al. (1990)
S. tricorniculata	*G. falcata*	Clamp (1991)
Sincothurnia branchiata	*A. leptodactylus*	Boshko (1995)
Stentor sp.	*P. (P.) l. leniusculus*	Cuellar et al. (2002)
Stylonichia sp.	*P. (G.) simulans*	Lahser, Jr. (1975)
Tetrahymena pyriformis	*A. leptodactylus, C. quadricarinatus*	Edgerton et al. (1996a), Fard et al. (2011)
Thuricola folliculata	*C. (C.) patzcuarensis*	Mayén-Estrada and Aladro-Lubel (2002)
Tokophrya cyclopum	*C. destructor*	O'Donoghue et al. (1990)
T. lemnarum	*A. leptodactylus*	Morado and Small (1995)
T. quadripartita	*A. leptodactylus, C. (C.) patzcuarensis*	Fernandez-Leborans and Tato-Porto (2000b), Mayén-Estrada and Aladro-Lubel (2001)
Trichophrya (=Dendrosoma) astaci	*A. leptodactylus*	Fernandez-Leborans and Tato-Porto (2000b), Morado and Small (1995)
T. cambari	*Cambarus* sp.	Fernandez-Leborans and Tato-Porto (2000b)
Trochilia sp.	*C. quadricarinatus*	Romero and Jiménez (2002)
Vaginicola ampulla	*C. cainii, C. destructor*	O'Donoghue et al. (1990)
Vaginicola sp.	*C. cainii, C. destructor*	O'Donoghue et al. (1990)
Vorticella alba	*C. albidus?*	Morado and Small (1995), Warren (1986)
V. calciformis	*C. cainii, C. destructor*	O'Donoghue et al. (1990)
V. campanula	*C. (C.) patzcuarensis*	Mayén-Estrada and Aladro-Lubel (2002), Warren (1986)
V. communis (=subsphaerica)	*C. (C.) patzcuarensis*	Mayén-Estrada and Aladro-Lubel (2002), Warren (1986)
V. convallaria (=similis)	*A. leptodactylus, C. cainii, C. destructor*	Fard et al. (2011), O'Donoghue et al. (1990), Warren (1986)
V. flexulosa	*C. cainii, C. destructor*	O'Donoghue et al. (1990)
V. fromenteli (=cucullus)	*C. (C.) patzcuarensis*	Mayén-Estrada and Aladro-Lubel (2002), Warren (1986)
V. infusionum (=utriculus =abbreviata)	*C. (C.) patzcuarensis*	Mayén-Estrada and Aladro-Lubel (2002), Warren (1986)
V. jaerae	*C. cainii, C. destructor*	O'Donoghue et al. (1990)
V. latifunda	*C. (C.) patzcuarensis*	Mayén-Estrada and Aladro-Lubel (2002), Warren (1986)
V. microstoma	*C. (C.) patzcuarensis, P. (G.) simulans*	Lahser, Jr. (1975), Mayén-Estrada and Aladro-Lubel (2002), Warren (1986)
V. natans	*C. (C.) patzcuarensis*	Mayén-Estrada and Aladro-Lubel (2002), Warren (1986)

Table 5. contd....

Table 5. contd.

Genus/species	Host(s)	Reference(s)
V. poznaniensis	"Crayfish"	Fernandez-Leborans and Tato-Porto (2000a), Warren (1986)
V. sertularium	*C. (C.) zempoalensis*	Lopez-Ochoterena and Gasea (1971)
Vorticella sp.	*C. cainii, C. destructor, C. quadricarinatus*	Herbert (1987), O'Donoghue et al. (1990)
V. striata	*C. (C.) patzcuarensis*	Mayén-Estrada and Aladro-Lubel (2002), Warren (1986)
Zoothamnium dichotomum	*C. (C.) patzcuarensis*	Mayén-Estrada and Aladro-Lubel (2002)
Z. ponticum	*C. (C.) zempoalensis*	Lopez-Ochoterena and Gasea (1971)
Z. procerius	*A. astacus*	Sprague and Couch (1971)
Z. simplex	*C. (C.) patzcuarensis*	Mayén-Estrada and Aladro-Lubel (2002)
Zoothamnium sp.	*A. leptodactylus, C. cainii, C. quadricarinatus, P. (P.) l. leniusculus, P. (S.) clarkii*	Cuellar et al. (2002), Fard et al. (2011), Herbert (1987), Vogelbein and Thune (1988)

Phylum Platyhelminthes – class Trematoda (Digenea)

Digenea, or trematodes, are normally endoparasitic in a range of hosts, including crayfish. Identification is based on a number of features, and as a general rule, they possess a pair of suckers and an incomplete digestive system (see Fig. 2). The taxonomy of the group is more or less stable, but with changes proposed to the classification of a number of genera and families made annually by researchers. No specific keys exist to identify larval stages of digeneans in crayfish, with identification being reliant on the skills of the diagnostician and a good understanding of the available literature. A greater emphasis has been placed on molecular tools to identify digeneans over the past few years, and although not specifically developed for identification of crayfish digeneans, they could and should be used to verify parasite identity. In addition, it is important that these tools are further refined and developed to confirm the some of the identifications reported in the literature and to assist with understanding phylogeography of these parasites translocated as a result of human intervention. For example, Mohamed et al. (2005) identify a number of digeneans in a sample of 1151 non-native *Procambarus clarkii* in Egypt. The origin of these digeneans and the role of crayfish as a host, are unclear with the authors reporting, e.g., *Metagonimoides oregonensis*, originally described from the USA, in the sample. The authors did, however, conduct lifecycle studies for each digenean found and were able to apparently obtain adult digeneans from vertebrate hosts. Edgerton et al. (2002a) included results from an unpublished report suggesting that representatives of the Digenea families Cathaemasiidae, Microphallidae and Plagiorchiidae occurred in Australian *Cherax cainii*; the application of molecular tools on new samples may help to elucidate the identification of these larval stages. It should be noted that Lotz and Font (1983) considered that the report of *Paralecithodendrium naviculum* by Williams (1967) was incorrect and should have been reported as *Ochoterenatrema diminutum*.

Digenea require at least two hosts to complete their lifecycle; typically adults occur in vertebrate hosts and developmental stages (including metacercariae, sporocysts, rediae) occurring on or in invertebrates such as crayfish. It is known that several digenean species in the genus *Alloglossidium*, as well as *Sogandaritrema progeneticum*, *Crepidostomum cornutum*, *Opecoelus variabilis* and *Astacotrema cirrigerum* can achieve sexual maturity in crayfish in a process known a progenensis (Lotz and Corkum 1983, Smythe and Font 2001). Digenean infections can be systemic (e.g., *C. cornutum*, *O. variabilis* or *O. isostomata*) or limited to specific organs such as the musculature (e.g., *A. cirrigerum*, *C. sinensis*, *C. obscurus*, *M. typicus*, *M. spinulosus*, *O. diminutum*, *P. westermani*, *P. siliculus* or *Renifer* sp.), gills

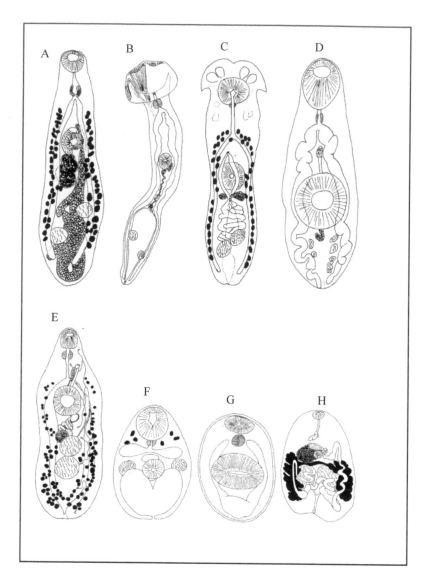

Figure 2. Examples of Digenea reported in crayfish. (A) Progenetic metacercaria of *Alloglosidium progeneticum* from *Procambarus* (*Pennides*) *spiculifer* (after Sullivan and Heard 1969). (B) Metacercaria of *Choanocotyle elegans* isolated from *Cherax* sp. (after Jue Sue and Platt 1998). (C) Progenetic metacercaria of *Crepidostomum cornutum* from *Cambarus sciotensis* (after Cheng 1957). (D) Metacercaria of *Gorgodera amplicava* from *Orconectes* (*Buannulifictus*) *palmericreolanus* (after Sogandares-Bernal 1965). (E) Adult *Opecoelus variabilis* from an atyid shrimp (after Cribb 1985). (F) *Ochoterenatrema diminutum* from *Orconectes* (*Procericambarus*) *rusticus* (after Williams 1967). (G) Metacercaria of *Astacatrematula macrocotyla* from *Pacifastacus* (*Pacifastacus*) *leniusculus trowbridgii* (after Macy and Bell 1968). (H) Progenetic metacercaria of *Sogandaritrema progeneticum* from *Cambarellus* (*Pandicambarus*) *puer* (after Sogandares-Bernal 1965). Not to scale.

(e.g., *A. macrocotyla*, *Maritrema* spp. or *P. jaenschi*), hepatopancreas (e.g., *C. elegans* or *Microphallus* spp.), antennal gland (e.g., *A. filiformis*, *Alloglossidium* spp.) or are found encysted on external surfaces of crayfish (e.g., *A. parvus*, *G. hominis*).

Digeneans of crayfish are found in the following families: Allocreadiidae (*Crepidostomum*), Cyathocotylidae (*Prohemistomum*), Choanocotylidae (*Choanocotyle*), Cladorchidae (*Allasostomoides*), Echinostomatidae (*Petasiger*), Gastrodiscidae (*Gastrodiscoides*), Gorgoderidae (*Gorgodera*), Gyrabascidae (*Cephalophallus*), Haematoloechidae (*Haematoloechus*), Heterophyidae (*Centrocestus*, *Heterophyes*, *Metagonimoides*, *Pygidiopsis*), Lecithodendriidae (*Ochoterenatrema*), Macroderodidae (*Alloglossidium*, *Macroderoides*), Microphallidae (*Maritrema*, *Microphallus*, *Quasimaritremopsis*), Opisthorchiidae

(*Clonorchis*), Opecoelidae (*Opecoelus, Plagioporus*), Orchipedidae (*Orchipedium*), Paragonimidae (*Paragonimus*), Superfamily Plagiorchioidea (*Allocorrigia*), Reniferidae (*Renifer*), Psilostomidae (*Astacatrematula, Astacotrema*), Troglotrematidae (*Macroorchis*) (Table 6).

Table 6. List of digeneans reported from crayfish.

Genus/species	Host(s)	Reference(s)
Allasostomoides (=*Allasostoma*) *parvus*	*Orconectes* (*Crockerinus*) *propinquus*	Beaver (1929), Brooks (1975), Suter and Richardson (1977), Watson and Rohde (1995)
Allocorrigia filiformis	*Procambarus* (*Scapulicambarus*) *clarkii*	Turner and Corkum (1977), Turner (2006)
Alloglossidium (=*Alloglossoides*) *cardicolum*	*Procambarus* (*Ortmannicus*) *acutus*	Corkum and Turner (1977), Turner (1999)
A. (=*Plagiorchis*) *corti* (=*ameirurensis*)	*Orconectes* (*Buannulifictus*) *palmeri longimanus, Orconectes* (*Procericambarus*) *acares, Procambarus* (*Tenuicambarus*) *tenuis*	McAllister et al. (2011), McCoy (1928)
A. (=*Alloglossoides*) *dolandi*	*Procambarus* (*Ortmannicus*) *acutissimus, P.* (*O.*) *acutus, Procambarus* (*Ortmannicus*) *epicyrtus, Procambarus* (*Scapulicambarus*) *howellae, Procambarus* (*Scapulicambarus*) *paeninsulanus, Procambarus* (*Scapulicambarus*) *troglodytes*	Turner and McKeever (1993), Turner (2007, 2009)
A. greeri	*Cambarellus* (*Dirigicambarus*) *shufeldtii*	Font (1994)
A. (=*Macroderoides*) *progeneticum* (=*progeneticus*)	*Procambarus* (*Pennides*) *spiculifer*	Carney and Brooks (1991), Sogandares-Bernal (1965), Sullivan and Heard (1969)
Astacatrematula (=*Sphaeridiotrema*) *macrocotyla*	*Pacifastacus* (*Pacifastacus*) *leniusculus trowbridgii*	Macy and Bell (1968)
Astacotrema (=*Distoma*) *cirrigerum*	*Astacus astacus, Astacus leptodactylus*	Warren (1903)
Centrocestus cuspidatus	*P.* (*S.*) *clarkii*	Mohamed et al. (2005)
Cephalophallus obscurus	*P.* (*P.*) *leniusculus trowbridgii*	Macy and Moore (1954)
Choanocotyle elegans	*Cherax* sp.	Jue Sue and Platt (1998)
Clonorchis sinensis	*P.* (*S.*) *clarkii*	Lun et al. (2005)
Crepidostomum cornutum (=*Bunodera cornuta*)	*C.* (*D.*) *shufeldtii, Cambarellus* (*Pandicambarus*) *puer, Cambarus sciotensis, O.* (*C.*) *propinquus, Orconectes* (*Gremicambarus*) *nais, Orconectes* (*Gremicambarus*) *virilis, Orconectes* (*Tragulicambarus*) *lancifer, Orconectes* (*Trisellescens*) *immunis, Procambarus* (*Giardiella*) *simulans, P.* (*S.*) *clarkii, P.* (*O.*) *acutus, Procambarus* (*Pennides*) *penni*	Ameel (1937), Cheng and James (1960a,b), Henderson (1938), Lefebvre and Poulin (2005), Sogandares-Bernal (1965)
Gastrodiscoides hominis	"Crayfish"	Chai et al. (2009)
Gorgodera amplicava	*Cambarus* sp., *Orconectes* (*Buannulifictus*) *palmeri creolanus, P.* (*S.*) *clarkii*	Sogandares-Bernal (1965)
G. cygnoides	"Crayfish"	Baker (2007)
Gorgoderina attenuata	"Crayfish"	Baker (2007)
G. vitelliloba	"Crayfish"	Baker (2007)
Haematoloechus sp.	"Crayfish"	Morrison (1966)
Heterophyes aequalis	*P.* (*S.*) *clarkii*	Mohamed et al. (2005)
Macroderoides typicus	*O.* (*T.*) *lancifer, P.* (*O.*) *acutus, P.* (*S.*) *clarkii*	Sogandares-Bernal (1965)

Table 6. contd....

Table 6. contd.

Genus/species	Host(s)	Reference(s)
Macroorchis spinulosus	*Cambaroides similis*	Chai et al. (1996)
Maritrema sp.	*Cambarus* sp.	Stafford (1931)
Maritrema (Atriospinosum) obstipum	*C. (D.) shufeldtii, P. (S.) clarkii*	Sogandares-Bernal (1965)
Metagonimoides oregonensis	*P. (S.) clarkii*	Mohamed et al. (2005)
Microphallus spp.	*O. (C.) propinquus, Orconectes (Procericambarus) rusticus*	Sargent et al. (2014)
M. fonti (=opacus) (=ovatus?)	*C. (P.) puer, O. (C.) propinquus, P. (S.) clarkii*	Caveny and Etges (1971), Osborn (1919), Overstreet et al. (1992), Sogandares-Bernal (1965), Stafford (1931)
M. minus	*P. (S.) clarkii*	Mohamed et al. (2005)
M. minutus	*Cherax destructor*	Johnston (1948)
Ochoterenatrema diminutum	*O. (P.) rusticus*	Lotz and Font (1983), Williams (1967)
Opecoelus variabilis	*Cherax depressus, Cherax dispar*	Cribb (1985)
Orchipedium (=Distoma) isostomata	All species of European crayfish	Dollfus et al. (1935)
Petasiger neocomense	*P. (S.) clarkii*	Mohamed et al. (2005)
Paragonimus kellicotti	*O. (C.) propinquus, O. (G.) virilis, Orconectes (Procericambarus) luteus, Orconectes (Procericambarus) punctimanus, O. (P.) rusticus, P. (O.) acutus, P. (S.) clarkii*	Fischer et al. (2011), Ishii (1966), Sogandares-Bernal (1965), Stromberg et al. (1978)
P. westermani	*C. similis*	Kim et al. (2009)
Plagioporus siliculus	*Pacifastacus* sp.	Sinitsin (1931)
Plagiorchis jaenschi	*C. destructor*	Johnston and Angel (1950)
Prohemistomum vivax	*P. (S.) clarkii*	Mohamed et al. (2005)
Pygidiopsis summa	*P. (S.) clarkii*	Mohamed et al. (2005)
Quasimaritremopsis (=Maritrema = Maritreminoides = Microphallus) medius	*?O. (C.) propinquus, ?O. (G.) virilis*	Deblock (1973)
Renifer (=Ochetosoma) sp.	*P. (S.) clarkii*	Sogandares-Bernal (1965)
Sogandaritrema (=Microphallus) progeneticum	*C. (D.) shufeldtii, C. (P.) puer, P. (S.) clarkii*	Lotz and Corkum (1983), Sogandares-Bernal (1962b, 1965)

Most reports of digeneans in crayfish tend to report presence/absence rather than describe any particular pathologies associated with the infection, perhaps reflecting the limited importance of this group of parasites as a population driver for crayfish. Infections of the antennal gland by *Alloglossidium cardicolum* lead to localised damage to the epithelium (Turner 1985). Digeneans in crayfish can be of concern for human health, particularly amongst cultures reliant on the consumption of raw or undercooked crayfish. Known or potential human-pathogenic digeneans utilising crayfish as hosts include *Clonorchis sinensis, Gastrodiscoides hominis, Heterophyes aequalis, Macroorchis spinulosus, Paragonimus kellocotti* and *P. westermani*.

Phylum Platyhelminthes – class Cestoda

Cestodes, or tapeworms are rarely found in crayfish with only two confirmed infections reported, namely *Hymenolepis (=Rodentolepis =Vampirolepis) diminuta* (Order Cyclophyllidea, Family Hymenolepididae) found in intestinal mucosa of *Cherax destructor* by O'Donoghue et al. (1990) and *Austramphilina elongata* (Order Amphilinidea), in the abdominal muscle of the same host. Reports of *Hymenolepis collaris* and *H. tenuirostris* from the body cavity of *Astacus astacus* by Hall (1929) were considered incorrect by Alderman and Polglase (1988) and Edgerton et al. (2002a). Further, attempts to transmit *Ophiotaenia testudo* through feeding of tapeworm eggs to crayfish were unsuccessful (Magath 1924); transmission of *H. diminuta* metacestodes from crayfish to laboratory rats was also unsuccessful (O'Donoghue et al. 1990). The lifecycle of *A. elongata* was completed by Rohde and Georgi (1983), who showed that eggs released from turtles released infective larvae that penetrated the cuticle of *C. destructor*. Transmission back to the vertebrate host occurred when turtles ate infected crayfish.

Phylum Platyhelminthes – order Temnocephalida

The taxonomic position of the temnocephalids has historically been controversial but recent studies have placed them within the Rhabdocoela (Phylum Platyhelminthes). Temnocephalids occur mainly on the Parastacidae, with *Temnocephala mexicana* being the only example of a temnocephalid on a Cambaridae (*Procambarus* (*Procambarus*) *digueti*). Other than a brief report of an unidentified *Temnocephala* sp. on farmed *Pacifastacus leniusculus* in Spain by Cuellar et al. (2002) and *Temnocephala minor* on the claws of *Astacus leptodactylus* in Turkey by Xylander (1997), no temnocephalids have been described from members of the Astacidae. Temnocephalids belong to a number of families including the Didymorchiidae (*Didymorchis*), Actinodactylellidae (*Actinodactylella*), Diceratocephalidae (*Diceratocephala*, *Decadidymus*), Temnocephalidae (*Dactylocephala*, *Temnocephala*, *Temnohaswellia*, *Temnomonticellia*, *Temnosewellia*, *Notodactylus*), and the Subfamily Craspedellinae (in the Temnocephalidae) (*Craspedella*, *Gelasinella*, *Heptacraspedella*, *Zygopella*) (Table 7). Taxonomy of the group is based on a number of features including the internal morphology of the reproductive organs (number of testes and arrangement of vagina and cirrus), the number and type of digitate processes (or "tentacles") and the presence or absence of dorsal scales (Fig. 3). Keys and monographs of the Temnocephala are provided by Cannon (1991), Cannon and Sewell (1995), Damborenea and Cannon (2001), Hickman (1967), Joffe et al. (1998), Martínez-Aquino et al. (2014), Sewell and Cannon (1998), Sewell et al. (2006), Sewell (2013) and Williams (1981).

As with a number of other commensals and pathogens of crayfish, temnocephalids have been transposed with their native host to new areas. Unidentified temnocephalids in the gills of *Cherax quadricarinatus* from Singapore and on *Cherax peknyi* from Indonesia that were imported into the UK were noted by Longshaw et al. (2012a); it is probable that the example from *C. quadricarinatus* was *Temnosewellia* cf. *minor* based on its morphological characteristics. The same species has been recorded on non-native *Cherax* species in several countries including *C. destructor* in Italy (Chiesa et al. 2015), *C. cainii* in Japan (Niwa and Ohtaka 2006, Oki et al. 1995), *C. cainii* in South Africa (identified incorrectly as *T. chaeropsis*) (Avenant-Oldewage 1993, Mitchell and Kock 1988); *Temnosewellia semperi* has been reported on *Cherax quadricarinatus* in China (Wen and Liu 2001). Another species of temnocephalid that has successfully translocated with its host is *Diceratocephala boschmai*, originally described from *Cherax* spp. in Indonesia. It has been recorded on non-native *C. destructor* in Thailand (Ngamniyom et al. 2014), and on *C. quadricarinatus* in South Africa (du Preez and Smit 2013) and Uruguay (Volonterio 2009); unidentified temnocephalids on the carapace of *C. quadricarinatus* in Ecuador most likely represent *D. boschmai* (Romero and Jiménez 2002).

Lifecycles are relatively simple with hermaphroditic adults laying eggs on the body surface of crayfish, which hatch to produce juveniles. Temnocephalids appear to have a limited impact on the overall health of their hosts, although there is some suggestion that their presence can reduce the overall marketability of farmed crayfish, thereby having an economic impact. Host switching is a concern with the risk that imported temnocephalids may infect native hosts. Xylander (1997) described *T. minor* from *A. leptodactylus* in Turkey; it appears that transfer occurred in artificial conditions from *Cherax* spp. in a mixed-culture tank. Whilst the risk is therefore low, it should not be underestimated and efforts should be made to disinfect crayfish prior to movement into a new area (Mitchell and Kock 1988). Several unidentified temnocephalids have been reported on crayfish including from *Pacifastacus leniusculus*, *Euastacus sulcatus*, *E. mirangudjin* and *E. gumar* (Coughran 2011a,b, Cuellar et al. 2002, Wild and Furse 2004); it is probable that there are a number of new, as yet undescribed species of these often overlooked symbionts on crayfish.

Table 7. List of temnocephalids, in alphabetical order, reported from crayfish.

Genus/species	Host(s)	Reference(s)
Actinodactylella blanchardii	*Engaeus fossor*	Suter and Richardson (1977), Watson and Rohde (1995)
Craspedella bribiensis	*Cherax robustus*	Sewell and Cannon (1998)
C. cooranensis	*Cherax depressus*	Sewell and Cannon (1998)
C. gracilis	*C. depressus*	Cannon and Sewell (1995)
C. joffei	*Cherax punctatus*	Sewell and Cannon (1998)
C. pedum	*Cherax quadricarinatus*	Cannon and Sewell (1995)
C. shorti	*C. depressus*	Cannon and Sewell (1995)
C. simulator	*Cherax cuspidatus, C. depressus, Cherax destructor, Cherax dispar*	Cannon and Sewell (1995)
C. spenceri	*C. depressus, C. destructor, C. dispar, C. punctatus*	Cannon and Jennings (1987), Cannon and Sewell (1995)
C. yabba	*C. depressus, C. dispar*	Cannon and Sewell (1995)
Dactylocephala (=Temnocephala) madagascarinensis	*Astacoides granulimanus, Astacoides madagascariensis, Astacoides* sp.	Cannon and Sewell (2001a)
Decadidymus gulosus	*C. quadricarinatus*	Cannon (1991)
Diceratocephala boschmai	*Cherax boschmani, Cherax communis, C. destructor, Cherax longipes, Cherax lorentzi, Cherax pallidus, C. quadricarinatus*	Cannon (1991), Ngamniyom et al. (2014)
Didymorchis astacopsis	*Euastacus* spp.	Haswell (1915)
D. cherapsis	*C. dispar, C. punctatus*	Cannon and Jennings (1987)
D. haswelli	*Parastacus saffordi*	Martínez-Aquino et al. (2014)
D. paranephropsis	*Paranephrops zealandicus*	Sewell (1998)
Gelasinella powellorum	*Euastacus spinifer*	Sewell and Cannon (1998)
Heptacraspedella peratus	*Euastacus bispinosus*	Cannon and Sewell (1995)
Notodactylus (=Temnocephala) handschini	*Cherax albertisii, C. boschmai, C. communis, C. longipes, C. lorentzi, "Cherax munida", C. pallidus, Cherax panaicus, C. quadricarinatus*	Cannon (1991)
Temnocephala axenos (=brasilensis = chilensis)	*Parastacus pugnax, Parastacus* sp., *Samastacus* sp.	Damborenea and Cannon (2001), Goetsch (1935)
T. mexicana	*Parastacus* sp., *Procambarus (Procambarus) digueti*	Damborenea and Cannon (2001)
Temnohaswellia alpina	*Euastacus rieki, Euastacus* sp.	Sewell et al. (2006)
T. breviumbella	*Euastacus bidawalus*	Sewell et al. (2006)
T. capricornia	*Euastacus monteithorum*	Sewell et al. (2006)
T. (=Temnocephala) comes (=pugna)	*Euastacus armatus?, Euastacus brachythorax, Euastacus clarkae, Euastacus dangadi, Euastacus dharawhalus, Euastacus gamilaroi, Euastacus gumar, Euastacus guwinus?, Euastacus hirsutus, Euastacus jagara, Euastacus maidae, Euastacus mirangudjin, Euastacus neohirsutus, Euastacus polysetosus, Euastacus setosus, Euastacus* sp., *Euastacus spinichelatus, E. spinifer, Euastacus sulcatus, Euastacus suttoni, Euastacus valentulus, Euastacus yanga*	Cannon (1993), Sewell et al. (2006)

Table 7. contd....

Table 7. contd.

Genus/species	Host(s)	Reference(s)
T. cornu	*E. jagara*	Sewell et al. (2006)
T. crotalum	*E. bispinosus, Euastacus kershawi, Euastacus neodiversus, Euastacus woiwuru, Euastacus yarraensis*	Sewell et al. (2006)
T. munifica	*Euastacus hystricosus*	Sewell et al. (2006)
T. (=T.) novaezealandiae	*P. zealandicus*	Williams (1994)
T. pearsoni	*Euastacus eungella*	Sewell et al. (2006)
T. (=T.) simulator (tetrica)	*E. armatus?, E. dangadi, E. gumar, E. neohirsutus, E. spinichelatus, E. sulcatus, E. suttoni, E. valentulus*	Cannon (1993), Sewell et al. (2006)
T. subulata	*Euastacus australasiensis*	Sewell et al. (2006)
T. umbella	*E. guwinus* cf. *dharawalus*	Sewell et al. (2006)
T. verruca	*E. armatus, E. bidawalus, E. brachythorax, Euastacus claytoni, Euastacus crassus, E. dharawalus, Euastacus gamilaroi, E. polysetosus, Euastacus reductus, E. spinifer, E. yanga, E. yarraensis*	Sewell et al. (2006)
Temnohaswellia sp.	*Euastacus urospinosus*	Sewell et al. (2006)
Temnomonticellia (=Temnocephala) aurantiaca	*Astacopsis* sp., *Engaeus cunicularis*	Hickman (1967), Jennings (1971)
T. (=T.) fulva	*Parastacoides tasmanicus* complex	Hickman (1967)
T. (=T.) pygmaea	*Astacopsis gouldi, Parastacoides tasmanicus* complex	Hickman (1967), Horwitz and Knott (1991)
T. (=T.) quadricornis	*A. franklinii, A. gouldi*	Hickman (1967), Jennings (1971)
T. (=T.) tasmanica	*A. franklinii, Engaeus fossor*	Hickman (1967), Jennings (1971), Suter and Richardson (1977)
Temnosewellia acicularis	*E. bidawalus, E. crassus*	Sewell et al. (2006)
T. acirra	*C. destructor*	Cannon and Sewell (2001b)
T. alba	*Euastacus balanensis, Euastacus fleckeri*	Sewell et al. (2006)
T. albata	*C. depressus, Euastacus robertsi*	Sewell et al. (2006)
T. aphyodes	*E. fleckeri*	Sewell et al. (2006)
T. apiculus	*E. kershawi*	Sewell et al. (2006)
T. urga	*Euastacus yigara*	Sewell et al. (2006)
T. argeta	*Cherax parvus, E. yigara*	Sewell et al. (2006)
T. argilla	*E. fleckeri*	Sewell et al. (2006)
T. aspinosa	*E. valentulus*	Sewell et al. (2006)
T. aspra	*E. balanensis*	Sewell et al. (2006)
T. bacrio	*E. maidae, E. sulcatus, E. valentulus*	Sewell et al. (2006)
T. bacrioniculus	*E. maidae, E. neohirsutus, E. setosus, E. sulcatus, E. valentulus*	Sewell et al. (2006)
T. batiola	*E. hystricosus, E. urospinosus*	Sewell et al. (2006)
T. belone	*E. brachythorax*	Sewell et al. (2006)
T. caliculus	*E. kershawi,?E. woiwuru,?E. yarraensis*	Sewell et al. (2006)
T. cestus	*C. dispar, E. urospinosus*	Sewell et al. (2006)

Table 7. contd....

Table 7. contd.

Genus/species	Host(s)	Reference(s)
T. (=Temnocephala) chaeopsis	Cherax cainii, Cherax preissii, Cherax cf. quinquecarinatus	Cannon and Sewell (2001b), Jennings (1971)
T. christineae	C. depressus, C. robustus	Cannon and Sewell (2001b)
T. (=T.) cita	Parastacoides tasmanicus complex	Hickman (1967), Horwitz and Knott (1991), Jennings (1971)
T. comythus	E. gumar, E. spinichelatus	Sewell et al. (2006)
T. coughrani	E. mirangudjin, E. sulcatus	Sewell et al. (2006)
T. cypellum	E. spinifer	Sewell et al. (2006)
T. (=T.) dendyi	Cherax albidus, Cherax bicarinatus, C. destructor, C. depressus, C. dispar, C. robustus	Cannon and Sewell (2001b), Jennings (1971)
T. (=T.) engaei	E. fossor	Jennings (1971), Suter and Richardson (1977)
T. (=T.) fasciata	E. australasiensis, E. clarkae, E. polysetosus, Euastacus sp., E. spinifer	Sewell et al. (2006)
T. fax	E. armatus, E. cf. crassus, E. hirsutus, E. yanga	Sewell et al. (2006)
T. flammula	E. neohirsutus	Sewell et al. (2006)
T. gingrina	E. dangadi, E. gumar, E. suttoni, E. sulcatus, E. valentulus	Sewell et al. (2006)
T. gracilis	E. guwinus cf. dharawalus	Sewell et al. (2006)
T. keras	E. kershawi, E. yarraensis	Sewell et al. (2006)
T. maculata	E. bispinosus	Sewell et al. (2006)
T. magna	E. armatus?	Sewell et al. (2006)
T. maxima	E. sulcatus	Sewell et al. (2006)
T. minima	E. sulcatus	Sewell et al. (2006)
T. (=T.) minor	C. albidus, C. bicarinatus, C. cainii, C. depressus, C. destructor, C. dispar, C. quadricarinatus	Cannon and Jennings (1987), Cannon and Sewell (2001b), Chiesa et al. (2015), Jennings (1971)
T. muscalingulata	E. armatus, E. crassus, E. neodiversus, E. rieki, E. woiwuru	Sewell et al. (2006)
T. phantasmella	Cherax rhynchotus	Cannon and Sewell (2001b)
T. possibilitas	E. bispinosus	Sewell et al. (2006)
T. punctata	C. cainii, Cherax cf. quinquecarinatus	Cannon and Sewell (2001b)
T. (=T.) rouxii	Cherax aruanus, C. quadricarinatus	Cannon (1991)
T. (=T.) semperi	C. quadricarinatus	Sewell et al. (2006), Wen and Liu (2001)
T. unguiculus	E. claytoni	Sewell et al. (2006)
Temnosewellia sp.	Cherax punctatus, E. neohirsutus	Cannon and Sewell (2001b), Sewell et al. (2006)
Zygopella deimata	C. cainii	Cannon and Sewell (1995)
Z. pista	C. cainii	Cannon and Sewell (1995)
Z. stenota	C. cf. bicarinatus, C. cainii, Cherax cf. quinquecarinatus	Cannon and Sewell (1995)

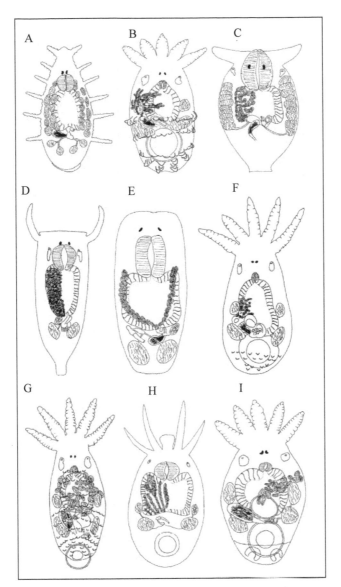

Figure 3. Representative examples of temnocephalids genera reported from crayfish hosts. (A) *Actinodactylella* (after Haswell 1893). (B) *Craspedella* (after Cannon and Sewell 1995). (C) *Decadidymus* (after Cannon 1991). (D) *Diceratocephala* (after Jones and Lester 1992). (E) *Didymorchis* (after Haswell 1915) (F) *Gelasinella* (after Sewell and Cannon 1998). (G) *Heptacraspedella* (after Cannon and Sewell 1995). (H) *Temnomonticellia* (after Hickman 1967). (I) *Zygopella* (after Cannon and Sewell 1995). Not to scale.

Phylum Nematoda

Nematodes associated with crayfish are generally considered commensal, occurring primarily on the gills and do not normally impact on host survival as they occur in low numbers. Occasionally, however, some of these commensals can occur in large numbers on their host (Schneider 1932). Due to either the methods used to sample crayfish or to lack of taxonomic specialisms, many nematodes of crayfish are not identified. These include unidentified nematodes in the gills of *A. pallipes* (Longshaw et al. 2012b), of *Procambarus* (*Scapulicambarus*) *clarkii* (Quaglio et al. 2006b), of *Cherax quadricarinatus* (Herbert 1987) and the gills of *Engaeus fossor* and *E. cisternarius* (Suter and Richardson 1977). In addition,

Unestam (1973) and Ljungberg and Monné (1968) report on the presence of nematode eggs in *Pacifastacus* (*Pacifastacus*) *leniusculus* and in the nerve cords and connective tissues of *Astacus astacus* respectively. Other nematodes reported in the gills include *Chromadorita leuckartii*, *Chrysonemoides* (=*Dorylaimus*) *holsaticus*, *Crocodorylaimus* (=*Dorylaimus*) *flavomaculatus*, *Eudorylaimus* (=*Dorylaimus*) *carteri* and *Eudorylaimus* (=*Dorylaimus*) *centrocercus*, *Monhystera* (?) *dispar*, *Mononchus truncatus* (=*macrostoma*), *Tobrilus* (=*Trilobus*) *gracilis* and *Tobrilus* (=*Trilobus*) *medius* on *Cambarus* sp. and *Austropotamobius* sp. (Schneider 1932), *Chromadorita* (=*Prochromadorella*) *astacicola* in *Astacus astacus*, *Cambarus* sp. and *Austropotamobius* sp. (Schneider 1932, Wiszniewski 1939), *Chromadorita* (=*Prochromadorella*) *viridis* in *A. astacus* (Wiszniewski 1939), *Dorylaimus* (?) sp. on *Procambarus* (*Girardiella*) *simulans* (Lahser, Jr. 1975), *Gammarinema* (=*Monhystera* = *Rhabditis*) *cambari* on *Cambarus* (*Puncticambarus*) *acuminatus* and *Procambarus* (*Ortmannicus*) *blandingii* (Allen 1933), *Gammarinema* (=*Monhystera*?) sp. on *Cherax albidus* (Jones and Lawrence 2001) and *Procambarus* (*Girardiella*) *simulans* (Lahser, Jr. 1975), *Rhabditis terricola* (=*teres*) and *R. inermis* in *Orconectes* (*Faxonius*) *limosus*, *Cambarus* sp. and *Austropotamobius* sp. (Schneider 1932, Wiszniewski 1939) and *Paractinolaimus* (=*Actinolaimus*) *macrolaimus* on *Astacus astacus*, *Cambarus* sp. and *Austropotamobius* sp. (Schneider 1932, Wiszniewski 1939). Wholly parasitic nematodes are rarely reported in crayfish but include the report of an unidentified nematode encapsulated within the intestine of *Procambarus* (*Scapulicambarus*) *clarkii*, the human pathogenic *Gnathostoma spinigerum* and the rat lungworm *Angiostrongylus cantonensis* in *Cambarus* sp. (Miyazaki 1954, Moravec 2007, Quaglio et al. 2006b). Woodhead (1950) erroneously proposed that crayfish were a host for the giant kidney worm *Dioctophyme* (=*Dioctophyma*) *renale*. *Orconectes* (*Crockerinus*) *propinquus* appear to be refractive to infections with the nematode *Drancunculus insignis* as attempts to transmit the infection to crayfish have proved unsuccessful (Crichton and Beverley-Burton 1977).

Phylum Annelida – class Polychaeta

Several species of *Stratiodrilus* (Family Histriobdellidae) have been described from the branchial chambers of parastacid crayfish, with no reports of mortalities or pathology associated with them. *Stratiodrilus haswelli* has been reported on *Astacoides madagascariensis* from Madagascar, *S. tasmanicus* on *Astacopsis franklinii* from Australia, *S. novaehollandiae* on *Euastacus* sp. (as *Astacopsis serratus*) as well as on *Cherax dispar* and *C. punctatus* from Australia, *S. pugnaxi* on *Parastacus* sp. and *Parastacus pugnax* from Chile and *S. vilae* on *Parastacus brasiliensis* and *P. defossus* (Amato 2001, Cannon and Jennings 1987, Moyano et al. 1993, Harrison 1928, Haswell 1922, Vila and Bahamonde 1985) (Fig. 4). In addition, Vila and Bahamonde (1985) report the presence of *S. aeglaphilus* and *S. circensis* (=*platensis*) on unspecified families of Parastacidae (see also Steiner and Amaral 1999). Keys to the described *Stratiodrilus*, excluding *S. vilae*, are provided by Vila and Bahamonde (1985).

Phylum Annelida – class Oligochaeta

Three species of ectoparasitic oligochaetes have been described from parastacid crayfish, namely *Astacopsidrilus notabilis* (=*Phreodrilus goddardi*), *A.* (=*Phreodrilus*) *fusiformis* and *A. jamiesoni* on *Euastacus* spp. (Brinkhurst 1971, Pinder and Brinkhurst 1997). In Europe, the main described species is *Hystricosoma chappuisi* occurring in the branchial chambers of *Astacus leptodactylus* and *Austropotamobius torrentium* (Boshko 1983, Subchev et al. 2007). Numbers are generally low but can reach up to 7000 individual worms per host (Boshko 1983).

Damage to the gills of *A. leptodactylus* by *Aeolosoma hempritichi* was reported by Fard et al. (2011). Other *Aeolosoma* species reported from crayfish include *A. quaternarium*, *A. tenebrarum*, *A. variegatum* and *Aeolosoma* sp. on *A. leptodactylus* and *Astacus astacus* (Boshko 1983, Wiszniewski 1939). Boshko (1983) reported *Dero obtusa*, *Nais barbata*, *Pristina aequiseta*, *Stylaria lacustris*, and *Vejdovskiella comata* from the branchial chambers of *A. leptodactylus* and *A. astacus*, suggesting that damage and deformation of the gills may occur when numbers of worms and their coccons were high.

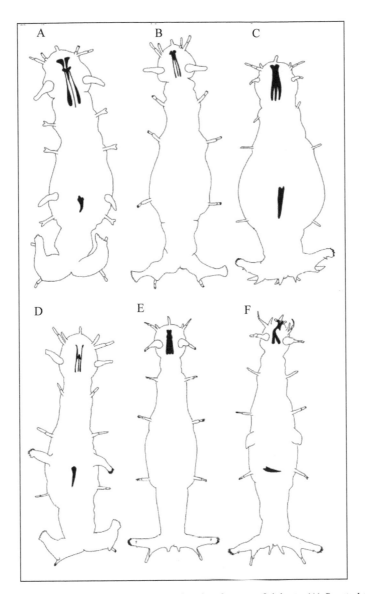

Figure 4. Line drawings of all the *Stratiodrilus* spp. reported to date from crayfish hosts. (A) *Stratiodrilus haswelli* (male) from *Astacoides madagascariensis* (after Harrison 1928). (B) *Stratiodrilus novaehollandiae* (female) from *Euastacus* sp. (after Harrison 1928). (C) *Stratiodrilus pugnaxi* (male) from *Parastacus pugnax* (after Vila and Bahamonde 1985). (D) *Stratiodrilus tasmanicus* (male) from *Astacopsis franklinii* (after Vila and Bahamonde 1985). (E) *Stratiodrilus vilae* (female) from *Parastacus brasiliensis* (after Amato 2001). (F) *Stratiodrilus vilae* (male) from *Parastacus brasiliensis* (after Amato 2001). Not to scale.

Phylum Annelida, class Clitellata – order Branchiobdellida

These annelids are normally considered to be ectocommensals or ectosymbionts of crayfish and whilst there has been some debate regarding their taxonomic position, it is generally accepted that they are derived oligochaetes (Apakupakul et al. 1999, Cardini and Ferraguti 2004, Erséus et al. 2008, Gelder and Siddall 2001, Gelder et al. 2012). Branchiobdellids are hermaphroditic and lay cocoons on a number of surfaces, including the carapace of their crayfish hosts. The impact of branchiobdellids on their crayfish host is subject to some debate with some authors suggesting that they may be pathogenic (Alderman and Polglase 1988, Hubault 1935) whilst others have suggested that they may in fact engage in a cleaning symbiosis with their hosts leading to improved growth rates (Brown et al. 2002, Keller 1992, Lee et al. 2009). Major reviews and taxonomic lists, including lists of synonyms can be found in Gelder (1996a,b, 2011, 2014), Gelder and Hall (1990), Gelder and Ohtaka (2002), Gelder et al. (1994), Goodnight (1940), Holt (1973a, 1986, 1989), Holt and Opell (1993), Subchev (2014) and Yamaguchi (1934). Four subfamilies of crayfish-infecting branchiobdellids are recognised including Branchiobdellidae (containing the genera *Ankyrodrilus, Branchiobdella, Cirrodrilus, Sinodrilus* and *Xironogiton*), Bdellodrilidae (containing the genera *Bdellodrilus, Cronodrilus, Hidejiodrilus* and *Uglukodrilus*), Cambarincolidae (containing the genera *Cambarincola, Ceratodrilus, Ellisodrilus, Forbesodrilus, Magmatodrilus, Oedipodrilus, Pterodrilus, Sathodrilus, Tettodrilus* and *Triannulata*) and the subfamily Xironodrilidae (containing the genus *Xironodrilus*) (see Table 8). An example of a generalised branchiobdellid is shown in Fig. 5a. Whilst they appear to be restricted to crayfish in the northern hemisphere and are generally restricted in their distribution, several are considered invasive, having been moved with their natural hosts to new areas

Table 8. List of branchiobdellids, in alphabetical order, reported from crayfish.

Genus/species	Host(s)	Reference(s)
Ankyrodrilus koronaeus	*Cambarus (Cambarus) angularis, Cambarus (Cambarus) bartonii bartonii, Cambarus (Cambarus) sciotensis, Cambarus (Erebicambarus) tenebrosus, Cambarus (Hiaticambarus) longulus, Cambarus (Puncticambarus) acuminatus*	Hobbs, Jr. et al. (1967), Williams et al. (2013)
A. legaeus	*C. (C.) angularis, C. (C.) bartonii bartonii, C. (C.) sciotensis, C. (E.) tenebrosus, Orconectes (Procericambarus) placidens*	Hobbs, Jr. et al. (1967), Holt (1973b), Gelder (1996a), Williams et al. (2013)
Bdellodrilus (=Branchiobdella) illuminatus	*Cambarellus (Cambarellus) montezumae, C. (C.) b. bartonii, Cambarus (Cambarus) bartonii cavatus, Cambarus (Cambarus) carinirostris, C. (C.) sciotensis, Cambarus (Hiaticambarus) longirostris, C. (H.) longulus, Cambarus (Jugicambarus) carolinus, C. (P.) acuminatus, Orconectes (Gremicambarus) virilis, Procambarus (Ortmannicus) acutus, Procambarus (Procambarus) digueti*	Gelder et al. (2001), Gelder and Williams (2011), Goodnight (1940), Hobbs, Jr. et al. (1967), Holt (1973a,b), Keenan et al. (2014), Williams et al. (2013)
Branchiobdella sp.	*Astacus leptodactylus salinus, Austropotamobius pallipes italicus, Austropotamobius pallipes pallipes*	Subchev (2012)
B. astaci	*Astacus astacus, Astacus leptodactylus, A. l. salinus, A. p. italicus, A. p. pallipes, Austropotamobius torrentium*	Subchev (2012, 2014)
B. balcanica sketi	*A. astacus, A. leptodactylus, A. torrentium, Orconectes (Faxonius) limosus*	Subchev (2014)
B. balcanica	*A. astacus, O. (F.) limosus*	Ďuriš et al. (2006), Füreder et al. (2009)
B. cheni	*Cambaroides dauricus, Cambaroides wladiwostokensis?*	Timm (1991)
B. digitata	*Cambaroides japonicus*	Yamaguchi (1934)
B. domina	*C. wladiwostokensis*	Timm (1991)
B. hexadonta	*A. astacus, A. leptodactylus, A. torrentium, A. p. italicus, A. p. pallipes, O. (F.) limosus*	Gherardi et al. (2002), Subchev et al. (2007), Subchev (2012, 2014)
B. italica	*A. astacus, A. p. italicus, A. p. pallipes, Procambarus (Scapulicambarus) clarkii*	Gherardi et al. (2002), Oberkofler et al. (2002), Subchev (2012)

Table 8. contd....

Table 8. contd.

Genus/species	Host(s)	Reference(s)
B. kobayashii	*Cambaroides similis*	Wang and Cui (2007), Yamaguchi (1934)
B. kozarovi	*A. astacus, A. leptodactylus, A. l. salinus*	Subchev (2012)
B. macroperistomium	*C. dauricus, Cambaroides schrenckii, C. similis*	Wang and Cui (2007)
B. minuta	*C. dauricus, C. schrenckii*	Timm (1991)
B. monodontus	*C. dauricus*	Wang and Cui (2007)
B. orientalis	*C. similis*	Wang and Cui (2007), Yamaguchi (1934)
B. papillosa	*A. torrentium*	Subchev (2014)
B. parasita	*A. astacus, A. leptodactylus, A. pallipes, A. torrentium, O. (F.) limosus, Pacifastacus (Pacifastacus) leniusculus leniusculus, P. (S.) clarkii*	Ďuriš et al. (2006), Subchev et al. (2007), Subchev (2012, 2014)
B. pentadonta	*A. astacus, A. leptodactylus, A. pallipes, A. torrentium, O. (F.) limosus, P. (S.) clarkii*	Ďuriš et al. (2006), Gelder et al. (2012), Subchev et al. (2007), Subchev (2012, 2014)
B. teresae	*C. similis*	Subchev (1986)
Cambarincola barbarae	*P. (S.) clarkii*	Gelder et al. (1994), Holt (1981)
C. bobbi	*C. (C.) b. bartonii, C. (E.) tenebrosus*	Holt (1988a), Holt and Opell (1993), Williams and Gelder (2011), Williams et al. (2013)
C. branchiophilus	*C. (C.) b. bartonii, C. (C.) sciotensis, C. (H.) longulus, C. (P.) acuminatus*	Hobbs, Jr. et al. (1967), Hoffman (1963), Holt and Opell (1993)
C. carcinophilus	*Procambarus (Austrocambarus) zapoapensis*	Holt (1973a)
C. chirocephalus	*C. (C.) b. bartonii, Orconectes (Crockerinus) propinquus, O. (G.) virilis, Orconectes (Procericambarus) longidigitus, Orconectes (Procericambarus) medius, O. (P.) placidens, Orconectes (Procericambarus) rusticus, P. (O.) acutus*	Gelder (1996a), Goodnight (1940), Holt and Opell (1993)
C. demissus	*Orconectes (Crockerinus) erichsonianus, O. (P.) rusticus, Cambarus (Puncticambarus) nerterius*	Hoffman (1963), Holt (1973b), Holt and Opell (1993)
C. dubius	*C. (E.) tenebrosus, Orconectes (Orconectes) inermis testii*	Holt (1973b), Holt and Opell (1993)
C. ellisi	*Procambarus (Girardiella) regiomontanus*	Holt (1973a)
C. fallax	*C. (C.) b. bartonii, C. (C.) sciotensis, C. (H.) longirostris, C. (H.) longulus, C. (J.) carolinus, C. (P.) acuminatus, Cambarus sp. – cataloochee morph, Orconectes (Crockerinus) obscurus, O. (P.) placidens, O. (P.) rusticus, P. (S.) clarkii, P. (P.) l. leniusculus*	Hobbs, Jr. et al. (1967), Hoffman (1963), Holt (1981), Holt and Opell (1993), Gelder (1996a), Gelder and Hall (1990), Gelder and Williams (2011), Gelder et al. (2001), Keller (1992), Williams et al. (2013)
C. floridanus	*Procambarus (Ortmannicus) fallax, Procambarus (Pennides) spiculifer*	Holt and Opell (1993), Williams et al. (2013)
C. goodnighti	*P. (O.) fallax, Procambarus (Scapulicambarus) paeninsulanus*	Holt and Opell (1993)

Table 8. contd....

Table 8. contd.

Genus/species	Host(s)	Reference(s)
C. gracilis	*Pacifastacus (Pacifastacus) leniusculus klamathensis, P. (P.) l. leniusculus, Pacifastacus (Pacifastacus) leniusculus trowbridgii, P. (S.) clarkii*	Gelder and Hall (1990), Holt (1981), Holt and Opell (1993), Williams et al. (2013)
C. heterognathus	*C. (C.) bartonii bartonii, C. (C.) sciotensis, C. (H.) longulus, C. (H.) longirostris, Cambarus* sp., *Orconectes (Procericambarus) forceps*	Gelder and Williams (2011), Hobbs, Jr. et al. (1967), Hoffman (1963), Holt and Opell (1993)
C. hoffmani	*Procambarus (Ortmannicus) caballeroi, Procambarus (Villalobosus) hoffmani*	Holt (1973a)
C. holostomus	*C. (C.) b. bartonii, C. (C.) b. cavatus, C. (C.) sciotensis, C. (H.) longirostris, C. (H.) longulus, C. (J.) carolinus, C. (P.) acuminatus, Cambarus* sp. – Cataloochee morph, *O. (C.) erichsonianus, O. (P.) forceps*	Gelder and Williams (2011), Hobbs, Jr. et al. (1967), Hoffman (1963), Holt and Opell (1993), Williams et al. (2013)
C. holti	*Cambarus (Depressicambarus) graysoni, Cambarus* sp.	Hoffman (1963), Holt and Opell (1993), Williams et al. (2013)
C. illinoisensis	*O. (G.) virilis*	Holt (1982), Holt and Opell (1993)
C. ingens	*C. (C.) b. bartonii, C. (C.) sciotensis, Cambarus (Hiaticambarus) chasmodactylus, C. (H.) longirostris, Cambarus* sp. - Cataloochee morph	Farrell et al. (2014), Gelder and Williams (2011), Hoffman (1963), Holt and Opell (1993), Williams et al. (2013)
C. jamapaensis	*Procambarus (Austrocambarus) mexicanus, Procambarus (Ortmannicus) cuevachicae, P. (V.) hoffmani*	Holt (1973a)
C. leoni	*Procambarus (Ortmannicus) orcinus, Troglocambarus maclanei*	Holt (1973b), Holt and Opell (1993)
C. leptadenus	*C. (E.) tenebrosus*	Holt (1973b), Holt and Opell (1993)
C. macrocephalus	*Pacifastacus (Hobbsastacus) connectens, Pacifastacus (Hobbsastacus) gambelii*	Hoffman (1963), Holt and Opell (1993)
C. macrodontus	*Cambarus (Lacunicambarus) diogenes, Cambarus (Lacunicambarus) ludovicianus, O. (C.) propinquus, O. (G.) virilis, O. (P.) medius, Orconectes (Procericambarus) menae, Orconectes (Trisellescens) immunis, Procambarus (Girardiella) hagenianus hagenianus, Procambarus (Girardiella) simulans, P. (O.) acutus, P. (S.) clarkii*	Goodnight (1940), Holt and Opell (1993), Williams et al. (2013)
C. manni	*P. (O.) fallax, P. (S.) paeninsulanus*	Holt and Opell (1993)
C. mesochoreus	*C. (C.) b. bartonii, Cambarus (Puncticambarus) robustus, O. (C.) obscurus, O. (F.) limosus, O. (G.) virilis, O. (T.) immunis, O. (P.) rusticus, Orconectes* sp., *P. (O.) acutus, P. (S.) clarkii*	Gelder et al. (1999, 2001), Hoffman (1963), Holt and Opell (1993), Williams et al. (2013)
C. meyeri	*C. (C.) b. bartonii*	Holt and Opell (1993), Williams et al. (2013)
C. micradenus	*Procambarus (Paracambarus) paradoxus*	Holt (1973a)

Table 8. contd....

Table 8. contd.

Genus/species	Host(s)	Reference(s)
C. (=Triannulata) okadai (=montanus)	P. (P.) l. klamathensis, P. (P.) l. leniusculus, P. (P.) l. trowbridgii, P. (S.) clarkii	Gelder and Ohtaka (2000a), Goodnight (1940), Holt (1981), Holt and Opell (1993), James et al. (2015), Williams et al. (2013)
C. olmecus	P. (A.) mexicanus, P. (O.) cuevachicae	Holt (1973a)
C. osceola	Fallicambarus (Creaserinus) fodiens, O. (G.) virilis, P. (S.) paeninsulanus	Hoffman (1963), Holt and Opell (1993), Williams et al. (2013)
C. ouachita	Orconectes sp.	Hoffman (1963), Holt and Opell (1993)
C. pamelae	P. (S.) clarkii	Gelder et al. (1994), Holt (1984a)
C. (=Branchiobdella) (=Astacobdella) philadelphicus	C. (C.) b. bartonii, C. (C.) carinirostris, (C.) sciotensis, Cambarus (Depressicambarus) latimanus, C. (E.) tenebrosus, ?C. (H.) chasmodactylus, C. (II.) longulus, C. (J.) carolinus, C. (P.) acuminatus, C. (P.) nerterius, Cambarus (Puncticambarus) reburrus, C. (P.) robustus, Cambarus sp., O. (C.) erichsonianus, O. (F.) limosus, O. (G.) virilis, O. (P.) placidens, Orconectes (Procericambarus) durelli, O. (T.) immunis, Procambarus (Ortmannicus) hayi, Procambarus (Pennides) spiculifer, P. (P.) rusticus	Farrell et al. (2014), Gelder (1996a), Gelder et al. (2001), Gelder and Williams (2011), Goodnight (1940), Hall (1915), Hobbs, Jr. et al. (1967), Holt (1973b), Williams et al. (2013)
C. restans	Orconectes sp.	Hoffman (1963), Holt and Opell (1993)
C. serratus	P. (H.) connectens	Holt (1981), Holt and Opell (1993)
C. sheltensis	Orconectes (Orconectes) australis australis	Holt (1973b), Holt and Opell (1993)
C. shoshone	P. (H.) connectens	Hoffman (1963), Holt and Opell (1993)
C. susanae	Procambarus (Austrocambarus) llamasi, P. (G.) regiomontanus, P. (O.) cuevachiae, Procambarus (Ortmannicus) gonopodcristatus	Holt (1973a)
C. toltecus	P. (A.) zapoapensis	Holt (1973a)
C. virginicus	C. (P.) acuminatus	Hoffman (1963), Holt and Opell (1993)
C. vitreus	C. (C.) b. bartonii, O. (C.) propinquus, O. (G.) virilis, O. (P.) longidigitus, Orconectes (Procericambarus) placidus, O. (P.) rusticus, O. (T.) immunis, Orconectes (Trisellescens) mississippiensis, P. (G.) simulans, P. (O.) acutus, P. (S.) clarkii	Gelder (2004), Goodnight (1940), Hoffman (1963), Holt and Opell (1993), Williams et al. (2009)
Cambarincola sp.	C. (C.) bartonii cavatus, C. (D.) graysoni, C. (E.) tenebrosus	Holt (1973b), Keenan et al. (2014), Williams et al. (2013)
Ceratodrilus ophiorhysis (=orphiorhysis)	P. (H.) connectens, P. (H.) gambelii	Holt (1988b), Holt and Opell (1993), Williams et al. (2013)
C. (=Cirrodrilus) thysanosomus	P. (H.) gambelii	Goodnight (1940), Holt and Opell (1993)
Cirrodrilus (=Stephanodrilus) aequiannulus	C. dauricus	Wang and Cui (2007)

Table 8. contd....

Table 8. contd.

Genus/species	Host(s)	Reference(s)
C. (=S.) anodontus	*C. dauricus*	Wang and Cui (2007)
C. (=S.) aomorensis	*C. japonicus*	Gelder and Ohtaka (2000b), Yamaguchi (1934)
C. (=S.) breviformis	*C. dauricus*	Wang and Cui (2007)
C. (=S.) chosen	*C. dauricus, C. similis, C. wladiwostokensis*	Timm (1991), Yamaguchi (1934)
C. (=S.) cirratus	*C. japonicus*	Yamaguchi (1934)
C. (=S.) ezoensis	*C. japonicus*	Yamaguchi (1934)
C. fimbriatus	*C. wladiwostokensis*	Timm (1991)
C. (=S.) heteroglandularis	*C. dauricus*	Wang and Cui (2007)
C. (=Cambarinicola) homodontus	*C. japonicus*	Yamaguchi (1934)
C. (=S.) inukaii	*C. japonicus*	Gelder and Ohtaka (2000b), Yamaguchi (1934)
C. (=S.) japonicus	*C. japonicus*	Yamaguchi (1934)
C. (=S.) kawamurai	*C. dauricus, C. similis*	Wang and Cui (2007), Yamaguchi (1934)
C. (=S.) liaoningensis	*C. dauricus*	Wang and Cui (2007)
C. (=S.) makinoi	*C. japonicus*	Yamaguchi (1934)
C. (=S.) megalodentatus	*C. japonicus*	Yamaguchi (1934)
C. (=S.) minimus	*C. dauricus*	Wang and Cui (2007)
C. (=Carcinodrilus) nipponicus	*C. japonicus*	Yamaguchi (1934)
C. (=S.) peristomalis	*C. dauricus*	Wang and Cui (2007)
C. pugnax	*C. wladiwostokensis*	Timm (1991)
C. (=S.) quadritentacularis	*C. schrenckii, C. dauricus*	Timm (1991)
C. (=S.) sapporensis	*C. japonicus*	Yamaguchi (1934)
C. (=S.) suzukii	*C. dauricus, C. similis*	Timm (1991), Yamaguchi (1934)
C. tsugarensis	*C. japonicus*	Gelder and Ohtaka (2000b)
C. (=S.) uchidai	*C. japonicus*	Yamaguchi (1934)
Cronodrilus ogygius	*Cambarus (Depressicambarus) englishi, P. (P.) spiculifer*	Gelder and Ferraguti (2001), Williams et al. (2013), Holt (1968a)
Ellisodrilus carronamus	*C. (E.) tenebrosus, Orconectes* sp.	Holt (1988a), Holt and Opell (1993)
E. clitellatus	*Cambarus (Jugicambarus) distans*	Holt and Opell (1993)
E. (=Pterodrilus) durbini	*O. (G.) virilis, Orconectes (Procericambarus) barrenensis, O. (P.) rusticus*	Goodnight (1940), Holt and Opell (1993)
Forbesodrilus (=Cambarincola) nanognathus (=nanagnathus)	*Procambarus (Austrocambarus) vazquezae*	Gelder (2011), Holt (1973a)
Hidejiodrilus (=Stephanodrilus) koreanus	*C. similis*	Gelder (2010)
Magmatodrilus (=Stephanodrilus) obscurus	*Pacifastacus (Hobbsastacus) fortis, Pacifastacus (Hobbsastacus) nigrescens*	Holt and Opell (1993), Williams et al. (2013)
Oedipodrilus anisognathus	*O. (C.) erichsonianus, O. (P.) forceps, Orconectes (Procericambarus) juvenilis, Orconectes* sp.	Gelder and Williams (2011), Holt (1988a), Holt and Opell (1993), Williams et al. (2013)
O. cuetzalanae	*Procambarus (Villalobosus) cuetzalanae*	Holt (1984b), Holt and Opell (1993)

Table 8. contd....

Table 8. contd.

Genus/species	Host(s)	Reference(s)
O. (=Cambarincola) macbaini	*C. (C.) b. bartonii, C. (C.) b. cavatus, C. (E.) tenebrosus, O. (C.) obscurus, O. (C.) propinquus, Orconectes (Gremicambarus) compressus, O. (G.) virilis, O. (P.) rusticus, Orconectes* sp., *O. (T.) immunis*	Holt (1973b, 1988a), Holt and Opell (1993), Williams et al. (2013)
O. oedipus	*O. (G.) compressus, O. (P.) placidens*	Gelder (1996a), Holt and Opell (1993)
Pterodrilus alcicornus	*C. (C.) b. bartonii, C. (C.) b. cavatus, C. (C.) sciotensis, C. (H.) chasmodactylus, C. (H.) longirostris, C. (H.) longulus, Cambarus (Jugicambarus) parvoculus, C. (P.) acuminatus, C. (P.) robustus, Cambarus (Puncticambarus) veteranus, Cambarus* sp. – Cataloochee morph, *Orconectes (Crockerinus) sanbornii, O. (P.) juvenilis*	Gelder and Williams (2011), Holt (1968b), Holt and Opell (1993), Williams et al. (2013)
P. annulatus	*O. (P.) placidens*	Gelder (1996a)
P. cedrus	*O. (P.) placidus, C. (E.) tenebrosus, O. (P.) juvenilis, O. (P.) rusticus*	Holt (1968b), Holt and Opell (1993)
P. choritonamus	*C. (E.) tenebrosus, O. (P.) placidus, Cambarus (Puncticambarus) extraneus*	Holt (1968b), Holt and Opell (1993)
P. distichus	*C. (C.) b. bartonii, C. (H.) chasmodactylus, C. (P.) robustus, O. (C.) obscurus, O. (C.) propinquus, O. (P.) juvenilis, O. (P.) rusticus, O. (T.) immunis*	Holt (1968b), Holt and Opell (1993), Williams et al. (2013)
P. hobbsi	*C. (C.) b. bartonii, C. (C.) b. cavatus, C. (C.) sciotensis, C. (D.) latimanus, Cambarus (Depressicambarus) striatus, Cambarus (Erebicambarus) rusticiformis, C. (E.) tenebrosus, Cambarus (Glarecola) friaufi, C. (H.) chasmodactylus, C. (H.) longirostris, C. (H.) longulus, C. (J.) distans, C. (J.) parvoculus, C. (P.) extraneus, C. (P.) robustus, C. (P.) veteranus, Camburus* sp., *O. (C.) erichsonianus, O. (P.) forceps, O. (P.) juvenilis, O. (P.) placidens, O. (P.) placidus, O. (P.) rusticus*	Gelder (1996a), Gelder and Williams (2011), Holt (1968b), Holt and Opell (1993), Williams et al. (2013)
P. mexicanus	*Orconectes (Buannulifictus) meeki meeki, O. (F.) limosus, Orconectes (Gremicambarus) nais, Orconectes (Procericambarus) hylas, Orconectes (Procericambarus) luteus, Orconectes (Procericambarus) nana, Orconectes (Procericambarus) neglectus neglectus, Orconectes (Procericambarus) ozarkae, Orconectes (Procericambarus) punctimanus, P. (A.) mexicanus*	Ellis (1919), Gelder (2004), Gelder et al. (2001), Holt (1968b, 1973a)
P. missouriensis	*O. (F.) limosus, O. (P.) luteus*	Gelder et al. (2001), Holt (1968b), Holt and Opell (1993)
P. robinae	*O. (P.) durelli*	Williams and Gelder (2011), Williams et al. (2013)

Table 8. contd....

Table 8. contd.

Genus/species	Host(s)	Reference(s)
P. simondsi	*C. (C.) b. bartonii, C. (D.) latimanus, Cambarus* sp.	Holt (1968b), Holt and Opell (1993)
Pterodrilus sp.	*C. (C.) b. bartonii, C. (D.) latimanus, C. (H.) longirostris, Cambarus (Jugicambarus) carolinus, Cambarus* sp.	Holt (1968b)
Sathodrilus attenuatus	*P. (P.) l. klamathensis, P. (P.) l. leniusculus*	Holt and Opell (1993), Kawai et al. (2004), Ohtaka et al. (2005), Williams et al. (2013)
S. carolinensis	*C. (D.) latimanus, Cambarus* sp.	Holt (1968a), Holt and Opell (1993)
S. chehalisae	*P. (P.) l. leniusculus, P. (P.) l. trowbridgii*	Holt and Opell (1993), Williams et al. (2013)
S. dorfus	*P. (P.) l. klamathensis*	Holt and Opell (1993)
S. (=Cambarincola) elevatus	*C. (L.) diogenes, C. (P.) robustus, O. (C.) obscurus, O. (C.) propinquus, O. (G.) virilis, O. (P.) rusticus, O. (P.) punctimanus, O. (T.) immunis*	Goodnight (1940), Holt (1978), Holt and Opell (1993)
S. hortoni	*C. (L.) diogenes, Cambarus* sp.	Holt and Opell (1993)
S. (=C.) inversus (=virgiliae)	*P. (H.) gambelii, P. (P.) l. klamathensis, P. (P.) l. leniusculus, P. (P.) l. trowbridgii*	Goodnight (1940), Holt and Opell (1993), Williams et al. (2013)
S. lobatus	*P. (P.) l. klamathensis, P. (P.) l. leniusculus*	Holt and Opell (1993), Williams et al. (2013)
S. megadenus	*C. (D.) latimanus*	Holt (1968a), Holt and Opell (1993)
S. nigrofluvius	Unknown	Holt (1989), Holt and Opell (1993)
S. norbyi	*P. (P.) l. klamathensis, P. (P.) l. leniusculus*	Holt and Opell (1993), Williams et al. (2013)
S. okaloosae	*Procambarus (Ortmannicus) evermanni, Procambarus (Pennides) versutus*	Holt and Opell (1993)
S. prostates	*P. (A.) zapoapensis, P. (O.) cuevachicae, P. (V.) hoffmani*	Holt (1973a)
S. rivigeae	*Orconectes (Buannulifictus) palmeri longimanus*	Holt (1988a), Holt and Opell (1993)
S. shastae	*P. (H.) fortis, P. (P.) l. leniusculus*	Gelder and Ferraguti (2001), Holt and Opell (1993)
S. veracruzicus	*Procambarus (Villalobosus) contrerasi, P. (V) hoffmani, Procambarus (Villalobosus) riojai*	Holt (1968a, 1973a)
S. villalobosi	*P. (P.) paradoxus, P. (V.) contrerasi, P. (V.) cuetzalanae, Procamabrus (Villalobosus) erichsoni, P. (V.) riojai*	Holt (1968a, 1973a, 1984b)
S. wardinus	*P. (P.) l. klamathensis*	Holt and Opell (1993)
Sinodrilus (=Branchiobdella) heterorchis	*C. dauricus*	Gelder and Brinkhurst (1990)
Tettodrilus friaufi	*C. (D.) graysoni, C. (D.) striatus, C. (E.) tenebrosus, Orconectes (Procericambarus) mirus, Orconectes (Trisellescens) rhoadesi*	Holt (1968a), Holt and Opell (1993)

Table 8. contd....

Table 8. contd.

Genus/species	Host(s)	Reference(s)
Triannulata magna	*P. (P.) l. klamathensis, P. (P.) l. leniusculus, P. (P.) l. trowbridgii*	Goodnight (1940), Holt (1981), Holt and Opell (1993), Williams et al. (2013)
Uglukodrilus (=*Adenodrilus*) *hemophagus*	*P. (P.) l. klamathensis, P. (P.) l. leniusculus*	Gelder and Ferraguti (2001), Holt (1977), Williams et al. (2013)
Xironodrilus appalachius	*C. (C.) b. bartonii, C. (H.) longirostris, Cambarus (Puncticambarus) chaugaensis, C. (P.) reburrus, Cambarus* sp. – Cataloochee morph	Brown et al. (2012), Gelder and Williams (2011), Goodnight (1943), Williams et al. (2013)
X. bashaviae	*C. (C.) b. bartonii*	Holt and Weigl (1979)
X. dentatus (=*pulcherrimus dentatus*)	*O. (B.) p. longimanus*	Goodnight (1940)
X. formosus	*C. (C.) b. bartonii, C. (C.) b. cavatus, C. (H.) longulus, Cambarus (Jugicambarus) setosus, C. (P.) acuminatus, O. (C.) propinquus, O. (G.) virilis, O. (P.) longidigitus, O. (P.) luteus, O. (P.) medius, O. (P.) neglectus, O. (P.) placidus, O. (P.) punctimanus, O. (P.) rusticus, O. (T.) immunis, O. (P.) placidens*	Ellis (1919), Gelder (1996a), Goodnight (1940), Hobbs, Jr. et al. (1967), Holt (1973b), Williams et al. (2013)
X. (=*Branchiobdella*) *pulcherrimus*	*C. (C.) b. bartonii, C. (C.) carinirostris, Cambarus (Jugicambarus) dubius, Orconectes (Crockerinus) obscurus*	Goodnight (1940), Hall (1915)
Xironodrilus sp.	*C. (H.) chasmodactylus*	DeWitt et al. (2013)
Xironogiton cassiensis	*P. (P.) l. leniusculus*	Holt (1974), Geasa (2014)
X. fordi	*P. (H.) connectens, P. (H.) gambelii*	Holt (1974)
X. kittitasi	*P. (P.) l. leniusculus*	Holt (1974), Gelder and Hall (1990), Williams et al. (2013)
X. occidentalis	*P. (H.) gambelii, P. (P.) l. klamathensis, P. (P.) l. leniusculus, P. (P.) l. trowbridgii*	Ellis (1919), Goodnight (1940), Williams et al. (2013)
X. victoriensis (=*oregonensis*) (=*instabilis*)	*P. (P.) l. leniusculus, P. (S.) clarkii, C. (C.) b. bartonii, C. (C.) b. cavatus, C. (C.) carinirostris, C. (C.) sciotensis, C. (H.) longulus, C. (J.) carolinus, C. (P.) robustus, O. (C.) obscurus, O. (F.) limosus, O. (P.) juvenilis, P. (H.) gambelii, P. (P.) l. klamathensis, P. (P.) l. trowbridgii*	Ellis (1919), Gelder and Hall (1990), Gelder et al. (2001, 2012), Goodnight (1940), Hobbs, Jr. et al. (1967), Holt (1973b), James et al. (2015), Oberkofler et al. (2002), Ohtaka et al. (2005), Vedia et al. (2014), Williams et al. (2013)

(Gelder 2014). These include *Cambarincola mesochoreus* originally described from *Orconectes* sp. in Indiana, USA (Hoffman 1963) which has been reported on a number of crayfish species in North America and Italy (Gelder et al. 1994, 1999, 2001), *C. okadai* and *Sathodrilus attenuatus* transferred from the USA to Japan on *Pacifastacus leniusculus* subsp. (Gelder and Ohtaka 2000a, Kawai et al. 2004, Ohtaka et al. 2005), *C. okadai* reported on non-native *P. leniusculus* in the United Kingdom (James et al. 2015) and *Xironogiton victoriensis* originally described in the USA but reported in mainland Europe (Ďuriš et al. 2006, James et al. 2015).

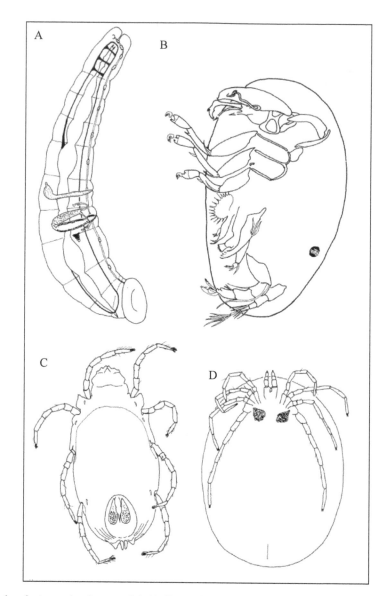

Figure 5. Examples of ectoparasites from crayfish. (A) Generalised view of a typical branchiobdellid worm (after Holt 1986). (B) Generalised entocytherid ostracod with right valve of shell removed (after Hobbs Jr. 1971). (C) Ventral view of a mite, *Astacopsiphagus parasiticus* (after Womersley 1941–1943). (D) Ventral view of a female *Astacocroton molle* (Arachnida) (after Haswell 1922). Not to scale.

Phylum Acanthocephala

Acanthocephalans are obligate endoparasites with complex lifecycles that usually involve a vertebrate final host and an arthropod as an intermediate host where they occur as cystacanths in the haemocoel. Although the higher phylogenetic relationships of acanthocephalans and other taxa are subject to some controversy, they appear to be related to the Rotifera (Fontaneto and Jondelius 2011, Gazi et al. 2012, Verweyen et al. 2011, Weber et al. 2013, Zrzavý et al. 1998). An up to date listing of extant acanthocephalans described to date can be found in Amin (2013). Five acanthocephalans have been reported from crayfish, including *Filicollis anatis* (Family Polymorphidae) in *Astacus astacus* (Golvan 1961), *Ibirhynchus* (=*Southwellina*) *dimorpha* (Family Polymorphidae) in *Procambarus* (*Scapulicambarus*) *clarkii* and *Cambarellus* (*Dirigicambarus*) *shufeldtii* (García-Varela et al. 2011, Richardson and Font 2006, Schmidt 1973), *Polymorphus biziuarae* (Family Polymorphidae) in *Cherax destructor* (Johnston and Edmonds 1948), *P. minutus* (=*boschadis*) in *Orconectes* (*Faxonius*) *limosus* and *Astacus astacus* (Golvan 1961, Unestam 1973) and *Neoechinorhynchus* (*Neoechinorhynchus*) *rutili* (subfamily Neoechinorhynchinae) in *Pacifastacus* (*Pacifastacus*) *leniusculus trowbridgii* by Merritt and Pratt (1964). Merritt and Pratt (1964) considered that crayfish were a paratenic host for *N. rutili* as only three out of 154 *P. (P.) l. trowbridgii* were infected. The importance of crayfish in the lifecycles of other acanthocephalans remains unknown.

Phylum Rotifera

Rotifers are normally found in the gill cavity of crayfish and are not normally considered as problematic for their hosts, being viewed as epizoic rather than parasitic (May 1989). Rotifers represented by seven families have been recorded on crayfish, namely Branchionidae (*Branchionius*), Notommatidae (*Cephalodella*) Dicranophoridae (*Dicranophorus*), Lecanidae (*Lecane*), Lepadellidae (*Lepadella*) Epiphanidae (*Mikrocodides*) and Philodinidae (*Embata*). Few recent records of rotifer infections in crayfish are noted, with most from the early twentieth century. No pathology has been recorded associated with rotifers in crayfish. Updated nomenclature of all rotifers is provided by Segers (2002, 2007).

Table 9. List of rotifers, in alphabetical order, reported from crayfish.

Genus/species	Host(s)	Reference(s)
Brachionius sp.	*Pacifastacus* (*Pacifastacus*) *leniusculus leniusculus, Procambarus* (*Girardiella*) *simulans*	Cuellar et al. (2002), Lahser, Jr. (1975)
Cephalodella (=*Diaschiza*) *jakubskii* (=*crassipes*)	*Astacus astacus, Austropotamobius torrentium*	Jersabek (2002), May (1989), Segers (2007)
Dicranophorus cambari	*A. torrentium, Orconectes* (*Faxonius*) *limosus*	Hauer (1959), Wulfert (1957)
D. hauerianus	*A. astacus, Astacus leptodactylus, O. (F.) limosus*	Boshko (1980), May (1989), Wiszniewski (1939)
D. siedleckii (=*hauerianus* var. *brachygnathus*)	*A. astacus, A. leptodactylus*	Wiszniewski (1939, 1953)
Embata (=*Callidina*) *parasitica*	*Cambarus* (*Ortmannicus*) *blandingii, Cambarus* (*Puncticambarus*) *acuminatus*	Allen (1933)
Lecane (=*Monostyla*) sp.	*P. (G.) simulans*	Lahser, Jr. (1975), Segers (2007)
Lepadella (*Xenolepadella*) *astacicola*	*A. astacus, A. leptodactylus, A. torrentium, O. (F.) limosus*	Bertani et al. (2011), Boshko (1980), Hauer (1959), Segers (2007), Wiszniewski (1939)
L. (X.) borealis	*O. (F.) limosus*	May (1989), Segers (2007)
L. (X.) branchicola	*A. astacus, A. leptodactylus*	Boshko (1980), Segers (2007), Wiszniewski (1939)
L. (X.) lata (=*raja*) (=var. *sinuata*)	*A. astacus, A. leptodactylus, O. (F.) limosus*	Boshko (1980), May (1989), Segers (2007), Wiszniewski (1939)
L. nana	*A. leptodactylus*	Boshko (1980)
L. (X.) parasitica	*A. astacus, A. torrentium*	Hauer (1959), Segers (2007), Wiszniewski (1939)
Mikrocodides sp.	*P. (G.) simulans*	Lahser, Jr. (1975)
Unidentified Rotifers	*Engaeus cisternarius, Engaeus fossor, P. (S.) clarkii*	Quaglio et al. (2006b), Suter and Richardson (1977)

Phylum Arthropoda – subclass Copepoda

Crayfish tunnels are utilised by a range of free-living copepods as refugia, including a number that also occur on the gills and carapace of crayfish (Reid 2001, Reid et al. 2006). Despite the large number of copepods described worldwide, few have been recorded on crayfish; those that have, tend to be viewed as commensal or accidental on their hosts. To date, no Copepoda have been reported as associated with the Parastacidae. Typically, lifecyles of these copepods are complex involving a series of moults between naupliar, copepodids and adult stages. Reports of copepods associated with crayfish include *Nitocra* (=*Nitokra*) *divaricata* (Family Ameiridae) on *Astacus astacus*, *Astacus leptodactylus* (Subchev and Stanimirova 1998) and *Austropotamobius torrentium* (Defaye 1996), *Nitocra divaricata caspica* on *A. leptodactylus caspius* (Huys et al. 2009), *Nitocra hibernica* on *A. leptodactylus* and *Pacifastacus* (*Pacifastacus*) *leniusculus leniusculus* (Huys et al. 2014), *Acanthocyclops* sp. (Family Cyclopidae) on *A. leptodactylus* (Huys et al. 2014), *Canthocamptus staphylinus* (Family Canthocamptidae) on *A. astacus* (Wiszniewski 1939), *Attheyella crassa* (Family Canthocamptidae) and *A. trispinosa* on *A. astacus* (Wiszniewski 1939), *A. carolinensis* on *Cambarus* (*Jugicambarus*) *distans* (Bowman et al. 1968) and *A. pilosa* on *Cambarus* (*Cambarus*) *bartonii bartonii*, *Cambarus* (*Cambarus*) *sciotensis*, *Cambarus* (*Erebicambarus*) *tenebrosus*, *Cambarus* (*Jugicambarus*) *asperimanus* and *Orconectes* (*Procericambarus*) *rusticus* (Bowman et al. 1968, Prins 1964). The reports of *A. carolinensis* on crayfish by Prins (1964) were misidentified and should be considered as *A. pilosa* according to Bowman et al. (1968). Additional reports of commensal copepods on crayfish include *Attheyella northumbrica* (=*dentata*) and *Bryocamptus minutus* (Family Canthocamptidae) on gills of unidentified crayfish (Reid 2001).

Phylum Arthropoda – class Ostracoda

Ostracods, obligate associates of crayfish, belong to the family Entocytheridae and further subdivided into two subfamilies, namely the subfamily Entocytherinae containing the genera *Ankylocythere*, *Ascetocythere*, *Cymocythere*, *Dactylocythere*, *Donnaldsoncythere*, *Entocythere*, *Geocythere*, *Harpagocythere*, *Hartocythere*, *Litocythere*, *Lordocythere*, *Okriocythere*, *Ornithocythere*, *Phymocythere*, *Plectocythere*, *Psittocythere*, *Rhadinocythere*, *Sagittocythere*, *Saurocythere*, *Thermastrocythere*, *Uncinocythere* and *Waltoncythere*, and the subfamily Notocytherinae containing the genera *Chelocythere*, *Elachistocythere*, *Herpetocythere*, *Hesperocythere*, *Laccocythere*, *Lichnocythere*, *Notocythere* and *Riekocythere* (Table 10, Fig. 5b). Members of the subfamily Notocytherinae occur on antipodean Parastacidae whilst members of the subfamily Entocytherinae occur predominately on the Cambaridae from the Americas. Major reviews of crayfish ostracods include those by Hart, Jr. (1962), Hart, Jr. and Hart (1967), Hart and Hart, Jr. (1974), Hobbs, Jr. (1971), Hobbs, Jr. and Hart, Jr. (1966), Hobbs, Jr. and Hobbs III (1970) and Mestre et al. (2014). *Ankylocythere sinuosa*, originally described from the USA, has been noted on established populations of *Procambarus* (Scapulicambarus) *clarkii* in the Iberian Penninsula (Aguilar-Alberola et al. 2011) and in Greater London, United Kingdom (Huys et al. 2014). No other ostracods were detected on *P.* (S.) *clarkii* during the survey of Huys et al. (2014) and the ostracods do not appear to have transferred to native crayfish species in the area. In addition, *Uncinocythere occidentialis* has been recorded in non-native *Pacifastacus* (*Pacifastacus*) *leniusculus leniusculus* from two sites in the Greater London area, United Kingdom (Huys et al. 2014) and from non-native *Pacifastacus* (*Pacifastacus*) *leniusculus trowbridgii* in Japan (Smith and Kamiya 2001). The same or similar species is likely to be present in most populations of signal crayfish in the United Kingdom (Longshaw et al. 2012a). Cuellar et al. (2002) reported the presence of *Eucypris virens* on the gills of *P. leniusculus* under culture conditions in Spain; however, the photograph provided in the manuscript of a lateral view of the ostracod does not appear to conform to that species and is more likely to be *U. occidentialis*. *Eucypris virens* has a more convex dorsal edge and a concave ventral edge in lateral view (Henderson 1990) compared with *U. occidentialis* which is clearly more rounded in lateral view (Hart et al. 1985), similar to the image provided by Cuellar et al. (2002). It should be noted that *E. virens* appears to be a species complex with more than 35 cryptic species in Europe which is normally recorded in grassy temporary pools (Koenders et al. 2012) and not as a commensal of crayfish.

Table 10. List of ostracods, in alphabetical order, reported from crayfish.

Genus/species	Crayfish host(s)	Reference(s)
Ankylocythere (=Entocythere) ancyla	*Cambarellus (Dirigicambarus) shufeldtii, Cambarellus (Pandicambarus) ninae, Cambarellus (Pandicambarus) puer, Cambarus (Depressicambarus) latimanus, Cambarus (Depressicambarus) striatus, Fallicambarus (Creaserinus) fodiens, Faxonella clypeata, Orconectes (Faxonius) limosus, Procambarus (Capillicambarus) incilis, Procambarus (Girardiella) simulans, Procambarus (Hagenides) advena, Procambarus (Hagenides) caritus, Procambarus (Hagenides) pygmaeus, Procambarus (Hagenides) talpoides, Procambarus (Hagenides) truculentus, Procambarus (Leconticambarus) barbatus, Procambarus (Leconticambarus) pubischelae pubischelae, Procambarus (Leconticambarus) pubischelae deficiens, Procambarus (Ortmannicus) acutus, Procambarus (Ortmannicus) enoplosternum, Procambarus (Ortmannicus) epicyrtus, Procambarus (Ortmannicus) litosternum, Procambarus (Ortmannicus) lunzi, Procambarus (Ortmannicus) pubescens, Procambarus (Ortannicus) seminolae, Procambarus (Scapulicambarus) clarkii, Procambarus (Scapulicambarus) howellae, Procambarus (Scapulicambarus) paeninsulanus, Procambarus (Scapulicambarus) troglodytes*	Andolshek and Hobbs, Jr. (1986), Baker (1969), Crawford, Jr. (1965), Hobbs, Jr. and Peters (1977), Huys et al. (2014), Peters and Pugh (1999)
A. barbouri	*Procambarus (Pennides) roberti*	Figueroa and Hobbs, Jr. (1974)
A. (=E.) bidentata	*Procambarus (Austrocambarus) llamasi, Procambarus (Austrocambarus) mexicanus, Procambarus (Austrocambarus) mirandai, Procambarus (Austrocambarus) oaxacaereddelli, Procambarus (Austrocambarus) pilosimanus, Procambarus (Austrocambarus) rodriguezi, Procambarus (Austrocambarus) ruthveni, Procambarus (Austrocambarus) vazquezae, Procambarus (Austrocambarus) veracruzanus, Procambarus (Austrocambarus) zapoapensis*	Hart, Jr. (1962), Hobbs, Jr. (1971, 1973)
A. burkeorum	*Cambarus (Lacunicambarus) ludovicianus, Cambarus (Lacunicambarus) diogenes, C. (D.) striatus*	Hobbs III (1971), Hobbs, Jr. and Peters (1991)
A. carpenteri	*C. (L.) diogenes*	Hobbs, Jr. and McClure (1983)
A. chipola	*C. (L.) diogenes, Procambarus (Hagenides) rogersiexpletus, Procambarus (Leconticambarus) kilbyi*	Hobbs III (1978)
A. (=E.) copiosa	*P. (S.) clarkii, Orconectes (Gremicambarus) virilis, Cambarus (Jugicambarus) batchi*	Hobbs, Jr. and Peters (1991), Huys et al. (2014)
A. (=E.) cubensis	*Procambarus (Austrocambarus) cubensis cubensis*	Hobbs, Jr. (1971)
A. ephydra	*Cambarus sp., Cambarus (Erebicambarus) tenebrosus, Orconectes (Crockerinus) erichsonianus, Orconectes (Gremicambarus) compressus, Orconectes (Procericambarus) juvenilis, Orconectes (Procericambarus) mirus*	Hart and Hart, Jr. (1971)
A. freyi	*Cambarus (Depressicambarus) reflexus, C. (D.) striatus, C. (L.) diogenes, Fallicambarus (Creaserinus) byersi, F. clypeata, P. (H.) talpoides, P. (L.) barbatus, Procambarus (Leconticambarus) shermani, Procambarus (Ortmannicus) acutissimus, P. (O.) lunzi, P. (O.) seminolae, P. (S.) howellae*	Hobbs III (1978), Andolshek and Hobbs, Jr. (1986)
A. (=E.) hamata	*Procambarus (Austrocambarus) atkinsoni, P. (A.) C. cubensis*	Hobbs, Jr. (1971)
A. harmani	*C. (P.) puer, Cambarus (Depressicambarus) sp., C. (L.) diogenes, C. (L.) ludovicianus, Fallicambarus (Fallicambarus) dissitus, Orconectes sp., Procambarus (Capillicambarus) hinei, P. (O.) acutus*	Hart and Hart, Jr. (1974)

Table 10. contd....

Table 10. contd.

Genus/species	Crayfish host(s)	Reference(s)
A. (=E.) heterodonta (=talirotunda)	Cambarellus (Cambarellus) chapalanus, Cambarellus (Cambarellus) montezuma, Cambarellus (Cambarellus) patzcuarensis, Cambarellus (Cambarellus) zempoalensis	Hart and Hart, Jr. (1974), Hobbs, Jr. (1971)
A. (=E.) hobbsi	C. (D.) reflexus, Cambarus (Puncticambarus) acuminatus, F. (C.) fodiens, F. clypeata, P. (H.) advena, P. (H.) caritus, Procambarus (Hagenides) rogersi rogersi, P. (H.) talpoides, Procambarus (Leconticambarus) alleni, P. (L.) kilbyi, P. (L.) P. pubischelae, Procambarus (Ortmannicus) blandingii, P. (O.) enoplosternum, Procambarus (Ortmannicus) leonensis, P. (O.) lunzi, Procambarus (Ortmannicus) pycnogonopodus, P. (O.) seminolae, Procambarus (Pennides) spiculifer, P. (S.) howellae, P. (S.) paeninsulanus, P. (S.) troglodytes	Andolshek and Hobbs, Jr. (1986), Hart and Hart, Jr. (1974)
A. hyba	Cambarus (Depressicambarus) sp., ?C. (D.) striatus, Orconectes sp., O. (C.) erichsonianus, O. (G.) compressus, O. (P.) mirus, Orconectes (Procericambarus) placidus, Orconectes (Procericambarus) rusticus, Procambarus sp., P. (H.) r. rogersi	Hart and Hart, Jr. (1974), Hobbs, Jr. and Walton (1963a)
A. krantzi	C. (L.) diogenes, F. (C.) byersi, F. clypeata, Procambarus (Acucauda) fitzpatricki, Procambarus (Leconticambarus) econfinae, Procambarus (Ortmannicus) bivittatus, Procambarus (Ortmannicus) evermanni	Hobbs III (1978)
A. maya	Procambarus sp., Procambarus (Austrocambarus) oaxacae oaxacae, P. (A.) o. reddelli, Procambarus (Procambarus) digueti	Hobbs, Jr. (1971, 1973)
A. prolata	Cambarus (Lacunicambarus) miltus	Hobbs, Jr. and Peters (1991)
A. (=E.) sinuosa (=cambari)	F. (C.) fodiens, O. (G.) virilis, Procambarus (Girardiella) regiomontanus, P. (S.) clarkii, P. (G.) simulans, P. (O.) acutus, Procambarus (Ortmannicus) cuevachicae, Procambarus (Ortmannicus) gonopodocristatus, Pacifastacus (Pacifastacus) leniusculus leniusculus, Procambarus (Paracambarus) paradoxus	Aguilar-Alberola et al. (2011), Hobbs, Jr. (1971), Holt (1973a), Huys et al. (2014), Lahser, Jr. (1975), Mestre et al. (2011, 2014), Peters and Pugh (1999), Young (1971)
A. spargosis	P. (H.) advena, P. (H.) pygmaeus, P. (L.) barbatus, P. (O.) enoplosternum, P. (O.) epicyrtus, P. (O.) litosternum, P. (O.) lunzi, P. (O.) seminolae, P. (S.) troglodytes	Andolshek and Hobbs, Jr. (1986)
A. tallapoosa	Cambarus (Depressicambarus) halli, C. (D.) latimanus, P. (P.) spiculifer	Hart and Hart, Jr. (1971)
A. talulus	P. (L.) alleni	Hart and Hart, Jr. (1974)
A. (=E.) telmoecea	C. (L.) diogenes, C. (D.) latimanus, C. (P.) acuminatus, F. clypeata, Procambarus (Ortmannicus) sp., P. (O.) acutus, P. (O.) enoplosternum, P. (O.) pubescens, P. (P.) spiculifer, P. (S.) howellae, P. (S.) paeninsulanus, P. (S.) troglodytes	Hobbs, Jr. and Peters (1977)
A. (=E.) tiphophila	Cambarus (Depressicambarus) truncatus, Cambarus (Jugicambarus) dubius, C. (L.) diogenes, F. clypeata, P. (O.) enoplosternum, P. (O.) pubescens, P. (P.) spiculifer, P. (S.) howellae, P. (S.) troglodytes	Andolshek and Hobbs, Jr. (1986), Hobbs, Jr. and Peters (1991)
A. toltecae	Procambarus (Ortmannicus) toltecae, Procambarus (Ortmannicus) villalobosi	Hobbs, Jr. (1971)
A. tridentata	C. (L.) diogenes, Orconectes (Trisellescens) immunis	Hart (1964), Hart and Hart, Jr. (1974)
A. villalobosi	P. (A.) llamasi, P. (A.) pilosimanus	Hobbs, Jr. (1971)

Table 10. contd....

Table 10. contd.

Genus/species	Crayfish host(s)	Reference(s)
Ascetocythere (=*Entocythere*) *asceta*	*Cambarus* (*Jugicambarus*) *carolinus*, *C.* (*J.*) *dubius*	Hart and Hart, Jr. (1974), Hobbs, Jr. et al. (1967)
A. batchi	*Cambarus* (*Jugicambarus*) *parvoculus*	Hobbs, Jr. and Walton (1968)
A. bouchardi	*Cambarus* (*Hiaticambarus*) *longirostris*, *C.* (*J.*) *parvoculus*	Hobbs, Jr. and Walton (1975)
A. coryphodes	*Cambarus* (*Cambarus*) *bartonii bartonii*, *Cambarus* (*Jugicambarus*) *monongalensis*	Hobbs, Jr. and Hart, Jr. (1966)
A. cosmeta	*Cambarus* (*Jugicambarus*) *asperimanus*, *C.* (*J.*) *carolinus*, *C.* (*J.*) *dubius*	Hobbs, Jr. and Hart, Jr. (1966), Hobbs, Jr. and Peters (1989)
A. didactylata	*Cambarus* (*Cambarus*) *bartonii cavatus*, *C.* (*J.*) *dubius*, *C.* (*J.*) *parvoculus*	Hobbs, Jr. and Hart, Jr. (1966), Hobbs, Jr. and Peters (1991)
A. hoffmani	*C.* (*J.*) *dubius*	Hobbs, Jr. and Hart, Jr. (1966)
Ascetocythere holti	*Cambarus* (*Depressicambarus*) *sphenoides*, *Cambarus* (*Jugicambarus*) *bouchardi*, ?*Cambarus* (*Jugicambarus*) *distans*, ?*Cambarus* (*Jugicambarus*) *obeyensis*	Hobbs, Jr. and Walton (1970)
A. hyperoche	*C.* (*C.*) *b. cavatus*, *C.* (*H.*) *longirostris*, *O.* (*P.*) *juvenilis*, *Orconectes* (*Procericambarus*) *spinosus*	Hobbs, Jr. and Hart, Jr. (1966), Hobbs, Jr. and Peters (1989)
A. jezerinaci	*C.* (*C.*) *b. bartonii*, *C.* (*J.*) *dubius*	Hobbs, Jr. and McClure (1983)
A. lita	*C.* (*D.*) *sphenoides*, *C.* (*J.*) *distans*, *Cambarus* (*Puncticambarus*) *buntingi*	Hobbs, Jr. and Hobbs III (1970), Hobbs, Jr. and Walton (1975)
A. myxoides	*C.* (*C.*) *b. bartonii*, *Cambarus* (*Cambarus*) *carinirostris*, *C.* (*J.*) *dubius*, *C.* (*J.*) *monongalensis*	Hobbs, Jr. and Hart, Jr. (1966), Hobbs, Jr. and Peters (1993)
A. ozalea	*C.* (*J.*) *dubius*	Hobbs, Jr. and Hart, Jr. (1966)
A. pseudolita	*C.* (*J.*) *distans*, *C.* (*P.*) *buntingi*	Hobbs, Jr. and Walton (1975)
A. riopeli	*C.* (*J.*) *distans*, *C.* (*J.*) *dubius*, *C.* (*P.*) *buntingi*, *Cambarus* (*Puncticambarus*) *robustus*, *O.* (*P.*) *juvenilis*, *O.* (*P.*) *rusticus*	Hobbs, Jr. and Peters (1993), Hobbs, Jr. and Walton (1976)
A. sclera	*C.* (*J.*) *dubius*, *C.* (*P.*) *robustus*, *O.* (*P.*) *juvenilis*	Hobbs, Jr. and Hart, Jr. (1966), Hobbs, Jr. and Peters (1993)
A. stockeri	*C.* (*C.*) *b. cavatus*, *C.* (*J.*) *dubius*, *C.* (*L.*) *diogenes*	Hobbs, Jr. and Peters (1989)
A. triangulata	*C.* (*D.*) *sphenoides*, *C.* (*J.*) *parvoculus*	Hobbs, Jr. and Walton (1975)
A. veruta	*C.* (*H.*) *longirostris*, *C.* (*J.*) *distans*	Hobbs, Jr. and Walton (1975)
Chelocythere kalganensis	*Cherax* sp.	Hart, Jr. and Hart (1967)

Table 10. contd....

Table 10. contd.

Genus/species	Crayfish host(s)	Reference(s)
Cymocythere clavata	*C. (C.) b. bartonii, C. (J.) asperimanus*	Crawford, Jr. (1965), Hart and Hart, Jr. (1974), Hobbs, Jr. and Hart, Jr. (1966)
C. (=Entocythere) cyma	*C. (C.) b. bartonii, Cambarus (Depressicambarus) sp., C. (D.) striatus, C. (H.) longirostris, C. (J.) carolinus, C. (L.) diogenes, Cambarus (Puncticambarus) extraneus, O. (C.) erichsonianus, P. (P.) spiculifer*	Hart and Hart, Jr. (1974), Hobbs, Jr. and Hart, Jr. (1966)
C. gonia	*C. (C.) b. bartonii, C. (C.) b. cavatus, Cambarus (Depressicambarus) deweesae, C. (D.) halli, C. (D.) striatus, Cambarus (Hiaticambarus) girardianus, C. (H.) longirostris, C. (L.) diogenes, C. (P.) extraneus, Cambarus (Tubericambarus) acanthura, O. (C.) erichsonianus, Orconectes (Procericambarus) forceps, O. (P.) juvenilis, O. (P.) spinosus, P. (O.) acutus, Procambarus (Ortmannicus) lophotus*	Hart and Hart, Jr. (1974), Hobbs, Jr. and Hart, Jr. (1966), Hobbs, Jr. and Peters (1993)
Dactylocythere amicula	*C. (C.) b. bartonii, Cambarus (Depressicambarus) sp., Cambarus (Jugicambarus) sp., C. (P.) robustus, O. (P.) juvenilis*	Hart, Jr. and Hart (1966)
D. amphiakis	*Cambarus sp., C. (E.) tenebrosus, Orconectes sp., Orconectes (Crockerinus) tricuspis*	Hart, Jr. and Hart (1966)
D. apheles	*C. (J.) carolinus*	Hobbs, Jr. and Walton (1976)
D. (=Entocythere) arcuata	*C. (E.) tenebrosus*	Hart, Jr. and Hobbs, Jr. (1961)
D. astraphes	*Cambarus (Depressicambarus) graysoni, C. (D.) sphenoides, C. (D.) striatus, Cambarus (Hiaticambarus) sp., C. (H.) girardianus C. (J.) parvoculus, Cambarus (Jugicambarus) unestami, O. (P.) spinosus, Orconectes (Trisellescens) alabamensis*	Hobbs, Jr. and Walton (1977)
D. banana	*C. (C.) b. bartonii, Cambarus (Hiaticambarus) longulus, C. (P.) acuminatus*	Hart and Hart, Jr. (1971)
D. brachydactylus	*C. (C.) b. bartonii, C. (H.) longirostris, ?O. (G.) virilis, O. (P.) forceps*	Hobbs, Jr. and Walton (1976)
D. brachystrix	*Cambarus sp., C. (C.) b. bartonii, O. (P.) spinosus*	Hobbs, Jr. and Walton (1966)
D. charadra	*Cambarus sp., C. (C.) b. bartonii, C. (H.) longirostris, C. (P.) robustus*	Hobbs III (1971)
D. (=E.) chalaza	*C. (C.) b. bartonii, C. (H.) longirostris, C. (H.) longulus, C. (J.) carolinus, C. (J.) dubius, C. (P.) acuminatus, C. (P.) robustus*	Hobbs, Jr. et al. (1967), Walton and Hobbs, Jr. (1971)
D. (=E.) chelomata	*C. (C.) b. bartonii, C. (H.) longirostris, O. (P.) juvenilis*	Hobbs, Jr. and Peters (1977)
D. coloholca	*Cambarus sp., C. (D.) sphenoides, C. (H.) longirostris, Cambarus (Jugicambarus) crinipes, C. (J.) distans, C. (J.) dubius, C. (L.) diogenes, O. (C.) erichsonianus*	Hobbs, Jr. and Peters (1993)
D. cooperorum	*Barbicambarus cornutus*	Hobbs, Jr. and Walton (1968)
D. corvus	*C. (D.) graysoni, Cambarus (Erebicambarus) rusticiformis, C. (E.) tenebrosus, Cambarus (Puncticambarus) cumberlandensis, O. (P.) placidus*	Hobbs, Jr. and Walton (1977)

Table 10. contd....

Table 10. contd.

Genus/species	Crayfish host(s)	Reference(s)
D. crawfordi	*Cambarus (Cambarus) ortmanni, C. (E.) tenebrosus, C. (L.) diogenes, F. (C.) fodiens, Orconectes (Crockerinus) sanbornii, O. (P.) rusticus*	Hobbs, Jr. and Peters (1993)
D. crena	*C. (D.) striatus, C. (H.) longirostris, C. (L.) diogenes, C. (T.) acanthura*	Hobbs, Jr. and Peters (1993), Hobbs, Jr. and Walton (1975)
D. cryptoteresis	*C. (L.) diogenes*	Hobbs, Jr. and Peters (1993)
D. (=E.) daphnoides	*C. (C.) b. bartonii, Cambarus (Cambarus) sciotensis, C. (D.) striatus, Cambarus (Hiaticambarus) chasmodactylus, C. (H.) longirostris, C. (J.) dubius, Cambarus (Puncticambarus) sp., C. (P.) robustus, Cambarus (Puncticambarus) veteranus, O. (C.) sanbornii, O. (G.) compressus, Orconectes (Procericambarus) hylas, O. (P.) juvenilis, O. (P.) placidus, Orconectes (Procericambarus) putnami, Orconectes (Procericambarus) quadruncus*	Hobbs, Jr. and Peters (1989, 1993)
D. demissa	*C. (D.) sphenoides, C. (J.) distans, C. (J.) parvoculus, Cambarus (Veticambarus) pristinus*	Hobbs, Jr. and Walton (1976)
D. enoploholca	*C. (H.) longirostris, Cambarus (Puncticambarus) sp.*	Hobbs, Jr. and Walton (1970)
D. exoura	*C. (C.) ortmanni, C. (E.) tenebrosus, C. (L.) diogenes, O. (P.) rusticus*	Hart, Jr. and Hart (1966)
D. (=E.) falcata	*C. (C.) b. bartonii, Cambarus (Depressicambarus) sp., C. (D.) latimanus, C. (H.) longirostris, C. (J.) parvoculus, C. (L.) diogenes, Cambarus (Puncticambarus) sp., C. (P.) acuminatus, O. (C.) erichsonianus, Orconectes (Crockerinus) propinquus, O. (P.) forceps, O. (P.) hylas, O. (P.) juvenilis, O. (P.) mirus, O. (P.) placidus, O. (P.) spinosus, P. (O.) acutus*	Hobbs, Jr. and Peters (1977)
D. guyandottae	*C. (J.) dubius, C. (L.) diogenes*	Hobbs, Jr. and Peters (1991)
D. isabelae	*Cambarus (Depressicambarus) sp.*	Hobbs, Jr. and Peters (1977)
D. jeanae	*C. (C.) b. bartonii, C. (H.) longulus, C. (L.) diogenes, C. (P.) acuminatus, Orconectes (Crockerinus) virginiensis, P. (O.) acutus*	Hobbs, Jr. (1967), Hobbs, Jr. and Peters (1977)
D. koloura	*Orconectes sp., O. (P.) rusticus*	Hart and Hart, Jr. (1971)
D. lepta	*C. (J.) dubius*	Hobbs, Jr. and Peters (1991)
D. (=E.) leptophylax	*C. (C.) b. bartonii, Cambarus (Depressicambrus) sp., C. (D.) latimanus, C. (J.) asperimanus, Procambarus (Pennides) raneyi, P. (P.) spiculifer, Procambarus (Pennides) versutus, O. (P.) forceps*	Hobbs, Jr. and Peters (1977)
D. macroholca	*Cambarus sp., Cambarus (Depressicambarus) sp., C. (D.) graysoni, C. (D.) striatus, C. (E.) rusticiformis, C. (E.) tenebrosus, Cambarus (Glarecola) friaufi, Cambarus (Hiaticambarus) sp., C. (H.) girardianus, Cambarus (Jugicambarus) sp., C. (J.) batchi, C. (J.) dubius, C. (L.) diogenes, Orcoenectes sp., Orconectes (Crockerinus) shoupi, O. (G.) compressus, O. (P.) forceps, O. (P.) juvenilis, O. (P.) placidus, O. (P.) putnami, O. (P.) rusticus, O. (P.) spinosus*	Hobbs, Jr. and Hobbs III (1970), Hobbs, Jr. and Peters (1993)

Table 10. contd....

Table 10. contd.

Genus/species	Crayfish host(s)	Reference(s)
D. (=E.) mecoscapha	*Cambarus (Depressicambarus)* sp., *C. (H.) girardianus,* *C. (H.) longirostris, O. (C.) erichsonianus, O. (G.) compressus,* *O. (P.) mirus, O. (P.) spinosus, P. (L.) barbatus, P. (O.) lophotus,* *P. (O.) lunzi*	Hobbs, Jr. and Peters (1991)
D. megadactylus	*C. (C.) b. bartonii, C. (P.) acuminatus, P. (O.) acutus*	Hart and Hart, Jr. (1971)
D. myura	*? C. (J.) carolinus, C. (J.) dubius*	Hobbs, Jr. and Peters (1993), Hobbs, Jr. and Walton (1970)
D. pachysphyrata	*Cambarus* sp.	Hobbs, Jr. and Walton (1966)
D. peedeensis	*Cambarus (Puncticambarus)* sp., *P. (O.) acutus*	Hobbs, Jr. and Peters (1977)
D. phoxa	*C. (C.) b. bartonii*	Hobbs, Jr. (1967)
D. prinsi	*Cambarus* sp., *C. (C.) b. bartonii, C. (J.) asperimanus*	Hobbs, Jr. and Peters (1977), Hobbs, Jr. and Walton (1968)
D. (=E.) prionata	*C. (C.) ortmanni, C. (D.) graysoni, C. (D.) striatus,* *C. (E.) tenebrosus, C. (L.) diogenes, O. (G.) compressus,* *Orconectes (Orconectes) australis packardi, O. (P.) putnami,* *O. (P.) rusticus*	Hobbs, Jr. and Peters (1993)
D. prominula	*C. (C.) b. bartonii, C. (D.) latimanus, C. (D.) striatus, C. (J.) distans,* *C. (J.) dubius, C. (H.) girardianus, C. (L.) diogenes, Cambarus* *(Puncticambarus)* sp., *O. (C.) erichsonianus, O. (P.) forceps,* *O. (P.) spinosus, P. (O.) lophotus*	Hobbs, Jr. and Walton (1977)
D. pughae	*C. (D.) sphenoides, Cambarus (Jugicambarus)* sp., *C. (L.) diogenes*	Hobbs, Jr. and Hobbs III (1970)
D. pygidion	*C. (J.) dubius*	Hobbs, Jr. and Peters (1991)
D. (=E.) runki	*C. (C.) b. bartonii, C. (H.) chasmodactylus, C. (H.) longirostris,* *C. (P.) robustus*	Hobbs, Jr. and Peters (1977)
D. sandbergi	*Orconectes* sp., *O. (G.) compressus, O. (P.) rusticus*	Hart and Hart, Jr. (1971)
D. scissura	*C. (C.) b. bartonii, Cambarus (Depressicambarus)* sp., *C. (J.) dubius*	Hobbs, Jr. and Walton (1975)
D. scotos	*C. (L.) diogenes*	Norden and Norden (1984)
D. speira	*Cambarus (Jugicambarus)* sp., *Orconectes* sp., *O. (P.) rusticus*	Hart and Hart, Jr. (1971)
D. spinata	*C. (J.) distans*	Hobbs, Jr. and Walton (1970)
D. spinescens	*Cambarus* sp., *C. (C.) b. bartonii, C. (H.) girardianus,* *O. (C.) erichsonianus*	Hobbs, Jr. and Walton (1977)
D. (=E.) steevesi	*Cambarus* sp., *C. (E.) tenebrosus, Orconectes* sp., *O. (O.) a. packardi, Orconectes (Orconectes) pellucidus,* *O. (P.) placidus*	Hart, Jr. and Hobbs, Jr. (1961)
D. (=E.) striophylax	*C. (C.) b. bartonii, C. (D.) striatus, C. (P.) acuminatus, Cambarus* *(Puncticambarus) spicatus, P. (O.) acutus, P. (O.) enoplosternum,* *P. (O.) pubescens, Procambarus (Pennides) petersi,* *P. (S.) troglodytes*	Andolshek and Hobbs, Jr. (1986), Hobbs, Jr. and Peters (1977)
D. susanae	*C. (E.) tenebrosus, O. (O.) a. packardi, Orconectes (Orconectes)* *inermis inermis*	Hobbs III (1971)

Table 10. contd....

Table 10. contd.

Genus/species	Crayfish host(s)	Reference(s)
D. (=E.) suteri	*C. (C.) b. bartonii, Cambarus (Depressicambarus)* sp., *C. (D.) latimanus, Cambarus (Depressicambarus) reduncus, C. (H.) longulus, C. (L.) diogenes, C. (P.) acuminatus, Cambarus (Puncticambrus)* sp., *C. (P.) spicatus, F. (C.) fodiens, O. (C.) erichsonianus, O. (F.) limosus, O. (P.) spinosus, P. (O.) acutus, P. (P.) spiculifer*	Hobbs, Jr. and Peters (1977)
D. (=E.) ungulata	*C. (C.) b. bartonii, C. (E.) tenebrosus, Cambarus (Jugicambarus)* sp., *C. (J.) distans, Orconectes* sp., *O. (C.) erichsonianus, Orconectes (Orconectes) australis australis, O. (O.) a. packardi, O. (O.) pellucidus, O. (P.) mirus, O. (P.) placidus, O. (P.) putnami, O. (P.) rusticus*	Hart, Jr. and Hobbs, Jr. (1961)
D. xystroides	*Cambarus (Depressicambarus)* sp., *C. (D.) striatus, C. (E.) tenebrosus, C. (H.) girardianus, Orconectes* sp., *O. (C.) erichsonianus, O. (G.) compressus, O. (P.) forceps, O. (P.) placidus, O. (P.) rusticus, O. (P.) spinosus, O. (T.) alabamensis*	Hobbs, Jr. and Walton (1963a)
Donnaldsoncythere ardis	*C. (C.) b. bartonii, C. (H.) longulus, C. (P.) acuminatus*	Hobbs, Jr. and Walton (1963b)
D. cayugaensis	*C. (P.) robustus*	Hobbs, Jr. and Walton (1966)
D. (=Entocythere) donnaldsonensis (=hiwasseensis = humesi = ileata = pennsylvanica = scalis = tuberosa)	*C. (C.) b. bartonii, C. (C.) b. cavatus, C. (C.) carinirostris, C. (C.) sciotensis, Cambarus (Depressicambarus)* sp., *C. (D.) latimanus, C. (D.) reduncus, C. (E.) tenebrosus, C. (H.) chasmodactylus, C. (H.) longirostris, C. (H.) longulus, C. (J.) asperimanus, C. (J.) dubius, C. (J.) monongalensis, C. (L.) diogenes, Cambarus (Puncticambarus)* sp., *C. (P.) acuminatus, Cambarus (Puncticambarus) nerterius, Cambarus (Puncticambarus) reburrus, C. (P.) robustus, O. (C.) sanbornii, O. (O.) a. australis, O. (O.) a. packardi, O. (O.) pellucidus, O. (P.) juvenilis, P. (O.) acutus, P. (P.) raneyi*	Hart, Jr. (1962), Hart, Jr. and Hobbs, Jr. (1961), Hobbs, Jr. and Peters (1977, 1991, 1993)
D. leptodrilus	*Cambarus* sp., *C. (C.) b. bartonii, C. (D.) reduncus, C. (H.) longulus, F. (C.) fodiens*	Hobbs, Jr. and Peters (1977)
D. truncata	*C. (C.) b. bartonii, C. (C.) sciotensis, C. (H.) longulus, C. (P.) acuminatus*	Hart and Hart, Jr. (1974)
Elachistocythere merista	*Cherax* sp.	Hart and Hart (1970)
Entocythere cambaria	*O. (C.) propinquus, Orconectes (Hespericambarus) difficilis, O. (P.) hylas, Orconectes (Procericambarus) medius, Orconectes (Procericambarus) neglectus neglectus, Orconectes (Procericambarus) ozarkae, Orconectes (Procericambarus) punctimanus, O. (P.) quadruncus, P. (O.) acutus*	Hart and Hart, Jr. (1974)
E. claytonhoffi	*P. (A.) llamasi, P. (A.) mexicanus, P. (A.) o. oaxacae, P. (A.) O. reddelli, P. (A.) pilosimanus, P. (A.) vazquezae, P. (A.) zapoapensis, P. (O.) cuevachicae, P. (O.) toltecae*	Hobbs, Jr. (1973)
E. costata	*F. (C.) fodiens, P. (O.) acutus, Procambarus (Ortmannicus) pearsei, Procambarus (Ortmannicus) plumimanus*	Hobbs, Jr. and Peters (1977)
E. dentata	*Cambarus (Depressicambarus) catagius, C. (D.) reduncus, C. (H.) chasmodactylus, C. (J.) dubius, C. (P.) robustus, Orconectes (Crockerinus) obscurus, P. (H.) advena, P. (L.) p. pubischelae*	Crawford, Jr. (1965), Hobbs, Jr. and Peters (1977)

Table 10. contd....

Table 10. contd.

Genus/species	Crayfish host(s)	Reference(s)
E. dorsorotunda	*P. (H.) advena, P. (H.) r. rogersi, P. (H.) talpoides, P. (L.) alleni, P. (L.) p. pubischelae, Procambarus (Ortmannicus) lecontei, P. (O.) seminolae, P. (S.) paeninsulanus, P. (H.) pygmaeus*	Andolshek and Hobbs, Jr. (1986)
E. elliptica	*C. (P.) puer, C. (C.) b. bartonii, Cambarus (Depressicambarus) sp., C. (D.) latimanus, C. (L.) diogenes, Cambarus (Puncticambarus) sp., Fallicambarus sp., F. (C.) fodiens, O. (H.) difficilis, O. (P.) spinosus, P. (C.) hinei, P. (G.) simulans, P. (H.) advena, P. (H.) r. rogersi, P. (L.) alleni, P. (L.) barbatus, P. (L.) kilbyi, Procambarus (Leconticambarus) latiplerum, P. (L.) shermani, P. (O.) acutissimus, P. (O.) acutus, P. (O.) bivittatus, P. (O.) enoplosternum, Procambarus (Ortmannicus) hybus, P. (O.) lecontei, P. (O.) leonensis, P. (O.) litosternum, P. (O.) pubescens, P. (O.) pycnogonopodus, P. (O.) seminolae, Procambarus (Ortmannicus) verrucosus, Procambarus (Ortmannicus) viaeviridis, P. (P.) spiculifer, Procambarus (Pennides) dupratzi, Procambarus (Pennides) vioscai vioscai, P. (P.) versutus, Procambarus (Scapulicambarus) okaloosae, P. (S.) paeninsulanus, P. (S.) troglodytes*	Andolshek and Hobbs, Jr. (1986)
E. harrisi	*C. (C.) b. bartonii, C. (D.) latimanus, C. (H.) longulus, C. (L.) diogenes, C. (P.) acuminatus, P. (G.) simulans, P. (O.) acutus*	Hobbs, Jr. and Peters (1977), Peters and Pugh (1999)
E. illinoisensis	*C. (C.) b. bartonii, Cambarus sp., Cambarus (Depressicambarus) sp., C. (D.) striatus, C. (E.) tenebrosus, C. (H.) longirostris, Cambarus (Jugicambarus) sp., Orconectes sp., Orconectes (Buannulifictus) meeki meeki, Orconectes (Buannulifictus) palmeri longimanus, O. (C.) propinquus, O. (G.) compressus, O. (G.) virilis, O. (P.) forceps, O. (P.) juvenilis, Orconectes (Procericambarus) longidigitus, O. (P.) ozarkae, O. (P.) placidus, O. (P.) putnami, O. (P.) rusticus, O. (P.) spinosus, O. (T.) alabamensis, O. (T.) immunis, P. (O.) acutus*	Hart and Hart, Jr. (1974)
E. internotalus	*C. (C.) b. bartonii, Cambarus (Depressicambarus) sp., C. (D.) halli, C. (D.) latimanus, C. (D.) reduncus, C. (D.) shufeldtii, C. (L.) diogenes, Cambarus (Puncticambarus) sp., C. (P.) acuminatus, C. (P.) spicatus, F. (C.) fodiens, Orconectes (Tragulicambarus) lancifer, Procambarus sp., P. (H.) advena, P. (L.) p. pubischelae, P. (O.) acutissimus, P. (O.) acutus, P. (O.) enoplosternum, P. (O.) pearsei, P. (O.) seminolae, P. (S.) paeninsulanus, Procambarus (Pennides) natchitochae, P. (P.) spiculifer, Procambarus (Pennides) vioscai vioscai, P. (S.) clarkii, P. (S.) howellae, P. (S.) troglodytes*	Hobbs, Jr. and Peters (1977)
E. kanawhaensis	*C. (C.) b. bartonii, C. (C.) sciotensis*	Hobbs, Jr. and Walton (1966)
E. lepta	*C. (C.) b. cavatus*	Hart and Hart, Jr. (1971)
E. mexicana	*Procambarus (Ortmannicus) caballeroi, Procambarus (Villalobosus) erichsoni, Procambarus (Villalobosus) hoffmani, Procambarus (Villalobosus) hortonhobbsi, Procambarus (Villalobosus) riojai*	Hobbs, Jr. (1971)
E. prisma	*F. clypeata, P. (H.) advena, P. (H.) caritus, P. (H.) pygmaeus, P. (H.) talpoides, Procambarus (Hagenides) truculentus, P. (L.) p. deficiens, P. (L.) barbatus, P. (O.) acutus, P. (O.) enoplosternum, P. (O.) litosternum, P. (O.) lunzi, P. (O.) seminolae, P. (S.) howellae, P. (S.) troglodytes*	Andolshek and Hobbs, Jr. (1986)
E. reddelli	*C. (P.) puer, F. (C.) fodiens, F. clypeata, P. (G.) simulans, P. (O.) acutus, P. (O.) viaeviridis, P. (S.) clarkii*	Hobbs, Jr. and Peters (1977), Hobbs, Jr. and Walton (1968)

Table 10. contd....

Table 10. contd.

Genus/species	Crayfish host(s)	Reference(s)
E. ruibali	*P. (A.) atkinsoni, P. (A.) C. cubensis*	Hobbs, Jr. (1971)
E. tyttha	*F. (C.) fodiens*	Hobbs, Jr. and Hobbs III (1970)
Geocythere acuta	*C. (L.) diogenes*	Hart and Hart, Jr. (1971)
G. (=Entocythere) geophila	*C. (L.) diogenes, P. (P.) spiculifer, P. (S.) paeninsulanus*	Hart and Hart, Jr. (1974)
G. gyralea	*C. (L.) diogenes, O. (C.) tricuspis, O. (G.) virilis*	Hart and Hart, Jr. (1974)
G. nessoides	*C. (L.) diogenes, C. (L.) ludovicianus, F. (C.) fodiens*	Hobbs, Jr. and Hobbs III (1970)
Harpagocythere baileyi	*C. (C.) b. bartonii, Cambarus (Jugicambarus) sp., C. (J.) asperimanus*	Hobbs, Jr. and Peters (1977)
H. georgiae	*C. (D.) latimanus, C. (J.) carolinus*	Hobbs III (1965)
II. tertius	*Cambarus* sp.	Hobbs, Jr. and Walton (1968)
Hartocythere (=Entocythere =Geocythere) torreya	*C. (L.) diogenes*	Hobbs III (1970)
Herpetocythere acanthoides	*Engaeus cisternarius, Euastacus australasiensis, Euastacus spinifer*	Hart, Jr. and Hart (1967), Suter and Richardson (1977)
H. australensis	*Euastacus hystricosus*	Hart, Jr. and Hart (1967)
H. bendora	*Euastacus crassus*	Hart, Jr. and Hart (1967)
H. gnoma	*Euastacus neohirsutus*	Hart, Jr. and Hart (1967)
H. labidioides	*E. hystricosus*	Hart, Jr. and Hart (1967)
H. mackenziei	*E. hystricosus*	Hart, Jr. and Hart (1967)
Hesperocythere klasteroides	*Cherax* sp.	Hart, Jr. and Hart (1967)
H. tallanalla	*Cherax* sp.	Hart, Jr. and Hart (1967)
H. xiphoides	*Cherax* sp.	Hart, Jr. and Hart (1967)
Laccocythere aotearoa	*Paranephrops planifrons*	Hart, Jr. and Hart (1971)
Lichnocythere synethes	*Euastacus setosus*	Hart, Jr. and Hart (1967)
L. tubrabucca	*Euastacus sp., E. hystricosus, Euastacus sulcatus*	Hart, Jr. and Hart (1967)
L. victoria	*E. australasiensis*	Hart, Jr. and Hart (1967)
Litocythere lucileae	*C. (J.) asperimanus*	Hobbs, Jr. and Walton (1968)
Lordocythere petersi	*C. (D.) sphenoides, C. (J.) dubius, Cambarus (Jugicambarus) nodosus, C. (L.) diogenes, C. (T.) acanthura*	Hobbs, Jr. and Hobbs III (1970), Hobbs, Jr. and Peters (1993)
Notocythere antichthon	*E. neohirsutus*	Hart, Jr. and Hart (1967)
N. blundelli	*Engaeus cymus*	Hart, Jr. and Hart (1967)
N. erica	*Engaeus affinis*	Hart, Jr. and Hart (1967)

Table 10. contd....

Table 10. contd.

Genus/species	Crayfish host(s)	Reference(s)
N. mirranatwa	*Cherax* sp., *Engaeus* sp., *Geocharax* sp., *Parastacoides* sp.	Hart, Jr. and Hart (1967)
N. rieki	*E. crassus*	Hart, Jr. and Hart (1967)
N. synomodites	*E. cymus*	Hart, Jr. and Hart (1967)
N. syssitos	*Astacopsis* sp., *Cherax* sp., *Cherax punctatus*, *Engaeus* sp., *Ombrastacoides leptomerus*	Hart, Jr. and Hart (1967)
N. tasmanica	*E. cisternarius*, *Parastacoides* sp.	Hart, Jr. and Hart (1967), Suter and Richardson (1977)
Okriocythere cheia	*C. (L.) diogenes*, *?F. (C.) fodiens*, *?P. (O.) acutus*	Hart (1964), Hobbs, Jr. and Peters (1977)
Ornithocythere aetodes	*C. (L.) diogenes*	Hobbs III (1970)
O. gypodes	*C. (L.) diogenes*	Hobbs III (1969, 1970)
O. popi	*C. (L.) diogenes*, *C. (L.) ludovicianus*	Hobbs III (1970)
O. rhea	*C. (L.) diogenes*	Hobbs III (1970)
O. waltonae	*C. (L.) diogenes*, *C. (P.) acuminatus*, *?F. (C.) fodiens*, *P. (O.) acutus*	Hobbs, Jr. (1967), Hobbs, Jr. and Peters (1977)
O. thomai	*C. (L.) diogenes*	Hobbs, Jr. and McClure (1983)
Phymocythere lophota	*C. (J.) monongalensis*	Hobbs, Jr. and Peters (1993)
P. (=Cymocythere = Entocythere) phyma	*Cambarus* sp., *C. (C.) b. bartonii*, *C. (C.) carinirostris*, *C. (C.) sciotensis*, *C. (J.) dubius*, *C. (P.) nerterius*, *C. (P.) robustus*, *Orconectes* sp., *O. (G.) virilis*	Hobbs, Jr. and Hart, Jr. (1966), Hobbs, Jr. and Peters (1993)
Plectocythere crotaphis	*C. (J.) carolinus*	Hobbs III (1965)
P. johnsonae	*C. (J.) carolinus*	Hobbs, Jr. and Hart, Jr. (1966)
P. kentuckiensis	*C. (J.) dubius*	Hobbs, Jr. and Peters (1991)
P. odelli	*C. (J.) cf. dubius*	Norden (1977)
Psittocythere psitta	*C. (J.) distans*	Hobbs, Jr. and Walton (1975)
Rhadinocythere (=Entocythere) serrata	*C. (L.) diogenes*	Hart and Hart, Jr. (1974)
Riekocythere cherax	*Cherax* cf. *crassimanus*	Hart, Jr. and Hart (1967)
R. xenika	*Cherax* sp.	Hart, Jr. and Hart (1967)
Sagittocythere (=Entocythere) barri	*C. (E.) tenebrosus*, *C. (J.) cf. distans*, *Orconectes* sp., *O. (O.) a. australis*, *O. (O.) a. packardi*, *O. (O.) i. inermis*, *Orconectes (Orconectes) inermis testii*, *O. (O.) pellucidus*	Hart, Jr. and Hobbs, Jr. (1961)
S. stygia	*O. (O.) pellucidus*	Hart, Jr. and Hart (1966)
Saurocythere rhipis	*C. (L.) diogenes*	Hobbs III (1969)

Table 10. contd....

Table 10. contd.

Genus/species	Crayfish host(s)	Reference(s)
Thermastrocythere riojai	*C. (L.) diogenes, O. (B.) m. meeki, O. (B.) p. longimanus, Orconectes (Buannulifictus) meeki brevis, Orconectes (Buannulifictus) palmeri palmeri, O. (C.) propinquus, Orconectes (Gremicambarus) nais, O. (G.) virilis, O. (H.) difficilis, Orconectes (Hespericambarus) hathawayi, O. (P.) hylas, O. (P.) longidigitus, Orconectes (Procericambarus) luteus, Orconectes (Procericambarus) macrus, O. (P.) medius, Orconectes (Procericambarus) nana, O. (P.) neglectus chaenodactylus, O. (P.) n. neglectus, O. (P.) ozarkae, Orconectes (Procericambarus) peruncus, O. (P.) punctimanus, O. (P.) quadruncus, O. (P.) rusticus, O. (T.) immunis, O. (T.) lancifer, P. (O.) acutus, P. (P.) dupratzi, P. (P.) natchitochae, P. (S.) clarkii*	Hobbs, Jr. and Hobbs III (1970)
Uncinocythere allenae	*O. (G.) compressus*	Hart and Hart, Jr. (1971)
U. (=Entocythere) ambophora	*Procambarus (Lonnbergius) acherontis, Procambarus (Lonnbergius) morrisi*	Hobbs, Jr. and Franz (1991)
U. (=E.) bicuspide (=uncinata)	*P. (A.) mexicanus, P. (P.) paradoxus, Procambarus (Villalobosus) contrerasi, P. (V.) erichsoni, P. (V.) hoffmani, P. (V.) hortonhobbsi, P. (V.) riojai, Procambarus (Villalobosus) zihuateutlensis*	Hobbs, Jr. (1971)
U. cassiensis	*Pacifastacus (Hobbsastacus) gambelli*	Hart, Jr. (1965)
U. (=E.) caudata	*Pacifastacus (Hobbsastacus) fortis, P. (H.) gambelii*	Kozloff (1955)
U. (=E.) clemsonella	*C. (C.) b. bartonii, Cambarus (Depressicambarus) sp., C. (D.) latimanus, P. (O.) lophotus*	Hart and Hart, Jr. (1974)
U. (=E.) columbia	*P. (H.) gambelii, Pacifastacus (Pacifastacus) leniusculus klamathensis, P. (P.) l. leniusculus, Pacifastacus (Pacifastacus) leniusculus trowbridgii*	Hart and Hart, Jr. (1974)
U. (=E.) cuadricuspide	*C. (C.) ortmanni, P. (O.) caballeroi, P. (P.) paradoxus, P. (V.) erichsoni, P. (V.) hoffmani, P. (V.) hortonhobbsi, P. (V.) zihuateutlensis*	Hobbs, Jr. (1971)
U. (=E.) dobbinae	*C. (C.) ortmanni, P. (O.) caballeroi, P. (P.) paradoxus, P. (V.) contrerasi, P. (V.) erichsoni, P. (V.) hoffmani, P. (V.) riojai, Procambarus (Villalobosus) teziutlanensis, Procambarus (Villalobosus) tlapacoyanensis*	Hobbs, Jr. (1971)
U. (=E.) equicurva (=lucifuga)	*Cambarellus (Pandicambarus) schmitti, C. (D.) latimanus, C. (D.) striatus, C. (D.) truncatus, C. (L.) diogenes, F. (C.) fodiens, F. clypeata, O. (P.) spinosus, P. (C.) hinei, P. (O.) acutus, P. (O.) enoplosternum, Procambarus (Ortmannicus) fallax, Procambarus (Ortmannicus) lucifugus alachua, Procambarus (Ortmannicus) pallidus, P. (O.) pubescens, P. (P.) spiculifer, P. (S.) howellae, P. (S.) paeninsulanus, P. (P.) petersi, P. (P.) versutus, P. (S.) troglodytes*	Andolshek and Hobbs, Jr. (1986), Hart, Jr. (1962)
U. (=E.) ericksoni	*P. (H.) gambelii, P. (P.) l. klamathensis, P. (P.) l. leniusculus, P. (P.) l. trowbridgii*	Hart, Jr. (1965)
U. holti	*O. (G.) virilis, O. (H.) hathawayi, O. (T.) immunis, O. (T.) lancifer, P. (H.) gambelii, P. (P.) dupratzi*	Hart, Jr. (1965)
U. (=E.) neglecta	*P. (H.) gambelii, Pacifastacus (Hobbsastacus) nigrescens, P. (P.) l. klamathensis*	Westervelt, Jr. and Kozloff (1959)
U. (=E.) occidentalis	*Pacifastacus sp., P. (P.) l. leniusculus, P. (P.) l. klamathensis, P. (P.) l. trowbridgii*	Kozloff and Whitman (1954), Smith and Kamiya (2001)
U. (=E.) pholetera	*Cambarus (Erebicambarus) hubrichti*	Hart, Jr. and Hobbs, Jr. (1961)

Table 10. contd....

Table 10. contd.

Genus/species	Crayfish host(s)	Reference(s)
U. (=E.) simondsi	*Cambarus* sp., *C. (C.) b. bartonii, Cambarus (Depressicambarus) sp., C. (D.) halli, C. (D.) latimanus, C. (H.) longirostris, C. (J.) dubius, C. (L.) diogenes, Cambarus (Puncticambarus) sp., C. (P.) nerterius, Orconectes sp., O. (C.) erichsonianus, Hobbseus cristatus, O. (G.) virilis, O. (O.) a. packardi, O. (P.) forceps, O. (P.) juvenilis, O. (P.) placidus, O. (P.) rusticus, O. (P.) spinosus, Procambarus (Ortmannicus) sp., P. (O.) acutus, P. (O.) acutissimus, Procambarus (Ortmannicus) lewisi, P. (O.) viaeviridis, P. (P.) spiculifer, P. (P.) versutus*	Hart and Hart, Jr. (1974), Hobbs, Jr. and Peters (1989)
U. spathe	*O. (C.) erichsonianus*	Hart and Hart, Jr. (1971)
U. stubbsi	*B. cornutus, C. (C.) b. bartonii, i C. (D.) latimanus, C. (D.) striatus, C. (E.) tenebrosus, C. (H.) girardianus, C. (L.) diogenes, C. (P.) robustus, Orconectes sp., O. (C.) erichsonianus, O. (C.) obscurus, O. (C.) propinquus, O. (C.) sanbornii, O. (C.) shoupi, Orconectes (Faxonius) indianensis, O. (G.) compressus, O. (G.) nais, O. (G.) virilis, O. (O.) pellucidus, Orconectes (Procericambarus) barrenensis, O. (P.) rusticus, O. (P.) spinosus, O. (P.) forceps, O. (P.) juvenilis, O. (P.) mirus, O. (P.) placidus, O. (P.) putnami, O. (P.) rusticus, O. (P.) spinosus, Orconectes (Rhoadesius) kentuckiensis, Orconectes (Rhoadesius) sloani, O. (T.) immunis, Orconectes (Trisellescens) rhoadesi, P. (L.) barbatus, P. (O.) lunzi, P. (P.) spiculifer*	Hobbs, Jr. and Walton (1966)
U. thektura	*O. (T.) immunis, P. (H.) gambelii*	Hart, Jr. (1965)
U. warreni	*Cambarus (Jugicambarus) cryptodytes*	Hobbs, Jr. and Walton (1968)
U. (=E.) xania	*Cambarus (Jugicambarus) setosus*	Hart, Jr. and Hobbs, Jr. (1961)
U. xena	*Orconectes* sp., *O. (P.) placidus*	Hart and Hart, Jr. (1971)
U. zancla	*B. cornutus, Cambarus sp., C. (C.) b. bartonii, C. (C.) b. cavatus, C. (C.) ortmanni, C.(D.) graysoni, C. (D.) striatus, C. (E.) rusticiformis, C. (E.) tenebrosus, Cambarus (Glarecola) brachydactylus, C. (G.) friaufi, C. (H.) girardianus, C. (J.) carolinus, C. (J.) crinipes, C. (J.) dubius, Cambarus (Jugicambarus) gentryi, C. (L.) diogenes, Cambarus (Puncticambarus) sp., C. (P.) cumberlandensis, C. (P.) robustus, Orconectes sp., O. (C.) shoupi, O. (G.) compressus, O. (O.) i. inermis, O. (P.) barrenensis, O. (P.) forceps, O. (P.) mirus, O. (P.) placidus, O. (P.) putnami, O. (P.) rusticus, O. (P.) spinosus, Orconectes (Trisellescens) rhoadesi, P. (P.) spiculifer*	Hobbs, Jr. and Peters (1993), Hobbs, Jr. and Walton (1963a)
Waltoncythere (=Aphelocythere) acuta	*C. (C.) b. bartonii, C. (J.) asperimanus*	Hobbs, Jr. and Peters (1977, 1979)

Sexes are seperate in the entocytherids and, following copulation, eggs are generally deposited on the gill setae of their crayfish hosts. Following egg hatching, at least seven moults occur giving rise to adult ostraocods. No reports of mortalities associated with ostracods have been reported, despite high numbers sometimes being recorded on crayfish.

Phylum Arthropoda – class Arachnida

Freshwater mites normally have a complex lifecycle involving a parasitic larval stage, followed by a quiescent protonymph, then a predatory deutonymph, a quiescent tritonymph and a predatory adult stage (Goldschmidt et al. 2002) (Fig. 5c,d). Haswell (1922) described a new genus and species of mite,

Astacocroton molle (Family Astacocrotonidae) from the gills of *Euastacus* sp. (as *Astacopsis serratus*) in exquisite detail. Adult males appeared to be transitory on the crayfish host, whilst blind adult females were considered to be obligate parasites; larval stages were not described. The obligate parasitic mite *Peza daps* (Family Pezidae), described from the gills of *Engaeus fultoni* in Australia was considered to be parasitic due to the absence of eyes and its habitat; only one single adult female was collected and described (Harvey 1990). A third antipodean species, *Astacopsiphagus parasiticus* (Subfamily Astacopsiphaginae) has been described from the gills of *Euastacus* sp. (as *Astacopsis serratus*) with three nymphal instar stages occurring on the crayfish host (Bartsch 1996, Viets 1931). European species of mites infecting the gills include *Limnohalacarus wackeri* v. *astacicola* (Subfamily Limnohalacarinae) and *Porohalacarus alpinus* (Subfamily Halacarinae) on *Astacus astacus* and *Orconectes* (*Faxionus*) *limosus* (Wiszniewski 1939, Zawal 1998), *Piona pusilla* (=*rotunda*) (Family Pionidae) on *A. astacus* and *Astacus leptodactylus* (Viets 1939, Wiszniewski 1939), *Porolohmannella* (=*Leptognatus* = *Trouessartella* = *Lohmannella*) *violacea* (Subfamily Limnohalacarinae) on *O.* (*F.*) *limosus* (Bartsch 2006, 2011, Zawal 1998), *Hygrobates* cf. *longipalpis* (Family Hygrobatidae) on farmed *Pacifastacus leniusculus* (Cuellar et al. 2002) and unidentified mites in the gills of *Austropotamobius pallipes* and *Procambarus* (*Ortmannicus*) *fallax* (Longshaw et al. 2012a,b). Mites are not generally considered detrimental to host survival and are not a cause for concern for human health.

Other fouling organisms and idiopathic conditions

In addition to harbouring a range of parasites and commensals, crayfish have been reported as hosts for fouling organisms that utilise the external surfaces without leading to a host response or to becoming problematic. These include cyanobacteria and algae (Lahser, Jr. 1975), corixid eggs (Abbott 1912a,b, Griffiths 1945, Meyer 1965), *Argulus* eggs (Wierzbicka and Smietana 1999), bryozoans (Ďuriš et al. 2006), zebra mussels (Brazner and Jensen 2000, Lamanova 1971, Ďuriš et al. 2006) and cladocerans (Huys et al. 2014).

Idiopathic conditions are rarely reported from crayfish partly due to a lack of clear case definitions. Amongst those reported include a cyst-like growth under the carapace of *Orconectes* (*Crockerinus*) *propinquus* (Dexter 1954), black to dark blue coloured spots on the carapace of *Cherax quadricarinatus* which rendered them unmarketable (Edgerton 2000, Edgerton and Owens 1999, Jiménez and Romero 1997), haemocytic enteritis (Edgerton 2000, Edgerton and Owens 1999), needle shaped crystals in the nephridial canal (Edgerton 2000), necrosis of and giant cells in the labyrinth epithelium (Edgerton 2000), as well as nodules, granulomas and iron granules in the hepatopancreas of *C. quadricarinatus* (Romero and Jiménez 2002).

Host parasite lists

The data in the following tables has been derived from the published literature; nomenclatural changes and relevant literature for each condition can be found in the preceding sections. The host species have been selected on the basis that they are either considered invasive, have been extensively farmed or are ecologically important.

Within the crayfish family Astacidae, data are provided for *Astacus astacus, A. leptodactylus, Austropotamobius pallipes* spp., *A. torrentium* and *Pacifastacus* (*Pacifastacus*) *leniusculus* spp. (Table 11). *Astacus astacus* is indigenous to mainland Europe but has been introduced to a number of countries including Norway, Sweden and the UK; *A. leptodactylus* is native to eastern Europe and the Middle East and has been introduced to northern and western Europe; *A. pallipes* and *A. torrentium* occur in Europe, but have undergone declines in several countries; *P. leniusculus* and its respective subspecies *leniusculus, klamathensis* and *trowbridgii* is native to British Columbia (Canada) and western USA. Due to interbreeding and anthropogenic movements the boundaries between subspecies have been diluted and the data on which species or subspecies occur in the different geographical regions is not well defined. *Pacifastacus leniusculus* has been widely introduced into other parts of the USA, Europe, Russia and Japan, initially for cultivation purposes. There are now established wild populations in many countries.

Table 11. Host/pathogen lists for selected members of the Astacidae through their distribution range.

Host	Parasites and pathogens
Astacus astacus	**Viruses**: *Aa*BV, IPNV, Picorna-like virus, WSSV; **Bacteria**: *Hafnia alvei, Pseudomonas aeruginosa, P. putida*; **Fungi**: *Achlya* sp., *Aphanomyces astaci, A. frigidophilus, Epicoccum nigrum, Fusarium avenaceum, F. solani, Leptolegnia* sp., *Mucor hiemalis, M. racemosus, Phytopythium* sp., *Ramularia astaci, Saprolegnia australis, S. ferax, S. hypogyna, S. littoralis, S. parasitica*, Saprolegniales I, II, III, *Scoliolegnia asterophora, Trichosporon beigelii*; **Microsporidia**: *Thelohania contejeani*; **Mesomycetozoea**: *Psorospermium* sp.; **Protista**: *Cothurnia astaci, C. curva, C. plachteri, C. sieboldii, C. variabilis, Discophrya astaci, Epistylis astaci, E. bimarginata, E. crassicollis, Opercularia crustaceorum, Podophrya astaci, Propyxidium asymmetrica, Zoothamnium procerius*; **Digenea**: *Astacotrema cirrigerum*; **Nematoda**: *Chromadorita astacicola, C. viridis, Paractinolaimus macrolaimus*, unidentified eggs; **Annelida**: *A. quaternarium, A. tenebrarum, A. variegatum* and *Aeolosoma* sp. *Dero obtusa, Nais barbata, Pristina aequiseta, Stylaria lacustris,* and *Vejdovskiella comata*; **Acanthocephala**: *Filicollis anatis, Polymorphus minutus*; **Mites**: *Limnohalacarus wackeri var. astacicola, Porohalacarus alpinus, Piona pusilla*; **Rotifera**: *Lepadella (Xenopadella) astacicola, L. (X.) branchicola, L. (X.) parasitica, L. (X.) lata, Cephalodella jakubskii, Dicranophorus siedleckii, D. hauerianus*; **Copepoda**: *Attheyella crassa, A. trispinosa, Canthocamptus staphylinus, Nitocra divaricata*; **Branchiobdellids**: *Branchiobdella astaci, B. balanica sketi, B. balanica, B. hexadonta, B. italica, B. kozarovi, B. parasita, B. pentadonta*
Astacus leptodactylus	**Virus**: WSSV; **Bacteria**: *Aeromonas hydrophila, Campylobacter* spp., *Escherichia coli, Listeria monocytogenes, Proteus morganii, P. vulgaris, Vibrio alginolyticus, V. harveyi, V. mimicus, V. vulnificus*; **Fungi**: *Acremonium* sp., *A. leptodactyli, Alternaria* sp., *Aphanomyces astaci, Aspegillus flavus, Fusarium* sp., *F. oxysporum, F. solani, Penicillium expansum, Saprolegnia* sp., *S. parasitica*; **Microsporidia**: *Thelohania contejeani*; **Mesomycetozoea**: *Psorospermium* sp.; **Protista**: *Acineta tuberosa, Chilodonella* sp., *Cothurnia astaci, C. bavarica, C. curva, C. sieboldii, C. variabilis, Cyclodonta bipartita, Discophrya astaci, Epistylis* sp., *E. astaci, E. chrysemidis, E. cambari, E. crassicollis, E. niagarae, Lernaeophrya capitata, Mantonella potamobii, Opercularia allensi, O. articularia, O. nutans, Podophrya fixa, Pyxicola annulata, Sincothurnia branchiata, Tetrahymena pyriformis, Tokophrya lemnarum, Trichophrya astaci, Vorticella covallaria, Zoothamnium* sp.; **Annelida**: *Aeolosoma hemipritchi, A. quaternarium, A. tenebrarum, A. variegatum, Aeolosoma* sp., *Dero obtusa, Hystricosoma chappuisi, Nais barbata, Pristina aequiseta, Stylaria lacustris* and *Vejdovskiella comata*; **Mites**: *Piona pusilla*; **Rotifer**: *Dicranophorus siedleckii, D. hauerianus, Lepadella (Xenopadella) astacicola, L. (X.) branchicola, L. (X.) lata, L. nana*; **Copepoda**: *Nitocra divaricata, N. hibernica, Acanthocyclops* sp.; **Branchiobdellids**: *Branchiobdella* sp., *B. astaci, B. balcanica sketi, B. hexadonta, B. kozarovi, B. parasita, B. pentadonta*
Austropotamobius pallipes spp.	**Virus**: *Ap*BV; WSSV; **Bacteria**: *Aeromonas hydrophila, Citrobacter* sp., *C. freundii, Hafnia alvei, Nocardia* sp., *Proteus morganii, P. vulgaris, Pseudomonas aeruginosa, P. flourescens, P. putida*; **Fungi**: *Acremonium* sp., *Alternaria* sp., *Aphanomyces astaci, A. frigidophilus, A. repetans, Cryptococcus gammari, Fusarium* sp., *F. oxysporum, F. solani, Plectosporium tabacinum, Geotrichum* sp., *Gliocladium* sp., *Penicillium* sp., *Saprolegnia* sp., *S. australis, S. parasitica, Trichoderma* sp.; **Microsporidia**: *Thelohania contejeani*; **Mesomycetozoea**: *Psorospermium* sp.; **Protista**: *Cothurnia* sp., *C. sieboldii, Epistylis* sp.; **Nematoda**: unidentified nematodes; **Mites**: unidentified mites; **Branchiobdellids**: *Branchiobdella* sp., *B. astaci, B. hexadonta, B. italica, B. parasita, B. pentadonta*
Austropotamobius torrentium	**Fungi**: *Aphanomyces astaci*; **Mesomycetozoea**: *Psorospermium* sp.; **Protista**: *Cothurnia plachteri, C. sieboldii, Discophrya astaci, Epistylis astaci, E. niagarae, Orbopercularia astacicola*; **Annelida**: *Hystricosoma chappuisi*; **Rotifer**: *Cephalodella jakubskii, Dicranophorus cambari, Lepadella (Xenopadella) astacicola, L. (X.) parasitica*; **Copepoda**: *Nitocra divaricata*; **Branchiobdellids**: *Branchiobdella astaci, B. balcanica sketi, B. hexadonta, B. papillosa, B. parasita, B. pentadonta*

Table 11. contd....

Table 11. contd.

Host	Parasites and pathogens
Pacifastacus (Pacifastacus) leniusculus spp.	**Viruses**: *P*IBV, WSSV; **Bacteria**: *Acinetobacter* sp., *Aeromonas hydrophila*, *Chryseobacterium* sp., *Citrobacter gillenii*, *C. murliniae/freundii*, *Hafnia alvei*, *Pseudomonas* sp., *P. guinea/peli*, *P. libanensis/gessardii*, *Spiroplasma* sp., *Vibrio alginolyticus*; **Fungi**: *Acremonium* sp., *Alternaria chlamydospora*, *Alternaria* sp., *Aphanomyces* sp., *A. astaci*, *A. repetans*, *Aspergillus* sp., *Circinella muscae*, *Dictyuchus* sp., *Fusarium* sp., *F. solani*, *Mucor* sp., *Penicillium* sp., *Saprolegnia* sp., *S. australis*, *S. parasitica*, *Trichoderma* sp.; **Microsporidia**: *Bacillidium* sp., *Cystosporogenes* sp., *Microsporidium* sp., *Thelohania contejeani*, *Vittaforma* sp., *V. corneae*; **Mesomycetozoa**: *Psorospermium* sp.; **Protista**: *Cothurnia tespa*, *C. transoceanica*, *C. ussurina*, *Lagenophrys leniusculus*; **Digenea**: *Astacatremulata macrocotyla*, *Cephalophallus obscurus*; **Acanthocephala**: *Neoechinorhynchus rutili*; **Nematoda**: unidentified eggs; **Ostracoda**: *Ankylocythere sinuosa*, *?Eucypris virens*, *Uncinocythere columbia*, *U. ericksoni*, *U. neglecta*, *U. occidentalis*; **Mites**: *Hygrobates* cf. *longipalpis*; **Copepoda**: *Nitocra hibernica*; **Branchiobdella**: *Branchiobdella parasita*, *Cambarincola fallax*, *C. gracilis*, *C. okadai*, *Sathodrilus attenuatus*, *S. chehalisae*, *S. dorfus*, *S. inversus*, *S. lobatus*, *S. norbyi*, *S. shastae*, *S. wardinus*, *Triannulata magna*, *Uglukodrilus hemophagus*, *Xironogiton cassiensis*, *X. instabilis*, *X. kittitasi*, *X. occidentalis*, *X. victoriensis*

Within the family Cambaridae, data are provided on *Orconectes (Faxonius) limosus*, *Orconectes (Gremicambarus) virilis*, *Orconectes (Procericambarus) juvenilis*, *Orconectes (Procericambarus) rusticus*, *Orconectes (Trisellescens) immunis*, *Procambarus (Ortmannicus) fallax* and *Procambarus (Scapulicambarus) clarkii* (Table 12). *Orconectes (F.) limosus* is native to the eastern states of the USA and Canada but has been introduced to many countries in Europe and to Russia with populations established in the wild; *O. (G.) virilis* is native to Canada and parts of the USA and has been introduced into several American states, to Mexico, France and Sweden; *O. (P.) juvenilis* is native to Kentucky, USA and has been recorded in France; *O. (P.) rusticus* is native to the Ohio River system but has been introduced to a number of areas in Canada and the USA; *O. (T.) immunis* is native to the USA and has been introduced into Germany where wild populations have been established; *P. (O.) fallax* is native to Florida and Georgia in the USA. The parthenogenic form of this species, *P. (O.) fallax* f. *virginalis* or Marmokrebs has been extensively used in the pet trade with established wild populations now recorded in Madagascar and Germany. Finally, *P. (S.) clarkii*, native to southern USA, has been widely introduced to other states in the USA, to southern and central America, Africa, Europe and Asia with wild populations established in many of these countries.

Table 12. Host/pathogen lists for selected members of the Cambaridae through their distribution range.

Host	Parasites and pathogens
Orconectes (Faxonius) limosus	**Virus**: WSSV; **Bacteria**: *Proteus morganii*, *P. vulgaris*; **Fungi**: *Aphanomyces astaci*, *Didymaria cambari*, *Phytophthora inundata-P. humicola*, *Saprolegnia australis*, *S. diclina*, *S. ferax*, *S. parasitica*, Saprolegniales II, III, IV; **Mesomycetozoa**: *Psorospermium orconectis*; **Protista**: *Cothurnia astaci*, *C. bavarica*, *C. curva*, *C. variabilis*, *Cyclodonta bipartita*, *Discophrya astaci*, *Epistylis nigarae*; **Acanthocephala**: *Polymorphus minutus*; **Nematoda**: *Rhabditis terricola*, *R. inermis*; **Mites**: *Limnohalacarus wackeri* var. *astacicola*, *Porohalacarus alpinus*, *Porolohmannella violacea*; **Ostracods**: *Ankylocythere ancyla*, *Dactylocythere suteri*; **Rotifer**: *Dicranophorus cambari*, *D. hauerianus*, *Lepadella (Xenolepadella) astacicola*, *L. (X.) borealis*, *L. (X.) lata*; **Branchiobdellida**: *Branchiobdella balcanica sketi*, *B. balcanica*, *B. hexadonta*, *B. parasita*, *B. pentadonta*, *Xironogiton instabilis*, *Cambarincola mesochoreus*, *C. philadelphicus*, *Pterodrilus missouriensis*
Orconectes (Gremicambarus) virilis	**Virus**: WSSV; **Bacteria**: *Acinetobacter* sp., *Arthrobacter* sp., *Erwinias* sp., *Phenylobacterium* sp., *Serratia rubidaea*; **Fungi**: *Aphanomyces astaci*, *Batrachochytrium dendrobatidis*; **Microsporidia**: *Thelohania* sp.; **Mesomycetozoa**: *Psorospermium* sp.; **Protista**: *Epistylis* sp.; **Digenea**: *Crepidostomum cooperi*, *Quasimaritremopsis*

Table 12. contd....

Table 12. contd.

Host	Parasites and pathogens
	medius, Paragonimus kellicotti; **Ostracoda**: *Ankylocythere copiosa, Dactylocythere brachydactylus, Entocythere illinoisensis, Geocythere gyralea, Phymocythere phyma, Thermastrocythere riojai, Uncinocythere holti, U. simondsi, U. stubbsi*; **Branchiobdellida**: *Bdellodrilus illuminatus, Cambarincola chirocephalus, C. illinoisensis, C. macrodontus, C. mesochoreus, C. osceola, C. philadelphicus, C. vitreus, Ellisodrilus durbini, Oedipodrilus macbaini, Sathodrilus elevatus, Xironodrilus formosus*
Orconectes (Procericambarus) juvenilis	**Ostracoda**: *Ankylocythere copiosa, A. sinuosa, Ascetocythere hyperoche, A. riopeli, A. sclera, Cymocythere gonia, Dactylocythere amicula, D. chelomata, D. daphanoides, D. falcata, D. macroholca, Donnaldsoncythere donnaldsonensis, Entocythere illinoisensis, Uncinocythere simondsi, U. stubbsi*; **Branchiobdellid**: *Oedipodrilus anisognathus, Pterodrilus alcicornus, P. cedrus, P. distichus, P. hobbsi, Xironogiton instabilis*
Orconectes (Procericambarus) rusticus	**Mesomycetozoa**: *Psorospermium* sp.; **Protista**: *Epistylis* sp.; **Digenea**: *Microphallus* sp., *Ochoterenatrema diminutum, Paragonimus kellicotti*; **Ostracoda**: *Ankylocythere hyba, Ascetocythere riopeli, Dactylocythere crawfordi, D. exoura, D. koloura, D. macroholca, D. prionata, D. sandbergi, D. speira, D. ungulata, D. xystroides, Entocythere illinoisensis, Thermastrocythere riojai, Uncinocythere simondsi, U. stubbsi, U. zancla*; **Branchiobdellida**: *Cambarincola chirocephalus, C. demissus, C. fallax, C. mesochoreus, C. philadelphicus, C. vitreus, Oedipodrilus macbaini, Pterodrilus cedrus, P. distichus, P. hobbsi, Sathodrilus elevatus, Xironodrilus formosus*; **Copepoda**: *Attheyella pilosa*
Orconectes (Trisellescens) immunis	**Fungi**: *Aphanomyces astaci*; **Mesomycetozoa**: *Psorospermium* sp.; **Digenea**: *Crepidostomum cornutum*; **Ostracoda**: *Ankylocythere tridentata, Entocythere illinoisensis, Thermastrocythere riojai, Uncinocythere holti, U. stubbsi, U. thektura*; **Branchiobdellida**: *Cambarincola macrodontus, C. mesochoreus, C. philadelphicus, C. vitreus, Oedipodrilus macbaini, Pterodrilus distichus, Sathodrilus elevatus, Xironodrilus formosus*
Procambarus (Ortmannicus) fallax	**Bacteria**: *Aeromonas sobria, Citrobacter freundii, Grimontia hollisae, Pasteurella multocida, Rickettsia*-like organism; **Fungi**: *Aphanomyces astaci*; **Mesomycetozoa**: *Psorospermium* sp.; **Protista**: Unidentified coccidian, unidentified ciliates; **Mites**: Unidentified mites; **Ostracoda**: *Uncinocythere equicurva*; **Branchiobdellida**: *Cambarincola floridanus, C. goodnighti, C. manni*
Procambarus (Scapulicambarus) clarkii	**Virus**: WSSV; **Bacteria**: *Acinetobacter* sp., *A. antitratum, A. calcoaceticus, A. lwoffi, Aeromonas hydrophila, A. liquefacieus, Arthrobacter* sp., *Bacillus* sp., *Citrobacter freundii, Corynebacterium* sp., *Enterobacter aerogenes, E. cloacae, Flavobacterium* sp., *F. dorminator, Francisella tularensis* var. *palaearctica, Micrococcus luteus, M. roseus, Oligella ureolytica, Pseudomonas* sp., *P. alcaligenes, P. mendocina, P. putrefaciens, P. stutzeri, Sphingobacterium multivorum, Spiroplasma eriocheiris, Staphylococcus epidermidis, Streptococcus* sp., *Vibrio* sp., *V. alginolyticus, V. cholerae, V. mimicus*; **Fungi**: *Absidia fusca, A. glauca, Acremonium* sp., *A. chrysogenum, A. kiliense, A. persicinum, Alternaria alternata, A. cheiranthi, A. chlamydospora, Aphanomyces astaci, A. repetans, Arthrinium* sp., *A. phaeospermum, Aspergillus* sp., *A. album, A. brasiliensis, A. clavatus, A. flavus, A. fumigatus, A. glaucus, A. niger, A. terreus, A. versicolor, Aureobasidium pullulans* var. *melanogenum, A. p.* var. *pullulans, Batrachochytrium dendrobatidis, Cephalotrichum microsporum, Chaetomella raphigera, Chaetomium* sp., *Cladosporium* sp., *C. chlorocephalum, C. cladosporoides, Clonostachys rosea, Coniella* sp., *Cryptococcus laurentii, Drechslera* sp., *Emericellopsis* sp., *Epicoccum nigrum, Erysiphe* sect. *Uncinula* sp., *Fusarium* sp., *F. dimerum, F. oxysporum, F. proliferatum, F. verticillioides, Gliocladium* sp., *Graphium* sp., *Hemicarpenteles ornatum, Hormisum* sp., *Khuskia oryzae, Microdochium bolleyi, Mortierella* sp., *M. turficola, Mucor* sp., *M. hiemalis, M. plumbleus, Oidiodendron flavum, Paecilomyces* sp., *P. farinosus, P. inflatus, P. lilacinum, Penicillium* sp., *P. verrucosum, Pestalotiopsis guepinii, Phoma* sp., *P. glomerata, Rhizopus* sp., *R. stolonifer, Rhodotorula* sp., *Saprolegnia parasitica, Scopulariopsis* sp., *Sordaria fimicola, Talaromyces flavus, Trichoderma* sp., *T. viridae, Ulocladium* sp.; **Mesomycetozoa**: *Psorospermium* sp.; **Protista**: *Acineta tuberosa, Colpoda* sp., *Cothurnia* sp., *Epistylis* sp., *Euplotes* sp., *Lagenophrys* sp., *Paramecium* sp., *Zoothamnium* sp.; **Digenea**: *Clonorchis sinensis, Crepidostomum cornutum,*

Table 12. contd.

Host	Parasites and pathogens
	Prohemistomum vivax, Petasiger neocomense, Gorgodera amplicava, Centrocestus cuspidatus, Heterophyes aequalis, Metagonimoides oregonensis, Pygidiopsis summa, Macroderoides typicus, Maritrema (Atriospinosum) obstipum, Microphallus fonti, M. minus, Sogandaritrema progeneticum, Paragonimus kellicotti, Allocorrigia filiformis, Renifer sp.; **Acanthocephala**: *Ibirhynchus dimorpha*; **Nematoda**: Unidentified nematodes; **Ostracoda**: *Ankylocythere copiosa, A. sinuosa, Entocythere internotalus, E. reddelli, Thermastrocythere riojai*; **Rotifera**: unidentified; **Branchiobdellids**: *Branchiobdella italica, B. parasita, B. pentadonta, Cambarincola barbarae, C. fallax, C. gracilis, C. macrodontus, C. mesochoreus, C. okadai, C. pamelae, C. vitreus, Xironogiton victoriensis*

Within the family Parastacidae, data are provided for *Cherax albidus, C. cainii, C. destructor* and *C. quadricarinatus* (Table 13). *Cherax albidus* is considered invasive within Australia; *C. cainii* is native to south-western Australia but has been translocated to other parts of Australia, Chile, China, Japan, Africa, New Zealand and the USA; *C. destructor*, native to Australia has been noted in Spain and Italy; *C. quadricarinatus* is native to northern Australia and Papua New Guinea but has been translocated for

Table 13. Host/pathogen lists for selected members of the Parastacidae through their distribution range.

Host	Parasites and pathogens
Cherax albidus	**Virus**: WSSV, picorna-like virus; **Bacteria**: *Acinetobacter lwoffi, Aeromonas* sp., *A. hydrophila, A. sobria, A. veroni, Alcaligenes* sp., *Bacillus* sp., *Citrobacter freundii,* Coliform-like spp., *Corynebacterium* sp., *Escherichia coli, Flavobacterium* sp., *Hafniaalvei, Kurthia* sp., *Micrococcus* sp., *Plesiomonas shigelloides, Proteus* sp., *Pseudomonas* sp., *Shewanella putrefaciens, Staphylococcus* sp., *Vibrio anguillarum, V. cholerae, V. mimicus*; **Microsporidia**: *Thelohania parastaci, Vavraia parastacida*; **Mesomycetozoea**: *Psorospermium* sp.; **Protista**: *Epistylis* sp., *Lagenophrys* sp., *L. willisi, Setonophrys communis, S. lingulata, S. occlusa, S. seticola, Vorticella alba*; **Nematoda**: *Gammarinema* sp.; **Temnocephalida**: *Temnosewellia dendyi, T. minor*
Cherax cainii	**Bacteria**: *Aeromonas hydrophila*; **Fungi**: *Saprolegnia* sp.; **Microsporidia**: *Vavraia parastacida*; **Mesomycetozoea**: *Psorospermium* sp.; **Protista**: *Epistylis* sp., *Lagenophrys deserti, Zoothamnium* sp.; **Temnocephalids**: *Temnosewellia chaeropsis, T. minor, T. punctata, Zygopella deimata, Z. pista, Z. stenota*
Cherax destructor	**Virus**: *Cd*SPV; **Bacteria**: *Pseudomonas* sp.; **Fungi**: *Aphanomyces astaci, Saprolegnia* sp.; **Microsporidia**: *Thelohania montirivulorum, T. parastaci, Vairimorpha cheracis*; **Protista**: *Lagenophrys willisi, Setonophrys spinosa, S. communis, S. lingulata, S. occlusa, S. seticola*; **Acanthocephala**: *Polymorphus biziurae*; **Digenea**: *Microphallus minutus, Plagiorchis jaenschi*; **Cestoda**: *Austramphilina elongata, Hymenolepis diminata*; **Temnocephala**: *Craspedella simulator, C. spenceri, Diceratocephala boschmani, Temnosewellia acirra, T. dendyi, T. minor*
Cherax quadricarinatus	**Virus**: CGV, *Cq*BV, *Cq*PlV, *Cq*PB, *Cq*RV, *Mr*NV, PMergDNV, SMV, YHD, WSSV; **Bacteria**: *Acinetobacter* sp., *Aeromonas hydrophila, A. sobria, Alcaligenes* sp., *Bacillus* sp., *Citrobacter* sp., *C. freundii, Corynebacterium* sp., *Coxiella cheraxi, Edwardsiella tarda, Enterobacter agglomerans, E. intermedium, Escherichia coli, Flavobacterium* sp., *Klebsiella pneumoniae, Micrococcus* sp., *M. luteus,* Mollicute-like prokaryont, *Moraxella* sp., *Mycobacterium chelonae, Pleisomonas shigelloides, Pseudomonas* sp., *P. cepacia, P. maltophila,* Rickettsia-like organism, *Shewanella putrefaciens, Staphylococcus cohnii, S. epidermidis, Vibrio cholerae, V. mimicus*; **Fungi**: *Achlya* sp., *Allomyces* sp., *Aphanomyces astaci, Lagenidium* sp., *Phytopythium* sp., *Saprolegnia* sp.; **Microsporidia**: *Thelohania* sp., *Vavraia parastacida*; **Mesomycetozoa**: *Psorospermium* sp.; **Protista**: *Acineta* sp., *Cothurnia* sp., *Epistylis* sp., *Lagenophrys* sp., *L. darwini, Setonophrys communis, Tetrahymena pyriformis, Trochilia* sp., *Vorticella* sp., *Zoothamnium* sp.; **Nematoda**: unidentified; **Temnocephalida**: *Craspedella pedum, Decadidymus gulosus, Diceratocephala bososchmai, Notodactylus handschini, Temnosewellia minor, T. rouxii, T. semperi*

aquaculture purposes to a large number of countries in Asia, North and South America, Africa and Europe. Established feral populations have been reported in central and south America, Asia and Africa.

Conclusions and future directions

This chapter has provided the first comprehensive list of pathogens, parasites and commensals of crayfish across their native and extended range that has been reported in the literature to date. It was apparent during the process of putting the lists together that there were a range of errors and mis-identifications in the literature that will need to be re-assessed using new and novel methods and re-sampling from animals across their range. In addition, there is an absence of data of the lifecycles of some parasites reported which should be collected. The lack of taxonomic rigour applied to some studies may lead to confusion when allegedly identical pathogens are reported on different continents thus making biogeographical interpretations difficult. Correct taxonomy of all fauna associated with crayfish is important to minimise the transfer of potential pathogens between watercourses, countries or continents. Incorrectly assuming that the same parasite occurs in the receiving water due to a lack of care in identification risks transfer of pathogenic forms to naïve populations.

It is recommended that efforts are made to identify disease agents, parasites and commensals when new species of crayfish are described or new populations of crayfish are examined. It is clear from the literature that for those crayfish species that have enjoyed extensive studies, they have a wider range and number of agents, e.g., approximately 70 different infections in the well studied *C. quadricarinatus* (Table 13), around 80 infections in *A. astacus* (Table 11) and around 150 infections in *P. (S.) clarkii* (Table 12)—it would disingenuous to suggest that this is because those species naturally have more infections (although some of these will have been obtained as a result of anthropogenic movements), rather it likely represents the potential diversity and range of infections present in all ≈640 extant crayfish species that have been described worldwide. If only those infections which are unique to those three species above are considered, the numbers of species reported in these hosts drop to around 50 for *C. quadricarinatus*, to around 70 for *A. astacus* and around 130 for *P. (S.) clarkii*. Taking the lowest estimate of 50 infections in *C. quadricarinatus* and multiplying by the number of extant crayfish species, there is the potential to report 32,000 infections in crayfish; this compares unfavourably with the estimated 900 parasites and pathogens reported in crayfish to date.

Examination of the health status of crayfish in their native range may provide information useful to control invasive species in their non-native range. The principle of biological control by infecting non-natives with species specific pathogens from their home range is embedded within the enemy release hypothesis which suggests that introduced species can be successful in a new area if released from control by their natural enemies, including disease. It is interesting to note that the bulk of viral infections reported in crayfish are from antipodean species with a distinct lack of information on viral infections in the Americas—this perhaps reflects the different interests of researchers in the two continents but leads to the promise that many potential pathogenic viruses of crayfish exist in, e.g., the Americas that might have a utility in controlling invasive species. However, caution needs to be exercised to ensure that transmission to non-target organisms does not occur.

Finally, there needs to be a concerted effort to move from studies with a narrow focus on a small range of disease-causing agents that are perceived to be problematic to taking a more holistic view that considers the impact of all infections in crayfish populations that have the potential to interact synergistically or independently leading to lethal and sub-lethal outcomes. Failure to do so will continue to stifle crayfish disease research long into the future with no advancement in the discipline of crayfish pathology in the widest sense and provide limited opportunities to study these important creatures in totality. The integration of different disciplines in the study of the health status of crayfish will lead to a greater understanding of the risk factors affecting crayfish populations, and ultimately to the protection of crayfish across their range through the removal or reduction of the threat of disease.

Acknowledgements

I would like to express my undying love and gratitude to my family who've allowed me the priviledge of hiding in my office at odd times of the day and night, during lost weekends, evenings, high days and holidays. This chapter is dedicated to Clare, Tom and Lottie.

References

Abbott, J.F. 1912a. An unusual symbiotic relation between a water bug and a crayfish. Am. Nat. 46: 553–556.

Abbott, J.F. 1912b. A new type of Corixidae (*Ramphocorixa balanodis*, n. gen., et sp.) with an account of its life history. Can. Entomol. 44: 113–121.

Aguilar-Alberola, J.A., F. Mesquita-Joanes, S. López, A. Mestre, J.C. Casanova, J. Rueda and A. Ribas. 2011. An invaded invader: high prevalence of entocytherid ostracods on the red swamp crayfish *Procambarus clarkii* (Girard, 1852) in the Eastern Iberian Peninsula. Hydrobiologia 688: 63–73.

Alderman, D.J. and J.L. Polglase. 1985a. Disinfection for crayfish plague. Aquac. Fish. Manag. 16: 203–205.

Alderman, D.J. and J.L. Polglase. 1985b. *Fusarium tabacinum* (Beyma) Gams. as a gill parasite in the crayfish, *Austropotamobius pallipes* Lereboullet. J. Fish Dis. 8: 249–252.

Alderman, D.J. and J.L. Polglase. 1988. Pathogens, parasites and commensals. pp. 167–212. *In*: D.M. Holdich and R.S. Lowery (eds.). Freshwater Crayfish: Biology, Management and Exploitation. Croom Helm, London.

Alderman, D.J., S.W. Feist and J.L. Polglase. 1986. Possible nocardiosis of crayfish, *Austropotamobius pallipes*. J. Fish Dis. 9: 345–347.

Allen, S. 1933. Parasites and commensals of North Carolina crayfishes. J. Elisha Mitchell Sci. Soc. 49: 119–121.

Amato, J.F.R. 2001. A new species of *Stratiodrilus* (Polychaeta, Histriobdellidae) from freshwater crayfishes of Southern Brazil. Iheringia, Série Zool. 90: 37–44.

Ambas, I., N. Buller and R. Fotedar. 2015. Isolation and screening of probiotic candidates from marron, *Cherax cainii* (Austin, 2002) gastrointestinal tract (GIT) and commercial probiotic products for the use in marron culture. J. Fish Dis. 38: 467–76.

Amborski, R.L., G. LoPiccolo, G.F. Amborski and J. Huner. 1975. A disease affecting the shell and soft tissues of Louisiana crayfish, *Procambarus clarkii*. Freshw. Crayfish 2: 299–316.

Ameel, D.J. 1937. The life history of *Crepidostomum cornutum* (Osborn). J. Parasitol. 23: 218–220.

Amin, O.M. 2013. Classification of the Acanthocephala. Folia Parasitol. 60: 273–305.

Anda, P., J. Segura del Pozo, J.M. Díaz García, R. Escudero, F.J. García Peña, M.C. López Velasco, R.E. Sellek, M.R. Jiménez Chillarón, L.P. Sánchez Serrano and J.F. Martínez Navarro. 2001. Waterborne outbreak of tularemia associated with crayfish fishing. Emerg. Infect. Dis. 7: 575–82.

Anderson, I.G. and H.C. Prior. 1992. Baculovirus infections in the mud crab, *Scylla serrata*, and a freshwater crayfish, *Cherax quadricarinatus*, from Australia. J. Invertebr. Pathol. 60: 265–273.

Andersson, M.G. and L. Cerenius. 2002. Analysis of chitinase expression in the crayfish plague fungus *Aphanomyces astaci*. Dis. Aquat. Org. 51: 139–147.

Andolshek, M.D. and H.H. Hobbs, Jr. 1986. The entocytherid ostracod fauna of Southeastern Georgia. Smithson. Contrib. to Zool. 424: 1–43.

Apakupakul, K., M.E. Siddall and E.M. Burreson. 1999. Higher level relationships of leeches (Annelida: Clitellata: Euhirudinea) based on morphology and gene sequences. Mol. Phylogenet. Evol. 12: 350–359.

Aquiloni, L., M.P. Martín, F. Gherardi and J. Diéguez-Uribeondo. 2011. The North American crayfish *Procambarus clarkii* is the carrier of the oomycete *Aphanomyces astaci* in Italy. Biol. Invasions 13: 359–367.

Aspán, A. and K. Söderhäll. 1991. Purification of prophenoloxidase from crayfish blood cells, and its activation by an endogenous serine proteinase. Insect Biochem. 21: 363–373.

Aspán, A., M. Hall and K. Söderhäll. 1990. The effect of endogeneous proteinase inhibitors on the prophenoloxidase activating enzyme, a serine proteinase from crayfish haemocytes. Insect Biochem. 20: 485–492.

Avenant-Oldewage, A. 1993. Occurrence of *Temnocephala chaeropsis* on *Cherax tenuimanus* imported into South Africa, and notes on its infestation of an indigenous crab. S. Afr. J. Sci. 89: 427–428.

Baker, D.G. 2007. Flynn's Parasites of Laboratory Animals. Blackwell Publishing Ltd., Oxford.

Baker, J.H. 1969. On the relationship of *Ankylocythere sinuosa* (Rioja, 1942) (Ostracoda, Entocytheridae) to the crayfish *Procambarus simulans simulans* (Faxon, 1884). Trans. Am. Microsc. Soc. 88: 293–294.

Ballesteros, I., M.P. Martín and J. Diéguez-Uribeondo. 2006. First isolation of *Aphanomyces frigidophilus* (Saprolegniales) in Europe. Mycotaxon 95: 335–340.

Ballesteros, I., M.P. Martín, L. Cerenius, K. Söderhäll, M.T. Telleria and J. Diéguez-Uribeondo. 2009. Lack of specificity of the molecular diagnostic method for identification of *Aphanomyces astaci*. Bull. Fr. Pêche Piscic. 385: 17–24.

Bangyeekhun, E., H.J. Ryynanen, P. Henttonen, J.V. Huner, L. Cerenius and K. Söderhäll. 2001. Sequence analysis of the ribosomal internal transcribed spacer DNA of the crayfish parasite *Psorospermium haeckeli*. Dis. Aquat. Org. 46: 217–222.

Bartsch, I. 1996. Halacarids (Halacaroidea, Acari) in freshwater. Multiple invasions from the Paleozoic onwards? J. Nat. Hist. 30: 67–99.

Bartsch, I. 2006. The freshwater mite *Porolohmannella violacea* (Kramer, 1879) (Acari: Halacaridae), description of juveniles and females and notes on development and distribution. Bonner Zool. Beiträge 55: 47–59.

Bartsch, I. 2011. North American freshwater Halacaridae (Acari): literature survey and new records. Int. J. Acarol. 37: 490–510.

Bateman, K.S., I. Tew, C. French, R.J. Hicks, P. Martin, J. Munro and G.D. Stentiford. 2012. Susceptibility to infection and pathogenicity of White Spot Disease (WSD) in non-model crustacean host taxa from temperate regions. J. Invertebr. Pathol. 110: 340–351.

Baumgartner, W.A., J.P. Hawke, K. Bowles, P.W. Varner and K.W. Hasson. 2009. Primary diagnosis and surveillance of white spot syndrome virus in wild and farmed crawfish (*Procambarus clarkii, P. zonangulus*) in Louisiana, USA. Dis. Aquat. Organ. 85: 15–22.

Bean, N.H., E.K. Maloney, M.E. Potter, P. Korazemo, B. Ray, J.P. Taylor, S. Seigler and J. Snowden. 1998. Crayfish: A newly recognized vehicle for *Vibrio* infections. Epidemiol. Infect. 121: 269–273.

Beaver, P.C. 1929. Studies on the development of *Allassostoma parvum* Stunkard. J. Parasitol. 16: 13–23.

Bertani, I., H. Segers and G. Rossetti. 2011. Biodiversity down by the flow: new records of monogonont rotifers for Italy found in the Po River. J. Limnol. 70: 321–328.

Bi, K., H. Huang, W. Gu, J. Wang and W. Wang. 2008. Phylogenetic analysis of Spiroplasmas from three freshwater crustaceans (*Eriocheir sinensis, Procambarus clarkia* and *Penaeus vannamei*) in China. J. Invertebr. Pathol. 99: 57–65.

Bochow, S., K. Condon, J. Elliman and L. Owens. 2015. First complete genome of an Ambidensovirus; *Cherax quadricarinatus* densovirus, from freshwater crayfish *Cherax quadricarinatus*. Mar. Genomics 24: 305–312.

Boshko, E. 1980. Rotifers (Rotatoria) in long clawed crayfish gill cavity from the Dnieper river basin (Ukrainian SSR). Vestn. Zool. 6: 41–46.

Boshko, E. 1981. The occurrence of *Psorospermium haeckeli* Hilgendorf in river crabs in the water bodies of the Dnieper (Ukrainian SSR). Vestn. Zool. 6: 73–76.

Boshko, E. 1983. Crayfish oligochaetes in the water bodies of the Ukraine. pp. 22–23. *In*: Proceedings of the Fourth All-Union Symposium, Tbilisi, 5–7 October 1983.

Boshko, E. 1995. New species of infusoria of the genera *Sincothurnia* and *Lagenophrys*. Zool. Zh. 74: 5–9.

Bouckenooghe, A.R. and B.J. Marino. 2001. *Psorospermium haeckelii*: a cause of pseudoparasitosis. South. Med. J. 94: 233–234.

Bowater, R.O., M. Wingfield, M. Fisk, M.L.C. Kelly, A. Reid, H. Prior and E.C. Kulpa. 2002. A parvo-like virus in cultured redclaw crayfish *Cherax quadricarinatus* from Queensland, Australia. Dis. Aquat. Org. 50: 79–86.

Bowman, T., R. Prins and B. Morris. 1968. Notes on the harpacticoid copepods *Attheyella pilosa* and *A. carolinensis*, associates of crayfishes in the Eastern United States. Proc. Biol. Soc. Washingt. 81: 571–586.

Brannelly, L., T. McMahon, M. Hinton, D. Lenger and C. Richards-Zawacki. 2015. *Batrachochytrium dendrobatidis* in natural and farmed Louisiana crayfish populations: prevalence and implications. Dis. Aquat. Org. 112: 229–235.

Brazner, J.C. and D.A. Jensen. 2000. Zebra mussel [*Dreissena polymorpha* (Pallas)] colonization of rusty crayfish [*Orconectes rusticus* (Girard)] in Green Bay, Lake Michigan. Am. Midl. Nat. 143: 250–256.

Brinkhurst, R.O. 1971. The aquatic oligochaeta known from Australia, New Zealand, Tasmania, and the adjacent islands. Univ. Queensl. Press 3: 99–128.

Brooks, D.R. 1975. A review of the genus *Allassostomoides* Stunkard 1924 (Trematoda: Paramphistomidae) with a redescription of *A. chelydrae* (MacCallum 1919) Yamaguti 1958. J. Parasitol. 61: 882–885.

Brown, B.L., R.P. Creed and W.E. Dobson. 2002. Branchiobdellid annelids and their crayfish hosts: are they engaged in a cleaning symbiosis? Oecologia 132: 250–255.

Brown, B.L., R.P. Creed, J. Skelton, M.A. Rollins and K.J. Farrell. 2012. The fine line between mutualism and parasitism: complex effects in a cleaning symbiosis demonstrated by multiple field experiments. Oecologia 170: 199–207.

Brown, P.B., M.R. White, D.L. Swann and M.S. Fuller. 1993. A severe outbreak of ectoparasitism due to *Epistylis* sp. in pond-reared orconectid crayfish. J. World Aquac. Soc. 24: 116–120.

Browning, J.S. and S.C. Landers. 2012. Exuviotrophic apostome ciliates from freshwater decapods in southern Alabama (USA) and a description of *Hyalophysa clampi* n. sp. (Ciliophora, Apostomatida). Eur. J. Protistol. 48: 207–214.

Cammà, C., N. Ferri, D. Zezza, M. Marcacci, A. Paolini, L. Ricchiuti and R. Lelli. 2010. Confirmation of crayfish plague in Italy: Detection of *Aphanomyces astaci* in white clawed crayfish. Dis. Aquat. Org. 89: 265–268.

Cannon, L.R.G. 1991. Temnocephalan symbionts of the freshwater crayfish *Cherax quadricarinatus* from northern Australia. Hydrobiol. 227: 341–347.

Cannon, L.R.G. 1993. New temnocephalans (Platyhelminthes): ectosymbionts of freshwater crabs and shrimps. Mem. Queensl. Museum 33: 17–40.

Cannon, L.R.G. and J.B. Jennings. 1987. Occurrence and nutritional relationships of four ectosymbiotes of the freshwater crayfish *Cherax dispar* Riek and *Cherax punctatus* Clark (Crustacea: Decapoda) in Queensland. Aust. J. Mar. Freshw. Res. 38: 419–427.

Cannon, L.R.G. and K.B. Sewell. 1995. Craspedellinae Baer, 1931 (Playthelminthes: Temnocephalida) ectosymbionts from the branchial chamber of Australian crayfish (Crustacea: Parasitica). Mem. Queensl. Museum 38: 397–418.

Cannon, L.R.G. and K.B. Sewell. 2001a. Observations on *Dactylocephala madagascariensis* (Vayssiere, 1892), a temnocephalan with twelve tentacles from Madagascar. Zoosystema 23: 11–18.

Cannon, L.R.G. and K.B. Sewell. 2001b. A review of *Temnosewellia* (Platyhelminthes) ectosymbionts of *Cherax* (Crustacea: Parastacidae) in Australia. Mem. Queensl. Museum 46: 385–399.

Cardini, A. and M. Ferraguti. 2004. The phylogeny of Branchiobdellida (Annelida, Clitellata) assessed by sperm characters. Zool. Anz. 243: 37–46.

Carney, J.P. and D.R. Brooks. 1991. Phylogenetic analysis of *Alloglossidium* Simer, 1929 (Digenea: Plagiorchiiformes: Macroderoididae) with discussion of the origin of truncated life cycle patterns in the genus. J. Parasitol. 77: 890–900.

Caveny, B.A. and F.J. Etges. 1971. Life history studies of *Microphallus opacus* (Trematoda: Microphallidae). J. Parasitol. 57: 1215–1221.

Cerenius, L., P. Henttonen, O.V. Lindqvist and K. Söderhäll. 1991. The crayfish pathogen *Psorospermium haeckeli* activates the prophenoloxidase activating system of freshwater crayfish *in vitro*. Aquaculture 99: 225–233.

Cerenius, L., S. Rufelt and K. Söderhäll. 1992. Effects of ampropylfos ((RS)-1-aminopropylphosphonic acid) on zoospore formation, repeated zoospore emergence and oospore formation in *Aphanomyces* spp. Pestic. Sci. 36: 189–194.

Chaga, O., M. Lignell and K. Söderhäll. 1995. The haemopoietic cells of the freshwater crayfish *Pacifastacus leniusculus*. Anim. Biol. 4: 59–70.

Chai, J.Y., W.M. Sohn, S. Huh, M.H. Choi and S.H. Lee. 1996. Redescription of *Macroorchis spinulosus* Ando, 1918 (Digenea: Nanophyetidae) encysted in the fresh water crayfish, *Cambaroides similis*. Korean J. Parasitol. 34: 1–6.

Chai, J.Y., E.H. Shin, S.H. Lee and H.J. Rim. 2009. Foodborne intestinal flukes in Southeast Asia. Korean J. Parasitol. 47 (suppl.): S69–S102.

Cheng, T.C. 1957. A study of the metacercaria of *Crepidostomum cornutum* (Osborn, 1903), (Trematoda: Allocreadiidae). Proc. Helminthol. Soc. Wash. 24: 107–109.

Cheng, T. and H. James. 1960a. The histopathology of *Crepidostomum* sp. infection in the second intermediate host, *Sphaerium striatinum*. Proc. Helminthol. Soc. Wash. 27: 67–68.

Cheng, T.C. and H.A. James. 1960b. Studies on the germ cell cycle, morphogenesis and development of the cercarial stage of *Crepidostomum cornutum* (Osborn, 1903) (Trematoda: Allocreadiidae). Trans. Am. Microsc. Soc. 79: 75–85.

Chiesa, S., M. Scalici, L. Lucentini and F. Marzano. 2015. Molecular identification of an alien temnocephalan crayfish parasite in Italian freshwaters. Aquat. Invasions 10: 209–216.

Chinain, M. and A. Vey. 1987. Infection caused by *Fusarium solani* in crayfish *Astacus leptodactylus*. Freshw. Crayfish 7: 195–202.

Clamp, J.C. 1987. Five new species of *Lagenophrys* (Ciliophora, Peritricha, Lagenophryidae) from the United States with observations on their developmental stages. J. Protozool. 34: 382–392.

Clamp, J.C. 1988. *Lagenophrys anticthos* n. sp. and *L. aegleae* Mouchet-Bennati, 1932 (Ciliophora, Peritricha, Lagenophryidae), ectocommensals of South American crustaceans. J. Protozool. 35: 164–169.

Clamp, J.C. 1991. Revision of the Family Lagenophryidae Bütschli, 1889 and description of the Family Usconophryidae n. fam. (Ciliophora, Peritricha). J. Protozool. 38: 355–377.

Clamp, J.C. 1992. Three new species of lagenophryid peritrichs (Ciliophora) ectocommensal on freshwater decapod crustaceans from Madagascar. J. Protozool. 39: 732–740.

Clamp, J.C. 1994. New species of *Lagenophrys* (Ciliophora, Peritrichia) from New Zealand and Australia. J. Eukaryot. Microbiol. 41: 343–349.

Claydon, K., B. Cullen and L. Owens. 2004a. Methods to enhance the intensity of intranuclear bacilliform virus infection in *Cherax quadricarinatus*. Dis. Aquat. Org. 60: 173–178.

Claydon, K., B. Cullen and L. Owens. 2004b. OIE white spot syndrome virus PCR gives false-positive results in *Cherax quadricarinatus*. Dis. Aquat. Org. 62: 265–268.

Cooper, A., R. Layton, L. Owens, N. Ketheesan and B. Govan. 2007. Evidence for the classification of a crayfish pathogen as a member of the genus *Coxiella*. Lett. Appl. Microbiol. 45: 558–563.

Corbel, V., Z. Zuprizal, C. Shi, J.-M. Arcier and J.-R. Bonami. 2001. Experimental infection of European crustaceans with white spot syndrome virus (WSSV). J. Fish Dis. 24: 377–382.

Corkum, K.C. and H.M. Turner. 1977. *Alloglossoides caridicola* gen. et sp. n. (Trematoda: Macroderoididae) from a Louisiana Crayfish. Proc. Helminthol. Soc. Wash. 44: 176–178.

Coughran, J. 2011a. Aspects of the biology and ecology of the Orange-Bellied Crayfish, *Euastacus mirangudjin* Coughran 2002, from northeastern New South Wales. Aust. Zool. 35: 750–756.

Coughran, J. 2011b. Biology of the Blood Crayfish, *Euastacus gumar* Morgan 1997, a small freshwater crayfish from the Richmond Range, northeastern New South Wales. Aust. Zool. 35: 685–697.

Crawford, E., Jr. 1965. Three new species of epizoic ostracods (Ostracoda, Entocytheridae) from North and South Carolina. Am. Midl. Nat. 74: 148–154.

Cribb, T.H. 1985. The life cycle and biology of *Opecoelus variabilis* sp. nov. (Digenea: Opecoelidae). Aust. J. Zool. 33: 715–728.

Crichton, V.F.J. and M. Beverley-Burton. 1977. Observations on the seasonal prevalence, pathology and transmission of *Dracunculus insignis* (Nematoda: Dracunculoidea) in the raccoon (*Procyon lotor* (L.)) in Ontario. J. Wildl. Dis. 13: 273–280.

Cuellar, M., I. Garcia-Cuenca and J. Fontanillas. 2002. Description de la zooépibiose de l'écrevisse signal (*Pacifastacus leniusculus*, Dana) en astaciculture. Bull. Fr. Pêche Piscic. 367: 959–972.

Damborenea, M.C. and L.R.G. Cannon. 2001. On neotropical *Temnocephala* (Platyhelminthes). J. Nat. Hist. 35: 1103–1118.

Davidson, E.W., J. Snyder, D. Lightner, G. Ruthig, J. Lucas and J. Gilley. 2010. Exploration of potential microbial control agents for the invasive crayfish, *Orconectes virilis*. Biocontrol Sci. Technol. 20: 297–310.

Deblock, S. 1973. Contribution to the study of Microphallidae Travassos, 1920 (Trematoda). 27. On some species described by S. Yamaguti in Japan: A. Invalidation of the genus *Maritreminoides* Rankin. Creation of genus satellites of the genus *Maritrema*: *Quasimaritrema*, Marit. Ann. Parasitol. Hum. Comp. 48: 543–557.

Defaye, D. 1996. Redescription of *Nitokra divaricata* Chappuis, 1923 (Copepoda, Harpacticoida) with first records from *Austropotamobius torrentium* Schrank. Acta Zool. Acad. Sci. Hung. 42: 145–155.

DeWitt, P.D., B.W. Williams, Z.-Q.Q. Lu, A.N. Fard and S.R. Gelder. 2013. Effects of environmental and host physical characteristics on an aquatic symbiont. Limnologica 43: 151–156.

Dexter, R.W. 1954. An unusual pearl from a freshwater mussel and a pearl-like growth from a crayfish. Ohio. J. Sci. 54: 241–242.

Diéguez-Uribeondo, J., J. Pinedo-Ruiz and L. Cerenius. 1993. Presence of *Psorospermium haeckeli* (Hilgendorf) in a *Pacifastacus leniusculus* (Dana) population of Spain. Freshw. Crayfish 9: 286–288.

Diéguez-Uribeondo, J., L. Cerenius and K. Söderhäll. 1994. *Saprolegnia parasitica* and its virulence on three different species of crayfish. Aquaculture 120: 219–228.

Diéguez-Uribeondo, J., T.S. Huang, L. Cerenius and K. Söderhäll. 1995. Physiological adaptation of an *Aphanomyces astaci* strain isolated from the warm-water crayfish *Procambarus clarkii*. Mycol. Res. 99: 574–578.

Diéguez-Uribeondo, J., J.M. Fregeneda-Grandes, L. Cerenius, E. Pérez-Iniesta, J.M. Aller-Gancedo, M.T. Tellería, K. Söderhäll and M.P. Martín. 2007. Re-evaluation of the enigmatic species complex *Saprolegnia diclina-Saprolegnia parasitica* based on morphological, physiological and molecular data. Fungal Genet. Biol. 44: 585–601.

Diéguez-Uribeondo, J., M.A. Garcia, L. Cerenius, E. Kozubíková, I. Ballesteros, C. Windels, J. Weiland, H. Kator, K. Söderhäll and M.P. Martín. 2009. Phylogenetic relationships among plant and animal parasites, and saprotrophs in *Aphanomyces* (Oomycetes). Fungal Genet. Biol. 46: 365–376.

Diler, Ö. and Y. Bolat. 2001. Isolation of *Acremonium* species from crayfish, *Astacus leptodactylus* in Egirdir Lake. Bull. Eur. Assoc. Fish Pathol. 21: 164–168.

Ding, Z., K. Bi, T. Wu, W. Gu, W. Wang and J. Chen. 2007. A simple PCR method for the detection of pathogenic spiroplasmas in crustaceans and environmental samples. Aquaculture 265: 49–54.

Ding, Z., J. Du, J. Ou, W. Li, T. Wu, Y. Xiu, Q. Meng, Q. Ren, W. Gu, H. Xue, J. Tang and W. Wang. 2012. Classification of circulating hemocytes from the red swamp crayfish *Procambarus clarkii* and their susceptibility to the novel pathogen *Spiroplasma eriocheiris in vitro*. Aquaculture 356-357: 371–380.

Ding, Z., W. Yao, J. Du, Q. Ren, W. Li, T. Wu, Y. Xiu, Q. Meng, W. Gu, H. Xue, J. Tang and W. Wang. 2013. Histopathological characterization and *in situ* hybridization of a novel spiroplasma pathogen in the freshwater crayfish *Procambarus clarkii*. Aquaculture 380-383: 106–113.

Ding, Z.F., S.Y. Xia, H. Xue, J.Q. Tang, Q. Ren, W. Gu, Q.G. Meng and W. Wang. 2015. Direct visualization of the novel pathogen, *Spiroplasma eriocheiris*, in the freshwater crayfish *Procambarus clarkii* (Girard) using fluorescence *in situ* hybridization. J. Fish Dis. 38: 787–794.

Dollfus, R.-P., J. Callot and C. Desportes. 1935. *Distoma isostoma* Rudolphi 1819, parasite d'*Astacus*, est une metacercaire d'*Orchipedum*. Ann. Parasitol. Hum. Comp. 13: 116–132.

Dörr, A.J.M., M. Rodolfi, M. Scalici, A.C. Elia, L. Garzoli and A.M. Picco. 2011. *Phoma glomerata*, a potential new threat to Italian inland waters. J. Nat. Conserv. 19: 370–373.

Dörr, A.J.M., A.C. Elia, M. Rodolfi, L. Garzoli, A.M. Picco, M. D'Amen and M. Scalici. 2012. A model of co-occurrence: segregation and aggregation patterns in the mycoflora of the crayfish *Procambarus clarkii* in Lake Trasimeno (central Italy). J. Limnol. 71: 135–143.

Du, H., Z. Xu, X. Wu, W. Li and W. Dai. 2006. Increased resistance to white spot syndrome virus in *Procambarus clarkii* by injection of envelope protein VP28 expressed using recombinant baculovirus. Aquaculture 260: 39–43.

Du, H., L. Fu, Y. Xu, Z. Kil and Z. Xu. 2007. Improvement in a simple method for isolating white spot syndrome virus (WSSV) from the crayfish *Procambarus clarkii*. Aquaculture 262: 532–534.

Du, H., W. Dai, X. Han, W. Li, Y. Xu and Z. Xu. 2008. Effect of low water temperature on viral replication of white spot syndrome virus in *Procambarus clarkii*. Aquaculture 277: 149–151.

Du, Z.-Q., J.-F. Lan, Y.-D. Weng, X.-F. Zhao and J.-X. Wang. 2013. BAX inhibitor-1 silencing suppresses white spot syndrome virus replication in red swamp crayfish, *Procambarus clarkii*. Fish Shellfish Immunol. 35: 46–53.

du Preez, L. and N. Smit. 2013. Double blow: Alien crayfish infected with invasive temnocephalan in South African waters. S. Afr. J. Sci. 109: 1–4.

Dunn, J.C., H.E. McClymont, M. Christmas and A.M. Dunn. 2009. Competition and parasitism in the native white clawed crayfish *Austropotamobius pallipes* and the invasive signal crayfish *Pacifastacus leniusculus* in the UK. Biol. Invasions 11: 315–324.

Ďuriš, Z., I. Horká, J. Kristian and P. Kozák. 2006. Some cases of macro-epibiosis on the invasive crayfish *Orconectes limosus* in the Czech Republic. Bull. Fr. Pêche Piscic. 380-381: 1325–1337.

Duvic, B. and K. Söderhäll. 1990. Purification and characterization of a beta-1,3-glucan binding protein from plasma of the crayfish *Pacifastacus leniusculus*. J. Biol. Chem. 265: 9327–9332.

Eaves, L.E. and P.J. Ketterer. 1994. Mortalities in red claw crayfish *Cherax quadricarinatus* associated with systemic *Vibrio mimicus* infection. Dis. Aquat. Org. 19: 233–237.

Edgerton, B. 1996. A new bacilliform virus in Australian *Cherax destructor* (Decapoda: Parastacidae) with notes on *Cherax quadricarinatus* bacilliform virus (=*Cherax* baculovirus). Dis. Aquat. Org. 27: 43–52.

Edgerton, B.F. 2000. A compendium of idiopathic lesions observed in redclaw freshwater crayfish, *Cherax quadricarinatus* (von Martens). J. Fish Dis. 23: 103–113.

Edgerton, B.F. 2002. Hazard analysis of exotic pathogens of potential threat to European freshwater crayfish. Bull. Fr. Pêche Piscic. 367: 813–820.

Edgerton, B.F. 2003. Further studies reveal that *Austropotamobius pallipes* bacilliform virus (*Ap*BV) is common in populations of native freshwater crayfish in south-eastern France. Bull. Eur. Assoc. Fish Pathol. 23: 7–12.

Edgerton, B. 2004a. Studies on the susceptibility of the European white-clawed freshwater crayfish, *Austropotambius pallipes* (Lereboullet), to white spot syndrome virus for analysis of the likelihood of introduction and impact on European freshwater crayfish populations. Freshw. Crayfish 14: 228–235.

Edgerton, B.F. 2004b. Susceptibility of the Australian freshwater crayfish *Cherax destructor albidus* to white spot syndrome virus (WSSV). Dis. Aquat. Org. 59: 187–193.

Edgerton, B. and L. Owens. 1997. Age at first infection of *Cherax quadricarinatus* by *Cherax quadricarinatus* bacilliform virus and *Cherax* giardiavirus-like virus, and production of putative virus-free crayfish. Aquaculture 152: 1–12.

Edgerton, B.F. and L. Owens. 1999. Histopathological surveys of the redclaw freshwater crayfish, *Cherax quadricarinatus*, in Australia. Aquaculture 180: 23–40.

Edgerton, B.F. and H.C. Prior. 1999. Description of a hepatopancreatic rickettsia-like organism in the redclaw crayfish *Cherax quadricarinatus*. Dis. Aquat. Org. 36: 77–80.

Edgerton, B.F., L. Owens, B. Glasson and S. DeBeer. 1994. Description of a small dsRNA virus from freshwater crayfish *Cherax quadricarinatus*. Dis. Aquat. Org. 18: 63–69.

Edgerton, B., L. Owens, L. Harris, A. Thomas and M. Wingfield. 1995. A health survey of farmed redclaw crayfish, *Cherax quadricarinatus* (von Martens), in tropical Australia. Freshw. Crayfish 10: 322–337.

Edgerton, B., P. O'Donoghue, M. Wingfield and L. Owens. 1996a. Systemic infection of freshwater crayfish *Cherax quadricarinatus* by hymenostome ciliates of the *Tetrahymena pyriformis* complex. Dis. Aquat. Org. 27: 123–129.

Edgerton, B.F., P. Paasonen, P. Henttonen and L. Owens. 1996b. Description of a bacilliform virus from the freshwater crayfish, *Astacus astacus*. J. Invertebr. Pathol. 68: 187–190.

Edgerton, B., R. Webb and M. Wingfield. 1997. A systemic parvo-like virus in the freshwater crayfish *Cherax destructor*. Dis. Aquat. Org. 29: 73–78.

Edgerton, B.F., R. Webb, I.G. Anderson and E.C. Kulpa. 2000. Description of a presumptive hepatopancreatic reovirus, and a putative gill parvovirus, in the freshwater crayfish *Cherax quadricarinatus*. Dis. Aquat. Org. 41: 83–90.

Edgerton, B.F., L.H. Evans, F.J. Stephens and R.M. Overstreet. 2002a. Synopsis of freshwater crayfish diseases and commensal organisms. Aquaculture 206: 57–135.

Edgerton, B.F., H. Watt, J.-M. Becheras and J.-R. Bonami. 2002b. An intranuclear bacilliform virus associated with near extirpation of *Austropotamobius pallipes* Lereboullet from the Nant watersed in Ardéche, France. J. Fish Dis. 25: 523–531.

Edgerton, B., P. Henttonen, J. Jussila, A. Mannonen, P. Paasonen, T. Taugbøl, L. Edsman and C. Souty-Grosset. 2004. Understanding the causes of disease in European freshwater crayfish. Conserv. Biol. 18: 1466–1474.

Edsman, L., P. Nyström, A. Sandström, M. Stenberg, H. Kokko, V. Tiitinen, J. Makkonen and J. Jussila. 2015. Eroded swimmeret syndrome in female crayfish *Pacifastacus leniusculus* associated with *Aphanomyces astaci* and *Fusarium* spp. infections. Dis. Aquat. Org. 112: 219–228.

Ellis, M.M. 1919. The branchiobdellid worms in the collections of the United States National Museum, with descriptions of new genera and new species. Proc. U.S. Natl. Mus. 55: 241–265.

Erséus, C., M.J. Wetzel and L. Gustavsson. 2008. ICZN rules—a farewell to Tubificidae (Annelida, Clitellata). Zootaxa 1744: 66–68.

Fang, D.-A., X.-M. Huang, Z.-Q. Zhang, D.-P. Xu, Y.-F. Zhou, M.-Y. Zhang, K. Liu, J.-R. Duan and W.-G. Shi. 2013. Molecular cloning and expression analysis of chymotrypsin-like serine protease from the redclaw crayfish (*Cherax quadricarinatus*): a possible role in the junior and adult innate immune systems. Fish Shellfish Immunol. 34: 1546–1552.

Fard, A.N., A.A. Motalebi, B.J. Jafari, M.A. Meshgi, D. Azadikhah and M. Afsharnasab. 2011. Survey on fungal, parasites and epibionts infestation on the *Astacus leptodactylus* (Eschscholtz, 1823), in Aras Reservoir West Azarbaijan, Iran. Iran. J. Fish. Sci. 10: 266–275.

Farrell, K., R. Creed and B. Brown. 2014. Reduced densities of ectosymbiotic worms (Annelida: Branchiobdellida) on reproducing female crayfish. Southeast. Nat. 13: 523–529.

Fernandez-Leborans, G. and M.L. Tato-Porto. 2000a. A review of the species of protozoan epibionts on crustaceans. I. Peritrich ciliates. Crustaceana 73: 643–683.

Fernandez-Leborans, G. and M.L. Tato-Porto. 2000b. A review of the species of protozoan epibionts on crustaceans. II. Suctorian ciliates. Crustaceana 73: 1205–1237.

Figueroa, A. and H.H. Hobbs, Jr. 1974. Three new crustaceans from La Media Luna, San Luis Potosí, Mexico. Smithson. Contrib. to Zool. 174: 1–18.

Fischer, P.U., K.C. Curtis, L.A. Marcos and G.J. Weil. 2011. Molecular characterization of the North American lung fluke *Paragonimus kellicotti* in Missouri and its development in Mongolian gerbils. Am. J. Trop. Med. Hyg. 84: 1005–11.

Font, W.F. 1994. *Alloglossidium greeri* n. sp. (Digenea: Macroderoididae) from the cajun dwarf crayfish, *Cambarellus schufeldti*, in Louisiana, U.S.A. Trans. Am. Microsc. Soc. 113: 86–89.

Fontaneto, D. and U. Jondelius. 2011. Broad taxonomic sampling of mitochondrial cytochrome c oxidase subunit I does not solve the relationships between Rotifera and Acanthocephala. Zool. Anz. 250: 80–85.

Füreder, L., M. Summerer and A. Brandstätter. 2009. Phylogeny and species composition of five European species of *Branchiobdella* (Annelida: Clitellata: Branchiobdellida) reflect the biogeographic history of three endangered crayfish species. J. Zool. 279: 164–172.

Gao, M., F. Li, L. Xu and X. Zhu. 2014. White spot syndrome virus strains of different virulence induce distinct immune response in *Cherax quadricarinatus*. Fish Shellfish Immunol. 39: 17–23.

García-Varela, M., G. Pérez-Ponce de León, F.J. Aznar and S.A. Nadler. 2011. Erection of *Ibirhynchus* gen. Nov. (Acanthocephala: Polymorphidae), based on molecular and morphological data. J. Parasitol. 97: 97–105.

Garzoli, L., D. Paganelli, M. Rodolfi, D. Savini, M. Moretto, A. Occhipinti-Ambrogi and A.M. Picco. 2014. First evidence of microfungal "extra oomph" in the invasive red swamp crayfish *Procambarus clarkii*. Aquat. Invasions 9: 47–58.

Gatta, C.L., E. Comunale and C.I. Menghi. 2009. *Psorospermium haeckelii*: pseudoparásito hallado en un vector. Rev. Argent. Microbiol. 41: 198.

Gazi, M., T. Sultana, G.S. Min, Y.C. Park, M. García-Varela, S.A. Nadler and J.K. Park. 2012. The complete mitochondrial genome sequence of *Oncicola luehei* (Acanthocephala: Archiacanthocephala) and its phylogenetic position within Syndermata. Parasitol. Int. 61: 307–316.

Geasa, N. 2014. Pathological and histopathological investigations of disease causing organisms in freshwater crayfish *Pacifastacus leniusculus*. African J. Sci. Issues 2: 87–93.

Gelder, S. 1996a. Description of a new branchiobdellidan species, with observations on three other species, and a key to the genus *Pterodrilus* (Annelida: Clitellata). Proc. Biol. Soc. Washingt. 109: 256–263.

Gelder, S.R. 1996b. A review of the taxonomic nomenclature and a checklist of the species of the Branchiobdellae (Annelida: Clitellata). Proc. Biol. Soc. Washingt. 109: 653–663.

Gelder, S. 2004. Endemic ectosymbiotic branchiobdellidans (Annelida: Clitellata) reported on three "export" species of North American crayfish (Crustacea: Astacoidea). Freshw. Crayfish 14: 221–227.

Gelder, S. 2010. Re-description of the branchiobdellidan *Hidejiodrilus koreanus* (Pierantoni, 1912) (Annelida: Clitellata), from the Republic of Korea, and the designation of a neotype and paraneotype specimens. Acta Zool. Bulg. 62: 21–26.

Gelder, S. 2011. Reassignment of a Central American species of the Branchiobdellida (Annelida: Clitellata) to *Forbesodrilus* n.g. Acta Zool. Bulg. 63: 119–123.

Gelder, S. 2014. Review of the geographic distribution of acceptable crustacean hosts for obligate, ectosymbiotic branchiobdellidans (Annelida: Clitellata). Freshw. Crayfish 20: 81–85.

Gelder, S.R. and R.O. Brinkhurst. 1990. An assessment of the phylogeny of the Branchiobdellida (Annelida: Clitellata), using PAUP. Can. J. Zool. 68: 1318–1326.

Gelder, S.R. and L.A. Hall. 1990. Description of *Xironogiton victoriensis* n. sp. from British Columbia, Canada, with remarks on other species and a Wagner analysis of *Xironogiton* (Clitellata: Branchiobdellida). Can. J. Zool. 68: 2352–2359.

Gelder, S.R. and A. Ohtaka. 2000a. Redescription and designation of lectotypes of the North American *Cambarincola okadai* Yamaguchi, 1933 (Annelida: Clitellata: Branchiobdellidae). Proc. Biol. Soc. Washingt. 113: 1089–1095.

Gelder, S.R. and A. Ohtaka. 2000b. Description of a new species and a redescription of *Cirrodrilus aomorensis* (Yamaguchi, 1934) with a detailed distribution of the branchiobdellidans (Annelida: Clitellata) in northern Honshu, Japan. Proc. Biol. Soc. Washingt. 113: 633–643.

Gelder, S.R. and M. Ferraguti. 2001. Diversity of spermatozoan morphology in two families of Branchiobdellida (Annelida: Clitellata) from North America. Can. J. Zool. 79: 1380–1393.

Gelder, S.R. and M.E. Siddall. 2001. Phylogenetic assessment of the Branchiobdellidae (Annelida, Clitellata) using 18S rDNA, mitochondrial cytochrome c oxidase subunit I and morphological characters. Zool. Scr. 30: 215–222.

Gelder, S. and A. Ohtaka. 2002. A review of the Oriental branchiobdellidans (Annelida: Clitellata) with reference to the rediscovered slide collection of Prof. Hideji Yamaguchi. Species Divers. 7: 333–344.

Gelder, S.R. and B.W. Williams. 2011. First distributional study of Branchiobdellida (Annelida: Clitellata) in the Great Smoky Mountains National Park, North Carolina and Tennessee, USA, with a redescription of *Cambarincola holostomus* Hoffman, 1963. Southeast. Nat. 10: 211–220.

Gelder, S.R., G.B. Delmastro and M. Ferraguti. 1994. A report on branchiobdellidans (Annelida: Clitellata) and a taxonomic key to the species in northern Italy, including the first record of *Cambarincola mesochoreus* on the introduced American red swamp crayfish. Boll. di Zool. Naples 61: 179–183.

Gelder, S.R., G.B. Delmastro and J.N. Rayburn. 1999. Distribution of native and exotic branchiobdellidans (Annelida: Clitellata) on their respective crayfish hosts in northern Italy, with the first record of native *Branchiobdella* species on an exotic North American crayfish. J. Limnol. 58: 20–24.

Gelder, S.R., H.C. Carter and D.N. Lausier. 2001. Distribution of crayfish worms or branchiobdellidans (Annelida: Clitellata) in New England. Northeast. Nat. 8: 79–92.

Gelder, S.R., J.F. Parpet and F. Quaglio. 2012. First report of two North American branchiobdellidans (Annelida: Clitellata) or crayfish worms on signal crayfish in Europe with a discussion of similar introductions into Japan. Ann. Limnol. 48: 315–322.

Gherardi, F., F. Cenni, G. Crudele and M. Mori. 2002. Infestation rate of branchiobdellids in *Austropotamobius pallipesitalicus* from a stream of central Italy: preliminary results. Bull. Fr. Pêche Piscic. 367: 785–792.

Gill, E.E. and N.M. Fast. 2006. Assessing the microsporidia-fungi relationship: Combined phylogenetic analysis of eight genes. Gene 375: 103–109.

Glockling, S.L., W.L. Marshall and F.H. Gleason. 2013. Phylogenetic interpretations and ecological potentials of the Mesomycetozoea (Ichthyosporea). Fungal Ecol. 6: 237–247.

Goetsch, W. 1935. Fauna Chilensis. II. Untersuchungen zur kentnis der zoologie und biogeographie Chiles. Biologie und regeneration von *Temnocephala chilensis*. Zool. Jahrb. 67: 195–212.

Goldschmidt, T., R. Gerecke and G. Alberti. 2002. *Hygrobates salamandrarum* sp. nov. (Acari, Hydrachnidia, Hygrobatidae) from China: the first record of a freshwater mite parasitizing newts (Amphibia, Urodela). Zool. Anz. 241: 297–304.

Golvan, Y.J. 1961. Le phylum des Acanthocephala. Troisieme note. La classe des Palaeacanthocephala (Meyer, 1931). Ann. Parasitol. 36: 76–91.

Goodnight, C. 1940. The Branchiobdellidae (Oligochaeta) of North American crayfishes. Illinois Biol. Monogr. 17: 5–75.

Goodnight, C. 1943. Report on a collection of branchiobdellids. J. Parasitol. 29: 100–102.

Goodrich, H.P. 1956. Crayfish epidemics. Parasitology 46: 480–483.

Graham, L. and R. France. 1986. Attempts to transmit experimentally the microsporidian *Thelohania contejeani* in freshwater crayfish (*Orconectes virilis*). Crustaceana 51: 208–211.

Griffiths, M.E. 1945. The environment, life history and structure of the water boatman, *Ramphocorixa acuminata* (Uhler) (Hemiptera, Corixidae). Univ. Kans. Sci. Bull. 30: 241–365.

Grimes, B.H. 1976. Notes on the distribution of *Hyalophysa* and *Gymnodinioides* on crustacean hosts in coastal North Carolina and a description of *Hyalophys atrageri* sp. n. J. Protozool. 23: 246–251.

Groff, J.M., T. McDowell, C.S. Friedman and R.P. Hedrick. 1993. Detection of a nonoccluded baculovirus in the freshwater crayfish *Cherax quadricarinatus* in North America. J. Aquat. Anim. Health 5: 275–279.

Gydemo, R. 1996. Signal crayfish, *Pacifastacus leniusculus*, as a vector for *Psorospermium haeckeli* to noble crayfish, *Astacus astacus*. Aquaculture 148: 1–9.

Halder, M. and W. Ahne. 1988a. *Astacus astacus* L. identified as IPNV-vector. Freshw. Crayfish 7: 303–308.

Halder, M. and W. Ahne. 1988b. Freshwater crayfish *Astacus astacus*—a vector for infectious pancreatic necrosis virus (IPNV). Dis. Aquat. Org. 4: 205–209.

Hall, M. 1915. Desriptions of a new genus and species of the discodrilid worms. Proc. U.S. Natl. Museum 48: 187–193.

Hall, M.C. 1929. Arthropods as intermediate hosts of helminths. Smithson. Misc. Collect. 81: 1–81.

Hall, M., R. Wang, R. van Antwerpen, L. Sottrup-Jensen and K. Söderhäll. 1999. The crayfish plasma clotting protein: a vitellogenin-related protein responsible for clot formation in crustacean blood. Proc. Natl. Acad. Sci. U.S.A. 96: 1965–1970.

Harlioğlu, M.M. 1999. The first record of *Epistylis niagarae* on *Astacus leptodactylus* in a crayfish rearing unit, Cip. Turkish. J. Zool. 23: 13–15.

Harrison, L. 1928. On the genus *Stratiodrilus* (Archiannelida: Histriobdellidae), with a description of a new species from Madagascar. Rec. Aust. Museum 16: 116–122.

Hart, C., Jr. 1962. A revision of the ostracods of the Family Entocytheridae. Proc. Acad. Nat. Sci. Philadelphia 114: 121–147.

Hart, C.W. 1964. Two new entocytherid ostracods from the vicinity of Washington, D. C. Proc. Biol. Soc. Washingt. 77: 243–246.

Hart, C., Jr. 1965. Three new entocytherid ostracods from the western United States, with new locality data for two previously described western entocytherids. Crustaceana 8: 190–196.

Hart, C., Jr. and H.H. Hobbs, Jr. 1961. Eight new troglobitic ostracods of the genus *Entocythere* (Crustacea, Ostracoda) from the eastern United States. Proc. Acad. Nat. Sci. Philadelphia 113: 173–185.

Hart, C., Jr. and D. Hart. 1966. Four new entocytherid ostracods from Kentucky, with notes on the troglobitic *Sagittocythere barri*. Not. Naturae. Acad. Nat. Sci. Philadelphia 388: 1–10.

Hart, C., Jr. and D. Hart. 1967. The entocytherid ostracods of Australia. Proc. Acad. Nat. Sci. Philadelphia 119: 1–51.

Hart, C., Jr. and D. Hart. 1971. A new ostracod (Entocytheridae, Notocytherinae) commensal on New Zealand crayfish. Proc. Biol. Soc. Washingt. 83: 579–584.

Hart, C.W., Jr., L.-A.C. Hayek, J. Clark and W.H. Clark. 1985. The life history and ecology of the entocytherid ostracod *Uncinocythere occidentalis* (Kozloff and Whitman) in Idaho. Smithson. Contrib. Zool. 419: 1–22.

Hart, D. and C. Hart. 1970. A new ostracod (Entocytheridae, Notocytherinae) on New Guinea crayfish. Zool. Meded. 44: 279–283.

Hart, D. and C. Hart, Jr. 1971. New entocytherid ostracods of the genera *Ankylocythere, Dactylocythere, Entocythere, Geocythere* and *Uncinocythere*: with a new diagnosis of the genus. Proc. Acad. Nat. Sci. Philadephia 123: 105–125.

Hart, D. and C. Hart, Jr. 1974. The ostracod family Entocytheridae. Proc. Acad. Nat. Sci. Philadelphia 18: 1–239.

Harvey, M. 1990. Pezidae, a new freshwater mite family from Australia (Acarina: Halacaroidea). Invertebr. Taxon 3: 771–781.

Haswell, W.A. 1893. On an apparently new type of the Platyhelminthes (Trematode?). Linn. Soc. N. S. W., Macleay Memorial Volume: 153–158.

Haswell, W. 1915. Studies on the Turbellaria. Part III. *Didymorchis*. Q. J. Microscop. Sc. Ser. 61: 161–169.

Haswell, W. 1922. *Astacocroton*, a new type of acarid. Proc. Linn. Soc. New South Wales 47: 329–343.

Hauck, A.K., M.R. Marshall, J.K.-K. Li and R.A. Lee. 2001. A new finding and range extension of bacilliform virus in the freshwater red claw crayfish in Utah, USA. J. Aquat. Anim. Health 13: 158–162.

Hauer, J. 1959. Raumparasitische rotatorien aus der kiemenhöhle des steinkrebses (*Potamobius torrentium* Schrank). Beiträge zur naturkundlichen Forsch Südwestdeutschl 18: 92–105.

Hayakijkosol, O. and L. Owens. 2011. Investigation into the pathogenicity of reovirus to juvenile *Cherax quadricarinatus*. Aquaculture 316: 1–5.

Hayakijkosol, O. and L. Owens. 2012. B2 or not B2: RNA interference reduces *Macrobrachium rosenbergii* nodavirus replication in redclaw crayfish (*Cherax quadricarinatus*). Aquaculture 326-329: 40–45.

Hayakijkosol, O., K. La Fauce and L. Owens. 2011. Experimental infection of redclaw crayfish (*Cherax quadricarinatus*) with *Macrobrachium rosenbergii* nodavirus, the aetiological agent of white tail disease. Aquaculture 319: 25–29.

Heath, I.B. and T. Unestam. 1974. Mycoplasma-like structures in the aquatic fungus *Aphanomyces astaci*. Science 183: 434–435.

Heidarieh, M., M. Soltani, F.M. Sedeh and N. Sheikhzadeh. 2013. Low water temperature retards white spot syndrome virus replication in *Astacus leptodactylus* crayfish. Acta Sci. Vet. 41: 1–6.

Henderson, H.E. 1938. The cercaria of *Crepidostomum cornutum* (Osborn). Trans. Am. Microsc. Soc. 57: 165–172.

Henderson, P.A. 1990. Freshwater ostracods. Synopses of the British Fauna (New Series). Number 42. 228 pp.

Henttonen, P., J.V. Huner and O.V. Lindqvist. 1994. Occurrence of *Psorospermium* sp. in several North American crayfish species, with comparative notes on *Psorospermium haeckeli* in the European crayfish, *Astacus astacus*. Aquaculture 120: 209–218.

Henttonen, P., J.V. Huner, P. Rata and O.V. Lindqvist. 1997. A comparison of the known life forms of *Psorospermium* spp. in freshwater crayfish (Arthropoda, decapoda) with emphasis on *Astacus astacus* L. (Astacidae) and *Procambarus clarkii* (Girard) (Cambaridae). Aquaculture 149: 15–30.

Herbert, B. 1987. Notes on diseases and epibionts of *Cherax quadricarinatus* and *C. tenuimanus* (Decapoda: Parastacidae). Aquaculture 64: 165–173.

Herbert, B.W. 1988. Infection of *Cherax quadricarinatus* (Decapoda: Parastacidae) by the microsporidium *Thelohania* sp. (Microsporida: Nosematidae). J. Fish Dis. 11: 301–308.

Hickman, V.V. 1967. Tasmanian Temnocephalidea. Proc. R. Soc. Tasmania 101: 227–250.

Hirsch, P.E., J. Nechwatal and P. Fischer. 2008. A previously undescribed set of *Saprolegnia* spp. in the invasive spiny-cheek crayfish (*Orconectes limosus*, Rafinesque). Fundam. Appl. Limnol. 172: 161–165.

Hirt, R.P., J.M. Logsdon, B. Healy, M.W. Dorey, W.F. Doolittle and T.M. Embley. 1999. Microsporidia are related to Fungi: Evidence from the largest subunit of RNA polymerase II and other proteins. Proc. Natl. Acad. Sci. 96: 580–585.

Hobbs III, H.H. 1965. Two new genera and species of the ostracod family Entocytheridae with a key to the genera. Proc. Biol. Soc. Washingt. 78: 159–164.

Hobbs III, H.H. 1969. A new genus and two new species of entocytherid ostracods from Alabama and Mississippi. Proc. Biol. Soc. Washingt. 82: 167–170.

Hobbs III, H.H. 1970. New entocytherid ostracods of the genus *Ornithocythere* and the description of a new genus. Proc. Biol. Soc. Washingt. 83: 171–182.

Hobbs III, H.H. 1971. New entocytherid ostracods of the genera *Ankylocythere* and *Dactylocythere*. Proc. Biol. Soc. Washingt. 84: 137–146.

Hobbs III, H.H. 1978. New species of ostracods from the Gulf Coastal Plain (Ostracoda: Entocytheridae). Trans. Am. Microsc. Soc. 97: 502–511.

Hobbs, H.H., Jr. 1967. A new genus and three new species of ostracods with a key to genus *Dactylocythere* (Ostracoda: Entocytheridae). Proc. U.S. Natl. Museum 122: 1–10.

Hobbs, H.H., Jr. 1971. The entocytherid ostracods from Mexico and Cuba. Smithson. Contrib. to Zool. 81: 1–55.

Hobbs, H.H., Jr. 1973. Three new troglobitic decapod crustaceans from Oaxaca, Mexico. Assoc. Mex. Cave Stud. Bull. 5: 25–38.

Hobbs, H.H., Jr. and M. Walton. 1963a. Three new ostracods (Ostracoda, Entocytheridae) from the Duck River Drainage in Tennessee. Am. Midl. Nat. 69: 456–461.

Hobbs, H.H., Jr. and M. Walton. 1963b. Four new species of the genus *Donnaldsoncythere* (Ostracoda, Entocytheridae) from Virginia with a key to the species of the genus. Trans. Am. Microsc. Soc. 82: 363–370.

Hobbs, H.H., Jr. and M. Walton. 1966. A new genus and six new species of entocytherid ostracods (Ostracoda, Entocytheridae). Proc. U.S. Natl. Museum 119: 1–12.

Hobbs, H.H., Jr. and C. Hart, Jr. 1966. On the entocytherid ostracod genera *Ascetocythere*, *Plectocythere*, *Phymocythere* (gen. nov.), and *Cymocythere*, with descriptions of new species. Proc. Acad. Nat. Sci. Philadelphia 118: 35–61.

Hobbs, H.H., Jr. and M. Walton. 1968. New entocytherid ostracods from the Southern United States. Proc. Acad. Nat. Sci. Philadelphia 120: 237–252.

Hobbs, H.H., Jr. and H.H. Hobbs III. 1970. New entocytherid ostracods with a key to the genera of the subfamily Entocytherinae. Smithson. Contrib. to Zool. 47: 1–19.

Hobbs, H.H., Jr. and M. Walton. 1970. New entocytherid ostracods from Tennessee and Virginia. Proc. Biol. Soc. Washingt. 82: 851–864.

Hobbs, H.H., Jr. and M. Walton. 1975. New entocytherid ostracods from Tennessee USA with a key to the species of the genus *Ascetocythere*. Proc. Biol. Soc. Washingt. 88: 5–20.

Hobbs, H.H., Jr. and M. Walton. 1976. New entocytherid ostracods from Kentucky and Tennessee. Proc. Biol. Soc. Washingt. 89: 393–404.

Hobbs, H.H., Jr. and M. Walton. 1977. New entocytherid ostracods of the genus *Dactylocythere*. Proc. Biol. Soc. Washingt. 90: 600–614.

Hobbs, H.H., Jr. and D. Peters. 1977. The entocytherid ostracods of North Carolina. Smithson. Contrib. to Zool. 247: 1–73.

Hobbs, H.H., Jr. and D.J. Peters. 1979. A substitute name for the homonym *Aphelocythere* Hobbs and Peters (Ostracoda, Entocytheridae). Proc. Biol. Soc. Washingt. 91: 1037.

Hobbs, H.H., Jr. and A. McClure. 1983. On a small collection of entocytherid ostracods with the descriptions of three new species. Proc. Biol. Soc. Washingt. 96: 770–779.

Hobbs, H.H., Jr. and D.J. Peters. 1989. New records of entocytherid ostracods infesting burrowing crayfishes, with the description of a new species, *Ascetocythere stockeri*. Proc. Biol. Soc. Washingt. 102: 324–330.

Hobbs, H.H., Jr. and D. Peters. 1991. Additional records of entocytherid ostracods infesting burrowing crayfishes, with descriptions of five new species. Proc. Biol. Soc. Washingt. 104: 64–75.

Hobbs, H.H., Jr. and R. Franz. 1991. A new troglobitic crayfish, *Procambarus* (*Lonnbergius*) *morrisi* (Decapoda: Cambaridae) from Florida. Proc. Biol. Soc. Washingt. 104: 55–63.

Hobbs, H.H., Jr. and D.J. Peters. 1993. New records of entocytherid ostracods infesting burrowing and cave-dwelling crayfishes, with descriptions of two new species. Proc. Biol. Soc. Washingt. 106: 455–466.

Hobbs, H.H., Jr., P.C. Holt and M. Walton. 1967. The crayfishes and their epizootic ostracod and branchiobdellid associates of the Mountain Lake, Virginia, region. Proc. U.S. Natl. Museum 123: 1–84.

Hochwimmer, G., R. Tober, R. Bibars-Reiter, E. Licek and R. Steinborn. 2009. Identification of two GH18 chitinase family genes and their use as targets for detection of the crayfish-plague oomycete *Aphanomyces astaci*. BMC Microbiology 9: 184.

Hoffman, R. 1963. A revision of the North American annelid worms of the genus *Cambarincola* (Oligochaeta: Branchiobdellidae). Proc. U.S. Natl. Museum 114: 271–371.

Holdich, D.M., J.D. Reynolds, C. Souty-Grosset and P.J. Sibley. 2009. A review of the ever increasing threat to European crayfish from non-indigenous crayfish species. Knowl. Manag. Aquat. Ecosyst. 394-395: 1–46.

Holt, P.C. 1968a. New genera and species of branchiobdellid worms (Annelida: Clitellata). Proc. Biol. Soc. Washingt. 81: 291–318.

Holt, P.C. 1968b. The genus *Pterodrilus* (Annelida: Branchiobdellida). Proc. U.S. Natl. Museum 125: 1–44.

Holt, P.C. 1973a. A summary of the branchiobdellid (Annelida: Clitellata) fauna of Mesoamerica. Smithson. Contrib. to Zool. 142: 1–40.

Holt, P.C. 1973b. Branchiobdellids (Annelida: Clitellata) from some eastern North American caves, with descriptions of new species of the genus *Cambarincola*. Int. J. Speleol. 5: 219–255.

Holt, P.C. 1974. The genus *Xironogiton* Ellis, 1919 (Clitellata: Branchiobdellida). Va. J. Sci. 25: 5–19.

Holt, P.C. 1977. A gill-inhabiting new genus and species of the Branchiobdellida (Annelida: Clitellata). Proc. Biol. Soc. Washingt. 90: 726–734.

Holt, P.C. 1978. The Reassignment of *Cambarincola elevatus* (Clitellata: Branchiobdellida) to the genus *Sathodrilus* Holt, 1968. Proc. Biol. Soc. Washingt. 91: 472–483.

Holt, P.C. 1981. A resume of members of the genus *Cambarincola* (Annelida: Branchiobdellida) from Pacific drainage of the USA. Proc. Biol. Soc. Washingt. 94: 675–695.

Holt, P.C. 1982. A new species of the genus *Cambarincola* (Clitellata: Branchiobdellida) from Illinois with remarks on the bursa of *Cambarincola vitreus* and the status of *Sathodrilus* Holt, 1968. Proc. Biol. Soc. Washingt. 95: 251–255.

Holt, P.C. 1984a. A new species of the genus *Cambarincola* (Clitellata: Branchiobdellida) from California. Proc. Biol. Soc. Washingt. 97: 544–549.

Holt, P.C. 1984b. On some branchiobdellids (Annelida: Clitellata) from Mexico with the description of new species of the genera *Cambarincola* and *Oedipodrilus*. Proc. Biol. Soc. Washingt. 97: 35–42.

Holt, P.C. 1986. Newly established families of the Order Branchiobdellida (Annelida: Clitellata) with a synopsis of the genera. Proc. Biol. Soc. Washingt. 99: 676–702.

Holt, P.C. 1988a. Four new species of cambarincolids (Clitellata: Branchiobdellida) from the southeastern United States with are description of *Oedipodrilus macbaini* (Holt, 1955). Proc. Biol. Soc. Washingt. 101: 794–808.

Holt, P.C. 1988b. The correct name of *Ceratodrilus orphiorhysis* Holt, 1960 (Annelida: Branchiobdellida). Proc. Biol. Soc. Washingt. 101: 308.

Holt, P.C. 1989. A new species of the cambarincolid genus *Sathodrilus* from Missouri, with the proposal of a replacement name for *Adenodrilus* Holt, 1977 (Clitellata: Branchiobdellida). Proc. Biol. Soc. Washingt. 102: 738–741.

Holt, P.C. and A. Weigl. 1979. A new species of *Xironodrilus* Ellis 1918 from North Carolina (Clitellata: Branchiobdellida). Brimleyana 1: 23–29.

Holt, P.C. and B.D. Opell. 1993. A checklist of and illustrated key to the genera and species of the Central and North American Cambarincolidae (Clitellata: Branchiobdellida). Proc. Biol. Soc. Washingt. 106: 251–295.

Horwitz, P. and B. Knott. 1991. The faunal assemblage in freshwater crayfish burrows in sedgeland and forest at Lightning Plains, western Tasmania. Pap. Proc. R. Soc. Tasmania. 125: 29–32.

Huang, C.H., L.R. Zhang, J.H. Zhang, L.C. Xiao, Q.J. Wu, D.H. Chen and J.K.-K. Li. 2001. Purification and characterization of White Spot Syndrome Virus (WSSV) produced in an alternate host: Crayfish, *Cambarus clarkii*. Virus Res. 76: 115–125.

Huang, T.S., L. Cerenius and K. Söderhäll. 1994. Analysis of the genetic diversity in crayfish plague fungus, *Aphanomyces astaci*, by random amplification of polymorphic DNA assay. Aquaculture 26: 1–10.

Huang, T.S., H. Wang, S.Y. Lee, M.W. Johansson, K. Söderhäll and L. Cerenius. 2000. A cell adhesion protein from the crayfish *Pacifastacus leniusculus*, a serine proteinase homologue similar to *Drosophila* masquerade. J. Biol. Chem. 275: 9996–10001.

Hubault, E. 1935. Une epizootie sur *Potamobius pallipes* Lereboullet. Ann. Parasitol. Hum. Comp. 2: 109–112.

Hüseyin, S. and B. Selcuk. 2005. Prevalence of *Epistylis* sp. Ehrenberg, 1832 (Peritrichia, Sessilida) on the narrow-clawed crayfish, *Astacus leptodactylus* (Eschscholtz, 1823) from Manyas Lake in Turkey. J. Anim. Vet. Adv. 4: 789–793.

Huys, R., J. Mackenzie-Dodds and J. Llewellyn-Hughes. 2009. Cancrincolidae (Copepoda, Harpacticoida) associated with land crabs: a semiterrestrial leaf of the ameirid tree. Mol. Phylogenet. Evol. 51: 143–56.

Huys, R., B. Oidtmann, M. Pond, H. Goodman and P.F. Clark. 2014. Invasive crayfish and their symbionts in the Greater London area: new data and the fate of *Astacus leptodactylus* in the Serpentine and Long Water Lakes. Ethol. Ecol. Evol. 26: 320–347.

Imhoff, E.M., R.J.G. Mortimer, M. Christmas and A.M. Dunn. 2010. Non-lethal tissue sampling allows molecular screening for microsporidian parasites in signal, *Pacifasticus leniusculus* (Dana), and vulnerable white-clawed crayfish, *Austropotamobius pallipes* (Lereboullet). Freshw. Crayfish 17: 145–150.

Ishii, Y. 1966. Differential morphology of *Paragonimus kellicotti* in North America. J. Parasitol. 52: 920–925.

James, J., J. Cable, G. Richardson, K.E. Davidson and A.S.Y. Mackie. 2015. Two alien species of Branchiobdellida (Annelida: Clitellata) new to the British Isles: a morphological and molecular study. Aquat. Inv. 10: 371–383.

Jennings, J.B. 1971. Parasitism and commensalism in the Turbellaria. Adv. Parasitol. 9: 1–32.

Jersabek, C. 2002. A case of considerable confusion in rotifer taxonomy: The *Cephalodella crassipes* complex. Arch. f. Hydrobiol. 139(suppl.): 265–274.

Jiménez, R. and X. Romero. 1997. Infection by intracellular bacterium in red claw crayfish, *Cherax quadricarinatus* (von Martens) in Ecuador. Aquac. Res. 28: 923–929.

Jiménez, R., R. Barniol, X. Romero and M. Machuca. 1998. A prokaryotic intracellular organism in the cuticular epithelium of cultured crayfish, *Cherax quadricarinatus* (von Martens), in Ecuador. J. Fish Dis. 21: 387–390.

Jiravanichpaisal, P., E. Bangyeekhun, K. Söderhäll and I. Söderhäll. 2001. Experimental infection of white spot syndrome virus in freshwater crayfish *Pacifastacus leniusculus*. Dis. Aquat. Org. 47: 151–157.

Jiravanichpaisal, P., K. Söderhäll and I. Söderhäll. 2004. Effect of water temperature on the immune response and infectivity pattern of white spot syndrome virus (WSSV) in freshwater crayfish. Fish Shellfish Immunol. 17: 265–275.

Jiravanichpaisal, P., S. Sricharoen, I. Söderhäll and K. Söderhäll. 2006. White spot syndrome virus (WSSV) interaction with crayfish haemocytes. Fish Shellfish Immunol. 20: 718–727.

Jiravanichpaisal, P., S.Y. Lee, Y.-A. Kim, T. Andrén and I. Söderhäll. 2007. Antibacterial peptides in hemocytes and hematopoietic tissue from freshwater crayfish *Pacifastacus leniusculus*: characterization and expression pattern. Dev. Comp. Immunol. 31: 441–455.

Jiravanichpaisal, P., S. Roos, L. Edsman, H. Liu and K. Söderhäll. 2009. A highly virulent pathogen, *Aeromonas hydrophila*, from the freshwater crayfish *Pacifastacus leniusculus*. J. Invertebr. Pathol. 101: 56–66.

Joffe, B.I., L.R.G. Cannon and E.R. Schockaert. 1998. On the phylogeny of families and genera within the Temnocephalida. Hydrobiologia 383: 263–268.

Johansson, M.W. and K. Söderhäll. 1985. Exocytosis of the prophenoloxidase activating system from crayfish haemocytes. J. Comp. Physiol. B 156: 175–181.

Johansson, M.W., M.I. Lind, T. Holmblad, P.O. Thörnqvist and K. Söderhäll. 1995. Peroxinectin, a novel cell adhesion protein from crayfish blood. Biochem. Biophys. Res. Commun. 216: 1079–1087.

Johnston, T.H. 1948. *Microphallus minutus*, a new trematode from the Australian water rat. Rec. South Aust. Museum 9: 93–100.

Johnston, T.H. and S.J. Edmonds. 1948. Australian Acanthocephala no. 7. Trans. R. Soc. South Aust. 72: 69–76.

Johnston, T.H. and L.M. Angel. 1950. The life history of *Plagiorchis jaenschi*, a new trematode from the Australian water rat. Trans. R. Soc. South Aust. 74: 49–58.

Jones, J.B. 1980. Freshwater crayfish *Paranephrops planifrons* infected with the microsporidian *Thelohania*. N. Zeal. J. Mar. Freshw. Res. 14: 45–46.

Jones, J.B. and C.S. Lawrence. 2001. Diseases of yabbies (*Cherax albidus*) in Western Australia. Aquaculture 194: 221–232.

Jones, T.C. and R.J.G. Lester. 1992. The life history and biology of *Diceratocephala boschmai* (Platyhelminthes; Temnocephalida), an ectosymbiont on the redclaw crayfish *Cherax quadricarinatus*. Hydrobiol. 248: 193–199.

Jue Sue, L. and T.R. Platt. 1998. Description and life-cycle of two new species of *Choanocotyle* n. g. (Trematoda: Plagiorchiida), parasites of Australian freshwater turtles, and the erection of the family Choanocotylidae. Syst. Parasitol. 41: 47–61.

Jussila, J., J. Makkonen and H. Kokko. 2011. Peracetic acid (PAA) treatment is an effective disinfectant against crayfish plague (*Aphanomyces astaci*) spores in aquaculture. Aquaculture 320: 37–42.

Kageyama, D., H. Anbutsu, M. Shimada and T. Fukatsu. 2007. *Spiroplasma* infection causes either early or late male killing in *Drosophila*, depending on maternal host age. Naturwissenschaften 94: 333–337.

Kane, J.R. 1965. The genus *Lagenophrys* Stein, 1852 (Ciliata, Peritricha) on Australasian Parastacidae. J. Protozool. 12: 109–122.

Kawai, T., T. Mitamura and A. Ohtaka. 2004. The taxonomic status of the introduced North American signal crayfish, *Pacifastacus leniusculus* (Dana, 1852) in Japan, and the source of specimens in the newly reported population in Fukushima Prefecture. Crustaceana 77: 861–870.

Keenan, S., M. Niemiller and B. Williams. 2014. Observations of an ectosymbiotic association between *Cambarus bartonii cavatus* (Decapoda: Cambaridae) and branchiobdellidans (Annelida: Clitellata) in Cruze Cave, Knox County, Tennessee, USA. Speleobiology Notes 6: 55–61.

Keller, N.S., M. Pfeiffer, I. Roessink, R. Schulz and A. Schrimpf. 2014. First evidence of crayfish plague agent in populations of the marbled crayfish (*Procambarus fallax forma virginalis*). Knowl. Manag. Aquat. Ecosyst. 414: 15.

Keller, T.A. 1992. The effect of the branchiobdellid annelid *Cambarincola fallax* on the growth rate and condition of the crayfish *Orconectes rusticus*. J. Freshw. Ecol. 7: 165–171.

Khamesipour, F., A.K. Shahraki, M. Moumeni, R.K. Boroujeni and M. Yadegari. 2013. Prevalence of *Listeria monocytogenes* in the crayfish (*Astacus leptodactylus*) by polymerase chain reaction in Iran. Int. J. Biosci. 3: 160–169.

Kim, E.-M., J.-L. Kim, S.-I. Choi, S.-H. Lee and S.T. Hong. 2009. Infection status of freshwater crabs and crayfish with metacercariae of *Paragonimus westermani* in Korea. Kor. J. Parasitol. 47: 425–426.

Klarberg, D.P., P. Henttonen and J.V. Huner. 2000. Occurrence of the enigmatic, unicellular *Psorospermium* organism in several cultured, sympatric populations of the freshwater crayfishes *Procambarus clarkii* and *Procambarus zonangulus*. J. World Aquac. Soc. 31: 264–273.

Koenders, A., K. Martens, S. Halse and I. Schön. 2012. Cryptic species of the *Eucypris virens* complex (Ostracoda, Crustacea) from Europe have invaded Western Australia. Biol. Invasions 14: 2187–2201.

Kopáček, P., L. Grubhoffer and K. Söderhäll. 1993a. Isolation and characterization of a hemagglutinin with affinity for lipopolysaccharides from plasma of the crayfish *Pacifastacus leniusculus*. Dev. Comp. Immunol. 17: 407–418.

Kopáček, P., M. Hall and K. Söderhäll. 1993b. Characterization of a clotting protein, isolated from plasma of the freshwater crayfish *Pacifastacus leniusculus*. Eur. J. Biochem. 213: 591–597.

Kozloff, E. 1955. Two new species of *Entocythere* (Ostracoda: Cytheridae), commensal on *Pacifastacus gambelii* (Girard). Am. Midl. Nat. 53: 156–161.

Kozloff, E. and D. Whitman. 1954. *Entocythere occidentalis* sp. nov., a cytherid ostracod commensal on western species of *Pacifastacus*. Am. Midl. Nat. 52: 159–163.

Kozubíková, E., S. Viljamaa-Dirks, S. Heinikainen and A. Petrusek. 2011a. Spiny-cheek crayfish *Orconectes limosus* carry a novel genotype of the crayfish plague pathogen *Aphanomyces astaci*. J. Invertebr. Pathol. 108: 214–216.

Kozubíková, E., T. Vrålstad, I. Filipová and A. Petrusek. 2011b. Re-examination of the prevalence of *Aphanomyces astaci* in North American crayfish populations in Central Europe by TaqMan MGB real-time PCR. Dis. Aquat. Org. 97: 113–125.

Kozubíková-Balcarová, E., O. Koukol, M.P. Martín, J. Svoboda, A. Petrusek and J. Díeguez-Uribeondo. 2013. The diversity of oomycetes on crayfish: morphological vs. molecular identification of cultures obtained while isolating the crayfish plague pathogen. Fungal Biol. 117: 682–691.

Krugner-Higby, L., D. Haak, P.T.J. Johnson, J.D. Shields, W.M. Jones III, K.S. Reece, T. Meinke, A. Gendron and J.A. Rusak. 2010. Ulcerative disease outbreak in crayfish *Orconectes propinquus* linked to *Saprolegnia australis* in Big Muskellunge Lake, Wisconsin. Dis. Aquat. Org. 91: 57–66.

La Fauce, K. and L. Owens. 2007. Investigation into the pathogenicity of *Penaeus merguiensis* densovirus (*Pmerg* DNV) to juvenile *Cherax quadricarinatus*. Aquaculture 271: 31–38.

Lahser, C., Jr. 1975. Epizoöites of crayfish I. Ectocommensals and parasites of crayfish of Brazos County, Texas. Freshw. Crayfish 2: 277–285.

Lamanova, A.I. 1971. Attachment by zebra mussels and acorn barnacles on crayfish. Hydrobiol. J. 6: 89–91.

Lan, J.-F., X.-C. Li, J.-J. Sun, J. Gong, X.-W. Wang, X.-Z. Shi, L.-J. Shi, Y.-D. Weng, X.-F. Zhao and J.-X. Wang. 2013. Prohibitin interacts with envelope proteins of white spot syndrome virus and prevents infection in the red swamp crayfish, *Procambarus clarkii*. J. Virol. 87: 12756–12765.

Langdon, J.S. 1991. Description of *Vavraia parastacida* sp. nov. (Microspora: Pleistophoridae) from marron, *Cherax tenuimanus* (Smith) (Decapoda: Parastacidae). J. Fish Dis. 14: 619–629.

Langdon, J.S. and T. Thorne. 1992. Experimental transmission *per os* of microsporidiosis due to *Vavraia parastacida* in the marron, *Cherax tenuimanus* (Smith), and yabby, *Cherax albidus* Clark. J. Fish Dis. 15: 315–322.

Lanz, H., V. Tsutsumi and H. Aréchiga. 2009. Morphological and biochemical characterization of *Procambarus clarki* blood cells. Dev. Comp. Immunol. 17: 389–397.

Lee, J.H., T.W. Kim and J.C. Choe. 2009. Commensalism or mutualism: conditional outcomes in a branchiobdellid-crayfish symbiosis. Oecologia 159: 217–224.

Lee, J.J., G.F. Leedale and P. Bradbury. 2000. An illustrated Guide to the Protozoa, second edition. Society of Protozoologists, Lawrence, Kansas, U.S.A. 1432 pp.

Lee, S.C., N. Corradi, E.J. Byrnes III, S. Torres-Martinez, F.S. Dietrich, P.J. Keeling and J. Heitman. 2008. Microsporidia evolved from ancestral sexual fungi. Curr. Biol. 18: 1675–1679.

Lee, S.Y. and K. Söderhäll. 2001. Characterization of a pattern recognition protein, a masquerade-like protein, in the freshwater crayfish *Pacifastacus leniusculus*. J. Immunol. 166: 7319–7326.

Lee, S.Y., B.L. Lee and K. Söderhäll. 2003. Processing of an antibacterial peptide from hemocyanin of the freshwater crayfish *Pacifastacus leniusculus*. J. Biol. Chem. 278: 7927–7933.

Lefebvre, F. and R. Poulin. 2005. Progenesis in digenean trematodes: a taxonomic and synthetic overview of species reproducing in their second intermediate hosts. Parasitology 130: 587–605.

Li, J., P. Du, Z. Li, Y. Zhou, W. Cheng, S. Wu, F. Chen and X. Wang. 2015. Genotypic analyses and virulence characterization of *Listeria monocytogenes* isolates from crayfish (*Procambarus clarkii*). Curr. Microbiol. 70: 704–709.

Lin, X. and I. Söderhäll. 2011. Crustacean hematopoiesis and the astakine cytokines. Blood 117: 6417–6424.

Lin, X., K. Söderhäll and I. Söderhäll. 2011. Invertebrate hematopoiesis: an astakine-dependent novel hematopoietic factor. J. Immunol. 186: 2073–2079.

Liu, H., K. Söderhäll and P. Jiravanichpaisal. 2009. Antiviral immunity in crustaceans. Fish Shellfish Immunol. 27: 79–88.

Liu, Y.T., C.I. Chang, J.R. Hseu, K.F. Liu and J.M. Tsai. 2013. Immune responses of prophenoloxidase and cytosolic manganese superoxide dismutase in the freshwater crayfish *Cherax quadricarinatus* against a virus and bacterium. Mol. Immunol. 56: 72–80.

Ljungberg, O. and L. Monné. 1968. On the eggs of an enigmatic nematode parasite encapsulated in the connective tissue of the European crayfish, *Astacus astacus* in Sweden. Bull. Off. Int. Epizoot. 69: 1231–1235.

Lo, C.-F., H.-C. Hsu, M.-F. Tsai, C.-H. Ho, S.-E. Peng, G.-H. Kou and D.V. Lightner. 1999. Specific genomic DNA fragment analysis of different geographical clinical samples of shrimp white spot syndrome virus. Dis. Aquat. Org. 35: 175–185.

Lom, J., F. Nilsen and I. Dyková. 2001. *Thelohania contejeani* Henneguy, 1892: dimorphic life cycle and taxonomic affinities, as indicated by ultrastructural and molecular study. Parasitol. Res. 87: 860–872.

Longshaw, M. 2011. Diseases of crayfish: a review. J. Invert. Path. 106: 54–70.

Longshaw, M., K.S. Bateman, P. Stebbing, G.D. Stentiford and F.A. Hockley. 2012a. Disease risks associated with the importation and release of non-native crayfish species in mainland Britain. Aquat. Biol. 16: 1–15.

Longshaw, M., P.D. Stebbing, K.S. Bateman and F.A. Hockley. 2012b. Histopathological survey of pathogens and commensals of white-clawed crayfish (*Austropotamobius pallipes*) in England and Wales. J. Invertebr. Pathol. 110: 54–59.

Lopez-Ochoterena, E. and E. Gasea. 1971. Protozoarios ciliados de Mexico. 17. Algunos aspectos biologicos de veinte especies epizoicas del crustaceo *Cambarellus montezumae zempoalensis* Villalobos. Revt. Lat.-Am. Microbiol. 13: 221–231.

Lotz, J.M. and K.C. Corkum. 1983. Studies on the life history of *Sogandaritrema progeneticus* (Digenea: Microphallidae). J. Parasitol. 69: 918–921.

Lotz, J. and W.F. Font. 1983. Review of the Lecithodendriidae (Trematoda) from *Eptesicus fuscus* in Wisconsin and Minnesota. Proc. Helminthol. Soc. Wash. 50: 83–102.

Lucić, A., I. Maguire and R. Erben. 2004. Occurrence of the pathogen *Psorospermium haeckeli* (hilgendorf) in astacid populations in Croatia. Bull. Fr. Pêche Piscic. 372-373: 375–385.

Lun, Z.-R., R.B. Gasser, D.-H. Lai, A.-X. Li, X.-Q. Zhu, X.-B. Yu and Y.-Y. Fang. 2005. Clonorchiasis: a key foodborne zoonosis in China. Lancet Infect. Dis. 5: 31–41.

Macy, R. and D. Moore. 1954. On the life cycle and taxonomic relations of *Cephalophallus obscurus* n. g., n. sp., an intestinal trematode (Lecithodendriidae) of mink. J. Parasitol. 40: 328–335.

Macy, R.W. and W.D. Bell. 1968. The life cycle of *Astacatrematula macrocotyla* gen. et sp. n. (Trematoda: Psilostomidae) from Oregon. J. Parasitol. 54: 319–323.

Maeda, M., T. Itami, E. Mizuki, R. Tanaka, Y. Yoshizu, K. Doi, C. Yasunaga-Aoki, Y. Takahashi and T. Kawarabata. 2000. Red swamp crawfish (*Procambarus clarkii*): an alternative experimental host in the study of white spot syndrome virus. Acta Virol. 44: 371–374.

Maestracci, V. and A. Vey. 1987. Fungal infection of gills in crayfish: histological, cytological and physiopathological aspects of the disease. Freshw. Crayfish 7: 187–194.

Magath, T. 1924. *Ophiotaenia testudo*, a new species from *Amyda spinifera*. J. Parasitol. 11: 44–49.

Makkonen, J., H. Kokko, P. Henttonen, M. Kivistik, M. Hurt, T. Paaver and J. Jussila. 2010. Fungal isolations from Saaremaa, Estonia: noble crayfish (*Astacus astacus*) with melanised spots. Freshw. Crayfish 17: 155–158.

Makkonen, J., J. Jussila, L. Koistinen, T. Paaver, M. Hurt and H. Kokko. 2013. *Fusarium avenaceum* causes burn spot disease syndrome in noble crayfish (*Astacus astacus*). J. Invertebr. Pathol. 113: 184–190.

Mann, H. 1940. Die brandfleckenkrankheit beim Sumpfkrebs (*Potamobius leptodactylus* Eschh.). Z. Parasitenkd. 11: 430–432.

Mann, H. and U. Pieplow. 1938. The necrotic spot disease in crayfish, and its causative agents. Zeitschrift. f. Fischerei. 38: 225–240.

Marino, F., T. Pretto, F. Tosi, S. Monaco, C. De Stefano, A. Manfrin and F. Quaglio. 2014. Mass mortality of *Cherax quadricarinatus* (Von Martens, 1868) reared in Sicily (Italy): crayfish plague introduced in an intensive farming. Freshw. Crayfish 20: 93–96.

Martínez-Aquino, A., F. Brusa and C. Damborenea. 2014. Checklist of freshwater symbiotic temnocephalans (Platyhelminthes, Rhabditophora, Temnocephalida) from the Neotropics. Zoosyst. Evol. 90: 147–162.

May, L. 1989. Epizoic and parasitic rotifers. Hydrobiologia 186/187: 59–67.

Mayén-Estrada, R. and M. Aladro-Lubel. 2000. First record of *Lagenophrys dennisi* (Ciliophora: Peritrichia) on the exoskeleton of crayfish *Cambarellus patzcuarensis*. J. Eukaryot. Microbiol. 47: 57–61.

Mayén-Estrada, R. and M.A. Aladro-Lubel. 2001. Epibiont peritrichids (Ciliophora: Peritrichida: Epistylidae) on the crayfish *Cambarellus patzcuarensis* in Lake Pátzcuaro, Michoacán, Mexico. J. Crustac. Biol. 21: 426–434.

Mayén-Estrada, R. and M.A. Aladro-Lubel. 2002. Distribution and prevalence of 15 species of epibiont peritrich ciliates on the crayfish *Cambarellus patzcuarensis* Villalobos, 1943 in Lake Pátzcuaro, Michoacán, Mexico. Crustaceana 74: 1213–1224.

McAllister, C.T., H.W. Robison and W.F. Font. 2011. Metacercaria of *Alloglossidium corti* (Digenea: Macroderoididae) from 3 species of crayfish (Decapoda: Cambaridae) in Arkansas and Oklahoma, U.S.A. Comp. Parasitol. 78: 382–386.

McCoy, O.R. 1928. Life history studies on trematodes from Missouri. J. Parasitol. 14: 207–228.

McMahon, T.A., L.A. Brannelly, M.W.H. Chatfield, P.T.J. Johnson, M.B. Joseph, V.J. McKenzie, C.L. Richards-Zawacki, M.D. Venesky and J.R. Rohr. 2013. Chytrid fungus *Batrachochytrium dendrobatidis* has nonamphibian hosts and releases chemicals that cause pathology in the absence of infection. Proc. Natl. Acad. Sci. U.S.A. 110: 210–215.

Merritt, S.V. and I. Pratt. 1964. The life history of *Neoechinorhynchus rutili* and its development in the intermediate host (Acanthocephala: Neoechinorhynchidae). J. Parasitol. 50: 394–400.

Mestre, A., J.S. Monros and F. Mesquita-Joanes. 2011. Comparison of two chemicals for removing an entocytherid (Ostracoda: Crustacea) species from its host crayfish (Cambaridae: Crustacea). Int. Rev. Hydrobiol. 96: 347–355.

Mestre, A., J. Monrós and F. Mesquita-Joanes. 2014. A review of the Entocytheridae (Ostracoda) of the world: updated bibliographic and species checklists and global georeferenced database, with insights into host. Crustaceana 87: 921–953.

Meyer, F.P. 1965. A pseudoparasitic infestation of crayfish. Prog. Fish-Cult. 27: 19.

Middleton, D., D.M. Holdich and N.A. Ratcliffe. 1996. Haemagglutinins in six species of freshwater crayfish. Comp. Biochem. Physiol. Part A 114: 143–152.

Mitchell, S.A. and D.J. Kock. 1988. Alien symbionts introduced with imported marron from Australia may pose a threat to aquaculture. S. Afr. J. Sci. 84: 877–878.

Miyazaki, I. 1954. Studies on *Gnathostoma* occurring in Japan (Nematoda: Gnathostomidae): II. Life history of *Gnathostoma* and morphological comparison of its larval forms. Kyushu Mem. Med. Sci. 5: 123–140.

Mohamed, A., M. Amin, O. Amer and A. Amin. 2005. Studies of the role of shellfish as a source for transmitting some parasites of zoonotic importance. J. Vet. Med. 7: 1–25.

Moodie, E.G., L.F. Le Jambre and M.E. Katz. 2003a. *Thelohania montirivulorum* sp. nov. (Microspora: Thelohaniidae), a parasite of the Australian freshwater crayfish, *Cherax destructor* (Decapoda: Parastacidae): fine ultrastructure, molecular characteristics and phylogenetic relationships. Parasitol. Res. 91: 215–228.

Moodie, E.G., L.F. Le Jambre and M.E. Katz. 2003b. *Thelohania parastaci* sp. nov. (Microspora: Thelohaniidae), a parasite of the Australian freshwater crayfish, *Cherax destructor* (Decapoda: Parastacidae). Parasitol. Res. 91: 151–165.

Moodie, E.G., L.F. Le Jambre and M.E. Katz. 2003c. Ultrastructural characteristics and small subunit ribosomal DNA sequence of *Vairimorpha cheracis* sp. nov. (Microspora: Burenellidae), a parasite of the Australian yabby, *Cherax destructor* (Decapoda: Parastacidae). J. Invertebr. Pathol. 84: 198–213.

Morado, J.F. and E.B. Small. 1995. Ciliate parasites and related diseases of Crustacea: a review. Rev. Fish. Sci. 3: 275–354.

Moravec, F. 2007. Some aspects of the taxonomy and biology of adult spirurine nematodes parasitic in fishes: a review. Folia Parasitol. 54: 239–257.

Morrison, E.O. 1966. Crayfish, possible secondary intermediate host for lung flukes (*Haematoloechus*) of bullfrogs in Jefferson County, Texas. Yearb. Am. Philosphical Soc. 1966: 361–362.

Moyano, H.I., F. Carrasco and S. Gacitúa. 1993. Sobre las especies Chilenas de *Stratiodrilus* Haswell, 1900 (Polychaeta, Histriobdellidae). Bol. Soc. Biol. Concepción, Chile. 64: 147–157.

Mrugała, A., E. Kozubíková-Balcarová, C. Chucholl, S. Resino, S. Vljamaa-Dirks, J. Vukić and A. Petrusek. 2015. Trade of ornamental crayfish in Europe as a possible introduction pathway for important crustacean diseases: crayfish plague and white spot syndrome. Biol. Invasions 17: 1313–1326.

Ngamniyom, A., T. Sriyapai and K. Silprasit. 2014. *Diceratocephala boschmai* (Platyhelminthes: Temnocephalida) from crayfish farms in Thailand: investigation of the topographic surface and analysis of 18S ribosomal DNA sequences. Turkish J. Zool. 38: 471–478.

Nguyen, K.Y.K., K. Sakuna, R. Kinobe and L. Owens. 2014. Ivermectin blocks the nuclear location signal of parvoviruses in crayfish, *Cherax quadricarinatus*. Aquaculture 420-421: 288–294.

Ninni, A.P. 1864. Sulla mortalità dei gamberi (*Astacus fluviatilis*) nel Veneto e più particolarmente nella provincia trevigiana. Atti. Imp. Roy. Inst. Veneto 3: 1203–1209.

Niwa, N. and A. Ohtaka. 2006. Accidental introduction of symbionts with imported freshwater shrimps. pp. 182–186. *In*: F. Koike, M.N. Clout, M. Kawamichi, M. DePoorter and K. Iwatsuki (eds.). Assessment and Control of Biological Invasion Risks. Shoukadoh Book Sellers, Kyoto, Japan.

Norden, A. 1977. A new entocytherid ostracod of the genus *Plectocythere*. Proc. Biol. Soc. Washingt. 90: 491–494.

Norden, A.W. and B.B. Norden. 1984. A new entocytherid ostracod of the genus *Dactylocythere*. Proc. Biol. Soc. Washingt. 98: 627–629.

O'Donoghue, P.J. and R.D. Adlard. 2000. Catalogue of protozoan parasites recorded in Australia. Mem. Queensl. Museum 45: 1–163.

O'Donoghue, P., I. Beveridge and P. Phillips. 1990. Parasites and ectocommensals of yabbies and marron in South Australia. S.A. Department Agric., Adelaide: 46 pp.

Oberkofler, B., F. Quaglio, L. Füreder, M.L. Fioravanti, S. Giannetto, C. Morolli and G. Minelli. 2002. Species of Branchiobdellidae (Annelida) on freshwater crayfish in south Tyrol (northern Italy). Bull. Fr. Pêche Piscic. 367: 777–784.

Ohtaka, A., S. Gelder, T. Kawai, K. Saito, K. Nakata and M. Nishino. 2005. New records and distributions of two North American branchiobdellidan species (Annelida: Clitellata) from introduced signal crayfish, *Pacifastacus leniusculus*, in Japan. Biol. Invasions 7: 149–156.

Ohtsuka, M., N. Kikuchi, T. Yamamoto, T. Suzutani, K. Nakanaga, K. Suzuki and N. Ishii. 2014. Buruli ulcer caused by *Mycobacterium ulcerans* subsp. *shinshuense*: a rare case of familial concurrent occurrence and detection of insertion sequence 2404 in Japan. J.A.M.A. Dermatology 150: 64–67.

Oidtmann, B. and R.W. Hoffman. 1999. Bacteriological investigations on crayfish. Freshwat. Crayfish 12: 288–302.

Oidtmann, B., S. Bausewein, L. Hölzle, R. Hoffmann and M. Wittenbrink. 2002. Identification of the crayfish plague fungus *Aphanomyces astaci* by polymerase chain reaction and restriction enzyme analysis. Vet. Microbiol. 85: 183–194.

Oki, I., S. Tamura, M. Takai and M. Kawakatsu. 1995. Chromosomes of *Temnocephala minor*, an ectosymbiotic turbellarian on Australian crayfish found in Kagoshima Prefecture, with karyological notes on exotic turbellarians found in Japan. Hydrobiologia 305: 71–77.

Orozova, P., I. Sirakov, V. Chikova, R. Popova, A.H. Al-Harbi, M. Crumlish and B. Austin. 2014. Recovery of *Hafnia alvei* from diseased brown trout, *Salmo trutta* L., and healthy noble crayfish, *Astacus astacus* (L.), in Bulgaria. J. Fish Dis. 37: 891–898.

Osborn, H.L. 1919. Observations on *Microphallus ovatus* sp. nov. from the crayfish and black bass of Lake Chautauqua, N.Y. J. Parasitol. 5: 123–127.

Overstreet, R.M., R.W. Heard and J.M. Lotz. 1992. *Microphallus fonti* sp. n. (Digenea: Microphallidae) from the red swamp crawfish in southern United States. Mem. Inst. Oswaldo Cruz 87: 175–178.

Owens, L. and C. McElnea. 2000. Natural infection of the redclaw crayfish *Cherax quadricarinatus* with presumptive spawner-isolated mortality virus. Dis. Aquat. Org. 40: 219–223.

Palm, M.E., W. Gams and H.I. Nirenberg. 1995. *Plectosporium*, a new genus for *Fusarium tabacinum*, the anamorph of *Plectosphaerella cucumerina*. Mycologia 87: 397–406.

Peters, D.J. and J.E. Pugh. 1999. On the entocytherid ostracods of the Brazos River Basin and adjacent coastal region of Texas. Proc. Biol. Soc. Washingt. 112: 338–351.

Pinder, A.M. and R.O. Brinkhurst. 1997. Review of the Phreodrilidae (Annelida: Oligochaeta: Tubificida) of Australia. Invertebr. Syst. 11: 443–523.

Poulos, B.T., C.R. Pantoja, D. Bradley-Dunlop, J. Aguilar and D.V. Lightner. 2001. Development and application of monoclonal antibodies for the detection of white spot syndrome virus of penaeid shrimp. Dis. Aquat. Org. 47: 13–23.

Prins, R. 1964. *Attheyella carolinensis* Chappuis (Copepoda: Harpacticoida) on freshwater crayfishes from Kentucky. Trans. Am. Microsc. Soc. 83: 370–371.

Quaglio, F., C. Morolli, R. Galuppi, C. Bonoli, F. Marcer, L. Nobile. G. de Luise and M.P. Tampieri. 2006a. Preliminary investigations of disease-causing organisms in the white-clawed crayfish *Austropotamobius pallipes* complex from streams of northern Italy. Bull. Fr. Pêch. Piscic. 380-381: 1271–1290.

Quaglio, F., C. Morolli, R. Galuppi, M.P. Tampieri, C. Bonoli, F. Marcer, G. Rotundo and G.S. Germinara. 2006b. Sanitary-pathological examination of red swamp crayfish (*Procambarus clarkii*, Girard 1852) in the Reno Valley. Freshw. Crayfish 15: 365–375.

Quaglio, F., R. Galuppi, F. Marcer, C. Morolli, C. Bonoli, B. Fioretto, M. Tampieri, S. Bassi, A. Lavazza, M. Gianaroli and F. Malagoli. 2008. Mortality episodes of white-clawed crayfish (*Austropotamobius pallipes* complex) in three streams of Modena province (Northern Italy). Ittiopatologia 5: 99–127.

Quilter, C.G. 1976. Microsporidian parasite *Thelohania contejeani* Henneguy from New Zealand freshwater crayfish. N.Z. J. Mar. Freshw. Res. 10: 225–231.

Ragan, M.A., C.L. Goggin, R.J. Cawthorn, L. Cerenius, A.V.C. Jamieson, S.M. Plourde, T.G. Rand, K. Söderhäll and R.R. Gutell. 1996. A novel clade of protistan parasites near the animal-fungal divergence. Proc. Natl. Acad. Sci. U.S.A. 93: 11907–11912.

Raissy, M., F. Khamesipour, E. Rahimi and A. Khodadoostan. 2014. Occurrence of *Vibrio* spp., *Aeromonas hydrophila*, *Escherichia coli* and *Campylobacter* spp. in crayfish (*Astacus leptodactylus*) from Iran. Iran. J. Fish. Sci. 13: 944–954.

Rantamaki, J., L. Cerenius and K. Söderhäll. 1992. Prevention of transmission of the crayfish plague fungus (*Aphanomyces astaci*) to the freshwater crayfish *Astacus astacus* by treatment with $MgCl_2$. Aquaculture 104: 11–18.

Regassa, L.B. and G.E. Gasparich. 2006. Spiroplasmas: evolutionary relationships and biodiversity. Front. Biosci. 11: 2983–3002.

Reid, J.W. 2001. A human challenge: discovering and understanding continental copepod habitats. Hydrobiologia 453-454: 201–226.

Reid, J.W., C.K. Noro, L. Buckup and J. Bisol. 2006. Copepod crustaceans from burrows of *Parastacus defossus* Faxon, 1898 in southern Brazil. Nauplius 14: 23–30.

Richardson, D.J. and W.F. Font. 2006. The Cajun dwarf crawfish (*Cambarellus shufeldtii*): an intermediate host for *Southwellina dimorpha* (Acanthocephala). J. Ark. Acad. Sci. 60: 192–193.

Rohde, K. and M. Georgi. 1983. Structure and development of *Austramphilina elongata* Johnston, 1931 (Cestodaria: Amphilinidea). Int. J. Parasitol. 13: 273–287.

Romero, X. and R. Jiménez. 2002. Histopathological survey of diseases and pathogens present in redclaw crayfish, *Cherax quadricarinatus* (Von Martens), cultured in Ecuador. J. Fish Dis. 25: 653–667.

Royo, F., G. Andersson, E. Bangyeekhun, J.L. Múzquiz, K. Söderhäll and L. Cerenius. 2004. Physiological and genetic characterisation of some new *Aphanomyces* strains isolated from freshwater crayfish. Vet. Microbiol. 104: 103–112.

Rudolph, E.H. 2013. *Parastacus pugnax* (Poeppig, 1835) (Crustacea, Decapoda, Parastacidae): conocimiento biológico, presión extractiva y perspectivas de cultivo. Lat. Am. J. Aquat. Res. 41: 611–632.

Rudolph, E., F. Retamal and A. Martínez. 2007. First record of *Psorospermium haecklii* Hilgendorf, 1883 in a South American parastacid, the burrowing crayfish *Parastacus pugnax* (Poeppig, 1835) (Decapoda, Parastacidae). Crustaceana 80: 939–946.

Rusaini, E. Ariel, G. Burgess and L. Owens. 2013. Investigation of an idiopathic lesion in redclaw crayfish *Cherax quadricarinatus* using suppression subtractive hybridization. J. Virol. Microbiol. 2013: ID 569032.

Saelee, N., C. Noonin, B. Nupan, K. Junkunlo, A. Phongdara, X. Lin, K. Söderhäll and I. Söderhäll. 2013. β-Thymosins and hemocyte homeostasis in a crustacean. PloS ONE 8: e60974.

SamCookiyaei, A., M. Afsharnasab, V. Razavilar, A.A. Motalebi, S. Kakoolaki, Y. Asadpor, M. Yahyazade and A. Nekuie Fard. 2012. Experimentally pathogenesis of *Aeromonas hydrophila* in freshwater Crayfish (*Astacus leptodactylus*) in Iran. Iran J. Fish. Sci. 11: 644–656.

Sargent, L., A. Baldridge and M. Vega-Ross. 2014. A trematode parasite alters growth, feeding behavior, and demographic success of invasive rusty crayfish (*Orconectes rusticus*). Oecologia 175: 947–958.

Scheer, D. 1979. *Psorospermium orconectis* n. sp., ein neuer parasit in *Orconectes limosus*. Arch. f. Protistenkd. 121: 381–391.

Schmidt, G.D. 1973. Resurrection of *Southwellina* Witenberg, 1932, with a description of *Southwellina dimorpha* sp. n., and a key to genera in Polymorphideae (Acanthocephala). J. Parasitol. 59: 299–305.

Schneider, W. 1932. Nematoden aus der kiemenhohle des flunkrebses. Arch. f. Hydrobiol. 24: 629–637.

Schrimpf, A., C. Chucholl, T. Schmidt and R. Schulz. 2013. Crayfish plague agent detected in populations of the invasive North American crayfish *Orconectes immunis* (Hagen, 1870) in the Rhine River, Germany. Aquat. Invasions 8: 103–109.

Scott, J.R. and R.L. Thune. 1986a. Ectocommensal protozoan infestations of gills of red swamp crawfish, *Procambarus clarkii* (Girard), from commercial ponds. Aquaculture 55: 161–164.

Scott, J.R. and R.L. Thune. 1986b. Bacterial flora of hemolymph from red swamp crawfish, *Procambarus clarkii* (Girard), from commercial ponds. Aquaculture 58: 161–165.

Segers, H. 2002. The nomenclature of the Rotifera: annotated checklist of valid family and genus-group names. J. Nat. Hist. 36: 631–640.

Segers, H. 2007. Annotated checklist of the rotifers (Phylum Rotifera), with notes on nomenclature, taxonomy and distribution. Zootaxa 1564: 1–104.

Sewell, K.B. 1998. The taxonomic status of the ectosymbiotic flatworm *Didymorchis paranephropis* Haswell. Mem. Queensl. Museum 42: 585–595.

Sewell, K.B. 2013. Key to the genera and checklist of species of Australian temnocephalans (Temnocephalida). Museum Victoria Sci. Reports 17: 1–13.

Sewell, K.B. and L.R.G. Cannon. 1994. Symbionts and biodiversity. Mem. Queensl. Museum 36: 33–40.

Sewell, K.B. and L.R.G. Cannon. 1998. New temnocephalans from the branchial chamber of Australian *Euastacus* and *Cherax* crayfish hosts. Proc. Linn. Soc. New South Wales 119: 21–36.

Sewell, K.B., L.R.G. Cannon and D. Blair. 2006. A review of *Temnohaswellia* and *Temnosewellia* (Platyhelminthes: Temnocephalida: Temnocephalidae), ectosymbionts from Australian crayfish *Euastacus* (Parastacidae). Mem. Queensl. Museum 52: 199–279.

Shi, X.-Z., X.F. Zhao and J.-X. Wang. 2014. A new type antimicrobial peptide astacidin functions in antibacterial immune response in red swamp crayfish *Procambarus clarkii*. Dev. Comp. Immunol. 43: 121–128.

Shi, Z., C. Huang, J. Zhang, D. Chen and J.R. Bonami. 2000. White spot syndrome virus (WSSV) experimental infection of the freshwater crayfish, *Cherax quadricarinatus*. J. Fish Dis. 23: 285–288.

Shi, Z., H. Wang, J. Zhang, Y. Xie, L. Li, X. Chen, B.F. Edgerton and J.R. Bonami. 2005. Response of crayfish, *Procambarus clarkii*, haemocytes infected by white spot syndrome virus. J. Fish. Dis. 28: 151–156.

Sinitsin, D. 1931. Studien uber die phylogenie der trematoden. IV. The life histories of *Plagioporus siliculus* and *Plagioporus virens*, with special reference to the origin of Digenea. Zeitschrift fur Wissenschaftliche Zool. 138: 409–456.

Smith, R. and T. Kamiya. 2001. The first record of an entocytherid ostracod (Crustacea: Cytheroidea) from Japan. Benthos Res. 56: 57–61.

Smith, V.J. and K. Söderhäll. 1986. Crayfish pathology: an overview. Freshw. Crayfish 6: 199–211.

Smythe, A.B. and W.F. Font. 2001. Phylogenetic analysis of *Alloglossidium* (Digenea: Macroderoididae) and related genera: life-cycle evolution and taxonomic revision. J. Parasitol. 87: 386–391.

Söderhäll, I. 2013. Recent advances in crayfish hematopoietic stem cell culture: a model for studies of hemocyte differentiation and immunity. Cytotechnology 65: 691–695.

Söderhäll, I., C. Wu, M. Novotny, B.L. Lee and K. Söderhäll. 2009. A novel protein acts as a negative regulator of prophenoloxidase activation and melanization in the freshwater crayfish *Pacifastacus leniusculus*. J. Biol. Chem. 284: 6301–6310.

Söderhäll, K. and R. Ajaxon. 1982. Effect of quinones and melanin on mycelial growth of *Aphanomyces* spp. and extracellular protease of *Aphanomyces astaci*, a parasite on crayfish. J. Invertebr. Pathol. 39: 105–109.

Söderhäll, K., J. Rantamäki and O. Constantinescu. 1993. Isolation of *Trichosporon beigelii* from the freshwater crayfish *Astacus astacus*. Aquaculture 116: 25–31.

Sogandares-Bernal, F. 1962a. Presumable microsporidiosis in the dwarf crayfishes *Cambarellus puer* Hobbs and *C. shufeldti* (Faxon) in Louisiana. J. Parasitol. 48: 493.

Sogandares-Bernal, F. 1962b. *Microphallus progeneticus*, a new apharyngeate progenetic trematode (Microphallidae) from the dwarf crayfish *Cambarellus puer*, in Louisiana. Tulane Stud. Zool. 9: 319–322.

Sogandares-Bernal, F. 1965. Parasites from Louisiana crayfishes. Tulane Stud. Zool. 12: 79–85.

Soowannayan, C. and M. Phanthura. 2011. Horizontal transmission of white spot syndrome virus (WSSV) between red claw crayfish (*Cherax quadricarinatus*) and the giant tiger shrimp (*Penaeus monodon*). Aquaculture 319: 5–10.

Soowannayan, C., G.T. Nguyen, L.N. Pham, M. Phanthura and N. Nakthong. 2015. Australian red claw crayfish (*Cherax quadricarinatus*) is susceptible to yellow head virus (YHV) infection and can transmit it to the black tiger shrimp (*Penaeus monodon*). Aquaculture 445: 63–69.

Sprague, V. 1950. *Thelohania cambari* n. sp., a microsporidian parasite of North American crayfish. J. Parasitol. 36: 46.

Sprague, V. 1966. Two new species of *Plistophora* (Microsporida, Nosematidae) in decapods, with particular reference to one in the blue crab. J. Protozool. 13: 196–199.

Sprague, V. and J.A. Couch. 1971. An annotated list of protozoan parasites, hyperparasites and commensals of decapod crustaceans. J. Protozool. 18: 526–537.

Sricharoen, S., J.J. Kim, S. Tunkijjanukij and I. Söderhäll. 2005. Exocytosis and proteomic analysis of the vesicle content of granular hemocytes from a crayfish. Dev. Comp. Immunol. 29: 1017–1031.

Stafford, E.W. 1931. Platyhelmia in aquatic insects and Crustacea. J. Parasitol. 37: 131.

Steiner, T.M. and A.C.Z. Amaral. 1999. The family Histriobdellidae (Annelida, Polychaeta) including descriptions of two new species from Brazil and a new genus. Cont. Zool. 68: 95–108.

Stentiford, G.D., J.-R. Bonami, V. Alday-Sanz and P. Alday-Sanz. 2009. A critical review of susceptibility of crustaceans to Taura syndrome, Yellowhead disease and White Spot Disease and implications of inclusion of these diseases in European legislation. Aquaculture 291: 1–17.

Strand, D.A., A. Holst-Jensen, H. Viljugrein, B. Edvardsen, D. Klaveness, J. Jussila and T. Vrålstad. 2011. Detection and quantification of the crayfish plague agent in natural waters: Direct monitoring approach for aquatic environments. Dis. Aquat. Org. 95: 9–17.

Stromberg, P.C., M.J. Toussant and J.P. Dubey. 1978. Population biology of *Paragonimus kellicotti* metacercariae in central Ohio. Parasitology 77: 13–18.

Subchev, M. 1986. On the korean branchiobdellids (Annelida, Clitellata) with a description of a new species: *Branchiobdella teresae* sp. n. Acta Zool. Bulg. 31: 60–66.

Subchev, M.A. 2012. *Branchiobdella* (Annelida: Clitellata) species found in crayfish collection of London Natural History Museum. Acta Zool. Bulg. 64: 319–323.

Subchev, M. 2014. The genus *Branchiobdella* Odier, 1823 (Annelida, Clitellata, Branchiobdellida): a review of its european species. Acta Zool. Bulg. 66: 5–20.

Subchev, M. and L. Stanimirova. 1998. Distribution of freshwater crayfishes (Crustacea: Astacidae) and the epibionts of the genus *Branchiobdella* (Annelida: Branchiobdellae), *Hystricosoma chappuisi* Michaelsen, 1926 (Annelida: Oligochaeta) and *Nitocrella divaricata* (Crustacea: Copepoda) in Bulgaria. Hist. Nat. Bulg. 9: 5–18.

Subchev, M., E. Koutrakis and C. Perdikaris. 2007. Crayfish epibionts *Branchiobdella* sp. and *Hystricosoma chappuisi* (Annedlida: Clitellata) in Greece. Bull. Fr. Pêch. Piscic. 387: 59–66.

Sullivan, J.J. and R.W. Heard. 1969. *Macroderoides progeneticus* n. sp., a progenetic trematode (Digenea: Macroderoididae) from the antennary gland of the crayfish, *Procambarus spiculifer* (LeConte). Trans. Am. Microsc. Soc. 88: 304–308.

Suter, P. and A. Richardson. 1977. The biology of two species of *Engaeus* (Decapoda: Parastacidae) in Tasmania. III. Habitat, food, associated fauna and distribution. Aust. J. Mar. Freshw. Res. 28: 95–103.

Tan, C.K. and L. Owens. 2000. Infectivity, transmission and 16S rRNA sequencing of a rickettsia, *Coxiella cheraxi* sp. nov., from the freshwater crayfish *Cherax quadricarinatus*. Dis. Aquat. Org. 41: 115–122.

Taylor, S., M.J. Landman and N. Ling. 2009. Flow cytometric characterization of freshwater crayfish hemocytes for the examination of physiological status in wild and captive animals. J. Aquat. Anim. Health 21: 195–203.

Thörnqvist, P.-O. and K. Söderhäll. 1993. *Psorospermium haeckeli* and its interaction with the crayfish defence system. Aquaculture 117: 205–213.

Thune, R.L., J.P. Hawke and R.J. Sebeling. 1991. Vibriosis in the red swamp crawfish. J. Aquat. Anim. Health 3: 188–191.

Tilmans, M., A. Mrugała, J. Svoboda, M. Engelsma, M. Petie, D.M. Soes, S. Nutbeam-Tuffs, B. Oidtmann, I. Roessink and A. Petrusek. 2014. Survey of the crayfish plague pathogen presence in the Netherlands reveals a new *Aphanomyces astaci* carrier. J. Invertebr. Pathol. 120: 74–79.

Timm, T. 1991. Branchiobdellida (Oligochaeta) from the farthest South-East of the U.S.S.R. Zool. Scr. 20: 321–331.

Topić Popović, N., R. Sauerborn Klobučar, I. Maguire, I. Strunjak-Perović, S. Kazazić, J. Barišić, M. Jadan, G. Klobučar and R. Čož-Rakovac. 2014. High-throughput discrimination of bacteria isolated from *Astacus astacus* and *A. leptodactylus*. Knowl. Managt. Aquatic Ecosyst. 413: 4.

Toumanoff, C. 1965. Infections bactériennes chez les écrevisses (Entérobactériacées). I - Protéoses. Bull. Fr. Piscic. 219: 41–65.

Turner, H.M. 1985. Pathogenesis of *Alloglossoides caridicola* (Trematoda) infection in the antennal glands of the crayfish *Procambarus acutus*. J. Wildl. Dis. 21: 459–461.

Turner, H.M. 1999. Distribution and prevalence of *Alloglossoides caridicola* (Trematoda: Macroderoididae), a parasite of the crayfish *Procambarus acutus* within the State of Louisiana, U.S.A., and into adjoining states. J. Helminthol. Soc. Washingt. 66: 86–89.

Turner, H.M. 2006. Distribution and prevalence of *Allocorrigia filiformis* (Trematoda: Dicrocoeliidae), a parasite of the crayfish *Procambarus clarkii*, within the state of Louisiana and the Lower Mississippi River Valley, U.S.A. Comp. Parasitol. 73: 274–278.

Turner, H.M. 2007. New hosts, distribution, and prevalence records for *Alloglossidium dolandi* (Digenea: Macroderoididae), a parasite of procambarid crayfish, within the coastal plains of Georgia, Alabama, and Mississippi, U.S.A. Comp. Parasitol. 74: 148–150.

Turner, H.M. 2009. Additional distribution and prevalence records for *Alloglossidium dolandi* (Digenea: Macroderoididae) and a comparison with the distribution of *Alloglossidium caridicolum*, parasites of procambarid crayfish, within the coastal plains of the Southeastern United States. Comp. Parasitol. 76: 283–286.

Turner, H.M. and K.C. Corkum. 1977. *Allocorrigia filiformis* gen. et sp. n. (Trematoda: Dicrocoeliidae) from the crayfish, *Procambarus clarkii* (Girard, 1852). Proc. Helminthol. Soc. Wash. 44: 65–67.

Turner, H.M. and S. McKeever. 1993. *Alloglossoides dolandi* n. sp. (Trematoda: Macroderoididae) from the crayfish *Procambarus epicyrtus* in Georgia. J. Parasitol. 79: 353–355.

Unestam, T. 1973. Significance of diseases on freshwater crayfish. Freshw. Crayfish 5: 135–150.

Unestam, T. 1976. Defence reactions in and susceptibility of Australian and New Guinean freshwater crayfish to European-crayfish-plague fungus. Aust. J. Exp. Biol. Med. Sci. 53: 349–59.

Vafopoulou, X. 2009. Mechanisms of wound repair in crayfish. Invertebr. Surviv. J. 6: 125–137.

Vedia, I., J. Oscoz and J. Rueda. 2014. An alien ectosymbiotic branchiobdellidan (Annelida: Clitellata) adopting exotic crayfish: a biological co-invasion with unpredictable consequences. Inl. Waters 5: 89–92.

Vennerström, P., K. Söderhäll and L. Cerenius. 1998. The origin of two crayfish plague (*Aphanomyces astaci*) epizootics in Finland on noble crayfish, *Astacus astacus*. Ann. Zool. Fennici 35: 43–46.

Verweyen, L., S. Klimpel and H.W. Palm. 2011. Molecular phylogeny of the Acanthocephala (class Palaeacanthocephala) with a paraphyletic assemblage of the orders Polymorphida and Echinorhynchida. PLoS One 6: e28285.

Vey, A. 1977. Studies on the pathology of crayfish under rearing conditions. Freshw. Crayfish 3: 311–319.

Vey, A. 1981. Les maladies des ecrevisses, leur reconnaissance et la surveillance sanitaire des populations astacicoles. Bull. Fr. Pêch. Piscic. 281: 223–236.

Vey, A., N. Boemare and C. Vago. 1975. Recherches sur les maladies bacteriennes de l'ecrevisse *Austropotamobius pallipes* Lereboullet. Freshw. Crayfish 2: 287–297.

Viets, K. 1931. Über eine an Krebskiemen parasitierende Halacaride aus Australien. Zool. Anz. 96: 115–120.

Viets, K. 1939. *Piona rotunda* (Kramer) (Hydrachnellae) in der Kiemenhöhle von Krebsen. Arch. Hydrobiol. i Rybactwa 12: 115–116.

Vila, I. and N. Bahamonde. 1985. Two new species of *Stratiodrilus, Stratiodrilus aeglaphilus* and *Stratiodrilus pugnaxi* (Annelida, Histriobdellidae) from Chile. Proc. Biol. Soc. Washingt. 98: 347–350.

Viljamaa-Dirks, S., S. Heinikainen, H. Torssonen, M. Pursiainen, J. Mattila and S. Pelkonen. 2013. Distribution and epidemiology of genotypes of the crayfish plague agent *Aphanomyces astaci* from noble crayfish *Astacus astacus* in Finland. Dis. Aquat. Org. 103: 199–208.

Vogelbein, W.K. and R.L. Thune. 1988. Ultrastructural features of three ectocommensal Protozoa attached to the gills of the red swamp crawfish, *Procambarus clarkii* (Crustacea: Decapoda). J. Protozool. 35: 341–348.

Vogt, G. and M. Rug. 1999. Life stages and tentative life cycle of *Psorospermium haeckeli*, a species of the novel DRIPs clade from the animal-fungal dichotomy. J. Exp. Zool. part A 283: 31–42.

Vogt, G., M. Keller and D. Brandis. 1996. Occurrence of *Psorospermium haeckeli* in the stone crayfish *Austropotamobius torrentium* from a population naturally mixed with the noble crayfish *Astacus astacus*. Dis. Aquat. Org. 25: 233–238.

Vogt, G., L. Tolley and G. Scholtz. 2004. Life stages and reproductive components of the Marmorkrebs (marbled crayfish), the first parthenogenetic decapod crustacean. J. Morphol. 261: 286–311.

Volonterio, O. 2009. First report of the introduction of an Australian temnocephalidan into the New World. J. Parasitol. 95: 120–123.

Vrålstad, T., A.K. Knutsen, T. Tengs and A. Holst-Jensen. 2009. A quantitative TaqMan® MGB real-time polymerase chain reaction based assay for detection of the causative agent of crayfish plague *Aphanomyces astaci*. Vet. Microbiol. 137: 146–155.

Vranckx, R. and M. Durliat. 1981. Encapsulation of *Psorospermium haeckeli* by the haemocytes of *Astacus leptodactylus*. Experientia 37: 40–42.

Walker, G., R.G. Dorrell, A. Schlacht and J.B. Dacks. 2011. Eukaryotic systematics: a user's guide for cell biologists and parasitologists. Parasitology 138: 1638–1663.

Walton, M. and H.H. Hobbs, Jr. 1971. The distribution of certain entocytherid ostracods on their crayfish hosts. Proc. Acad. Nat. Sci. Philadelphia 123: 87–103.

Wang, H.-Z. and Y.-D. Cui. 2007. On the studies of microdrile Oligochaeta and Aeolosomatidae (Annelida) in China: brief history and species checklist. Acta Hydrobiol. Sin. 31: 87–98.

Wang, J., H. Huang, Q. Feng, T. Liang, K. Bi, W. Gu, W. Wang and J.D. Shields. 2009. Enzyme-liked immunosorbent assay for the detection of pathogenic spiroplasma in commercially exploited crustaceans from China. Aquaculture 292: 166–171.

Wang, W., L. Rong, W. Gu, K. Du and J. Chen. 2003. Study on experimental infections of *Spiroplasma* from the Chinese mitten crab in crayfish, mice and embryonated chickens. Res. Microbiol. 154: 677–680.

Wang, W., W. Gu, Z. Ding, Y. Ren, J. Chen and Y. Hou. 2005. A novel *Spiroplasma* pathogen causing systemic infection in the crayfish *Procambarus clarkii* (Crustacea: Decapod), in China. FEMS Microbiol. Lett. 249: 131–137.

Wang, W., W. Gu, G.E. Gasparich, K. Bi, J. Ou, Q. Meng, T. Liang, Q. Feng, J. Zhang and Y. Zhang. 2010. *Spiroplasma eriocheiris* sp. nov., a novel species associated with mortalities in *Eriocheir sinensis*, Chinese mitten crab. Int. J. Syst. Evol. Microbiol. 61: 703–708.

Warren, A. 1986. A revision of the genus *Vorticella* (Ciliophora, Peritrichida). Bull. Br. Museum (Nat. Hist. Zool.) 50: 1–57.

Warren, A. and J. Paynter. 1991. A revision of *Cothurnia* (Ciliophora: Peritrichida) and its morphological relatives. Bull. Br. Museum (Nat. Hist. Zool.) 57: 17–59.

Warren, E. 1903. On the anatomy and development of *Distomum cirrigerum*, v. Baer. J. Cell. Sci. s2-47: 273–301.

Watson, N.A. and K. Rohde. 1995. Ultrastructure of spermiogenesis and spermatozoa in the platyhelminths *Actinodactylella blanchardi* (Temnocephalida, Actinodactylellidae), *Didymorchis* sp. (Temnocephalida, Didymorchidae) and *Gieysztoria* sp. (Dalyelliida, Dalyelliidae), with implications for the phylogeny of the Rhabdocoela. Invertebr. Reprod. Dev. 27: 145–158.

Weber, M., A.R. Wey-Fabrizius, L. Podsiadlowski, A. Witek, R.O. Schill, L. Sugár, H. Herlyn and T. Hankeln. 2013. Phylogenetic analyses of endoparasitic Acanthocephala based on mitochondrial genomes suggest secondary loss of sensory organs. Mol. Phylogenet. Evol. 66: 182–189.

Wen, R.-S. and L. Liu. 2001. Description of *Temnocephala semperi*, an ectosymbition (sic) on *Cherax quadricarinatus*. J. Jiaying Univ. 6: XX.

Westervelt, C., Jr. and E. Kozloff. 1959. *Entocythere neglecta* sp. nov., a cytherid ostracod commensal on *Pacifastacus nigrescens* (Stimpson). Am. Midl. Nat. 61: 239–244.

Wierzbicka, J. and P. Smietana. 1999. The food of *Branchiobdella* Odier, 1823 [Annelida] dwelling on crayfish and the occurrence of the fish parasite *Argulus* Muller, 1785 [Crustacea] on the carapace of *Pontastacus leptodactylus* [Esch.]. Acta Ichthy. Pisc. 29: 93–99.

Wild, C. and J. Furse. 2004. The relationship between *Euastacus sulcatus* and temnocephalan spp. (Platyhelminthes) in the Gold Coast hinterland, Queensland. Freshw. Crayfish 14: 236–245.

Williams, B.W. and S.R. Gelder. 2011. A re-description of *Cambarincola bobbi* Holt, 1988, a description of a new species of *Pterodrilus*, and observations of sympatric species of crayfish worms (Annelida: Clitellata: Branchiobdellida) from the Cumberland River Watershed in Tennessee. Southeast. Nat. 10: 199–210.

Williams, B.W., S.R. Gelder and H. Proctor. 2009. Distribution and first reports of Branchiobdellida (Annelida: Clitellata) on crayfish in the Prairie Provinces of Canada. West. North. Am. Nat. 69: 119–124.

Williams, B.W., S.R. Gelder, H.C. Proctor and D.W. Coltman. 2013. Molecular phylogeny of North American Branchiobdellida (Annelida: Clitellata). Mol. Phylogenet. Evol. 66: 30–42.

Williams, J.B. 1981. Classification of the Temnocephaloidea (Platyhelminthes). J. Nat. Hist. 15: 277–299.

Williams, J.B. 1994. Unicellular adhesive secretion glands and other cells in the parenchyma of *Temnocephala novaezealandiae* (Platyhelminthes, Temnocephaloidea): intercell relationships and nuclear pockets. N.Z. J. Zool. 21: 167–178.

Williams, R. 1967. Metacercariae of *Prosthodendrium naviculum* Macy, 1936 (Trematoda: Lecithodendriidae) from the crayfish, *Orconectes rusticus* (Girard). Proc. Pennsylvania Acad. Sci. 41: 38–41.

Wiszniewski, J. 1939. O faunie jamy skrzelowej raków rzecznych ze szczególnym uwzględnieniem wrotków. Arch. Hydrobiol. i Rybactwa 12: 124–155.

Wiszniewski, J. 1953. O wrotkach-komensalach niektórych skorupiaków. Pol. Arch. Hydrobiol. 1: 25–41.

Womersley, H. 1941–1943. On *Astacopsiphagus parasiticus* Vietz 1931 (Acarina-Halacaridae) parasitic in the gill chambers of *Euastacus sulcatus* Clark M.S. Rec. S. Aust. Mus. 7: 401–403.

Wong, F.Y.K., K. Fowler and P.M. Desmarchelier. 1995. Vibriosis due to *Vibrio mimicus* in Australian freshwater crayfish. J. Aquat. Anim. Health 7: 284–291.

Woodhead, A.E. 1950. Life history cycle of the giant kidney worm, *Dioctophyma renale* (Nematoda), of Man and many other mammals. Trans. Am. Microsc. Soc. 69: 21–46.

Wu, X.G., H.T. Xiong, Y.Z. Wang and H.H. Du. 2012. Evidence for cell apoptosis suppressing white spot syndrome virus replication in *Procambarus clarkii* at high temperature. Dis. Aquat. Org. 102: 13–21.

Wulfert, K. 1957. Ein neues Rädertier aus der Kiemenhöhle von *Cambarus affinis*. Zool. Anz. 158: 26–30.

Xie, X., H. Li, L. Xu and F. Yang. 2005. A simple and efficient method for purification of intact white spot syndrome virus (WSSV) viral particles. Virus Res. 108: 63–67.

Xu, Z., H. Du, Y. Xu, J. Sun and J. Shen. 2006. Crayfish *Procambarus clarkii* protected against white spot syndrome virus by oral administration of viral proteins expressed in silkworms. Aquaculture 253: 179–183.

Xylander, W.E.R. 1997. Epidermis and sensory receptors of *Temnocephala minor* (Plathelminthes, Rhabdocoela, Temnocephalida): an electron microscopic study. Zoomorphology 117: 147–154.

Yamaguchi, H. 1934. Studies on Japanese Branchiobdellidae with some revisions on the classification. J. Fac. Sci. Hokkaido Imp. Univ. 3: 177–219.

Yan, D.C., S.L. Dong, J. Huang and J.S. Zhang. 2007. White spot syndrome virus (WSSV) transmission from rotifer inoculum to crayfish. J. Invertebr. Pathol. 94: 144–148.

Young, W. 1971. Ecological studies of the Entocytheridae (Ostracoda). Am. Midl. Nat. 85: 399–409.

Zawal, A. 1998. Water mites (Hydracarina) in the branchial cavity of crayfish *Orconectes limosus* (Raf. 1817). Acta Hydrobiol. 40: 49–54.

Zeng, Y. 2013. Procambarin: a glycine-rich peptide found in the haemocytes of red swamp crayfish *Procambarus clarkii* and its response to white spot syndrome virus challenge. Fish Shellfish Immunol. 35: 407–12.

Zhang, H.W., C. Sun, S.S. Sun, X.F. Zhao and J.X. Wang. 2010. Functional analysis of two invertebrate-type lysozymes from red swamp crayfish, *Procambarus clarkii*. Fish Shellfish Immunol. 29: 1066–1072.

Zhang, Y., J.-F. Ning, X.Q. Qu, X.L. Meng and J.P. Xu. 2012. TAT-mediated oral subunit vaccine against white spot syndrome virus in crayfish. J. Virol. Methods 181: 59–67.

Zhu, F., Z.G. Miao, Y.H. Li, H.H. Du and Z.R. Xu. 2009. Oral vaccination trials with crayfish, *Procambarus clarkii*, to induce resistance to the white spot syndrome virus. Aquac. Res. 40: 1793–1798.

Zrzavý, J., S. Mihulka, P. Kepka and A. Bezděk. 1998. Phylogeny of the Metazoa based on morphological and 18S ribosomal DNA evidence. Cladistics 14: 249–285.

Zuo, D., D.-L. Wu, C.-A. Ma, H.-X. Li, Y.-H. Huang, D.-L. Wang and Y.-L. Zhao. 2015. Effects of white spot syndrome virus infection and role of immune polysaccharides of juvenile *Cherax quadricarinatus*. Aquaculture 437: 235–242.

Environmental Drivers for Population Success: Population Biology, Population and Community Dynamics

Ed Willis Jones,[1] *Michelle C. Jackson*[2] *and Jonathan Grey*[3,*]

Introduction

A good grasp of the environmental drivers for population success is particularly pertinent for two quite opposing reasons when considering crayfish. For the purpose of conserving particular crayfish species, then a clear understanding of environmental tolerances and thresholds is important when defining habitat requirements and implementing management strategies for maintaining crayfish habitat of suitable quality to best promote sustainable populations. Our understanding of the environmental factors that may limit their distribution, particularly beyond the very local scales often studied, is still often inadequate (Dyer et al. 2013). This would be particularly relevant in the case of 'ark site' selection, a key component of the crayfish conservation strategy being implemented within Europe (Schulz et al. 2002), since ark sites are typically a last ditch attempt to save a population by translocating it away from perceived threats. An ark site must represent 'the ideal', the optimal suite of environmental parameters in which to preserve and hopefully propagate population numbers to potentially use for recolonization of former sites when appropriate. In contrast, understanding when and under what environmental conditions populations of crayfish falter or fail might reveal chinks in the armour of particularly pernicious crayfish that are gaining repute as invasive species. Exploiting such chinks could be an avenue toward control measures for some of those species introduced either intentionally or accidentally into water bodies where they are now wreaking considerable ecological and economic damage.

[1] School of Biological and Chemical Sciences, Queen Mary University of London, Mile End Road, London, E1 4NS.
  Email: e.willis-jones@qmul.ac.uk
[2] Centre for Invasion Biology, Department of Zoology and Entomology, University of Pretoria, Hatfield, South Africa.
  Email: mjackson@zoology.up.ac.za
[3] Lancaster Environment Centre, Lancaster University, Lancaster, LA1 4YQ.
* Corresponding author: j.grey@lancaster.ac.uk

A further juxtaposition is revealed in the review of environmental drivers for crayfish population success, again related to whether the crayfish is considered a native or an invasive species. In general, the more recent studies have tended to focus on the environmental tolerances of invasive crayfish, whereas studies of native species requirements are typically much older. Given that we are now experiencing a period of unprecedented climate change, and from the perspective of species future management, be it for conservation of species or stocks to exploit, long-term and large-scale studies, and modelling approaches are critical for providing a better understanding of crayfish population dynamics in relation to environmental factors (e.g., Dyer et al. 2013, Zimmerman and Palo 2012).

Why are crayfish abundant in some areas (or at certain times) and not others?

Clearly, combining observations of distribution patterns with appropriate environmental data (including any sympatric species), and drawing associations between particular species and habitat conditions, will allow for assessment of those factors affecting population change, be it success or failure (e.g., Svobodová et al. 2012, Richman et al. 2015). It can be extended to use in modelling potential distributions of species under future environmental change scenarios (Feria and Faulkes 2011, Dyer et al. 2013), and also identifying invasion potential (Olden et al. 2011). To distil the wealth of information on environmental drivers for population success from the literature and try to find generalities for the three diverse families of crayfish is clearly challenging because various crayfish inhabit lakes, streams, temporary wetlands, swamps and strong burrowers can be found in terrestrial ecosystems (Crandall and Buhay 2008). There are also some species which are specialist cave dwellers which often play a key role in ecosystem processes, such as carrion breakdown (Huntsman et al. 2011). Species-specific responses to environmental variables have been reported between two sympatric congeners in small streams (e.g., Flinders and Magoulick 2007) let alone between epigean and hypogean, or between aquatic and semi-aquatic species. A study of habitat use by five sympatric Australian freshwater crayfish species of the Parastacidae family from a diversity hotspot in The Grampians National Park, Australia, by Johnston and Robson (2009) exemplifies this complexity. While their distribution was directly related to habitat type, and the environmental and physicochemical variables that characterised habitats, *Engaeus lyelli*, *Gramastacus falcata* and *G. insolitus* occurred predominantly in floodplain wetlands and flooded vegetation habitats, *Euastacus bispinosus* occurred only in flowing channels with soft-sediments, and *Cherax destructor* was found in all catchments and habitat types studied. Hence, even within a relatively small area, these crayfish species varied in their degree of habitat specialisation: strongly generalist (*C. destructor*) to occupying only a specific habitat type (*E. bispinosus*). Some species appeared specialised for seasonal wetlands (*G. insolitus* and *G. falcata*), and overlap in site occupancy also varied: *G. insolitus* and *G. falcata* distributions were strongly associated, whereas *C. destructor* appeared to occur opportunistically across habitats, both alone and co-occurring with all the other species. To add to this complexity, the relative importance of environmental drivers will also change with ontogeny (e.g., habitat use—DiStefano et al. (2003), Brewer et al. (2009); or tolerance to pH—France (1984)). However, the distributional patterns of crayfish within aquatic ecosystems, and by inference their likely sustainable populations, typically can be related to the prevalent physico-chemical features provided that they are evaluated at appropriate temporal and spatial scales.

Two recent and relatively large-scale studies have focussed on chemical elements. Svobodová et al. (2012) compared 1008 sites accounting for ~90% coverage of the Czech Republic for two native species (*Astacus astacus* and *Austropotamobius torrentium*) and one invader (*Orconectes limosus*). Their statistical analyses suggested only weak positive relationships for the native species with oxygen concentration and pH, and negative associations with parameters indicating nutrient enrichment (ammonium, nitrite, BOD_5); the invader was apparently more tolerant of these latter nutrient enriched conditions. However, species are generally introduced into waters close to human habitation (e.g., Jackson and Grey 2013) which are also typically of lower environmental quality, so we should be wary of confounding these two issues. Similarly, using a systematic literature review of the chemistry of waterbodies inhabited by *Austropotamobius pallipes*, Haddaway et al. (2015) examined those significant variables potentially influencing crayfish distribution, several of which appeared to have a threshold effect. They found that crayfish presence was associated with high dissolved oxygen, low conductivity, ammonium, sodium, and phosphate, and to a lesser extent low sulphate, nitrate, and total suspended solids.

A more physical habitat traits approach was taken by Flinders and Magoulick (2003) when they assessed crayfish community structure through quantitative sampling of riffle, run and pool habitats in 15 intermittent and 21 permanent streams in northern Arkansas and southern Missouri. For the four crayfish species collected: *Orconectes marchandi*, *O. punctimanus*, *O. ozarkae* and *Cambarus hubbsi*, overall crayfish densities were significantly greater in intermittent streams than in permanent streams, especially for populations of *O. marchandi* and *O. punctimanus*. Moreover, there was a significant relationship between crayfish relative density and abiotic environmental variables for permanent streams, but not for intermittent streams. In permanent streams, the percentage of gravel substrate, substrate diversity, and mean current velocity were among the most important factors in determining crayfish density, and the authors suggest that taxa mobility and predation risk are the most likely explanations for their observed patterns of crayfish density. Meanwhile, Dyer et al. (2013) included physical as well as chemical variables within a Maximum Entropy Species Distribution modelling approach to predict the current and future distributions of four crayfish species endemic to Oklahoma and Arkansas. Key factors providing 65–87% of their model's predictive power were soil composition, elevation, winter precipitation, and temperature. In a final example of modelling approaches, Zimmerman and Palo (2012) adopted a temporal approach and examined *Astacus astacus* capture data over 27 years from a Swedish river to explore the relative impact of different climate factors and density dependence on variability of catch sizes. Time series of catch per unit effort (CPUE, assumed to reflect population size) were analysed in relation to the North Atlantic Oscillation (NAO) index, regional weather factors and water flow. Both density dependence and climatic factors played a significant role in crayfish population fluctuations with varying time lags. The authors demonstrated that for example, water flow showed a 2 year lag to the CPUE, and reasoned that high flow in the river might be affecting adult survival, while an as yet unidentified factor related to NAO and causing a 6 year lag to the CPUE might act on the juvenile stages of the population. While in the paragraphs above we have cited studies tending to focus purely on physical, or chemical, a combination of those two, it is without doubt that the biotic component is intrinsically linked and will also have a bearing upon crayfish distribution and population success. Recent studies in this vein have for example aimed to link crayfish distributions to the other macroinvertebrates on which they feed (e.g., Grandjean et al. 2011, Trouilhé et al. 2012, Jandry et al. 2014).

Environmental tolerances

Hydrography and habitat availability

Habitat heterogeneity is, perhaps unsurprisingly, important in maintaining high crayfish secondary production and diversity (e.g., 'channel units' of riffles, runs, and pools in rivers; Brewer et al. (2009), or cobble area in lentic systems; Olsson and Nyström (2009)), but their production is often heavily reliant upon relatively small areas within aquatic waterbodies. Structurally complex, vegetated edge habitats or backwaters offer refugia from abiotic and biotic factors (Garvey et al. 1994, Lodge and Hill 1994, Smith et al. 1996), although such areas frequently comprise <15% of the available stream benthos (Flinders and Magoulick 2007). It is these habitats which are also some of the first to be impacted upon by natural or anthropogenic change (Bunn and Arthington 2002), and so alteration of natural flow regimes may impact heavily upon future population recruitment (Fig. 1).

Refugia availability throughout ontogeny is important for population success and many crayfish occupy and defend shelters (across the families Cambaridae, Astacidae, Parastacidae; Holdich 2002, Gherardi and Daniels 2004, Alonso and Martínez 2006). Young crayfish need protection until they grow large enough to reach a size-refuge from fish predation (Stein 1977, Englund and Krupa 2000), as well as from conspecifics intent on cannibalism (Olsson and Nyström 2009). Despite their relatively large size amongst the invertebrate community, adult crayfish are still prey for many fish, mammals and birds, and are more vulnerable at certain times during the years for example when moulting or when the females are berried, and we will pick up on this later. In habitats subject to water flow, there is typically a negative

Aquatic biodiversity and natural flow regimes

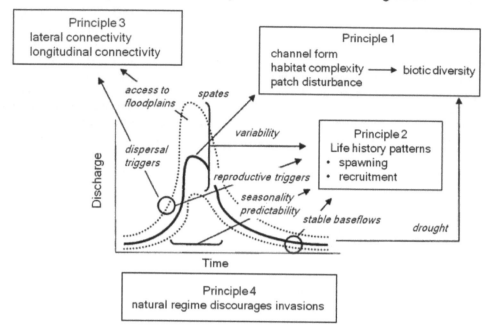

Figure 1. The natural flow regime of a river influences crayfish population success (particularly at reproductive and recruitment stages) via several interrelated mechanisms operating over different spatial and temporal scales. Redrawn and adapted from Bunn and Arthington (2002).

association between water velocity and crayfish density (e.g., Light 2003), so refugia are critical for all age-classes for protection against adverse flow conditions (Smith et al. 1996). Therefore, not only refuge availability, but competition with other species for shelters could significantly affect the fitness of crayfish (Bovbjerg 1970).

Extremes of the hydroperiod, both flooding and droughting conditions in wetland or stream habitats, can be a severe disturbance for all aquatic animals, but the severity of a disturbance is context dependent on the available substrate and the particular crayfish species traits. A reduction in hydroperiod in seasonal Florida wetlands led to a decrease in the density of the burrowing species, *Procambarus alleni*, and concomitant decreases in juvenile abundance and survival, while there was an increase in the number of smaller-bodied, dispersing adults (Acosta and Perry 2001). Taking an experimental approach in those Florida wetlands, Dorn and Volin (2009) demonstrated that *P. alleni* exhibited different burrowing abilities in marl and sand substrates leading to differential survival rates relative to a congener, *P. fallax*. Taylor (1983) found that drought differentially affected two resident species of crayfish. *Procambarus spiculifer*, a non-burrowing species, experienced reductions in adult and overall population densities, reduced adult body size, a shift in reproductive timing and increased juveniles within the population, whereas the burrowing species, *Cambarus latimanus*, showed no significant changes due to drought. Moreover, these drought-induced changes were maintained in one population of *P. spiculifer* for the duration of an 8-y study and resulted in the extinction of two sub-populations (Taylor 1988). Similarly, differential tolerance to drought and related temperature-oxygen concentration parameters resulted in ecological isolation within the natural range of *Orconectes virilis* and *O. immunis*; the former species preferred streams and lake margins whereas the latter inhabited ponds and sloughs (Bovbjerg 1970). Burrowing species have clearly evolved a survival strategy to cope with all but the most extreme of natural hydroperiod fluctuations, but any human activities exacerbating droughting (e.g., abstraction) may render the strategy insufficient for population persistence (Acosta and Perry 2001, Bunn and Arthington 2002). Indeed, interspecific differences in tolerance to stream drying and desiccation have been postulated as a key driver in success of some invasive crayfish over their native counterparts (e.g., in a laboratory study, the native crayfish *Orconectes eupunctus* was less tolerant of desiccation than the invasive *Orconectes neglectus chaenodactylus*; Larson et al. (2009)).

If there is a degree of connectivity with other waterbodies, so again context dependent, then crayfish may maintain populations by dispersal or emigration, even across land to avoid sub-optimal conditions (e.g., Acosta and Perry 2001). Cossette and Rodríguez (2004) noted that to avoid seasonal increases in water level and temperature, *Cambarus bartonii* moved from lotic into nearby lentic habitats, and this was especially apparent for 1+ individuals compared to adults. Counter to this is either the presence of natural barriers or anthropogenic obstructions such as dams that will limit the potential for dispersal (Joy and Death 2001). Indeed, limiting dispersal is proposed by various people as a management tool to curb invasive crayfish population success and spread (e.g., Kerby et al. 2005), but given that removal of anthropogenic barriers at least is currently in vogue to restore connectedness, rehabilitate geomorphology, and reinstate migration (see O'Connor et al. 2015), it is unlikely to be taken up.

Tolerance to some key chemical parameters

Human modification of the environment via cultural eutrophication, deforestation, agricultural improvement, abstraction and irrigation to name but a few all impact heavily upon water chemistry parameters and are increasingly identified as threats to crayfish populations (Richman et al. 2015). Several reviews of water chemistry parameters influencing crayfish distribution exist, e.g., Haddaway et al. (2015) who noted that magnesium, potassium, sodium, sulphate, nitrate, and total suspended solids may be tolerated at moderate to high concentrations in isolation by *A. pallipes* populations. However, suites of chemical conditions may act synergistically *in situ* and must be considered as multiple stressors alongside other environmental drivers like temperature. We will not dwell on the details of species-specific tolerance limits for the ≈640 species of freshwater crayfish here, but several parameters are worth examining further and some of these we return to later with particular reference to population success or failure.

pH

Crustacea such as crayfish have a relatively inflexible exoskeleton primarily composed of chitin to which calcium is added in the form of calcite crystals and calcium carbonate, making the skeleton robust. Hence, crayfish are predominantly found in waters with relatively high pH, e.g., *Austropotamobius torrentium* at pH 8.5 (Renz and Breithaupt 2000), *Orconectes* spp. at pH 7.4 (Capelli and Magnuson 1983), and between pH 7.2 and 8.9 for *Pacifastacus fortis* (Light et al. 1995). These preferences are supported by laboratory studies on crayfish growth and survival. For example, DiStefano et al. (1991) found high mortality in *Cambarus bartonii* maintained at pH < 3, Siewert and Buck (1991) found high mortality in *Orconectes virilis* at pH < 5, and Haddaway et al. (2013) reported decreasing survival (and growth) in juvenile *A. pallipes* with a reduction from pH 8.6 (94%) to pH 6.5 (25%). Similarly, as a result of the experimental acidification of a soft-water lake in the Experimental Lakes Area, Canada, populations of *O. virilis* declined considerably (Schindler and Turner 1982) or collapsed (France 1987). Mortality under extreme pH conditions has been attributed to disturbed ion regulation (Morgan and McMahon 1982), for example through increased excretion (Mauro and Moore 1987), and to exacerbation of the toxicity of other ions (Malley and Chang 1985, France 1987).

Of particular pertinence here is the impact of pH on growth and moulting, both of which will influence survival to reach adulthood, and hence reproduction (Haddaway et al. 2013). Berrill et al. (1985) compared the tolerance of three common Ontario crayfish species to low pH under natural and laboratory conditions. Both transplant and laboratory experiments indicated that exposure to a pH range of 5.4–6.1 in soft water was toxic to attached juvenile stages of *Orconectes rusticus* and *O. propinquus* but not to females carrying broods. In contrast, juveniles of *Cambarus robustus* moulted and survived in soft water at pH 4. Further laboratory experiments have established that newly independent crayfish are more sensitive than older juveniles and adults (France 1984), and that post-moult crayfish are more sensitive than inter-moult (Malley 1980). Calcium uptake across the gills of *O. virilis* is reduced at pH < 5.8 and inhibited at pH < 4, resulting in a decrease in the rate of re-hardening after moulting (Malley 1980). As there are widespread reports of declining availability of calcium in many freshwaters across the globe, this is a particular concern for the conservation of crayfish populations, indeed of all crustacea (e.g., Edwards et al. 2009, Jeziorski et al. 2015).

While acid conditions are clearly a stressor for crayfish, life-stage vulnerability and species-specific physiology likely account for differences reported from the field, and we should not ignore the fact that one such stressor may well interact with another or others to reveal some interesting complexities. For instance, Seiler and Turner (2004) evaluated the individual and population-level effects of acidification on crayfish (*Cambarus bartonii*) in 24 study reaches of nine Pennsylvania headwater streams using field experiments and survey data. Median base-flow pH varied from 4.4 to 7.4 with substantial variation found both among and within streams. They used bioassays to evaluate the relationship between stream pH and crayfish growth rates and showed that rates were always higher in circumneutral reaches than in acidic reaches. However, crayfish originating from acidic water grew less when transplanted into neutral water compared to crayfish originating in neutral water, thereby providing some evidence for a cost of acclimation to acidity. Stream surveys, perhaps counterintuitively at first glance, showed that crayfish density was six-fold higher in reaches with the lowest pH relative to circumneutral reaches. So, although individual crayfish suffered lower growth in acidified streams, increased acidity appeared to cause an increase in crayfish population size, and shifts in size structure, possibly by relieving predation pressure by fish which were less abundant in the acidic reaches.

Tolerance to nutrients, metals and salinity

The all-encompassing term water pollution is often touted as predictor of particular crayfish species' distributions. In the UK, indeed throughout Western Europe where there is a relative paucity of native species, they are all classically described as inhabitants of clean streams, rivers and ponds, and sensitive to nutrient enrichment (Favaro et al. 2010, Svobodová et al. 2012, reviewed for *A. pallipes* by Haddaway et al. 2015). Around 30 to 50% of threatened species from crayfish diversity hotspots like Australasia, Mexico, and the USA are specifically identified as at risk from pollution (Richman et al. 2015). Crayfish abundance has also been negatively associated with metal-contaminated waters and mining, particularly from surveys on Canadian lakes (e.g., Edwards et al. 2009, Iles and Rasmussen 2005, Keller et al. 1999, Richman et al. 2015). However, some studies suggest that crayfish are highly resistant to environmental metal contamination (Del Ramo et al. 1987, Roldan and Shivers 1987, Khan and Nugegoda 2007), and due to their rapid bioaccumulation and long retention times exhibited, fulfil the criteria described for bio-indicator species by Phillips and Rainbow (1993), Rainbow (1995), and Suárez-Serrano et al. (2010). Bioaccumulation and the effects of heavy metals on crayfish were reviewed by Kouba et al. (2010), and hence will not be covered again here in great detail.

Long-term exposure to nutrients associated with cultural eutrophication such as ammonium, nitrate and nitrite is toxic and reduces crayfish immunity, increasing the chance of infection (Meade and Watts 1995, Yildiz and Benli 2004); exposure to heavy metals such as aluminium has a similar effect (Alexopoulos et al. 2003, Ward et al. 2006). In a study of aluminium toxicity, Alexopoulos et al. (2003) found that despite its insolubility at circumneutral pH, freshly neutralized aluminium is toxic to a variety of freshwater organisms, and it has been included in a list of stressors as partly responsible for the severe declines (63–96%) of both native and non-native crayfish populations across south-central Ontario, Canada, by Edwards et al. (2009). According to Ward et al. (2006) studying *P. leniusculus* under lab conditions, aqueous aluminium impairs the functioning of the gills and ultimately produces a decrease in immunocompetence, and that exposure to episodic pulses of aqueous aluminium over the short term increases the risk of infection by impairing the ability of haemocytes to recognise and/or remove bacteria. However, the route of exposure is clearly important as it is for many contaminants. For example, crayfish can accumulate, store and excrete aluminium from contaminated food with only localised toxicity (Woodburn et al. 2011), while Simon and Boudou (2001) reported that crayfish take up mercury (Hg) and methylmercury (MeHg) from both water and food, with a marked tendency to accumulate MeHg. Moreover, temperature has been identified to have an interactive effect with metal toxicity; increasing temperatures by only 4°C (so within the potential range as predicted by various climate change scenarios; IPCC 2014) significantly increased the toxicity and reduced the oxygen consumptive ability of crayfish when exposed to metals like cadmium, copper, and zinc (Khan et al. 2006).

The tolerance of various crayfish species to salinity has mostly been studied from a conservation (e.g., Holdich et al. 1997, Pinder et al. 2005, Meineri et al. 2014) or from a physiological/aquaculture production perspective (e.g., Meade et al. 2002). Salinization is a considerable problem associated with intensive agriculture, particularly in crayfish diversity hotspots like Australia (Pinder et al. 2005, Richman et al. 2015). Maximal weight gain, growth efficiency, moulting success, and overall survival is typically better at low salinities between 0–5‰ for species such as *P. clarkii, P. leniusculus, A. leptodactylus, A. pallipes* and the *Cherax* species. However, experimental exposure to higher salinities indicate that juvenile *P. leniusculus* and *A. leptodactylus* at least are capable of surviving full strength seawater for short periods and this is of concern for limiting spread between river via estuaries in systems where they are considered invasive (Holdich et al. 1997). A further example of environmental drivers interacting, however, is provided by Morrissy (1978) who found that the range contraction of *Cherax caini* to the lower salinity reaches within some salinized rivers in Australia was greater than would have been predicted by the osmoregulatory ability of the adults. He suggested that changes which had occurred in parallel with salinization, such as eutrophication and the spread of exotic fish, might also be important.

Oxygen concentration and total suspended solids

Episodic or permanent hypoxia is often a consequence of cultural eutrophication and organic enrichment, and can be exacerbated by suspended solids when they contain organic matter and/or reduce light available for photosynthesis (Willis-Jones 2013). The results of a survey and literature review by Haddaway et al. (2015) found that *A. pallipes* in Europe typically inhabited waterbodies supporting relatively high dissolved oxygen and hence reflecting their classical association with 'clean' environments, but the species can be found across a wide range of concentrations and is tolerant of environmental hypoxia (3 mg O_2 l^{-1}) for 'prolonged' periods (Demers et al. 2006). Its congeners, *A. italicus* and *A. torrentium* have also been shown to tolerate such conditions for at least 12 days under experimental conditions (Demers et al. 2006). However, it is difficult to extrapolate from acute to chronic exposure. A number of other crayfish species such *Parastacus defossus* and *Procambarus clarkii* have been observed living in naturally hypoxic or periodically anoxic conditions associated with muddy and turbid habitats and are, to an extent, physiologically adapted to such (reviewed by McMahon 2002). Caine (1978) noted that Floridian species inhabiting surface streams were intolerant of hypoxia, whereas several cave-dwelling (troglobitic) species were tolerant, thereby demonstrating adaptive traits to their local environment. A differential ability to cope with hypoxia can lead to competitive exclusion between species; Bovbjerg (1970) reporting that *Orconectes virilis* was excluded from ponds that were subject to hypoxia whereas *O. immunis* remained.

The impact of suspended solids on crayfish is relatively understudied. Again, Haddaway et al. (2015) summarise knowledge from field surveys of *A. pallipes* that have accounted for total suspended solids and suggest that populations are generally found in systems with a lower concentration than the average of European and global river means, i.e., relatively clean systems again. But all flowing waters are naturally subject to periods of high turbidity, associated with the seasons and so crayfish must be able to cope with acute exposure. This is supported by a lab study by Rosewarne et al. (2014) who experimentally exposed *A. pallipes* to concentrations that might be encountered in a former quarry proposed as an ark site. Total suspended solids of >500 mgL^{-1} resulted in gill fouling in all exposed individuals, whilst 250 mgL^{-1} was associated with fouling in 92% of exposed individuals. However, Rosewarne et al. (2014) did not find any evidence of decreased survival over a 45 day period of exposure to 1000 mgL^{-1}, indicative of at least short-term tolerance for extremely high turbidity. It would be interesting to compare this to the tolerance of species like *P. clarkii* which, as for oxygen, probably have adaptive traits for coping with suspended solids given that they originate from sluggish, turbid rivers and bayous. Since many crayfish, if not all, bioturbate to varying degrees and hence modify the environment around them (engineering their ecosystems), they might actually induce some density dependent population control through their own actions.

Population success

Any population can be considered successful if it is able to sustain itself indefinitely over numerous generations. Consequently, achieving high recruitment and production is essential for long-term population success. In this section, the influence of environmental variables on these particular facets of crayfish population biology will be explored.

Recruitment is defined as the addition of mature individuals to a population, which is predominantly achieved through the onward growth of young individuals. Therefore, any environmental variable that affects the production or survival of the young will alter recruitment and hence population success. The first stage of recruitment is reproduction, and with the notable exception of the marbled crayfish ('Marmorkrebs'), *Procambarus fallax* f. *virginalis*, which is one of the few decapods known to reproduce through parthenogenesis (Scholtz et al. 2003, Vogt et al. 2004), all crayfish species reproduce sexually and require internal fertilization of the eggs. The success of such mating is determined chiefly by two things: the number of viable eggs produced by the female, and the viability of the sperm from the male.

Egg production in female crayfish is tightly linked with body size, with larger females typically able to assign more energy to reproduction and hence produce more eggs (Rhodes and Holdich 1982, Brewis and Bowler 1985, Oluoch 1990, Tropea et al. 2012). For example, *Astacus astacus*, maintains a constant egg size across all size classes with the total number of eggs produced increasing linearly with body size (Skurdal et al. 2011, Fig. 2).

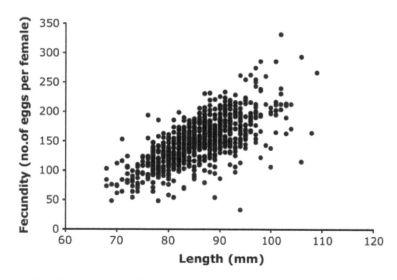

Figure 2. Variation in fecundity as a function of noble crayfish total length (mm). Data from Steinsfjorden (reproduced from Skurdal et al. 2011).

Similar relationships have been found for many crayfish species, although the relationship is not always linear; one study on *Austropotamobius pallipes*, found that clutch size plateaued once females exceeded carapace lengths of 33 mm (Brewis and Bowler 1985). Given the potential for female body size to affect reproduction efficiency, it becomes clear that the environmental factors that determine adult growth rate must be equally important for reproduction. Indeed, two major environmental drivers of adult growth, diet and temperature, have been found to have significant effects on egg production. Færøvig and Hessen (2003) found that in *A. astacus*, food quality (assessed by stoichiometry—C: N: P ratios) was important for somatic and reproductive growth and particularly that poor food quality (high C: P), or food shortage, could lead to reduced egg production. Thus, egg production may be highly sensitive to low availability of phosphorus but given the current trends in cultural eutrophication around the globe it is unlikely to be a limiting factor in all but the most oligotrophic, 'pristine' of ecosystems. Similarly, increased vitamin E intake has been linked with elevated egg production in *Astacus leptodactylus* (Harlioğlu et al. 2002).

Conversely, pollutants or contaminants may affect egg viability; for instance, mercury is known for its inhibitory effects on ovarian maturation in *Procambarus clarkii* (Reddy et al. 1997). With regard to temperature, several studies have found that low temperatures reduce adult growth rate and potential and consequently increase the time taken to reach sexual maturity, thereby reducing the number of eggs each female is able to produce (Oluoch 1990, Whitmore and Huryn 1999, Parkyn et al. 2002, Scalici and Gherardi 2007, Scalici et al. 2008). For example, a comparison of populations of the New Zealand species *Paranephrops planifrons* in streams situated on pastoral land and in native forest found that the greater number of annual degree days above 10°C in the pastoral streams (i.e., open-canopy) was linked with faster growth rates and a decrease in the age of female sexual maturity from 2 to 1 years (Parkyn et al. 2002). Similarly, *P. clarkii* introduced outside of its native sub-tropical range into the comparatively warmer, tropical locale of Lake Naivasha, Kenya, was found to be significantly larger at sexual maturity and hence capable of greater reproductive capacity (Oluoch 1990).

Aside from the total number of eggs produced, the viability of the eggs is of equal importance for reproductive success, since a large number of eggs with insufficient resources will result in a reduction of reproductive efficiency rather than an increase. Consequently, if a female is able to invest more resources in each egg and hence produce larger eggs, the offspring that she produces are likely to be more successful (Bernardo 1996). Therefore, it is possible that the non-linear relationships sometimes observed between body size and egg number may be in part due to greater investment in egg size in larger females. Indeed, studies on *Pacifastacus leniusculus, Austropotamobius torrentium* and *Cambaroides japonicus* have found positive relationships between female body size and egg size (Mason 1979, Nakata and Goshima 2006, Maguire et al. 2005); however, in all cases, the large amount of variation in egg size between females of similar sizes meant that the observed relationships were very weak. Furthermore, similar studies on other populations of *P. leniusculus* and other species have found that the variation between individuals of the same size was so great that no relationship could be found between body size and egg size/weight (Harlioğlu and Türkgülü 2000, Sáez-Royuela et al. 2006, Berber and Mazlum 2009, Tropea et al. 2012).

Marked variation in egg size suggests that environmental conditions may affect this aspect of egg production more directly than total egg number which, as demonstrated above, is predominantly affected indirectly through body size. Support for this comes from several studies that have found strong relationships between particular environmental variables and egg size (Huner and Lindqvist 1991, Rodríguez-González et al. 2006, 2009), with diet often highlighted as being of particular importance. Huner and Lindqvist (1991) found that poor nutrition in a pond culture of *Cherax tenuimanus* led to smaller eggs, whilst variations in the egg size of *Cherax quadricarinatus* have been linked with dietary lipid and protein content (Rodríguez-González et al. 2006, 2009). Population density and size structure has also been identified as a driver of egg size in a study of *Orconectes virilis* where an exploited population produced a greater number of smaller eggs than an unexploited population (Momot and Gowing 1977), indicating that the reduction in mean adult size and density due to exploitation reduced competitive interactions and hence increased the viability of offspring from smaller eggs.

Whilst egg number and size are the main determinants of reproductive potential in crayfish, sufficient quantities of viable sperm from the male are of course required to achieve this potential. For example, Aquiloni et al. (2009) found that irradiation of male *P. clarkii* led to sufficient reduction in sperm viability such that their reproductive output was reduced by 43%, suggesting that this method could be used to control invasive crayfish populations. Consequently, any significant impact of environmental conditions on sperm production has the potential to affect reproductive efficiency. One study on *C. quadricarinatus* found that both body size and temperature have the potential to affect sperm number and viability, with larger males able to produce more sperm and a more adherent spermatophore, suggesting larger males may be more successful reproducers, whilst sperm production was optimal at 27–29°C (Bugnot and López Greco 2009). Another study on *Austropotamobius italicus* found that male chelae size and asymmetry were negatively correlated with sperm number and longevity respectively (Galeotti et al. 2012), indicating that limitations in resources can lead to physiological trade-offs that can ultimately result in reduced reproductive efficiency.

Breeding in most crayfish species is predominantly synchronous in order that offspring emergence coincides with maximum resource abundance and to give greater protection from predation (Momot 1984). Consequently the factors that drive the timing of spawning are critical for population success. The exact conditions that give rise to mating are often highly species specific, although by comparing studies on several different species it is possible to determine which environmental variables are more generally important. In higher latitudes, food availability is closely tied with the seasons and so it is possible for animals to predict the future availability of food on the basis of environmental variables that indicate their current position in the seasonal cycle, such as temperature, day length or water flow regime. Indeed, several crayfish species have been found to use specific states of one or more of these variables as triggers for reproduction. For example, *P. clarkii* in Mediterranean wetlands has been found to closely match spawning time to the local flood regime, such that in areas with a short flooding period there is one reproductive peak in spring, whilst in areas with longer flooding periods they can achieve a second reproductive peak in the autumn (Ilhéu and Bernardo 1997, Gutiérrez-Yurrita and Montes 1999). Reproduction in *C. quadricarinatus* on the other hand is tightly controlled by photoperiod, with exposure to a winter-like photoperiod sufficient to cause a threefold increase in the rate of spawning under laboratory conditions (Karplus et al. 2003). In contrast, *A. astacus* reproduction is apparently unaffected by photoperiod and instead reliant on a period of decreasing temperature to trigger mating in the autumn (Westin and Gydemo 1986).

At lower latitudes, a reduction of 'seasonality' removes such constraints and if sufficient resources are available, then reproduction may progress for much of the year, if not all year round. Unfortunately, few crayfish species are native to tropical regions and so there is a distinct paucity of information on the triggers of reproduction in these species; however, the cosmopolitan species *P. clarkii* has been introduced to Lake Naivasha, Kenya, where the annual temperature range is 15.9–20.6°C which is entirely within the normal reproductive range of this species (Liu et al. 2013). Consequently, since the Lake rarely dries out completely, observations of this population can inform as to whether any other factors are important triggers of reproduction. Oluoch (1990) observed that this *P. clarkii* population was indeed capable of year round reproduction, indicating that without the constraints of low temperature or water level there was little else to restrict the timing of reproduction in this species and, thus, potentially other species as well. It must be noted though that there were still irregular peaks in reproductive activity throughout the year (Oluoch 1990). While another driver may still act to coordinate reproduction amongst the majority of the population, it could be physiological or genetic in nature rather than environmental.

Aside from coordinating offspring emergence with the occurrence of maximal resources, the timing of reproduction has other important implications for population success. One such implication is evident from the Lake Naivasha and Mediterranean studies on *P. clarkii* mentioned above (Oluoch 1990, Gutiérrez-Yurrita and Montes 1999), which is that species that have relatively loose controls over the timing of reproduction (i.e., a broad temperature range and an ability to synchronize with the local water regime) can be highly adaptable to new environments, since this enables them to maximise use of the available resources (Gutiérrez-Yurrita and Montes 1999). Consequently, species with highly adaptable reproduction can establish successful populations when they are introduced to new water bodies, contributing no doubt to why *P. clarkii* is such a successful and highly fecund invasive species across the globe, from equatorial Kenya to the temperate UK, and with expectations for further spread (Liu et al. 2011). Yet knowledge of a nuisance species' ability to adjust its reproductive cycle to the timing of flooding could be useful, with for example Gutiérrez-Yurrita and Montes (1999) suggesting a driver such as water regime could be manipulated for control of the invasive species in particular environments (in this case Doñana National Park, Spain) by managing water levels accordingly to disrupt reproduction. A further implication of the environmental control for the timing of reproduction is that for species in higher latitudes, the restricted nature of the growth season means that offspring produced at the beginning of the spring have a distinct advantage over those produced later since they can grow faster and attain a larger body size (Scalici and Gherardi 2007). Consequently if conditions enable a second reproductive peak to occur in the autumn, the offspring produced may become a distinctly less successful subset of the population (Scalici and Gherardi 2007).

The final factor that affects total recruitment is the survival of the offspring until they reach maturity. This is particularly important since several authors have identified differential growth and survival between age classes as the main driver of fluctuations in population size and fecundity (Momot 1967, Skurdal et al. 2011). Consequently the factors that affect the survival of juvenile crayfish are of great importance to population success.

Perhaps the most obvious driver of juvenile survival is predation pressure, since greater losses to predation will necessarily reduce recruitment success. For example, a study by Kellogg and Dorn (2012) found that predation by natural densities of sunfish could reduce recruitment of *Procambarus alleni* by over 99% and *P. fallax* by 62% in artificial wetland mesocosms. However, it is important to note that in natural systems without sunfish, *P. alleni* dominates to the near total exclusion of *P. fallax* (Kellogg and Dorn 2012). Therefore, although predation can severely limit the recruitment success of crayfish, it can in some cases actually improve population success if it affects an otherwise stronger competitor to a greater degree. Apart from interspecific predation, intraspecific cannibalism will clearly limit recruitment success. The occurrence of cannibalism often increases alongside population density, such that recruitment success can be highly density dependant, thereby creating a self-imposed population limit, as has been observed in *Orconectes virilis* (Momot and Gowing 1977, Momot et al. 1978).

Given the potential for predation to have such a dramatic effect on recruitment success, it is evident that the availability of refugia in which to hide from predators could be important in reducing the size of this effect. Indeed, several studies have linked habitat complexity and thus refuge availability with recruitment in crayfish (Momot and Gowing 1983, Olsson and Nyström 2009, Clark et al. 2013). Olsson and Nyström (2009) found that in an artificial stream experiment juvenile *P. leniusculus* survival was highest in streams with rockier habitat (cobbles) whilst Clark et al. (2013) found similar results for small *Orconectes obscurus* in a natural stream environment (Fig. 3).

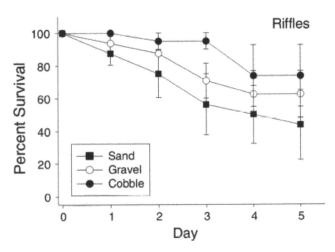

Figure 3. Survival of juvenile crayfish (mean ± 1 SE) in relation to habitat complexity (sand, gravel and cobble plots) within riffles of the Mahoning River, Ohio, USA. Reproduced from Clark et al. (2013).

Momot and Gowing (1983) on the other hand found that emergent vegetation was a key refuge for juvenile *O. virilis* and that reduction in the extent of this vegetation led to reduced cohort production. These studies therefore demonstrate that the exact nature of the refugia may well vary between species and/or water bodies; however, their presence is critical for recruitment success and their abundance can influence the carrying capacity of the water body (Momot and Gowing 1983).

Movement and dispersal

The ability to move both within and between water bodies is of great importance for crayfish population success as it enables the discovery of new food sources, refuges and mates and facilitates the establishment of new populations. Such movement can be either active (i.e., walking or tail flipping) or passive (i.e., drift or human mediated), with both types thought to be important for the success of all crustacean populations (Hänfling et al. 2011). Active dispersal results from intentional movements by individual crayfish, which often tend to be irregular, with many species combining periods of movement with periods of residence in a series of ephemeral home areas (Gherardi et al. 2000, Robinson et al. 2000). Furthermore, there is often great variation in activity between individuals within a population such that at any given moment one crayfish might be in a nomadic phase whilst another may be sedentary (Gherardi et al. 2000, Robinson et al. 2000). For example in one study of *A. pallipes* movement in North Yorkshire, UK, some individuals moved over 300 m in 10 days whilst others barely moved at all (Robinson et al. 2000). Despite this, when viewed at the population level, active movement and dispersal can often be seen to be driven or constrained by environmental conditions such as light, temperature, water regime or season, and indeed, the size and connectedness of the waterbody.

Temperature is the most common driver of movement due to the fact that crayfish are poikilothermic and so their metabolic activity and hence ability to move is tightly linked with the ambient temperature (Lehtikoivunen and Kivivuori 1994). The consequence of this for dispersal is that crayfish will be most active at higher temperatures and so dispersal and colonisation will typically be faster in the summer and in warmer climates. For example, the invasive *P. leniusculus* in the UK has been found to travel the greatest distances during the summer, with the timing and extent of movements closely coupled with fluctuations in temperature (Bubb et al. 2004, Johnson et al. 2014, Fig. 4).

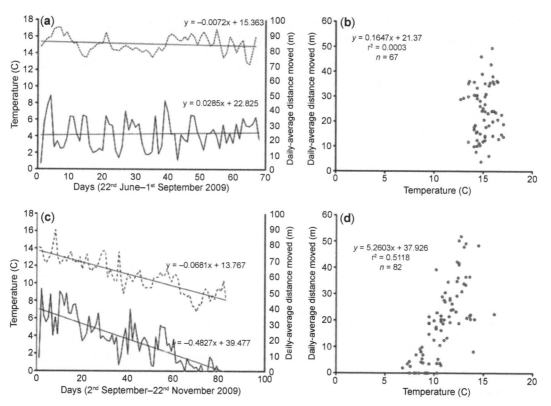

Figure 4. Effect of temperature on crayfish dispersal. Time-series of the daily-averaged distance moved by crayfish (solid line) and the water temperature (dashed line) in (a) summer and (c) autumn. Best-fit lines represent the regression of each variable on time. Scatter plots of daily-averaged distance moved by crayfish versus water temperature for (b) summer and (d) winter. Reproduced from Johnson et al. (2014).

Similarly, in Italy the invasive *P. clarkii* is typically most active at night. However, during the winter it has been observed to become relatively more active during the day, presumably because the low winter night-time temperatures render metabolic activity too low for such activity (Gherardi et al. 2000). This idea is supported by observations from NE Portugal, where the spread of *P. clarkii* along the River Maçãs has been rapid in all sections except in the upper reaches where the colder winter temperatures are believed to have limited it and allowed *P. leniusculus* to dominate instead (Bernardo et al. 2011).

Variations in temperature during the year not only drive changes in absolute activity, but also changes in the pattern of movement with the seasons. The reason for this is that the relationship between temperature and potential activity means that over the course of the year the relative risk of different behaviours and locations changes in accordance with the temperature. For example, *O. limosus* was found to migrate in spring and summer out from a reservoir and into an inflowing brook where there were fewer predators and competitors; however, in autumn the declining temperatures made the brook less hospitable and so the crayfish would migrate back downstream and overwinter in the reservoir where the temperature regime was more stable (Buřič et al. 2009). Similarly, in Lake Tahoe, *P. leniusculus* was found to migrate into deeper water at the onset of colder winter temperatures, which was believed to afford them greater protection from winter storms that cause high crayfish mortality for those remaining in shallower littoral areas (Flint 1977). Therefore, it is evident that temperature also has the potential to drive the large-scale spatial distribution of crayfish over the annual cycle.

The study by Gherardi et al. (2000) demonstrates the potential for crayfish activity to respond to light conditions, with *P. clarkii* typically being more active in the dark. A pattern of nocturnal behaviour is very common amongst crayfish and has been observed across a diverse range of taxa, such as *P. leniusculus*, *A. astacus* and *Orconectes nais* (Rice and Armitage 1974, Styrishave et al. 2007). It is generally believed to be predator avoidance tactic since the predators large enough to prey on crayfish are typically heavily reliant on vision for prey detection. As a consequence, restricting risky activities such as long distance movement to the hours of darkness dramatically increases the chance of survival (Flint 1977, Hill and Lodge 1994, Styrishave et al. 2007). Nocturnal behaviour is linked in most species with physiological circadian rhythms such that oxygen consumption and heart rate are usually higher overnight; however, species such as *P. leniusculus* are physiologically capable of reasonably high levels of activity during the day (Styrishave et al. 2007). Indeed, while the majority of movement occurs at night, many species are also active to a lesser degree during the day (Hill and Lodge 1994, Gherardi et al. 2000, Robinson et al. 2000), indicating a flexibility in diurnal behaviour to cater for variation in other environmental factors, such as predation or temperature.

The water regime is another environmental factor that can control crayfish dispersal and movement. This is particularly important in rivers where it has been noted for many species that rates of active upstream movement are slower that for downstream because the water flow impedes upstream movement (Bubb et al. 2004, Kerby et al. 2005, Bernardo et al. 2011). Furthermore, water flow is not constant throughout the year, nor between rivers or even river reaches and so upstream dispersal is often most restricted during high flow periods such as the winter/wet season and in upland streams. For example, in the Portuguese River Maçãs *P. leniusculus* was observed to spread downstream at 2.8 km yr^{-1} versus 1.7 km yr^{-1} upstream, with the major part of the upstream spread undertaken during low current velocity conditions in late spring and summer (Bernardo et al. 2011). Artificial and natural barriers in rivers affecting connectedness of the system will clearly impact upon crayfish dispersal, particularly by restricting or preventing active upstream movement. In one study of 32 river sections in southern California it was found that the invasive *P. clarkii* was frequently absent upstream of large barriers such as waterfalls or culverts (Kerby et al. 2005). In addition, a further mark-recapture survey found that barriers significantly reduced their movement between pools within the river (Kerby et al. 2005). Consequently, it has been suggested that river sections bounded by downstream barriers (either natural or artificial) could be the best systems to try to protect from alien crayfish invasion.

Passive dispersal is by definition directly controlled by the environmental conditions. There are two major ways in which passive crayfish dispersal can occur. The most widespread is through being washed downstream during high flow events such as a flood. The viability of this as a dispersal mechanism has been investigated in a number of species with mixed results. One study found downstream dispersal of *P. leniusculus* increased from 2.8 to 6.7 km yr^{-1} following a high flow event (Bernardo et al. 2011) whilst another on the same species in a different river identified no significant difference in passive dispersal (Bubb et al. 2004); similarly contradictory results have been found for *P. clarkii*, again for different rivers (e.g., Kerby et al. 2005). However, this type of dispersal is inherently dangerous and is likely to result in higher mortality, counter to population success. Indeed, Robinson et al. (2000) found two out of five crayfish they were tracking to be dead after a flood event. Taken together these studies indicate that this type of dispersal is unlikely to be responsible for large-scale population movements, but it may be important in accelerating the colonization of downstream reaches.

The second type of passive dispersal is human mediated, since several crayfish species have been selected by humans and transported around the world for the aquaculture and aquarium trades. The most notable amongst these are *P. leniusculus, P. clarkii, O. virilis, O. rusticus, O. limosus, A. leptodactylus* and *C. destructor* since these species have managed to attain a truly global distribution as a direct result of human activities and are well known to have widespread negative impacts on the systems they invade, including local extirpation of native crayfish species (Holdich 1999, Harlioğlu and Harlioğlu 2006). Crayfish introductions to non-native systems have frequently been un-intentional (Holdich 1999); however, given the various capabilities of these species for dispersal and recruitment described above, it is little wonder that they have managed to establish and expand populations in the majority of places where they have been given the opportunity. Consequently, for those species which we have distributed beyond their native range and we continue to disperse, then human mediated passive dispersal is of great importance for their overall success as it enables the establishment of new populations in otherwise unreachable locations. However, the success of these invasive species often comes at the expense of any native crayfish species already inhabiting those systems (Twardochleb et al. 2013), and so this dispersal mechanism can lead to differential population success amongst species. Of course, intentional deployment of this dispersal mechanism can be used to introduce threatened species to 'ark sites' where there are no invasives, thereby preventing their immediate extinction (Peay and Füreder 2011).

Survival and mortality

A successful population is one that can sustain and expand itself indefinitely. This is predominantly achieved through recruitment and dispersal; however, in order for these processes to occur the adult breeding population needs to survive for long enough to undertake them. Consequently, adult survival and mortality and the environmental factors that drive them are of great importance for population success.

There are two major drivers of adult mortality: predation and disease. The major predators of adult crayfish are often large predatory fish, such as sunfish or bass in North America and pike in Europe (Elvira et al. 1996, Neveu 2001, Kellogg and Dorn 2012) although mammalian and avian predators such as otters and herons are also capable of consuming them in reasonable quantities (Beja 1996, Montesinos et al. 2008). The effect of such predation on crayfish populations can be sizeable, especially if appropriate shelter is unavailable due to competition or low abundance of refugia. For instance, Garvey et al. (1994) found that in a mixed crayfish species assemblage of *O. rusticus, O. propinquus* and *O. virilis*, as is found in several North American lakes, the native *O. virilis* could be excluded from shelters and was consequently consumed at a very high rate by predatory fish, leading the authors to conclude that this process could lead to the local extinction of this species. The importance of predation and refugia for adult population success is further demonstrated by another study on *O. rusticus*, which found that adult crayfish density was disproportionately high in habitats that provide shelter, such as cobbles, especially when predator density was high, and that mortality was lowest in those habitats (Kershner and Lodge 1995). Therefore, it is evident that predation of adult crayfish has the potential to affect population size and distribution, such that crayfish are likely to be most successful in habitats that provide ample shelter or lack high densities of predators, e.g., temporary water bodies.

Humans can also be a significant predator of adult crayfish as they are trapped for both food and to try to manage invasive populations. Whilst crayfish can be caught in large numbers through intensive trapping, it has often been found that this has little effect on the actual population size if regular trapping is not maintained since many of these exploited or managed species have very fast reproduction and growth rates (Gherardi et al. 2011). Furthermore, the intrinsic bias of traps to catch the largest individuals in a population can actually reduce the top-down population control of intraspecific cannibalism and thus increase recruitment success, thereby maintaining the population size (Gherardi et al. 2011). However, it has been suggested that timing trapping to coincide with the period of greatest activity during the mating season will not only maximize the catch per unit effort, but will also reduce overall reproductive success of the crayfish as there will be fewer adults available for reproduction (Rogowski et al. 2013). Therefore, predation by humans, especially when conducted in conjunction with other predators may have the potential to negatively impact crayfish population success. Aspects of predation for population success will be picked up again when considering crayfish from a community perspective.

Disease is the other major environmental driver of crayfish mortality. Most crayfish diseases, much like diseases in other organisms, do not occur in the majority of individuals and so even if they cause the death of their victim they will not have much effect at the population level (e.g., psorospermiasis and porcelain disease; Longshaw 2011). However, there are a small number of diseases that can become very widespread in a population and result in a population crash or even local extinction. This is particularly common when diseases are introduced to populations that have had no previous exposure to the pathogen in their evolutionary history. The best known example of this is the so-called crayfish plague (*Aphanomyces astaci*) which is non-lethal to its native American hosts such as *P. leniusculus* and *P. clarkii* but is acutely pathogenic to all five native European crayfish species (Longshaw 2011). Since its introduction to Europe in the mid-nineteenth century crayfish plague is recognised to have caused the collapse of numerous crayfish populations and is considered a major threat to their conservation (Edgerton et al. 2004). Therefore, disease has the potential to be a major driver of population success and this potential is only likely to increase as we continue to introduce alien crayfish species and their associated diseases around the world.

Ontogeny

Ontogeny is the development of an individual throughout its lifetime. Crayfish undergo direct development and so their progress through each ontogenctic stage can be easily tracked through their pattern of moulting and growth. As animals grow, it is often the case that their environmental requirements change or that they are able to utilise new resources, and consequently the duration and success of each ontogenetic stage can vary between populations depending on the local environmental conditions and available resources.

As previously discussed in the recruitment section, the growth of both juveniles and adults can be influenced by the ambient temperature, such that populations in warmer (micro)climates can grow faster and bigger and therefore breed sooner and hence be more resilient as a population (Oluoch 1990, Parkyn et al. 2002). Additionally, diet quality also has been identified as a major driver of crayfish growth (Færøvig and Hessen 2003). Further to this, it has been found in several species that juvenile crayfish can exhibit different diet preferences to adults, with juveniles typically consuming more invertebrates and adults feeding mainly on detritus (Parkyn ct al. 2001). The reason for this is generally believed to be that juveniles need a greater protein intake for growth than adults (Momot 1995) and indeed it has been found in *C. quadricarinatus* that as juveniles grow, their dietary enzyme production shifts from mainly proteases towards mainly carbohydrascs (Figueiredo and Anderson 2003), their physiology therefore perfectly mirroring the observed shift in dietary preference. The implication of this ontogenetic shift in diet is that progression through certain growth stages could be impaired if the appropriate food source is under-represented in the local environment; however, studies on this effect havc determined that due to the typically omnivorous nature of crayfish, they are usually able to find something to eat and so limitation of ontogenetic stages by diet is unlikely to be a regular occurrence (Parkyn et al. 2001, Bondar et al. 2005).

Growth and survival of both juveniles and adults can be heavily influenced by the local habitat types, with the availability of refugia identified as particularly critical. However, the type of refugia required can change with the ontogenetic stage, since as crayfish grow larger, they will require more spacious refugia. Evidence for this comes from Rabeni (1985) who found that *Orconectes punctimanus* juveniles mainly utilised stands of vegetation in shallow water whilst adults were strongly associated with large substrate particles such as cobbles. Similarly, juveniles of the congeneric *O. luteus* utilised most types of refugia but adults were restricted to areas with larger particle sizes. In addition, other studies have shown that adult *O. rusticus* prefer cobbled habitat despite vegetated habitats offering similar levels of protection against predation (Kershner and Lodge 1995), which is believed to be because the dense vegetation restricts movement and foraging, whilst juvenile *O. virilis* were found to be heavily reliant on the refugia provided by vegetation rather than other habitat types (Momot and Gowing 1983). It is therefore evident that for many crayfish species their habitat requirements change as they grow and so if certain habitat types are in short supply then the success of the associated ontogenetic stage is likely to be affected, as was seen in the Momot and Gowing (1983) study where reduction of the macrophyte cover was associated with reduced juvenile survival.

Community dynamics and species interactions

It is estimated that somewhere between 1.5 and 30 million different species live on Earth today with recent estimates of eukaryote diversity alone ranging from 5 ± 3 million (Costello et al. 2013) to 8.7 ± 1.3 million (Mora et al. 2011). Those species living in the same geographical area are a part of the same ecological community and, thus, there is the potential for thousands of interactions between species, creating vast and complex ecological networks (Woodward et al. 2008, Ings et al. 2009). Such interspecific interactions control community structure and vary dynamically over space and time, with implications for the success of crayfish populations in both their native and invaded habitats. Different processes, including those operating over a long time scale such as natural selection, and short-term ecological interactions such as grazing pressure, can influence the number and the identity of species in communities. Crayfish success therefore will vary depending on a number of community features, such as the availability of niche space or the presence of predators. Crayfish also play a key role in structuring communities via trophic and non-trophic interactions and although they are denizens of the benthos, it is important to consider their wider influence beyond the benthos to pelagic (plankton) communities for example. Here, we describe the role of crayfish in this regard and relate it to their population success.

Patterns of community assembly

Freshwater ecosystems are often fragmented or physically isolated in a terrestrial landscape (e.g., lakes and ponds), or linear in nature (e.g., streams and rivers). The habitats they comprise are diverse and may be both above and below ground, permanent and temporary, and the water may be still or flowing. Freshwater communities are shaped by several interacting factors: species respond to abiotic stimuli and individuals interact with conspecifics and with individuals of other species. The strength of these factors in relation to one another will determine community structure (Mutshinda et al. 2009). Species interactions and the variation in species responses to their environment ensure that community structure is dynamic, both spatially and temporally. A fundamental goal of ecological research is to understand the mechanisms which determine community structure and control ecosystem stability, and this is becoming increasingly important with accelerating rates of global change (Raupach et al. 2007, Jackson and Grey 2013, Marcott et al. 2013). Indeed, we have a vested interest in understanding the determinants of community structure of freshwater habitats because ultimately it will influence biodiversity and the provisioning of essential ecosystem goods and services upon which we rely (Ormerod et al. 2010, Strayer and Dudgeon 2010).

Crayfish play an important role across a range of freshwater habitats around the world and, in systems where they are not naturally present, other crustacean species such as crabs and shrimps tend to occupy a similar functional role. Their ability to occupy a broad range of habitats within a particular ecosystem, and their role within wider the community and the reciprocal influence of community dynamics upon crayfish population success, will vary between those habitats. Juvenile and adult crayfish may perform different functional roles due to variation in body size and dietary requirements. Juveniles are typically cryptic and their ecological roles usually resemble those of other smaller invertebrate species, while the relatively much larger-bodied adults may dominate the invertebrate community in terms of biomass and, in this respect, have similar ecological roles to fish. More importantly, as invaders, the impact of crayfish is often intensified with crayfish size (Usio et al. 2009) and larger crayfish will be less susceptible to predation by native fish since fish are gape-limited predators. The size structure of crayfish populations will, therefore, have implications for their success and for community structure.

Keystone and engineer species

Secondary production and biomass in many freshwater systems is dominated by fish, rendering them important in determining community structure via top-down control or perhaps indirectly via ecosystem engineering. However, freshwater crayfish can be just as, or indeed more, significant in influencing food web structure and hence wider ecosystem functioning. Where crayfish and fish coexist in freshwater systems, a degree of interaction will occur to control community structure through them both achieving relatively large size and having relatively greater longevity, and we will revisit this toward the end of the chapter. In systems that do not support fish but where crayfish may thrive, such as ephemeral ponds, then crayfish may dominate invertebrate (and non-fish vertebrate) biomass. For example, *Orconectes limosus* comprised 49% of macroinvertebrate biomass in a lake in Germany (Haertel-Borer et al. 2005) and various *Pacifastacus* species have been shown to contribute >90% to the total biomass in streams in America (Haggerty et al. 2002). Indeed, their dominance of biomass can be especially important where they are introduced species, with an 'invasive effect' exacerbated relative to their already strong food web effects in native systems (Nyström et al. 1999, Rodríguez et al. 2005). For this reason, the interactions crayfish have with other species are important in shaping community structure, although it should be noted that crayfish densities vary considerably in both space and time in natural ecosystems and may exert seemingly little influence upon ecosystem functioning when very low (Nyström et al. 2006). Nevertheless, some crayfish species have been considered as keystone species (Nyström et al. 1996), exerting an influence on community structure and ecosystem processes that is disproportionate to their biomass. They are also often referred to as ecosystem engineers (or geomorphic agents). These are defined as organisms that create, modify or maintain habitats (or microhabitats) by causing physical state changes in biotic and abiotic materials that, directly or indirectly, modulate the availability of resources to other species (Jones et al. 1994, 1997).

Crayfish are ably equipped as ecosystem engineers with their large chelae, and burrowing and tail-flipping behaviours. It is perhaps not surprising then that a considerable amount of research has focused on their ability to move and re-organize sediment structure, and particularly for the environmental damage that invasive species may be causing to waterways and the banks that bound them. *Pacifastacus leniusculus* is capable of sediment displacement in the order of 1.7 kg m^{-2}d^{-1} under experimental conditions, with the majority (78%) of this volume change associated with small scale ($\leq$1 median grain diameter) movements of surface grains (Johnson et al. 2010). However, individual crayfish were able to move material up to 38 mm in diameter that had a submerged weight six times that of their own bodies. From a physical standpoint, by modifying the arrangement of sediment grains on surface substrates, crayfish may counteract the low-flow physical consolidation of river beds and reduce the entrainment stresses required to move river bed material (Statzner et al. 2003, Johnson et al. 2010, Harvey et al. 2014).

From an ecological view, altering grain size and structure changes the habitat and/or prey availability for the local benthic community (Statzner et al. 2003, Hayes 2012). Taking this further, bioturbation or re-suspension of fine sediments increases turbidity in the water column, and eventually results in deposition of fine sediments elsewhere in the system (e.g., Usio and Townsend 2002, Harvey et al. 2014, Fig. 5).

While nutrient enrichment of the water column or reduction in penetration of photosynthetically active radiation and their subsequent effects to primary production have been considered as a consequence of bioturbation (e.g., Bilotta and Brazier 2008), the less obvious, indirect effects such as oxygen deficits associated with increased microbial respiration or release of reduced chemical species back into an oxygenated water column have not (Willis-Jones 2013, Fig. 6).

Macrophytes play a vital role in aquatic ecosystems by influencing physical water movement, water chemistry, and biological interactions by providing both dietary resource and physical refugia, and determining key environmental parameters such as light extinction, temperature, substrate, oxygen concentration, and nutrient availability (Carpenter and Lodge 1986, Dorn and Wojdak 2004). Alterations to macrophyte species composition and/or density brought about either by direct consumption or indirectly through alteration of the rooting substrate or light climate will, therefore, have repercussions for the whole ecosystem (Nyström et al. 1999, Momot 1995, Usio et al. 2009). Indeed, when *P. clarkii* was introduced

268 *Biology and Ecology of Crayfish*

Figure 5. Photograph of experimental pond mesocosms to measure direct and indirect effects of crayfish bioturbation (ecosystem engineering); Willis-Jones (2013). Turbidity is evident in the pond with a high density of *Procambarus clarkii* (front left) while the water column remains clear where they are absent (front right). The probes are Unisense oxygen electrodes.

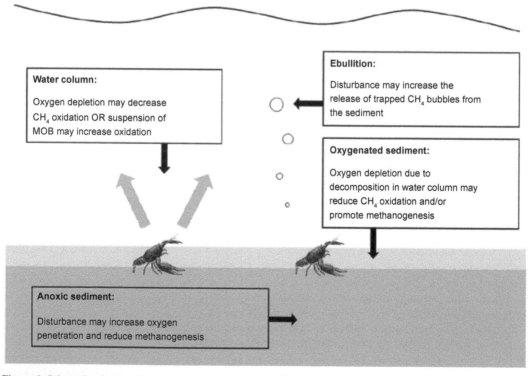

Figure 6. Schematic of potential interactions between crayfish bioturbation and the biogeochemical cycling of methane. Reproduced from Willis-Jones (2013).

to a clear, mesotrophic, macrophyte-dominated lake in Spain, within three years it had experienced a 90% decline in macrophyte surface cover and switched to a turbid state (Rodríguez et al. 2003). Turbid lakes that are dominated by phytoplankton are often maintained perpetually in this state by the presence of crayfish which graze, clip and up-root submerged macrophytes, preventing them from (re)establishing (Lodge and Lorman 1987). *Procambarus clarkii* was also proposed as responsible for the elimination of native submerged plants in Lake Naivasha, Kenya (Smart et al. 2002), and conspecifics in their native range in California, America have been shown to eradicate *Potamogeton pectinatus* from freshwater marshes (Feminella and Resh 1989).

Certain crayfish species have been shown to exert a dietary preference for submerged macrophytes (e.g., Smart et al. 2002). Using an enclosure-exclosure experiment to control the presence of native *Orconectes rusticus* at natural densities in the Great Lakes of the USA, Lodge and Lorman (1987) found that crayfish reduced submerged macrophyte shoot number and total biomass by up to 64%. Additionally, in some lakes in that region, native crayfish have eliminated submerged macrophytes completely or reduced diversity by up to 80% (Wilson et al. 2004). This herbivory and ecosystem engineering by crayfish influences the timing and scale of nutrient turnover by macrophytes as well as the macrophyte biomass entering the detrital food chain (Lodge 1991). Loss of macrophyte richness and density leads to a decrease in important refugia and dietary resource for fish, amphibians and invertebrates and, by inference, less food for crayfish (Wilson et al. 2004). For example, Rodríguez et al. (2005) discovered that the loss of macrophytes following the crayfish invasion in Spain was directly related to the exclusion of 71% of macroinvertebrate genera, 83% of amphibian species and 75% of duck species. Hence, as a consequence of their actions, by restructuring or in many cases by removing macrophytes from habitats, crayfish are also removing a valuable refuge and resource for themselves (Harper et al. 2002, Garvey et al. 2003, Grey and Jackson 2012); this is likely to result in decreased population success.

Habitat links

As discussed earlier, crayfish can be highly mobile species and may move considerable distances, for example migrating seasonally to deeper water to breed (Momot and Gowing 1972) or diurnally to forage for food, and by doing so they create links between otherwise isolated habitats. Ruokonen et al. (2012) found crayfish with littoral carbon stable isotope values in the deep profundal zone of lakes, indicating that the crayfish fed in the littoral zone and migrated to deeper water on a daily basis. By processing organic matter and mixing and re-suspending sediments, crayfish can release bound nutrients and make those available again for primary production up in the water column, and impact upon pelagic community dynamics well away from the benthos where they reside (Covich et al. 1999). As integrators between otherwise separated habitats within aquatic ecosystems then, crayfish create and shift energy pathways, and play a key role in community functioning.

The exchange of organisms and energy across ecosystem boundaries (ecosystem subsidies) has major implications for food web structure and dynamics; food webs in lakes, streams and riparian habitats are often heavily influenced and shaped by aquatic-terrestrial links (Grey et al. 2001, Knight et al. 2005). Inputs of allochthonous leaf litter to streams from the riparian zone can amount to 20–2000 g dry mass m^{-2}year^{-1}, providing considerable energy and exerting strong bottom-up effects on stream food webs (Bartels et al. 2012). Terrestrial invertebrates that fall into streams are consumed by fish and there is a reciprocal flow of energy in the form of adult aquatic insects emerging into the riparian habitat for terrestrial predators (Knight et al. 2005). As large-bodied omnivores capable of shredding and ingesting leaf litter, as well as preying upon other macroinvertebrates, crayfish play an important role in ecotonal coupling and can be considered as a major processer of energy between terrestrial and aquatic food webs. Terrestrial leaf litter is a major component in the diet of many crayfish populations and the amount introduced into waterbodies can have implications for crayfish population success; Kobayashi et al. (2011) found that the abundance of invasive *Procambarus clarkii* populations in ponds in Japan increased with leaf litter inputs. Larson et al. (2011) found that the contribution of leaf litter to the diet of crayfish was inversely correlated with lake size because smaller lakes have an increased prevalence of allochthonous resources due to a larger perimeter to surface area ratio. Since the relative importance of leaf litter in the diet of crayfish will have implications for their effect on other aquatic resources, it is likely that their influence on community dynamics will also be dependent on lake size. In smaller lakes, the benthic and littoral communities of which crayfish are a part are relatively more important to ecosystem functioning than the pelagic community (which dominates in larger systems) and, therefore, crayfish effects will be more important in smaller systems.

There is a reciprocal flow of energy when crayfish become prey for partially terrestrial species, including wading birds. This adds aquatic-derived energy to the terrestrial food chain, through secondary production and excretion. For example, invasive crayfish have promoted the population growth of many threatened species across Spain and Portugal, such as otter, Egyptian mongoose, night heron and white stork (Correia 2001, Tablado et al. 2010), but have also been implicated in sustaining and hence facilitating other invaders such as mink (Melero et al. 2014). As an extreme example of ecotonal coupling, crayfish have even been known to leave the water and graze directly upon terrestrial plants. A study of the invasive *Procambarus clarkii* in Lake Naivasha, Kenya, demonstrated using stable isotopes that when macrophyte density was low in the lake, individuals could be found that had left the lake and were residing in hippo foot prints or burrows above the waterline, and grazing on terrestrial plants at night which contributed a significant proportion of their biomass (Grey and Jackson 2012). Such adaptable behavior must allow populations to persist when conditions are sub-optimal. In a similar manner, DiStefano et al. (2009) discovered that *Orconectes williamsi* and *O. meeki meeki* in America occupy the hyporheic zones of intermittent streams during periods of drought. Of course, while we have thus far only really considered the freshwater species of crayfish in this chapter, there are some species of crayfish that are terrestrial specialists. In South Carolina, America, *Distocambarus crockeri*, a primary burrower, is actually negatively associated with aquatic habitats and, instead, it is found in soils with seasonal perched water tables in ridge-tops (Welch and Eversole 2006). In fact, contrary to the view that all crayfish are aquatic, burrowing species comprise 15% of the total United States cambarid crayfish fauna (Welch and Eversole 2006), some with important roles in terrestrial community dynamics, although this is not well documented. Burrowing crayfish are often associated with open, treeless communities (Hobbs and Rewolinski 1985, Hobbs and Whiteman 1991) and population success will also depend on land use (disturbance) and distance from groundwater seeps (Loughman 2010). The abandoned burrows of terrestrial crayfish provide habitat for other non-crayfish species, offering important refugia for organisms such as salamanders, snakes and toads (Welch and Eversole 2006, Loughman 2010). Burrowing behaviour has strong local ecosystem effects through soil mixing (Welch and Eversole 2006). Indeed, the disturbance caused by burrowing animals, for example, crayfish of the genus *Fallicambarus*, is important in maintaining the plant community in some habitats (Brewer 1999). These community level contributions of primary burrowers suggests that, like aquatic crayfish, partially terrestrial crayfish may also play a considerable role in ecosystem functioning.

Crayfish in food webs

Omnivory

Omnivorous and opportunistic in their feeding habits (Correia 2003, Stenroth et al. 2006, Grey and Jackson 2012), crayfish play an important role in many aquatic ecosystems (Stenroth and Nyström 2003, Creed and Reed 2004, Creed et al. 2010). Crayfish are usually considered to be opportunistic scavengers, with their diet ranging from algae, detritus and dung to invertebrates and carrion (Grey and Harper 2002, Parkyn et al. 2001, Correia 2003, Ilhéu et al. 2007). They are also occasional active predators of fish eggs and small fish (e.g., Savino and Miller 1991). However, despite their omnivorous nature, crayfish can be selective consumers (Nyström et al. 1999), and individuals within a population may specialize on particular food items (see Fig. 7). For instance, they will selectively consume smaller size-classes of mussels (MacIsaac 1994) and demonstrate a preference for submerged macrophytes over emergent plants (Nyström and Strand 1996). Their classification as a generalist arises from their ability to readily switch diets when a preferred resource becomes limited (Grey and Jackson 2012).

As we have seen, some crayfish species can occupy a wide range of habitats and tolerate high levels of disturbance; consequently a single species may be exposed to a diverse range in dietary items across temporal and spatial scales (Correia 2003, Grey and Jackson 2012, Ruokonen et al. 2012). There is some

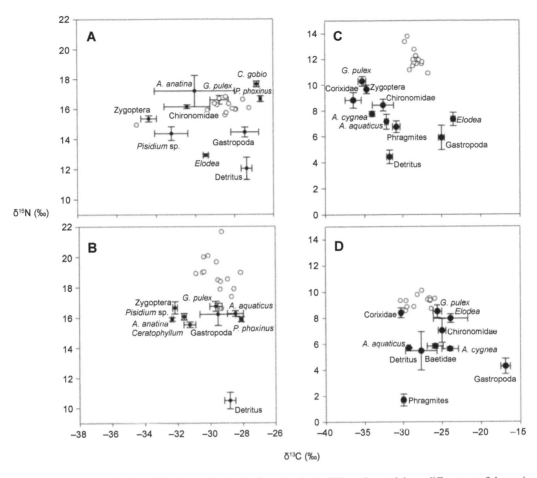

Figure 7. Stable isotope bi-plots of freshwater food webs from sites in the UK, each containing a different crayfish species demonstrating interspecific variation in position within the food web, and degree of intraspecific variation in diet selection. Open symbols represent individual (A) *Pacifastacus leniusculus*, (B) *Orconectes virilis*, (C) *Procambarus clarkii*, and (D) *Astacus leptodactylus*. Closed symbols represent putative resources (means ± 1 SE, n = 3 to 10) which were present and sufficiently abundant during collection to be analysed for stable isotopes.

evidence of seasonal variation in crayfish diet, with fluctuations in assimilated diet correlating with resource availability. For example, Correia (2003) found that introduced *P. clarkii* in Portugal consumed more plant material in the summer months reflecting seasonal growth. Furthermore, as we noted earlier, ontogenetic and sex differences in diet have also been documented (Johnson and Nack 2010) with some studies indicating a decrease in the consumption of animal material with increasing size (Parkyn et al. 2001, Correia 2003). Other studies, however, have found no evidence of intraspecific variation in diet (Bondar et al. 2005).

While terrestrial leaf litter and other plant detritus appears to contribute a significant proportion of crayfish diet in many populations (Parkyn et al. 2001, Correia 2003), and crayfish abundance can be promoted by elevated leaf litter subsidies (Kobayashi et al. 2011), it is not a particularly high quality resource. It might simply be consumed while sorting through detritus for more profitable animal prey and/or when more beneficial resources are limited. Stable isotope studies indicate that most of the energy for reproduction and growth is provided by animal material, even when it is consumed in smaller amounts than allochthonous leaf litter (in terms of biomass; Whitledge and Rabeni 1997, Parkyn et al. 2001, Hollows et al. 2002). Indeed, despite leaf litter comprising >75% of the gut contents of *Orconectes luteus* and *O. punctimanus*, Whitledge and Rabeni (1997) found that invertebrates contributed between 29% and

50% to assimilated diet. Slow moving and sessile invertebrates, such as cased caddisflies, clams and snails, are important in this respect (Parkyn et al. 2001, Stenroth and Nyström 2003, zu Ermgassen and Aldridge 2011).

Food web structure

Crayfish play a pivotal role in freshwater food webs because, as relatively large-bodied consumers, there are numerous food web links in which they interact positively or negatively with others species. Crayfish have a disproportional influence on food webs due to their omnivory which spreads the flow of energy throughout a food web by diffusing the effects of consumption across many trophic levels. This alters predator-prey relationships, disrupts trophic cascades, and has implications for food web connectance and stability. The relationship between omnivory and stability has created some debate amongst ecologists, with early theory suggesting that omnivory will decrease community stability (May 1973) by increasing food web connectance and thus, reducing the strength of individual links. However, others advocate that parameters and processes such as prey refugia and adaptive feeding behaviour of the omnivore weaken interactions and, therefore, maintain stability (McCann and Hastings 1997, Emmerson and Yearsley 2004). Indeed, omnivory is common in nature and most food webs, particularly in freshwater ecosystems, are complex, typically comprising hundreds to thousands of links and only two degrees of separation between each species (Williams et al. 2002, Woodward et al. 2008). Consequently, as large invertebrate omnivores, capable of maintain high population biomass and production rates, crayfish play an important role in determining the structure of food webs (Dorn and Wojdak 2004).

The broad diet of crayfish allows them to control or eradicate vulnerable prey while persisting on alternative resources from other trophic levels. Crayfish have direct food web links with invertebrates, fish, amphibians, macrophytes and periphyton, which can comprise up to four different trophic levels; from primary producer to tertiary consumer. For instance, in a pond experiment *Orconectes virilis* consumed fish eggs, tadpoles, snails, algae and macrophytes, causing significant declines in all their abundances (Dorn and Wojdak 2004). There is extensive evidence that crayfish, particularly as invaders, cause considerable reductions in the abundance and biomass of their preferred diet items (Stenroth and Nyström 2003, Cruz et al. 2008, Klose and Cooper 2012), sometimes resulting in local extinctions. This effect can cascade through the food web, particularly if the extinct species was highly connected (Dunne et al. 2002). Declines in abundance of numerous species across multiple trophic levels may also result in a simplified or collapsed food web (Stenroth and Nyström 2003, Geiger et al. 2005).

Food webs in ecosystems with abundant crayfish populations are structurally different from those without crayfish because their simultaneous feeding across multiple trophic levels diversifies trophic linkages (Polis and Strong 1996, Stenroth and Nyström 2003). Due to their predatory and grazing activity, crayfish essentially concentrate energy pathways through food webs, reducing the number of strong food web interactions. Consequently, omnivorous crayfish increase food web connectance which spreads and dilutes the flow of energy through food webs while decreasing the strength of most food web links. Crayfish may also represent a resource, consumed by larger organisms within food webs and, without shelter, they make easy prey for many fish, birds, mammals and reptiles (Correia 2001, Nyström et al. 2006, Ogada 2006, Bašić et al. 2015). During moulting, crayfish are more vulnerable to predation, and cannibalism may also occur (Brewis and Bowler 1983). Food web interactions can also include competition for shared resources; interspecific competition for food occurs not only between crayfish species, but also with other taxa, especially benthic fishes (Dorn and Mittelbach 1999, Reynolds 2011). In mainland Africa, where there are no native crayfish, introduced species, which include *Procambarus clarkii* in Kenya and *Cherax destructor* in South Africa, often compete with native crabs (Foster and Harper 2007). Crayfish also overlap in range, and compete, with native freshwater crabs in southern Europe (Barbaresi and Gherardi 1997, Cumberlidge et al. 2009) and invasive Chinese mitten crabs in Europe and America (Rudnick and Resh 2005, Jackson and Grey 2013).

Trophic cascades

In simple terms, as omnivores straddling trophic levels, crayfish serve as conduits of energy in complex food webs, may link terrestrial and aquatic food webs and, within lakes, link littoral, profundal and pelagic zones. The impact they have on their food sources can cascade to lower trophic levels via trophic interactions and this influences the flow of energy throughout the food web with implications for ecosystem functioning. Ecosystem functioning embraces a range of processes, including ecosystem metabolism, primary and secondary production, nutrient cycling and energy flux, and crayfish often play a key role in regulating a number of these processes, including primary productivity, decomposition and energy pathways.

Energy flows from the bottom-up through food webs. In contrast to this, the early descriptions of trophic cascades involved top-down control (Carpenter et al. 1985), whereby a predator influences prey abundance and, in turn, this effects primary producer abundance (e.g., more planktivorous fish = less zooplankton = more phytoplankton). Trophic cascades are well documented in aquatic ecosystems (Carpenter 1987, Baum and Worm 2009) with important roles in determining community dynamics. Cascades are facilitated by vulnerable prey, strong predator-prey interactions and low diversity. As both grazers and predators, crayfish have a number of roles in cascading interactions. Firstly, predators of crayfish can control their abundance and, therefore, reduce the effect of their omnivorous feeding across multiple trophic levels. For example, predatory fish reduce crayfish abundance which decreases crayfish predation pressure on primary consumers (Dorn et al. 2006). Secondly, as a predator, crayfish can instigate top-down effects on lower trophic levels and ecosystem functioning due to a crayfish - primary consumer - primary producer cascade; the mechanistic nature, strength and importance of this effect, however, has been debated. Classic food chain theory advocates that crayfish reduce the abundance of primary consumers by directly consuming them, and thus, fewer herbivores signifies a reduced impact on the primary producers. For instance, crayfish control the abundance of grazing invertebrates, such as snails, through direct consumption in many ecosystems. When crayfish abundance is high, this can cause significant declines in snail abundance which reduces net grazing pressure on epiphyton and periphyton. Consequently, the reduction in snail abundance has a cascading effect on primary productivity; by reducing the biomass of grazing invertebrates, crayfish indirectly promote algal growth and, therefore, boost productivity. This has been demonstrated in American streams whereby the presence of invasive crayfish results in an increase in stream periphyton productivity by 4–7 times compared to crayfish-free controls (Charlebois and Lamberti 1996). An alternative mechanism is a trait-mediated trophic cascade, whereby the mere presence of crayfish causes primary consumers to partake in predator avoidance behaviour, reducing grazing efficiency and, thus, increasing primary production without a decrease in secondary production. Snails can exhibit predator avoidance behavior in the presence of crayfish by moving closer to the water surface. This can instigate the trait-mediated cascade, resulting in elevated periphyton abundance in the usual habitat of snails, and a decrease in periphyton abundance in their refuge near the water surface (Bernot and Turner 2001). Pelagic secondary production can also be regulated by crayfish, driven by their negative effects on fish recruitment. There is evidence that crayfish reduce the abundance of zooplanktivorous fish, promoting zooplankton abundance which, in some cases, can reduce primary productivity of the phytoplankton community as a result of increased grazing pressure by zooplankton (Dorn and Wojdak 2004). Alternatively, crayfish can reduce the abundance of sessile filter feeders such as freshwater mussels or surface grazers such as snails (zu Ermgassen and Aldridge 2011, Strayer and Malcom 2012, Jackson et al. 2014, Fig. 8), which has the potential to promote phytoplankton or periphyton populations.

Crayfish have both direct and indirect influences on leaf litter breakdown, rendering them key species in the process of decomposition and the cycling of nutrients. A trophic cascade, whereby crayfish consume other invertebrate species such as gammarids and asellids which shred and consume leaf litter, reduces net leaf litter breakdown. However, many species of crayfish will also consume leaf litter directly which decouples the trophic cascade and ultimately results in an overall increase in breakdown rates (Usio 2000, Jackson et al. 2014, Fig. 8). By efficiently processing detritus, crayfish open up the detrital food chain to more predators (Geiger et al. 2005), promoting the flux of terrestrial nutrients through the aquatic food web.

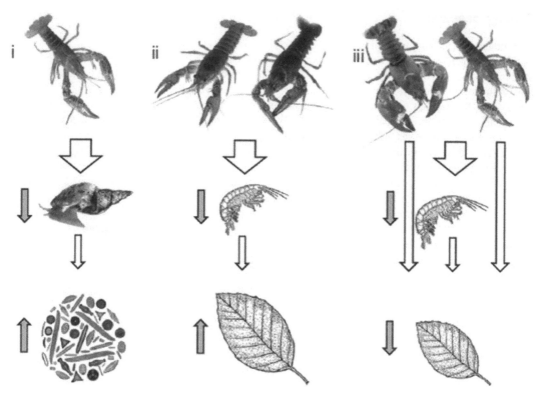

Figure 8. Schematic illustrating the trophic cascades (and potential to decouple these) instigated by invasive crayfish within experimental pond mesocosms—(i) *Procambarus clarkii*, (ii) *Orconectes virilis* and *Astacus leptodactylus*, and (iii) *Pacifastacus leniusculus* and *P. clarkii*. White arrows depict the direction of the cascade and grey arrows indicate the effect of the cascade on biomass compared to a control (zero crayfish). Reproduced from Jackson et al. (2014).

Bottom-up control of food webs, which goes by the theory that many prey can feed many predators, also exists but is less well documented than top-down cascades. As prey items, there is some evidence that crayfish influence higher trophic levels. Migrations of water birds in the USA often coincide with crayfish harvests and Fleury and Sherry (1995) found that water bird abundance and crayfish aquaculture production was positively correlated. Melero et al. (2014) found a strong positive relationship between the proportion of crayfish in mink diet and mink population density, and a negative relationship between the proportion of crayfish in mink diet and mink home range size, with crayfish contribution to mink diet reflecting their abundance in the ecosystem. It is now clear that introduced species like *Procambarus clarkii* in Portugal have become an important resource for many mammals and birds, playing a key role in food web interactions, and in structuring community dynamics in aquatic habitats and their associated riparian and terrestrial zones (Correia 2001).

Interactions with fish

Earlier in the chapter we made reference to crayfish, especially the larger bodied adults, having a more similar ecological role to fish rather than other macroinvertebrates. Consequently, we focus this last section from the community perspective on their interactions with fish. The mechanisms of interaction between fish and crayfish might be direct, such as competition or predation, or indirect through trophic cascades or habitat alteration. Here, we concentrate on direct links, and the implications for fish community structure and crayfish population success. Predator-prey links are reciprocal in many instances, with crayfish as both predator and prey. Competitive interactions focus on both shelter and food. These interactions may be further complicated by other controlling factors, such as disturbance, ontogenetic shifts, and species invasions. In the latter case, many species of crayfish and fish have been introduced outside their native range around the world and, therefore, crayfish-fish interactions can be native-native, native-invasive or invasive-invasive.

Crayfish as a resource

Crayfish are an important resource for many fish species in lakes and streams. A wide range of fish, including, but not limited to, barbel (Bašić et al. 2015), bass (Hill and Lodge 1995, Garvey et al. 2003, Hein et al. 2006), trout (Gowing and Momot 1979, Nyström et al. 2006), eels (Hicks 1997, Harrod and Grey 2006) and perch (Garvey et al. 2003, Nyström et al. 2006), consume a variety of crayfish species in their native habitats. Some species of fish have strong top-down effects on crayfish abundance through their predator-prey interaction. For instance, Hill and Lodge (1995) found native and invasive *Orconectes* spp. mortality was 50% higher in the presence of native largemouth bass (*Micropterus salmoides*) and, due to the negative impacts crayfish have on macrophytes, bass abundance was positively correlated with macrophyte abundance in the Great Lakes, America. Invasive crayfish populations can be successfully controlled by native fish (Hein et al. 2007) and there are also instances of invasive fish consuming invasive crayfish. Invasive *P. clarkii*, for instance, are consumed by invasive pike (*Esox lucius*) and largemouth bass, in Portugal (Elvira et al. 1996) and Kenya (Britton et al. 2010), respectively. Invasive fish may also have negative effects on native crayfish. For example, native *Paranephrops zealandicus* in New Zealand are negatively associated with the presence of introduced brown trout (*Salmo trutta*; Usio and Townsend 2000), and Edwards et al. (2009) in part attributed severe and widespread declines in both native and non-native crayfish to indiscriminate introductions of smallmouth bass (*M. dolomieu*). In Ohio, America, size-selective predation by fish on smaller native *Orconectes* species facilitated the invasion success of a larger, introduced crayfish; *Orconectes rusticus* (Mather and Stein 1993). Crayfish can also make smaller fish more vulnerable to piscivores by ousting them from refugia (Rahel and Stein 1988).

The impact of fish predation is related to the size distribution of both prey and predator. Adult crayfish often reach a large enough size that they are no longer consumed by gape-limited predators. Smallmouth bass (*Micropterus dolomieu*), for instance, consume *Orconectes propinquus* in ascending order of size (Stein 1977) and larger crayfish often have a size refuge from fish (Mather and Stein 1993). Habitat complexity may also be important in determining fish-crayfish predator-prey links. Stein (1977) found that the bass selected for small-sized crayfish in sand substrate and medium-sized crayfish in cobble substrate because cobbles provided a refuge for the smaller individuals.

Crayfish as a consumer

Crayfish have many different ecological relationships with fish, and interactions certainly are not restricted to the traditional idea of fish as predators and crayfish as prey. Crayfish may be reciprocal predators of fish eggs or the smaller fish species, especially if injured, and this behaviour has been particularly well documented from invasive crayfish populations, perhaps disproportionately so. In the United Kingdom, invasive *P. leniusculus* presence is negatively associated with fish abundance, particularly that of brown trout (*Salmo trutta*), bullhead (*Cottus gobio*) and stone loach (*Noemacheilus barbatulus*) (Guan and Wiles 1997, Peay et al. 2009). A combination of competition for shelter and direct predation by crayfish are blamed for the population declines (Guan and Wiles 1997). In contrast, Stenroth and Nyström (2003) found that *P. leniusculus* had no effect on brown trout fry in experimental enclosures in Swedish streams. In North America, the decline of game fish populations has been attributed to crayfish invasions, and although the mechanisms behind this are unclear, egg predation by crayfish is one hypothesis (Hobbs et al. 1989). Fish recruitment has been shown experimentally to be negatively affected by invasive crayfish due to changes in fish behaviour and the direct consumption of fish eggs. In ponds in North America, bluegill sunfish (*Lepomis macrochirus*) failed to nest successfully in the presence of invasive crayfish (*Orconectes virilis*) due to the invaders directly feeding on eggs (Dorn and Mittelbach 2004). Similarly, there is evidence that *Orconectes* spp. consume the eggs of lake sturgeon (*Acipenser fulvescens*), lake trout (*Salvelinus namaycush*), rainbow trout (*Oncorhynchus mykiss*), Arctic charr (*Salvelinus alpinus*) and fathead minnow (*Pimephales promelas*) (Savino and Miller 1991, Fitzsimons et al. 2002, Corkum and

Cronin 2004, Caroffino et al. 2010, Setzer et al. 2011). Evidence for crayfish consumption of juvenile and adult fish is more limited. However, Guan and Wiles (1997) reported that population declines in bullhead and stone loach were due to crayfish predation and several other studies report the occurrence of fish in crayfish diet (Lorman and Magnuson 1978, Correia 2003, Ilhéu et al. 2007). Ilhéu et al. (2007) examined the stomach contents of *P. clarkii* from isolated shallow pools in their invasive range in Portugal and found that fish consumption increased with fish density in pools, and that the most prominent fish in the crayfish diet was mosquitofish (*Gambusia affinis*), the most abundant fish species present.

Non-trophic interactions

Crayfish have been associated in the decline of fish populations through competition for shelter and food, and through the destruction of macrophytes, an important fish habitat. In artificial tanks, juvenile Atlantic salmon (*Salmo salar*) from Scotland were outcompeted by introduced *P. leniusculus* for shelters, which may make them more vulnerable to other predators such as birds under natural conditions (Griffiths et al. 2004). Again, under experimental conditions, both burbot (*Lota lota*) (Hirsch and Fischer 2008) and bullhead (Bubb et al. 2009) moderated their use of shelters when invasive crayfish were present. The loss of macrophyte beds which can be severe in the presence of crayfish, further exacerbates shelter competition by removing this important juvenile fish refuge.

Interactions among crayfish species

Numerous species of crayfish coexist naturally, particularly in North America, where it is not uncommon to find various species in the same stream. These species have evolved together and often occupy distinct ecological niches as a result. However, a small number of crayfish species have been translocated and introduced to waterbodies way beyond their native ranges for human food, fish forage, the aquarium trade, biocontrol, and for bait (Hobbs et al. 1989, Strayer 2010), and a few occupy a truly global distribution (e.g., Capinha et al. 2011). Native crayfish have not evolved with these newly introduced congeners and, therefore, interactions between them often have negative consequences for the native. Invasive species of crayfish regularly out-compete native crayfish because they are not subject to the same biotic (or abiotic) factors within the existing community that control population size such as selective predation and natural enemies (Hill and Lodge 1999). Introduced crayfish commonly exhibit faster growth rates and achieve larger sizes than their native counterparts, which gives them a further advantage with respect to increased fecundity and success in refugia competition scenarios (Alonso and Martínez 2006). Additionally, invasive crayfish may carry diseases and parasite burdens to which native species can be vulnerable (e.g., Alderman et al. 1984, Holdich et al. 2009, Longshaw et al. 2012).

Many aquatic environments have been invaded numerous times by numerous different species (Cohen and Carlton 1998, Leppäkoski and Olenin 2000, Jackson and Grey 2013), giving rise to the Invasion Meltdown Model which predicts that the disruption caused by the establishment of one invasive species can facilitate the success of further invaders (Simberloff and Von Holle 1999). However, sympatric invasive species may compete for resources and hence, have a detrimental impact on one another's success (Lohrer and Whitlatch 2002, Jackson et al. 2012). Despite the fact that invasive crayfish are now coexisting in various places around the world (e.g., Nakata et al. 2005, Bernardo et al. 2011), relatively few studies have examined interactions among them, or the consequences of multiple invasions of different crayfish for the recipient community. Empirical evidence from species other than crayfish supports both facilitative and negative interactions between sympatric invaders in aquatic environments (e.g., Jensen et al. 2002, Wonham et al. 2005). The occurrence of several stressors, such as multiple invaders, has the potential to moderate or amplify their impacts on the ecosystem (Darling and Côté 2008). Alternatively, if two sympatric invasive species have similar independent impacts, their combined effect might be additive. For example, invasive rusty crayfish (*Orconectes rusticus*) and invasive Chinese mystery snails (*Bellamya chinensis*) both independently reduce native snail biomass by consumptive and competitive interactions, respectively (Johnson et al. 2009). However, more functionally similar invaders, such as two or more crayfish species, are expected to interact and compete for shared resources causing a more complex combined impact.

There have been reports of co-existing populations of invasive crayfish species, including *P. leniusculus* and *P. clarkii* in Portugal (Bernardo et al. 2011) and Japan (Nakata et al. 2005), while *P. leniusculus* and *O. virilis*, and *P. clarkii* and *A. leptodactylus* are, if not yet co-existing, then in very close proximity in connected waterbodies in the UK (Ellis et al. 2012, Jackson et al. 2014). In other instances there has been serial replacement of invasive crayfish due to superior competition; Hill and Lodge (1999) described how the established invasive *Orconectes propinquus* in North America has been replaced by invading *O. limosus* by this means. Invasive crayfish are unlikely to facilitate one another's establishment; competition is far more probable. However, their interactions within the community may mitigate or amplify one another's impact on ecosystem structure and functioning (see Fig. 8 for different cascading impacts to primary production and leaf litter decomposition; Jackson et al. 2014).

Concluding remarks

Within this chapter, we have tried to draw together some of the inherently complex information summarizing our knowledge of the environmental drivers for crayfish population success (or failure). Many early studies have investigated only one environmental parameter in relation to one particular crayfish species, whereas we know now that abiotic factors can (and often do) interact in concert, that species-specificity (and ontogeny) limits general conclusions across the full range of ≈640 species, and that interactions are further complicated by the wider biotic community of which crayfish may be only a small part. However, mining of this seemingly bewildering wealth of information can lead to distillation of patterns, a good example of this being the paper by Richman et al. (2015) identifying the multiple drivers of decline in the global status of freshwater crayfish. They used the IUCN Categories and Criteria to evaluate extinction risk and found that 32% of all crayfish species are currently threatened with extinction. The level of extinction risk identified differed between crayfish families, with more threatened species in the Parastacidae and Astacidae as compared to the Cambaridae. There was also clear geographical variation in the dominant threats affecting hotspots of crayfish diversity. The majority of threatened US and Mexican species are affected by urban development, pollution, damming and water management, whereas the majority of threatened Australian species are affected by climate change, harvesting, agriculture and invasive species. Identifying the drivers for population declines can hopefully be turned around and used to define management criteria and strategies for promoting population success, especially when resources are limited and priorities must be set (Peters and Lodge 2013). Given that one of the threats to crayfish population success is other introduced crayfish, that the occurrence of multiple invasive species is seemingly on the increase, and that different species will adjust to climate change with varying degrees of success (Gherardi et al. 2013), then understanding the parameters that might promote and inhibit populations is of paramount importance in conserving native stocks.

References

Acosta, C.A. and S.A. Perry. 2001. Impact of hydropattern disturbence on crayfish population dynamics in the seasonal wetlands of Everglades National Park, USA. Aquat. Conserv. Mar. Freshw. Ecosyst. 11: 45–57.

Alderman, D.J., J.L. Polglase, M. Frayling and J. Hogger. 1984. Crayfish plague in Britain. J. Fish Dis. 7: 401–405.

Alexopoulos, E., C.R. McCrohan, J.J. Powell, R. Jugdaohsingh and K.N. White. 2003. Bioavailability and toxicity of freshly neutralized aluminium to the freshwater crayfish *Pacifastacus leniusculus*. Arch. Environ. Contam. Toxicol. 45: 509–514.

Alonso, F. and R. Martínez. 2006. Shelter competition between two invasive crayfish species: a laboratory study. Bull. Français la Pêche la Piscic. 1121–1132.

Aquiloni, L., A. Becciolini, R. Berti, S. Porciani, C. Trunfio and F. Gherardi. 2009. Managing invasive crayfish: use of X-ray sterilisation of males. Freshw. Biol. 54: 1510–1519.

Barbaresi, S. and F. Gherardi. 1997. Italian freshwater decapods: exclusion between the crayfish *Austropotamobius pallipes* (Faxon) and the crab *Potamon fluviatile* (Herbst). Bull. Fr. La Pech. La Piscic. 347: 731–747.

Bartels, P., J. Cucherousset, K. Steger, P. Eklov, L.J. Tranvik and H. Hillebrand. 2012. Reciprocal subsidies between freshwater and terrestrial ecosystems structure consumer resource dynamics. Ecology 93: 1173–1182.

Bašić, T., J.R. Britton, M.C. Jackson, P. Reading and J. Grey. 2015. Angling baits and invasive crayfish as important trophic subsidies for a large cyprinid fish. Aquat. Sci. 77: 153–160.

Baum, J.K. and B. Worm. 2009. Cascading top-down effects of changing oceanic predator abundances. J. Anim. Ecol. 78: 699–714.

Beja, P. 1996. An analysis of otter Lutra lutra predation on introduced American crayfish *Procambarus clarkii* in Iberian streams. J. Appl. Ecol. 33: 1156–1170.

Berber, S. and Y. Mazlum. 2009. Reproductive efficiency of the narrow-clawed crayfish, *Astacus leptodactylus*, in several populations in Turkey. Crustaceana 82: 531–542.

Bernardo, J. 1996. The particular maternal effect of propagule size, especially egg size: patterns, models, quality of evidence and interpretations. Integr. Comp. Biol. 36: 216–236.

Bernardo, J.M., A.M. Costa, S. Bruxelas and A. Teixeira. 2011. Dispersal and coexistence of two non-native crayfish species (*Pacifastacus leniusculus* and *Procambarus clarkii*) in NE Portugal over a 10-year period. Knowl. Manag. Aquat. Ecosyst. 401: 28.

Bernot, R.J. and A.M. Turner. 2001. Predator identity and trait-mediated indirect effects in a littoral food web. Oecologia 129: 139–146.

Berrill, M., L. Hollett, A. Margosian and J. Hudson. 1985. Variation in tolerance to low environmental pH by the crayfish *Orconectes rusticus*, *O. propinquus*, and *Cambarus robustus*. Can. J. Zool. 63: 2586–2589.

Bilotta, G.S. and R.E. Brazier. 2008. Understanding the influence of suspended solids on water quality and aquatic biota. Water Res. 42: 2849–2861.

Bondar, C.A., K. Bottriell, K. Zeron and J.S. Richardson. 2005. Does trophic position of the omnivorous signal crayfish (*Pacifastacus leniusculus*) in a stream food web vary with life history stage or density? Can. J. Fish. Aquat. Sci. 62: 2632–2639.

Bovbjerg, R.V. 1970. Ecological isolation and competitive exclusion in two crayfish (*Orconectes virilis* and *Orconectes immunis*). Ecology 51: 225–236.

Brewer, J.S. 1999. Effects of competition, litter, and disturbance on an annual carnivorous plant (*Utricularia juncea*). Plant Ecol. 140: 159–165.

Brewer, S.K., R.J. DiStefano and C.F. Rabeni. 2009. The influence of age-specific habitat selection by a stream crayfish community (*Orconectes* spp.) on secondary production. Hydrobiologia 619: 1–10.

Brewis, J.M. and K. Bowler. 1983. A study of the dynamics of a natural population of the freshwater crayfish, *Austropotamobius pallipes*. Freshw. Biol. 13: 443–452.

Brewis, J.M. and K. Bowler. 1985. A study of reproductive females of the freshwater crayfish *Austropotamobius pallipes*. Hydrobiologia 121: 145–149.

Britton, R.R., D.M. Harper, D.O. Oyugi and J. Grey. 2010. The introduced *Micropterus salmoides* in an equatorial lake: a paradoxical loser in an invasion meltdown scenario? Biol. Invasions 12: 3439–3448.

Bubb, D.H., T.J. Thom and M.C. Lucas. 2004. Movement and dispersal of the invasive signal crayfish *Pacifastacus leniusculus* in upland rivers. Freshw. Biol. 49: 357–368.

Bubb, D.H., O.J. O'Malley, A.C. Gooderham and M.C. Lucas. 2009. Relative impacts of native and non-native crayfish on shelter use by an indigenous benthic fish. Aquat. Conserv. Mar. Freshw. Ecosyst. 19: 448–455.

Bugnot, A.B. and L.S. López Greco. 2009. Sperm production in the red claw crayfish *Cherax quadricarinatus* (Decapoda, Parastacidae). Aquaculture 295: 292–299.

Bunn, S.E. and A.H. Arthington. 2002. Basic principles and ecological consequences of altered flow regimes for aquatic biodiversity. Environ. Manage. 30: 492–507.

Buřič, M., P. Kozák and A. Kouba. 2009. Movement patterns and ranging behavior of the invasive spiny-cheek crayfish in a small reservoir tributary. Fundam. Appl. Limnol. 174: 329–337.

Caine, E. 1978. Comparative ecology of epigean and hypogean crayfish (Crustacea: Cambaridae) from northwestern Florida. Am. Midl. Nat. 99: 315–329.

Capelli, G.M. and J.J. Magnuson. 1983. Morphoedaphic and biogeographic analysis of crayfish distribution in Northern Wisconsin. J. Crustac. Biol. 3: 548–564.

Capinha, C., B. Leung and P. Anastácio. 2011. Predicting worldwide invasiveness for four major problematic decapods: An evaluation of using different calibration sets. Ecography. 34: 448–459.

Caroffino, D.C., T.M. Sutton, R.F. Elliott and M.C. Donofrio. 2010. Predation on early life stages of lake sturgeon in the Peshtigo River, Wisconsin. Trans. Am. Fish. Soc. 139: 1846–1856.

Carpenter, S.R. 1987. Regulation of lake primary productivity by food web structure. Ecology 68: 1863–1876.

Carpenter, S.R. and D.M. Lodge. 1986. Effects of submersed macrophytes on ecosystem processes. Aquat. Bot. 26: 341–370.

Carpenter, S.R., J.F. Kitchell and J.R. Hodgson. 1985. Cascading trophic interactions and lake productivity. Bioscience 35: 634–639.

Charlebois, P.M. and G.A. Lamberti. 1996. Invading crayfish in a Michigan stream: direct and indirect effects on periphyton and macroinvertebrates. J. North Am. Benthol. Soc. 15: 551–563.

Clark, J.M., M.W. Kershner and J.J. Montemarano. 2013. Habitat-specific effects of particle size, current velocity, water depth, and predation risk on size-dependent crayfish distribution. Hydrobiologia 716: 103–114.

Cohen, A.N. and J.T. Carlton. 1998. Accelerating invasion rate in a highly invaded estuary. Science 279: 555–558.

Corkum, L.D. and D.J. Cronin. 2004. Habitat complexity reduces aggression and enhances consumption in crayfish. J. Ethol. 22: 23–27.

Correia, A.M. 2001. Seasonal and interspecific evaluation of predation by mammals and birds on the introduced red swamp crayfish *Procambarus clarkii* (Crustacea, Cambaridae) in a freshwater marsh (Portugal). J. Zool. 255: 533–541.

Correia, A.M. 2003. Food choice by the introduced crayfish *Procambarus clarkii*. Ann. Zool. Fennici 40: 517–528.

Cossette, C. and M.A. Rodríguez. 2004. Summer use of a small stream by fish and crayfish and exchanges with adjacent lentic macrohabitats. Freshwat. Biol. 49: 931–944.

Costello, M.J., R.M. May and N.E. Stork. 2013. Can we name Earth's species before they go extinct? Science 339: 413–416.

Covich, A.P., M.A. Palmer and T.A. Crowl. 1999. The role of benthic invertebrate species in freshwater ecosystems - zoobenthic species influence energy flows and nutrient cycling. Bioscience 49: 119–127.

Crandall, K.A. and J.E. Buhay. 2008. Global diversity of crayfish (Astacidae, Cambaridae, and Parastacidae - Decapoda) in freshwater. Hydrobiologia 595: 295–301.

Creed, R.P. and J.M. Reed. 2004. Ecosystem engineering by crayfish in a headwater stream community. J. North Am. Benthol. Soc. 23: 224–236.

Creed, R.P., A. Taylor and J.R. Pflaum. 2010. Bioturbation by a dominant detritivore in a headwater stream: litter excavation and effects on community structure. Oikos 119: 1870–1876.

Cruz, M.J., P. Segurado, M. Sousa and R. Rebelo. 2008. Collapse of the amphibian community of the paul do boquilobo natural reserve (central Portugal) after the arrival of the exotic american crayfish *Procambarus clarkii*. Herpetol. J. 18: 197–204.

Cumberlidge, N., P.K.L. Ng, D.C.J. Yeo, C. Magalhães, M.R. Campos, F. Alvarez, T. Naruse, S.R. Daniels, L.J. Esser, F.Y.K. Attipoe, F.L. Clotilde-Ba, W. Darwall, A. McIvor, J.E.M. Baillie, B. Collen and M. Ram. 2009. Freshwater crabs and the biodiversity crisis: importance, threats, status, and conservation challenges. Biol. Conserv. 142: 1665–1673.

Darling, E.S. and I.M. Côté. 2008. Quantifying the evidence for ecological synergies. Ecol. Lett. 11: 1278–1286.

Del Ramo, J., J. Díaz-Mayans, A. Torreblanca and A. Núñez. 1987. Effects of temperature on the acute toxicity of heavy metals (Cr, Cd, and Hg) to the freshwater crayfish, *Procambarus clarkii* (Girard). Bull. Environ. Contam. Toxicol. 38: 736–741.

Demers, A., C. Souty-Grosset, M.C. Trouilhé, L. Füreder, B. Renai and F. Gherardi. 2006. Tolerance of three European native species of crayfish to hypoxia. Hydrobiologia 560: 425–432.

DiStefano, R.J., R.J. Neves, L.A. Helfrich and M.C. Lewis. 1991. Response of the crayfish *Cambarus-bartonii-bartonii* to acid exposure in southern Appalachian streams. Can. J. Zool. 69: 1585–1591.

DiStefano, R.J., J.J. Decoske, T.M. Vangilder and L.S. Barnes. 2003. Macrohabitat partitioning among three crayfish species in two Missouri streams, U.S.A. Crustaceana 76: 343–362.

DiStefano, R.J., D.D. Magoulick, E.M. Imhoff and E.R. Larson. 2009. Imperiled crayfishes use hyporheic zone during seasonal drying of an intermittent stream. J. North Am. Benthol. Soc. 28: 142–152.

Dorn, N.J. and G.G. Mittelbach. 1999. More than predator and prey: a review of interactions between fish and crayfish. Vie et Milieu 49: 229–237.

Dorn, N.J. and G.G. Mittelbach. 2004. Effects of a native crayfish (*Orconectes virilis*) on the reproductive success and nesting behavior of sunfish (*Lepomis* spp.). Can. J. Fish. Aquat. Sci. 61: 2135–2143.

Dorn, N.J. and J.M. Wojdak. 2004. The role of omnivorous crayfish in littoral communities. Oecologia 140: 150–159.

Dorn, N.J. and J.C. Volin. 2009. Resistance of crayfish (*Procambarus* spp.) populations to wetland drying depends on species and substrate. J. North Am. Benthol. Soc. 28: 766–777.

Dorn, N.J., J.C. Trexler and E.E. Gaiser. 2006. Exploring the role of large predators in marsh food webs: evidence for a behaviorally-mediated trophic cascade. Hydrobiologia 569: 375–386.

Dunne, J.A., R.J. Williams and N.D. Martinez. 2002. Network structure and biodiversity loss in food webs: robustness increases with connectance. Ecol. Lett. 5: 558–567.

Dyer, J.J., S.K. Brewer, T.A. Worthington and E.A. Bergey. 2013. The influence of coarse-scale environmental features on current and predicted future distributions of narrow-range endemic crayfish populations. Freshw. Biol. 58: 1071–1088.

Edgerton, B.F., P. Henttonen, J. Jussila, A. Mannonen, P. Paasonen, T. Taugbøl, L. Edsman and C. Souty-Grosset. 2004. Understanding the causes of disease in European freshwater crayfish. Conserv. Biol. 18: 1466–1474.

Edwards, B.A., D.A. Jackson and K.M. Somers. 2009. Multispecies crayfish declines in lakes: implications for species distributions and richness. J. North Am. Benthol. Soc. 28: 719–732.

Ellis, A., M.C. Jackson, I. Jennings, J. England and R. Phillips. 2012. Present distribution and future spread of louisiana red swamp crayfish *Procambarus clarkii* (Crustacea, Decapoda, Astacida, Cambaridae) in Britain: implications for conservation of native species and habitats. Knowl. Manag. Aquat. Ecosyst. 406: 05.

Elvira, B., G.G. Nicola and A. Almodovar. 1996. Pike and red swamp crayfish: a new case on predator-prey relationship between aliens in central Spain. J. Fish Biol. 48: 437–446.

Emmerson, M. and J.M. Yearsley. 2004. Weak interactions, omnivory and emergent food-web properties. Proc. Biol. Sci. 271: 397–405.

Englund, G. and J.J. Krupa. 2000. Habitat use by crayfish in stream pools: influence of predators, depth and body size. Freshw. Biol. 43: 75–83.

Færøvig, P.J. and D.O. Hessen. 2003. Allocation strategies in crustacean stoichiometry: the potential role of phosphorus in the limitation of reproduction. Freshw. Biol. 48: 1782–1792.

Favaro, L., T. Tirelli and D. Pessani. 2010. The role of water chemistry in the distribution of *Austropotamobius pallipes* (Crustacea Decapoda Astacidae) in Piedmont (Italy). Comptes Rendus Biologies 333: 68–75.

Feminella, J. and V. Resh. 1989. Submersed macrophytes and grazing crayfish: an experimental study of herbivory in a California freshwater marsh. Ecography 12: 1–8.

Feria, T.P. and Z. Faulkes. 2011. Forecasting the distribution of Marmorkrebs, a parthenogenetic crayfish with high invasive potential, in Madagascar, Europe, and North America. Aquat. Invasions 6: 55–67.

Figueiredo, M.S.R.B. and A.J. Anderson. 2003. Ontogenetic changes in digestive proteases and carbohydrases from the Australian freshwater crayfish, redclaw *Cherax quadricarinatus* (Crustacea, Decapoda, Parastacidae). Aquac. Res. 34: 1235–1239.

Fitzsimons, J.D., D.L. Perkins and C.C. Krueger. 2002. Sculpins and crayfish in lake trout spawning areas in lake ontario: estimates of abundance and egg predation on lake trout eggs. J. Great Lakes Res. 28: 421–436.

Fleury, B. and T. Sherry. 1995. Long-term population trends of colonial wading birds in the southern United States: the impact of crayfish aquaculture on Louisiana populations. Auk 112: 613–632.

Flinders, C.A. and D.D. Magoulick. 2003. Effects of stream permanence on crayfish community structure. Am. Midl. Nat. 149: 134–147.

Flinders, C.A. and D.D. Magoulick. 2007. Habitat use and selection within ozark lotic crayfish assemblages: spatial and temporal variation. J. Crustac. Biol. 27: 242–254.

Flint, R.W. 1977. Seasonal activity, migration and distribution of the crayfish, *Pacifastacus Ieniusculus*, in Lake Tahoe. Am. Midl. Nat. 97: 280–292.

Foster, J. and D. Harper. 2007. Status and ecosystem interactions of the invasive Louisianan red swamp crayfish *Procambarus clarkii* in East Africa. pp. 91–101. *In*: F. Gherardi (ed.). Biological Invaders in Inland Waters: Profiles, Distribution, and Threats. Springer, Netherlands.

France, R. 1984. Comparative tolerance to low pH of three life stages of the crayfish *Orconectes virilis*. Can. J. Zool. 62: 2360–2363.

France, R. 1987. Reproductive impairment of the crayfish *Orconectes-virilis* in response to acidification of lake-223. Can. J. Fish. Aquat. Sci. 44: 97–106.

Galeotti, P., G. Bernini, L. Locatello, R. Sacchi, M. Fasola and D. Rubolini. 2012. Sperm traits negatively covary with size and asymmetry of a secondary sexual trait in a freshwater crayfish. PLoS One 7: e43771.

Garvey, J.E., R.A. Stein and H.M. Thomas. 1994. Assessing how fish predation and interspecific prey competition influence a crayfish assemblage. Ecology 75: 532–547.

Garvey, J.E., J.E. Rettig, R.A. Stein, D.M. Lodge and S.P. Klosiewski. 2003. Scale-dependent associations among fish predation, littoral habitat, and distributions of crayfish species. Ecology 84: 3339–3348.

Geiger, W., P. Alcorlo, A. Baltanás and C. Montes. 2005. Impact of an introduced Crustacean on the trophic webs of Mediterranean wetlands. Biol. Inv. 7: 49–73.

Gherardi, F. and W.H. Daniels. 2004. Agonism and shelter competition between invasive and indigenous crayfish species. Can. J. Zool. 82: 1923–1932.

Gherardi, F., S. Barbaresi and G. Salvi. 2000. Spatial and temporal patterns in the movement of *Procambarus clarkii*, an invasive crayfish. Aquat. Sci. 62: 179–193.

Gherardi, F., J.R. Britton, K.M. Mavuti, N. Pacini, J. Grey, E. Tricarico and D.M. Harper. 2011. Learning to face biological invasions in Africa: a lesson from Lake Naivasha (Kenya). Biol. Cons. 144: 2585–2596.

Gherardi, F., A. Coignet, C. Souty-Grosset, D. Spigoli and L. Aquiloni. 2013. Climate warming and the agonistic behaviour of invasive crayfishes in Europe. Freshw. Biol. 58: 1958–1967.

Gowing, H. and W.T. Momot. 1979. Impact of brook trout (*Salvelinus fontinalis*) predation on the crayfish *Orconectes virilis* in three Michigan lakes. J. Fish. Res. Board Canada 36: 1191–1196.

Grandjean, F., J. Jandry, E. Bardon, A. Coignet, M.C. Trouilhé, B. Parinet, C. Souty-Grosset and M. Brulin. 2011. Use of Ephemeroptera as bioindicators of the occurrence of white-clawed crayfish (*Austropotamobius pallipes*). Hydrobiologia 671: 253–258.

Grey, J. and D.M. Harper. 2002. Using stable isotope analyses to identify allochthonous inputs to Lake Naivasha mediated via the hippopotamus gut. Isot. Env. Health. St. 38: 245–250.

Grey, J. and M.C. Jackson. 2012. "Leaves and eats shoots": direct terrestrial feeding can supplement invasive red swamp crayfish in times of need. PLoS One 7: e42575.

Grey, J., R.I. Jones and D. Sleep. 2001. Seasonal changes in the importance of the source of organic matter to the diet of zooplankton in Loch Ness, as indicated by stable isotope analysis. Limnol. Oceanogr. 46: 505–513.

Griffiths, S.W., P. Collen and J.D. Armstrong. 2004. Competition for shelter among over-wintering signal crayfish and juvenile Atlantic salmon. J. Fish Biol. 65: 436–447.

Guan, R.Z. and P.R. Wiles. 1997. Ecological impact of introduced crayfish on benthic fishes in a British lowland river. Conserv. Biol. 11: 641–647.

Gutiérrez-Yurrita, P.J. and C. Montes. 1999. Bioenergetics and phenology of reproduction of the introduced red swamp crayfish, *Procambarus clarkii*, in Doñana National Park, Spain, and implications for species management. Freshw. Biol. 42: 561–574.

Haddaway, N.R., R.J.G. Mortimer, M. Christmas and A. Dunn. 2013. Effect of pH on growth and survival in the freshwater crayfish *Austropotamobius pallipes*. Freshwat. Crayfish 19: 53–62.

Haddaway, N.R., R.J.G. Mortimer, M. Christmas and A.M. Dunn. 2015. Water chemistry and endangered white-clawed crayfish: a literature review and field study of water chemistry association in *Austropotamobius pallipes*. Knowl. Manag. Aquat. Ecosyst. 416: 01.

Haertel-Borer, S.S., D. Zak, R. Eckmann, U. Baade and F. Hölker. 2005. Population density of the crayfish, *Orconectes limosus*, in relation to fish and macroinvertebrate densities in a small mesotrophic lake—implications for the lake's food web. Int. Rev. Hydrobiol. 90: 523–533.

Haggerty, S.M., D.P. Batzer and C.R. Jackson. 2002. Macroinvertebrate assemblages in perennial headwater streams of the Coastal Mountain range of Washington, U.S.A. Hydrobiologia 479: 143–154.

Hänfling, B., F. Edwards and F. Gherardi. 2011. Invasive alien Crustacea: dispersal, establishment, impact and control. BioControl 56: 573–595.

Harlioğlu, M.M. and I. Türkgülü. 2000. The relationship between egg size and female size in freshwater crayfish, *Astacus leptodactylus*. Aquac. Int. 8: 95–98.

Harlioğlu, M.M. and A.G. Harlioğlu. 2006. Threat of non-native crayfish introductions into Turkey: global lessons. Rev. Fish Biol. Fish. 16: 171–181.

Harlioğlu, M.M., K. Köprücü and Y. Özdemir. 2002. The effect of dietary vitamin E on the pleopodal egg number of *Astacus leptodactylus* (Eschscholtz, 1823). Aquac. Int. 10: 391–397.

Harper, D.M., A.C. Smart, S. Coley, S. Schmitz, A.C.G. De Beauregard, R. North, C. Adams, P. Obade and M. Kamau. 2002. Distribution and abundance of the Louisiana red swamp crayfish *Procambarus clarkii* Girard at Lake Naivasha, Kenya between 1987 and 1999. Hydrobiologia 488: 143–151.

Harrod, C. and J. Grey. 2006. Isotopic variation complicates analysis of trophic relations within the fish community of Plußsee: a small, deep, stratifying lake. Arch. Hydrobiol. 167: 281–299.

Harvey, G.L., A.J. Henshaw, T.P. Moorhouse, N.J. Clifford, H. Holah, J. Grey and D.W. Macdonald. 2014. Invasive crayfish as drivers of fine sediment dynamics in rivers: field and laboratory evidence. Earth Surf. Process. Landforms 39: 259–271.

Hayes, R.B. 2012. Consequences for lotic ecosystems of invasion by signal crayfish. Ph.D. Thesis, Queen Mary University of London, London.

Hein, C.L., B.M. Roth, A.R. Ives and M.J. Vander Zanden. 2006. Fish predation and trapping for rusty crayfish (*Orconectes rusticus*) control: a whole-lake experiment. Can. J. Fish. Aquat. Sci. 63: 383–393.

Hein, C.L., M.J. Vander Zanden and J.J. Magnuson. 2007. Intensive trapping and increased fish predation cause massive population decline of an invasive crayfish. Freshw. Biol. 52: 1134–1146.

Hicks, B.J. 1997. Food webs in forest and pasture streams in the Waikato region, New Zealand: a study based on analyses of stable isotopes of carbon and nitrogen, and fish gut contents. New Zeal. J. Mar. Freshw. Res. 31: 651–664.

Hill, A.M. and D.M. Lodge. 1994. Diel changes in resource demand: competition and predation in species replacement among crayfishes. Ecology 75: 2118–2126.

Hill, A.M. and D.M. Lodge. 1995. Multi-trophic-level impact of sublethal interactions between bass and omnivorous crayfish. J. North Am. Benthol. Soc. 14: 306–314.

Hill, A.M. and D.M. Lodge. 1999. Replacement of resident crayfishes by an exotic crayfish: the roles of competition and predation. Ecol. Appl. 9: 678–690.

Hirsch, P.E. and P. Fischer. 2008. Interactions between native juvenile burbot (*Lota lota*) and the invasive spinycheek crayfish (*Orconectes limosus*) in a large European lake. Can. J. Fish. Aquat. Sci. 65: 2636–2643.

Hobbs, H.H. and S.A. Rewolinski. 1985. Notes on the burrowing crayfish *Procambarus* (Girardiella) *gracilis* (Bundy) (Decapoda, Cambaridae) from Southeastern Wisconsin, U.S.A. Crustaceana 48: 26–33.

Hobbs, H.H. and M. Whiteman. 1991. Notes on the burrows, behavior, and color of the crayfish *Fallicambarus (F.) devastator* (Decapoda: Cambaridae). Southwest. Nat. 36: 127–135.

Hobbs, H.H., J.P. Jass and J.V. Huner. 1989. A review of global crayfish introductions with particular emphasis on two north american species (Decapoda, Cambaridae). Crustaceana 56: 299–316.

Holdich, D.M. 1999. The negative effects of established crayfish introductions. pp. 31–47. *In*: F. Gherardi and D.M. Holdich (eds.). Crayfish in Europe as Alien Species: How to Make the Best of a Bad Situation. Balkema, Rotterdam.

Holdich, D.M. 2002. Distribution of crayfish in europe and some adjoining countries. Bull. Français la Pêche la Piscic. 611–650.

Holdich, D.M., M.M. Harlioglu and I. Firkins. 1997. Salinity adaptations of crayfish in british waters with particular reference to *Austropotamobius pallipes*, *Astacus leptodactylus* and *Pacifastacus leniusculus*. Estuarine, Coast. Shelf Sci. 44: 147–154.

Holdich, D.M., J.D. Reynolds, C. Souty-Grosset and P.J. Sibley. 2009. A review of the ever increasing threat to European crayfish from non-indigenous crayfish species. Knowl. Manag. Aquat. Ecosyst. 394-395: 11.

Hollows, J.W., C.R. Townsend and K.J. Collier. 2002. Diet of the crayfish *Paranephrops zealandicus* in bush and pasture streams: Insights from stable isotopes and stomach analysis. New Zeal. J. Mar. Freshw. Res. 36: 129–142.

Huner, J.V. and O.V. Lindqvist. 1991. Special problems in freshwater crayfish egg production. pp. 235–266. *In*: A. Wenner and A. Kuris (eds.). Crustacean Egg Production. Crustacean Issues, Vol. 7. Balkema, Rotterdam.

Huntsman, B.M., M.P. Venarsky and J.P. Benstead. 2011. Relating carrion breakdown rates to ambient resource level and community structure in four cave stream ecosystems. J. North Am. Benthol. Soc. 30: 882–892.

Iles, A.C. and J.B. Rasmussen. 2005. Indirect effects of metal contamination on energetics of yellow perch (*Perca flavescens*) resulting from food web simplification. Freshw. Biol. 50: 976–992.

Ilhéu, M. and J.M. Bernardo. 1997. Life history and population biology of red swamp crayfish, *Procambarus clarkii*, in a Mediterranean reservoir. Freshwat. Crayfish 11: 54–59.

Ilhéu, M., J.M. Bernardo and S. Fernandes. 2007. Predation of invasive crayfish on aquatic vertebrates: the effect of *Procambarus clarkii* on fish assemblages in Mediterranean temporary streams. pp. 543–558. *In*: F. Gherardi (ed.). Biological Invaders in Inland Waters: Profiles, Distribution, and Threats. Springer, Netherlands.

Ings, T.C., J.M. Montoya, J. Bascompte, N. Blüthgen, L. Brown, C.F. Dormann, F. Edwards, D. Figueroa, U. Jacob, J.I. Jones, R.B. Lauridsen, M.E. Ledger, H.M. Lewis, J.M. Olesen, F.J.F. Van Veen, P.H. Warren and G. Woodward. 2009. Ecological networks—Beyond food webs. J. Anim. Ecol. 78: 253–269.

IPCC. 2014. Climate Change 2014: Impacts, Adaptation, and Vulnerability. Part A: Global and Sectoral Aspects. *In*: C.B. Field, V.R. Barros, D.J. Dokken, K.J. Mach, M.D. Mastrandrea, T.E. Bilir, M. Chatterjee, K.L. Ebi, Y.O. Estrada, R.C. Genova, B. Girma, E.S. Kissel, A.N. Levy, S. MacCracken, P.R. Mastrandrea and L.L. White (eds.). Contribution of Working Group II to the Fifth Assessment Report of the Intergovernmental Panel on Climate Change. Cambridge University Press, Cambridge, United Kingdom and New York, NY, USA.

Jackson, M.C. and J. Grey. 2013. Accelerating rates of freshwater invasions in the catchment of the River Thames. Biol. Invasions 15: 945–951.

Jackson, M.C., I. Donohue, A.L. Jackson, J.R. Britton, D.M. Harper and J. Grey. 2012. Population-level metrics of trophic structure based on stable isotopes and their application to invasion ecology. PLoS One 7: e31757.

Jackson, M.C., T. Jones, M. Milligan, D. Sheath, J. Taylor, A. Ellis, J. England and J. Grey. 2014. Niche differentiation among invasive crayfish and their impacts on ecosystem structure and functioning. Freshw. Biol. 59: 1123–1135.

Jandry, J., B. Parinet, M. Brulin and F. Grandjean. 2014. Ephemeroptera communities as bioindicators of the suitability of headwater streams for restocking with white-clawed crayfish, *Austropotamobius pallipes*. Ecol. Indic. 46: 560–565.

Jensen, G.C., P.S. McDonald and D.A. Armstrong. 2002. East meets west: competitive interactions between green crab *Carcinus maenas*, and native and introduced shore crab *Hemigrapsus* spp. Mar. Ecol. Prog. Ser. 225: 251–262.

Jeziorski, A., A.J. Tanentzap, N.D. Yan, A.M. Paterson, M.E. Palmer, J.B. Korosi, J.A. Rusak, M.T. Arts, W.B. Keller, R. Ingram, A. Cairns and J.P. Smol. 2015. The jellification of north temperate lakes. Proc. Biol. Sci. 282: 20142449.

Johnson, J.H. and C.C. Nack. 2010. Ontogenetic variation in food consumption of rusty crayfish (*Orconectes rusticus*) in a Central New York stream. J. Freshw. Ecol. 25: 59–64.

Johnson, M.F., S.P. Rice and I. Reid. 2010. Topographic disturbance of subaqueous gravel substrates by signal crayfish (*Pacifastacus leniusculus*). Geomorphology 123: 269–278.

Johnson, M.F., S.P. Rice and I. Reid. 2014. The activity of signal crayfish (*Pacifastacus leniusculus*) in relation to thermal and hydraulic dynamics of an alluvial stream, UK. Hydrobiologia 724: 41–54.

Johnson, P.T.J., J.D. Olden, C.T. Solomon and M.J. Vander Zanden. 2009. Interactions among invaders: Community and ecosystem effects of multiple invasive species in an experimental aquatic system. Oecologia 159: 161–170.

Johnston, K. and B.J. Robson. 2009. Habitat use by five sympatric Australian freshwater crayfish species (Parastacidae). Freshw. Biol. 54: 1629–1641.

Jones, C.G., J.H. Lawton and M. Shachak. 1994. Organisms as ecosystem engineers. Oikos 69: 373–386.

Jones, C.G., J.H. Lawton and M. Shachak. 1997. Positive and negative effects of organisms as physical ecosystem engineers. Ecology 78: 1946–1957.

Joy, M.K. and R.G. Death. 2001. Control of freshwater fish and crayfish community structure in Taranaki, New Zealand: dams, diadromy or habitat structure? Freshw. Biol. 46: 417–429.

Karplus, I., H. Gideon and A. Barki. 2003. Shifting the natural spring-summer breeding season of the Australian freshwater crayfish *Cherax quadricarinatus* into the winter by environmental manipulations. Aquaculture 220: 277–286.

Keller, W., J.H. Heneberry and J.M. Gunn. 1999. Effects of emission reductions from the Sudbury smelters on the recovery of acid- and metal-damaged lakes. J. Aquat. Ecosyst. Stress Recovery 6: 189–198.

Kellogg, C.M. and N.J. Dorn. 2012. Consumptive effects of fish reduce wetland crayfish recruitment and drive species turnover. Oecologia 168: 1111–1121.

Kerby, J.L., S.P.D. Riley, L.B. Kats and P. Wilson. 2005. Barriers and flow as limiting factors in the spread of an invasive crayfish (*Procambarus clarkii*) in southern California streams. Biol. Conserv. 126: 402–409.

Kershner, M. and D. Lodge. 1995. Effects of littoral habitat and fish predation on the distribution of an exotic crayfish, *Orconectes rusticus*. J. North Am. Benthol. Soc. 14: 414–422.

Khan, M.A.Q., S.A. Ahmed, B. Catalin, A. Khodadoust, O. Ajayi and M. Vaughn. 2006. Effect of temperature on heavy metal toxicity to juvenile crayfish, *Orconectes immunis* (Hagen). Environ. Toxicol. 21: 513–520.

Khan, S. and D. Nugegoda. 2007. Sensitivity of juvenile freshwater crayfish *Cherax destructor* (Decapoda: Parastacidae) to trace metals. Ecotoxicol. Environ. Saf. 68: 463–469.

Klose, K. and S.D. Cooper. 2012. Contrasting effects of an invasive crayfish (*Procambarus clarkii*) on two temperate stream communities. Freshw. Biol. 57: 526–540.

Knight, T.M., M.W. McCoy, J.M. Chase, K.A. McCoy and R.D. Holt. 2005. Trophic cascades across ecosystems. Nature 437: 880–883.

Kobayashi, R., Y. Maezono and T. Miyashita. 2011. The importance of allochthonous litter input on the biomass of an alien crayfish in farm ponds. Popul. Ecol. 53: 525–534.

Kouba, A., M. Buřič and P. Kozák. 2010. Bioaccumulation and effects of heavy metals in crayfish: a review. Water Air Soil Pollut. 211: 5–16.

Larson, E.R., D.D. Magoulick, C. Turner and K.H. Laycock. 2009. Disturbance and species displacement: different tolerances to stream drying and desiccation in a native and an invasive crayfish. Freshw. Biol. 54: 1899–1908.

Larson, E.R., J.D. Olden and N. Usio. 2011. Shoreline urbanization interrupts allochthonous subsidies to a benthic consumer over a gradient of lake size. Biol. Lett. 7: 551–554.

Lehtikoivunen, S.M. and L.A. Kivivuori. 1994. Effect of temperature acclimation in the crayfish *Astacus astacus* L. on the locomotor activity during a cyclic temperature change. J. Therm. Biol. 19: 299–304.

Leppäkoski, E. and S. Olenin. 2000. Non-native species and rates of spread: Lessons from the brackish Baltic Sea. Biol. Invasions 2: 151–163.

Light, T. 2003. Success and failure in a lotic crayfish invasion: the roles of hydrologic variability and habitat alteration. Freshw. Biol. 48: 1886–1897.

Light, T., D.C. Erman, C. Myrick and J. Clarke. 1995. Decline of the shasta crayfish (*Pacifastacus fortis* Faxon) of northeastern California. Conserv. Biol. 9: 1567–1577.

Liu, S., S. Gong, J. Li and W. Huang. 2013. Effects of water temperature, photoperiod, eyestalk ablation, and non-hormonal treatments on spawning of ovary-mature red swamp crayfish. N. Am. J. Aquac. 75: 228–234.

Liu, X., Z. Guo, Z. Ke, S. Wang and Y. Li. 2011. Increasing potential risk of a global aquatic invader in Europe in contrast to other continents under future climate change. PLoS One 6: e18429.

Lodge, D.M. 1991. Herbivory on freshwater macrophytes. Aquat. Bot. 41: 195–224.

Lodge, D.M. and J.G. Lorman. 1987. Reductions in submerged macrophyte biomass and species richness by the crayfish *Orconectes rusticus*. Can. J. Fish. Aquat. Sci. 44: 591–597.

Lodge, D.M. and A.H. Hill. 1994. Factors governing species composition, population size, and productivity of cool-water crayfishes. Nord. J. Freshw. Res. 69: 111–136.

Lohrer, A.M. and R.B. Whitlatch. 2002. Interactions among aliens: apparent replacement of one exotic species by another. Ecology 83: 719–732.

Longshaw, M. 2011. Diseases of crayfish: a review. J. Invertebr. Pathol. 106: 54–70.

Longshaw, M., K.S. Bateman, P. Stebbing, G.D. Stentiford and F.A. Hockley. 2012. Disease risks associated with the importation and release of non-native crayfish species into mainland Britain. Aquat. Biol. 16: 1–15.

Lorman, J.G. and J.J. Magnuson. 1978. The role of crayfishes in aquatic ecosystem. Fisheries 3: 8–10.

Loughman, Z.J. 2010. Ecology of *Cambarus dubius* (upland burrowing crayfish) in North-Central West Virginia. Southeast. Nat. 9: 217–230.

MacIsaac, H.J. 1994. Size-selective predation on zebra mussels (*Dreissena polymorpha*) by crayfish (*Orconectes propinquus*). J. North Am. Benthol. Soc. 13: 206–216.

Maguire, I., G.I.V. Klobučar and R. Erben. 2005. The relationship between female size and egg size in the freshwater crayfish *Austropotamobius torrentium*. Bull. Français la Pêche la Piscic. 777–785.

Malley, D.F. 1980. Decreased survival and calcium-uptake by the crayfish *Orconectes virilis* in low pH. Can. J. Fish. Aquat. Sci. 37: 364–372.

Malley, D.F. and P.S.S. Chang. 1985. Effects of aluminum and acid on calcium uptake by the crayfish *Orconectes virilis*. Arch. Environ. Contam. Toxicol. 14: 739–747.

Marcott, S.A., J.D. Shakun, P.U. Clark and A.C. Mix. 2013. A reconstruction of regional and global temperature for the past 11,300 years. Science 339: 1198–201.

Mason, J.C. 1979. Significance of egg size in the freshwater crayfish, *Pacifastacus leniusculus* (Dana). Freshw. Crayfish 4: 83–92.

Mather, M.E. and R.A. Stein. 1993. Direct and indirect effects of fish predation on the replacement of a native crayfish by an invading congener. Can. J. Fish. Aquat. Sci. 50: 1279–1288.

Mauro, N.A. and G.W. Moore. 1987. Effects of environmental pH on ammonia excretion, blood pH, and oxygen uptake in fresh water crustaceans. Comp. Biochem. Physiol. Part C Comp. Pharmacol. 87: 1–3.

May, R.M. 1973. Stability and complexity in model ecosystems. Monogr. Popul. Biol. 6: 1–235.

McCann, K. and A. Hastings. 1997. Re-evaluating the omnivory–stability relationship in food webs. Proc. R. Soc. B Biol. Sci. 264: 1249–1254.

McMahon, B.R. 2002. Physiological adaption to environment. pp. 327–376. *In*: D.M. Holdich (ed.). Biology of Freshwater Crayfish. Blackwell Science, Oxford, England.

Meade, M.E. and S.A. Watts. 1995. Toxicity of ammonia, nitrite, and nitrate to juvenile Australian crayfish, *Cherax quadricarinatus*. J. Shellfish Res. 14: 341–346.

Meade, M.E., J.E. Doeller, D. Kraus and S.A. Watts. 2002. Effects of temperature and salinity on weight gain, oxygen consumption rate, and growth efficiency in juvenile red-claw crayfish *Cherax quadricarinatus*. J. World Aquac. Soc. 33: 188–198.

Meineri, E., H. Rodriguez-Perez, S. Hilaire and F. Mesleard. 2014. Distribution and reproduction of *Procambarus clarkii* in relation to water management, salinity and habitat type in the Camargue. Aquat. Conserv. Mar. Freshw. Ecosyst. 24: 312–323.

Melero, Y., S. Palazón and X. Lambin. 2014. Invasive crayfish reduce food limitation of alien American mink and increase their resilience to control. Oecologia 174: 427–434.

Momot, W. and H. Gowing. 1977. Production and population-dynamics of crayfish *Orconectes-virilis* in 3 michigan lakes. J. Fish. Res. Board Canada 34: 2041–2055.

Momot, W., H. Gowing and P. Jones. 1978. The dynamics of crayfish and their role in ecosystems. Am. Midl. Nat. 99: 10–35.

Momot, W.T. 1967. Population dynamics and productivity of the crayfish, *Orconectes virilis*, in a marl lake. Am. Midl. Nat. 78: 55–81.

Momot, W.T. 1984. Crayfish production: A reflection of community energetics. J. Crustac. Biol. 4: 35–54.

Momot, W.T. 1995. Redefining the role of crayfish in aquatic ecosystems. Rev. Fish. Sci. 3: 33–63.

Momot, W.T. and H. Gowing. 1972. Differential seasonal migration of the crayfish, *Orconectes virilis* (Hagen), in marl lakes. Ecology 53: 479–483.

Momot, W.T. and H. Gowing. 1983. Some factors regulating cohort production of the crayfish, *Orconectes virilis*. Freshw. Biol. 13: 1–12.

Montesinos, A., F. Santoul and A.J. Green. 2008. The diet of the night heron and purple heron in the Guadalquivir Marshes. Ardeola 55: 161–167.

Mora, C., D.P. Tittensor, S. Adl, A.G.B. Simpson and B. Worm. 2011. How many species are there on earth and in the ocean? PLoS Biol. 9: e1001127.

Morgan, D.O. and B.R. McMahon. 1982. Acid tolerance and effects of sublethal acid exposure on iono-regulation and acid-base status in two crayfish *Procambarus clarkii* and *Orconectes rusticus*. J. Exp. Biol. 97: 241–252.

Morrissy, N.M. 1978. The past and present distribution of marron, *Cherax tenuimanus* (Smith), in Western Australia. Fish. Res. Bull. (West. Aust.) 22: 1–38.

Mutshinda, C.M., R.B. O'Hara and I.P. Woiwod. 2009. What drives community dynamics? Proc. Biol. Sci. 276: 2923–2929.

Nakata, K. and S. Goshima. 2006. Asymmetry in mutual predation between the endangered japanese native crayfish *Cambaroides japonicus* and the north american invasive crayfish *Pacifastacus leniusculus*: a possible reason for species replacement. J. Crustac. Biol. 26: 134–140.

Nakata, K., K. Tsutsumi, T. Kawai and S. Goshima. 2005. Coexistence of two North American invasive crayfish species, *Pacifastacus leniusculus* (Dana, 1852) and *Procambarus clarkii* (Girard, 1852) in Japan. Crustaceana 78: 1389–1394.

Neveu, A. 2001. Can resident carnivorous fishes slow down introduced alien crayfish spread? Efficacy of 3 fish species versus 2 crayfish species in experimental design. Bull. Français la Pêche la Piscic. 683–704.

Nyström, P. and J.A. Strand. 1996. Grazing by a native and an exotic crayfish on aquatic macrophytes. Freshw. Biol. 36: 673–682.

Nyström, P., C. Bronmark and W. Graneli. 1996. Patterns in benthic food webs: a role for omnivorous crayfish? Freshw. Biol. 36: 631–646.

Nyström, P., C. Bronmak and W. Graneli. 1999. Influence of an exotic and a native crayfish species on a littoral benthic community. Oikos 85: 545–553.

Nyström, P., P. Stenroth, N. Holmqvist, O. Berglund, P. Larsson and W. Granéli. 2006. Crayfish in lakes and streams: individual and population responses to predation, productivity and substratum availability. Freshw. Biol. 51: 2096–2113.

O'Connor, J.E., J.J. Duda and G.E. Grant. 2015. 1000 dams down and counting. Science 348: 496–497.

Ogada, M.O. 2006. Effects of the Louisiana crayfish invasion and other human impacts on the African Clawless otter in the Ewaso Ng'iro ecosystem. Ph.D. thesis, Kenyatta University, Nairobi, Kenya.

Olden, J.D., M.J. Vander Zanden and P.T.J. Johnson. 2011. Assessing ecosystem vulnerability to invasive rusty crayfish (*Orconectes rusticus*). Ecol. Appl. 21: 2587–2599.

Olsson, K. and P. Nyström. 2009. Non-interactive effects of habitat complexity and adult crayfish on survival and growth of juvenile crayfish (*Pacifastacus leniusculus*). Freshw. Biol. 54: 35–46.

Oluoch, A.O. 1990. Breeding biology of the Louisiana red swamp crayfish *Procambarus clarkii* Girard in Lake Naivasha, Kenya. Hydrobiologia 208: 85–92.

Ormerod, S.J., M. Dobson, A.G. Hildrew and C.R. Townsend. 2010. Multiple stressors in freshwater ecosystems. Freshw. Biol. 55: 1–4.

Parkyn, S.M., K.J. Collier and B.J. Hicks. 2001. New Zealand stream crayfish: functional omnivores but trophic predators? Freshw. Biol. 46: 641–652.

Parkyn, S.M., K.J. Collier and B.J. Hicks. 2002. Growth and population dynamics of crayfish *Paranephrops planifrons* in streams within native forest and pastoral land uses. New Zeal. J. Mar. Freshw. Res. 36: 847–862.

Peay, S. and L. Füreder. 2011. Two indigenous European crayfish under threat—how can we retain them in aquatic ecosystems for the future? Knowl. Manag. Aquat. Ecosyst. 401: 33.

Peay, S., N. Guthrie, J. Spees, E. Nilsson and P. Bradley. 2009. The impact of signal crayfish (*Pacifastacus leniusculus*) on the recruitment of salmonid fish in a headwater stream in Yorkshire, England. Knowl. Manag. Aquat. Ecosyst. 394-395: 12.

Peters, J.A. and D.M. Lodge. 2013. Habitat, predation, and coexistence between invasive and native crayfishes: prioritizing lakes for invasion prevention. Biol. Invasions 15: 2489–2502.

Phillips, D.J.H. and P.S. Rainbow. 1993. Biomonitoring of trace aquatic comtaminants. Elsevier Applied Science, London, England.

Pinder, A.M., S.A. Halse, J.M. McRae and R.J. Shiel. 2005. Occurrence of aquatic invertebrates of the wheatbelt region of Western Australia in relation to salinity. Hydrobiologia 543: 1–24.

Polis, G.A. and D.R. Strong. 1996. Food web complexity and community dynamics. Am. Nat. 147: 813–846.

Rabeni, C.F. 1985. Resource partitioning by stream-dwelling crayfish: the influence of body size. Am. Midl. Nat. 113: 20–29.

Rahel, F.J. and R.A. Stein. 1988. Complex predator-prey interactions and predator intimidation among crayfish, piscivorous fish, and small benthic fish. Oecologia 75: 94–98.

Rainbow, P.S. 1995. Biomonitoring of heavy metal availability in the marine environment. Mar. Pollut. Bull. 31: 183–192.

Raupach, M.R., G. Marland, P. Ciais, C. Le Quéré, J.G. Canadell, G. Klepper and C.B. Field. 2007. Global and regional drivers of accelerating CO_2 emissions. Proc. Natl. Acad. Sci. U.S.A. 104: 10288–10293.

Reddy, P.S., S.R. Tuberty and M. Fingerman. 1997. Effects of cadmium and mercury on ovarian maturation in the red swamp crayfish, *Procambarus clarkii*. Ecotoxicol. Environ. Saf. 37: 62–65.

Renz, M. and T. Breithaupt. 2000. Habitat use of the crayfish *Austropotamobius torrentium* in small brooks and in Lake Constance, Southern Germany. Bull. Français la Pêche la Piscic. 356: 139–154.

Reynolds, J.D. 2011. A review of ecological interactions between crayfish and fish, indigenous and introduced. Knowl. Manag. Aquat. Ecosyst. 401: 10.

Rhodes, C.P. and D.M. Holdich. 1982. Observations on the fecundity of the freshwater crayfish, *Austropotamobius pallipes* (Lereboullet) in the British Isles. Hydrobiologia 89: 231–236.

Rice, P.R. and K.B. Armitage. 1974. The effect of photoperiod on oxygen consumption of the crayfish *Orconectes nais* (Faxon). Comp. Biochem. Physiol. Part A Physiol. 47: 261–270.

Richman, N.I., M. Böhm, S.B. Adams, F. Alvarez, E.a. Bergey, J.J.S. Bunn, Q. Burnham, J. Cordeiro, J. Coughran, K.A. Crandall, K.L. Dawkins, R.J. DiStefano, N.E. Doran, L. Edsman, A.G. Eversole, L. Fureder, J.M. Furse, F. Gherardi, P. Hamr, D.M. Holdich, P. Horwitz, K. Johnston, C.M. Jones, J.P.G. Jones, R.L. Jones, T.G. Jones, T. Kawai, S. Lawler, M. Lopez-Mejia, R.M. Miller, C. Pedraza-Lara, J.D. Reynolds, A.M.M. Richardson, M.B. Schultz, G.A. Schuster, P.J.

Sibley, C. Souty-Grosset, C.A. Taylor, R.F. Thoma, J. Walls, T.S. Walsh and B. Collen. 2015. Multiple drivers of decline in the global status of freshwater crayfish (Decapoda: Astacidea). Philos. Trans. R. Soc. B Biol. Sci. 370: 20140060.

Robinson, C.A., T.J. Thom and M.C. Lucas. 2000. Ranging behaviour of a large freshwater invertebrate, the white-clawed crayfish *Austropotamobius pallipes*. Freshw. Biol. 44: 509–521.

Rodríguez, C.F., E. Bécares and M. Fernández-Aláez. 2007. Shift from clear to turbid phase in Lake Chozas (NW Spain) due to the introduction of American red swamp crayfish (*Procambarus clarkii*). Hydrobiol. 506-509: 421–426.

Rodríguez, C.F., E. Bécares, M. Fernández-Aláez and C. Fernández-Aláez. 2005. Loss of diversity and degradation of wetlands as a result of introducing exotic crayfish. Biol. Invasions 7: 75–85.

Rodríguez-González, H., M. García-Ulloa, A. Hernández-Llamas and H. Villarreal. 2006. Effect of dietary protein level on spawning and egg quality of redclaw crayfish *Cherax quadricarinatus*. Aquaculture 257: 412–419.

Rodríguez-González, H., H. Villarreal, M. García-Ulloa and A. Hernández-llamas. 2009. Dietary lipid requirements for optimal egg quality of redclaw crayfish, *cherax quadricarinatus*. J. World Aquac. Soc. 40: 531–539.

Rogowski, D.L., S. Sitko and S.A. Bonar. 2013. Optimising control of invasive crayfish using life-history information. Freshw. Biol. 58: 1279–1291.

Roldan, B.M. and R.R. Shivers. 1987. The uptake and storage of iron and lead in cells of the crayfish (*Orconectes propinquus*) hepatopancreas and antennal gland. Comp. Biochem. Physiol. C. 86: 201–214.

Rosewarne, P.J., J.C. Svendsen, R.J.G. Mortimer and A.M. Dunn. 2014. Muddied waters: suspended sediment impacts on gill structure and aerobic scope in an endangered native and an invasive freshwater crayfish. Hydrobiologia 722: 61–74.

Rudnick, D. and V. Resh. 2005. Stable isotopes, mesocosms and gut content analysis demonstrate trophic differences in two invasive decapod crustacea. Freshw. Biol. 50: 1323–1336.

Ruokonen, T.J., M. Kiljunen, J. Karjalainen and H. Hämäläinen. 2012. Invasive crayfish increase habitat connectivity: a case study in a large boreal lake. Knowl. Manag. Aquat. Ecosyst. 407: 08.

Sáez-Royuela, M., J.M. Carral, J. Celada, J.R. Pérez and A. González. 2006. Pleopodal egg production of the white-clawed crayfish *Austropotamobius pallipes* Lereboullet under laboratory conditions: relationship between egg number, egg diameter and female size. Bull. Français la Pêche la Piscic. 380-381: 1207–1214.

Savino, J.F. and J.E. Miller. 1991. Crayfish (*Orconectes-virilis*) feeding on young lake trout (*Salvelinus-namaycush*)—Effect of rock size. J. Freshw. Ecol. 6: 161–170.

Scalici, M. and F. Gherardi. 2007. Structure and dynamics of an invasive population of the red swamp crayfish (*Procambarus clarkii*) in a Mediterranean wetland. Hydrobiologia 583: 309–319.

Scalici, M., A. Belluscio and G. Gibertini. 2008. Understanding population structure and dynamics in threatened crayfish. J. Zool. 275: 160–171.

Schindler, D.W. and M.A. Turner. 1982. Biological, chemical and physical responses of lakes to experimental acidification. Water. Air. Soil Pollut. 18: 259–271.

Scholtz, G., A. Braband, L. Tolley, A. Reimann, B. Mittmann, C. Lukhaup, F. Steuerwald and G. Vogt. 2003. Parthenogenesis in an outsider crayfish. Nature 421: 806–806.

Schulz, R., T. Stucki and C. Souty-Grosset. 2002. Roundtable Session 4A. Management: reintroductions and restocking. Bull. Fr. Pêche Piscic. 367: 917–922.

Seiler, S.M. and A.M. Turner. 2004. Growth and population size of crayfish in headwater streams: individual- and higher-level consequences of acidification. Freshw. Biol. 49: 870–881.

Setzer, M., J.R. Norrgård and T. Jonsson. 2011. An invasive crayfish affects egg survival and the potential recovery of an endangered population of Arctic charr. Freshw. Biol. 56: 2543–2553.

Siewert, H.F. and J.P. Buck. 1991. Effects of Low pH on survival of crayfish (*Orconectes virilis*). J. Freshw. Ecol. 6: 87–91.

Simberloff, D. and B. Von Holle. 1999. Positive interactions of nonindigenous species: invasional meltdown? Biol. Invasions 1: 21–32.

Simon, O. and A. Boudou. 2001. Simultaneous experimental study of direct and direct plus trophic contamination of the crayfish *Astacus astacus* by inorganic mercury and methylmercury. Environ. Toxicol. Chem. 20: 1206–1215.

Skurdal, J., D.O. Hessen, E. Garnås and L.A. Vøllestad. 2011. Fluctuating fecundity parameters and reproductive investment in crayfish: driven by climate or chaos? Freshw. Biol. 56: 335–341.

Smart, A.C., D.M. Harper, F. Malaisse, S. Schmitz, S. Coley and A.C. Gouder De Beauregard. 2002. Feeding of the exotic Louisiana red swamp crayfish, *Procambarus clarkii* (Crustacea, Decapoda), in an African tropical lake: Lake Naivasha, Kenya. Hydrobiologia 488: 129–142.

Smith, G.R.T., M.A. Learner, F.M. Slater and J. Foster. 1996. Habitat features important for the conservation of the native crayfish *Austropotamobius pallipes* in Britain. Biol. Conserv. 75: 239–246.

Statzner, B., O. Peltret and S. Tomanova. 2003. Crayfish as geomorphic agents and ecosystem engineers: effect of a biomass gradient on baseflow and flood-induced transport of gravel and sand in experimental streams. Freshw. Biol. 48: 147–163.

Stein, R.A. 1977. Selective predation, optimal foraging, and the predator-prey interaction between fish and crayfish. Ecology 58: 1237–1253.

Stenroth, P. and P. Nyström. 2003. Exotic crayfish in a brown water stream: effects on juvenile trout, invertebrates and algae. Freshw. Biol. 48: 466–475.

Stenroth, P., N. Holmqvist, P. Nyström, O. Berglund, P. Larsson and W. Graneli. 2006. Stable isotopes as an indicator of diet in omnivorous crayfish (*Pacifastacus leniusculus*): the influence of tissue, sample treatment, and season. Can. J. Fish. Aquat. Sci. 63: 821–831.

Strayer, D.L. 2010. Alien species in fresh waters: ecological effects, interactions with other stressors, and prospects for the future. Freshw. Biol. 55: 152–174.

Strayer, D.L. and D. Dudgeon. 2010. Freshwater biodiversity conservation: recent progress and future challenges. J. North Am. Benthol. Soc. 29: 344–358.

Strayer, D.L. and H.M. Malcom. 2012. Causes of recruitment failure in freshwater mussel populations in southeastern New York. Ecol. Appl. 22: 1780–1790.

Styrishave, B., B.H. Bojsen, H. Witthøfft and O. Andersen. 2007. Diurnal variations in physiology and behaviour of the noble crayfish *Astacus astacus* and the signal crayfish *Pacifastacus leniusculus*. Mar. Freshw. Behav. Physiol. 40: 63–77.

Suárez-Serrano, A., C. Alcaraz, C. Ibáñez, R. Trobajo and C. Barata. 2010. *Procambarus clarkii* as a bioindicator of heavy metal pollution sources in the lower Ebro River and Delta. Ecotoxicol. Environ. Saf. 73: 280–286.

Svobodová, J., K. Douda, M. Štambergová, J. Picek, P. Vlach and D. Fischer. 2012. The relationship between water quality and indigenous and alien crayfish distribution in the Czech Republic: patterns and conservation implications. Aquat. Conserv. Mar. Freshw. Ecosyst. 22: 776–786.

Tablado, Z., J.L. Tella, J.A. Sánchez-Zapata and F. Hiraldo. 2010. The paradox of the long-term positive effects of a north american crayfish on a european community of predators. Conserv. Biol. 24: 1230–1238.

Taylor, R.C. 1983. Drought-induced changes in crayfish populations along a stream continuum. Am. Midl. Nat. 110: 286–298.

Taylor, R.C. 1988. Population dynamics of the crayfish *Procambarus spiculifer* observed in different-sized streams in response to two droughts. J. Crustac. Biol. 8: 401–409.

Tropea, C., M. Arias, N.S. Calvo and L.S. López Greco. 2012. Influence of female size on offspring quality of the freshwater crayfish *Cherax quadricarinatus* (Parastacidae: Decapoda). J. Crustac. Biol. 32: 883–890.

Trouilhé, M.-C., G. Freyssinel, J. Jandry, M. Brulin, B. Parinet, C. Souty-Grosset and F. Grandjean. 2012. The relationship between Ephemeroptera and presence of the white-clawed crayfish (*Austropotamobius pallipes*). Case study in the Poitou-Charentes region (France). Fundam. Appl. Limnol. 179: 293–303.

Twardochleb, L.A., J.D. Olden and E.R. Larson. 2013. A global meta-analysis of the ecological impacts of nonnative crayfish. Freshw. Sci. 32: 1367–1382.

Usio, N. 2000. Effects of crayfish on leaf processing and invertebrate colonisation of leaves in a headwater stream: decoupling of a trophic cascade. Oecologia 124: 608–614.

Usio, N. and C.R. Townsend. 2000. Distribution of the New Zealand crayfish *Paranephrops zealandicus* in relation to stream physical chemistry, predator fish and invertebrate prey. New Zeal. J. Mar. Freshw. Res. 34: 557–567.

Usio, N. and C.R. Townsend. 2002. Functional significance of crayfish in stream food webs: roles of omnivory, substrate heterogeneity and sex. Oikos 98: 512–522.

Usio, N., R. Kamiyama, A. Saji and N. Takamura. 2009. Size-dependent impacts of invasive alien crayfish on a littoral marsh community. Biol. Conserv. 142: 1480–1490.

Vogt, G., L. Tolley and G. Scholtz. 2004. Life stages and reproductive components of the marmorkrebs (marbled crayfish), the first parthenogenetic decapod Crustacean. J. Morphol. 261: 286–311.

Ward, R.J.S., C.R. McCrohan and K.N. White. 2006. Influence of aqueous aluminium on the immune system of the freshwater crayfish *Pacifasticus leniusculus*. Aquat. Toxicol. 77: 222–228.

Welch, S.M. and A.G. Eversole. 2006. The occurrence of primary burrowing crayfish in terrestrial habitat. Biol. Conserv. 130: 458–464.

Westin, L. and R. Gydemo. 1986. Influence of light and temperature on reproduction and moulting frequency of the crayfish, *Astacus astacus* L. Aquaculture 52: 43–50.

Whitledge, G.W. and C.F. Rabeni. 1997. Energy sources and ecological role of crayfishes in an Ozark stream: insights from stable isotopes and gut analysis. Can. J. Fish. Aquat. Sci. 54: 2555–2563.

Whitmore, N. and A.D. Huryn. 1999. Life history and production of *Paranephrops zealandicus* in a forest stream, with comments about the sustainable harvest of a freshwater crayfish. Freshw. Biol. 42: 467–478.

Williams, R.J., E.L. Berlow, J.a. Dunne, A.-L. Barabási and N.D. Martinez. 2002. Two degrees of separation in complex food webs. Proc. Natl. Acad. Sci. U.S.A. 99: 12913–12916.

Willis-Jones, W.E. 2013. The indirect impacts of bioturbation by red swamp crayfish, *Procambarus clarkii*. M.Sc. Thesis, Queen Mary University of London, London.

Wilson, K.A., J.J. Magnuson, D.M. Lodge, A.M. Hill, T.K. Kratz, W.L. Perry and T.V. Willis. 2004. A long-term rusty crayfish (*Orconectes rusticus*) invasion: Dispersal patterns and community change in a north temperate lake. Can. J. Fish. Aquat. Sci. 61: 2255–2266.

Wonham, M.J., M. O'Connor and C.D.G. Harley. 2005. Positive effects of a dominant invader on introduced and native mudflat species. Mar. Ecol. Prog. Ser. 289: 109–116.

Woodburn, K., R. Walton, C. McCrohan and K. White. 2011. Accumulation and toxicity of aluminium-contaminated food in the freshwater crayfish, *Pacifastacus leniusculus*. Aquat. Toxicol. 105: 535–542.

Woodward, G., G. Papantoniou and R.B. Lauridsen. 2008. Trophic trickles and cascades in a complex food web: impacts of a keystone predator on stream community structure and ecosystem processes. Oikos 117: 683–692.

Yildiz, H.Y. and A.C.K. Benli. 2004. Nitrite toxicity to crayfish, *Astacus leptodactylus*, the effects of sublethal nitrite exposure on hemolymph nitrite, total hemocyte counts, and hemolymph glucose. Ecotoxicol. Environ. Saf. 59: 370–375.

Zimmerman, J.K.M. and R.T. Palo. 2012. Time series analysis of climate-related factors and their impact on a red-listed noble crayfish population in northern Sweden. Freshw. Biol. 57: 1031–1041.

zu Ermgassen, P.S.E. and D.C. Aldridge. 2011. Predation by the invasive American signal crayfish, *Pacifastacus leniusculus* Dana, on the invasive zebra mussel, *Dreissena polymorpha* Pallas: The potential for control and facilitation. Hydrobiologia 658: 303–315.

Field Sampling Techniques for Crayfish

Eric R. Larson[1],* and *Julian D. Olden*[2]

Introduction

Why do we study crayfish? Answers may range from their cultural or economic value (Jones et al. 2006) to ecological importance (Usio and Townsend 2004) to the high conservation need of many species (Taylor et al. 2007). Alternatively, one justification for the emergence of crayfish as model organisms (e.g., Crandall 2000) has been their ubiquity and ease of collection relative to a rewarding range of biological insights. For example, Thomas Henry Huxley (1884) in his introduction to the study of zoology framed his book around crayfish in part because the "[the crayfish] is readily obtained." Yet those of us who need to quantitatively sample crayfish recognize that "readily obtained" does not necessarily translate into representative of broader populations or communities. Huxley (1884) might be countered by observations like those of Rabeni et al. (1997), who note that no crayfish sampling method is without biases that may misrepresent attributes ranging from relative abundance to size and age structure of populations. Ultimately, quantitative sampling for crayfish presents a number of challenges to confound even the most experienced of field biologists.

Writing a comprehensive review of field sampling methods for crayfish must accommodate both the diversity of crayfish themselves and the diversity of researchers interested in them. Crayfish occur in habitats ranging from large lakes to wadeable streams, to difficult to sample environments like caves and terrestrial burrows. These different habitats demand different sampling tools and approaches. Further, widely varying research objectives justify a need to sample crayfish, from studies of evolution or phylogeography (Trontelj et al. 2005) to bioassessments of freshwater habitats (Reynolds and Souty-Grosset 2012) to surveys of diseases like the crayfish plague (Holdich and Reeve 1991) or commensals like crayfish worms (Williams et al. 2009). Accordingly, a wide variety of field sampling methodologies for crayfish have been developed and tested over the past century. We are unaware, however, of any comprehensive review that has attempted to make sense of how to sample for crayfish, or at a minimum provide a thorough bibliography to serve as a foundation for researchers seeking to work with and sample crayfish in the field (but see Parkyn 2015 for a synthesis of the most common crayfish sampling techniques reported in the journal *Freshwater Crayfish*).

[1] Department of Natural Resources and Environmental Sciences, University of Illinois, Urbana, IL 61801 USA.
[2] School of Aquatic and Fishery Sciences, University of Washington, Seattle, WA 98195 USA.
  Email: olden@uw.edu
* Corresponding author: erlarson@illinois.edu

In our review, we've attempted to represent insights on field sampling for crayfish from a breadth of researchers in different countries and continents working on a wide range of organisms and questions. However, we should first identify the experiences and accompanying biases we bring to the task of writing a review on field sampling methods for crayfish. We are freshwater ecologists interested in animal distributions and ecological processes at landscape scales. Accordingly, our perspectives may skew toward sampling and monitoring for studying trends in crayfish populations over relatively large areas, whether for conservation of native species or management of invasive crayfishes (e.g., Olden et al. 2006, 2009, 2011, Larson and Olden 2008, 2013, Larson et al. 2012). Further, we work predominantly in North America, and may be most familiar with crayfish literature from the United States and Canada. Consequently, we may give unintentional short shrift to biologists interested in collecting or sampling crayfish for other reasons, in habitats other than the streams and lakes we most typically work in, or using novel sampling tools and techniques we have not encountered. These caveats aside, our review of the literature is extensive and our hope is that we've provided a useful roadmap for researchers starting out on the task of quantitative field sampling for crayfish.

We also have not set out to write a book chapter on field sampling in general. For that need, we direct readers to texts on sampling design and methods for the ecological sciences or freshwater fisheries (e.g., Morrison et al. 2008, Bonar et al. 2009, Magurran and McGill 2011, Zale et al. 2013). Further, we note that many crayfish of conservation concern may be difficult to detect on the landscape, and recent research and publications on sampling for rare or cryptic species may be worth consulting (e.g., Thompson 2004, MacKenzie et al. 2006). We instead focus on two areas: a review of approaches for sampling crayfish in disparate habitats, and a summary of some important considerations for planning a field sampling program for crayfish. The first of these sections outlines crayfish sampling by habitat type: lentic or large lotic environments; wadeable streams and rivers; terrestrial environments; and caves. The second of these sections emphasizes methods to evaluate accuracy of field sampling techniques for crayfish via sampling efficiency; precision via power analysis; reliability of occupancy estimates by quantifying detection probability; and the use of mark-recapture methods to not only monitor crayfish populations but also evaluate other sampling approaches. We conclude with a brief section on the value and potential benefits of improving transferability and transparency of results between different studies and regions by standardizing crayfish sampling methodologies.

Sampling for crayfish by habitat type

Crayfish in lentic and large lotic environments

Crayfish are ecologically and economically important organisms in many large waterbodies globally, from tropical to temperate lakes (Harper et al. 2002, Jansen et al. 2009) and major rivers (Larson et al. 2010b). These habitats provide unique challenges for crayfish sampling relative to more tractable wadeable streams and rivers, where a diversity of sampling methods of different merits are available (Rabeni et al. 1997). Conversely, lentic or large lotic environments typically necessitate sampling by one of three approaches: baited traps; visual searches or collection by divers or snorkelers; and throw traps. Of these, baited traps are perhaps the most commonly applied and also potentially the most problematic, with known biases favoring large adults and males over other members of the population (e.g., Brown and Brewis 1978, Capelli and Magnuson 1983). Related to baited trapping, installation and recovery of habitat structures or "bundles" has been suggested as an alternative or complementary approach that may sample under-represented members of the population or community (Parkyn et al. 2011). Dive or snorkel surveys are perhaps more accurate than trapping in representing population attributes or abundance (Lamontagne and Rasmussen 1993), but seem less commonly used owing to the required technical expertise (e.g., SCUBA certification) and the time- and labor-intensive nature of this method. Moreover, environments characterized by poor light conditions and low water clarity may limit the application of underwater surveys. Surveys of some lentic environments like freshwater marshes and wetlands have also been conducted with throw traps, a less commonly applied and tested methodology (but see Dorn et al. 2005).

Figure 1. Examples of some sampling techniques for crayfish commonly used in lentic or large lotic environments. Baited trapping for crayfish with modified Gee minnow traps, showing retrieval of a trap (A) and two separated trap pieces with collected signal crayfish *Pacifastacus leniusculus* (B). Visual searches for crayfish by snorkelers (C) with a collected crayfish (D). All photographs by the authors.

Baited traps applied to sample for crayfish vary by design and dimensions between studies and countries, from the Swedish trappy commonly used in Europe (e.g., Edsman and Söderbäck 1999) to modified (i.e., enlarged openings) Gee minnow traps in North America (e.g., Capelli and Magnuson 1983) to baited hoop nets used in Australia (e.g., de Graaf et al. 2010, Fulton et al. 2012). In many cases these traps are fundamentally similar. For example, Harlioĝlu (1999) gives dimensions of the Swedish trappy as an 0.5 m long mesh cylinder of 0.2 m diameter with funnels at either end with 4.5 cm diameter openings, whereas Larson and Olden (2013) give dimensions for modified Gee minnow traps as an 0.42 m long mesh cylinder of 0.21 m diameter with 6.0 cm openings. Both Stuecheli (1999) and Huner and Espinoza (2004) demonstrated how choice of trap opening diameter can have implications for the number, size, and sex of crayfish caught. These studies found that smaller diameter openings tend to trap smaller crayfish and larger openings larger crayfish; that larger diameter openings may disproportionally favor male crayfish (Stuecheli 1999); and that different diameter openings may select for different crayfish species (Huner and Espinoza 2004). There has been persistent interest over time in "building a better crayfish trap," with numerous alternative designs proposed (e.g., Slater 1995, Mangan et al. 2009), although many of these do not seem to have gained wide popularity or application. An exception may be the triangular or pyramid trap designs used in commercial harvest of crayfish (e.g., Huner and Espinoza 2004). Many studies have noted that crayfish trap design—including attributes like trap shape, entrance funnel diameter or slope, or size of trap mesh—can have implications for crayfish recruitment to and retention in traps (Westman et al. 1978a, Fjälling 1995, Mangan et al. 2009). Yet most contemporary studies seem to use fairly standard and commercially available crayfish traps, a potential asset for transferability of results between studies and regions (see below). For additional comparisons of performance of different crayfish trap designs, see also Bean and Huner (1979), Whisson and Campbell (2000), and Khanipour and Melnikov (2007).

Studies trapping to sample for crayfish often differ by the bait used. A cursory review of this literature provides wide ranging preferences for bait types, including fish (Taugbøl et al. 1997), beef liver (Capelli and Magnuson 1983), commercial crayfish baits (Cange et al. 1986), canned cat food (Edwards et al. 2009), dried dog food (Larson and Olden 2013), and mashed potato (Usio et al. 2006). Thomas Henry Huxley (1884) even offers an opinion, suggesting that aspiring zoologists might collect their crayfish with "hoop-nets baited with frogs." Quantitative evaluations of the effect of bait type on crayfish catch have produced contradictory results. Somers and Stechey (1986) found no significant differences in overall crayfish catch-per-unit effort (CPUE; the number of crayfish per trap) between traps baited with beef liver, chicken, fish, or dry dog food, but found interactions between bait type and the species and size of crayfish collected. Alternatively, Kutka et al. (1992) found that fish or a commercial crayfish bait (dependent on species) outperformed dry dog food, whereas Rach and Bills (1987) and Romaire and Osorio (1989) found that commercial crayfish baits outperformed fish. To further complicate matters, Huner et al. (1990) found little difference between traps baited with commercial crayfish pellets and fish but differences between fish species (see also Taugbøl et al. 1997), and Cange et al. (1986) found significant but very small differences in CPUE between 18 commercial crayfish baits and a fish control. Many of these studies share the finding that factors like crayfish species or water temperature can influence the efficacy of different crayfish baits (Somers and Stechey 1986, Kutka et al. 1992, Beecher and Romaire 2010). Romaire and Osorio (1989) observed no difference in effectiveness of trapping in response to bait quantity (150 g, 225 g, or 300 g). As there is unlikely to be a universal answer to "which crayfish bait is best?", we might suggest as above for trap design that any standardization of bait use between studies and regions would be invaluable for improving transferability of results, such as comparisons of relative abundance of crayfish as CPUE between native and non-native or invasive ranges (Larson et al. 2010a, Larson and Olden 2013). Other considerations for researchers in choosing a crayfish bait may include cost, availability, and ease of use, transport or disposal (Rach and Bills 1987). Finally, a meta-analysis (i.e., Gurevitch et al. 2001) on the influence of crayfish identity (species to family), habitat type, water temperature, or other factors on performance of different baits might find resolution in seemingly contradictory literature.

Baited crayfish traps are typically deployed overnight, with some reported durations spanning between 15 to 26 hours (e.g., Capelli and Magnuson 1983, Lewis 1997, Edwards et al. 2009, Larson and Olden 2013). Researchers should anticipate that CPUE will be higher for traps deployed overnight than during the day owing to the generally nocturnal nature of crayfish (but see Romaire and Osorio 1989). Lewis (1997) demonstrated CPUE for a reservoir population of the signal crayfish *Pacifastacus leniusculus* increased following sunset and declined following sunrise. Related, as crayfish have been noted to escape traps over time (Westman et al. 1978a), Harlioğlu (1999) suggested that traps should be checked and emptied at several intervals over the night. This suggestion has not to our knowledge been commonly adopted in North America, although some trapping implementations in other regions retrieve traps after short durations (Policar and Kozák 2005, de Graaf et al. 2010, Fulton et al. 2012). We suggest escape of crayfish from traps may be of greater interest to commercial harvesters seeking to maximize their catch or managers attempting to eradicate invasive crayfish (e.g., Hein et al. 2006), and of less interest to ecologists more concerned with whether CPUE results of any systematically applied methodology (e.g., duration) reliably correspond with presence or true abundance of crayfish populations (see below). Further, returning to and emptying traps multiple times over short durations may often be infeasible for studies at large spatial scales where study sites are separated by large distances (e.g., Edwards et al. 2009, Larson and Olden 2013). Trapping for crayfish is a passive sampling approach that is not only dependent on the overall abundance of crayfish but also their behavior (i.e., activity), and crayfish behavior will unavoidably interact with trapping results in potentially complex ways (Dorn et al. 2005). Beyond the potential for crayfish to exit traps, large and aggressive crayfish will defend traps as habitat and exclude smaller individuals thus reducing overall CPUE (Ogle and Kret 2008), and the presence of predatory fish may not only influence CPUE of crayfish trapping via overall crayfish abundance but also by changing crayfish behavior (Collins et al. 1983). In such cases, independent validation by alternative sampling methods (e.g., visual searches by divers) is recommended to parse the role of behavior relative to actual crayfish presence or density on CPUE (Collins et al. 1983, Somers and Green 1993).

Timing of trapping and associated environmental variables (e.g., temperature, lunar phase) may affect CPUE or species and size distributions of collected crayfish. A number of studies have documented a consistent relationship between water temperature and CPUE from trapping, with CPUE increasing with warmer temperatures (Somers and Stechey 1986, Araujo and Romaire 1989, Somers and Green 1993). Effects of other seasonal or temporal concerns like lunar phase or weather events have received more equivocal support. Araujo and Romaire (1989) used multivariate statistics to analyze patterns in crayfish CPUE among environmental axes, and found that temperature explained 85% of variation in crayfish CPUE with trivial variation explained by other factors like lunar phase or weather events (e.g., rain). Somers and Stechey (1986) similarly found minimal effect of lunar phase on crayfish CPUE, but noted that larger crayfish were more likely to be collected on moonlit nights. Season may also affect crayfish catch independent of factors like temperature. For example, some events in crayfish life histories (e.g., molting, reproduction) may affect CPUE or render some individuals (e.g., berried females) extremely difficult to detect (Malley and Reynolds 1979, Somers and Green 1993, Richards et al. 1996). Attempts to account for the effect of life history on estimates of relative abundance or population parameters (e.g., sex ratios) will need to accommodate the particular species or population of crayfish being studied (see chapters on growth and reproduction, behavior, and ecology in the current volume).

Given that temperature exerts such a strong influence on CPUE from baited trapping, most trapping studies for crayfish in temperate regions occur in warm months of summer or early autumn (e.g., Somers and Green 1993). Researchers sampling at these times in lentic systems have noted that crayfish are usually found in shallow, warm littoral habitats above the thermocline or metalimnion (Abrahamsson and Goldman 1970, Capelli and Magnuson 1976, Momot and Gowing 1972), although Lamontagne and Rasmussen (1993) reported that steeper lake beds supported crayfish at greater depths. Beyond temperature, crayfish may avoid soft or flocculent sediments common in deeper waters (Elser et al. 1994) or perhaps other limnological gradients like dissolved oxygen (e.g., anoxic hypolimnion). Some commonly used depths in crayfish trapping studies of lentic environments have included 1 m by both Puth and Allen (2004) and Roth et al. (2007), 1–3 m by Capelli and Magnuson (1983), 0.5–6 m by Larson and Olden (2013), and 0.5 to 8 m by Edwards et al. (2009). However, crayfish can occasionally use considerably deeper waters in both lentic and large lotic environments. For example, Flint (1977) reported seasonal migrations of *P. leniusculus* in Lake Tahoe, California and Nevada, USA, from shallow waters in summer to deeper waters (e.g., 40 m) over winter. Working with the same species in a large reservoir in Oregon, USA, Lewis (1997) collected crayfish to depths of 100 m but found 98% of the population in waters shallower than 70 m with peak relative abundance at 10–20 m.

Many studies have reported that crayfish in lentic or large lotic waters are most often associated with firm and rocky substrates (Capelli and Magnuson 1983, Garvey et al. 2003). Rock substrate often provides shelter to crayfish from both aquatic and terrestrial predators as well as cannibalism by other crayfish (Nyström et al. 2006, Olsson and Nyström 2009). Studies of temperate lakes have found lower crayfish densities in open or macrophyte-dominated habitats relative to rocky substrates (e.g., Kershner and Lodge 1995, Pilotto et al. 2008). Conversely, crayfish in lentic wetlands lacking rocky substrates may occur in extremely high densities in aquatic macrophytes (Dorn et al. 2005), and some crayfish species will prefer fine silt substrates over firmer, coarser substrates for burrowing (Dorn and Volin 2009). Habitat preferences or tolerances by a crayfish species may change between studies and study regions, as well. For example, *P. leniusculus* in its native range is more likely to occur in lakes with firm and rocky riparian zone substrates (Larson and Olden 2013), but Usio et al. (2006) documented that an invasive population of this crayfish burrowed into undercut banks in a wetland system in Japan where rocky substrates were absent. When a researcher is studying a crayfish species with known substrate preferences (e.g., *Orconectes rusticus*; Kershner and Lodge 1995), trapping where these substrates predominate is likely to improve probability of detection or increase CPUE. Such targeted trapping has been used to assist in the eradication of this invasive species from a lake where introduced in northern Wisconsin, USA (Hein et al. 2006). Conversely, for species where substrate preferences are not well-known or in studies seeking to evaluate habitat selection, researchers should sample representatively among available habitat types with a random or stratified sampling design (e.g., Lewis 1997).

There is little good guidance on the selection of the number of traps to use at a given site in a crayfish sampling effort. Several studies have attempted to quantify the area over which crayfish may recruit to baited traps. Abrahamsson and Goldman (1970) adapted a methodology from Cukerzis (1959) that related trap CPUE to adjacent density estimates from SCUBA surveys within 0.725 m^2 enclosures, calculating that a mean CPUE of 18.5 adult crayfish per trap in areas with known densities of 1.4 adult crayfish per m^2 corresponded with a sampling area of 13.0 m^2. Lewis (1997) used a more complicated field methodology and calculation in a similar study system and with the same species as Abrahamsson and Goldman (1970) to instead estimate capture ranges from mean 92 m^2 to 116 m^2 depending on habitat (substrate) types. Acosta and Perry (2000) used marked crayfish stocked at known distances from traps in a Florida, USA wetland to estimate an effective trap area of 56 m^2. Researchers regularly attempt to maintain independence by separating individual crayfish traps by a minimum distance, such as the 3 m used in linear trap lines of a standard sampling protocol in Ontario, Canada (David et al. 1994, Edwards et al. 2009) or the 10–30 m between traps used by Larson and Olden (2013). The overall intensity of trapping effort has varied by study and region, from a fairly typical range of 12–24 traps per individual temperate lake in the USA (Capelli and Magnuson 1983, Puth and Allen 2004, Roth et al. 2007), to 54 traps per lake in Canada (David et al. 1994, Edwards et al. 2009), to 100–120 traps per lake or river reach in Sweden (Nyström et al. 2006, Zimmerman and Palo 2011). Statistical simulation or power analysis (see below) based on empirical data from studies that sample with a high effort (e.g., Edwards et al. 2009) would be useful to evaluate the effect of trapping intensity on common attributes of interest such as species richness or mean CPUE (i.e., how does randomly omitting some trap data influence raw results or subsequent analyses). Experimental studies can also be used to evaluate the effect of trapping effort on reliability of CPUE as a measure of relative abundance. Recently, Zimmerman and Palo (2011) compared CPUE from trapping effort of 15 and 120 traps in reaches of a Swedish river to population estimates from mark-recapture, and found $R^2 = 0.23$ for the lower sampling effort and $R^2 = 0.46$ for the higher sampling effort. Notably, the lower trapping effort over-estimated abundance relative to the mark-recapture estimates (see below).

Trapping for crayfish is biased in representing different components of crayfish populations and communities. Researchers have consistently documented that baited trapping systematically under-represents small, juvenile, and female crayfish in favor of large, aggressive, male crayfish (Brown and Brewis 1978, Capelli and Magnuson 1983, Rabeni et al. 1997, Ogle and Kret 2008, Chucholl 2011). Because of these biases, some researchers like Lodge et al. (1986) have used CPUE of adult male crayfish as an index of population relative abundance instead of CPUE from all collected crayfish. Other studies have found baited trapping favors some crayfish species over others present in the community (Huner and Espinoza 2004, Price and Welch 2009, Parkyn et al. 2011). Diameter of entrances to traps may be varied to represent crayfish of different sizes, sexes or species (Lewis 1997, Stuecheli 1999), but such modifications are still unlikely to overcome the systematic misrepresentation of crayfish population or community structure by baiting trapping. We recommend against exclusive reliance on baited trapping for studies of crayfish life histories, although trapping might complement other sampling methodologies by sampling older and larger individuals (Rabeni et al. 1997, Harper et al. 2002, Price and Welch 2009, Chucholl 2011). Researchers have also recently proposed deploying artificial habitat structures or bundles along with baited traps to better represent components of the population neglected by trapping. Habitat structures consist of rock or wood bundles that are deployed and then colonized by crayfish prior to recovery. Some examples of habitat structures range from a traditional Māori design used in New Zealand (Kusabs and Quinn 2009) to an application by Warren et al. (2009) to sample crayfish from silty lowland streams of the Mississippi River delta, USA. Warren et al. (2009) deployed habitat structures for 14 days, whereas Kusabs and Quinn (2009) deployed their design for up to a month before recovery. Parkyn et al. (2011) evaluated performance of habitat bundles as an alternative or complement to baited crayfish traps in Missouri, USA, and found that the bundles expanded the representation of species collected and better represented female and juvenile crayfish in populations. Fjälling (2011) recently proposed a new "enclosure trap" specifically for trapping juvenile crayfish, which colonize enclosed substrate bundles that are deployed prior to release of eggs by female crayfish and recovered over monthly intervals. Peay (2003) similarly discussed use of un-baited refuge traps that work on the basis of crayfish colonizing traps for habitat.

How well do baited traps represent the overall abundance of crayfish populations in sampled habitats? Studies that have compared CPUE from traps as a measure of relative abundance to either known population sizes or alternative estimates of relative abundance have reported performance ranging from poor (e.g., Collins et al. 1983, Dorn et al. 2005) to excellent (e.g., Capelli and Magnuson 1983, Olsen et al. 1991). Capelli and Magnuson (1983) reported $R^2 = 0.97$ between CPUE from trapping and crayfish density estimates from visual diver surveys for lakes of Wisconsin, USA. Olsen et al. (1991) similarly compared CPUE from traps to visual survey estimates in the same lake district and found $R^2 = 0.71$ for all species and $R^2 = 0.95$ for *O. rusticus*. Olsen et al. (1991) also found that CPUE from traps over-estimated crayfish abundance in lakes of moderate or intermediate abundance. This finding of high CPUE at moderate abundances is a common occurrence in many fisheries systems, where it is termed "hyperstability" and given as a reason why fisheries catch data may fail to anticipate or detect declining populations (Harley et al. 2001). Hockley et al. (2005) and Jones et al. (2008) outline related concerns for using angler-reported CPUE to monitor population status of harvested crayfish in Madagascar, although Zimmerman and Palo (2011) are more optimistic with respect to the potential for angler-reported CPUE to represent population status in a reintroduced and recovering population of *Astacus astacus* in Sweden. Hyperstability is not the only way that CPUE may misrepresent true population abundance, as "hyperdepletion" occurs when CPUE under-estimates abundance. Dorn et al. (2005) found CPUE under-estimated abundance at high crayfish densities, particularly for smaller individuals, in an enclosure study in a Florida, USA wetland. Collins et al. (1983) similarly found CPUE could under-estimate true crayfish abundance when fish populations suppressed CPUE through crayfish anti-predator behavior. Collins et al. (1983) could only produce an $R^2 = 0.19$ between visual density estimates and CPUE when information on predatory fish abundance was withheld from regression models, although not all fish affected crayfish CPUE to the same extent. Agreement between CPUE and crayfish density observed from visual surveys improved the most when information on rock bass *Ambloplites rupestris* abundance was included in models, whereas other fish species like yellow perch *Perca flavescens* did not negatively affect crayfish trapability.

Given the limitations of baited trapping outlined in preceding paragraphs, why then have so many researchers used trapping in field studies of crayfish distributions or population processes? We offer that passive sampling for crayfish by baited trapping may often be more convenient and expedient than labor-intensive active sampling approaches, and this may allow researchers to sample larger areas or more sites at greater frequencies. All field sampling methodologies involve some tradeoffs between rigor at individual locations and the number of locations that can be sampled (Jones 2011). Further, the use of baited trapping may facilitate comparison between studies and study regions if aspects like trap design and baits used are standardized, whereas approaches like visual searching by divers could be vulnerable to observer effects or other biases that have not been well-evaluated for studies of crayfish (see below). We'd like to counter the many known weaknesses and limitations of baited trapping for crayfish with a number of examples of the utility of this approach in studies of crayfish conservation and management. Baited trapping has proven successful in documenting replacement of native crayfish species by invasive crayfishes (Lodge et al. 1986, Westman et al. 2002, Olden et al. 2011), as well as documenting the spread of invasive crayfishes in new habitats (Wilson et al. 2004, Olden et al. 2006) and the decline of invasive crayfish populations following control and removal efforts (Hein et al. 2007). Baited trapping has also succeeded in documenting native crayfish population declines through time (Edwards et al. 2009), and demonstrated crayfish habitat preferences and community associations between lakes over large areas (Capelli and Magnuson 1983, Garvey et al. 2003). Accordingly, we emphasize that baited trapping has a valuable role in studies of crayfish ecology and management, but urge researchers to be aware of the many known biases and limitations of this approach.

Visual surveys of crayfish density or abundance by divers or snorkelers (hereafter just divers) may often be preferable to relative abundance estimated as CPUE from baited traps (Lamontagne and Rasmussen 1993). Visual surveys of crayfish by divers have a reasonably long history dating back at least to Cukerzis (1959), subsequently adapted and popularized by Abrahamsson and Goldman (1970). Such visual surveys for sampling crayfish have received some methodological evaluations in the field (Lamontagne and Rasmussen 1993, Pilotto et al. 2008), albeit less than the preponderance of method papers on baited trapping for crayfish (see above). Visual surveys for crayfish typically take the form of intensive searching for crayfish in either quadrats (squares) of set area, line transects of a set length and width, or timed searches of a set duration (Quinn and Janssen 1989, Lamontagne and Rasmussen 1993, Mueller 2002, Magoulick 2004, Pilotto et al. 2008). Small quadrats may occasionally include mesh or wire screening or walls to inhibit crayfish escape by swimming behaviors when disturbed from shelter (Abrahamsson and Goldman 1970, Pilotto et al. 2008). Researchers interested in using active visual searches for crayfish, whether in lentic and large lotic systems or in other habitats (see below), should consult a number of references on distance sampling and line transect surveys (Buckland et al. 1993, Ensign et al. 1995, Thomas et al. 2010). Specific to crayfish, Lamontagne and Ramussen (1993) provide a particularly useful analysis, which compared timed searches (10 minutes) to quadrat samples of different dimensions (1 m^2 and 10 m^2) and performed power analysis to evaluate the degree of replication needed to detect differences in crayfish density or relative abundance (see section on precision and power analysis). These authors found that quadrat samples were seemingly more accurate than brief timed counts; that fewer replicate samples are needed with larger quadrat samples because larger quadrat samples reduce variance in density estimates; and that low crayfish densities demand potentially restrictive high sampling effort to produce precise estimates. Pilotto et al. (2008) also recommended quadrat sampling over CPUE from timed searches (2 hours) from a similar study, although neither of the preceding two studies evaluated line transect (i.e., set distance and width) surveys commonly used for this purpose (e.g., Quinn and Janssen 1989, Mueller 2002, Magoulick 2004), which might be anticipated to perform more like quadrat samples than timed searches.

Researchers using visual searches for crayfish in lentic or large lotic environments may conduct surveys at night when crayfish are active by using dive lamps (e.g., Davies and Ramsey 1989), but it is seemingly more common to conduct surveys during the day by systematically overturning and searching substrates (e.g., rock, woody debris) that crayfish are likely to use as shelter. In some cases crayfish are only counted (e.g., timed counts in Lamontagne and Rasmussen 1993), although researchers will often want to physically capture crayfish to make accurate measures of size or accurate determinations of sex. Crayfish can be collected using hand nets or more elaborate tools like a "suction gun" (Davies and Ramsey 1989) or "diver-operated dredge-seive" (Odelstrom 1983). Active searches by divers will better represent all ages, sizes, and sexes of the crayfish population than baited trapping (Davies and Ramsey 1989, France et al. 1991, Chucholl 2011), although active searching should still not be interpreted as perfectly accurate relative to actual population attributes (see below). A further benefit of active searches by divers is that this approach permits a very fine grain for identifying crayfish habitat use within study sites (Kershner and Lodge 1995, Garvey et al. 2003), as opposed to baited trapping where recruitment distances to traps may be uncertain or very large and hence the habitat type actually sampled by a trap location may be ambiguous. While we have noted (above) that baited trapping has succeeded in representing crayfish population declines over time, so too have visual searches by divers been found useful in monitoring crayfish population responses to perturbations such as experimental acidification of lakes (Davies 1989).

Visual searches by divers may better represent crayfish population sex or size structure (e.g., France et al. 1991) or relative abundance (e.g., Collins et al. 1983) when compared to baited trapping, but we note that few studies have tested performance of visual searches for crayfish against either known or independently estimated (e.g., mark-recapture) population attributes (but see Pilotto et al. 2008). Other disciplines that regularly use visual searches by divers (e.g., marine biology) have reported a number of limitations and biases of this approach, such as failing to detect or under-estimating the abundance of small or juvenile individuals in populations or cryptic species in communities (e.g., Willis 2001, Edgar et al. 2004). Such biases likely manifest in visual diver searches for crayfish, albeit at a considerably lower magnitude than for baited trapping. For example, juvenile or age-0 crayfish can be difficult to detect even in dive searches and often constitute small proportions of resulting length-frequency distributions for entire populations (e.g., France et al. 1991). We recommended that more studies critically evaluate the accuracy and precision of dive survey estimates for crayfish using tools like mark-recapture population estimates (Edgar et al. 2004, Pilotto et al. 2008), estimates made in enclosures on stocked crayfish of known densities (Dorn et al. 2005, Larson et al. 2008), or possibly through application of toxicants like rotenone to provide post-survey measures of absolute population size (Fisher 1987, Willis 2001). Visual searches by divers are also likely vulnerable to "observer effects" in which individuals conducting sampling differ in their capacity to observe or capture the organisms of interest, a form of bias that may be more severe for visual searches than for standardized baited trapping and that may affect transferability of results between studies (reviewed in Elphick 2008). Finally, the ease of visual surveys by divers varies with habitat depth. Surveys by snorkelers with little technical proficiency are possible in shallow habitats, whereas surveys by SCUBA-certified divers are necessary in deeper habitats, and some deep habitats will be altogether infeasible to sample by visual survey. Consequently, there may be some cases where deep water habitat use by crayfish will need to be evaluated with baited trapping even if shallower waters are visually surveyed by divers (Lewis 1997), although the general preference by crayfish for shallow littoral waters does make visual survey by divers an excellent sampling tool for many applications (Lamontagne and Rasmussen 1993).

Many studies on crayfish sampling methodologies have been conducted in temperate lakes of North America or Europe (e.g., Somers and Stechey 1986, Lamontagne and Rasmussen 1993, Pilotto et al. 2008). Less research is available on sampling for crayfish in large water bodies in tropical regions or in wetlands that may differ considerably from temperate lakes by either habitat attributes or members of the crayfish community and their behaviors (Acosta and Perry 2000, Harper et al. 2002). As one important exception, Dorn et al. (2005) provided a good methods paper on sampling for crayfish from such a wetland habitat, contrasting an active throw trap (1 m^2) against baited trapping. Dorn et al. (2005) suggested that this throw trap is most equivalent to active area-based (e.g., quadrat) sampling in lentic and lotic environments, and demonstrated that the throw trap provides more accurate and precise estimates of crayfish density, relative to know densities of crayfish stocked in enclosures, than baited trapping. Dorn et al. (2005) noted that the good performance of the throw trap may be restricted to macrophyte-dominated wetlands or similar habitats, and less effective over hard or uneven substrates (see Kobza et al. 2004). The throw trap of Dorn et al. (2005) seems to have received few subsequent implementations outside of sampling by the same authors in the Everglades of Florida, USA, although Harper et al. (2002) used a seemingly similar "cage" sampler of 1 m^3 volume in Lake Naivasha, Kenya for the introduced red swamp crayfish *Procambarus clarkii*. Researchers working in macrophyte-dominated wetland habitats have commonly supplemented baited trapping with near-shore kick sampling or seining (Usio et al. 2006; see section on sampling wadeable streams and rivers) or searches of macrophtyes with hand nets (Harper et al. 2002).

The methods outlined above are not the only approaches used to sample crayfish from lentic or large lotic environments, but they are either the most common (baited trapping, visual surveys by divers) or the most rigorously evaluated (throw traps; Dorn et al. 2005). Some alternative methods include the use of "baited sticks" in Europe (Policar and Kozák 2005) and a range of crayfish sampling or harvesting tools in Australia such as baited cameras, drop nets, scoop nets, and snares (de Graaf et al. 2010, Fulton et al. 2012). We also recognize that incidental capture of crayfish during sampling for fish can be of high value in lentic and large lotic environments. Crayfish regularly become entangled in fish sampling gear like gill nets, where they are often discarded as by-catch without identification or enumeration. We emphasize that valuable information can and should be acquired when crayfish are incidentally collected with fish sampling. For example, Jansen et al. (2009) reported spread of invasive *O. rusticus* in Lake of the Woods, Canada, between 1976 and 2006 based on by-catch from gill nets for fish sampling, and Mueller (2001) discovered the first occurrence of *P. clarkii* in Washington State, USA, during routine fish community sampling conducted by the state fisheries agency. Similarly, Paragamian (2010) used CPUE from gill nets to report a negative relationship between a declining predatory fish species, burbot *Lota lota*, and the crayfish *P. leniusculus* between the years 1994 and 2006, suggesting that increasing *P. leniusculus* abundance was related to ongoing release from predation. The conservation and management of crayfish would benefit greatly if management agencies retained, identified, enumerated, and reported crayfish collected under other monitoring programs, whether for bioassessments based on benthic macroinvertebrate communities or regular surveys of fish populations. Finally, a number of crayfish management programs depend on either angler surveys or angler-reported CPUE to monitor population trends for exploited crayfish. We refer interested readers to Hockley et al. (2005), Jones et al. (2008), de Graaf et al. (2010) and Zimmerman and Palo (2011) for some considerations in designing and implementing angler or community-based crayfish monitoring programs.

Crayfish in wadeable streams and rivers

Wadeable streams and rivers may be the easiest habitats to quantitatively sample for crayfish. Relative to lentic and large lotic environments, wadeable streams and rivers provide tractable sizes where wetted widths and depths rarely inhibit access by researchers to different areas or habitats. Further, a number of methods from either field sampling for smaller benthic macroinvertebrates (e.g., Hess or Surber samplers; Surber 1937, Hess 1941) or generally larger and more mobile freshwater fish (e.g., seining or backpack electrofishing, e.g., Hayes and Baird 1996) have been adapted to collect crayfish in wadeable streams and rivers. Some common sampling approaches used for crayfish in wadeable streams and rivers include collection by hand nets, dip nets, or D-frame nets, electrofishing, kick sampling, quadrat sampling, and seining (Rabeni et al. 1997, Englund 1999, DiStefano et al. 2003, Flinders and Magoulick 2005, Price and Welch 2009, Gladman et al. 2010, Wooster et al. 2012). We decline to revisit here those methods outlined in detail in the section (above) on sampling for crayfish in lentic and large lotic environments, but note that both baited trapping and visual surveys by snorkelers or divers are also commonly applied in wadeable streams and rivers. For example, much of our previous discussion on baited trapping also relates to wadeable streams and rivers (see Peay et al. 2009 for an application), although we caution that researchers should anticipate that directional stream flows will affect how crayfish recruit to baits (Keller et al. 2001) and high stream flows carry risks of both inhibiting crayfish entrance into baited traps as well as dislodging traps downstream. Similarly, visual surveys by divers either in line transects (Charlebois and Lamberti 1996, Magoulick 2004) or quadrats (Pintor and Sih 2011) have been used in wadeable streams and rivers, and many of the considerations and recommendations from lentic and large lotic environments are directly relevant albeit with methodological adaptations to stream flow (i.e., crayfish escape behavior by swimming may be modified by flow).

Figure 2. Examples of some sampling techniques for crayfish commonly used in wadeable streams and rivers. One researcher operates a backpack electrofishing unit while a second nets stunned organisms (A). "Kick sampling" is conducted in which one researcher holds a seine net, and a second overturns potential shelter substrates, flushing crayfish downstream (B). Two researchers use the contained quadrat sampler design of Rabeni (1985), overturning substrates to a depth of 15 cm and flushing crayfish downstream into a long net end (C). All photographs by the authors.

A handful of studies provide useful evaluations of sampling methods for crayfish in wadeable streams and rivers (e.g., Price and Welch 2009, de Graaf et al. 2010, Gladman et al. 2010), although researchers interested in quantitative sampling for crayfish would do well to start with Rabeni et al. (1997). Rabeni et al. (1997) assessed baited trapping, direct observation at night, electrofishing, hand netting at night, and quadrat sampling for the crayfish *Paranephrops planifrons* in a New Zealand stream. Rabeni et al. (1997) ultimately dismissed direct observation at night and baited trapping as not useful for their species and system, and found size biases for other approaches: the quadrat sampler (see below) favored smaller individuals, hand collection with dip nets at night favored larger individuals, and electrofishing was intermediate. Electrofishing followed by quadrat sampling provided the most accurate population estimates relative to mark-recapture values, whereas hand netting and direct observation led to very few individuals collected or observed. Finally, Rabeni et al. (1997) suggested that mark-recapture combining several sampling methodologies and repeated site visits (see Ricker 1975) is more effective than serial depletion of individuals during a single site visit (see Zippin 1958) for estimating crayfish population size in a stream reach. Regrettably, few studies like Rabeni et al. (1997) have evaluated performance of a diversity of crayfish sampling gears for wadeable streams and rivers in other regions and on other species. Exceptions include Price and Welch (2009), who compared baited trapping, dip netting, electrofishing, and seining in a diversity of habitat types of eastern USA lowlands, and Gladman et al. (2010), who evaluated hand netting, electrofishing, kick sampling into a D-frame net, and Surber sampling for detecting invasive populations of *P. leniusculus* in Scotland. Price and Welch (2009) generally supported Rabeni et al. (1997) in favoring electrofishing followed by seining while finding biases for dip netting and baited traps consistent with past results, but noted that effectiveness of sampling approaches differed by crayfish species. Gladman et al. (2010) suggested a combination kick sampling (3-minute effort, 250 mm width net) and subsequent electrofishing protocol as the best for detecting *P. leniusculus* presence in riffles. As with field sampling for lentic and large lotic environments (above), there may be many cases where a combination of sampling approaches for crayfish in wadeable streams and rivers will be necessary to address the question of interest. For example, studies intensively quantifying production of crayfish in streams or rivers may need to combine a variety of sampling methodologies for different habitat types and age or size classes (Mason 1975, Evans-White et al. 2003).

Electrofishing in streams typically uses a backpack with a battery or generator unit for power and a cathode tail and anode wand (direct current) or two equal-sized electrode wands (alternating current) for delivering the electrical field to the water. A single researcher using an electrofishing backpack is usually accompanied by assistants netting organisms from the water. Wider streams and rivers may necessitate multiple backpack electrofishing units. Decisions on whether to use direct or alternating current and the voltage setting are dependent on both environmental conditions (e.g., water conductivity) and biological attributes of the sampled organisms (e.g., size, organismal conductivity relative to water conductivity). Importantly, the effect of the electric field is a function of the size of the organism, and consequently electrofishing is less effective for small and juvenile crayfish (Alonso 2001). Researchers sampling for fish often blocknet stream segments to prevent escape by flight behaviors (Zale et al. 2013), but the lesser swimming ability of crayfish may minimize this concern—Alonso (2001) notes that crayfish often respond to electrofishing by becoming immobilized on the stream benthos. "Electric seines" have been used as alternatives to backpack electrofishing (Angermeier et al. 1991), albeit not to our knowledge for crayfish. Researchers interested in using electrofishing to sample for crayfish in wadeable streams and rivers should consult general sources like Cowx (1990) or Zale et al. (2013). Westman et al. (1978b) were early proponents of electrofishing for crayfish in wadeable streams and rivers. These authors reported generally good performance of the approach albeit with predictable difficulties capturing smaller crayfish; suggested that electrofishing might be most effective sampling at night when crayfish are likely to be more active out of benthic substrates or burrows; and noted that electrofishing has limitations in habitats where dense riparian vegetation or large woody debris obstructs capture of crayfish. Beyond Rabeni et al. (1997) and Gladman et al. (2010), Alonso (2001) evaluated electrofishing for crayfish via serial depletion for *Austropotamobius pallipes* in three Spanish streams. Alsono (2001) reported that catchability of crayfish increased with body size but small crayfish (<40 mm total length) still composed nearly 1/3 of the total catch. Alonso (2001) reported that an individual electrofishing pass generally led to the collection of 60%

of the estimated population, with 3–4 serial depletions capturing over 90% of the estimated population. Electrofishing settings for crayfish in wadeable streams and rivers have varied from low (30–50 V; Alonso 2001) to higher (300–600 V; Westman et al. 1978b, Burskey and Simon 2010) voltage, with Westman et al. (1978b) recommending direct current. Some other examples of electrofishing to sample crayfish populations in wadeable streams and rivers include Bernardo et al. (1997), Usio and Townsend (2001), and Burskey and Simon (2010).

It is our perception that relatively few studies sampling for crayfish in wadeable streams or rivers use electrofishing, often favoring snorkel surveys (e.g., Pintor and Sih 2011), quadrat sampling (e.g., DiStefano et al. 2003), or various seining or kick sampling methodologies (e.g., Flinders and Magoulick 2005). The cost of backpack electrofishing gear may be prohibitive to some researchers, while the technical expertise to operate it and associated safety risks to researchers may be a further deterrent (Cowx 1990, Alonso 2001, Zale et al. 2013). Electrofishing can be harmful to freshwater organisms including crayfish (Alonso 2001), although any sampling approach for crayfish in wadeable streams and rivers will carry some risk of injury or mortality to study specimens. For example, quadrat sampling or kick sampling will often crush or damage a minority of collected crayfish during the process of disturbing the substrate (see below). Electrofishing may also fail to collect crayfish that are burrowed deeply into substrates or stream banks, prompting some researchers to suggest sampling at night when crayfish are likely to be more active (Westman et al. 1978b; see Evans-White et al. 2003 for the same concern for seining for crayfish). Recently, Rogowski et al. (2013) reported that electrofishing was ineffective for sampling crayfish in a turbid desert river of the southwestern USA, and expressed concern that burrowing behavior and shelter use may result in many individuals never being collected by multiple depletion passes. Approaches like quadrat sampling (see below) that actively disturb substrate or shelter habitats may be more successful at collecting crayfish during the day. Further, some researchers may prefer to study crayfish populations as densities (individuals per m^2) rather the number of overall crayfish collected from a length of river or stream reach. For example, such density-specific estimates are useful for relating crayfish habitat selection to substrate type or presence or absence of aquatic macrophytes (DiStefano et al. 2003, Flinders and Magoulick 2005). While more evaluations of electrofishing for sampling crayfish in wadeable streams and rivers would be useful, this method merits serious consideration from researchers owing to its favorable performance in past comparative studies of crayfish sampling approaches (Rabeni et al. 1997, Price and Welch 2009, Gladman et al. 2010; but see Rogowski et al. 2013).

Quadrat sampling is a common approach for crayfish in wadeable streams and rivers, at least in areas of the central and southern USA. Popularized by Rabeni (1985), the quadrat sampler is a 1 m^2 metal frame structure that is 0.4 m tall with net mesh walls on upstream and side surfaces and a downstream tapered bag or "cod-end" that is 0.50 m long (see DiStefano et al. 2003, Larson et al. 2008 for illustrations). The quadrat sampler is placed on the stream bed and excess mesh netting from the front and side walls is buried into the substrate to inhibit crayfish escape around quadrat margins. Researchers then disturb the substrate within the quadrat sampler for 3–5 min to a depth of 0.15 m and flush crayfish, with the aid of streamflow, into the downstream end. Quadrat sampling can typically be conducted in wadeable streams and rivers by researchers in hip or chest waders, although deeper habitats might necessitate snorkeling or SCUBA diving (Rabeni 1985). DiStefano et al. (2003) describes the quadrat sampling methodology in detail and demonstrates its use in identifying crayfish habitat associations (e.g., riffle, pool, vegetation, etc.) and temporal trends over eight years (1991–1998) in two rivers of Missouri, USA. DiStefano et al. (2003) found that the quadrat sampler was moderately precise, with good statistical power for detecting spatial differences between habitat types or study sites but less power for detecting temporal trends. Brewer et al. (2009) applied DiStefano et al.'s (2003) methodology to intensively estimate crayfish production in different stream habitats of the same region. Larson et al. (2008) evaluated quadrat sampler estimates of crayfish density relative to known densities of marked crayfish stocked in stream enclosures. Larson et al. (2008) reported that mean density estimates from the quadrat sampler were 69% of the known stocked density of three crayfish per m^2, a performance similar to a single pass by backpack electrofishing for crayfish (see Alonso 2001 above) and equivalent to performance of quadrat sampling for benthic fish (Fisher 1987, Peterson and Rabeni 2001). However, there was a wide range in crayfish density estimates between stream segments, from under-estimates to positively biased over-estimates, and power analysis suggested that high replication might be necessary to produce very precise estimates of density. This finding is similar to Lamontagne and Rasmussen's (1993) assessment and power analysis for snorkel surveys of quadrat samples in lakes. Further, although quadrat sampling performs well for smaller or juvenile crayfish, it likely under-estimates larger and adult crayfish and may be difficult to use in habitats with large or difficult to move obstruction like boulders or large woody debris (Rabeni et al. 1997).

Other quadrat sampler designs have been advocated, such as Walton and Cook's (2010)'s "microhabitat" quadrat sampler, which is 0.25 m² with a larger 1 m² net mesh skirt for minimizing crayfish escapes. We caution from Lamontagne and Rasmussen (1993) that smaller quadrats will likely produce larger variance around crayfish density estimates and consequently reduce power in studies, but appreciate that there may be contexts where sampling smaller microhabitats may be desirable (see also Usio 2007). Regardless, similar inflations in variance should be anticipated for crayfish density estimates (or other parameters of interest) produced from benthic sampling tools that have been developed to collect smaller macroinvertebrates from generally smaller areas (e.g., Surber 1937, Hess 1941). It is also possible to sample quadrats without the metal frame and mesh net design used by Rabeni (1985) and subsequent applications; for example, Pintor and Sih (2011) sampled 1 m² areas with snorkel surveys where walled quadrats were not applied. DiStefano et al. (2009) used a PVC pipe frame to delineate a 1 m² area to sample for crayfish from the hyporheic zone of an intermittent stream that had dried entirely (i.e., substrate also excavated to 0.15 m), with density estimates compared to those from the contained quadrat sampler design applied when the same stream was flowing earlier in the year. To our knowledge, Rabeni's (1985) design has not been compared for accuracy and precision of crayfish density estimates to quadrat samples that do not contain the 1 m² area (or other size) with a metal frame and mesh net walls, or to more relaxed area-based sampling approaches like kick sampling (see below). Rabeni's (1985) contained design is anticipated to minimize loss to crayfish flight (i.e., "tail flipping"), but does come with tradeoffs in transportation difficulty (e.g., bulky or awkward to transport) and time spent sampling.

Accordingly, other less intensive area-based sampling approaches have been developed for crayfish in wadeable streams and rivers. One example is "kick sampling," in which a set area (or time) is disturbed by kicking over or through substrates with crayfish swimming or being flushed downstream into a net or seine (e.g., Mather and Stein 1993, Smith et al. 1996, Gladman et al. 2010). Flinders and Magoulick (2005) provide a good demonstration of quantitative kick sampling (or "kick netting," "kick seining") for a diverse crayfish community in Arkansas and Missouri, USA. Adapting their methodology from Mather and Stein (1993), Flinders and Magoulick (2005) disturbed stream substrates over a 1 m² area with crayfish flushed downstream into a 1.5 m by 1.0 m seine net held by two researchers. Estimated densities of six collected crayfish species were then related to habitat attributes measured in these 1 m² sample units (e.g., substrate type, depth, current velocity, etc.), and biological attributes recorded from collected crayfish were also reported (crayfish length, weight, reproductive status, etc.). Advantages of this approach over quadrat sampling (above) include easier transport of the sampling gear, generally faster application of the sampling methodology to each 1 m² area (e.g., due to disturbing substrates to shallower depths), and possible implementation with only a single researcher with adjustment of the downstream net size (e.g., even relatively small D-frame nets are feasible; see Gladman et al. 2010). Smith et al. (1996) used a timed rather than area-based kick sampling approach to evaluate habitat preferences for *Austropotamobius pallipes* in Britain; these authors measured CPUE as crayfish collected per duration of kick sampling, and observed that kick sampling selected for smaller crayfish relative to more focused hand-collection by overturning large substrates (see below; also Gladman et al. 2010). As discussed above, there may be many research questions where combining several field sampling approaches for crayfish is advisable. For example, crayfish life history studies that seek to quantify age structure or growth rates through length-frequency histograms might benefit from combining several sampling approaches that differentially favor large or small individuals in the population.

Finally, as emphasized by Huxley (1884), collecting crayfish from wadeable streams and rivers is generally not difficult, particularly if collecting crayfish is the only objective (e.g., for subsequent laboratory observation or experiments). As demonstrated by Smith et al. (1996) and also Usio (2007), hand collection of crayfish is effective where crayfish are abundant and can be conducted from the shoreline, while wading, or by snorkelers or divers. Rabeni et al. (1997) cautioned against approaches like hand collection or observation at night for representing crayfish abundance or population attributes, but some researchers have succeeding in applying these methods to crayfish field sampling. For example, Ludlam and Magoulick (2009) needed crayfish density estimates in association with a manipulative exclosure experiment of grazer effects in stream pools drying under drought, but felt that active sampling approaches like quadrat sampling or electrofishing might disturb their small, isolated study sites. As a consequence, these authors estimated crayfish densities using line transect surveys (0.92 m width, ≥15 m length transects) at night by either snorkeling or walking along shorelines when stream pools were too shallow. Similarly contradicting recommendations of Rabeni et al. (1997), Usio (2007) used hand-collection for the endangered crayfish *Cambaroides japonicus* in small streams of Japan to evaluate selection of microhabitats at smaller grain sizes (see Wiens 1989 and below) than kick or quadrat sampling allow. As these examples demonstrate, there is no "cook book" method that will meet every crayfish sampling need, and we encourage researchers to adapt the field sampling approaches outlined above to their own systems and research questions. While we advocate throughout this book chapter for sampling standardization wherever possible, and we believe many approaches like electrofishing or quadrat sampling are widely applicable, the variety of habitats that crayfish occupy inevitably leads to unique sampling challenges. Researchers will need to apply some creativity when crayfish behavior or habitat selection provides sampling difficulties that are not easily accommodated by the more routine methodologies characterized here.

One area where sampling for crayfish in lotic environments differs from lentic environments is the importance of stream or river flow. As per sampling in lentic environments (see above), researchers in temperate regions should anticipate that crayfish will be easier to collect under warmer conditions in streams and rivers due to increased activity levels (e.g., Bubb et al. 2004). However, the flow regime of a stream or river is a further aspect of environmental seasonality that will need to be accommodated by researchers sampling for crayfish. High river flows, whether from predictable (e.g., snowmelt in spring) or unpredictable events (e.g., storms), will often suppress crayfish catch through behavioral responses, such as becoming less active or burrowing to deeper substrates, and will affect the performance and/or safety of different crayfish sampling methodologies. Severe floods may even affect local population size through mortality or downstream displacement of crayfish (Momot 1966, Parkyn and Collier 2004). Sampling for crayfish in wadeable streams and rivers will often be easiest and safest at low or base flow conditions, although extremely low flows due to seasonal or supraseasonal droughts may also drive crayfish into hyporheic habitats via burrowing, to migrate to permanent refugia, or lead to crayfish mortality and population declines (Adams and Warren 2005, DiStefano et al. 2009). Further, many of the sampling methods outlined above for crayfish in wadeable streams and rivers are dependent on or affected by stream flow velocity. For example, quadrat and kick sampling both depend in part on stream flow to displace exposed or swimming crayfish downstream. These sampling methods may be less effective in either extremely high flows where the gear is difficult or unsafe to use or in slow-flowing and stagnant waters where crayfish will not be driven by current into downstream nets. In such stagnant habitats or study systems, approaches like actively pulling seines upstream through the sample area of interest may be necessary (Evans-White et al. 2003, Price and Welch 2009). In these cases the distinction between "lentic" and "lotic" habitats is blurred, and some of the methodologies discussed in the preceding section may merit consideration (e.g., baited trapping, throw traps), just as some implementations discussed for sampling wadeable streams and rivers may be applied to non-flowing freshwater habitats (e.g., beach seining; Poulin et al. 2007).

Where multiple crayfish species co-occur, researchers should anticipate that habitat preferences may vary by species. Flinders and Magoulick (2005) found that some crayfish species preferred fast-flowing riffles and runs, others preferred slow-flowing pools and backwaters, and some crayfish were generalists that could be found across these habitat types. In the same system, Flinders and Magoulick (2003) documented that some crayfish occurred in highest densities in intermittent streams, whereas other crayfish species were only found in permanent streams and rivers. Similarly well-defined habitat preferences have been documented for a diverse crayfish community in lotic and lentic environments of Australia (Johnston and Robson 2009a). Accordingly, researchers may often need to representatively sample available habitat types via random or stratified designs to best represent crayfish populations or habitat preferences within communities. This same recommendation applies to the well-documented differences in habitat selection between life stages of crayfish within the same species. A number of researchers have noted that small or juvenile crayfish often occur in shallow habitats, perhaps to avoid fish predators, whereas larger or adult crayfish often occur in deeper habitats, potentially to avoid terrestrial predators (Rabeni 1985, Englund and Krupa 2000, Flinders and Magoulick 2007). Similarly, juvenile crayfish may often select coarse and rocky substrates to avoid predation or cannibalism by other crayfish (Usio and Townsend 2000, Nyström et al. 2006), whereas large crayfish may have more flexibility in selecting shelter habitats, such as burrowing deeply into stream banks (Guan 2000, Usio et al. 2006). As noted in the section on sampling lentic and large lotic habitats (above), berried or ovigerous female crayfish (carrying eggs or early instar juveniles) will often become less active and select shelter habitats, reducing their detectability in populations by both active and passive sampling techniques and skewing observed sex ratios towards male-dominance (Mason 1970, Somers and Green 1993, Richards et al. 1996). Malley and Reynolds (1979) further outline life history considerations in freshwater sampling programs with the crayfish *Orconectes virilis* as an example. Researchers will often need to anticipate and account for heterogeneity in habitat preferences by different life stages when studying crayfish in wadeable streams and rivers.

We conclude by emphasizing that wadeable streams and rivers are tremendously heterogenous environments (Palmer and Poff 1997). As a consequence, freshwater organisms in wadeable streams and rivers often have patchy or clumped distributions as responses to a variety of abiotic, biotic and spatial factors (Jackson et al. 2001). Wadeable streams and rivers may be easier to work in comprehensively than other habitats like deep lentic waters, but these systems pose considerable challenges to understanding the distribution, behavior, and ecology of freshwater organisms through space and time. We note as well that many of the sampling options outlined above for crayfish in wadeable streams and rivers differ by both the extent of area they are sampling and the grain of the sampling unit itself (Wiens 1989). Backpack electrofishing is typically applied over fairly large or long stream reaches, and may be incapable of identifying finer scale habitats used by crayfish (e.g., for shelter or foraging) within these sampled extents. Approaches like quadrat sampling, kick sampling, or hand collection may be most appropriate for identifying crayfish associations with fine-grain habitat features, but may suffer low statistical power owing to inflated variance caused by the patchy distributions of crayfish, rendering them potentially less useful for relating crayfish presence or population size to larger scale environmental factors (e.g., land use in contributing watersheds). We note that not only have few studies explicitly considered multi-scale predictors of crayfish habitat use or presence/abundance in wadeable streams and rivers (but see Usio 2007, Wooster et al. 2012), but we are aware of no studies that have evaluated how the extents and grains of areas sampled affect our ability to understand the spatial extents and grains over which community and ecosystem attributes influence crayfish presence and abundance. As crayfish in wadeable streams and rivers are simultaneously responding to both habitat and prey availability at fine spatial scales and to factors like wide-ranging predator abundance or water chemistry/quality over large spatial scales, we urge researchers to be scale conscious in choosing (or combining) sampling methods for crayfish in these environments. Similarly, the "riverscape" concept advocated by Fausch et al. (2002) seeks to better accommodate and account for the complex spatially continuous nature of stream and river networks. To our knowledge, crayfish have not yet been explicitly studied in such a riverscape context. Better relating crayfish presence or abundance to the spatially continuous habitat features of streams and rivers may be a productive area of future work that may also require some innovation in field sampling methodologies relative to those outlined above (e.g., spatially continuous single-pass electrofishing; Bateman et al. 2005).

Burrowing or terrestrial crayfish

All crayfish can burrow under some circumstances, but many crayfish species are obligate burrowers with minimal requirements for permanent surface waters, and some species are effectively terrestrial organisms (e.g., Welch and Eversole 2006a). Hobbs (1942, 1981) categorized crayfish as primary burrowers if they spend the majority of their life in or near burrows and are infrequently found in open surface water, as secondary burrowers if they use surface water seasonally during wet periods of the year, and as tertiary burrowers if they live predominantly in surface water but build simple burrows for reproduction or in response to abiotic disturbance like drought (see Horwitz and Richardson 1986 and Welch and Eversole

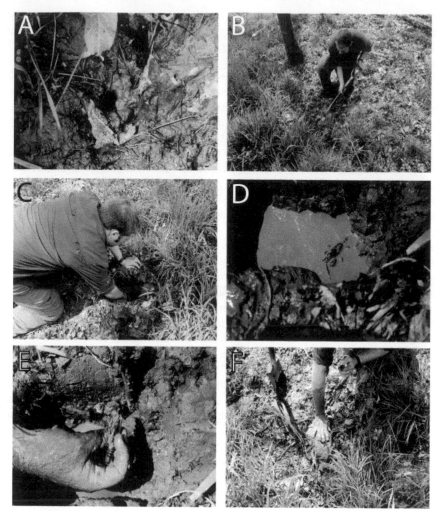

Figure 3. A researcher excavates a terrestrial crayfish burrow. The burrow (A) is first excavated with a shovel (B) and then by hand (C), with plunging (see main text) to dislodge the crayfish (D). The species in this case is *Cambarus monongalensis* from West Virginia, U.S.A. (E). The burrow is subsequently re-filled (F). All photographs by Zachary J. Loughman.

2006b for additional ecological classifications of burrowing). Our focus here is on primary burrowing crayfish, but the methods for sampling burrows are also applicable to secondary and tertiary burrowers under many circumstances.

Primary burrowing crayfish are common in eastern North America and Australia, and can be found as well in South America and Madagascar (e.g., Jones et al. 2007, Noro and Buckup 2010). An estimated 30% of critically imperiled crayfish in North America are primary burrowers despite representing just 15% of this region's crayfish species (per Hopper and Huryn 2012), and a number of burrowing crayfish species are recognized as threatened in Australia (Bryant and Jackson 1999). Moore et al. (2013) reported that only 2 of 61 (3%) primary burrowing crayfish species in the USA and Canada had published life history studies by 2012. One reason for this dearth of information on primary burrowing crayfish (hereafter just "burrowing crayfish") is undoubtedly the difficulty in sampling these organisms, although Australian researchers have perhaps been most persistent in acquiring ecological knowledge on burrowing crayfishes (e.g., Lake and Newcombe 1975, Richardson and Swain 1980, Horwitz et al. 1985). Burrowing crayfish may be sparse on the landscape or difficult to detect due to little prior biological information (Welch and Eversole 2006a), and sampling approaches for these crayfish range from passive trapping that is not always effective to active burrow excavations that can be labor intensive and destructive of crayfish habitat (Ridge et al. 2008). Perhaps as a consequence, some researchers have developed techniques to study the behavior and ecology of burrowing crayfish *ex situ* in laboratory or mesocosm experiments (Dorn and Volin 2009, Stoeckel et al. 2011). However, more field-based studies of the distribution and ecology of burrowing crayfish would be invaluable in aiding the conservation and management of these poorly known organisms, as well as improving our understanding of their potentially important role in ecosystems. For example, burrowing crayfish can move considerable volumes of soil (Welch et al. 2008), and create fossorial habitat used by a wide diversity of other organisms (Lake 1977, Pintor and Soluk 2006, Johnston and Robson 2009b, Loughman 2010).

Perhaps the simplest and most effective way to collect burrowing crayfish is to physically excavate burrows, most often by digging by hand or with a shovel (Simon 2004). Loughman (2010) describes a typical burrow excavation process as first identifying a burrow as active (occupied) when recently exhumed mud or organized pellets are evident at burrow portals, and then excavating to a resting chamber, which when breached is filled with water and plunged vigorously to dislodge the crayfish or prompt its movement to the surface. The "plunging" process described by Loughman (2010) can be done by hand or with an actual plunger (Simon 2001). Johnston and Fiegel (1997) demonstrate the use of burrow excavations for a life history study of an endemic burrowing crayfish in pitcher-plant bogs of Mississippi, USA; they subsampled the study site using quadrat samples in transects, excavated all burrows in 1 m² quadrats, and measured attributes of both burrows (e.g., chimney height, burrow entrance width, number of connections, etc.) and crayfish (e.g., carapace length and width, chelae length and width, sex and reproductive status, etc.). In this study, monthly collecting for over a year produced only 87 crayfish, low sample sizes relative to studies in lotic or lentic environments (see above) where a half dozen baited crayfish traps can regularly produce similar numbers of organisms. Depths of excavated burrows can be shallow or fairly deep, with 0.5 m depths common (Grow and Merchant 1980) but burrows up to 3 m deep reported for some species (Hogger 1988). Beyond the labor-intensive nature of excavating crayfish burrows, this approach inevitably disturbs or destroys crayfish habitat and often renders certain field studies, like mark-recapture population estimates, impossible. For this reason, Norrocky (1984) developed a passive trapping approach for sampling burrowing crayfish (see below). The destructive nature of active excavation of crayfish burrows can also restrict when and where researchers are able to work on burrowing crayfish. Johnston and Robson (2009b) sought to study commensal burrow relationships between a burrowing and generally non-borrowing crayfish in an Australian national park, but park rangers restricted burrow excavation due to its damaging nature to a single ranger-observed day in 2004 and a single ranger-observed day in 2007. We anticipate that many researchers proposing burrow excavations for sensitive species or in sensitive ecosystems might encounter similar restrictions, or even rejections of collecting permits, from government agencies.

Passive trapping of burrowing crayfish was first proposed by Norrocky (1984) as an alternative to burrow excavations. The Norrocky burrowing crayfish trap is a cylinder with a hinged, one-way door that is inserted at the entrance of a burrow. The hinged door allows crayfish to enter the trap but inhibits retreat back into the burrow. The Norrocky burrowing crayfish trap was found ineffective by Welch and Eversole (2006c), who proposed instead a "burrowing crayfish net" consisting of a 20 x 150 cm dimension piece of avian mist netting folded into 20 x 20 cm segments and inserted into a burrow entrance. Crayfish seeking to exit burrows become entangled in this netting, which can then be removed for the live capture of crayfish. Welch and Eversole (2006c) compared catch rates of both trap types, and reported that the burrowing crayfish net caught five times as many crayfish as the Norrocky burrowing crayfish trap. These authors also noted that the burrowing crayfish net is lighter and easier to transport than the rigid cylindrical tubes of the Norrocky burrowing crayfish trap. Ridge et al. (2008) compared efficacy of both the Norrocky burrowing crayfish trap and the burrowing crayfish net to direct excavations of burrows. These authors found comparable performance of the two crayfish traps, but both crayfish traps collected considerably fewer crayfish than direct excavations of burrows. Despite this, Ridge et al. (2008) suggested the burrowing crayfish net for non-destructive sampling of crayfish burrows owing to its ease of transport, and suggested that its poorer performance in their study relative to Welch and Eversole (2006c) may have been due to differences in crayfish morphology between the two regions. These authors proposed that spinier or more tuberculate crayfish may become more readily entangled in the burrowing crayfish net than smoother crayfish. Interestingly, Ridge et al. (2008) found that both trapping methods were influenced by habitat quality, with more crayfish collected from less disturbed habitats, whereas the efficacy of burrow excavations did not vary with habitat quality. Most recently, Hopper and Huryn (2012) developed a "reverse pitfall trap" for burrowing crayfish. In their study system, these authors found the Norrocky burrowing crayfish trap to regularly be fouled by mud extruded by burrowing crayfish, impairing its effectiveness. The reverse pitfall trap encases the burrow opening with a modified two gallon bucket with a 7 cm diameter hole cut its bottom and a funnel inserted into the burrow opening. Crayfish exiting the burrow via the provided funnel are then trapped in the bucket, where a shallow depth of water is provided to prevent desiccation and a lid is closed to prevent predation. Hopper and Huryn (2012) reported higher catch success than past implementations of either the Norrocky burrowing crayfish trap or the burrowing crayfish net. Yet we note that in all cases (including above), burrowing crayfish catch by passive traps is exceedingly low. Hopper and Huryn (2012) collected only 12 crayfish total for eighteen traps deployed and checked on weekly intervals over four weeks. Such low catches likely contribute to the paucity of available studies on burrowing crayfish ecology and life history from the field.

Burrowing crayfish can be studied by means other than active excavation or passive trapping of burrows. Burrowing crayfish will leave burrows at night to forage and disperse, and consequently observation and collection of active crayfish at night has been used to study these organisms (Hobbs 1981). Williams et al. (1974) used active searching at night to find burrowing crayfish, which were often located or perched at the entrances of burrows. These authors used a clever approach to prevent crayfish from retreating back into burrows when spotted: active burrows had thin metal sheets inserted into the base of the chimney during the day, then retracted to allow crayfish access to chimney top. When crayfish were observed at the chimney top at night, often extruding mud or pellets, the metal sheets were slid across the tunnel shaft to prevent escape back into the burrow. Loughman et al. (2013) have proposed baited lines to lure burrowing crayfish to the surface at night, finding high catch rates pooled across multiple species (91.5% success on 50 attempts). Other researchers like Taylor and Anton (1998) and Loughman et al. (2012) have opted to use lentic or lotic sampling approaches, such as baited trapping, to sample for burrowing crayfish when these organisms are briefly or seasonally using surface waters like vernal ponds. Some researchers choose to census the distribution and attributes of burrows on the landscape in place of directly capturing and studying burrowing crayfish themselves. For example, March and Robson (2006) surveyed the distribution and attributes of crayfish burrows to study the effects of riparian land use, ranging from native forest to cattle pasture, on burrowing crayfish distribution and density. Studying burrows in place of sampling crayfish directly requires knowledge of which crayfish are present at a study site, and is likely less useful for distributional work of poorly known species or initial census of new study sites. Regardless, directly studying burrow attributes can often be an interesting and important surrogate for acquiring knowledge on burrowing crayfish behavior and ecology. One emerging method of characterizing burrow size and complexity is to inject burrows with substances like gypsum, polyurethane foam or polyester resin, and then excavate the burrow casts and quantify their attributes (Lawrence et al. 2001, Welch et al. 2008, Noro and Buckup 2010). Such studies can be used to quantify volumes of soil moved and excavated by burrowing crayfish and measure attributes of burrow complexity like number of surface openings, number of nodes, linear burrow distance, and maximum depth (Welch et al. 2008). Noro and Buckup (2010) recommended polyester resin for the creation of burrow casts, noting that gypsum produced good casts but was fragile and that polyurethane foam reacted so rapidly that the foam hardened prior to completely filling the burrow system.

Welch and Eversole (2006a) use burrowing crayfish to provide an important cautionary example of the role of prior assumptions and researcher bias in the design of field sampling studies. While many burrowing crayfish in Australia (i.e., *Engaeus* species) are recognized as semi-terrestrial organisms owing to their ability to persist for long durations in burrows above the water table (Horwitz and Richardson 1986), burrowing crayfish in North America have typically been assumed to require at least one burrow shaft reaching the water table (Hobbs 1981). With this in mind, Welch and Eversole (2006a) sought to sample for a rare burrowing crayfish in South Carolina, USA, focusing at first on floodplain or wetland habitats where such crayfish were expected to occur. After failing to find the burrowing crayfish, these authors widened their sampling extent over several iterations, ultimately discovering that the species occurred almost exclusively in dry upland habitats with well-drained, non-hydric soils (i.e., a terrestrial organism). With this example in mind, we urge researchers initiating field studies of poorly known species, including the vast majority of burrowing crayfish, to carefully and critically consider the biological assumptions underlying the selection of sampling extent and focal habitats to survey. Further, due to the difficulty in collecting large numbers of burrowing crayfish, site occupancy (presence/absence) may be a more appropriate or tractable response variable for population monitoring than relative abundance, and researchers in many instances may be well-advised to account for the role of imperfect detection of organisms in their estimates of occupancy (MacKenzie et al. 2003, MacKenzie 2005). Loughman et al. (2012) demonstrate a sampling program for monitoring burrowing crayfish occupancy in large river bottomlands of West Virginia, USA, that accounts for potential incomplete detection through repeated site visits (see section on detection probability below). Additional guidance on sampling design and statistical approaches for studying rare and difficult to detect species is given by Thompson (2004), including adaptive sampling and noninvasive genetic sampling. On this latter subject, we speculate that emerging approaches like monitoring of animal populations by environmental DNA (e.g., Jerde et al. 2011, Thomsen et al. 2012, Tréguier et al. 2014) might be applicable to burrowing crayfish, perhaps by siphoning water from burrows and sequencing DNA to identify associated species.

We conclude by emphasizing that sampling approaches outlined above for burrowing crayfish can also be relevant for species more typically associated with lentic or lotic habitats. Even predominantly aquatic crayfish may leave water under some circumstances to disperse or forage on land (Furse et al. 2004, Grey and Jackson 2012), and all crayfish will burrow in response to stressful conditions like severe drought. As such, researchers working on predominantly aquatic crayfish may still need to excavate and describe burrows (Guan 2000) or occasionally search for crayfish dispersing overland (Claussen et al. 2000). In many cases, burrows of more aquatic crayfish will not resemble those of primary burrowers. For example, DiStefano et al. (2009) studied responses of two crayfish species to complete drying of an intermittent stream, finding that these tertiary burrowers sought refuge in the hyporheic zone rather than migrating downstream to permanent surface water in a reservoir. Yet owing to the coarse, rocky substrates at this study site, no conventional burrows were found, but rather crayfish were migrating to the hyporheic zone either by moving gravel and pebble substrate that collapsed behind them or through interstitial spaces between larger rocks. This finding was somewhat surprising as coarser substrates are often unsuitable for crayfish burrowing (Dorn and Volin 2009). Yet in general, researchers in streams or lakes with fine, silty substrates will often encounter tertiary burrowers making shallow, simple burrows that should be easier to excavate than the deep, complex burrows discussed above (Berrill and Chenoweth 1982, Guan 2000). Direct sampling of overland dispersing or foraging crayfish is, to our knowledge, a largely unexplored area. Many anecdotal accounts report crayfish dispersing overland, laboratory studies have estimated potential dispersal distances by relating movement rates to desiccation tolerances (Claussen et al. 2000), and some genetics studies have tested for, and generally failed to find, evidence of overland connectivity between crayfish populations (Hughes and Hillyer 2003, Bentley et al. 2010). Night observations and hand collection used to study burrowing crayfish on land (e.g., Williams et al. 1974) could be applied to sample for overland dispersing aquatic crayfish, although the infrequency of such overland dispersal events may make them difficult to design a sampling protocol for. Another alternative to evaluate potential crayfish overland movement might be application of drift fences with pitfall traps commonly used to study dispersing amphibians (e.g., Searcy et al. 2013). Regardless, we suggest the occasional use of terrestrial habitats by predominantly aquatic crayfish as an area of further research need and sampling design innovation.

Cave-dwelling or stygobitic crayfish

Like burrowing crayfish, cave-dwelling or stygobitic crayfish are both highly imperiled and generally poorly known. Stygobitic crayfish occur in karst regions of the central and eastern USA as well as a few locations in Cuba and Mexico (Hobbs et al. 1977), and include a number of species of high conservation concern (e.g., Graening et al. 2006). Researchers might be interested in sampling stygobitic crayfish to census populations of rare species (Graening et al. 2006), investigate patterns of evolution and historical phylogeography (Buhay et al. 2007), characterize regional cave faunas (Schneider and Culver 2004), trace

Figure 4. Researchers conduct visual searches with hand nets for crayfish in a cave stream. Photograph by Chuck Sutherland.

groundwater contamination (Dickson et al. 1979), or evaluate energy sources supporting cave food webs (Streever 1996, Opsahl and Chanton 2006). Many of the sampling methods outlined above for lotic or lentic environments are applicable to stygobitic crayfish. Stygobitic crayfish can be collected with baited traps (Schneider and Culver 2004, Opsahl and Chanton 2006), but hand collecting or visual searching is perhaps most common (Graening et al. 2006, Huntsman et al. 2011). Purvis and Opsahl (2005) describe an interesting technique for trapping groundwater wells to census cave populations of crayfish; their design modifies polycarbonate drink containers into baited traps that can be lowered into wells in karst landscapes or cave systems that are not directly entered by researchers. Conversely, a representative visual searching approach is described by Graening et al. (2006), who report results of 30 years of annual surveys for the Benton Cave crayfish *Cambarus aculabrum*—a species listed under the US Endangered Species Act. These surveys were conducted by one to three observers moving slowly upstream through caves, snorkeling where necessary, and visually searching with headlamps or dive lamps. For the four caves where *C. aculabrum* is known to occur, Graening et al. (2006) reported a maximum annual observed census of 56 individuals, demonstrating that the small sample sizes common to studies of burrowing crayfish will often be applicable to stygobitic crayfish, as well. Perhaps owing to the intensive focus on single sites or cave systems and the low relative abundance of many stygobitic crayfish, mark-recapture has been an important tool to investigate the biology of these organisms (Cooper 1975, Streever 1996, Venarsky et al. 2012). As for burrowing crayfish (above), we propose that the emerging tool of environmental DNA (Jerde et al. 2011, Thomsen et al. 2012, Tréguier et al. 2014) could be useful for monitoring presence of populations of cave crayfish that might otherwise be difficult to detect or where sampling or collecting might be anticipated to harm small populations (Graening et al. 2006). We conclude by cautioning that research in caves carries its own logistical issues and safety concerns, and we encourage researchers initiating sampling studies for stygobitic crayfish to consult with experienced researchers or organizations like the National Speleological Society (USA) for guidance on safety.

Evaluating crayfish sampling approaches

As noted in our introduction, we have not written a book chapter on general field sampling or sampling design. Many good guides to these topics already exist (e.g., Morrison et al. 2008, Magurran and McGill 2011, Zale et al. 2013), and we also suggest that even experienced ecologists or fisheries biologists would often benefit from consulting with a statistician or biometrician prior to implementing a new field project. Yet we do wish to briefly emphasize the ways that field studies risk misleading results when some important sampling considerations are overlooked. How accurately does a sampling gear represent a population or community attribute of interest? And is the magnitude or direction of such (likely) gear biases known? How many sampling replicates are needed for a researcher to successfully detect a pattern of interest, such as differences in habitat preference or trends in population size over time? And does the ability of gear types or observers to detect organisms vary by habitat type, season, or community interactions? As reviewed in the preceding sections, studies that have rigorously evaluated such questions for field sampling of crayfish are few and far between. We have certainly learned some valuable lessons with respect to issues of accuracy, precision, and imperfect detection when sampling for crayfish in the field. We summarize those lessons below, while emphasizing areas where additional methods studies could improve our ability to understand crayfish populations and communities in the field. Further, because mark-recapture is a tool that is commonly applied not only to understand populations but also to evaluate our sampling approaches, we provide a brief guide to crayfish mark-recapture studies and their conclusions.

Evaluating accuracy with sampling efficiency

Accuracy is the difference between a scientist's measurement and the true value of the object of interest. If 300 crayfish actually occur in a 10 m reach of stream, a single electrofishing pass that collects 250 of them is more accurate than a single seine pull that only collects 50. Accuracy is of understandably high value to field biologists, but we caution that the assumption that any sampling approach is perfectly accurate is a fantasy. As outlined in the sections above, no gear or sampling protocol has been found to perfectly represent the true abundance of crayfish in the field. A single electrofishing pass may collect only 60% of an estimated population in a block-netted stream reach (Alonso 2001), whereas a 1 m^2 quadrat sample may collect only about 70% of the crayfish actually inhabiting that area (Larson et al. 2008). Field sampling estimates that are inaccurate but randomly distributed around true values would be considered unbiased, but such estimates are typically biased relative to true values in consistent or systematic ways. As an example, baited trapping for crayfish will reliably favor larger and male individuals in the population and be biased against smaller, juvenile, or female individuals. A researcher that naively assumed baited trapping was either accurate or unbiased in its inaccuracy would draw misleading conclusions with respect to crayfish age or size structure or sex ratios in a population, as well as the true abundance or size of the population. This misinterpretation is unlikely to occur because the biases of baited trapping for crayfish have been so thoroughly documented and widely disseminated (e.g., Brown and Brewis 1978), but we note that other gear types for sampling crayfish simply carry different size, sex, or even species biases (e.g., Price and Welch 2009). Field biologists should recognize and accept that some sampling bias is inevitable, and that what is most important is understanding how bias manifests and responding appropriately in either analyzing or interpreting results (e.g., Bayley and Dowling 1990, Bayley and Peterson 2001).

"Sampling efficiency" is the term given by fisheries biologists in North America to describe how accurately a gear represents the actual presence or abundance of targeted organisms in a given area (e.g., Bayley and Dowling 1993, Peterson and Rabeni 2001). Some researchers have interpreted "efficiency" as meaning not only the number of organisms collected but also the time or effort required to collect them (e.g., Ridge et al. 2008), which we believe is better described as catch-per-unit effort (CPUE), an attribute that may not only vary between implementations of the same gear but also between multiple gear types (e.g., the hypothetical seine haul above has lower CPUE than a single backpack electrofishing pass assuming both take the same amount of time). A researcher might justifiably favor a gear with higher CPUE over a gear with lower CPUE with respect to collecting a larger number of individuals over less time, but this doesn't inherently mean either or both of the gear types accurately reflect population or community attributes of interest. Evaluating the efficiency of a sampling gear is challenged by the unknown "true" value of interest for wild populations. It may seem trite relative to more profound mysteries of the universe, but the exact number of crayfish occurring in a reach of river is unknowable with absolute certainty. Consequently, researchers use a variety of techniques to estimate or approximate sampling efficiency: stocking a known number of individuals in an enclosed area that is subsequently sampled (Dorn et al. 2005, Larson et al. 2008); repeatedly sampling an area and measuring and modeling the decline in organisms captured to estimate the actual population ("depletion"; Alonso 2001), or in some cases applying a toxicant to the area following sampling to dislodge or expose individuals that were missed (Fisher 1987, Bayley and Peterson 2001).

Larson et al. (2008) stocked crayfish at a density of three per m^2 in block-netted riffles of a stream, and then evaluated how density estimates from the quadrat sampler compared to the known true density.

Dorn et al. (2005) stocked known densities of crayfish in enclosed areas of a wetland, and then compared how faithfully two sampling techniques represented known density gradients. Such stocking studies offer certainty in knowing the exact value of the population attribute of interest, but inevitably come with costs of biological realism and a limited extent of sampled area. Alonso (2001) demonstrated multi-pass depletion from backpack electrofishing to estimate "true" abundance of stream-dwelling crayfish, which can then be used to evaluate efficiency of a single pass or alternative sampling approaches. Similarly, applying a toxicant after sampling has often been used to evaluate efficiency in studies of sampling gears for stream fish (Fisher 1987, Bayley and Peterson 2001), but would likely fail to reveal unobserved crayfish that are deeply burrowed into substrates or stream banks. Results of such evaluations of sampling efficiency are often used by fisheries biologists to correct their estimates of interest; for example, by adjusting measures of species richness to account for imperfect detection of organisms (Bayley and Peterson 2001). Few studies have truly evaluated efficiency of sampling for crayfish, and we are unaware of any application of results of these studies to subsequent field sampling programs (i.e., correcting field estimates by known sampling efficiencies; but see Williams et al. 2014 for estimation of correction factors between quadrat sampler and kick sampling density estimates for crayfish in wadeable Missouri, USA streams). Rather, such efficiency studies for crayfish more often seem to be used to compare and choose among different gear types (e.g., Dorn et al. 2005) or simply evaluate a single gear for accuracy (e.g., Larson et al. 2008). Further, studies of sampling efficiency for crayfish have typically been constrained to a single species and habitat type, leaving unexplored the ways that sampling efficiency may vary by habitat attributes, study species, or season (e.g., Peterson et al. 2004).

There are many field sampling situations for crayfish where estimating or interpreting efficiency may be infeasible. A researcher working on crayfish in small, wadeable streams might aspire to produce abundance estimates that represent with high accuracy the true abundance of crayfish. Conversely, researchers working on large lotic or lentic environments cannot know or collect the exact number of total crayfish that occur in such a study system, and often use measures of relative abundance like CPUE from baited traps that do not necessarily have a meaningful efficiency relationship to total abundance. In these circumstances, estimates of population size—often with fairly large error bounds—can be produced by mark-recapture (see below), and population estimates from mark-recapture studies have been used to evaluate relative abundance measures from sampling methods like baited trapping (e.g., Rabeni et al. 1997, Zimmerman and Palo 2011). Again, such evaluations of accuracy in sampling for crayfish are rare. More typically, researchers evaluate how one sampling approach for crayfish compares to a second approach irrespective of knowing how either compares to the true value of interest. Examples include contrasting density estimates of quadrat samples to CPUE from baited traps in lakes (e.g., Collins et al. 1983), or comparing the abundance, size distribution and species of crayfish collected by a variety of gears (e.g., Price and Welch 2009). These studies obviously provide important insights, such as demonstrating how passive and active sampling approaches can produce disparate estimates of relative abundance as a consequence of the dependency of passive approaches on organismal behavior (Collins et al. 1983). Yet researchers sampling for crayfish in many habitat types will have to accept that "true" values of variables of interest are unknown and likely unknowable, particularly for studies over large spatial extents (e.g., many separate sample sites) where labor- and time-intensive approaches like mark-recapture population estimates may be impractical. Yet even accepting these limitations, we emphasize that imperfect estimates of relative abundance can provide invaluable information. As noted in sections above, even a sampling gear as undeniably biased as baited trapping has succeeded in detecting native crayfish declines over time (Edwards et al. 2009) and documenting replacement by invasive crayfish (Westman et al. 2002). Reliable inferences of population or community change over time and space are not dependent on a perfectly accurate gear type, but rather knowing that the biases of the gear are also consistent over time and space. Consequently, although we suggest that more studies of sampling efficiency or accuracy of crayfish sampling gears may be valuable, we also propose that more might be gained by developing and implementing standardized sampling protocols for crayfish sampling gears where biases are already well-known (e.g., baited trapping, quadrat sampling, kick sampling or seining).

Evaluating precision with power analysis

Precision is how repeatable a measurement on the same object or subject is over time or space. An accurate sampling method is not necessarily precise, and a precise sampling method is not necessarily accurate. Returning to our previous hypothetical example (above), a single pass of seining a 10 m reach of stream for crayfish was less accurate than a single pass of electrofishing. But perhaps these estimates of abundance could be repeated multiple times at the same location after returning all crayfish to the stream reach, and assuming that crayfish behavior or mortality was unaffected by previous sampling and immigration/ emigration was minimal. If a single pass of seining was then repeated and collected 49 crayfish (relative to 50 previously) but the single pass of electrofishing was repeated and collected 150 crayfish (relative to 250 previously), we might conclude – preferably with more replicates – that seining is more precise than electrofishing. If the relationship between estimated and true abundance for the two gear types was consistent regardless of true abundance and across sites or habitat types, it would be preferable to favor seining over backpack electrofishing for many crayfish sampling applications. For example, if a researcher wanted to evaluate how abundance of crayfish varied between different streams, the greater precision of seining would allow for more statistical power (see below) in detecting true differences in abundance despite being inaccurate in representing absolute abundance. Conversely, a researcher that wanted to estimate absolute biomass or production of crayfish in a single stream might value accuracy over precision, endeavoring to collect all crayfish possible and finding seining less adequate for this goal. We use this hypothetical example to emphasize that accuracy and precision are often independent, and researchers may value one over the other dependent on their study objectives.

While the preceding seining vs. electrofishing scenario was hypothetical, examples abound of sampling approaches that are accurate but imprecise or the opposite. For crayfish, Rabeni et al. (1997) wrote "it was possible to obtain quite small confidence intervals using a particular technique even though the technique was highly biased," identifying hand collecting of crayfish in particular as being inaccurate by extremely precise. Perhaps this precision has contributed to the preference by some researchers for hand collection (e.g., Smith et al. 1996, Usio 2007) over sampling approaches identified as more accurate by Rabeni et al. (1997). One important reason for researchers to be conscientious of precision in designing field sampling programs for crayfish is its influence on statistical power. Statistical power is the ability to identify a difference or effect as significant or present where it actually exists (i.e., to avoid a Type II statistical error). Researchers are often well-advised to conduct a power analysis prior to designing and implementing a field sampling program to identify the number of replicates needed to answer their question; conversely, *post hoc* power analysis can be used to evaluate whether an effect of a given size could even have been detected with the replication used (e.g., Acosta and Perry 2002, Pintor and Sih 2009). Statistical power is dependent on the effect size itself; larger effects are easier to detect. Statistical power is also dependent on sampling replication; more replicates increase statistical power, and identifying the minimal number of replicates necessary to identify a given effect size is typically the focus of power analysis. Statistical power is affected by variance of the sampled population; high variance in quadrat sampling estimates of mean crayfish density caused low potential power for detecting differences in crayfish density between habitats or sites in Larson et al. (2008). Finally, power varies by significance level, although most researchers conducting a power analysis are likely to use the customary $\alpha = 0.05$. For further explanation and advice on statistical power and power analysis, we refer readers to statistics textbooks for biologists like Quinn and Keough (2002).

Only a few researchers have published results of power analyses for crayfish sampling, but these studies offer important cautionary advice with respect to the amount of replication needed to identify patterns of crayfish population and community processes in the field. For quadrat samples of varying sizes (1 m^2 and 10 m^2) in lakes, Lamontagne and Rasmussen (1993) found high power where crayfish densities were high, but suggested that high replication (e.g., > 75 1 m^2 quadrat samples) would be necessary to identify differences where crayfish densities were low. Larson et al. (2008) similarly suggested that extremely high precision (e.g., confidence intervals 10% of mean density) in quadrat samples of stream dwelling crayfish might require implausibly high replication (i.e., 82 quadrat samples), whereas more moderate precision (e.g., confidence intervals 40% of mean density) were attainable with more realistic albeit still high replication (i.e., 17 quadrat samples). DiStefano et al. (2003) reported reasonably good power in detecting differences in crayfish density between different habitat types in streams, but found lower power in detecting trends in crayfish abundance over time through five years of sampling. These examples demonstrate that researchers who have not considered power in designing field studies for crayfish may often lack the replication necessary to detect their effects or responses of interest. These studies also only address power for quadrat samples, whether by divers in lakes or the contained designed of Rabeni (1985) for wadeable streams and rivers. Studies reporting results of power analysis on other crayfish sampling approaches, such as baited trapping, appear particularly rare (but see Acosta and Perry 2002). How many traps per area or traps per lake are necessary to detect differences in relative abundance between sites? To detect all of the species (richness) present in a lake? We suggest that power analysis and statistical simulation of rich existing datasets (e.g., Edwards et al. 2009) would be valuable for answering methodological questions for crayfish field sampling in additional habitats and with additional gear types.

Accounting for detection probability in occupancy estimates

Much of our preceding discussion on field sampling for crayfish is focused on abundance or relative abundance as the response of interest. However, there may be instances where occupancy (presence/absence) is the preferred variable for purposes like documenting species distributions or monitoring trends through time over large landscapes. Although estimates of both abundance and occupancy are affected by incomplete detection of organisms (Royle et al. 2005), occupancy estimates are particularly sensitive to bias caused by imperfect detection (MacKenzie et al. 2003). Accordingly, much effort has recently been dedicated by wildlife biologists to develop sampling and statistical techniques to study occupancy while correcting for incomplete detection probabilities (MacKenzie et al. 2006, Jones 2011). This fits within a broader movement to better understand and account for the zero-inflated data common in ecological field studies (Martin et al. 2005). In general, these approaches typically use repeated visits to sampling sites to search for organism presence, and then these repeated visits are modeled similarly to mark-recapture population estimates (see below) to estimate occupancy while accounting for imperfect detection probability. This is done as a two step hierarchical process in which both occupancy (presence or absence) and detection when present (detected or not detected) are modeled, with detection evaluated by presumably independent estimates from the repeated site visits and the occupancy state between visits assumed closed (i.e., organism presence at a site is assumed to not change between repeated visits). Site covariates are often included in models to estimate not only occupancy but also detection probability; in the case of crayfish, it is not difficult to think of some site attributes that could affect detection probability. As noted in sections above, crayfish in temperate regions will often be easier to trap or collect when water temperatures are warm. Accordingly, water temperature could be a site covariate expected to influence detection of crayfish. Sampling too early in the year when water temperatures are cold might result in failure to detect a crayfish population where present, and repeated visits to the same site over a seasonal or temperature gradient could be used to identify the role of water temperature in detecting crayfish presence. Some additional site or environmental covariates that could affect crayfish detection where truly present might include lunar cycle (Araujo and Romaire 1989), stream or river flow with respect to floods or droughts (Parkyn and Collier 2004), or behavioral responses of crayfish to predatory fish (Collins et al. 1983).

Few studies have considered detection probability in estimating crayfish occupancy. Loughman et al. (2012) used repeated site visits to account for detection probability in sampling burrowing crayfish of West Virginia, USA. These authors did not include any specific environmental covariates that might influence detection probability, finding that best supported models included a constant detection probability over time-varying detection probabilities. More recently, Pearl et al. (2013) accounted for detection probability in studying and modeling occupancy for one native and two non-native crayfish species in western Oregon, USA. These authors did not make repeated site visits, but rather treated multiple crayfish traps per site as independent observation of occupancy. Covariates considered as potentially affecting crayfish detection included year sampled, date of the year, whether traps were baited or unbaited (approximately half of traps were unbaited), and habitat attributes including whether the substrate was silty, if the habitat was manmade, whether the habitat was lotic or lentic, and if the habitat had permanent connectivity to other nearby surface waters. These same habitat attributes were also considered as covariates for crayfish occupancy. Pearl et al. (2013) modeled occupancy as two-species single-season models that allowed for evaluation of how species mutually affected their co-occurrence for each of the two native and non-native species pairs. Pearl et al. (2013) found that whether or not traps were baited strongly affected crayfish detection probability, perhaps predictably given the known influence of bait on crayfish recruitment to traps (see above). Pearl et al. (2013) found per-trap detection probabilities of 0.11 to 0.44 depending on crayfish species, although we would anticipate from these results that single-visit per-site detection probabilities at typical levels of trap replication (e.g., 12–24 traps per lake in the USA; see above) would have considerably higher detection probabilities. Even ten traps per site visit would be anticipated to typically result in species detection at the lowest probability of only 0.11 per-trap reported by Pearl et al. (2013). Finally, these authors found that detection probabilities were higher for *P. leniusculus* in lotic than lentic habitats, whereas sampling date positively and manmade habitats negatively affected detection probability for the ringed crayfish *Orconectes neglectus*, although sample sizes were relatively low for this invasive species (Pearl et al. 2013).

Correction for detection probability in occupancy studies appears far more common in the field of terrestrial wildlife biology than in freshwater fisheries or ecology (but see Falke et al. 2012 and Sethi and Benolkin 2013 for good examples of applications to freshwater research). There are reasons why researchers may abstain from accounting for detection probability in field studies of occupancy. The repeated site visits often used to evaluate detection probability obviously come with costs of time and effort. A researcher might be faced with the option of visiting 100 sites once for a crayfish sampling survey, or visiting 25 sites four times each to make robust estimates of detection probability. Tyre et al. (2003) offer guidance on choosing among this tradeoff between statistical power (visiting more sites) and error (incorrectly inferring species absence due to false negatives); these authors generalize that when error rates of false negatives are ≤50% that researchers should sample more sites, whereas when error rates are >50% repeated site visits should be used. It may be that most crayfish researchers assume false negative errors rates ≤50%, a premise that could be evaluated with more studies explicitly evaluating detection probability when sampling for crayfish. Detection probability may also be evaluated without making repeated site visits. For example, detection probability might be evaluated for one gear type at a single sampling visit by also using a second gear type; studies of crayfish occupancy or abundance in lakes using baited traps could be complemented by divers searching quadrats or line transects (e.g., Collins et al. 1983). Yet while such pairing of sampling approaches may account for detection errors attributable to gear or habitat types, they will not account for detection errors with a temporal signature (e.g., water temperature). Further, studies comparing detection probability between different gear types still necessitate increased sampling effort per each individual site visit, likely reducing the overall number of site visits possible. Finally, Welsh et al. (2013) critically evaluated modeling approaches for occupancy accounting for detection probability, noting that these models often have multiple solutions, problems with convergence, and bias caused by the dependency of detection on abundance. These authors conclude that in some cases "trying to adjust occupancy models for non-detection can be as misleading as ignoring non-detection completely."

Despite the limitations and criticisms of occupancy approaches accounting for detection probability outlined above, we do suggest that this increasingly prominent methodology may be appropriate for some crayfish field sampling applications. Approaches accounting for imperfect detection probability provide a well-established sampling and modeling framework that could be applied to answer important methodological questions for crayfish sampling. How does crayfish detection probability vary by time? Between habitat types? And which gear types are best able to detect crayfish presence with the lowest rates of false negatives (e.g., Schloesser et al. 2012)? We also suggest that many rare crayfish species that occur at low abundances or patchily on the landscape may be best monitored with occupancy accounting for detection probability. For example, both primary burrowing (e.g., Loughman et al. 2012) and cave-dwelling crayfishes (e.g., Graening et al. 2006) can occur at low densities in difficult to sample habitats, and it is precisely these kind of species where occupancy estimation approaches accounting for detection probability are often recommended (MacKenzie et al. 2006). When monitoring populations of rare or endangered species, it may often be important to parse whether apparent trends are the product of actual population processes or instead noise from imperfect detection and false negatives. Occupancy estimation with detection probability offers a means of doing this that is more demanding than most sampling typically applied for crayfish studies (above) but still affords a potentially broader landscape or multi-site focus than even more labor-intensive mark-recapture studies (below). As a result, we anticipate that many more papers specific to detection probability issues and crayfish sampling may be published in the near future.

Evaluating sampling techniques with mark-recapture

Mark-recapture has a distinguished history as one of the most important tools in fish and wildlife conservation and management (Pollock et al. 1990). Among the earliest published applications of mark-recapture to studying crayfish populations was by Johnson (1971), who used this technique to produce an estimate of total population size of *P. leniusculus* in a small lake of Washington State, USA. Over the following decades, many other researchers have used mark-recapture to characterize crayfish population size, demographic rates, or trends through time or between habitat types (Norrocky 1991, Parkyn et al. 2002, Jones and Coulson 2006, Nowicki et al. 2008). Researchers interested in applying mark-recapture techniques to studying crayfish populations should consult general sources like Pollock et al. (1990), or Nowicki et al.'s (2008) specific recommendations and limitations for monitoring crayfish with mark-recapture. Our interest here is in briefly reviewing mark-recapture for crayfish as a means of evaluating other field sampling protocols and gears for these organisms. Owing to the difficulty of knowing absolute crayfish population sizes in many study systems (see above), mark-recapture estimates of population size may be the only avenue to evaluating accuracy of measures of relative abundance in representing populations. For example, Zimmerman and Palo (2011) used population estimates from mark-recapture to evaluate the accuracy of CPUE from baited traps for several rivers in Sweden. Rabeni et al. (1997) depended on mark-recapture population estimates for evaluating a number of crayfish sampling gears in a wadeable stream, and preferred this approach over population estimates from serial depletion during a single site visit. Even researchers who are not interested in producing estimates of total crayfish population size may borrow some tools or techniques from mark-recapture studies in evaluating crayfish sampling gears. Larson et al. (2008) used a known population size of stocked, marked crayfish to evaluate accuracy and precision of the quadrat sampler for estimating densities of stream-dwelling crayfish. Consequently, we also briefly comment on methods for marking crayfish, whether for mark-recapture population estimates or other applications.

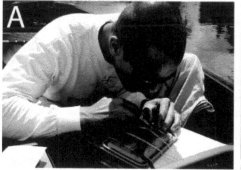 

Figure 5. A researcher marks a red swamp crayfish *Procambarus clarkii* with visible implant alphanumeric tags (A). The tag is being injected into the abdomen of the crayfish while it is immobilized with rubber bands on a plastic board (B). All photographs by the authors.

At its simplest, mark-recapture refers to a process of estimating population size (and other demographic parameters, e.g., Jones and Coulson 2006) by marking individual organisms that have been sampled from a population, returning them to the population, revisiting the site after a time interval, and then sampling again to evaluate the proportion of marked organisms to all individuals collected. From this proportion of marked to unmarked individuals a total population size estimate is possible, contingent on assumptions that the population is effectively closed to immigration, emigration, mortality, or birth between sampling intervals. Different models to account for different sampling structures or violations of assumptions (e.g., closure) are reviewed in sources like Pollock et al. (1990), and probability of capture can also be evaluated by use of repeated site visits (see section above on detection probability in occupancy studies). A good summary of mark-recapture for studying crayfish is presented by Nowicki et al. (2008), who used an ongoing annual monitoring program for adult *Austropotamobius pallipes* in an Italian river system to discuss and synthesize strengths and limitations of mark-recapture techniques for crayfish. Nowicki et al. (2008) found that the adult component of populations of *A. pallipes* appeared closed except during the winter, likely due to high mortality during this season, and early summer, possibly due to emigration from study sites. Consequently, Nowicki et al. (2008) proposed that populations of this crayfish be considered closed for seasons including spring and late summer to early autumn. Nowicki et al. (2008) suggested these two seasons as primary sampling periods for rigorous studies of crayfish ecology, with five capture sessions per sampling period separated by an interval of two weeks. This sampling interval was found to reduce a strong but short-term negative effect of crayfish capture on recapture. Nowicki et al. (2008) proposed a simpler sampling program for basic estimates of crayfish population size, suggesting that a single sampling period in late summer and autumn with as few as three capture sessions could produce fairly precise population estimates. Nowicki et al. (2008) argued that their methodological recommendations should be generalizeable to crayfish with ecology similar to *A. pallipes*, which we interpret as meaning similar habitat types (i.e., wadeable streams and rivers) in temperate climates.

We are not confident that Nowicki et al.'s (2008) recommendations will extrapolate to crayfish with different habitat specializations or to tropical regions with markedly different temperature seasonality and precipitation regimes. Researchers working in tropical regions might refer to Jones and Coulson (2006), who used mark-recapture to study demography of a harvested freshwater crayfish in Madagascar. One important conclusion from Jones and Coulson (2006) is that repeated handling through recapture did affect crayfish mortality, a factor that should be incorporated in models or by minimizing recapture events (see Nowicki et al. 2008 above). Regardless, simple mark-recapture protocols for estimating crayfish population size like those advocated by Nowicki et al. (2008) could be extremely valuable for evaluating other gears that instead represent crayfish relative abundance; this could be done not only for baited trapping (Zimmerman and Palo 2011) but also techniques like kick sampling, quadrat sampling, or visual searches by divers (e.g., Rabeni et al. 1997), not to mention sampling approaches for cave-dwelling or terrestrial crayfish. While mark-recapture may be an excellent tool for studying or monitoring one to few populations of crayfish, even in its simplest implementations it is unmanageable for studying crayfish at dozens to hundreds of sites or locations as in Edwards et al. (2009) or Larson and Olden (2013). As all field sampling approaches come with tradeoffs between rigor at individual sites and the number and distribution of sites that can be visited (Jones 2011), researchers should choose sampling approaches that best fit their needs and study questions. This may mean mark-recapture studies at one to few study sites, occupancy estimation accounting for detection probability at an intermediate number of sites, and estimates of occupancy or relative abundance without evaluating detection probabilities at higher numbers of sites. Importantly, we recommend here that lessons learned from mark-recapture or detection probability studies with respect to field sampling techniques may be extremely informative in choosing sampling approaches when scaling-up crayfish surveys to larger landscapes.

Related to mark-recapture studies, a wide variety of different approaches have been recommended for marking individual crayfish. These range from systems of physically marking crayfish like punching holes in uropods (Guan 1997) or branding by freezing or heat (Abrahamsson 1965) to more recent tagging systems including visible implant elastomer (VIE; Clark and Kershner 2006), visible implant alphanumeric (VIalpha; Jerry et al. 2001) and passive integrated transponder (PIT; Black et al. 2010) tags. Haddaway et al. (2011) provided an excellent review and summary of different crayfish tagging methods. This review included comparing the costs of marking or tagging systems, whether they're internally or externally located, how they're detected (visual, dissection, electromagnet, etc.), retention rates of tags or marks, mortality rates of crayfish, and effects of marks or tags on crayfish growth. Combining this summary with an experimental comparison between electric cauterization of crayfish and VIE tags, Haddaway et al. (2011) ultimately concluded that VIE tags are among the best options for marking crayfish due to their low cost and lack of detectable effects on crayfish growth or survival. In our experience VIE tags are an excellent tool for marking crayfish, but there are many good options for crayfish researchers to consider.

Nowicki et al.'s (2008) mark-recapture study discussed above used the simple uropod ablation system of Guan (1997), and PIT tags have the advantage of being detected by a hand held reader that can now detect tags up to 18 cm distant and allows for systematic searching and detection of tagged crayfish in the field even when sheltered under substrate or in burrows (e.g., Bubb et al. 2002). Beyond such PIT tag systems, we also note that extremely valuable work on crayfish behavior and movement has been conducted using larger radiotelemetry tags; although not under the umbrella of field sampling for crayfish, we refer readers interested in this subject to sources like Robinson et al. (2000) and Webb and Richardson (2004) and the book chapters here on crayfish ecology and behavior.

Conclusions

The preceding book chapter demonstrates that we are not lacking for published studies on the subject of field sampling for crayfish. Researchers initiating new studies on crayfish in the field generally do not need to re-invent the wheel; many good resources exist that can provide guidance on crayfish sampling for a range of purposes in a diversity of habitat types. While we might like to see additional rigorous evaluations of the accuracy and precision of some common crayfish sampling techniques, and we recommend above a number of sampling and modeling tools applicable to this end, the development of new approaches to field sampling for crayfish does not seem to be a pressing need. Rather, we conclude by emphasizing the importance of finding order and agreement in a diverse and generally uncoordinated crayfish sampling literature. Field sampling approaches for crayfish have developed with little coordination between researchers and regions, resulting in a disparate array of preferences that hinder transferability of study results for crayfish ecology and conservation. For example, it may be easier to characterize the ecology of widespread invasive crayfishes between native and introduced ranges if researchers working on these organisms are sampling similarly. Likewise, many crayfish of conservation concern span multiple political jurisdictions, whether within or between nations, and agreement among researchers in different regions to use the same sampling protocols could facilitate better assessments of conservation status and trends.

The field of ecology is increasingly interested in data transparency, sharing, and accessibility. Such efforts are not only dependent on the willingness or ability of researchers to archive their data in publicly accessible places (Hampton et al. 2012), but also for researchers to agree on standard sampling methods and protocols for similar habitats and species (Bonar and Hubert 2002). As an example, the American Fisheries Society has recently published recommended standard methods for sampling freshwater fish in North American, and is developing a depository where data collected under these standardized sampling methods can be accessed and shared between researchers and management agencies (Bonar et al. 2009). The preceding book chapter demonstrates that we are far from such universally agreed on standard sampling protocols for field work on crayfish. Widely varying preferences can be found even for an approach as seemingly simple as baited trapping for crayfish, with different researchers preferring different trap designs, bait types, and trapping intensity. For the immediate future, we might suggest that researchers initiating new field studies of crayfish start by consulting with sampling and monitoring protocols developed for their region, a nearby region, or for a region they may wish to compare their results to. We are aware of suggested crayfish sampling protocols for Sweden (Edsman and Söderbäck 1999), the United Kingdom (Peay 2003), provinces of Canada (David et al. 1994) and states of the USA (Simon 2004, Maxted and Vander Zanden 2007). These existing crayfish sampling guides are also good candidates for the first places to look in synthesizing and developing standardized sampling protocols for application across state, provincial, or national borders.

In the longer term, arriving at a set of standardized sampling practices for crayfish will likely require coordination among an international community of researchers. We have not translated our own sampling preferences into recommended standard protocols here in part because we appreciate that our limited taxonomic and geographic experiences may not incorporate the needs of researchers working in different systems or on different species. Instead, we propose that standardized guidelines for crayfish field sampling might best arise through the coordination of a professional society like the International Association of Astacology (IAA). The IAA has commendably advanced our understanding of crayfish through its biennial international meeting and corresponding publication of proceedings as the journal

Freshwater Crayfish. Owing to the international scope of IAA, it might be possible for this professional society to convene experts from across continents and countries for a working group with the specific goal of producing recommended standardized protocols for crayfish field sampling. These researchers would ideally make recommendations of best practices for all habitat types used by crayfish, from lentic and lotic waters to caves and terrestrial burrows, and include perspectives on sampling for crayfish from tropical to temperate regions. Further, like the American Fisheries Society example discussed above, it may be possible for IAA or a cooperating museum or university to develop an online depository or archive for crayfish field sampling data. This would allow managers and researchers to better follow distributional changes of crayfish species over space as well trends in individual populations over time. If such a coordinated international effort is ultimately infeasible, we would still urge regional groups of scientists to attempt to develop continental-scale standardized practices, perhaps facilitated through groups that meet to evaluate crayfish conservation status for North America (Taylor et al. 2007) or Europe (Souty-Grosset et al. 2006).

We do not naively assume that a single standardized sampling protocol for crayfish will be useful for every researcher in every circumstance. Inevitably, researchers will need to exercise their own judgment to resolve some unique sampling challenges that arise. And there may be some areas where new advancements in field sampling may yet transform how we study crayfish in the field; for example, our suggestion (above) that environmental DNA might be a powerful tool for monitoring rare crayfish populations or difficult to sample environments. This emerging approach has recently received the first of likely many subsequent applications to and evaluations for crayfish (Tréguier et al. 2014). Yet as Bonar and Hubert (2002) outline, field biologists often fall back to a routine set of excuses for postponing adopting standardized sampling protocols: that these protocols suppress their own judgment or expertise, inhibit creativity, aren't applicable to their "special" or "different" study system or species, and create breaks with historical data sets or monitoring programs using local or idiosyncratic sampling approaches. We side with Bonar and Hubert (2002) that the benefits of developing and implementing standardized sampling protocols outweigh these perceived disadvantages. Standardization of techniques has been critical to the advancement of science, and many disciplines that field ecology is inherently dependent on, ranging from meteorology to water quality and chemistry monitoring, use data collection and reporting protocols that are standardized between research groups and political jurisdictions. We propose here that the greatest future gains in studying crayfish in the field will not necessarily come from developing new sampling techniques, but rather from improving our ability to compare and synthesize research results between different regions, and that this will be dependent on convincing crayfish researchers to standardize sampling and data reporting. As other chapters in this book attest, many of our greatest challenges in crayfish management are international in scope, and we believe that moving toward regionally or internationally recognized best practices for crayfish field sampling may be a critical foundation for improving our understanding and conservation of these organisms.

References

Abrahamsson, S.A.A. 1965. A method of marking crayfish *Astacus astacus* (Linne) in population studies. Oikos 16: 228–231.

Abrahamsson, S.A.A. and C.R. Goldman. 1970. Distribution, density and production of the crayfish *Pacifastacus leniusculus* Dana in Lake Tahoe, California-Nevada. Oikos 21: 83–91.

Acosta, C.A. and S.A. Perry. 2000. Effective sampling area: a quantitative method for sampling crayfish populations in freshwater marshes. Crustaceana 73: 425–431.

Acosta, C.A. and S.A. Perry. 2002. Spatio-temporal variation in crayfish production in disturbed marl prairie marshes of the Florida Everglades. J. Freshwater Ecol. 17: 641–650.

Adams, S.B. and M.L. Warren. 2005. Recolonization by warmwater fishes and crayfishes after severe drought in upper coastal plain hill streams. Trans. Am. Fish. Soc. 134: 1173–1192.

Alonso, F. 2001. Efficiency of electrofishing as a sampling method for freshwater crayfish populations in small creeks. Limnetica 20: 59–72.

Angermeier, P.L., R.A. Smogor and S.D. Steele. 1991. An electric seine for collecting fish in streams. N. Am. J. Fish. Manage. 11: 352–357.

Araujo, M.A. and R.P. Romaire. 1989. Effects of water quality, weather and lunar phase on crawfish catch. J. World Aquac. Soc. 20: 199–207.

Bateman, D.S., R.E. Gresswell and C.E. Torgersen. 2005. Evaluating single-pass catch as a tool for identifying spatial pattern in fish distribution. J. Freshwat. Ecol. 20: 335–345.

Bayley, P.B. and D.C. Dowling. 1990. Gear efficiency calibrations for stream and river sampling. Aquatic Ecology Technical Report 90/8, Illinois Natural History Survey, Champaign, Illinois.

Bayley, P.B. and D.C. Dowling. 1993. The effect of habitat in biasing fish abundance and species richness estimates when using various sampling methods in streams. Polish Archives of Hydrobiology 40: 5–14.

Bayley, P.B. and J.T. Peterson. 2001. An approach to estimate probability of presence and richness of fish species. Trans. Am. Fish. Soc. 130: 620–633.

Bean, R.A. and J.V. Huner. 1979. An evaluation of selected crawfish traps and trapping methods. Freshwater Crayfish 4: 141–151.

Beecher, L.E. and R.P. Romaire. 2010. Evaluation of baits for harvesting Procambarid crawfishes with emphasis on bait type and bait quantity. Journal of Shellfish Research 29: 13–18.

Bentley, A.I., D.J. Schmidt and J.M. Hughes. 2010. Extensive intraspecific genetic diversity of a freshwater crayfish in a biodiversity hotspot. Freshwater Biology 55: 1861–1873.

Bernardo, J.M., M. Ilheu and A.M. Costa. 1997. Distribution, population structure and conservation of *Austropotamobius pallipes* in Portugal. Bull. Fr. Peche. Piscic. 347: 617–624.

Berrill, M. and B. Chenoweth. 1982. The burrowing ability of nonburrowing crayfish. Am. Midl. Nat. 108: 199–201.

Black, T.R., S.S. Herleth-King and H.T. Mattingly. 2010. Efficacy of internal PIT tagging of small-bodied crayfish for ecological study. Southeast. Nat. 9: 257–266.

Bonar, S.A. and W.A. Hubert. 2002. Standard sampling of inland fish: benefits, challenges, and a call for action. Fisheries 27: 10–16.

Bonar, S.A., W.A. Hubert and D.W. Willis. 2009. Standard Methods for Sampling North American Freshwater Fishes. American Fisheries Society, Bethesda, Maryland.

Brewer, S.K., R.J. DiStefano and C.F. Rabeni. 2009. The influence of age-specific habitat selection by a stream crayfish community (*Orconectes* spp.) on secondary production. Hydrobiologia 619: 1–10.

Brown, D.J. and J.M. Brewis. 1978. A critical look at trapping as a method of sampling a population of *Austropotamobius pallipes* (Lereboullet) in a mark and recapture study. Freshwater Crayfish 4: 159–164.

Bryant, S. and J. Jackson. 1999. Tasmania's Threatened Fauna Handbook. Threatened Species Unit, Parks and Wildlife Service, Department of Primary Industries, Water, and Environment, Tasmania.

Bubb, D.H., M.C. Lucas, T.J. Thom and R. Rycroft. 2002. The potential use of PIT telemetry for identifying and tracking crayfish in their natural environment. Hydrobiologia 483: 225–230.

Bubb, D.H., T.J. Thom and M.C. Lucas. 2004. Movement and dispersal of the invasive signal crayfish *Pacifastacus leniusculus* in upland rivers. Freshwater Biology 49: 357–368.

Buckland, S.T., D.R. Anderson, K.P. Burnham and J.L. Laake. 1993. Distance Sampling: Estimating Abundance of Biological Populations. Chapman and Hall, London.

Buhay, J.E., G. Moni, N. Mann and K.A. Crandall. 2007. Molecular taxonomy in the dark: evolutionary history, phylogeography, and diversity of cave crayfish in the subgenus *Aviticambarus*, genus *Cambarus*. Molecular Phylogenetics and Evolution 42: 435–448.

Burskey, J.L. and T.P. Simon. 2010. Reach- and watershed-scale associations of crayfish within an area of varying agricultural impacts in west-central Indiana. Southeastern Naturalist 9: 199–216.

Cange, S.W., D. Pavel, C. Burns, R.P. Romaire and J.W. Avault, Jr. 1986. Evaluation of eighteen artificial crayfish baits. Freshwater Crayfish 6: 270–273.

Capelli, G.M. and J.J. Magnuson. 1976. Reproduction, moulting, and distribution of *Orconectes propinquus* (Girard) in relation to temperature in a northern mesotrophic lake. Freshwater Crayfish 2: 415–564.

Capelli, G.M. and J.J. Magnuson. 1983. Morphoedaphic and biogeographic analysis of crayfish distribution in northern Wisconsin. Journal of Crustacean Biology 3: 548–564.

Charlebois, P.M. and G.A. Lamberti. 1996. Invading crayfish in a Michigan stream: direct and indirect effects on periphyton and macroinvertebrates. Journal of the North American Benthological Society 15: 551–563.

Chucholl, C. 2011. Population ecology of an alien "warm water" crayfish (*Procambarus clarkii*) in a new cold habitat. Knowledge and Management of Aquatic Ecosystems 401: 29.

Clark, J.M. and M.W. Kershner. 2006. Size-dependent effects of visible implant elastomer on crayfish (*Orconectes obscures*) growth, mortality, and tag retention. Crustaceana 79: 275–284.

Claussen, D.L., R.A. Hopper and A.M. Sanker. 2000. The effects of temperature, body size, and hydration state on the terrestrial locomotion of the crayfish *Orconectes rusticus*. Journal of Crustacean Biology 20: 218–223.

Collins, N.C., H.H. Harvey, A. Tierney and D.W. Dunham. 1983. Influence of predator density on trapability of crayfish in Ontario lakes. Canadian Journal of Fisheries and Aquatic Sciences 40: 1820–1828.

Cooper, J.E. 1975. Ecological and behavioral studies in Shelta Cave, Alabama, with emphasis on decapods crustaceans. Ph.D. Dissertation, University of Kentucky, Lexington, Kentucky.

Cowx, I.G. 1990. Developments in Electric Fishing. Blackwell Science Publications, Oxford.

Crandall, J.A. 2000. Crayfish as model organisms. Freshwater Crayfish 13: 3–12.

Cukerzis, J. 1959. Zählmetoden zur Bestimmung, sowie Aufzuchtverfahren und Schonmassnahemen zer Hebung der Edelkrebsbestände Litauens. Ubersetzung Nr. 140, Statens Naturvetenskapliga Forskningsgrad, Stockholm.

David, S.M., J.M. Somers and R.A. Reid. 1994. Long-term trends in the relative abundance of crayfish from acid sensitive, softwater lakes in south central Ontario: a data summary for the first 5 years, 1988–1992. Ontario Ministry of the Environment, Dorset, Ontario.

Davies, I.J. 1989. Population collapse of the crayfish *Orconectes virilis* in response to experimental whole-lake acidification. Canadian Journal of Fisheries and Aquatic Sciences 46: 910–922.

Davies, I.J. and D.J. Ramsey. 1989. A diver operated suction gun and collection bucket for sampling crayfish and other aquatic macroinvertebrates. Canadian Journal of Fisheries and Aquatic Sciences 46: 923–927.

de Graaf, M., S. Beatty and B.M. Molony. 2010. Evaluation of the recreational marron fishery against environmental change and human interaction. Fisheries Research Report No. 211, Western Australia Department of Fisheries, North Beach, Western Australia.

Dickson, G.W., L.A. Briese and J.P. Giesy. 1979. Tissue metal concentrations in two crayfish species cohabiting a Tennessee cave stream. Oecologia 44: 8–12.

DiStefano, R.J., C.M. Gale, B.A. Wagner and R.D. Zweifel. 2003. A sampling method to assess lotic crayfish communities. Journal of Crustacean Biology 23: 678–690.

DiStefano, R.J., D.D. Magoulick, E.M. Imhoff and E.R. Larson. 2009. Imperiled crayfishes use hyporheic zone during seasonal drying of an intermittent stream. Journal of the North American Benthological Society 28: 132–152.

Dorn, N.J. and J.C. Volin. 2009. Resistance of crayfish (*Procambarus* spp.) populations to wetland drying depends on species and substrate. Journal of the North American Benthological Society 28: 766–777.

Dorn, N.J., R. Urgelles and J.C. Trexler. 2005. Evaluating active and passive sampling methods to quantify crayfish density in a freshwater wetland. Journal of the North American Benthological Society 24: 346–356.

Edgar, G.J., N.S. Barrett and A.J. Morton. 2004. Biases associated with the use of underwater visual census techniques to quantify the density and size-structure of fish populations. Journal of Experimental Marine Biology and Ecology 308: 269–290.

Edsman, L. and B. Söderbäck. 1999. Standardised sampling method for crayfish—the Swedish protocol. Freshwater Crayfish 12: 705–713.

Edwards, B.A., D.A. Jackson and K.M. Somers. 2009. Multispecies crayfish declines in lakes: implications for species distributions and richness. Journal of the North American Benthological Society 28: 719–732.

Elphick, C.S. 2008. How you count counts: the importance of methods research in applied ecology. Journal of Applied Ecology 45: 1313–1320.

Elser, J.J., C. Junge and C.R. Goldman. 1994. Population structure and ecological effects of the crayfish *Pacifastacus leniusculus* in Castle Lake, California. Great Basin Naturalist 54: 162–169.

Englund, G. 1999. Effects of fish on the local abundance of crayfish in stream pools. Oikos 87: 48–56.

Englund, G. and J.J. Krupa. 2000. Habitat use by crayfish in stream pools: influence of predators, depth and body size. Freshwater Biology 43: 75–83.

Ensign, W.E., P.L. Angermeier and C.A. Dolloff. 1995. Use of line transect methods to estimate abundance of benthic stream fishes. Canadian Journal of Fisheries and Aquatic Sciences 52: 213–222.

Evans-White, M.A., W.K. Dodds and M.R. Whiles. 2003. Ecosystem significance of crayfishes and stonerollers in a prairie stream: functional differences between co-occurring omnivores. Journal of the North American Benthological Society 22: 423–441.

Falke, J.A., L.L. Bailey, K.D. Fausch and K.R. Bestgen. 2012. Colonization and extinction in dynamic habitats: an occupancy approach for a Great Plains stream fish assemblage. Ecology 93: 858–867.

Fausch, K.D., C.E. Torgersen, C.V. Baxter and H.W. Li. 2002. Landscapes to riverscapes: Bridging the gap between research and conservation of stream fishes. BioScience 52: 483–498.

Fisher, W.L. 1987. Benthic fish sampler for use in riffle habitats. Transactions of the American Fisheries Society 116: 768–772.

Fjälling, A. 1995. Crayfish traps employed in Swedish fisheries. Freshwater Crayfish 8: 201–214.

Fjälling, A.B. 2011. The enclosure trap, a new tool for sampling juvenile crayfish. Knowledge and Management of Aquatic Ecosystems 401: 09.

Flinders, C.A. and D.D. Magoulick. 2003. Effects of stream permanence on crayfish community structure. American Midland Naturalist 149: 134–147.

Flinders, C.A. and D.D. Magoulick. 2005. Distribution, habitat use and life history of stream-dwelling crayfish in the Spring River drainage of Arkansas and Missouri with a focus on the imperiled Mammoth Spring crayfish (*Orconectes marchandi*). American Midland Naturalist 154: 358–374.

Flinders, C.A. and D.D. Magoulick. 2007. Effects of depth and crayfish size on predation risk and foraging profitability of a lotic crayfish. Journal of the North American Benthological Society 26: 767–778.

Flint, R.W. 1977. Seasonal activity, migration and distribution of the crayfish, *Pacifastacus leniusculus*, in Lake Tahoe. American Midland Naturalist 97: 280–292.

France, R., J. Holmes and A. Lynch. 1991. Use of size-frequency data to estimate the age composition of crayfish populations. Canadian Journal of Fisheries and Aquatic Sciences 48: 2324–2332.

Fulton, C.J., D. Starrs, M.P. Ruibal and B.C. Ebner. 2012. Counting crayfish: active searching and baited cameras trump conventional hoop netting in detecting *Euastacus armatus*. Endangered Species Research 19: 39–45.

Furse, J.M., C.H. Wild and N.N. Villamar. 2004. In-stream and terrestrial movements of *Euastacus sulcatus* in the Gold Coast hinterland: developing and testing a method of assessing freshwater crayfish movements. Freshwater Crayfish 14: 213–220.

Garvey, J.E., J.E. Rettig, R.A. Stein, D.M. Lodge and S.P. Klosiewski. 2003. Scale-dependent associations among fish predation, littoral habitat, and distributions of crayfish species. Ecology 84: 3339–3348.

Gladman, Z.F., W.E. Yeomans, C.E. Adams, C.W. Bean, D. McColl, J.P. Olszewska, C.W. McGillivray and R. McCluskey. 2010. Detecting North American signal crayfish (*Pacifastacus leniusculus*) in riffles. Aquatic Conservation: Marine and Freshwater Ecosystems 20: 588–594.

Graening, G.O., M.E. Slay, A.V. Brown and J.B. Koppelman. 2006. Status and distribution of the endangered Benton Cave crayfish, *Cambarus aculabrum* (Decapoda: Cambaridae). Southwestern Naturalist 51: 376–381.

Grey, J. and M.C. Jackson. 2012. 'Leaves and eats shoots': Direct terrestrial feeding can supplement invasive red swamp crayfish in times of need. Plos One 7(8): e42575.

Grow, L. and H. Merchant. 1980. The burrow habitat of the crayfish, *Cambarus diogenes diogenes*. American Midland Naturalist 103: 231–237.

Guan, R.-Z. 1997. An improved method for marking crayfish. Crustaceana 70: 641–652.

Guan, R.-Z. 2000. Abundance and production of the introduced signal crayfish in a British lowland river. Aquaculture International 8: 59–76.

Gurevitch, J., P.S. Curtis and M.H. Jones. 2001. Meta-analysis in ecology. Advances in Ecological Research 32: 199–247.

Haddaway, N.R., R.J.G. Mortimer, M. Christmas and A.M. Dunn. 2011. A review of marking techniques for Crustacea and experimental appraisal of electric cauterisation and visible implant elastomer tagging for *Austropotamobius pallipes* and *Pacifastacus leniusculus*. Freshwater Crayfish 18: 55–67.

Hampton, S.E., J.J. Tewksbury and C.A. Strasser. 2012. Ecological data in the information age. Frontiers in Ecology and the Environment 10: 59.

Harley, S.J., R.A. Myers and A. Dunn. 2001. Is catch-per-unit-effort proportional to abundance? Canadian Journal of Fisheries and Aquatic Sciences 58: 1760–1772.

Harlioğlu, M.M. 1999. The efficiency of the Swedish trappy in catching freshwater crayfish *Pacifastacus leniusculus* and *Astacus leptodactylus*. Turkish Journal of Zoology 23: 93–98.

Harper, D.M., A.C. Smart, S. Coley, S. Schmitz, A.-C. Gouder de Beauregard, R. North, C. Adams, P. Obade and M. Kamau. 2002. Distribution and abundance of the Louisiana red swamp crayfish *Procambarus clarkii* Girard at Lake Naivasha, Kenya between 1987 and 1999. Hydrobiologia 488: 143–151.

Hayes, J.W. and B.B. Baird. 1996. Estimating relative abundance of juvenile brown trout in rivers by underwater census and electrofishing. New Zealand Journal of Marine and Freshwater Research 28: 243–255.

Hein, C.L., B.M. Roth and M.J. Vander Zanden. 2006. Fish predation and trapping for rusty crayfish (*Orconectes rusticus*) control: a whole-lake experiment. Canadian Journal of Fisheries and Aquatic Sciences 63: 383–393.

Hein, C.L., M.J. Vander Zanden and J.J. Magnuson. 2007. Intensive trapping and increased fish predation cause massive population decline of an invasive crayfish. Freshwater Biology 52: 1134–1146.

Hess, A.D. 1941. New limnological sampling equipment. Limnological Society of America Special Publication 6: 1–5.

Hobbs, H.H., Jr. 1942. The crayfishes of Florida. University of Florida Biological Science Series 3: 1–179.

Hobbs, H.H., Jr. 1981. The crayfishes of Georgia. Smithsonian Contributions to Zoology 318: 1–549.

Hobbs, H.H., Jr., H.H. Hobbs III and M.A. Daniel. 1977. A review of the troglobitic Decapod crustaceans of the Americas. Smithsonian Contributions to Zoology 244: 1–177.

Hockley, N.J., J.P.G. Jones, F.B. Andriahajaina, A. Manica, E.H. Ranambitsoa and J.A. Randriamboahary. 2005. When should communities and conservationists monitor exploited resources? Biodiversity and Conservation 14: 2795–2806.

Hogger, J.B. 1988. Ecology, population biology and behavior. pp. 114–144. *In*: D.M. Holdich and R.S. Lowery (eds.). Freshwater Crayfish: Biology, Management and Exploitation. Croom Helm Ltd., Kent, United Kingdom.

Holdich, D.M. and I.D. Reeve. 1991. Distribution of freshwater crayfish in the British Isles, with particular reference to crayfish plague, alien introductions and water quality. Aquatic Conservation: Marine and Freshwater Ecosystems 1: 139–158.

Hopper, J.D. and A.D. Huryn. 2012. A new, non-destructive method for sampling burrowing crayfish. Southeastern Naturalist 11: 43–48.

Horwitz, P.H.J. and A.M.M. Richardson. 1986. An ecological classification of the burrows of Australian freshwater crayfish. Australian Journal of Marine and Freshwater Research 37: 237–242.

Horwitz, P.H.J., A.M.M. Richardson and P.M. Cramp. 1985. Aspects of the life history of the burrowing crayfish, *Engaeus leptorhynchus*, at Rattrays Marsh, North East Tasmania. Tasmanian Naturalist 82: 1–5.

Hughes, J.M. and M.K. Hillyer. 2003. Patterns of connectivity among populations of *Cherax destructor* (Decapoda: Parastacidae) in western Queensland, Australia. Marine and Freshwater Research 54: 587–596.

Huner, J.V. and J. Espinoza. 2004. Effect of trap inner funnel diameter on crayfish catch. Freshwater Crayfish 14: 59–69.

Huner, J.V., V.A. Pfister, R.P. Romaire and T.J. Baum. 1990. Effectiveness of commercially formulated and fish baits in trapping Cambarid crayfish. Journal of the World Aquaculture Society 21: 288–294.

Huntsman, B.M., V.P. Vernarsky and J.P. Benstead. 2011. Relating carrion breakdown rates to ambient resource level and community structure in four cave stream ecosystems. Journal of the North American Benthological Society 30: 882–892.

Huxley, T.H. 1884. The Crayfish: An Introduction the Study of Zoology. D. Appleton and Company, New York.

Jackson, D.A., P.R. Peres-Neto and J.D. Olden. 2001. What controls who is where in freshwater fish communities—The roles of biotic, abiotic and spatial factors. Canadian Journal of Fisheries and Aquatic Sciences 58: 157–170.

Jansen, W., N. Geard, T. Mosindy, G. Olson and M. Turner. 2009. Relative abundance and habitat association of three crayfish (*Orconectes virilis, O. rusticus,* and *O. immunis*) near an invasion front of *O. rusticus*, and long-term changes in their distribution in Lake of the Woods, Canada. Aquatic Invasions 4: 627–649.

Jerde, C.L., A.R. Mahon, W.L. Chadderton and D.M. Lodge. 2011. "Sight-unseen" detection of rare aquatic species using environmental DNA. Conservation Letters 4: 150–157.

Jerry, D.R., T. Stewart, I.W. Purvis and L.R. Piper. 2001. Evaluation of visual implant elastomer and alphanumeric internal tags as a method to identify juveniles of the freshwater crayfish, *Cherax destructor.* Aquaculture 193: 149–154.

Johnson, E.A. 1971. Biological studies on the crayfish, *Pacifastacus leniusculus trowbridgii* (Stimpson), in Fern Lake, Washington. M.S. Thesis, University of Washington, Seattle, Washington.

Johnston, C.E. and C. Fiegel. 1997. Microhabitat parameters and life-history characteristics of *Fallicambarus gordoni* Fitzpatrick, a crayfish associated with pitcher-plant bogs in southern Mississippi. Journal of Crustacean Biology 17: 687–691.

Johnston, K. and B.J. Robson. 2009a. Habitat use by five sympatric Australian freshwater crayfish species (Parastacidae). Freshwater Biology 54: 1629–1641.

Johnston, K. and B.J. Robson. 2009b. Commensalism used by freshwater crayfish species to survive drying in seasonal habitats. Invertebrate Biology 128: 269–275.

Jones, J.P.G. 2011. Monitoring species abundance and distribution at the landscape scale. Journal of Applied Ecology 48: 9–13.

Jones, J.P.G. and T. Coulson. 2006. Population regulation and demography in a harvested freshwater crayfish from Madagascar. Oikos 11: 602–611.

Jones, J.P.G., F.B. Andriahajaina, E.H. Ranambinintsoa, N.J. Hockley and O. Ravoahangimalala. 2006. The economic importance of freshwater crayfish harvesting in Madagascar and the potential of community-based conservation to improve management. Oryx 40: 168–175.

Jones, J.P.G., F.B. Andriahajaina, N.J. Hockley, K.A. Crandall and O.R. Ravoahangimalala. 2007. The ecology and conservation status of Madagascar's endemic freshwater crayfish (Parastacidae; *Astacoides*). Freshwater Biology 52: 1820–1833.

Jones, J.P.G., M.M. Andriamarovololona, N. Hockley, J.M. Gibbons and E.J. Milner-Gulland. 2008. Testing the use of interviews as a tool for monitoring trends in the harvesting of wild species. Journal of Applied Ecology 45: 1205–1212.

Keller, T.A., A.M. Tomba and P.A. Moore. 2001. Orientation in complex chemical landscapes: spatial arrangement of chemical sources influences crayfish food-finding efficiency in artificial streams. Limnology and Oceanography 46: 238–247.

Kershner, M.W. and D.M. Lodge. 1995. Effects of littoral habitat and fish predation on the distribution of an exotic crayfish, *Orconectes rusticus.* Journal of the North American Benthological Society 14: 414–422.

Khanipour, A.A. and V.N. Melnikov. 2007. Determination of suitable trap type for the Caspian crayfish, *Astacus leptodactylus eichwaldi*, in Anzali coastal area, Iran. Iranian Journal of Fisheries Sciences 6: 59–76.

Kobza, R.M., J.C. Trexler, W.F. Loftus and S.A. Perry. 2004. Community structure of fishes inhabiting aquatic refuges in a threatened karst wetland and its implications for ecosystem management. Biological Conservation 116: 153–165.

Kusabs, I.A. and J.M. Quinn. 2009. Use of a traditional Māori harvesting method, the tau kōura, for monitoring kōura (freshwater crayfish, *Paranephrops planifrons*) in Lake Rotoiti, North Island, New Zealand. New Zealand Journal of Marine and Freshwater Research 43: 713–722.

Kutka, F.J., C. Richards, G.W. Merick, P.W. Devore and M.E. McDonald. 1992. Bait preference and trapability of two common crayfishes in northern Minnesota. The Progressive Fish Culturist 54: 250–254.

Lake, P.S. 1977. Pholeteros—the faunal assemblage found in crayfish burrows. Australian Society for Limnology Newsletter 15: 57–60.

Lake, P.S. and K.J. Newcombe. 1975. Observations on the ecology of the crayfish *Parastacoides tasmanicus* (Decapoda: Parastacidae) from south-western Tasmania Australian Zoologist 18: 197–214.

Lamontagne, S. and J.B. Rasmussen. 1993. Estimating crayfish density in lakes using quadrats: maximizing precision and efficiency. Canadian Journal of Fisheries and Aquatic Sciences 50: 623–626.

Larson, E.R. and J.D. Olden. 2008. Do schools and golf courses represent emerging pathways for crayfish invasions? Aquatic Invasions 3: 465–468.

Larson, E.R. and J.D. Olden. 2013. Crayfish occupancy and abundance in lakes of the Pacific Northwest, USA. Freshwater Science 32: 94–107.

Larson, E.R., R.J. DiStefano, D.D. Magoulick and J.T. Westhoff. 2008. Efficiency of a quadrat sampling technique for estimating riffle-dwelling crayfish density. North American Journal of Fisheries Management 28: 1036–1043.

Larson, E.R., J.D. Olden and N. Usio. 2010a. Decoupled conservatism of Grinnellian and Eltonian niches in an invasive arthropod. Ecosphere 1: art16.

Larson, E.R., C.A. Busack, J.D. Anderson and J.D. Olden. 2010b. Widespread distribution of the non-native northern crayfish (*Orconectes virilis*) in the Columbia River basin. Northwest Science 84: 108–111.

Larson, E.R., C.L. Abbott, N. Usio, N. Azuma, K.A. Wood, L.-M. Herborg and J.D. Olden. 2012. The signal crayfish is not a single species: cryptic diversity and invasions in the Pacific Northwest range of *Pacifastacus leniusculus.* Freshwater Biology 57: 1823–1838.

Lawrence, C.S., J.I. Brown and J.E. Bellanger. 2001. Morphology and occurrence of yabby (*Cherax albidus* Clark) burrows in Western Australian farm dams. Freshwater Crayfish 13: 253–264.

Lewis, S.D. 1997. Life history, population dynamics, and management of signal crayfish in Lake Billy Chinook, Oregon. M.S. Thesis, Oregon State University, Corvallis, Oregon.

Lodge, D.M., T.K. Kratz and G.M. Capelli. 1986. Long-term dynamics of three crayfish species in Trout Lake, Wisconsin. Canadian Journal of Fisheries and Aquatic Sciences 43: 993–998.

Loughman, Z.J. 2010. Ecology of *Cambarus dubius* (Upland Burrowing Crayfish) in north-central West Virginia. Southeastern Naturalist 9: 217–230.

Loughman, Z.J., S.A. Welsh and T.P. Simon. 2012. Occupancy rates of primary burrowing crayfish in natural and disturbed large river bottomlands. Journal of Crustacean Biology 32: 557–564.

Loughman, Z.J., D.A. Foltz and S.A. Welsh. 2013. Baited lines: an active nondestructive collection method for burrowing crayfish. Southeastern Naturalist 12: 809–815.

Ludlam, J.P. and D.D. Magoulick. 2009. Spatial and temporal variation in the effects of fish and crayfish on benthic communities during stream drying. Journal of the North American Benthological Society 28: 371–382.

MacKenzie, D.I. 2005. What are the issues with presence-absence data for wildlife managers? Journal of Wildlife Management 69: 849–860.

MacKenzie, D.I., J.D. Nichols, J.E. Hines, M.G. Knutson and A.B. Franklin. 2003. Estimating site occupancy, colonization and local extinction when a species is detected imperfectly. Ecology 84: 2200–2207.

MacKenzie, D.I., J.D. Nichols, J.A. Royle, K.H. Pollock, L.L. Bailey and J.E. Hines. 2006. Occupancy Estimation and Modeling: Inferring Patterns and Dynamics of Species Occurrence. Academic Press, San Diego, California.

Magoulick, D.D. 2004. Effects of predation risk on habitat selection by water column fish, benthic fish and crayfish in stream pools. Hydrobiologia 527: 209–221.

Magurran, A.E. and B.J. McGill. 2011. Biological Diversity: Frontiers in Measurement and Assessment. Oxford University Press, Oxford, United Kingdom.

Malley, D.F. and J.B. Reynolds. 1979. Sampling strategies and life history of non-insectan freshwater invertebrates. Journal of the Fisheries Research Board of Canada 36: 311–318.

Mangan, B.P., A.D. Ciliberto and M.T. Homewood. 2009. A versatile and economical trap for capturing wild crayfish. Journal of Freshwater Ecology 24: 119–124.

March, T.S. and B.J. Robson. 2006. Association between burrow densities of two Australian freshwater crayfish (*Engaeus sericatus* and *Geocharax gracilis*: Parastacidae) and four riparian land uses. Aquatic Conservation: Marine and Freshwater Ecosystems 16: 181–191.

Martin, T.G., B.A. Wintle, J.R. Rhodes, P.M. Kuhnert, S.A. Field, S.J. Low-Choy, A.J. Tyre and H.P. Possingham. 2005. Zero tolerance ecology: improving ecological inference by modeling the source of zero observation. Ecology Letters 8: 1235–1246.

Mason, J.C. 1970. Maternal-offspring behavior of the crayfish, *Pacifastacus trowbridgii* (Stimpson). American Midland Naturalist 84: 463–473.

Mason, J.C. 1975. Crayfish production in a small woodland stream. Freshwater Crayfish 2: 449–479.

Mather, M.E. and R.A. Stein. 1993. Direct and indirect effects of fish predation on the replacement of a native crayfish by an invading congener. Canadian Journal of Fisheries and Aquatic Sciences 50: 1279–1288.

Maxted, J. and M.J. Vander Zanden. 2007. Protocol for Wisconsin Crayfish Samping. University of Wisconsin, Madison, Wisconsin.

Momot, W.T. 1966. Upstream movement of crayfish in an intermittent Oklahoma stream. American Midland Naturalist 75: 150–159.

Momot, W.T. and H. Gowing. 1972. Differential seasonal migration of the crayfish, *Orconectes virilis* (Hagen), in marl lakes. Ecology 53: 479–483.

Moore, M.J., R.J. DiStefano and E.R. Larson. 2013. An assessment of life history studies for United States and Canadian crayfishes: identifying biases and knowledge gaps to improve conservation and management. Freshwater Science 32: 1276–1287.

Morrison, M.L., W.M. Block, M.D. Strickland, B.A. Collier and M.J. Peterson. 2008. Wildlife Study Design. Springer, New York City, New York.

Mueller, K.W. 2001. First records of the red swamp crayfish, *Procambarus clarkii* (Girard, 1852) (Decapoda, Cambaridae) from Washington State, U.S.A. Crustaceana 74: 1003–1007.

Mueller, K.W. 2002. Habitat associations of introduced smallmouth bass and native signal crayfish of Lake Whatcom, Washington. Journal of Freshwater Ecology 17: 13–18.

Noro, C.K. and L. Buckup. 2010. The burrows of *Parastacus defossus* (Decapoda: Parastacidae), a fossorial freshwater crayfish from southern Brazil. Zoologia 27: 341–346.

Norrocky, M.J. 1984. Burrowing crayfish trap. Ohio Journal of Science 84: 65–66.

Norrocky, M.J. 1991. Observations on the ecology, reproduction and growth of the burrowing crayfish *Fallicambarus* (Creaserinus) *fodiens* (Decapoda: Cambaridae) in North-central Ohio. American Midland Naturalist 125: 75–86.

Nowicki, P., T. Tirelli, R.M. Sartor, F. Bona and D. Pessani. 2008. Monitoring crayfish using a mark-recapture method: potentials, recommendations, and limitations. Biodiversity and Conservation 17: 3513–3530.

Nyström, P., P. Stenroth, H. Holmqvist, O. Berglund, P. Larsson and W. Granéli. 2006. Crayfish in lakes and streams: individual and population responses to predation, productivity and substratum availability. Freshwater Biology 51: 2096–2113.

Odelstrom, T. 1983. A portable hydraulic diver-operated dredge-sieve for sampling juvenile crayfish. Description and experience. Freshwater Crayfish 5: 270–274.

Ogle, D.H. and L. Kret. 2008. Experimental evidence that captured rusty crayfish (*Orconectes rusticus*) exclude uncaptured rusty crayfish from entering traps. Journal of Freshwater Ecology 23: 123–129.

Olden, J.D., J.M. McCarthy, J.T. Maxted, W.W. Fetzer and M.J. Vander Zanden. 2006. The rapid spread of rusty crayfish (*Orconectes rusticus*) with observations on native crayfish declines in Wisconsin (U.S.A.) over the past 130 years. Biological Invasions 8: 1621–1628.

Olden, J.D., J.W. Adams and E.R. Larson. 2009. First record of *Orconectes rusticus* (Girard, 1852) (Decapoda, Cambaridae) west of the Great Continental Divide in North America. Crustaceana 82: 1347–1351.

Olden, J.D., M.J. Vander Zanden and P.T.J. Johnson. 2011. Assessing ecosystem vulnerability to invasive rusty crayfish (*Orconectes rusticus*). Ecological Applications 21: 2587–2599.

Olsen, T.M., D.M. Lodge, G.M. Capelli and R.J. Houlihan. 1991. Mechanisms of impact of an introduced crayfish (*Orconectes rusticus*) on littoral congeners, snails, and macrophytes. Canadian Journal of Fisheries and Aquatic Sciences 48: 1853–1861.

Olsson, K. and P. Nyström. 2009. Non-interactive effects of habitat complexity and adult crayfish on survival and growth of juvenile crayfish (*Pacifastacus leniusculus*). Freshwater Biology 54: 35–46.

Opsahl, S.P. and J.P. Chanton. 2006. Isotopic evidence for methane-based chemosynthesis in the Upper Floridian Aquifer food web. Oecologia 150: 89–96.

Palmer, M.A. and N.L. Poff. 1997. The influence of environmental heterogeneity on patterns and processes in streams. Journal of the North American Benthological Society 16: 169–173.

Paragamian, V.L. 2010. Increase in abundance of signal crayfish may be due to decline in predators. Journal of Freshwater Ecology 25: 155–157.

Parkyn, S.M. 2015. A review of current techniques for sampling freshwater crayfish. pp. 205–220. *In*: T. Kawai, Z. Faulkes and G. Scholtz (eds.). Freshwater Crayfish: A Global Overview. CRC Press, Boca Raton, Florida.

Parkyn, S.M. and K.J. Collier. 2004. Interaction of press and pulse disturbance on crayfish populations: flood impacts in pasture and forest streams. Hydrobiologia 527: 113–124.

Parkyn, S.M., K.J. Collier and B.J. Hicks. 2002. Growth and population dynamics of crayfish *Paranephrops planifrons* in streams within native forest and pastoral land uses. New Zealand Journal of Marine and Freshwater Research 36: 847–861.

Parkyn, S.M., R.J. DiStefano and E.M. Imhoff. 2011. Comparison of constructed microhabitat and baited traps in Table Rock Reservoir, Missouri, USA. Freshwater Crayfish 18: 69–74.

Pearl, C.A., M.J. Adams and B. McCreary. 2013. Habitat and co-occurrence of native and invasive crayfish in the Pacific Northwest, USA. Aquatic Invasions 8: 171–184.

Peay, S. 2003. Monitoring the white-clawed crayfish *Austropotamobius pallipes*. Conserving Natura 20000 Rivers Monitoring Series No. I, English Nature, Peterborough, United Kingdom.

Peay, S., N. Guthrie, J. Spees, E. Nilsson and P. Bradley. 2009. The impact of signal crayfish (*Pacifastacus leniusculus*) on the recruitment of salmonid fish in a headwater stream in Yorkshire, England. Knowledge and Management of Aquatic Ecosystems 12: 394–395.

Peterson, J.T. and C.F. Rabeni. 2001. Evaluating the efficiency of a one-square-meter quadrat sampler for riffle-dwelling fish. North American Journal of Fisheries Management 21: 76–85.

Peterson, J.T., R.F. Thurow and J.W. Guzevich. 2004. An evaluation of multipass electrofishing for estimating the abundance of stream-dwelling salmonids. Transactions of the American Fisheries Society 133: 462–475.

Pilotto, F., G. Free, G. Crosa, F. Sena, M. Ghiani and A.C. Cardoso. 2008. The invasive crayfish *Orconectes limosus* in Lake Varese: estimating abundance and population size structure in the context of habitat and methodological constraints. Journal of Crustacean Biology 28: 633–640.

Pintor, L.M. and D.A. Soluk. 2006. Evaluating the non-consumptive, positive effects of a predator in the persistence of an endangered species. Biological Conservation 130: 584–591.

Pintor, L.M. and A. Sih. 2009. Differences in growth and foraging behavior of native and introduced populations of an invasive crayfish. Biological Invasions 11: 1895–1902.

Pintor, L.M. and A. Sih. 2011. Scale dependent effects of native prey diversity, prey biomass and natural disturbance on the invasion success of an exotic predator. Biological Invasions 13: 1357–1366.

Policar, T. and P. Kozák. 2005. Comparison of trap and baited stick catch efficiency for noble crayfish (*Astacus astacus* L.) in the course of the growing season. Bull. Fr. Peche Piscic. 376-377: 675–686.

Pollock, K.H., J.D. Nichols, C. Brownie and J.E. Hines. 1990. Statistical inference for capture-recapture experiments. Wildlife Monographs 107: 1–97.

Poulin, B., G. Lefebvre and A.J. Crivelli. 2007. The invasive red swamp crayfish as a predictor of Eurasian bittern density in the Camargue, France. Journal of Zoology 273: 98–105.

Price, J.E. and S.M. Welch. 2009. Semi-quantitative methods for crayfish sampling: Sex, size and habitat bias. Journal of Crustacean Biology 29: 208–216.

Purvis, K. and S.P. Opsahl. 2005. A novel technique for invertebrate trapping in groundwater wells identifies new populations of the troglobitic crayfish, *Cambarus cryptodytes*, in southwest Georgia, USA. Journal of Freshwater Ecology 20: 361–365.

Puth, L.M. and T.F.H. Allen. 2004. Potential corridors for the rusty crayfish, *Orconectes rusticus*, in northern Wisconsin (USA) lakes: lessons for exotic invasions. Landscape Ecology 20: 567–577.

Quinn, G.P. and M.J. Keough. 2002. Experimental Design and Data Analysis for Biologists. Cambridge University Press, Cambridge, United Kingdom.

Quinn, J.P. and J. Janssen. 1989. Crayfish competition in southwestern Lake Michigan: a predator mediated bottleneck. Journal of Freshwater Ecology 5: 75–85.

Rabeni, C.F. 1985. Resource partitioning by stream-dwelling crayfish: the influence of body size. American Midland Naturalist 113: 20–29.

Rabeni, C.F., K.F. Collier, S.M. Parkyn and B.J. Hicks. 1997. Evaluating techniques for sampling stream crayfish (*Paranephrops planifrons*). New Zealand Journal of Marine and Freshwater Research 31: 693–700.

Rach, J.J. and T.D. Bills. 1987. Comparison of three baits for trapping crayfish. North American Journal of Fisheries Management 7: 601–603.

Reynolds, J.D. and C. Souty-Grosset. 2012. Management of Freshwater Biodiversity: Crayfish as Bioindicators. Cambridge University Press, Cambridge, United Kingdom.

Richards, C., F.J. Kutka, M.E. McDonald, G.W. Merrick and P.W. Devore. 1996. Life history and temperature effects on catch of northern orconectid crayfish. Hydrobiologia 319: 111–118.

Richardson, A.M.M. and R. Swain. 1980. Habitat requirements and distribution of *Engaeus cisternarius* and three subspecies of *Parastacoides tasmanicus* (Decapoda: Parasticade), burrowing crayfish from an area of south-western Tasmania. Australian Journal of Marine and Freshwater Research 31: 475–484.

Ricker, W.E. 1975. Computation and interpretation of biological statistics of fish populations. Bulletin of the Fisheries Research Board of Canada 191: 1–382.

Ridge, J., T.P. Simon, D. Karns and J. Robb. 2008. Comparison of three burrowing crayfish capture methods based on relationships with species morphology, seasonality, and habitat quality. Journal of Crustacean Biology 28: 466–472.

Robinson, C.A., T.J. Thom and M.C. Lucas. 2000. Ranging behavior of a large freshwater invertebrate, the white-clawed crayfish *Austropotamobius pallipes*. Freshwater Biology 44: 509–521.

Rogowski, D.L., S. Sitko and S.A. Bonar. 2013. Optimising control of invasive crayfish using life-history information. Freshwater Biology 58: 1279–1291.

Romaire, R.P. and V.H. Osorio. 1989. Effectiveness of crawfish baits as influenced by habitat type, trap-set time, and bait quantity. The Progressive Fish Culturist 51: 232–237.

Roth, B.M., J.C. Tetzlaff, M.L. Alexander and J.F. Kitchell. 2007. Reciprocal relationships between exotic rusty crayfish, macrophytes, and *Lepomis* species in northern Wisconsin lakes. Ecosystems 10: 74–85.

Royle, J.A., J.D. Nichols and M. Kéry. 2005. Modelling occurrence and abundance of a species when detection is imperfect. Oikos 110: 353–359.

Schloesser, J.T., C.P. Paukert, W.J. Doyle, T.D. Hill, K.D. Steffensen and V.H. Travnichek. 2012. Heterogenous detection probabilities for imperiled Missouri River fishes: implications for large-river monitoring programs. Endangered Species Research 16: 211–224.

Schneider, K. and D.C. Culver. 2004. Estimating subterranean species richness using intensive sampling and rarefaction curves in a high density cave region in West Virginia. Journal of Cave and Karst Studies 66: 39–45.

Searcy, C.A., E. Gabbai-Saldate and H.B. Shaffer. 2013. Microhabitat use and migration distance of an endangered grassland amphibian. Biological Conservation 158: 80–87.

Sethi, S.A. and E. Benolkin. 2013. Detection efficiency and habitat use to inform inventory and monitoring efforts: juvenile coho salmon in the Knik River basin, Alaska. Ecology of Freshwater Fish 22: 398–411.

Simon, T.P. 2001. Checklist of the crayfish and freshwater shrimp (Decapoda) of Indiana. Proceedings of the Indiana Academy of Science 110: 104–110.

Simon, T.P. 2004. Standard operating procedures for the collection and study of burrowing crayfish in Indiana. I. Methods for the collection of burrowing crayfish in streams and terrestrial habitats. Occassional Papers of the Indiana Biological Survey Aquatic Research Center 2: 1–18.

Slater, F.M. 1995. A simple crayfish trap. Freshwater Crayfish 10: 194–195.

Smith, G.R.T., M.A. Learner, F.M. Slater and J. Foster. 1996. Habitat features important for the conservation of the native crayfish *Austropotamobius pallipes* in Britain. Biological Conservation 75: 239–246.

Somers, K.M. and D.P.M. Stechey. 1986. Variable trapability of crayfish associated with bait type, water temperature and lunar phase. American Midland Naturalist 116: 36–44.

Somers, K.M. and R.H. Green. 1993. Seasonal patterns in trap catches of the crayfish *Cambarus bartoni* and *Orconectes virilis* in six south-central Ontario lakes. Canadian Journal of Zoology 71: 1136–1145.

Souty-Grosset, C., D.M. Holdich, P.Y. Noël, J.D. Reynolds and P. Haffner. 2006. Atlas of Crayfish in Europe. Muséum National d'Histoire Naturelle, Paris, France.

Stoeckel, J.A., B.S. Helms and E. Cash. 2011. Evaluation of a crayfish burrowing chamber design with simulated groundwater flow. Journal of Crustacean Biology 31: 50–58.

Streever, W.J. 1996. Energy economy hypothesis and the troglobitic crayfish *Procambarus eryhtrops* in Sim's Sink Cave, Florida. American Midland Naturalist 135: 357–366.

Stuecheli, K. 1999. Trapping bias in sampling crayfish with baited funnel traps. North American Journal of Fisheries Management 11: 236–239.

Surber, E.W. 1937. Rainbow trout and bottom fauna production in one mile of stream. Transactions of the American Fisheries Society 66: 193–202.

Taugbøl, T., J. Skurdal, A. Burba, C. Munoz and M. Sáez-Royuela. 1997. A test of crayfish predatory and nonpredatory fish species as bait in crayfish traps. Fisheries Management and Ecology 4: 127–134.

Taylor, C.A. and T.G. Anton. 1998. Distribution and ecological notes on some of Illinois' burrowing crayfish. Transactions of the Illinois State Academy of Science 92: 137–145.

Taylor, C.A., G.A. Schuster, J.E. Cooper, R.J. DiStefano, A.G. Eversole, P. Hamr, H.H. Hobbs III, H.W. Robison, C.E. Skelton and R.E. Thoma. 2007. A reassessment of the conservation status of crayfishes of the United States and Canada after 10+ years of increased awareness. Fisheries 32: 372–389.

Thomas, L., S.T. Buckland, E.A. Rexstad, J.L. Laake, S. Strindberg, S.L. Hedley, J.R.B. Bishop, T.A. Marques and K.P. Burnham. 2010. Distance software: design and analysis of distance sampling surveys for estimating population size. Journal of Applied Ecology 47: 5–14.

Thompson, W. 2004. Sampling Rare or Elusive Species: Concepts, Designs, and Techniques for Estimating Population Parameters. Island Press, Washington D.C.

Thomsen, P.F., J. Kielgast, L.I. Iversen, C. Wiuf, M. Rasmussen, M.T.P. Gilbert, L. Orlando and E. Willerslev. 2012. Monitoring endangered freshwater biodiversity using environmental DNA. Molecular Ecology 21: 2565–2573.

Tréguier, A., J.M. Paillisson, T. Dejean, A. Valentini, M.A. Schlaepfer and J.M. Roussel. 2014. Environmental DNA surveillance for invertebrate species: advantages and technical limitations to detect invasive crayfish *Procambarus clarkii* in freshwater ponds. Journal of Applied Ecology 51: 871–879.

Trontelj, P., Y. Machino and B. Sket. 2005. Phylogenetic and phylogeographic relationships in the crayfish genus *Austropotamobius* inferred from mitochondrial COI gene sequences. Molecular Phylogenetics and Evolution 34: 212–226.

Tyre, A.J., B. Tenhumberg, S.A. Field, D. Niejalke, K. Parris and H.P. Possingham. 2003. Improving precision and reducing bias in biological surveys: Estimating false-negative error rates. Ecological Applications 13: 1790–1801.

Usio, N. 2007. Endangered crayfish in northern Japan: Distribution, abundance and microhabitat specificity in relation to stream and riparian environment. Biological Conservation 134: 517–526.

Usio, N. and C.R. Townsend. 2000. Distribution of the New Zealand crayfish *Paranephrops zealandicus* in relation to stream physico-chemistry, predatory fish, and invertebrate prey. New Zealand Journal of Marine and Freshwater Research 34: 557–567.

Usio, N. and C.R. Townsend. 2001. The significance of crayfish *Paranephrops zealandicus* as shredders in a New Zealand headwater stream. Journal of Crustacean Biology 21: 354–359.

Usio, N. and C.R. Townsend. 2004. Roles of crayfish: Consequences of predation and bioturbation for stream invertebrates. Ecology 85: 807–822.

Usio, N., H. Nakajima, R. Kamiyama, I. Wakana, S. Hiruta and N. Takamura. 2006. Predicting the distribution of invasive crayfish (*Pacifastacus leniusculus*) in a Kusiro Moor marsh (Japan) using classification and regression trees. Ecological Research 21: 271–277.

Venarsky, M.P., A.D. Huryn and J.P. Benstead. 2012. Re-examining extreme longevity of the cave crayfish *Orconectes australis* using new mark-recapture data: a lesson on the limitations of iterative size-at-age models. Freshwater Biology 57: 1471–1481.

Walton, C.F., Jr. and S.B. Cook. 2010. The microhabitat quadrat sampler—a 0.25-m2 quadrat sampler for microhabitat of crayfishes. Journal of Freshwater Ecology 25: 313–315.

Warren, M.L., Jr., A.L. Sheldon and W.R. Haag. 2009. Constructed microhabitat bundles for sampling fishes and crayfishes in coastal plain streams. North American Journal of Fisheries Management 29: 330–342.

Webb, M. and A.M.M. Richardson. 2004. A radio telemetry study of movement in the giant Tasmanian freshwater crayfish, *Astacopsis gouldi*. Freshwater Crayfish 14: 197–204.

Welch, S.M. and A.G. Eversole. 2006a. The occurrence of primary burrowing crayfish in terrestrial habitat. 130: 458–464.

Welch, S.M. and A.G. Eversole. 2006b. An ecological classification of burrowing crayfish based on habitat affinities. Freshwater Crayfish 15: 155–161.

Welch, S.M. and A.G. Eversole. 2006c. Comparison of two burrowing crayfish trapping methods. Southeastern Naturalist 5: 27–30.

Welch, S.M., J.L. Waldron, A.G. Eversole and J.C. Simoes. 2008. Seasonal variation and ecological effects of Camp Shelby burrowing crayfish (*Fallicambarus gordoni*) burrows. American Midland Naturalist 159: 378–384.

Welsh, A.H., D.B. Lindenmayer and C.F. Donnelly. 2013. Fitting and interpreting occupancy models. PLoS One 8(1): e52015.

Westman, K., M. Pursiainen and R. Vilkman. 1978a. A new folding trap model which prevents crayfish from escaping. Freshwater Crayfish 4: 235–242.

Westman, K., O. Sumari and M. Puriainen. 1978b. Electric fishing in sampling crayfish. Freshwater Crayfish 4: 251–256.

Westman, K., R. Savolainen and M. Julkunen. 2002. Replacement of the native crayfish *Astacus astacus* by the introduced species *Pacifastacus leniusculus* in a small, enclosed Finnish lake: a 30-year study. Ecography 25: 53–73.

Whisson, G.J. and L. Campbell. 2000. Catch efficiency of five freshwater crayfish traps in south-west Western Australia. Freshwater Crayfish 13: 58–66.

Wiens, J.A. 1989. Spatial scaling in ecology. Functional Ecology 3: 385–397.

Williams, B.W., S.R. Gelder and H. Proctor. 2009. Distribution and first reports of Branchiobdellida (Annelida: Clitellata) on crayfish in the prairie provinces of Canada. Western North American Naturalist 69: 119–124.

Williams, D.D., N.E. Williams and H.B.N. Hynes. 1974. Observations on the life history and burrow construction of the crayfish *Cambarus fodiens* (Cottle) in a temporary stream in southern Ontario. Canadian Journal of Zoology 52: 365–370.

Williams, K., S.K. Brewer and M.R. Ellersieck. 2014. A comparison of two gears for quantifying abundance of lotic-dwelling crayfish. Journal of Crustacean Biology 34: 54–60.

Willis, T.H. 2001. Visual census methods underestimate density and diversity of cryptic reef fishes. Journal of Fish Biology 59: 1408–1411.

Wilson, K.A., J.J. Magnuson, D.M. Lodge, A.M. Hill, T.K. Kratz, W.L. Perry and T.V. Willis. 2004. A long-term rusty crayfish (*Orconectes rusticus*) invasion: dispersal patterns and community change in a north temperate lake. Canadian Journal of Fisheries and Aquatic Sciences 61: 2255–2266.

Wooster, D., J.L. Snyder and A. Madsen. 2012. Environmental correlates of signal crayfish, *Pacifastacus leniusculus* (Dana, 1852), density and size at two spatial scales in its native range. Journal of Crustacean Biology 32: 741–752.

Zale, A.V., D.L. Parrish and T.M. Sutton. 2013. Fisheries Techniques. American Fisheries Society, Bethesda, Maryland.

Zimmerman, J.K.M. and R.T. Palo. 2011. Reliability of catch per unit effort (CPUE) for evaluation of reintroduction programs—a comparison of the mark-recapture method with standardized trapping. Knowledge and Management of Aquatic Ecosystems 401: 07.

Zippin, C. 1958. The removal method of population estimation. Journal of Wildlife Management 22: 82–90.

Laboratory Methods for Crayfish

Matt Longshaw[1],* and *Paul Stebbing*[2]

Introduction

In the previous chapter, Larson and Olden raised the point there were many good reasons for studying and collecting crayfish. They provided a review of the literature and detailed methods for field sampling of crayfish across their range. Rightly, they stop short of describing methods for the transportation, holding and sampling of crayfish on the laboratory side, which we cover below. This chapter is not intended as a formal review of the literature, rather it reflects a combination of published information and our own experiences in the laboratory. The reader should use this chapter as a way of considering more widely potential pitfalls and improvements in transporting animals and obtaining samples, but not necessarily adhere strictly to the sampling methods described. Differences in laboratory capabilities, availability of consumables and overall cost as well as storage capacity for samples will ultimately determine what is feasible.

One of the crucial elements that we would like to see as crayfish studies move forward is the integration of sampling to the extent that even if individual researchers are not working in a particular field that they collect samples for use by others. Imagine the strength of a study whereby appropriate numbers of crayfish are collected in a field survey and the correct morphological taxonomy of the crayfish is applied and morphometric data is collected, where tissue samples are stored for genetic analysis, where reproductive status is assessed, where the overall health of the population is characterised, where pathogens and parasites of individual crayfish are described, and where contemporaneous environmental samples are collected to allow for a holistic interpretation of the individual and population success. The power of such a multi-disciplinary approach becomes more worthwhile when one considers samples collected from rare or protected animals or from previously undescribed species. It doesn't follow that any one individual needs to do all the analysis on all the samples collected but it would mean that a network of like-minded individuals could willingly exchange samples and information for the greater good of astacology. Of critical importance to ensure validity of the studies is to have a robust method for the labelling of individuals and groups of samples to allow data collected for the different purposes to be comparable to each other and to allow integration of data.

[1] Benchmark Animal Health Ltd. Bush House, Edinburgh Technopole, Milton Bridge, EH26 0BB.
[2] Cefas Weymouth Laboratory, Barrack Road, The Nothe, Weymouth, Dorset, DT4 8UB.
  Email: paul.stebbing@cefas.co.uk
* Corresponding author: matt.longshaw@bmkanimalhealth.com

This chapter does not cover aspects such as experimental study design nor does it consider numbers of crayfish that need to be sampled from an epidemiological or statistical perspective. These are outside the scope of the chapter and the reader is referred to a range of papers and books that readily describe these in appropriate detail (Magurran 1988, 2003, 2010, Legendre and Legendre 2012, Southwood and Henderson 2000, Raidal et al. 2004, Sutherland 2006, MacDonald 2009, Dytham 2010, Ruxton and Colegrave 2010, Bate and Clark 2014).

Readers are actively encouraged to ensure they adhere to local and national regulations and framework agreements on the welfare, collection, handling, transport and holding of wild and captive crayfish (Fotedar and Evans 2011, Crook 2013), to ensure they follow local and national legislation on the use, release and disposal of waste material including carcasses, water and chemicals following good biosecurity rules at all times and to comply with international rules on transfer of genetic material between countries if applicable and adhering to the Convention on Biological Diversity (https://www.cbd.int/convention/). In addition, care must be taken to minimise any the risk of personal injury and reduce, where possible, potential risks associated with contaminated animals and environmental samples.

Transportation

Aquatic crustaceans, including many species of crayfish, are transported live globally for a number of reasons, e.g., for human consumption, aquaculture and the aquarium trade. The process of transporting live aquatic animals can be stressful, especially as a result of changing and/or adverse environmental conditions, such as high levels of ammonia, temperature fluctuations, exposure to air and physical trauma through handling processes. To combat these factors and to minimise mortality and morbidity during transport there have been a number of technological developments including specially designed tanks and containers, modified transport vehicles and a greater understanding of the effects of environmental stressors on the health status of the organisms during the transport process (Fotedar and Evans 2011).

Transportation of live crayfish is an essential element of many scientific studies, e.g., moving animals from site of capture to experimental facilities for behavioural studies or for dissection if examining the disease status of the donor population. Although crayfish are characterised by a relatively high tolerance to the stress of live transport, measures to maximise survival rates are similar to those for other crustacean species. Transportation can involve moving animals over considerable distances, which in some cases can take days and may involve multiple modes of transport across international borders. Many countries have laws relating to the keeping and transportation of crayfish. It is therefore essential that the laws relevant to the country or countries in which the animals are being transported are understood and adhered to. Many national or regional government web-pages will contain the relevant information and should be referred to as part of the process of planning the transportation. This section focuses on the transportation of live crayfish rather than tissue samples, and includes some broad guidelines on the best practice of transporting live crayfish to maximise health and survival.

When transporting live animals ensuring they arrive at their destination in good health and unstressed is the ultimate aim of the process, especially where the continued survival post transportation of the animals is required. Stress during transport can lead to susceptibility of animals to diseases, thus potentially jeopardising the survival of the stock. All species are defined by tolerance limits to certain environmental conditions (see Table 1 for examples of tolerance levels to some environmental parameters); where possible these should be referred to when developing plans for transportation. However, some individual variability of tolerance to stress will be observed as a result of factors such as overall health status, moult stage, age, sex and size. In addition natural genetic variation in tolerance to stressors may be observed between individuals and populations. It is likely for there to be some mortalities and/or morbidity during or post the transport process, but measures can be taken to keep these to a minimum. There are some key stressors to consider and try and reduce when arranging the transportation of animals:

1. **Physical damage or disturbance** can be caused during the capture and transport process, causing stress and potentially increasing susceptibility to diseases.
2. **Overcrowding** may result in increased aggressive interaction and potential physical damage (e.g., lost appendages) or cannibalism.

3. **Adverse environmental conditions** such as increased ammonia levels, decreased dissolved oxygen will ultimately impact on the animals' survival during and post transport. Prolonged chronic exposure to adverse environmental conditions can also impact on the animals' immunity.
4. **Temperature fluctuations** is a factor to which crustaceans are particularly susceptible.
5. **Desiccation or prolonged exposure to air** will cause stress or potential mortality. Although crayfish are facultative air breathers, moisture is still required to facilitate the respiratory process.

Table 1. Comparative tolerances of crayfish (adapted from Nyström 2002).

Species	Temperature range (°C)	Minimum oxygen (mg L⁻¹)	Salinity range (ppt)
Astacidae			
Astacus astacus	1–28	3.2	0
Astacus leptodactylus	4–32	4	0–14
Austropotamobius pallipes	1–28	2.7	0
Pacifastacus leniusculus	1–33	-	0–21
Parastacidae			
Cherax destructor	1–35	1.0	0–12
Cherax tenuimanus	4–28	3.7	0.3–12
Cherax quadricarinatus	10–30	0.5	0–12
Cambaridae			
Orconectes rusticus	2.5–33		
Procambarus clarkii	<35	0.4	0–12

Pre-transport planning

Prior to transporting crayfish or any live animals, it is important that a clear plan is developed. The plan should consider the following points:

1. Where the animals are coming from and where they are going
2. How the animals will be transported
3. How long the animals will be in transit for
4. Who is responsible for the animals at the different stages of the transport process
5. Any legal requirements are met
6. A biosecurity plan (this is discussed in a separate section below)

Suitable methods for the collection of crayfish from the wild have been discussed in the previous chapter, and will not be discussed further here, but ensuring animals are not physically damaged during the process is important to avoid issues during transportation or future storage. Likewise, ensuring animals are not maintained in stressful condition prior to packaging for transportation is also important. Storage is covered further on in this chapter and the basic principles apply here.

Ensuring a clear destination for the animals, with suitable facilites which are prepared in advance will further reduce transit time and stress. Making sure that the receivers are also aware of how many animals they will be receiving and when will aid in setting up of the receiving storage and ensuring that someone is available to take receipt of the animals at destination. Having animals arrive on a weekend, for example, when no one is working is likely to lead to mortalities.

Suitable conditions for how the animals are to be transported is discussed in more detail below. In essence maintaining environmental conditions within the species tolerance range and keeping levels of morbidity and mortality during transit to a minimum is a race against the time it takes to transport the animals—the longer the transit time the more effort required to maintain suitable conditions for the duration. Ensuring suitable arrangements are made for the transportation, especially where multiple modes of transport are required is important to reduce the transit time. This can be logistically complicated, for example when transporting animals internationally, and will require particular care in the planning phase. How long the animals are in transit for will be dictated by the distance being transported and modes of transport, this will likewise influence the conditions in which the animals will be transported.

Having named individuals responsible for the animals at the different stages of transit will aid in reducing transit time and ensuring care is taken of the animals during the process, if required. This is particularly important with long distance transits. At the very least an individual should be responsible for dispatching the container with the animals in and receiving the animals at the point of destination. If transit time is long then additional husbandry may be required during transit for which an individual should be made responsible for. If multiple modes of transport are being used then an individual will need to be responsible for the transition of the animals. In some cases this may be handled by a haulage or courier company.

Legal requirements for the transportation of live animals vary between countries and regions. Ensuring a clear understanding and adherence to the legal requirements is essential in ensuring successful and timely delivery of the animals. It is best practice, and normally a legal requirement, that the container is clearly labelled with the address of the point of destination (obviously required if a haulage or courier is transporting the animals), a contact telephone number for a person responsible for the animals, what is inside the container, including species and total number, the date of dispatch, and normally a "Handle with care" label.

Selecting the correct container and maintaining the environment

The container in which the animals are placed in for transportation will have an influence on how all of the stressors described above (i.e., physical damage or disturbance, overcrowding, adverse environmental conditions, temperature fluctuations and desiccation or air exposure) can be controlled for. In some cases, where animals are being transported short distances for human consumption, they are held in mesh sacks. Added layers of sacking with finer meshing can be added for additional biosecurity, especially when transporting invasive species. Regular hosing or dousing of the sacks with freshwater water, or covering the sacks with damp cloth is sufficient to maintain respiratory function. While suitable for transporting for human consumption and/or over short distances, this process is not ideal for transporting animals for use in biological studies, as physical damage can easily occur as a result of overcrowding and environmental conditions, while temperature fluctuations and desiccation are not easily controlled for.

Guidance for the transportation of many aquatic crustacean species recommend animals are chilled (2–10°C) during transport to maximise health and survival. With more elaborate operations this is achieved by transporting the animals in chiller units or specially designed tanks or viviers. However, it is unlikely that such systems are available to the average astacologist, and therefore other methods of chilling can be used, such as the use of pre-chilled sawdust or wood shavings, hessian or freezer packs. The chilling process induces lethargy in the animals, which reduced respiratory rates and physical interaction between individuals. Care should be taken, however, to reduce the temperature slowly as rapid temperature changes cause stress and can result in loss of appendages. Likewise, at point of destination temperature should be increased slowly to ambient (or the temperature at which the animals will be maintained).

Ensuring animals are not overcrowded in the transport container will reduce the risk of stress through fighting or cannibalism, and will prevent damage to appendages. Knowing the approximate number of animals to be transported will allow for a suitably sized container to be obtained. Containers should be water tight with a lid which can be secured; cool boxes are ideal for this purpose. The secure lid will aid in maintaining a constant environment for the animals during transport, as well as providing an additional level of biosecurity. Furthermore, rigid containers will prevent animals being crushed during transport while aiding in handling. This can be of particular importance if the animals will be transported as airfreight. However, when being transported in rigid containers animals can sustain damage if knocked about within the container during transit. It is therefore important to ensuring that the animals are packed securely within the container to avoid such damage. Cool boxes are also ideal for maintaining temperature, where materials used to chill the animals can be easily placed in the bottom of the container or, in the case of chiller blocks, taped to the inside of the lid. Care should be taken to ensure that the animals do not come into direct contact with the chilled material, as this can cause damage and undue stress. Packing material should be used as a barrier to prevent direct contact. The packing material can also provide shelter within the container helping to reduce stressful interaction between individuals. Short length of plastic piping can provide suitable hides for the animals, as can large quantities of pond weed.

Transporting crayfish submersed can lead to water quality issues over time, such as decreased oxygen levels, which can result in mortalities. Maintaining water quality and oxygen levels of submersed animals requires specialist equipment, it is therefore preferable to move animals in damp conditions. Damp conditions can be easily maintained for short periods using wet paper, cloth or pond weed (taken from the same location as the animals) included in the container with the animals. The damp material used to maintain humidity can be used as a barrier between the animals and the chilled materials used to maintain temperature. Over longer distances, changing the damp material (and possibly the chilled material) may be required to maintain environmental conditions.

In some cases, such as for the aquarium trade, crayfish are transported submersed, using oxygen saturated water. The animals are also purged prior to transportation to reduce biological oxygen demand, as well as being chilled to reduce respiratory rate.

Biosecurity

Biosecurity is of utmost importance at all stages of the transport process. Reducing the risk of transferring pathogens and invasive species (termed here biological pests) during the transportation process is important in maintaining biodiversity at both the site of origin of the species being transported, the point of destination and maintaining the health of the animals and any other stock which are being held in the facility which the animals are destined for. There are several potential pathways/vectors by which the movement of biological pests can be moved via during the transportation process. Primarily these are:

1. The outside of the container, which may become contaminated during the packing process
2. The packing material used (if taken from the environment from which the animals originated) may contained invasive species
3. Water used to dampen material to maintain humidity could contain pathogens or invasive species
4. The species being transported may carry diseases or are invasive in their own right

Biosecurity is most effective when a precautionary approach is taken. Therefore, assuming that the animals being transported carry disease, or the packing material contains invasive species is the best default position. Simple measures can be taken to reduce the risk posed by these potential pathways/vectors. For examples:

1. The outside of the container can be checked carefully for any hitchhiking plants or animals and cleaned using a disinfectant. Similarly, the container and many other materials used during transport which is to be kept (e.g., chiller blocks) should be carefully checked and cleaned after use.
2. Packing material can be carefully disposed of as biological waste at the point of destination and not disposed of with general waste.
3. Water can be should be disposed of in such a way as to avoid it entering any natural water systems. This may involve chemically treating the water prior to disposal, or at the very least ensuring that the water will go through a tertiary sewage treatment system.
4. The crayfish should not be released or allowed to escape during any stages of the transport process, even if native to the region. Using containers which are sealable are ideal as biosecurity measures. At point of destination, ideally the crayfish should be kept separate from other stock, even if from the sample population, as diseases status of the population may have changed.

There are most likely other pathways and vectors which will be specific to the transport process that you as an astacologist are planning, and which should be carefully considered, with mitigation measures, during the planning process. This section only provides examples, rather than a definitive list, but highlights the need to consider biosecurity as an essential part of the process.

Holding of crayfish species

This section will discuss some of the basic principles of holding and husbandry of crayfish under experimental conditions based on the authors' experience and on published literature. It will not cover experimental design, or any specific keeping requirements in relation to experimentation, or the breeding of crayfish. As crayfish are diverse so are their keeping requirements, so it is important to understand the specific environmental tolerance limits and conditions required for the species being kept prior to setting up a holding facility (see Table 1 for some environmental parameters for a number of species). As the author's experience is primarily with Astacidae crayfish then this section, does have a bias towards these species, but many of the principles apply to other species as well.

Holding crayfish, as with their transport, requires the minimisation of stressors to ensure good health and survival. Keeping physical interaction between animals to a minimum, will help in avoiding physical damage. Likewise, avoiding overcrowding will reduce physical interaction as well as reducing the risk from disease. Ensuring environmental conditions are maintained within the tolerance parameters of the species will further prevent stress to the animals. Holding facilities are unlike the natural environment, but an attempt to replicate the natural environment, as much as practicable will helped to minimise stress and maintain the health of the crayfish.

Where to place the holding facility?

Consideration should be given to where holding tanks are to be set up prior to obtaining stock animals. Ideally tanks should be placed in a location where the environmental parameters (e.g., temperature and light) can be controlled or at least maintained at a constant level. Climate control rooms where the temperature and light can be easily manipulated are ideal, but not always available. Means of controlling the light/dark ratio and temperature are important as many of the natural process of crayfish are governed by these variables. How natural process may be disrupted as a result of holding facility design are key considerations when designing experiments, especially those related to behaviour or reproduction. Excessive light, noise or physical disturbance can all be the cause of stress, or disruption to natural behaviour, which will affect the health and wellbeing of the animals. Where possible, tanks/aquaria should be maintained in locations where they will not be disturbed from regular human traffic or other noise/vibrations such as from heavy machinery, generators, air conditioning units and the like. Other practical considerations such as sufficient space, a source of freshwater (running water if a flow-through system is to be used) and a drain will all limit where a system can be placed.

Biosecurity should also be a key consideration when deciding on the location of holding tanks. This could potentially have legislative requirements in relation to the keep of non-native species, where access needs to be limited to licence holders. Therefore, facilities will need to have means to limit access. Again legal requirements relating to the keep of non-native species may dictate where effluent from the system goes and the inclusion of measures to prevent escape of crayfish from the system. In relation to the location of the holding tanks legal requirements may specify animals are held in enclosed rooms with a lip on the door, doors that open inwards, and no drains at ground level. As a precautionary measure it is practical to consider the disease status of the animals, with measures taken to prevent transfer of disease from or to the stock animals. Having separate sets of nets and other equipment used as part of the husbandry process for use with different stock reduces the risk of transfer of disease, likewise having foot baths containing disinfectant at the entrance to the holding facility will further reduce the risk from disease.

Holding tank set up

Many crayfish species are sensitive to water quality. Maintain good water quality can be achieved by having a flow-through systems, where clean water flows into the holding tank and then flows to waste, although this is not always practical and uses significant amounts of water. Recirculating units, consisting of a large sump containing water which is then pumped into the holding tank and then back into the sump, increases the volume of water in which the animals are kept, helps to aerate the water and is less wasteful. Pumps can be fitted with filters to remove suspended solids in addition to nitrates and nitrates. Recirculating systems will need regular water changes however, for example 50% water change once every 2 weeks and a 100% water change every 2 months, depending on the volume of total volume (tanks + sump) of water used and the number of animals being held. Still water tanks can also be used, although water changes will need to be more frequent in addition to requiring aeration by air pumps. Maintaining calcium levels can be important depending on the hardness of the water source used. Calcium blocks can be used to buffer the water in the tanks or sumps.

It has been shown that increased complexity of habitat reduces the amount of aggressive interaction and thus cannibalism in crayfish (Corkrum and Cronin 2004). Ensuring there are surplus hiding places and shelters for the number of animals being held will greatly reduce mortality rates. In addition, having a variation in the diameter of the hiding places to accommodate the different sizes of animals being held will reduce stress. PVC tubing cut into short lengths is ideal as it comes in a variety of diameters and it sinks to the bottom of the tank. Other systems such as bricks with holes are also effective. Substrate type helps to naturalise the habitat while further reducing suspended solids. Fine gravel or sand also encourages natural burrowing behaviour. Plants and other natural features can be included, but in the author's experience are eaten or quickly uprooted and shredded which can block outflows.

If using glass tanks then sides should be covered to reduce reflections to reduce stress. Colour backdrops have been shown to reduce stress in some fish species (Sebire personal communication), and may also reduce stress while creating a more natural environment for the crayfish.

Crayfish are excellent escape artists, and can easily climb out of tanks using shelters located near to the sides of tanks or airlines trailing into the tank. Having lids or covers on tanks can prevent escape, and in some cases are a legal requirement for the keeping of invasive species. Transparent lids allow for behaviour not to be disrupted as a result of prolonged dark periods. Deep sided tanks where water levels are maintained <50% of the total high can also help to reduce the risk of escape.

General husbandry

Feeding stock animals with a surplus of food can reduce aggressive interaction as well as ensuring that all animals have consumed enough food (Corkrum and Cronin 2004). As crayfish are omnivorous they will eat a wide range of food types. Food types which can easily be removed from tanks aid with general husbandry. Potatoes, carrots and other root vegetables are ideal for leaving in the tank as they do not breakdown and can be easily removed by hand or net. Other food types such as fish food pellets can also be used, but are prone to breaking down thus increasing suspended solids. Providing variety in the food offered to the crayfish will help to maintain health. Tinned ham or pieces of fish provide variety and a source of protein, although they breakdown and therefore increase the frequency of required cleaning. A feeding regime used by the author's when keeping signal crayfish (*Pacifastacus leniusculus*) is for the animals to be fed with excess 1 cm^3 chunks of carrot and/or potatoes once every two days, with old food being removed the same day that new food is added. Once per week, meat items (1 cm^3) are added to excess and removed the following day. In the author's experience, this has been an effective feeding regime in maintaining stock crayfish.

Checking water quality on a regular basis is important in assessing if water changes are being made frequently enough or if addition filtration is required. Using basic water quality testing kits available from aquarium shops are sufficient, although more accurate means are available. Measuring nitrates, nitrites and ammonia levels provide a good assessment of water quality although additional tests for certain parameters, e.g., calcium, may be considered depending on the quality of the source water.

Checking animals and holding tanks on a daily basis is important. Ensuring that flow levels are sufficient, air lines are not snagged or filters are clean is essential in maintaining the holding systems and the health of the stock. Looking for unusual behaviour/morbidity or clinical signs of disease and removing these animals from the stock tanks can help to prevent the spread of disease; likewise dead animals should be removed. Testing for disease in the removed animals can help to identify problems, although available tests may be limited depending on the state of the animal.

Keeping accurate records of all activity relating to husbandry of the animals and who has conducted the particular activity, can help to ensure that checks are made regularly as well as providing useful knowledge of the life history of the animals and identify issues. Records should include: the source of the animals, when they arrived in the holding facility, total number of animals, any observation of the animals when they arrived (e.g., any gravid females), water quality test results, any mortalities and the date when they occurred, the date of feeds and what they were fed, the date of water changes and tank cleaning.

Dissections/sample collection

External examinations and morphometrics

Recent advances in molecular methods allow for the detection of minute traces of DNA in environmental samples (eDNA) without the need to directly sample or visualise the target animal. Such approaches have utility in areas that are difficult to sample of access and have been used to detect invasive species as well as potential disease causing agents (Strand et al. 2011, Longshaw et al. 2012, Hartikainen et al. 2014, Kolby et al. 2015). These approaches, whilst useful, may require some development of species- or genus-specific primers as well as refinement to ensure detection of the target animal amplicon (e.g., Figiel and Bohn 2015, Tréguier et al. 2014). For crayfish, the likely samples to detect eDNA may well be water in which the target crayfish reside, sub-samples of which can also be used to assess levels of contaminants, metals, oxygen, etc... Under experimental conditions, it may be possible to collect faecal matter from animals to provide additional information. The application of analysing faecal matter from wild crayfish may not be as easy due to the issues around collection of such samples under field conditions and the limits of detection of current methods. However, as with eDNA methods for water, the approach may provide information such as diet choices, distributional data of crayfish and data on overall health status. It may be possible to sample casts left after crayfish have moulted for molecular analysis as well as for the presence of external parasites, commensals, etc... this is, however, time limited with degradation of DNA and loss of ectocommensals likely to occur soon after the crayfish discards its outer shell.

Prior to conducting any external examinations and/or taking of morphometric data, the animals may be anaesthetised (with recovery) or killed (covered in detail below). However, it is important to note that for some techniques, such as histology, the time between death of the animal and taking of samples should be minimised. External observations should include an assessment of the overall appearance of the animal – colouration, general behaviour (active, lethargic, aggressive, placid, etc...), moult status, external damage (including missing or damaged limbs, damage to antenna, eyes, etc...), evidence of regeneration, lesions and/or melanised areas on carapace, presence of ectocommensals/ectoparasites (ensuring these are collected for later identification) and possibly swabs or scrapes of carapace. The reproductive state as well as the sex of the individuals should be noted.

At a minimum, the following external measurements should be taken (see Fig. 1): chelae width, depth and length (ensuring left and right are measured separately to account for dimorphic growth), carapace length (from tip of rostrum to back of carapace), carapace depth and width, post-orbital/carapace length (from posterior edge of the eye socket to the posterior edge of the carapace) and abdomen length (from the posterior edge of the carapace to the posterior edge of the telson) and abdomen width as well as total weight (Deniz (Bök) et al. 2010, Endrizzi et al. 2013, Chybowski 2014).

Anaesthesia (with and without recovery)/sedatives/humane killing

There is now a greater emphasis on the humane treatment and welfare of experimental and wild animals, including invertebrates (Crook 2013), particularly around use of anaesthetics (with and without recovery) and in the development of humane methods of stunning and killing them. A number of methods have been proposed, some of which are in common use. It is important to stress however, that any methods used should take account of prevailing laws and guidelines; as such, we are not advocating one method over another, rather that the responsibility for anaesthetising or killing of crayfish rests with the reader. Ross and Ross (2008) provide a good summary of legislative and safety considerations with use of anaesthetic agents,

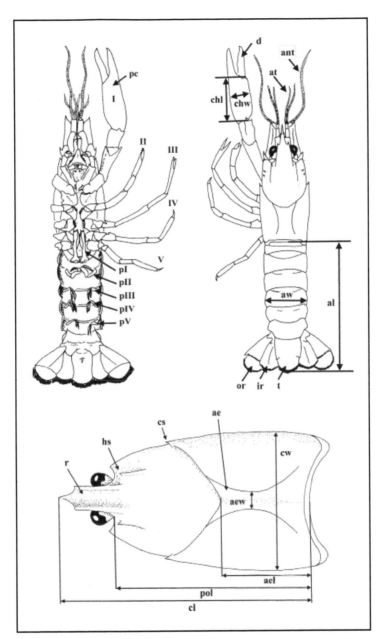

Figure 1. Diagrammatic crayfish showing major appendages and measurements required. Top left, ventral view; top right, dorsal view; bottom, carapace. Where ae = areola, ael = areola length, aew = areola width, ant = antenna, at = antennule, al = abdomen length, aw = abdomen width, chl = chela length, chw = chela width, cl = carapace length, cw = carapace width, cs = cervical spine, d = dactyl, hs = hepatic spine, ir = inner ramus of uropod, or = outer ramus of uropod, pc = propodus of chela, pol = postorbital length, r = rostrum, I-V = pereiopods I-V, pI-pV = pleopods I-V. Not to scale. After Hobbs and Jass (1988).

albeit with a strong focus on teleosts. Different crayfish species and indeed the same crayfish species at different physiological states may react differently to anaesthetics so care should be taken when applying anaesthetic agents, particularly if the animal is expected to recover. If studies are to be conducted on animals that includes assessment of ectocommensals and/or ectoparasites, then workers need to be aware that some anaesthetic methods may cause loss of these ectocommensals leading to an underestimate of parasite burdens (e.g., Mestre et al. 2011).

Some researchers advocate the use of ice to slow down the metabolism of crayfish, which will allow for the subsequent sampling (with or without recovery) of individuals. Recently developed technology such as electrical stunning of crayfish is considered to be a humane and rapid method for killing crayfish. A number of chemicals have been trialled, being either added to the water, dispersed in air or injected into the animal including lidocaine-HCl, ketamine-HCl, halothane, chloroform, carbonated water, clove oil (and its synthetic equivalent AQUI-S®) (Kleinholz 1947, Obradović 1986, Brown et al. 1996, McRae et al. 1999, Bondar et al. 2005, Ross and Ross 2008, Xie et al. 2010, Fotedar and Evans 2011).

Non-lethal sampling

A limited number of non-lethal/non-invasive methods have been developed for crayfish for assessment of metabolic status, disease screening, for assessment of stable isotopes or for DNA analysis. Sampling of water has been discussed in a previous section and will not be covered further here. Swabs or scrapes of the carapace may have a utility in identification of some external pathogens but results should be treated with caution as the sample may well contain environmental contaminants that are of no consequence to crayfish.

A non-lethal method for sampling abdominal muscle tissues was described by Imhoff et al. (2010). Amplification of parasite DNA from the tissues was successful and survival, growth and moulting of sampled crayfish was 100% over the six weeks of the study. Similarly, sampling of haemolymph can be a useful method for collecting tissue samples along with removal of 1–2 legs (Simčič et al. 2012), of part of the telson or soft abdominal cuticle (Oidtmann et al. 2006) and other components of the crayfish exoskeleton (Li et al. 2011), all of which have proved useful in assessing health status and for molecular studies.

Tissue sampling

It is important to determine the purpose of tissue sampling before starting. Whilst best practice might suggest collecting all tissues in every fixative/preservative possible with a view to doing everything that is currently technically feasible as well as collecting tissues that may have a utility in the future as technologies develop, it is recognised that there is a balance to be reach between cost, time and storage capacity. Therefore, there is a need to be pragmatic but we would suggest that, at a minimum, tissues are preserved to allow for histology, electron microscopy and molecular/genetic studies to be conducted. It is possible to collect matched tissue samples from individual crayfish to allow for comparison of results from these three methods. Clearly, the size of the animals to be sampled will impact on the amounts of tissue that can be collected and so some hard decisions may need to be made on which of the tissues and fixatives are paramount to the study, sacrificing one preservation method over another. Details of specific laboratory techniques and methods are not covered in this chapter and readers are advised to determine these from published literature pertinent to their interests or from methods available within their own facilities.

Following the external assessment of the crayfish for morphometrics and overall appearance as described above and ensuring that the animal is suitably sedated and killed, relevant, study-specific tissues need to be collected. At the risk of being repetitive, for each tissue collected, it should be possible to collect material for histology/swabs/smears/imprints/tissue squashes, electron microscopy and for molecular/genetic studies. Haemolymph can be collected and either smeared onto a slide for subsequent fixation and staining or stored in eppendorf-type tubes for additional analysis. The tail should be separated from the cephalothorax by cutting across at the point where the two meet. Cutting through the underside of the tail will expose the central nerve cord and allow for sampling of abdominal muscle tissue as well as the nerve cord. One or two pleopods and pieces of claw muscle should be sampled. Opening the main body by cutting along the top of the carapace exposes the underlying organs and tissues. Tissues to be sampled include heart, hepatopancreas, gonad, gill and epidermis. Study-specific tissues might also include eyes and antennal gland. Each organ should be examined and note made of any abnormalities, including colour changes and lesions.

Conclusions

It has been over a decade since Edgerton and Jussila (2004) wrote a manuscript proposing in essence that a fully integrated trans-European project should be instigated to more widely understand and study crayfish pathology. The principle was laudable but didn't appear to gain much traction with the wider astacology community. At the start of this chapter we proposed that future studies should strongly consider taking a similar integrated approach, albeit more widely than pathology. We hope that this chapter provides the impetus for some workers to adopt this holistic approach, especially if combining their studies with field work as described by Larson and Olden in the previous chapter. We recognise that sacrifices may need to be made with regards to what is measured and what is collected, especially when material, finances and assistance is sparse but such limitations should not be seen as an easy option to avoid collection of samples for fellow researchers.

As technologies advance, it is probable that fewer environmental and tissue samples will need to be taken to understand and describe various aspects of crayfish biology. In order to assist with and to contribute to these future developments, efforts should be made to appropriately store tissues in order to provide a repository (or tissue bank) for the future, being mindful of the issues of space and appropriateness of methods for long term storage of tissues. It is of course the desire of any "hands-on" biologist that such methods do not become the primary source of information regarding our chosen animals, rather that they are complimentary to the more "traditional" methods described herein of catching, holding, handling and observing crayfish. We hope that future generations of biologists do not lose the pleasure to be gained from these critical skills.

References

Bate, S.M. and R.A. Clark. 2014. The Design and Statistical Analysis of Animal Experiments. Cambridge University Press, Cambridge, 324 pp.

Bondar, C.A., K. Bottriell, K. Zeron and J.S. Richardson. 2005. Does trophic position of the omnivorous signal crayfish (*Pacifastacus leniusculus*) in a stream food web vary with life history stage or density? Can. J. Fish. Aquat. Sci. 62: 2632–2639.

Brown, P.B., M.R. White, J. Chaille, M. Russell and C. Oseto. 1996. Evaluation of three anesthetic agents for crayfish (*Orconectes virilis*). J. Shellfish Res. 15: 433–435.

Chybowski, Ł. 2014. Morphometric differentiation in four populations of signal crayfish, *Pacifastacus leniusculus* (Dana), in Poland. Arch. Polish Fish. 22; 229–233.

Corkum, L.D. and D.J. Cronin. 2004. Habitat complexity reduces aggression and enhances consumption in crayfish. J. Ethol. 22: 23–27.

Crook, R. 2013. The welfare of invertebrate animals in research: can science's next generation improve their lot? J. Postdr. Res. 1: 9–20.

Deniz (Bök), T., M.M. Harlıoğlu and M.C. Deval. 2010. A study on the morphometric characteristics of *Astacus leptodactylus* inhabiting the Thrace region of Turkey. Knowl. Manag. Aquat. Ecosyst. 397: article 5.

Dytham, C. 2010. Choosing and Using Statistics: A Biologists Guide, 3rd edition. Wiley-Blackwell, Oxford, 320 pp.

Edgerton, B. and J. Jussila. 2004. Crayfish pathology in Europe: past, present and a programme for the future. Bull. Fr. Pêche Piscic. 372-373: 473–482.

Endrizzi, S., M.C. Bruno and B. Maiolini. 2013. Distribution and biometry of native and alien crayfish in Trentino (Italian Alps). J. Limnol. 72: 343–360.

Figiel, Jr., C.R. and S. Bohn. 2015. Laboratory experiments for the detection of environmental DNA of crayfish: examining the potential. Freshwat. Crayf. 21: 159–163.

Fotedar, S. and L. Evans. 2011. Health management during handling and live transport of crustaceans: a review. J. Invertebr. Pathol. 106: 143–152.

Hartikainen, H., G.D. Stentiford, K.S. Bateman, C. Berney, S.W. Feist, M. Longshaw, B. Okamura, D. Stone, G. Ward, C. Wood and D. Bass. 2014. Mikrocytids: a novel radiation of parasitic protists with a broad invertebrate host range and distribution. Curr. Biol. 24: 807–812.

Hobbs III, H. and J.P. Jass. 1988. The crayfishes & shrimp of Wisconsin (Cambaridae, Palaemonidae). Milwaukee Public Museum, 177 pp.

Imhoff, E.M., R. Mortimer, M. Christmas and A.M. Dunn. 2010. Non-lethal tissue sampling allows molecular screening for microsporidian parasites in signal, *Pacifasticus leniusculus* (Dana), and vulnerable white-clawed crayfish, *Austropotamobius pallipes* (Lereboullet). Freshw. Crayfish 17: 145–150.

Kleinholz, L. 1947. A method for removal of the sinus gland from the eyestalks of crustaceans. Biol. Bull. 93: 52–55.

Kolby, J.E., K.M. Smith, S.D. Ramirez, F. Rabemananjara, A.P. Pessier, J.L. Brunner, C.S. Goldberg, L. Berger and L.F. Skerratt. 2015. Rapid response to evaluate the presence of amphibian chytrid fungus (*Batrachochytrium dendrobatidis*) and ranavirus in wild amphibian populations in Madagascar. PLoS One 10: e0125330.

Legendre, P. and L. Legendre. 2012. Numerical Ecology. Elsevier Science Ltd., London, 1006 pp.

Li, Y., W. Wang, X. Liu, W. Luo, J. Zhang and Y. Gul. 2011. DNA extraction from crayfish exoskeleton. Indian J. Exp. Biol. 49: 953–957.

Longshaw, M., S.W. Feist, B. Oidtmann and D.M. Stone. 2012. Applicability of sampling environmental DNA for aquatic diseases. Bull. Eur. Assoc. Fish Pathol. 32: 69–76.

MacDonald, J. 2009. Handbook of Biological Statistics. Sparky House Publishing, Baltimore, 287 pp.

Magurran, A. 1988. Ecological Diversity and Its Measurement. Princeton University Press, Princeton, 179 pp.

Magurran, A. 2003. Measuring Biological Diversity. Wiley-Blackwell, Oxford, 256 pp.

Magurran, A. 2010. Biological Diversity: Frontiers in Measurement and Assessment. Oxford University Press, Oxford, 345 pp.

McRae, T., K.L. Horsley and B. McKenzie. 1999. Evaluation of anaesthetic agents for crayfish. Freshw. Crayfish 12: 221–232.

Mestre, A., J.S. Monros and F. Mesquita-Joanes. 2011. Comparison of two chemicals for removing an entocytherid (Ostracoda: Crustacea) species from its host crayfish (Cambaridae: Crustacea). Int. Rev. Hydrobiol. 96: 347–355.

Obradović, J. 1986. Effects of anaesthetics (halothane and MS-222) on crayfish, *Astacus astacus*. Aquaculture 52: 213–217.

Oidtmann, B., S. Geiger, P. Steinbauer, A. Culas and R.W. Hoffmann. 2006. Detection of *Aphanomyces astaci* in North American crayfish by polymerase chain reaction. Dis. Aquat. Org. 72: 53–64.

Raidal, S.R., G. Cross, S. Fenwick, P.K. Nicholls, B. Nowak, K. Ellard and F. Stephens. 2004. Aquatic animal health: exotic disease training manual. Fisheries Research and Development Corp. and Murdoch University, 154 pp.

Ross, L.G. and B. Ross. 2008. Anaesthetic and Sedative Techniques for Aquatic Animals. Wiley-Blackwell, Oxford, 222 pp.

Ruxton, G.D. and N. Colegrave. 2010. Experimental Design for the Life Sciences. Oxford University Press, Oxford, 200 pp.

Simčič, T., F. Pajk, A. Brancelj and A. Vrezec. 2012. Preliminary multispecies test of a model for non-lethal estimation of metabolic activity in freshwater crayfish. Acta Biol. Slov. 55: 15–27.

Southwood, T.R.E. and P.A. Henderson. 2000. Ecological Methods. Blackwell Science, Oxford, 575 pp.

Strand, D.A., A. Holst-Jensen, H. Viljugrein, B. Edvardsen, D. Klaveness, J. Jussila and T. Vrålstad. 2011. Detection and quantification of the crayfish plague agent in natural waters: Direct monitoring approach for aquatic environments. Dis. Aquat. Org. 95: 9–17.

Sutherland, W. 2006. Ecological Census Techniques: A Handbook. Cambridge University Press, Cambridge, 432 pp.

Tréguier, A., J.M. Paillisson, T. Dejean, A. Valentini, M.A. Schlaepfer and J.-M. Roussel. 2014. Environmental DNA surveillance for invertebrate species: advantages and technical limitations to detect invasive crayfish *Procambarus clarkii* in freshwater ponds. J. Appl. Ecol. 51: 871–879.

Xie, H.-M., H.-X. Bian, Y. Yang, W.-C. Zhang, C.-M. Lin and X.-L. Gao. 2010. Clove oil anesthesia for the improvement of survival rate of crayfish. Food Sci. 31: 247–250.

The Management of Invasive Crayfish

Paul Stebbing

"No problem can withstand the assault of sustained thinking"

- Voltaire

Introduction

For millions of years geographical features have provided environmental barriers that have given rise to isolated ecosystems essential for species to evolve. With the advent of humans, these natural barriers to the migration of species, which previously resulted in such a wealth of biodiversity, have been rendered ineffective by the gradual 'opening up' of the globe. This process has resulted in species being moved outside of the native range and introduced into new areas. Sometimes these species have become established and thrived, and can have considerable impact on the invaded environment.

It is likely that the process of humans moving animals and plants with them as they migrated started many thousands of years ago, with humans moving species that were of use to them outside of their native ranges and introducing them into new areas from before the Neolithic era, circa 10,200 B.C. (Webb 1985). According to Grandjean et al. (1997) it is possible that movements of humans from continental Europe into the British Isles after the last Ice Age may have resulted in the introduction of the white-clawed crayfish (*Austropotamobius pallipes*), the only crayfish found in Britain and Ireland considered native. White-clawed crayfish may have been introduced deliberately as a food source or possible accidently with the movement of other aquatic animals or vegetation.

The development of boats, and the ability of humans to cross huge tracts of open water, as demonstrated by Thor Heyerdahl's KonTiki expedition, and with the discovery of the Americas by the Vikings and Columbus, made transoceanic movement of species possible. This process most likely saw the deliberate movements of domesticated animals for farming and aquaculture, but also the accidental movement of hitchhikers, either with cargo, or as bio-fouling species attached to the hulls of boats. The black rat (*Rattus rattus*), for example, has spread extensively as stowaways from its native range of tropical Asia, and is now found in every continent. With the growth of globalisation, as well as an emphasis on free trade and commercialisation, opportunities for the accidental or deliberate introduction of non-native species have

Cefas Weymouth Laboratory, Barrack Road, The Nothe, Weymouth, Dorset, UK, DT4 8UB.

increased significantly, especially over recent years (Cohen and Carlton 1998, Ruiz et al. 2000, MacIsaac et al. 2001).

Many countries rely on introduced species as key elements to their food supply. In the United States for example, 98% of the food supply comes from introduced, non-native species such as wheat, rice, domestic cattle and poultry with a combined value of more than $500 billion a year. However, some non-native species do have dramatic environmental and economic impacts. These invasive non-native species (INNS) are considered the biggest driver of global biodiversity loss after climate change (Mack et al. 2000, WWF 2014). INNS have both ecological and social impacts, replacing native species, altering invaded environments and introducing novel diseases, in addition to reducing natural hazard prevention, impacting on ecosystem services and increasing threats to human health (Hatcher et al. 2012, Pimentel et al. 2005, Vilà et al. 2010, Williams et al. 2010). In the EU, the cost of managing INNS, has been estimated at €12 billion per year (Shine et al. 2009); and in the United States $137 billion per annum (Pimentel et al. 2000). This was considered by Perring et al. (2002) to be a conservative estimate as it deals only with a subset of INNS. It is, however, impossible to put a value on the actually biological and economic damage that INNS incur as INNS are considered one of the main proximate causes of extinctions worldwide (Glowka et al. 1994), an effect that cannot be ascribed a cost.

Normally these immense ecological and economic problems are a result of the introduction of only a few INNS, with as many as 80–90% of established non-native species having minimal detectable effects (Williamson 1996). This is more clearly illustrated by the 'Three Tens' rule of Williamson (1996), where approximately 10% of imported species become introduced, 10% of introduced species become established, and 10% of those established become invasive. An exception to this rule are crayfish, with a much larger percentage of introduced crayfish becoming invasive.

Crayfish are globally recognised as some of the most widely distributed and invasive aquatic species. Invasive species of crayfish have been associated with the decline of native species, through disease transmission and competitive exclusion, and in some cases domination of the biomass. Invasive crayfish have much wider environment influence than native crayfish, potentially impacting on the whole ecosystem, for example: (1) negative effects on wider invertebrate communities; (2) competitive interactions with native fish; and (3) impacts on river morphology through burrowing and sediment mobilisation. Not only do they have significant environmental and economic impacts, but also compromise progress towards compliance with environmental targets, such as the European Union Water Framework Directive. The implementation of management programmes to eradicate and control invasive crayfish species is, therefore, of global high importance.

This chapter provides a brief introduction to some of the main principles of managing INNS, and an attempt has been made to link these principles into the more specific management of invasive crayfish. A summary of different management frameworks and methods used to control or eradicate invasive crayfish species is provided, and how these methods could constitute a wider management strategy. It should be noted that this is not an exhaustive list and there is likely to be some methods which are not mentioned or are currently under development. Although many of the examples used are from the UK the principles and methods discussed will be applicable elsewhere.

Principles of INNS management

The control of any INNS is complex, in some cases requiring the application of multiple disciplines to be effective (e.g., ecology, behavioural ecology, pathology, toxicology, population ecology, molecular biology, sociology, economics, project and resource management). It is therefore useful to identify the main risks associated with INNS to aid in the development of targeted management strategies, especially where limited resources are available. Within this context the nature of the risks (and ultimately the subsequent management approaches employed) relate to geographically or politically discreet areas (referred to here after as the control area). The control area in question may vary considerably in size and/or political status, for example a country, state, county, district, region, river catchment, or a single isolated pond. The identification of the control area to which the management strategy is applied is an essential primary stage. The risks posed by INNS to the control area can then be identified, and a management strategy developed aimed at tackling the various risks. The following are examples of risks that may be considered within a management strategy. The examples are broad, high level categories, which could be split down or made more specific depending on the level of detail required or the control area in question:

1. *Introduction risks*: The risk of INNS not yet present within the control area entering the control area. For example, the marbled crayfish (*Procambarus fallax* f. *virginalis*) is a recognised invasive species of crayfish, but (at the time of writing) is not established in the UK. However, the popularity of the species in the aquarium trade poses a major threat to its potential introduction as demonstrated by established populations being found elsewhere in Europe. The inclusion of this issue into a management programme could help to prevent introductions from occurring therefore avoiding problems associated with the species.

2. *Recent introductions or those of limited distribution*: INNS that are of a recognised or potential risk which have recently been introduced or currently have limited distribution within the control area. Immediate action to eradicate the species will prevent their further spread, and potentially further environmental and/or economic impact. The cost of eradication and remediation can increase exponentially as the species continues to spread. For example, the white river crayfish (*Procambarus acutus*) is found in one single enclosed fishery in the UK. It is thought to have been present in the fishery for only a short period. A rapid response to this population could see its removal with comparatively limited environmental or economic impact before it spreads into other water bodies and becomes more difficult to manage.

3. *Established species*: INNS which have become established and potentially wide spread across the control area. The eradication of the species from the control area could be a significant undertaking, possibly as a result of multiple populations found in different and potentially highly variable habitats across the control area. However, the complete removal of the species could be of significant environmental and economic benefit. Containment of the population(s) to reduce the rate of further spread may be the only management measure realistically available. For example, the signal crayfish (*Pacifastacus leniusculus*) is widely distributed throughout much of the UK and Europe. It is found in almost every major catchment and in both lotic and lentic systems of varying sizes. Eradication of the species on a UK scale would be a considerable undertaking, although limiting their further spread, or eradicating populations at a localised geographical level may still be short term viable options.

The United Nations Convention on Biological Diversity (CBD) (Secretariat of the Convention on Biological Diversity 2004) have set out a hierarchal approach to the management of INNS in the form of responses to identified risks:

1. The prevention of introductions between and within states;
2. Early detection and rapid response to new introductions;
3. Containment and long-term control measures, where eradication is not possible or resources are not available.

This final point could be reworded as: 'eradication and, where this is not possible or resources not available, containment and long-term control measures are applied'.

For prevention, early detection, rapid response, eradication, containment and long-term control measures to be effective they require effort, a strategy (or management programme) and funding. For long term effectiveness, infrastructure needs to be in place; for example, to support rapid response there needs to be a contingency plan, an effective method of eradication/control established, a team recognised and retained to undertake the response, in addition to funds available to purchase required equipment and materials. Therefore the effective management of INNS needs commitment and appropriate resourcing. The essential element of any INNS management programme are tried and tested methods to eradicate or control the target species. Therefore, as part of the effort to control invasive species there is a need for research to develop suitable methods of management. Methods should be socially and ethically acceptable, efficient, non-polluting and non-threatening to native species, human or domestic animals or crops. Clearly these are ideals and very few methods will meet all of these criteria, but attempts should be made meet them where possible.

Prevention of introduction

Complete eradication of an established INNS, pessimistically, is rarely possible (Kolar and Lodge 2001, Mack 2000) and can be economically (and potentially environmentally) very costly, or methods may just not exist which can be realistically used. Therefore prevention is recognised as the most effective management approach (Caffrey et al. 2014, Caplat and Coutts 2011). On the premise that 'prevention is better than cure', the CBD Aichi Biodiversity Targets for 2020 and new EU legislation on the prevention and management of INNS both focus on identifying and managing the risk associated with pathways and vectors of INNS introduction and spread (European Commission 2013). In relation to control areas, pathway management can vary considerably in form depending on scale, from cross boarder import controls, to biosecurity measures implemented at a single water body. While it is highly unlikely that all potential introductions will be prevented, measures to reduce propagule pressure along the transit route will decrease the risk of introductions and establishment occurring. Understanding the potential pathways that may result in the introduction (or spread) of invasive crayfish is an essential step in the management process. Reducing the risk of introduction is essential before attempted eradication as reintroduction may occur without adequate management.

Pathways of introduction/spread will vary considerably and need to be assessed on a case by case basis for each control area. Likewise the means by which the pathway can be managed will be equally as diverse. An example of possible introductions already mentioned is the marbled crayfish and its potential introduction into the UK from other countries via the aquarium trade. There are possible scenarios where crayfish may be introduced through accidental or deliberate actions. For example, the translocation of crayfish inadvertently caught with fish in one water body destined for restocking in another, the deliberate introduction of crayfish into a water body for weed control, or to seed populations for future food harvesting.

Early detection and rapid response

Early detection is an integrated system of active or passive surveillance to identify the presence of new INNS, or spread of an existing established INNS as early after entry into the control area as possible. This will facilitate the rapid response to eradicate and control the population in the early stages of establishment when success is still feasible and less costly. For early detection and rapid response, there needs to be effective means of detecting species at low densities, either soon after introduction or in early stages of establishment. Methods of detection/sampling crayfish are covered in detail by Larson and Olden within this book, but suffice to say that there are few methods that can be applied consistently across large areas that are suitable for the detection of low density populations. As mentioned by Larson and Olden, methods are being developed which may help with early detection, such as environmental DNA or eDNA sampling. This is an approach where a sample of water or sediment is taken from a water course, and through molecular analysis, DNA released from any individuals of the species being monitored for is detected if present. There are drawbacks to the application of this method to crayfish. Crayfish are unlikely to release large quantities of DNA into the environment, in comparison to fish species where DNA will be released in mucus, scales and faecal castes for example, making detection, especially of a low biomass populations very difficult. Levels of DNA released from a crayfish population is also likely to vary dramatically through time, not just in relation to increasing biomass of a population, but also as a result of breeding, spawning and moulting events. Despite this, with further development, eDNA does present an exciting and very viable means of effectively and quickly monitoring a large number of water bodies for low density populations.

A potential means for making monitoring methods for early detection attempts more targeted is through focusing efforts at locations within the control area where crayfish are most likely to arrive. Identifying 'hot spot' of introduction is normally based on the quantification of likely pathways of introduction and the relative proximity to suitable habitat to the pathway. Targeting high risk locations is a viable method to reduce the resource demand on surveillance programmes, but does need reviewing regularly as pathways and there relative risk will change over time, especially in response to pathway management measures.

Once a population of unwanted crayfish has been detected this will ideally initiate a rapid response process. Rapid response is a systematic effort to contain, control or eradicate an INNS while it is still in the early stages of establishment or soon after its initial detection. It may be implemented in response to a new introduction of an unwanted species, or to the spread of a previously established species. Ideally, as with all eradication or control attempts, there should be preliminary assessments and subsequent monitoring to determine effectiveness. Normally the main focus of the response is rapidity and efficiency in the removal of the species, limiting the potential for further spread and the need for further action. The development of contingency plans for the rapid response to certain high priority species will aid in this process. In some cases eradication may not be possible and control measures are alternatively implemented.

Eradication, containment and long-term control measures of established species

Although eradication is often the primary intention of rapid response, within this context it is referring to the eradication of established INNS rather than new introductions, therefore the species in question is likely to have a wide distribution within the control area. Although the eradication of a single population may be feasible, within this context there will need to be procedures in place to ensure that reintroduction do not occur in cleared areas. Therefore an eradication strategy would also need to consist of a robust biosecurity and monitoring programme. Where eradication is not considered feasible, containment and long-term control measures may be implemented focusing on preventing further spread of the species, with a focus on biosecurity. There may also be a desire to reduce the impact of INNS, which is normally achieved through reduction in population size if eradication is not feasible.

Assessing feasibility

It is important that feasibility is assessed, as however desirable eradication or control may be, a failed attempt can be counterproductive. There has been a relative hiatus in the management of invasive crayfish as a result of attempted eradication, containment or control programmes which have been perceived as failures. In some cases the causes of these failures where due to a lack of clearly defined end points, or limited feasibility assessments. Bomford and O'Brien (1995) suggested six criteria for assessing feasibility:

1. Rate of removal exceeds rate of increase at all population densities. Any control method needs to remove the population more quickly than the rate of replacement (through immigration or reproduction).
2. Immigration is prevented or reduced. If animals can migrate or be released from captivity into the control area, eradication will be transient.
3. All reproductive animals within the target population must be at risk. All reproductive or potentially reproductive animals within the population must be potentially susceptible to the control mechanism. Therefore all of a target population must be treated.
4. Animals can be detected at low densities. Without a successful mechanism by which the target species can be detected at low densities, there is no way to determine the level of success of eradication efforts or if it has been achieved.
5. Discounted cost-benefit analysis in favour of eradication. The overall cost of the eradication/control programme needs to be carefully compared to costs incurred by the presence of the species, including damage to resources, remediation and control.
6. Suitable socio-political environment. Social and political factors can play an important role in determining if eradication programmes should proceed, even when technical and economic criteria are met.

Although ideally all 6 of these criteria should be met, as discussed previously the detection of low density populations of crayfish (point 4 above) is difficult, but all the other criteria can be met. A common flaw in the assessment of the effectiveness of control/eradication methods, is that immigration (point 2) has not been prevented, or the whole target population has not been treated (point 3). This may have resulted in the current view that there is no definitive methodology for the control/eradication (i.e., point 1 above) for invasive crayfish (see Holdich et al. 1999, Gherardi et al. 2011). However, there are several methods that have shown considerable potential, in addition to novel approaches that require further investigation. A key element of Integrated Pest Management is the use of multiple tools to achieve the effective management of the target species, but it is relatively rare to see attempts to eradicate or control invasive crayfish populations where multiple methods are applied simultaneously or in a staged process (in addition to complying with the 6 criteria above). Where multiple methods have been applied then successes have been observed.

Perception of management

INNS management is poised at a crossroads where effective management programmes require the input of ecologists, social scientists, resource managers, and economists (Simberloff et al. 2012). This make the development and implementation of management programmes complicated and open to scrutiny from many perspectives. This has added to a general pessimism concerning the prospects of effective management of INNS (Simberloff 2009); often leading to a lack of action, especially when dealing with established INNS. Failed or unsuccessful attempt at control and/or eradication can lead to scepticism, leading to risk aversion exactly when considered risk-taking is required (Parkes and Panetta 2009). Often failures in eradication, containment or control attempts can lead to abandonment of the problem, when a properly planned, sustained programme may have addressed the problem. Setting realistic and attainable objectives in relation to the feasibility of controlling INNS will aid significantly in overcoming some of this pessimism, as well as providing a measure against which the relative success of the management programme can be determined. Some attempted management programme trials for invasive crayfish have set unrealistic objectives or poorly defined end points, resulting in the misinterpretation of the results and incorrect conclusions being drawn in relation to the management programmes effectiveness.

Management methods

Within this section methods of managing crayfish are discussed. The methods have been categorised as (1) mechanical; (2) physical; (3) biological; (4) biocidal; (5) autocidal and alternative, and (6) legislative management. Reference is made within each section to the application of the management methods in relation to the CBD hierarchal approach to the management of INNS: (1) the prevention of introductions between and within states, (2) early detection and rapid response to introduction and (3) containment and long-term control measures.

Mechanical management

Mechanical control encompasses the physical removal of crayfish from water bodies using traps, nets, electrocution or removal by hand. Humans have proven to be very effective at driving species to extinction over the years by their physical removal. The effects of over exploitation on many economically valuable species have been well documented, and examples exist where species are under threat from local and ultimately global extinction by the effects of physical removal (e.g., Cheung et al. 2005). For physical removal to be effective at eradicating, or at least reducing population size significantly, the methods used need to comply with the feasibility criteria, specifically the rate of removal needs to be higher than the rate of immigration and reproduction. There are no apparent reasons why invasive crayfish population cannot be over exploited through physical removal to the point of extinction.

There is a history of trapping crayfish from the wild for human consumption, predominantly in Northern Europe and the Southern States of America. A wide variety of trap types have been used. All follow the same basic principle of a submersed container with multiple entrances that facilitate ingress of animals while limiting egress from a central chamber with an attractant, normally comprising a food based bait, located in the main chamber (Bean and Huner 1979, Westman et al. 1979, Fjälling 1995, Campbell and

Whisson 2000). The most commonly used trap type is a cylindrical funnel trap, or Swedish 'Trappy' trap, with entrances at either end of a central chamber (Fjälling 1995). Trapping traditionally takes place in the summer months when crayfish are most active, and therefore more readily trapped.

Trapping is often considered to be inherently biased to the removal of dominant large adult males (Gherardi et al. 2011). For this reason not all of the target population may be at risk from trapping, at least at the beginning of a trapping operation. However, the sex ratio of catches and the size of animals caught depend on a number of factors. Seasonal variation in the sex ratio of catches have been observed, with more females being caught immediately after the release of juveniles and in the breeding season. In addition, smaller animals are more likely to enter traps in the absence of larger crayfish (Peay and Hiley 2001). With the exploitation of a population, larger animals are removed first, therefore, with prolonged trapping smaller animals will become more readily trapped, in addition to the sex ratio of catches becoming more equal. The process of 'trapping down' a population will increase the number of life stages susceptible to trapping making eradication more likely as more life stages become susceptible to the method. However, this does mean that for reliable results trapping requires a considerable amount of effort. Effort is a combination several elements of the trapping process: the number of traps used, the type of trap used, the type of bait used, the frequency at which traps are emptied, how long traps are deployed for and how long the trapping programme is run for. It is due to a lack of trapping effort in addition to not meeting feasibility requirements that some studies have concluded that the eradication of a crayfish population using trapping alone is not feasible.

It has been noted that some species subject to exploitation compensate with high breeding and survival rates due to an increased availability of resources (Bomford and O'Brien 1995). This has been noted in crayfish, where the removal of large dominant males has been suggested to reduce pressure on juveniles thereby giving rise to larger populations (Gherardi et al. 2011). This was observed in populations of noble crayfish (*Astacus astacus*) where the removal of larger animals reduced competition on smaller animals resulting in the development of much denser populations (Skurdal and Qvenild 1986). This may result in a reduction in the average size of animals in the population with earlier maturation (Freeman et al. 2010). These observations may be a result of the smaller animals becoming more active and therefore appearing more abundant (or trappable) as a result of the removal of the larger animals rather than a population level compensatory response.

Trapping programmes on riverine systems found removal of large adult males from a section acted as a drain on adjacent areas, with large adult males from adjacent untrapped areas moving into the available space formed by the depletion of the population in trapped areas. This process resulted in a perceived enhancement of the population in adjacent areas (Ibbotson et al. 1997, Holdich et al. 1999, Moorhouse and Macdonald 2010). Although again this may be a result of smaller animals becoming more active (and therefore trappable) in the absence of larger animals. This process does highlight the level of immigration/ migration in crayfish populations and the need to ensure all of the population is affected by a control method to be feasible.

The use of commercially available trap types in the management of invasive crayfish populations has been used to suppress populations, but eradication was not achieved (Bills and Marking 1988, Roqueplo et al. 1995, Frutiger et al. 1999, Holdich et al. 1999). These observations may be a result of a number of factors: (1) a lack of effort, (2) trapping only being conducted during summer months when catches would be dominated by large animals, or (3) trapping may not conducted over a long enough period (or with enough effort) to see the reduction in larger animals and the trapping of smaller animals.

Trap design effects the catch composition, such as the quantities caught, the size of the animals and possibly sex ratio of the catch. For example, commercial 'Trappy' traps have a diamond shaped mesh with a maximum diameter of 3.5 cm, the purpose of which is to allow small animals to escape from the trap to maintain a viable fishery. Commercial trap designs are those that are most readily available and therefore most commonly used in control/eradication studies, possibly being one reason for small animals not being caught so readily in some studies, in addition to large adult competitive interference. Trap retention was an issue with the majority commercially available traps tested by Westman (1991), with crayfish being able to enter and exit some trap designs at will, especially smaller animals. Modification of the entrance to crayfish traps to a slit-like aperture increased retention considerably (Westman 1991). A reduction in the diameters of trap entrances have also resulted in a more equal sex ratio in catches, suggesting that modifying traps can remove the perceived bias to the removal of large adult males, especially during the initial stages of trapping (Stuecheli 1991). The retention rate of traps can also be improved by decreasing the mesh size (Peay and Hiley 2001), also resulting in traps catching a wider size range of animals (Wright and Williams 2000).

Traps with a large internal 'volume' appeared to catch the highest quantities of crayfish and have the best retention (Bean and Huner 1979, Fjälling 1995, Campbell and Whisson 2000). The increased volume of the traps may negate the prior occupancy effect of larger animals deterring smaller animals from entering a trap as the additional volume makes encounters less frequent. The improvements in retention may be as a result of the animals not being able to relocating entrances more easily, and therefore escape.

Several long-term trapping programmes have been implemented in an attempt to control/eradicate crayfish populations. For example, trapping commenced on the River Lark, England in 2001 and is ongoing (West 2011). The project used homemade traps with smaller mesh size and larger holding areas than commercial 'Trappy' traps. A wide range of trap styles where used in an attempt to capture as many life stages of the invasive signal crayfish as possible. Trapping was also conducted upstream of the control area to reduce immigration. Although trapping effort has varied throughout the study, there has been a total reduction of 70% in the size of catches. This has resulted in observed recovery of the immediate ecosystem, such as river banks and fish populations. Another long term trapping programme on the River Clyde in Scotland has seen a significant reduction in total numbers of signal crayfish caught from 10,625 in 2001–2002 to 5335 in 2006–2007 using commercial 'Trappy' traps (Reeve 2004, Freeman et al. 2010).

Electrofishing has been used as a method of sampling fish for a number of years. While fish respond to an electric current by moving towards the source, crayfish are stunned by the process becoming completely motionless and easily removed by net. Electrofishing can therefore be employed to stun and capture crayfish, but is only effective against those out in the open, but not those in cover or burrows at the time of treatment. Electrofishing equipment mounted on boats is sometimes used to harvest crayfish in the USA, demonstrating its effectiveness as a means of removing crayfish (Huner 1988). Electrofishing is considered effective at removing all sizes of crayfish (Westman et al. 1978), but catch vary with time of day (Laurent 1995), so is best suited for use at night when crayfish are most active (Westman et al. 1978). Electrofishing has been used to remove large numbers of crayfish over a broad range of sizes (Sinclair and Ribbens 1999), removing a large number of animals over a broad size range, but no depletion in total catch numbers per run were observed.

The use of electrocution in the control of invasive crayfish is being examined by other means in addition to the traditional use of electrofishing. The system reported by Peay and Mckimm (2011) is effectively a large, fixed position electrofishing kit, with lengths of cathode spread across a river bed or which can be driven into the bankside. These are used to send 96 kw pulses through the water (in comparison to 0.5 kw of a normal electrofishing kit) killing the crayfish rather than stunning, even when in cover.

There are a number of drawbacks to using electrofishing or electrocution methods. Health and safety is a major issue, not only to those applying the method, but members of the public, livestock and pets. In addition only trained teams should conduct such exercises, making the method less accessible in comparison to trapping. The application of electricity is also limited in its applications to shallow, clear water, with clement weather, therefore only small water systems can be treated during summer months. Methods, such as the system reported by Peay and Mckimm (2011), would also have to be coupled with fish removals and possibly a period of recovery to allow re-colonisation by macro-invertebrates affected by the treatment process adding to costs.

Mechanical removal of crayfish from a water body is labour intensive, and can require extended periods of time to be effective depending on the amount of effort applied, thus potentially incurring considerable cost (Gherardi et al. 2011). To date there have been no successful eradications of invasive crayfish populations using mechanical methods. While a reduction in population size have been observed with the use of consistent trapping pressure, as control programmes progress it takes more time and expense to locate and remove animals (i.e., the economics of diminishing returns). Hence removal rates are lower at low population densities (Bomford and O'Brien 1995). It is important therefore to maintain the rate of removal despite very low numbers being caught. Further refining methods of trapping could increase efficacy considerably, increasing the possibility of eradication.

The mechanical removal of crayfish from a population could potentially be part of a rapid response process, if large numbers can be removed over a short period, reducing propagule pressure and therefore limiting the risk of further spread. This would only be considered if other means of eradication where not considered feasible. Commercial trappers in the UK have been reported to use up to 120 traps per acre (about 1 trap every 34 m²), emptied every 24 hour period for 3 months resulting in a massive reduction in the affected populations size. However, even though mechanical removal is one of the most widely used and accessible means of controlling and containing invasive crayfish populations, it has yet to be demonstrated as a standalone approach in fully eradicating any crayfish population. However, with further structured research and the development of technology, eradication of crayfish populations using mechanical removal may be a viable option for the future. In the interim, combining trapping with other approaches may be better suited for rapid response and total eradication.

Physical management

Physical control includes: (1) de-watering and the removal of suitable habitat, exposing animals to conditions that will cause mortality, e.g., desiccation or predation; and (2) the inclusion of barriers to prevent dispersion of the population. Holdich and Reeve (1991) reported a management programme implemented on ponds infested with invasive crayfish where the ponds were drained and left to dry. The ponds were refilled shortly after, but crayfish that had survived desiccation in burrows then reappeared the following year. Peay and Hiley (2001) reported that a lake in Wales was drained and dug out, but females carrying eggs were found 3 months after the process. Kozák and Policar (2003) drained ponds in Poland, and despite over wintering at temperatures below – 20°C, crayfish still emerged the following spring when the ponds were refilled. Given the robustness of the majority of invasive crayfish species, it is unsurprising that they are able to survive in only damp conditions for such prolonged periods of time. Although these methods do appear to be effective at reducing populations size, they are circumstantial in their application.

There are reported cases in America where whole ponds containing invasive crayfish have been filled in to prevent further spread (Peay pers. comm.). Although this is an extreme approach, resulting in the complete destruction of the pond, it is cost effective and removes the possibility for the population to spread further.

Large barriers such as waterfalls have been shown to be effective at limiting the movement of crayfish (Kerby et al. 2005, Peay and Hiley 2001). Dams have been used to limit the spread of invasive crayfish in an upland stream in Spain (Dana et al. 2011), with a series of three dams which were less than 3 m high constructed in a specific manner to prevent crayfish movement. A crayfish barrier has recently (2011) been installed between the headwaters of the River Clyde and River Annan in Scotland in an attempt to control the spread of signal crayfish (C. Bean pers. comm.). An electrical barrier was developed by Unestam et al. (1972), and deployed with some success, but due to the large power requirements, the cost of implementation was prohibitive. Despite reported successes of physical barriers Frings et al. (2013) demonstrated that crayfish use their swimming capability to pass barriers, concluding that crayfish are able to pass all barriers. A barrier deployed in a river between Sweden and Norway to prevent the movement of crayfish was found not to be effective (Johnsen et al. 2008). However, this does not detract from the fact that barriers can be effective mechanism by which the natural movement of crayfish can be halted (or at least delayed), containing populations for a period. Peay and Hiley (2001) examined the potential benefits of catch pits on the outflow of an infested lake to prevent further spread into an adjoining river, but there are no results on how successful the catch pit is at preventing movements.

De-watering and habitat modification provide a range of potential management methods, suitable for both rapid response and more long term management. Although de-watering or habitat removal/infilling do pose extreme measures, with environmental impact, they can have significant impact on crayfish populations. Barriers provide a long term control measure, which are comparatively cheap, with low running costs (i.e., maintenance). Despite this there is potential for flooding, human activities, natural dispersal or predators to result in the movement of crayfish beyond such barriers. Given that barriers can be circumvented by some many means, they will only ever delay what some may consider as the inevitable. This could lead to the pessimistic view that 'they do not work', but barrier are still an important tool to consider in the management of invasive crayfish. The potential impact that the construction of barriers may have on other species (e.g., migratory fish) and the ability of the waterway to be navigated would have to be considered, potentially limiting their application, although barriers that are passable by fish have been developed (Frings et al. 2013).

Biological management

Invasive species can be successful in a new environment if they are introduced in the absence of constraints such as pathogens or predators that would normally keep population numbers under control. This idea is wholly ensconced in the enemy release hypothesis (ERH) (Clay 2003, Hilker et al. 2005), which states that the abundance or impact of a non-indigenous species may be related to the absence of natural enemies of these species in their introduced range, compared with those occurring in their indigenous range (Colautti et al. 2004). The premise of biological control is to utilise pathogens or predators that are detrimental to survival of the target organism. In principle, biological control may be viewed by the general public as a more 'natural' approach to the control of pest species, particularly due to growing concerns surrounding over-reliance on harmful chemicals that may not be species specific.

There are a number of species that naturally predate on crayfish, however, fish are possibly the most suitable as they can be easily transported are comparatively easy to obtain and will stay within the target water body. Eels (*Anguilla anguilla*), burbot (*Lota lota*), perch (*Perca fluviatilis*), pike (*Esox lucius*), chub (*Squalius cephalus*), trout (e.g., *Salmo trutta*) and (*Oncorhynchus mykiss*), tench (*Tinca tinca*) and carp (*Cyprinus carpio*) are all recognised predators of crayfish. A number of studies have examined the impact of fish predation on crayfish populations (e.g., Westman 1991). The presence of predatory fish have resulted in a reduction in population size, reduced individual growth, and altered behaviour, such as increased utilisation of shelter (Blake and Hart 1995). There is size selective predation depending on the species of fish, for example pike predate on all sizes whereas perch, carp and tench predate on smaller animals (Neveu 2001). Aquiloni et al. (2010) found that eel gape size limited the maximum size of the animals predated on. They also found that eels could enter into burrows, but were not found to be voracious feeders. Eels have been attributed as the main cause of decline in crayfish populations in a study by Frutiger and Müller (2002). In concurrence, West (2011) concluded that the dramatic increase of crayfish in the River Lark (England) was as a direct result of the alleviation of predatory control from eels. The declining eel stocks in many GB rivers may help to explain the expansion of invasive crayfish, again reiterating the advantages of healthy waterways as protection against invasions. This is illustrated by a study where fish were removed from a lake in Finland resulting in a dramatic increase in the crayfish population, highlighting the natural control of fish on crayfish (Westman 1991).

Although the introduction of predatory fish does apply some level of control to invasive crayfish populations, there are potential issues. The fish may predate on non-target species, a particular issue once the target population of crayfish has been reduced. In addition, the introduced fish may impact on the environment (e.g., carp causing turbidity), and migrate away from the area of control if used in an open water system. Despite these issues, if applied under the correct circumstances fish may be a viable means of supressing crayfish populations. Invasive crayfish are recognised predators of fish, including eggs and juveniles. Reduction of predator pressure on fish populations as a result of trapping, for example, may result in an increase in fish recruitment and therefore predator pressure on the crayfish. This is a potential benefit of trapping which has not been fully investigated, although recovery of fish populations have been noted as a result of trapping exercises on a number of occasions.

Matt Longshaw (in this book) has already provided a comprehensive review of crayfish pathogens and alluded to their potential use as biological control method. To date there does not appear to be any examples of successful commercial scale control of aquatic animals using pathogens, although the utility of pathogens to control aquatic plants or their pests has been demonstrated (Freeman 1977, Thanabalu et al. 1992, Mcfadyen 1998).

A number of viruses may present potentially viable options for the control of invasive crayfish, not only as a result of mortality but also due to their species specificity. Examples include, *Cherax quadricarinatus* bacilliform virus, *Cq*BV which has been shown to induce mortality, particularly under farmed conditions, suggesting that other similar viruses may have some utility in biological control (Edgerton et al. 1995). *Pacifastacus leniusculus* bacilliform virus (*Pl*BV) which infects the hepatopancreas and midgut (Hedrick et al. 1995, Hauck et al. 2001) has also been associated with mortality events. White spot virus syndrome (WSSV) is an exception to this rule and has a wide host range, including crayfish; potentially being able to infect all decapod crustaceans (Stentiford et al. 2009). The virus has been transmitted under experimental and field conditions to several crayfish species (Shi et al. 2000, Jiravanichpaisal et al. 2001, Claydon et al. 2004, Du et al. 2007, 2008, Baumgartner et al. 2009, Stentiford et al. 2009, Davidson et al. 2010, Soowannayan and Phanthura 2011). Davidson et al. (2010) suggested that WSSV could be used as a biological control agent against invasive crayfish as it was considered to be highly pathogenic to the host and was readily transmitted between crayfish through cannibalism. While the use of WSSV appears at first to show promise, it should be recognised that the virus is not host specific and thus would readily infect non-target decapods. The release of this virus into the wild may lead to the untenable position of a pathogenic strain affecting native hosts at the same time as resistance building up in the target species. The use of viruses as biological control agents would require the production of large quantities, which without a crustacean specific cell line would be very difficult.

Bacterial infections of crayfish tend to be opportunistic and non-specific with few infections leading to significant mortalities, particularly under wild conditions, unless underlying factors such as prevailing environmental conditions or other interacting pathogens are present (Alderman and Polglase 1988, Edgerton et al. 2002, Longshaw 2011). Furthermore, several of the bacteria isolated are known to be a source of gastroenteritis in humans (Longshaw et al. 2012) so do not present viable options for control.

One bacterial-like group that shows some promise as a biological control agent is the spiroplasmas. These are small helical bacteria normally associated with plants, arthropods and ticks where they have been shown to act as direct mortality drivers, as sex distorters or as male killing agents (e.g., Gazla and Carracedo 2009). Most importantly, methods exist for the culture of large volumes of bacteria potentially needed for use as a control agent (Nunan et al. 2005, Ding et al. 2007). Two spiroplasmas have been reported in crayfish. *Procambarus clarkii* are susceptible to "crayfish weakness disease", and a *Spiroplasma* from the male gonads of signal crayfish (Longshaw et al. 2012) which appears to limit sperm production. Studies would be required to be conducted on transmissibility, host specificity and impact on crayfish reproduction before release into the environment. Furthermore, methods for delivery of the pathogen and upscale production of the bacterium would need to be considered before possible use as a control agent.

Many fungi have been isolated from crayfish throughout their range, in some cases associated with disease and mortalities, but tend to be opportunistic, invading crayfish tissues through breaches in the cuticle. *Aphanomyces astaci* originated in North America and through the anthropogenic movements of crayfish, became established in numerous countries in Europe where it has been implicated in the decline of several indigenous crayfish species (Alderman 1993, Bohman et al. 2006, Harlioglu 2008). There are at least three molecular clades of *A. astaci* (Huang et al. 1994, Lilley et al. 1997, Kozubíková et al. 2011). The possibility that different strains of *A. astaci* exist, each potentially being able to infect different hosts with different pathogenicity leads to the possibility of isolating strains that would specifically pathogenic to invasive crayfish (Kozubíková et al. 2011). However, until suitable experimental challenges are completed that demonstrate the host specificity the use of *A. astaci* as a biological control agent is not viable.

One major genus of Microsporidia, *Thelohania*, and several minor genera occur in crayfish, although the taxonomy of the genus in crustacean hosts is confused (Brown and Adamson 2006). *Thelohania contejeani*, the causative agent of the chronic 'porcelain disease' infects many different invasive and indigenous crayfish species and is a recognised mortality driver in crayfish (Dunn et al. 2009). Morphologically similar, but undescribed microsporidians have been reported from New Zealand and Canada in crayfish (Quilter 1976, Graham and France 1986). In both cases, mortalities were associated with the infections, although at low levels. Microsporidians have been utilised as control agents in insects, albeit as long-term regulators of populations and thus may prove useful in the control of non-native crayfish.

Although there is a growing understanding of pathogens in invasive crayfish species, and potential candidates for use in their control have been identified, there is still considerable effort required to develop this field of work for them to become viable control options. As yet no pathogen has been found which is virulent enough for us as an eradication tool, with the exception of WSSV, but the potential repercussion of its use do not make it a viable option. Pathogens do present an excellent tool for species control, however, as long as a species specific pathogen can be found to which resistance is not developed, something that to date has alluded scientists.

Currently biological management methods are only suitable for long term management strategies, but with the further development and investigation in the use of pathogens, viable rapid response methods may become a reality.

Biocidal management

Normally in response to the use of chemical control agents, people think of horror stories surrounding such compounds as DDT and TBT, with environmental impacts on non-target species and the development of resistance raising major concerns surrounding their use. Recent developments as a direct response to these concerns have resulted in the development of compounds which are far more specific in their modes of action. Although the fact still remains that no biocide has been found yet that is specific to a species of crayfish (Peay and Hiley 2001), and without dedicated effort to discover a selective crayfish biocide the economics make it unlikely that such an approach is feasible. There have been several reviews conducted on potential biocides (e.g., Holdich et al. 1999). Given the lack of specificity, focus has been given to chemicals that are not persistent in the environment, are readily available and inexpensive (Gherardi et al. 2011). This has resulted in two main chemicals being used in field trials: Pyblast (R) (3.0% pytherins plus piperonylbutoxide and alcohol ethoxylate), and BETAMAX VET (R) (Cypermethrin, a synthetic pyrethroid).

Pyblast has been used by Peay et al. (2006) in an attempt to eradicate a signal crayfish population in ponds in Scotland. Pyblast is toxic to fish, crustaceans and insects, but has a low toxicity to mammals and birds and is harmless to plants. Therefore fish were removed from the treatment site and stored for reintroduction. The treatment resulted in effectively complete extermination of life in the treated pond. While Pyblast rapidly breaks down in sunlight and there is no harmful residue, allowing some treated waters to recover rapidly, in deep and/or turbid ponds there is little UV penetration and therefore chemical degradation. There have been crayfish detected at the site subsequent to the treatment (see Gherardi et al. 2011) so further monitoring is ongoing. Pyblast has also been trailed in the control of red swamp crayfish with some success (Cecchinelli et al. 2011).

BETAMAX VET is highly toxic to aquatic crustaceans, originally developed for the treatment of sea lice, and so was selected to control signal crayfish in ponds, after the first discovery of their invasions, in Norway (Sandodden and Johnsen 2010). BETAMAX VET is a synthetic pyretheroid and therefore is very similar in its effects to Pyblast. Two treatment of the ponds were conducted with the chemical being dispersed both on the surface and bottom of the ponds. The ponds were subsequently drained and no crayfish were discovered. Post-treatment surveillance is still ongoing at the treated sites.

Biocides present a viable option for rapid response to new invasions, and to situations where the potential threat posed by a populations of invasive crayfish is outweighed by the relative short term environmental impact and the cost of treatment. However, this will limit their useful application to specific locations and circumstances, e.g., enclosed bodies of water. Further research into the possible application of biocides to the control of invasive crayfish is required to try and identify more specific compounds and reduce the potential for environmental impact. Currently biocides present the only tried and tested means of eradication and their further refinement will allow for their application to a wider range of environmental conditions.

Autocidal and alternative management

Autocidal control methods are those that limit the ability of the target population to produce viable or fertile offspring. Autocidal management has been developed and most commonly applied to pest insect control whereby an all-male population of the target species is mass reared; the population is then exposed to a radiation source, which induces genetic aberrations in the individual males, rendering them either sterile or unable to produce viable progeny. The males are released en masse; whereupon they mate with wild females, producing non-viable progeny or no progeny at all. The method is associated with population control to a level where eradication of the pest is possible. For example, successful elimination of the screw worm (*Cochliomyia hominivorax*) from North America was achieved using male sterilisation methods (Knipling 1960). The technique has been employed successfully against a number of other pest species such as Mediterranean fruit fly (*Ceratitis capitata*), melon fly (*Ceratitis capitata*), pink bollworm (*Pectinophoragos sypiella*), codling moth (*Cydia pomonella*) and tsetse fly (*Glossina austeni*) (Wyss 2000, Hendrichs et al. 2005, Klassen and Curtis 2005). The technique is species specific, with no significant environmental impact, and is inversely density dependent, becoming more effective over time as the population of fertile males decreases. This makes it an ideal method for eradication of invasive alien species (Franz and Robinson 2011).

The sterile male technique has been subject to some preliminary studies as a method for the control of invasive alien crayfish (Aquiloni et al. 2009). Exposure of male red swamp crayfish to X-rays was found to reduce the size of testes and limit spermatogenesis, while not compromising the survival or mating ability of the males. The number of aborted eggs in the clutches sired by treated males was higher than those observed in untreated males. There are, however, some distinct drawbacks of the process, such as the catching or mass rearing of male crayfish in sufficient numbers, the transport of the animals to a facility where they can be treated and transport back to the water for release. There is also a requirement for highly trained staff to operate the irradiation equipment. In addition to the technical difficulties there may be public concerns about the release of irradiated animals into the food chain. Despite these drawbacks, which could lead to a prohibitive cost to such operations, the general principle of male sterilization by irradiation has been demonstrated.

The mass rearing of crayfish to the extent required to release sufficient numbers of sterilised males seems unlikely given current knowledge. As the sex ratio in crayfish is normally 1:1, and the technology to breed only males does not currently exist, then farming sufficient numbers would be very time consuming and costly. However, commercial crayfish trappers would be able to supply substantial quantities of male crayfish, but this may be an expensive means of acquiring suitable numbers and would require the transportation of large numbers of animals potentially over long distances. An alternative approach would be to combine the male sterilisation method with a trapping programme, where males are sterilised as they are taken out of the water and any females or smaller males caught are destroyed. The trapping would lower the female population density, while providing male animals for sterilisation and return to the remaining population. Removal of the male gonopods has proven to be an effective means of rendering a male affectively sterile (Stebbing et al. 2014). The males' survivability or ability to find, compete and copulate with females is not affected by this process, but the male is unable to deposit spermatophores effectively, resulting in a reduced rate or no fertilisation of eggs. The method is cheap to apply (only requiring a pair of scissors), can be applied by anybody without any need for specialist training or equipment, and does not involve the transportation of animals away from their point of capture. When applied to large adult males the technique has been estimated to remain effective for approximately 3 years (assuming 1 moult per year). How this method will affect crayfish populations is unclear. There is a significant gap in our understanding of crayfish population dynamics, factors such as how many females a male can mate with in a breeding season, and if males form harems will influence how effective this method will be. While population modelling has been conducted to provide an indication of how the treatment may work in a simulated population, full field trials of the method are required.

Animals respond to chemical cues, called semio-chemicals, which alter their behaviour. Semio-chemicals in the form of pheromones are commonly used in the management of insect (specifically lepidopteran and coleopteran) pest populations. The use of pheromones to control insect pest populations can be divided into two strategies: Mating disruption ('sexual confusion') whereby pheromone dispensers release plumes of pheromones, so that the natural release of pheromone by females to attract males is masked. In this situation, males will be unable to locate females and mating will not occur, leading to a decline in population size over time. 'Attract and kill' whereby the pheromone dispenser lures males to a trap, removing them from the population and preventing them from mating, controlling the population in the subsequent generation (El-Sayed et al. 2009).

There are several studies that have demonstrated the presence of sex pheromones in crayfish (e.g., Ameyaw-Akumfi and Hazlett 1975, Stebbing et al. 2003, Berry and Breithaupt 2008). Field trials of a signal crayfish (putative) pheromone traps was successful at attracting sexually mature males during the breeding season, but did not attract as many individuals as a normal food baited trap (Stebbing et al. 2004). Similar results were obtained by Aquiloni et al. (2010) when using a similar method to control red swamp crayfish.

In addition to the possible application of sex pheromones, other semio-chemicals may prove to be useful in the management of invasive crayfish, for example brood pheromones and disturbance semio-chemicals. Crayfish have an extended period of brood care during which the female carries an egg mass, and later the juveniles up to stage three of their life cycle during which they will become more independent from their mother (Little 1976, Holdich 1992). Evidence suggests the female produces a pheromone which she starts to release after egg deposition which allows the young to discriminate between the mother and non-brooding adults and therefore avoid cannibalisation by other adults (Little 1976). With further knowledge of how this pheromone functions may provide opportunities to disrupt parental care.

Zulandt-Schneider and Moore (2000) classified disturbance chemicals into context specific categories: avoidance chemicals are released directly from a repellent stimulus, e.g., a predator; alarm chemicals are released from a damaged conspecific; while stress chemicals are released from a stressed but undamaged conspecific. Zulandt-Schneider and Moore (2000) showed that red swamp crayfish can detect stressed and damaged conspecifics. An alarm pheromone is then released in the urine due to the presence of predator odours, warning nearby conspecifics of impending danger. Urine from stressed individuals caused conspecifics to walk significantly faster and retreat from the source of the signal. This exhibits the use of avoidance chemical signals against certain predators. Hazlett (1994) demonstrated that virile crayfish also responded to alarm pheromones released by damaged and stressed, as well as crushed, congenerics. Similar effects have been found in signal crayfish (Stebbing et al. 2010). Such chemicals could be used to deter crayfish away from certain areas and prevent further spread.

Biosecurity measures implemented to reduce the risk of the introduction and/or spread of invasive species are a key element of any management programme. There is a need to ensure that biosecurity measures are practical, effective and environmentally sound where possible. Anderson et al. (2015) examined the use of hot water as a means of cleaning equipment to remove a range of invasive plant and animal species. Submersion of adult signal crayfish for 5 minutes in water heated to 40°C resulted in 100% mortality. Adults were used as a proxy for juvenile crayfish as part of the study, with the assumption that adults will be more resilient to heat than the juveniles, therefore presenting an upper limit for the temperature and time of exposure required for 100% efficacy. This methods offers an effective means by which items, e.g., fishing nets, can be cleaned in a manner which will result in 100% mortality of crayfish (and other invasive organisms).

Autocidal and other alternative means of management present potential means of controlling or eradicating invasive crayfish populations, in addition to reducing the risk of their further spread. Autocidal techniques have proven very effective in the control of pest insects and there is no reason why, with further development they should not prove equally as effective on invasive crayfish. Male sterilisation techniques could significantly enhance the impact of mechanical removal, while maintaining population dynamics, making such methods more viable. Sex pheromones, if purified and concentrated could not only be used to enhance mechanical removal, but also be deployed to detect populations at low densities. Heated water offers an environmentally sound means of cleaning equipment safely and has been demonstrated as effective against a range of invasive species. Although the method does have its drawbacks, such as application in remote locations, it does offer a much needed biosecurity measure.

Legislative management

Legislation can be a very effective method of preventing both the introduction and spread of invasive crayfish by anthropogenic means, therefore constituting an important element in any management strategy. Legislation, however, is only as effective as its enforcement, and needs strong political drive to finance enforcement activities. Education can play an important role in relation to the implementation and enforcement, requiring sustained effort to be effective, with the legislation acting as a deterrent. Complicated legislation can create confusion amongst users, making education difficult, but also more necessary. A simplistic approach to how invasive crayfish, or other INNS, are regulated is ideal making for easily understood and enforceable regulation. In England and Wales the Prohibition of keeping of Live Crayfish Order 1996 made it an offence to keep any invasive crayfish except under licence. An exception was made for the keeping of signal crayfish in areas where that species was already widely established. This exemption resulted in the division of the control area into what were termed either 'go areas', where the signal crayfish was not subject to control, or 'no-go areas', where the keeping of this species could only be carried out under licence. In addition, two general licences were issued: A licence to keep live crayfish in hotels, markets and restaurants for human consumption, and a licence to keep tropical crayfish for ornamental purposes in heated indoor aquaria. The combination of different species, geographical zones and end uses affecting how the legislation applies, has caused confusion amongst enforcers and the general public alike making the legislation less robust, resulting in illegal activities through ignorance rather than deliberate actions. Education and public awareness are key elements in enforcement of legislation and obtaining public cooperation, not just in relation to legislative control but with other aspects of INNS management, for example, monitoring.

Careful consideration is required as to which stage of the process of introduction and/or spread is regulated. Ideally bottle necks in pathways should be identified and regulated. For example, a major route of introduction is through the importation of live animals, either for human consumption or the aquarium trade. As this trade is likely to cross political borders, these provide a bottle neck at which the trade can be controlled. Sweden, for example, has effectively closed it borders to the importation of live crayfish. Most introductions of ornamental animals into England and Wales originate from third countries (i.e., countries that are not within the European Community), all which have to enter via Border Inspection Posts (BIPs) at British airports. The bottle necking of trade allows import controls to be easily implemented, with the removal from consignments of any species that cannot be legally kept or released in England and Wales. However, this legislation targets end users and the keeping of crayfish, rather than their importation, where it would be more effective to target those trading in the animals, preventing introduction of the animals rather than attempting to control once the animals have entered the control area. The aquarium trade is now seen as the most probable route of introduction for new crayfish entering Europe. A more robust approach to legislating against the trade of invasive crayfish species would be of significant benefit in preventing further introductions.

The identification of loop holes in legislations and their avoidance is of particularly importance. Despite efforts in England and Wales to prevent introductions through the control of keeping crayfish, which has largely prevented the introduction of crayfish from third countries, there has continued to be a trade in ornamental crayfish from European sources with suppliers transporting invasive crayfish into the control area with relative impunity as a result of the comparative open borders to trade with the rest of Europe. These suppliers arguably do not commit a keeping offence, and are therefore able to act without fear of sanction under this legislation. This is an effective loop-hole in the legislation and its subsequent enforcement. Likewise the general licence for keeping live animals for human consumption in England and Wales in hotels, markets and restaurants has resulted in the release and establishment of invasive crayfish with the inevitable consequence that some are bought and released into ponds lakes and rivers. Unlike the ornamental trade, which typically sells low volumes of stock at high prices, the wholesale or retail of live crayfish for food poses a significant risk because animals are sold cheaply and in numbers that if released could result in the establishment of a new population.

Once a species has entered the control area, legislation regulating the species is much more difficult to enforce. In England and Wales the principal legislation relating to the control of introduction of INNS is the Wildlife and Countryside Act 1981 (WCA). The Act made it an offence to release, or to allow to escape, into the wild, any non-indigenous organism except under licence. The WCA is important in that it enshrines the policy principle that INNS should not be released into the wild, but as a practical tool for the enforcement of this policy for aquatic animals, the Act is not adequate. Unlike organisms released into terrestrial habitats it is very difficult to detect when an aquatic animal has been released or allowed to escape to the wild. As a result, in most cases it would be impractical to pursue a case against someone for a breach of the WCA, because of the lack of evidence regarding the actual offence. Banning transport and sale of live crayfish, which is the case in Arizona, USA, can make the enforcement process more achievable, with anyone found in the possession of live crayfish committing an offence. The division of England and Wales into 'go' and 'no-go' has not prevented the spread of signal crayfish between these two areas, despite the difference in licence requirements, there has still been movements of signal crayfish into the 'no-go' areas with relative impunity, as not only are those responsible for the introduction often difficult to identify, but even in the cases where this has happened, those responsible for supplying the animals are not regulated against. It is therefore important for legislation intended to limit the spread of invasive crayfish to be effective for both those committing the offence of introduction be liable, but also the supplier of the animals. This will act as a deterrent at both ends of the supply chain.

EC Regulation 708/2007 concerning alien and locally Absent Species in Aquaculture, and The Alien and Locally absent Species in Aquaculture Regulations 2011 (ASR) has effectively rendered the farming of crayfish species difficult and unlikely to be cost effective as all operations are required to be fully isolated and operate a double level of escape prevention. Typically this requires the site to operate indoor tank based systems, where the tanks have to be lidded and screened to prevent the escape of any crayfish (at all life stages) and where the building in which the tanks are housed is also escape-proofed (e.g., no open drains, doors must close onto raised walls, no holes in the walls). The ASR is a good example of how legislation can be implemented in such a way as to limit the risk of activities in relation to a certain pathway. However, the ASR only controls one pathway. Inconsistent regulation among multiple vectors that may introduce the same species are considered a weak link, where a consistency in regulation across all pathways and vectors is ideally required for legislative control to be truly effective. While legislation is one of the main tools in preventing the introduction of INNS, and therefore invasive crayfish, and it is recognised that there have been some effective regulations implemented to control invasive crayfish, there needs to be more harmonisation and robustness covering all potential pathways.

Towards a coherent management strategy

The continue impact and spread of invasive crayfish is well documented and recognised as an issues of increasing concern. The development of comprehensive and sustainable management strategies to prevent further introduction and to eradicate and/or control existing threats are therefore essential to combat the issue.

Anthropogenic activities are considered the main cause for the introduction and spread of invasive crayfish. Educating and raising awareness of the potential dangers of moving invasive crayfish can help significantly in reducing the risk of both deliberate and accidental introductions. Not only is it important for people to understand the potential environmental impact that introducing invasive crayfish may have, but also to provide tools as to how the risk of transfer can be reduced. While there has been some development of biosecurity measures, such as the Check, Clean, Dry campaigns of New Zealand and Britain, there is still a need to see more practical measures that members of the public can easily apply. Ensure that the biosecurity message and methods are easily understood and to apply is essential in ensuring maximum up-take. While taking a light-touch approach to the management of certain pathways can be the best approach, there needs to be enforced legislative control for others. This tends to be the cases in relation to activates where there financial gain to be made.

Despite the lack of current methods for the complete eradication of invasive crayfish populations, there are many methods that have either shown promise in the field and need minor modifications to be more effective, or are currently under development and need field testing to demonstrate effectiveness. As research in to control and eradication progresses there is a need to be ensure that future studies are assessed for feasibility and carefully planned so as to provide more realistic evaluation of the methods potential use. The guidelines set out within this chapter will hopefully provide at least a solid foundation from which these assessments can be made from. It remains likely that management methods will be situational and therefore the application of control methods will have to be determined on a case-by-case basis. While a 'silver bullet' for the eradication of invasive crayfish, which meets all criteria set out within this chapter, seems unlikely there is still a need for further research into management methods to complement and enhance management strategies. With a wider range of tools at the disposal for invasive crayfish management, strategies will be able to tackle a wider range of risks and scenarios, forming a more robust and comprehensive strategy.

A number of control methods have already been developed which may prove effective at eradication if combined into an integrated management approach, where multiple management methods are used in conjunction with one another. The combination of control methods for successful pest management is exemplified in the Intergrated Pest Management approach of terrestrial insects. In many cases the sum total is greater than its parts, where the combination of methods have a compounding effect on each other, resulting in a greater overall effect on the population. There have been several attempts at using multiple methods for the management of invasive crayfish populations. In Switzerland, for example, extensive trapping in addition to the introduction of predatory fish (eel and pike) significantly reduced the size of a population of red swamp crayfish by a factor of 10 over 3 years (Hefti and Stucki 2006). In Wisconsin, USA a lake containing an extensive invasive crayfish population was treated using a combination of intensive trapping with a change in legislation relating to the capture of predatory fish from the lake (Hein et al. 2006, Hansen et al. 2013). After 5 years of intensive trapping and fisheries management practices, a decline in crayfish of up to 95% was seen. Similarly a combination of control mechanisms was applied to a population of signal crayfish in Spain (Dana et al. 2011). Trapping, manual removal and electrofishing resulted in a sharp decline in the target population's size over a 4 year period, with a catch rate (crayfish per worker per day) of 30 in the first year decreasing to 10 in the 4th year. One of the key features of these combined approaches is they target multiple life stages, potentially resulting in a greater level of control than if single mechanisms were applied. In addition all of the examples provided where conducted over a long time period and effected the entire target populations. Despite the effective control of populations using a combination of mechanisms, there have been few recorded attempts at using multi-disciplinary approaches. While population modelling can help to assess the effectiveness of applying multiple methods of control further research is required to test multi-method approaches in the field. The further refinement of existing methods to make them more effective and less labour intensive, alongside the development of novel methods, and the investigation into integrated management strategies could lead to the extirpation of crayfish populations.

The development of control methods continues to be hindered by a lack of knowledge of a species life history and population dynamics, which are essential in their development, assessment and refinement. Understanding key biological aspects of the target species will greatly aid the development of methods of control. It will also provide a greater understanding of how management strategies will effect populations in the long term, enabling the further refinement of management strategies, in addition to assessing the feasibility of eradication. A greater understanding of population dynamics will facilitate the development of more robust and complex population models. These population models can then be used to effectively design and develop long term management strategies.

Management strategies are multidisciplinary requiring the input from a range of different experts. Collaboration and information sharing is therefore essential for their development and sustained application. Variation in approaches to the management of invasive crayfish across contiguous geographical areas can lead to discrepancies which may allow crayfish to move easily from one political state to another. It is therefore important that management strategies provide harmonisation in the methods and approaches used to assist in the global control of invasive crayfish. Lessons learnt in relation to the management of invasive crayfish should be shared as much as possible to help the further development of strategies and methods of control. It is clear that ecosystems and their incumbent populations face a growing threat from invasive crayfish; solutions to this problem exist which will require, as stated by Voltaire at the start of the chapter, "sustained thinking" and a clear desire by all to find appropriate and long-lasting solutions.

Acknowledgements

To my two children, Hugo and Emilie, both of which were born during the process of writing this chapter and editing this book. Also to my wonderful wife, Marion, who has given me so much.

References

Alderman, D.J. 1993. Crayfish plague in Britain, the first twelve years. Freshw. Crayfish 9: 266–272.

Alderman, D.J. and J.L. Polglase. 1988. Pathogens, parasites and commensals. pp. 167–212. *In*: D.M. Holdich and R.S. Lowery (eds.). Freshwater Crayfish: Biology, Management and Exploitation. Croom Helm, London.

Ameyaw-Akumfi, C. and B.A. Hazlett. 1975. Sex recognition in the crayfish *Procambarus clarkii*. Science 190: 1225–1226.

Anderson, L.G., A.M. Dunn, P.J. Rosewarne and P.D. Stebbing. 2015. Invaders in hot water: a simple decontamination method to prevent the accidental spread of aquatic invasive non-native species. Biol. Invasions 17: 2287–2297.

Aquiloni, L., A. Beccioloni, R. Berti, S. Porciani, C. Trunfio and F. Gherardi. 2009. Managing invasive crayfish: use of X-ray sterilisation of males. Freshwater Biol. 54: 1510–1519.

Aquiloni, L., E. Tricarico and F. Gherardi. 2010. Crayfish in Italy: distribution, threats and management. Int. Aquat. Res. 2: 1–14.

Baumgartner, W.A., J.P. Hawke, K. Bowles, P.W. Varner and K.W. Hasson. 2009. Primary diagnosis and surveillance of white spot syndrome virus in wild and farmed crawfish (*Procambarus clarkii, P. zonangulus*) in Louisiana, USA. Dis. Aquat. Org. 85: 15–22.

Bean, R.A. and J.V. Huner. 1979. An evaluation of selected crawfish traps and trapping methods. Freshw. Crayfish 4: 141–151.

Berry, F.C. and T. Breithaupt. 2008. Development of behavioural and physiological assays to assess discrimination of male and female odours in crayfish, *Pacifastacus leniusculus*. Behaviour 145: 1427–1446.

Bills, T.D. and L. Marking. 1988. Control of nuisance populations of crayfish with traps and toxicants. The Prog. Fish Cult. 50: 103–106.

Blake, M.A. and P.J.B. Hart. 1995. The vulnerability of juvenile signal crayfish to perch and eel predation. Freshwater Biol. 33: 233–244.

Bohman, P., F. Nordwall and L. Edsman. 2006. The effect of the large-scale introduction of signal crayfish on the spread of crayfish plague in Sweden. Bull. Fr. Pêch. Piscic. 380-381: 1291–1302.

Bomford, M. and P. O'Brien. 1995. Eradication or control for vertebrate pests? Wildlife Soc. B. 23: 249–255.

Brown, A.M.V. and M.L. Adamson. 2006. Phylogenetic distance of *Thelohania butleri* Johnston, Vernick, and Sprague, 1978 (Microsporidia; Thelohaniidae), a parasite of the smooth pink shrimp *Pandalus jordani*, from its congeners suggests need for major revision of the genus *Thelohania* Henneguy, 1892. J. Eukaryot. Microbiol. 53: 445–455.

Caffrey, J., J.-R. Baars, J. Barbour, P. Boets, P. Boon, K. Davenport, J. Dick, J. Early, L. Edsman, C. Gallagher, J. Gross, P. Heinimaa, C. Horrill, S. Hudin, P. Hulme, S. Hynes, H. MacIsaac, P. McLoone, M. Millane, T. Moen, N. Moore, J. Newman, R. O'Conchuir, M. O'Farrell, C. O'Flynn, B. Oidtmann, T. Renals, A. Ricciardi, H. Roy, R. Shaw, O. Weyl, F. Williams and F. Lucy. 2014. Tackling invasive alien species in Europe: the top 20 issues. Manag. Biol. Inv. 5: 1–20.

Campbell, L. and G.J. Whisson. 2000. Catch efficiency of five freshwater crayfish traps in south-west Western Australia. Freshw. Crayfish 13: 58–66.

Caplat, P. and S.R. Coutts. 2011. Integrating ecological knowledge, public perception and urgency of action into invasive species management. Environ. Manage. 48: 878–881.

Cecchinelli, E., L. Aquiloni, G. Maltagliati, G. Orioli, E. Tricarico and F. Gherardi. 2011. Use of natural pyrethrum to control the red swamp crayfish *Procambarus clarkii* in a rural district of Italy. Pest Manag. Sci. 68: 839–844.

Cheung, W.W.L., T.J. Pitcher and D. Pauly. 2005. A fuzzy logic expert system to estimate intrinsic extinction vulnerabilities of marine fishes to fishing. Biol. Conserv. 124: 97–111.

Clay, K. 2003. Parasites lost. Nature 421: 585–586.

Claydon, K., B. Cullen and L. Owens. 2004. OIE white spot syndrome virus PCR gives false-positive results in *Cherax quadricarinatus*. Dis. Aquat. Org. 62: 265–268.

Cohen, A.N. and J.T. Carlton. 1998. Accelerating invasion rate in a highly invaded estuary. Science 279: 555–558.

Colautti, R.I., A. Ricciardi, I.A. Grigorovich and H.J. Macisaac. 2004. Is invasion success explained by the enemy release hypothesis? Ecol. Lett. 7: 721–733.

Dana, E.D., J. García-De-Lomas, R. Gozález and F. Ortega. 2011. Effectiveness of dam construction to contain the invasive crayfish *Procambarus clarkii* in a Mediterranean mountain stream. Ecol. Eng. 37: 1607–1613.

Davidson, E.W., J. Snyder, D. Lightner, G. Ruthig, J. Lucas and J. Gilley. 2010. Exploration of potential microbial control agents for the invasive crayfish, *Orconectes virilis*. Biocontrol Sci. Techn. 20: 297–310.

Ding, Z., K. Bi, T. Wu, W. Gu, W. Wang and J. Chen. 2007. A simple PCR method for the detection of pathogenic spiroplasmas in crustaceans and environmental samples. Aquaculture 265: 49–54.

Du, H., L. Fu, Y. Xu, Z. Kil and Z. Xu. 2007. Improvement in a simple method for isolating white spot syndrome virus (WSSV) from the crayfish *Procambarus clarkii*. Aquaculture 262: 532–534.

Du, H., W. Dai, X. Han, W. Li, Y. Xu and Z. Xu. 2008. Effect of low water temperature on viral replication of white spot syndrome virus in *Procambarus clarkii*. Aquaculture 277: 149–151.

Dunn, J.C., H.E. Mcclymont, M. Christmas and A.M. Dunn. 2009. Competition and parasitism in the native white clawed crayfish *Austropotamobius pallipes* and the invasive signal crayfish *Pacifastacus leniusculus* in the UK. Biol. Inv. 11: 315–324.

Edgerton, B., L. Owens, L. Harris, A. Thomas and M. Wingfield. 1995. A health survey of farmed red-claw crayfish, *Cherax quadricarinatus* (von Martens), in tropical Australia. Freshw. Crayfish 10: 332–338.

Edgerton, B.F., L.H. Evans, F.J. Stephens and R.M. Overstreet. 2002. Synopsis of freshwater crayfish diseases and commensal organisms. Aquaculture 206: 57–135.

El-Sayed, A.M., D.M. Suckling, J.A. Byers, E.B. Jang and C.H. Wearing. 2009. Potential of "lure and kill" in long-term pest management and eradication of invasive species. J. Eco. Ent. 102: 815–835.

European Commission. 2013. Regulation of the European Parliament and of the Council on the Prevention and Management of the Introduction and Spread of Invasive Alien Species.

Fjälling, A. 1995. Crayfish traps employed in Swedish fisheries. Freshw. Crayfish 8: 201–214.

Franz, G. and A.S. Robinson. 2011. Molecular technologies to improve the effectivess of the sterile insect technique. Genetica 139: 1–5.

Freeman, M.A., J.F. Turnbull, W.E. Yeomans and C.W. Bean. 2010. Prospects for management strategies of invasive crayfish populations with an emphasis on biological control. Aquat. Conserv. 20: 211–223.

Freeman, T.E. 1977. Biological control of aquatic weeds with plant pathogens. Aquat. Bot. 3: 175–184.

Frings, R.M., S.C.K. Vaeßen, H. Groß, S. Roger, H. Schüttrumpf and H. Hollert. 2013. A fish-passable barrier to stop the invasion of non-indigenous crayfish. Biol. Conserv. 159: 521–529.

Frutiger, A. and R. Müller. 2002. Der Rote SumpfkrebsimSchübelweiher (GemeindeKüsnacht ZH, Schweiz). Auswertung der Maßnahmen 1998–2001 und Erkenntnisse. *Dübendorf: EAWAG*, 26 pp.

Frutiger, A., S. Borner, T. Büsser, R. Egge, R. Müller, S. Müller and H.R. Wasmer. 1999. How to control unwanted populations of *Procambarus clarkii* in central Europe? Freshw. Crayfish 12: 714–726.

Gazla, I.N. and M.C. Carracedo. 2009. Effect of intracellular *Wolbachia* on interspecific crosses between *Drosophila melanogaster* and *Drosophila simulans*. Genet. and Mol. Res. 8: 861–869.

Gherardi, F., L. Aquiloni, J. Diéguez-Uribeondo and E. Tricarico. 2011. Managing invasive crayfish: is there a hope? Aqu. Sci. 73: 185–200.

Glowka, L., F. Burhenne-Guilmin, H. Synge, J.A. McNeely and L. Gündling. 1994. A guide to the convention on biological diversity. International Union for the Conservation of Nature (IUCN), Gland, Switzerland.

Graham, L. and R. France. 1986. Attempts to transmit experimentally the microsporidian *Thelohania contejeani* in freshwater crayfish (*Orconectes virilis*). Crustaceana 51: 208–211.

Grandjean, F., C. Souty-Grosset and D.M. Holdich. 1997. Geographical variation of mitochondrial DNA between European populations of the white-clawed crayfish *Austropotamobius pallipes*. Freshwater Biol. 37: 493–501.

Hansen, G.J.A., C.L. Hein, B.M. Roth, J.V. Zanden, J.W. Gaeta, A.W. Latzka and S.R. Carpenter. 2013. Food web consequences of long-term invasive crayfish control. Can. J. Fish. Aquat. Sci. 70: 1109–1122.

Harlioglu, M.M. 2008. The harvest of the freshwater crayfish *Astacus leptodactylus* Eschscholtz in Turkey: harvest history, impact of crayfish plague, and present distribution of harvested populations. Aquat. Int. 16: 351–360.

Hatcher, M.J., J.T.A. Dick and A.M. Dunn. 2012. Disease emergence and invasions. Funct. Ecol. 26: 1275–1287.

Hauck, A.K., M.R. Marshall, J.K.K. Li and R.A. Lee. 2001. A new finding and range extension of bacilliform virus in the freshwater red claw crayfish in Utah, USA. J. Aquat. Anim. Health 13: 158–162.

Hazlett, B.A. 1994. Alarm responses in the crayfish *Orconectes virilis* and *Orconectes propinquus*. J. Chem. Ecol. 20: 1525–1535.

Hedrick, R.P., T.S. Mcdowell and C.S. Friedman. 1995. Baculoviruses found in two species of crayfish from California. Aquaculture 95, Abstracts: 135.

Hefti, D. and P. Stucki. 2006. Crayfish management for Swiss waters. Bull. Fr. Pêch. Piscic. 380-381: 937–950.

Hein, C.L., B.M. Roth, A.R. Ives and M.J. Vander Zanden. 2006. Fish predation and trapping for rusty crayfish (*Orconectes rusticus*) control: a whole-lake experiment. Can. J. Fish. Aquat. Sci. 63: 383–393.

Hendrichs, J., M. Vreysen, W. Enkerlin and J. Cayol. 2005. Strategic options in using sterile insects for area-wide integrated pest management. pp. 563–600. *In*: V.A. Dyck, J. Hendrichs and A.S. Robinson (eds.). Sterile Insect Technique: Principles and Practice in Area-Wide Integrated Pest Management. Springer, USA.

Hilker, F.M., M.A. Lewis, H. Seno, M. Langlais and H. Malchow. 2005. Pathogens can slow down or reverse invasion fronts of their hosts. Biol. Inv. 7: 817–832.

Holdich, D.M. 1992. Crayfish nomenclature and terminology: recommendations for uniformity. Finn. Fish. Res. 14: 157–159.

Holdich, D.M. and I.D. Reeve. 1991. Alien crayfish in the British Isles. Report for the Natural Environment Research Council, Swindon.

Holdich, D.M., R. Gydemo and W.D. Rogers. 1999. A review of possible methods for controlling alien crayfish populations. pp. 245–270. *In*: F. Gherardi and D.M. Holdich (eds.). Crayfish in Europe as Alien Species. How to Make the Best of a Bad Situation. A.A. Balkema, Rotterdam.

Huang, T.S., L. Cerenius and K. Söderhäll. 1994. Analysis of the genetic diversity in crayfish plague fungus, *Aphanomyces astaci*, by random amplification of polymorphic DNA assay. Aquaculture 26: 1–10.

Huner, J.V. 1988. *Procambarus* in North America and elsewhere. pp. 239–261. *In*: D.M. Holdich and R.S. Lowery (eds.). Freshwater Crayfish: Biology, Management and Exploitation. Croom Helm (Chapman and Hall), London.

Ibbotson, A.T., G. Tapir, M.T. Furse, J.M. Winder, J. Blackburn, P. Scarlett and J. Smith. 1997. Impact of the signal crayfish *Pacifastacus leniusculus* and its associated crayfishery on the River Thames. Report for the Environment Agency, Thames Division, Reading, UK.

Jiravanichpaisal, P., E. Bangyeekhun, K. Söderhäll and I. Söderhäll. 2001. Experimental infection of white spot syndrome virus in freshwater crayfish *Pacifastacus leniusculus*. Dis. Aquat. Org. 47: 151–157.

Johnsen, S.I., T. Jansson, J.K. Høye and T. Taugbøl. 2008. Vandaringssperre for signalkrep i Buåa, Edaakommun, Sverige-Ov ervåkingavsignalkrepsogkrepsepestsituasjonen. NINA Rapport 356, 15 s. (ISBN 978-82-426-1920-4).

Kerby, J., S. Riley, P. Wilson and L.B. Kats. 2005. Barriers and flow as limiting factors in the spread of an invasive crayfish (*Procambarus clarkii*) in southern California streams. Biol. Conserv. 126: 402–409.

Klassen, W. and C. Curtis. 2005. History of the sterile insect technique. pp. 3–36. *In*: V.A. Dyck, J. Hendrichs and A.S. Robinson (eds.). Sterile Insect Technique: Principles and Practice in Area-Wide Integrated Pest Management. Springer, USA.

Knipling, E.F. 1960. The eradication of the screwworm fly. Sci. Am. 203: 4–48.

Kolar, C.S. and D.M. Lodge. 2001. Progress in invasion biology: predicting invaders. Trends Ecol. Evol. 16: 199–204.

Kozák, P. and T. Policar. 2003. Practical elimination of signal crayfish (*Pacifastacus leniusculus*) from a pond. pp. 200–208. *In*: D.M. Holdich and P.J. Sibley (eds.). Management and Conservation of Crayfish. Proceedings of a Conference Held on 7th November, 2002. Environment Agency, Bristol.

Kozubíková, E., S. Viljamaa-Dirks, S. Heinikainen and A. Petrusek. 2011. Spiny-cheek crayfish *Orconectes limosus* carry a novel genotype of the crayfish plague pathogen *Aphanomyces astaci*. J. Invertebr. Pathol. 108: 214–216.

Laurent, P.J. 1995. Eradication of unwanted crayfish species for astacological management purposes. Freshw. Crayfish 8: 121–133.

Lilley, J.H., L. Cerenius and K. Söderhäll. 1997. RAPD evidence for the origin of crayfish plague outbreaks in Britain. Aquaculture 157: 181–185.

Little, E.E. 1976. Ontogeny of maternal behaviour and brood pheromone in crayfish. J. Comp. Phys. 112: 133–142.

Longshaw, M. 2011. Diseases of crayfish: a review. J. Invertebr. Pathol. 106: 54–70.

Longshaw, M., K.S. Bateman, P.D. Stebbing, G.D. Stentiford and F.A. Hockley. 2012. Disease risks associated with the importation and release of non-native crayfish species in mainland Britain. Aquat. Biol. 16: 1–15.

MacIsaac, H.J., I.A. Grigorvich and A. Ricciardi. 2001. Reassessment of species invasions concepts: the Great Lakes basin as a model. Biol. Inv. 3: 405–416.

Mack, R., D. Simberloff, W.M. Lonsdale, H. Evans, M. Clout and F.A. Bazzaz. 2000. Biotic invasions: causes, epidemiology, global consequences, and control. Ecol. Appl. 10: 689–710.

Mack, R.N. 2000. Biotic invasions: causes, epidemiology, global consequences, and control. Ecol. Appl. 10: 689–710.

Mcfadyen, R.E.C. 1998. Biological control of weeds. Annu. Rev. Ent. 43: 369–393.

Moorhouse, T.P. and D.W. Macdonald. 2010. Immigration rates of signal crayfish (*Pacifastacus leniusculus*) in response to manual control measures. Freshwater Biol. 56: 993–1001.

Neveu, A. 2001. Confrontation expérimentale entre des poisons omnivoresatochtones (11 espèces) et des écrevissesétrangèresintroduites (2 espèces). Bull. Fr. Pêche Piscic. 361: 705–735.

Nunan, L.M., D.V. Lightner, M.A. Oduori and G.E. Gasparich. 2005. *Spiroplasma penaei* sp. nov., associated with mortalities in *Penaeus vannamei*, Pacific white shrimp. Int. J. Syst. Evol. Micr. 55: 2317–2322.

Parkes, J.P. and D.F. Panetta. 2009. Eradication of invasive species: progress and emerging issues in the 21st century. pp. 47–62. *In*: M.N. Clout and P.A. Williams (eds.). Invasive Species Management. Oxford University Press, Oxford.

Peay, S. and P.D. Hiley. 2001. Eradication of alien crayfish. Phase II. Environment Agency Technical Report W1–037/TR1. Bristol: Environment Agency, 118 pp.

Peay, S. and R. McKimm. 2011. Electric treatment of signal crayfish in a small stream – field trial 2011. Project Report September 2011.

Peay, S., P.D. Hiley, P. Collen and I. Martin. 2006. Biocide treatment of ponds in Scotland to eradicate signal crayfish. Bull. Fr. Pêche Piscic. 380-381: 1363–1379.

Perring, C., M. Williamson, E.B. Barbier, D. Delfino, S. Dalmazzone, J. Shogren, P. Simmons and A. Watkinson. 2002. Biological invasion risks and the public good: an economic perspective. Conserv. Ecol. 6(1): 1.

Pimentel, D., L. Lach, R. Zuniga and D. Morrison. 2000. Environmental and economic costs of non-indigenous species in the United States. Bioscience 50: 53–65.

Pimentel, D., R. Zuniga and D. Morrison. 2005. Update on the environmental and economic costs associated with alien-invasive species in the United States. Ecol. Eco. 52: 273–288.

Quilter, C.G. 1976. Microsporidian parasite *Thelohania contejeani* Henneguy from New Zealand freshwater crayfish. N.Z. J. Mar. Freshw. Res. 10: 225–231.

Reeve, I.D. 2004. The removal of the North American signal crayfish (*Pacifastacus leniusculus*) from the River Clyde. Scottish Natural Heritage Commissioned Report, No. 020 (ROAME No. F00LI12).

Roqueplo, C., P.J. Laurent and A. Neveu. 1995. *Procambarus clarkii* Girard (écrevisse rouge des marais de Louisiana). Synthèse sur les problèmes poses par cetteespèceet sur les essais pour contrôlerses populations. L 'Astaciculteur de France 45: 2–17.

Ruiz, G.M., P.W. Fofonoff, J.T. Carlton, M.J. Wonham and A.H. Hines. 2000. Invasion of coastal marine communities in North America: apparent patterns, processes, and biases. Ann. Rev. Ecol. Syst. 31: 481–531.

Sandodden, R. and S.I. Johnsen. 2010. Eradication of introduced signal crayfish *Pacifastacus leniusculus* using the pharmaceutical BETAMAX VET. Aqua. Inv. 5: 75–81.

Secretariat of the Convention on Biological Diversity. 2004. Guidelines on biodiversity and tourism development: international guidelines for activities related to sustainable tourism development in vulnerable terrestrial, marine, and coastal ecosystems and habitats of major importance for biological diversity and protected areas, including fragile riparian and mountain ecosystems. Secretariat of the Convention on Biological Diversity, Montreal.

Shi, Z., C. Huang, J. Zhang, D. Chen and J.R. Bonami. 2000. White spot syndrome virus (WSSV) experimental infection of the freshwater crayfish, *Cherax quadricarinatus*. J. Fish Dis. 23: 285–288.

Shine, C., M. Kettunen, P.T. Brink, P. Genovesi and S. Gollasch. 2009. Technical Support to EU Strategy on Invasive Alien Species (IAS). Recommendations on policy options to minimise the negative impacts of invasive alien species on biodiversity in Europe and the EU. Final report.

Simberloff, D. 2009. We can eliminate invasions or live with them. Successful management projects. Biol. Inv. 11: 149–157.

Simberloff, D., J.L. Martin, P. Genovesi, V. Maris, D.A. Wardle, J. Aronson, F. Courchamp, B. Galil, E. Garcia-Berthoum, M. Pascal, P. Pyšek, R. Sousa, E. Tabacchi and M. Vila. 2012. Impacts of biological invasions: what's what and the way forward. Trends Ecol. Evo. 28(1): 58–66.

Sinclair, C. and J. Ribbens. 1999. Survey of American Signal Crayfish, *Pacifastus leniusculus*, distribution in the Kirkcudbrightshire Dee, Dumfries and Galloway, and assessment of the use of electrofishing as an eradication technique for Crayfish populations. Scottish National Heritage, Commissioned Report (F99AC601).

Skurdal, J. and T. Qvenild. 1986. Growth, maturity and fecundity of *Astacus astacus* in Lake Steinsfjorden, SE Norway. Freshw. Crayfish 6: 182–186.

Soowannayan, C. and M. Phanthura. 2011. Horizontal transmission of white spot syndrome virus (WSSV) between red claw crayfish (*Cherax quadricarinatus*) and the giant tiger shrimp (*Penaeus monodon*). Aquaculture 319: 5–10.

Stebbing, P.D., G.J. Watson, M.G. Bentley, D. Fraser, R. Jennings and S.P. Rushton. 2003. Reducing the threat: the potential use of pheromones to control invasive signal crayfish. Bull. Fr. Pêche Piscic. 370–371: 219–224.

Stebbing, P.D., G.J. Watson, M.G. Bentley, D. Fraser, R. Jennings, S.P. Rushton and P.J. Sibley. 2004. Evaluation of the capacity of pheromones for control of invasive non-native crayfish: Part 1. English Nature Research Report No. 578. Peterborough: English Nature.

Stebbing, P.D., G.J. Watson and M.G. Bentley. 2010. The response to disturbance chemicals and predator odours of juvenile and adult signal crayfish *Pacifastacus leniusculus* (Dana). Mar. Freshw. Behav. Phy. 43: 183–195.

Stebbing, P.D., M. Longshaw and A. Scott. 2014. Review of methods for the management on non-indigenous crayfish, with particular reference to Great Britain. J. Ethol. Ecol. Evo. 26: 204–231.

Stentiford, G.D., J.-R. Bonami and P. Alday-Sanz. 2009. A critical review of susceptibility of crustaceans to Taura syndrome, Yellowhead disease and White Spot Disease and implications of inclusion of these diseases in European legislation. Aquaculture 291: 1–17.

Stuecheli, K. 1991. Trapping bias in sampling crayfish with baited funnel traps. N. Am. J. Fish. Manage. 11: 236–239.

Thanabalu, T., J. Hindley, S. Brenner, C. Oei and C. Berry. 1992. Expression of the mosquitocidal toxins of *Bacillus sphaericus* and *Bacillus thuringiensis* subsp. *israelensis* by recombinant *Caulobacter crescentus*, a vehicle for biological control of aquatic insect larvae. Appl. Env. Microbiol. 58: 905–910.

Unestam, T., C.G. Nestell and S. Abrahamsson. 1972. An electrical barrier for preventing migration of freshwater crayfish in running water. A method to stop the spread of crayfish plague. Report of the Institute of Freshwater Research, Drottningholm 52: 199–203.

Vilà, M., C. Basnou, P. Pyšek, M. Josefsson, P. Genovesi, S. Gollasch, W. Nentwig, S. Olenin, A. Roques, D. Roy and P.E. Hulme. 2010. How well do we understand the impacts of alien species on ecosystem services? A pan-European, cross-taxa assessment. Front. Ecol. Environ. 8: 135–144.

Webb, D.A. 1985. What are the criteria for presuming native status? Watsonia 15: 231–236.

West, R.J. 2011. A review of signal crayfish trapping on the River Lark at Barton Mills, Suffolk, from 2001 to 2009. Lark Angling and Preservation Society, Suffolk UK.

Westman, K. 1991. The crayfish fishery in Finland—its past, present and future. Finn. Fish. Res. 12: 187–216.

Westman, K., O. Sumari and M. Pursiainen. 1978. Electric fishing in sampling Crayfish. Freshw. Crayfish 4: 251–255.

Westman, K., M. Pursiainen and R. Vilkman. 1979. A new folding trap model which prevents crayfish from escaping. Freshw. Crayfish 4: 235–242.

Williams, F., R. Eschen, A. Harris, D. Djeddour, C. Pratt, R.S. Shaw, S. Varia, J. Lamontagne-Godwin, S.E. Thomas and S.T. Murphy. 2010. The economic cost of invasive non-native species on Great Britain. CABI Project No. VM10066. CABI Europe-UK.

Williamson, M. 1996. Biological Invasions. Chapman and Hall, London.

Wright, R. and M. Williams. 2000. Long term trapping of signal crayfish at Wixoe on the River Stour, Essex. pp. 81–88. *In*: W.D. Rogers and J. Brickland (eds.). Proceedings of the Crayfish Conference held on 26th/27th April 2000 in Leeds. Environment Agency, Bristol.

WWF. 2014. Living Planet Report 2014: species and spaces, people and places. WWF, Gland, Switzerland.

Wyss, J.H. 2000. Screwworm eradication in the Americas. Ann. N.Y. Acad. Sci. 916: 186–193.

Zulandt-Schneider, R.A. and P.A. Moore. 2000. Urine as a source of conspecific disturbance signals in the crayfish *Procambarus clarkii*. J. of Exp. Biol. 203: 765–771.

Index

egg-sperm binding 91
Ejaculate size 90, 91, 152
ejaculatory duct 77, 78
Elachistocythere 214, 221
electrical stunning 334
electrocution 342, 344
electrofishing 296–298, 300, 301, 307–309, 344, 353
Elizabethkingia meningosepticum 176
Ellisodrilus 204, 208, 230
Embata 213
embryo attachment 96
Emericellopsis 183, 230
endangered 10, 15, 17, 18, 31, 32, 34, 43, 44, 50, 149, 300, 306, 311
Engaeus 7, 19, 20, 82, 99, 102, 103, 109, 120, 157, 190, 191, 198, 199, 201, 213, 223, 224, 227, 252, 304
Engaeus affinis 223
Engaeus cisternarius 99, 102, 109, 201, 213, 223, 224
Engaeus cymus 223, 224
Engaeus fossor 198–201, 213
Engaeus fultoni 191, 227
England 1, 13, 15, 32, 36, 46, 51, 149, 344, 346, 351, 352
Enterobacter 176, 177, 230, 231
Entocythere 214, 215, 217, 218, 221, 223–225, 230, 231
Entocytherinae 214
Entomophthoromycota 180
environmental data 252
environmental DNA (eDNA) 305, 306, 315, 332, 340
environmental drivers 251, 252, 255, 257, 258, 265, 277
environmental tolerances 251–253, 330
Epicoccum 183, 228, 230
epicuticle 63–65
epidermis 63–65, 334
Epiphanidae 213
Epistylis 186, 187, 189, 228–231
eradication 33, 50, 291, 339–345, 348, 349, 352, 353
Erwinia 176, 177, 229
Erymidae 3
escape responses 123, 132, 160
Escherichia 177, 179, 228, 231
Euastacus armatus 19, 71, 103, 109, 191, 198
Euastacus australasiensis 103, 109, 199, 223
Euastacus balanensis 199
Euastacus bidawalus 198
Euastacus bispinosus 20, 95, 103, 109, 198, 252
Euastacus brachythorax 198
Euastacus clarkae 9, 10, 198
Euastacus claytoni 199
Euastacus crassus 181, 191, 199, 223
Euastacus dangadi 9, 198
Euastacus dharawhalus 198
Euastacus eungella 199
Euastacus fleckeri 199
Euastacus gamilaroi 198, 199
Euastacus gumar 198
Euastacus guwinus 198
Euastacus hirsutus 198
Euastacus hystricosus 199, 223
Euastacus jagara 198
Euastacus kershawi 181, 199
Euastacus maidae 198
Euastacus mirangudjin 198
Euastacus monteithorum 198

Euastacus neodiversus 199
Euastacus neohirsutus 9, 198, 223
Euastacus polysetosus 198
Euastacus reductus 199
Euastacus rieki 198
Euastacus robertsi 199
Euastacus setosus 198, 223
Euastacus spinichelatus 198
Euastacus spinifer 103, 109, 181, 198, 223
Euastacus sulcatus 20, 197, 198, 223
Euastacus suttoni 198
Euastacus urospinosus 199
Euastacus valentulus 198
Euastacus woiwuru 199
Euastacus yanga 198
Euastacus yarraensis 199
Euastacus yigara 199
Eucypris virens 214, 229
Eudorylaimus
    carteri 202
    centrocercus 202
Euplotes 186, 187, 190, 230
Europe 4, 6, 11–15, 18, 21, 23, 32, 35, 36, 39, 40, 45–54, 94, 161, 171, 173, 185, 186, 202, 211, 214, 227, 229, 232, 251, 256, 257, 264, 265, 272, 289, 295, 315, 337, 339, 342, 347, 351
Everglades 295
evolutionary history 3, 6, 39, 44, 104, 107, 141, 265
evolve 31, 34, 98, 152, 337
ex situ 303
exoskeleton 62–66, 88, 175, 255, 334
extended parental care 154
external fertilization 96, 97
external measurements 332
external sperm storage 76, 84, 97, 105
extinction risk 55, 277
extinctions 18, 20, 22, 33, 44, 55, 171, 254, 264, 265, 267, 272, 277, 338, 342
eyes 1, 62–64, 133, 157, 174, 227, 332, 334
eyestalks 63, 146

F

facultative 83, 327
Fallicambarus (Creaserinus)
    byersi 215
    fodiens 189, 207, 215
Fallicambarus (Fallicambarus) dissitus 215
Fallicambarus fodiens 71, 100, 108, 120
Fallicambarus gordoni 71, 108, 119
false negatives 311
Faxonella clypeata 84, 108, 215
fecundity 101, 103, 107, 120, 258, 260, 276
feeding 63, 64, 70, 99, 104, 117–119, 123–125, 132, 135, 137, 140–142, 144, 154, 155–157, 159, 173, 174, 197, 265, 270, 272, 273, 275, 331
female abdomen 104
female choice 107, 152
female control 96, 97
female investment 104
female odour 84
female receptivity 99
female reproductive state 100

Milton Keynes UK
Ingram Content Group UK Ltd.
UKHW050454071024
449327UK00015B/374